D1159645

PHYSICAL AND CHEMICAL EQUILIBRIUM FOR CHEMICAL ENGINEERS

PHYSICAL AND CHEMICAL EQUILIBRIUM FOR CHEMICAL ENGINEERS

Second Edition

NOEL DE NEVERS
University of Utah
Department of Chemical Engineering
Salt Lake City, Utah

A JOHN WILEY & SONS, INC., PUBLICATION

Published by John Wiley & Sons, Inc., Hoboken, New Jersey
Published simultaneously in Canada

Library of Congress Cataloging-in-Publication Data:

De Nevers, Noel, 1932-
Physical and chemical equilibrium for chemical engineers / Noel de Nevers. – 2nd ed.
p. cm.
Includes index.
ISBN 978-0-470-92710-6 (hardback)
1. Thermodynamics. I. Title.
TP155.2.T45D4 2012
660′.2969–dc23

2011046731

Printed in the United States of America

10 9 8 7 6 5 4 3 2 1

CONTENTS

PREFACE

This book is intended for university juniors in chemical or environmental engineering. It explains the fundamentals of physical and chemical equilibrium and how these relate to practical problems in chemical and environmental engineering. The student will find that our understanding of equilibrium is based on thermodynamics. Nature attempts to minimize Gibbs energy; this book shows some of the details of that minimization.

Traditionally, this material is taught to chemical engineers as a second course in thermodynamics, following a fundamental thermodynamics course, substantially identical to the introductory thermodynamics course taught in mechanical engineering. This book assumes that its readers have completed such a course. A one-chapter review of that material is presented.

Physical and chemical equilibria present textbook authors with great opportunities to exercise their mathematical formalisms, but these formalisms often obstruct intuitive understanding of equilibrium. Furthermore, this topic introduces some material that is counterintuitive, and several properties, such as fugacity and activity, that are not easily related intuitively to the common experience of the student. As a result, most B.S. graduates in chemical engineering have a poor intuitive understanding of the relations between widely used equilibrium estimating methods (*K* values, relative volatility, equilibrium constants, liquid-liquid distribution coefficients) and the fundamental thermodynamics behind them. I certainly had little understanding of that topic when I received my B.S., Ch.E. In this book I have placed as many of those formalisms as I could in appendixes, and have added as much descriptive material as possible to try to help the student develop the intuitive connection between the working equilibrium tools of the chemical engineer and the thermodynamic basis for those tools. All of the material in this book can be presented in more mathematically compact and abstract form than it is here. I have deliberately preferred explanatory value to mathematical elegance. I have not sacrificed rigor, although the rigorous treatments are often in the appendixes.

I have been guided by three pedagogical maxims: (1) "The three rules of teaching are: from the known to the unknown, from the simple to the complex, one step at a time" (author unknown to me); (2) "If you don't understand something at least two ways, you don't understand it." (Alan Kay); and (3) "The purpose of computing is insight, not numbers!" (Richard Hamming). I have devoted more space and effort to determining numerical values of pertinent quantities than do most authors. I believe students need to develop a feel for how big? how fast? how hot? and how much?

In many areas of the book the treatment in the text is simple, with a more complex treatment outlined or discussed in one of the problems. Students are encouraged to at least read through all the problems, to see where more complex and complete treatments are either described or referred to. In many places in the book there are digressions, not directly applicable to the main flow of the text, and problems not directly related to chemical or environmental engineering. Some of these show interesting related technical issues. I include these because I think they help students build mental bridges to other parts of their personal experiences. The more the students are able to integrate the new information in this book into their existing knowledge base by such connections, the more likely they are to retain it and be able to use it.

Currently most of the industrial calculations of the type shown in this book are done by large computer programs. Most of the "real-world" calculations have no analytical solutions; they must be done numerically. I have not introduced the algorithms for those calculations, or supplied a CD allowing their use, because I consider it much more important for students to learn the physical basis of those calculations than to learn to use the programs. I use spreadsheets for numerical solutions and encourage the students to do so, because spreadsheets show the details plainly and their programming is totally intuitive.

In preparing the second edition, I have corrected the abundant typos and errors I know about from the first edition, simplified the notation and some language from the first edition, deleting some things that I liked but that the students apparently found confusing, and changed some names to match current usage (e.g. Gibbs free energy to Gibbs energy). I have added sections on Minimum and Maximum Work, Adsorption, Hydrates and Equilibrium in Biochemical reactions and The Bridgman Table. There are also some new problems and examples.

I thank my friend and colleague Geoff Silcox for his many suggestions and comments.

I will be very grateful to readers who point out to me typographic errors, incorrect equation numbers, incorrect figure numbers, or errors of any kind. Such errors will be corrected in subsequent editions or printings.

Noel de Nevers

Salt Lake City, Utah

ABOUT THE AUTHOR

Noel de Nevers received a B.S. from Stanford in 1954, and M.S. and Ph.D. degrees from the University of Michigan in 1956 and 1959, all in chemical engineering.

He worked for the research arms of the Chevron Oil Company from 1958 to 1963 in the areas of chemical process development, chemical and refinery process design, and secondary recovery of petroleum. He has been on the faculty of the University of Utah from 1963 to the present in the Department of Chemical Engineering becoming emeritus in 2002.

He has worked for the National Reactor Testing Site, Idaho Falls, Idaho, on nuclear problems, for the U.S. Army Harry Diamond Laboratory, Washington DC, on weapons, and for the Office of Air Programs of the U.S. EPA in Durham, NC, on air pollution.

He was a Fulbright student of Chemical Engineering at the Technical University of Karlsruhe, Germany, in 1954–1955, a Fulbright lecturer on Air Pollution at the Universidad del Valle, in Cali, Colombia, in the summer of 1974, and at the Universidad de la República, Montevideo Uruguay and the Universidad Naciónal Mar del Plata, Argentina in the Autumn of 1996.

His areas of research and publication are in fluid mechanics, thermodynamics, air pollution, technology and society, energy and energy policy, and explosions and fires. He regularly consults on air pollution problems, explosions, fires and toxic exposures.

In 2005 his textbook, *Fluid Mechanics for Chemical Engineers*, Third Edition, was issued by McGraw-Hill.

In 1993 he received the Corcoran Award from the Chemical Engineering Division of the American Society for Engineering Education for the best paper ("'Product in the way' Processes") that year in *Chemical Engineering Education*.

In 2000 his textbook, *Air Pollution Control Engineering*, Second Edition, was issued by McGraw-Hill, and reprinted by the Waveland Press in 2010.

In addition to his serious work he has three "de Nevers's Laws" in the latest "Murphy's Laws" compilation, and won the title "Poet Laureate of Jell-O Salad" at the Last Annual Jell-O Salad Festival in Salt Lake City in 1983. He is the official discoverer of Private Arch in Arches National Park.

NOMENCLATURE

SYMBOL	DESCRIPTION	SI DIMENSION	ENGLISH ENGINEERING DIMENSION
A	Helmholz energy $= U - TS$	J	Btu
a	Helmholz energy per unit mass or mol $= u - Ts$	J/mol or J/kg	Btu/lbmol or Btu/Ibm
A, B, C	constants in various equations	various	various
a_i	activity of species $i = f_i/f_i^{\circ}$		
B	second virial coefficient	L/mol	ft^3/lbmol
$\bar{b}_r$	partial molar second virial coefficient	L/mol	ft^3/lbmol
C	number of components in phase rule		
C_P	molar or mass heat capacity at constant pressure	J/(mol or kg)K	Btu/(lbm or lbmol) · °R
C_V	molar or mass heat capacity at constant volume	J/(mol or kg)K	Btu/(lbm or lbmol) · °R
E	Energy of all types	J	Btu
EOS	equation of state		
e	reaction coordinate (Chapter 12)		
f_i	fugacity of species i in a mixture	Pa	psia
G	Gibbs energy $= H - TS$	J	Btu
g	molar or specific Gibbs energy, $h - Ts$	J/(mol or kg)	Btu/(lbmol or lbm)
G^E	excess Gibbs energy	J	Btu
g^E	molar or mass excess Gibbs energy	J/(mol or kg)	Btu/(lbmol or lbm)
H_i	Henry's law constant for species i	atm	atm
H	enthalpy	J	Btu
h	molar or specific enthalpy	J/(mol or kg)	Btu/(Ibmol or lbm)
K	equilibrium constant, VLE, y_i/x_i		
K	distribution coefficient, LLE, $x_i^{(1)}/x_i^{(2)}$		
K	equilibrium constant, chemical reactions		
K_b	boiling-point elevation constant	°C/molal	seldom used
K_f	freezing-point depression constant	°C/molal	seldom used
K_p	equilibrium constant based on pressures	various	various
K_ϕ	equilibrium constant correction using Lewis–Randall rule		
$K_{\hat{\phi}}$	equilibrium constant correction for nonideal solutions		
K_{sp}	solubility product	various	various
k_{ij}	binary interaction coefficient		

SYMBOL	DESCRIPTION	SI DIMENSION	ENGLISH ENGINEERING DIMENSION
M	molecular weight	g/mol	lbm/lbmol
M, N	numbers used in derivation of phase rule		
m	mass	kg	lbm
m	molality	mol/kg	not used
NBP	normal boiling point	°C or K	°F or °R
n_e	mols of electrons transferred (electrochemical reactions)		
n_i	number of mols of species i	mol	lbmol
p_i	vapor pressure of species i	Pa	psia
P	pressure	Pa	psia
P	number of phases in phase rule		
Po	power	W	Btu/s or hp
P_c	critical pressure	Pa	atm
P_r	reduced pressure $= P/P_c$		
Q	heat quantity	J	Btu
R	universal gas constant (see inside back cover)	J/mol K	Btu/lbmol °R
r	radius	m	ft
S	entropy	J/K	Btu/°R
s	molar or specific entropy	J/(mol or kg)·K	Btu/(lbmol or lbm) · °R
stp	standard temperature and pressure (1 atm, 20°C)		
T	absolute temperature	K	°R
T_c	critical temperature	K	°R
T_r	reduced temperature $= T/T_c$		
t	time	s	s
U	internal energy	J	Btu
u	molar or specific internal energy	J/(mol or kg)	Btu/(lbmol or lbm)
V	volume	m^3	ft^3
v	molar or specific volume	m^3/(mol or kg)	ft^3/(lbmol or lbm)
V	number of degrees of freedom in phase rule		
VLE	vapor–liquid equilibrium		
W	work	J	Btu
x, y, z	distances in coordinate directions	m	ft
x, y, z	stand for any variable	various	various
x	mols formed by reaction	mol	lbmol
x_i	mol fraction of species i in liquid or solid		
$[X]$	concentration or activity of species X	various	various
$x_a^{(1)}$	value of variable x of species a in phase 1	various	various
x°	pure species or reference state value of x	various	various
$\bar{x}_i$	partial molar value of any variable x_i	various/mol	various/mol
x^*	ideal solution or ideal gas value of x	various	various
y_i	mol fraction of species i in vapor or gas		
z	compressibility factor		
α	volume residual	m^3/mol	ft^3/lbmol
α	relative volatility		
δ	solubility parameter	$(cal/cm^3)^{0.5}$ or $(MPa)^{0.5}$	not used
δ	cross term in virial equation for mixtures	L/mol	ft^3/lbmol
γ_i	activity coefficient of species i		
Γ	free energy parameter in Chapter 11		
θ	fractional surface coverage in adsorption		
μ_i	chemical potential of species i	J/mol	Btu/lbmol

SYMBOL	DESCRIPTION	SI DIMENSION	ENGLISH ENGINEERING DIMENSION
ν	stoichiometric coefficient (Chapter 12)		
ρ	density	kg/m^3	lbm/ft^3
σ	surface tension	N/m	lbf/ft
Φ	volume fraction in liquids		
ϕ	fugacity coefficient		
$\hat{\phi}_i$	partial fugacity coefficient		
ω	accentric factor		

Superscripts

0 and 1	base and correction terms in Pitzer-type EOS and functions derived from it
()	designates phase number or name
o	indicates standard state property.
–	superbar indicates partial molar property
*	on the symbol for a property (e.g. h^*) refers to the value of the property in the ideal gas state. On an equation e.g (Eq.7.22)* indicates and equation that is only applicable to ideal gases.
†	on an equation e.g. (Eq. 7.23)† indicates that the equation is only applicable to ideal solutions.

Subscripts

Letter	designates species *a*, *b*, *c*, *d*
Double	designates interaction between two molecules of the same (*ii*) or different (*ij*) types
Number	various, e.g., first and second states

1

INTRODUCTION TO EQUILIBRIUM

1.1 WHY STUDY EQUILIBRIUM?

The four basic tools used by chemical and environmental engineers are

1. Material balances
2. Energy balances
3. Equilibrium relations
4. Rate equations

You may be surprised that the second law of thermodynamics is not on the list. We will see later in this book that the second law of thermodynamics plays a key role in phase and chemical equilibrium. In fact, the principal use of the second law for chemical and environmental engineers is its indirect ulilization in computing the equilibrium states in process of technical interest.

Figure 1.1(a) shows four of the five practically-identical ammonia synthesis plants at the Donaldsonville, LA plant of CF Industries. Each of these produces 1500 tons/day of ammonia, for use in fertilizer. These five plants produce a total of about 5 billion pounds per year of ammonia, equivalent to 16 pounds per year for each person in the United States. That fertilizer contributes in a major way to the abundance, variety, and low cost of food in the United States. Such plants are vitally important to the human race. They produce the fixed nitrogen used in fertilizers throughout the world. Roughly 80 pounds of a variety of fertilizers are produced per year for each person on earth. If we lost this supply of synthetic fertilizers and then all stopped eating meat, we would be able to feed about 80% of the world's current population; the rest would starve [1]. Part (b) of the same figure shows a very simplified flow diagram of such an ammonia synthesis plant.

The overall reaction in the synthesis section of these plants is

$$3H_2 + N_2 \Leftrightarrow 2NH_3 \tag{1.A}$$

in which the $\Leftrightarrow$ symbol indicates a chemical equilibrium. (Every equation in this book has a number. Those, like this one, that are a specific description of a reaction or that are parts of examples or in other ways specific to some situation are number-letter combinations, like (1.A). Those that are not specific, but general, have number-number combinations, like (1.1).)

To design a new plant of this type or to analyze or understand this kind of plant we would use the four tools listed above, beginning with a material balance, determining the flow rates and compositions of all the process streams. We would need to know the chemical equilibrium in the reactor to estimate the fraction of the feed that is converted in one pass through the reactor according to Eq. 1.1 to determine the recycle flow rate and the total flow rate to the reactor. Then we would need to know the physical equilibrium in the separator to know the temperature and pressure in the separator required to separate most of that ammonia as a liquid from the unreacted synthesis gas, which is recycled as a gas. An energy balance would determine how much the exit temperature of the reactor exceeds the feed temperature, and how much heat must be removed from the feed plus recycle stream to get it to the proper temperature for the separator. Finally,

Physical and Chemical Equilibrium for Chemical Engineers, Second Edition. Noel de Nevers.

(a)

Feed preparation section

Synthesis section

Product ammonia plus unreacted feed is recycled to the chiller, separator and reactor

A small bleed stream removes impurities

Recirculating compressor

Feed preparation section reacts air, methane and water to produce a high-pressure, 3/1 mixture (by mol) of hydrogen and nitrogen, and rejects a waste stream of carbon dioxide

Separator removes ammonia as a liquid, and passes feed plus recycle to the reactor

Reactor partly converts nitrogen and hydrogen to ammonia

Chiller cools reactor exit stream plus feed stream enough to condense most of the ammonia

Ammonia product is removed as a liquid

(b)

FIGURE 1.1 (a) An aerial view of main part of the Donaldsonville, LA fertilizer complex of CF Industries. (Courtesy of CF Industries.) (b) Very simplified flow diagram of an ammonia synthesis plant. Only the synthesis section is discussed in the text. The feed preparation section is more complex and expensive than the synthesis section. The seemingly illogical placement of the chiller and separator so that they process the fresh feed plus the recycle is dictated by the fact that some feed impurities dissolve in the liquid ammonia, and thus are prevented from entering the reactor. The reactor converts only about 15% of the feed on each pass [2].

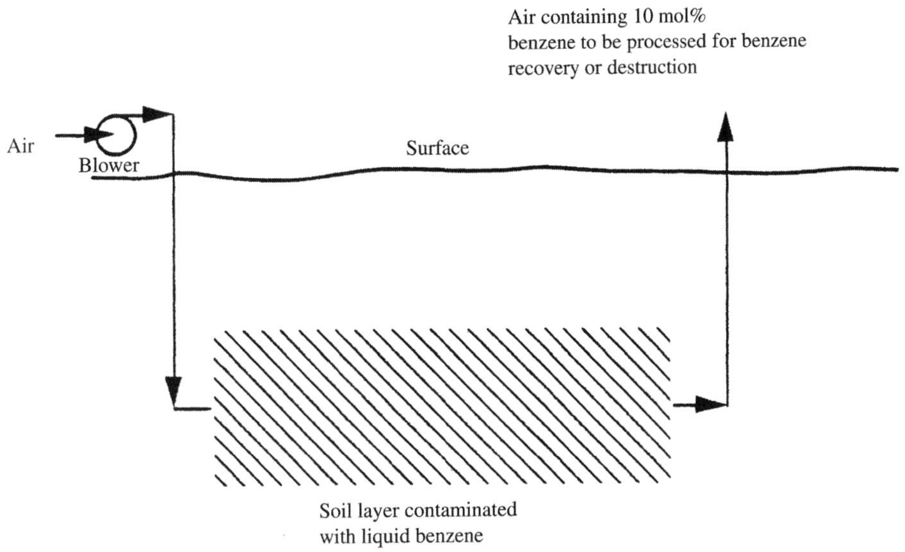

Air containing 10 mol%
benzene to be processed for benzene
recovery or destruction

Air

Blower

Surface

Soil layer contaminated
with liquid benzene

FIGURE 1.2 A common environmental engineering problem, removal of liquid benzene contamination from soils, to protect groundwater.

we would use the rate equations of fluid mechanics to choose the right pipe sizes, the rate equations of heat transfer to choose the right type and size for the heat exchanger, the rate equations for diffusion and chemical reactions to know how large the reactor must be, and how much catalyst it must contain, and the rate equations of mass transfer to design the separator.

In typical reactors of this type [2], only about 15% of the feed is converted to ammonia on one pass through the reactor. We would like to convert more, to limit the number of passes through the reactor and thus limit the cost of recirculating the unreacted gases. But the chemical equilibrium in Eq. 1.A limits the amount converted per pass. We will say much more about chemical equilibrium and about this particular chemical reaction in Chapter 12. Similarly, we would like to remove all of the ammonia in the chiller-separator combination, but we can remove only about 80% of it. Here the limitation is phase equilibrium, which we will discuss in the next several chapters.

All four basic tools are needed to understand such a process. Chemical engineers take courses in all these fields and become skilled in the use of all these tools. Figure 1.2 shows a common environmental engineering problem: Liquid benzene has leaked into an underground soil stratum. To remove it ("remediate the site"), air is pumped through the stratum. The benzene evaporates into the air, and is brought to the surface where it is treated to recover or destroy the benzene. The same four tools are used for this problem: the material balance to determine how much is to be removed, the equilibrium relationship to determine the maximum amount that a unit mass or unit volume of air can remove, the fluid flow rate equations to select the sizes of the pumps and lines, and the diffusion equations to estimate how close to equilibrium we would expect the exit gas to be. Here, again, we see that although we would like a high concentration of benzene in the air leaving the contaminated soil, we can only get a maximum of about 10 mol%, limited by the phase equilibrium between the liquid benzene in the soil layer and the air we pass through it.

The order of application of the four tools is not necessarily the same in both cases, but the tools are the same. Of the four tools, this book concerns only equilibrium.

The role of equilibrium in these processes is summarized in Figure 1.3. This shows that we want to go from where we are to somewhere else (e.g., make ammonia or remove the benzene contaminant), but that equilibrium acts like a brick wall between here and there, allowing us to get only part way. We use separation and recycle (Figure 1.1) or large amounts of one stream (Figure 1.2) to overcome this difficulty. To know the dimensions of our problem, we

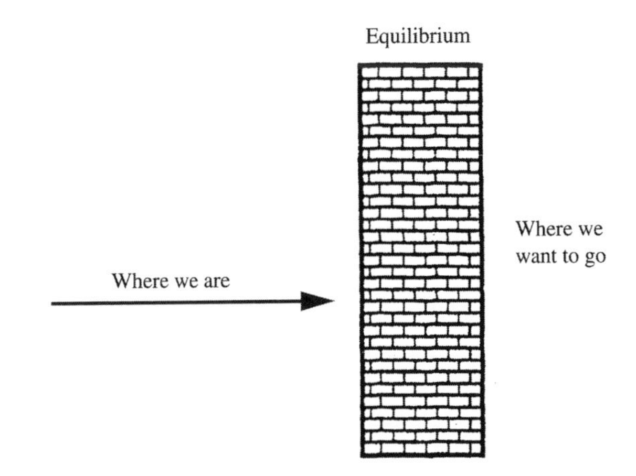

FIGURE 1.3 Equilibrium acts as a brick wall between where we are and where we want to go. Knowing where the wall is allows us to find ways around it, when the direct route is impossible.

must know where the equilibrium limits are. This book is about that.

1.2 STABILITY AND EQUILIBRIUM

A system is said to be at equilibrium if there is no change with time in any of the measurable properties of the system. For some systems we need to consider long time periods. For example, a piece of iron or steel left on the ground will turn completely to rust. If we look for changes in one day, we will see practically none. For most naturally occurring solids we need to view the changes on a geological time scale rather than the human time scale to determine if they are at equilibrium. For most of the problems of common engineering interest the changes are much more rapid, and we will consider a system to be at equilibrium if we can detect no change in it in a few hours or days. For large systems the changes may be small enough that we do not recognize them. For example, the deep oceans of the world have practically a uniform salt content. But it is not an equilibrium salt content. If we turned off the slow currents that mix the world's oceans and waited a few million years, we would see a different distribution of salt, with the concentration increasing slowly with depth [3] (see Chapter 14). Similarly, if we turned off the winds that mix the atmosphere and waited a few million years, we would see significant chemical concentration gradients in the atmosphere; the winds that mix the atmosphere keep its composition practically uniform (except for its water vapor content, which is generally less than a few mol%, but which varies significantly with time and place).

Left to themselves, all systems in the world move toward a state of phase and chemical equilibrium. How fast natural systems move in the direction of phase and chemical equilibrium depends on mass transfer rates (for phase equilibrium) and chemical reaction rates (for chemical equilibrium). Thermodynamics tell us little about these rates. So knowing the equilibrium state tells us in which direction the system will go, and how far the system is from its equilibrium state but not how fast it will move in that direction. The systems of greatest interest to us are mostly not left to themselves. The earth receives vast amounts of energy from the sun and that energy moves many systems on earth away from equilibrium; when the external energy source is removed, the relentless march toward equilibrium begins again. Our foods, fuels, electricity, and autos are all far from equilibrium, mostly based indirectly on solar energy. The equilibrium state for our bodies is to be converted mostly to water and carbon dioxide. By using the sun's energy, captured by plants to form foods, we can stay away from this equilibrium state for a long (and we hope interesting and useful) life (Figure 1.4). The oceans and atmosphere are far from equilibrium, mixed by solar-driven winds and currents. But for many systems of great practical interest we may safely assume that the system is at or very near equilibrium.

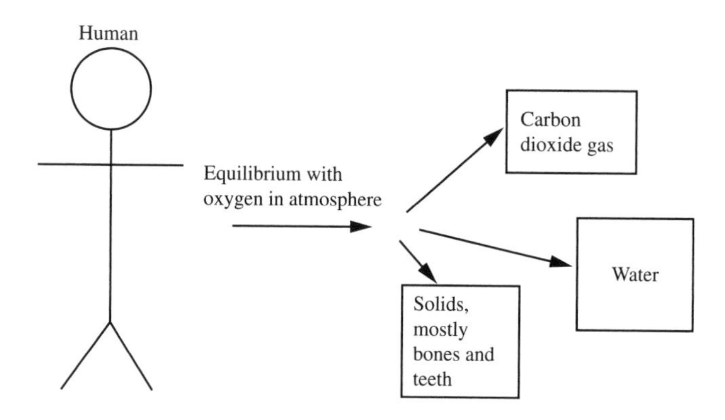

FIGURE 1.4 Equilibrium is not always desirable. If we bring a human to equilibrium with the oxygen in the atmosphere we will produce mostly water, carbon dioxide, and a solid residue made mostly of bones and teeth. We work hard at preventing this equilibrium, mostly by using the energy of the sun, concentrated in our foods.

The fact that there is no measurable change does not mean that the system is static. If we could see the atoms and molecules of, for example, steam and water, we would see that if the two phases are at equilibrium, there is a steady interchange of water molecules between the two phases. However, at equilibrium the flow of atoms or molecules in one direction is exactly equal to the flow in the other direction (as many water molecules per second pass from the water to the steam as from the steam to the water). Similarly, in all chemical reactions at equilibrium the concentrations of the reactants and products are not changing with time. That does not mean that the reaction has stopped. It means that the forward and reverse reactions are occurring at exactly the same rate, so that the net reaction rate (algebraic sum of the forward and backward reaction rates) is zero. We do not need this molecular view to make ordinary engineering calculations, but later we will see that it helps to form an intuitive picture of the relations we will use. So in this book we will occasionally refer to what is occurring at the molecular level to help us understand what is going on in engineering-scale systems.

The several kinds of equilibrium are most easily described in terms of mechanical models (Figure 1.5). A ball resting in a deep cup is in a stable equilibrium; if it is displaced a small amount from its rest position and then released, it will return to its original position (the bottom of the cup). This is a case of equilibrium with the surroundings. If the cup were suddenly removed, the ball would fall freely until it encountered the next obstacle, and then take up a new position of equilibrium with respect to its new surroundings. Such stability is also known in the field of chemical equilibria. In an aqueous solution at room temperature, the equilibrium product of the concentrations of hydrogen and hydroxyl ions is $\approx 10^{-14}$ (mol/L)^2. If we disturb this equilibrium by adding some acid or base, the concentrations will quickly readjust so that this product is again $\approx 10^{-14}$ (mol/L)^2.

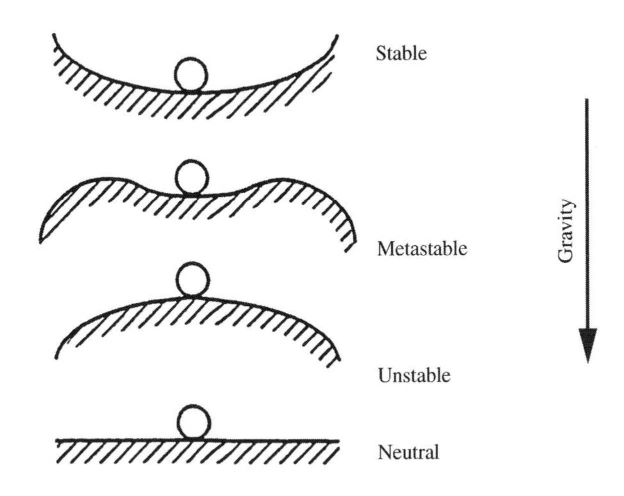

FIGURE 1.5 Mechanical models of stable, metastable, unstable, and neutral equilibrium.

If the cup in the above example is shallow then a very small displacement may result in the ball returning to the bottom of the cup, while a moderate-sized displacement may get it out of the cup and allow it to fall to a lower location. This is a metastable equilibrium. The typical chemical example of such metastable equilibrium is a stoichiometric mixture of hydrogen and oxygen at room temperature. We may change the temperature over a finite range without any significant change in chemical composition. However, if we raise the temperature of a small part of the mixture to a moderately high temperature, for example, by a spark, the system will convert to water explosively. Other examples are the supersaturated solution, which will crystallize if a small seed crystal is introduced, and the superheated liquid, which boils explosively when a boiling chip is introduced.

If a small ball were balanced exactly on the top of a very large ball, then any displacement of any measurable size in any direction would cause it to roll down the surface of the larger ball. This is an unstable equilibrium. We may think of this as the limiting case of a metastable equilibrium, in which the indentation in which the ball rests in the second part of the figure becomes shallower and shallower, eventually becoming flat and then curved upward. This situation exists in many nucleation phenomena. For example, as the temperature of a superheated liquid droplet is increased, eventually a "critical superheat temperature" is reached at which the drop boils spontaneously. At this temperature the drop is unstable and its own internal vibrations are apparently enough to cause it to boil.

A piece of steel in contact with air and water is also an example of this situation. The steel is actually rusting, so in mechanical analogy we would says that this is the equivalent of the small ball on top of the large ball. The rusting process is very slow, equivalent to the small ball rolling down the surface of the large one, but very, very slowly.

Neutral equilibrium is represented by a cylindrical pencil resting on a perfectly flat table. If a pencil at rest is given a small displacement, it does not return to its original position of equilibrium but remains at the new one. A corresponding phase equilibrium situation would be a mixture of ice and water.

If one adds a little heat to such a system, some of the ice will melt. When the heating process is stopped, the system does not go back to its original ratio of ice to water but remains at the new disturbed ice/water ratio.

1.3 TIME SCALES AND THE APPROACH TO EQUILIBRIUM

In the process shown in Figure 1.1, we would expect the stream leaving the reactor to be close to equilibrium, but not at equilibrium. To get all the way to equilibrium would require too large and expensive a reactor. It is more economical to use a smaller reactor and increase the amount recycled. In the process in Figure 1.2, the air flow is slow enough that if the benzene is well dispersed in the stratum, then the benzene concentration of the air leaving the stratum would be very close to equilibrium. Some natural processes, like flames, come very close to equilibrium. Others, like geologic processes, do not. Generally, small, fast processes come close to equilibrium; slow, large ones do not.

In spite of this, we most often compute the equilibrium conditions for processes, because these set the limits of what is possible. Then we must decide on the basis of economics how close to equilibrium we want to come and how much we are willing to pay, which generally means how big the reactor or separator must be.

1.4 LOOKING AHEAD, GIBBS ENERGY

In Chapter 4 we will show, on the basis of rigorous thermodynamics, that all natural systems try to lower their Gibbs energy:

$$\begin{aligned} &\text{Gibbs energy} &&= H - TS \\ &\text{Gibbs energy per mol or per unit mass} &&= h - Ts \end{aligned} \tag{1.1}$$

The symbols are all defined in the table of nomenclature. This says that natural processes proceed toward the lowest Gibbs energy consistent with the constraints imposed on them (e.g., the temperature, the pressure, the starting materials). At equilibrium natural systems are, in effect, at the bottom of a Gibbs energy basin, in which for any infinitesimal change in any direction the change in Gibbs energy is zero, and for any finite change the Gibbs energy increases. The result is borrowed from Chapter 4:

For any differential equilibrium change, chemical or physical or both, at constant T and P,

$$dG_{sys} = 0 \tag{4.8}$$

This whole book is simply the working out of the details of this statement. We will refer back to this idea often. As the details become complex, remember that they all rest on this one idea.

In dealing with phase or chemical equilibrium we may think of the situations in Figure 1.5, with the downward direction (the direction of gravity) replaced by the direction of *decreasing* Gibbs energy[1].

1.5 UNITS, CONVERSION FACTORS, AND NOTATION

In this book both English and SI units are used. As much as possible we use those units most commonly used in the United States in that particular area of engineering. Historically, scientists have used SI or metric (often the cgs version of metric), while U.S. engineers have used the English engineering system. In most practical equilibrium calculations we can ignore the effect of gravity, so that the confusion over weight and mass generally does not arise. Similarly, we have few accelerated systems, so that force and mass seldom appear in the same equation. For that reason, students have less trouble with units in this course than they do, for example, in fluid mechanics, where they first encounter the problems with force, weight, and mass.

Some students do have trouble with concentration units. Phase and chemical equilibrium inevitably lead to mixtures, and we need suitable ways of describing those mixtures. In chemical equilibrium calculations, as in almost all of chemistry, the normal unit of mass is the mol (sometimes called a gram mol),

$$\begin{pmatrix}\text{mols of} \\ \text{substance } x\end{pmatrix} = 6.023 \times 10^{23} \begin{pmatrix}\text{atoms or molecules} \\ \text{of substance } x\end{pmatrix} \tag{1.2}$$

So, for example, a mol of water (H_2O) is 6.023×10^{23} molecules of water. The molecular weight of water is $M = 18$ g/mol so that the mass of a mol of water is 18 g. (A better name for this quantity would be the molecular mass, because the gram is a unit of mass. But molecular weight is the common name.)

In U.S. engineering work the unit of mass is the pound mass, written lbm. We regularly use the pound mol written lbmol:

$$\begin{pmatrix}\text{lbmol of} \\ \text{substance } x\end{pmatrix} = 453.66 \cdot 6.023 \times 10^{23} \begin{pmatrix}\text{atoms of} \\ \text{substance } x\end{pmatrix} \tag{1.3}$$

Thus, a pound mol of water is 453.6 mol (1 lbm = 453.6 g), and we may write that for water

$$M_{water} = \text{molecular weight of water} = 18 \text{ g/mol} = 18 \text{ lbm/lbmol} \tag{1.4}$$

The relation between mass and mols is

$$\text{mols of } i = \frac{\text{mass of } i}{\text{molecular weight of } i} \qquad n_i = \frac{m_i}{M_i} \tag{1.5}$$

where n_i is the number of mols of i, m_i is the mass of i, and M_i is the molecular weight of i. With this definition, we can further define

$$(\text{mol fraction of } i) = x_i = \frac{\text{mols of } i \text{ in mixture}}{\text{total mols of all substances in mixture}} = \frac{n_i}{\sum_{\text{all substances}} n_j} \tag{1.6}$$

The mol fraction is dimensionless; all the mol fractions in any mixture sum to 1.00. The mol fraction of i is equivalent to the fraction of the molecules (or atoms) in the mixture that are of species i. This is the most widely used concentration unit in equilibrium calculations. By common convention the mol fraction of i in solids and liquids is given the symbol x_i, while that in the gas phase is given the symbol y_i.

One also regularly sees concentrations by mass (or weight); for example,

$$\text{mass fraction of } i = x_i = \frac{\text{mass of } i \text{ in mixture}}{\text{total mass of all substances in mixture}} = \frac{m_i}{\sum_{\text{all substances}} m_j} \tag{1.7}$$

and the symbol x_i is often used for this as well.

The concentrations of solutes in dilute solutions (of gas, liquid, or solid) are regularly expressed in parts per million (ppm). In the United States, ppm is almost always by *volume* or *mol* if it is concentration in a *gas*, and by *mass* or *weight* if it is a concentration in a *liquid* or *solid*. (For a liquid or a solid with a specific gravity of 1.00, like water or dilute solutions in water, ppm is the same as mg/kg, which is also widely used.) This mixed meaning for ppm is a source of confusion when both liquid or solid and gas concentrations appear in the same

[1] For most of the past 100 years the quantity we now call the Gibbs energy was called the Gibbs free energy. When you encounter this older name recognize that the two names describe the same quantity.

problem. (The same is true of parts per billion (ppb), which equals μg/kg for a solid or liquid material with specific gravity of 1.00.)

Example 1.1 One kg of sugar solution is made of 990 g of water, $M = 18$ g/mol, and 10 g of dissolved sugar (sucrose, $C_{12}H_{22}O_{11}$), $M = 342.3$ g/mol. What is the sucrose concentration, expressed in mass fraction, mol fraction, molality, and ppm? The mass fraction is

$$(\text{mass fraction of sucrose}) = x_i(\text{by mass})$$

$$= \frac{m_i}{\sum_{\text{all substances}} m_j} = \frac{10g}{10g + 990g} = 0.01 = 1\% \quad (1.B)$$

This is also the weight fraction. We would say that this is 1 wt% sugar (the common expression) or 1 mass% sugar (which we rarely hear).

The mol fraction is

$$(\text{mol fraction of sucrose}) = x_i(\text{by mol})$$

$$= \frac{n_i}{\sum_{\text{all substances}} n_j} = \frac{\dfrac{10\text{g}}{342.3\ \text{g/mol}}}{\dfrac{10\text{g}}{342.3\ \text{g/mol}} + \dfrac{990\text{g}}{18\text{g/mol}}}$$

$$= \frac{0.0292\ \text{mol}}{0.0292\ \text{mol} + 55.0\ \text{mol}} = 5.31 \times 10^{-4} = 0.0531\% \quad (1.C)$$

Mol% is 100 times the mol fraction, so we would say that this is 0.0531 mol%. For dilute solutions, like this one, we could also say that

$$\begin{pmatrix}\text{mol fraction}\\ \text{of solute}\end{pmatrix} \approx \begin{pmatrix}\text{mass fraction}\\ \text{of solute}\end{pmatrix} \cdot \frac{M_{\text{solvent}}}{M_{\text{solute}}}$$

$$\approx 0.01 \frac{18}{342.3} = 5.26 \times 10^{-4} = 0.0526\% \quad (1.D)$$

where ≈ means approximately equal (see Problem 1.2).

The *molality,* a concentration unit widely used in equilibrium calculations, is defined as

$$\text{molality} = \frac{\text{mols of solute}}{\text{kg of solvent}} = \frac{\dfrac{10\text{g}}{342.3\ \text{g/mol}}}{0.99\ \text{kg solvent}}$$

$$= 0.0295\ \text{molal} \quad (1.E)$$

For solutions of solids and liquids (but not gases) ppm almost always means ppm by mass, so 1% = 10,000 ppm. ■ (The ■ symbol indicates the end of an example.)

The concentrations used in Example 1.1 do not depend on the density of the mixture and do not change if we change that density by changing the temperature or pressure of the mixture. The mass and mol concentrations and molarity, which are also widely used, do depend on the density.

Example 1.2 The density at 20°C of 1.0 wt% sucrose solutions in water is 1.038143 g/cm^3 [4]. Using this value, find the mass and mol concentrations and molarity of the solution in Example 1.1.

In Example 1.1 the mass was chosen to be 1.00 kg, so that

$$V = \frac{m}{\rho} = \frac{1\ \text{kg}}{1.038\ \text{g/cm}^3} \cdot \frac{1000\ \text{g}}{\text{kg}} \cdot \frac{\text{m}^3}{10^6\text{cm}^3}$$

$$= 0.9634 \times 10^{-3}\text{m}^3 = 0.9634\text{L} \quad (1.F)$$

The mass concentration is

$$(\text{mass concentration of sucrose}) = \frac{\text{mass of sucrose}}{\text{volume of solution}}$$

$$= \frac{10\ \text{g}}{0.9634\ \text{L}} = 10.38 \frac{\text{g}}{\text{L}} \quad (1.G)$$

The mol concentration is

$$(\text{mol concentration of sucrose}) = \frac{\text{mols of sucrose}}{\text{volume of solution}}$$

$$= \frac{\dfrac{10\ \text{g}}{342.3\ \text{g/mol}}}{0.9634\ \text{L}} = 0.0303 \frac{\text{mol}}{\text{L}} \quad (1.H)$$

This is also the definition of the *molarity,* so this is an 0.0303 molar solution of sucrose in water. ■

These three concentration measures (or their English engineering unit equivalents) are widely used in process calculations in the United States. They are seldom used in equilibrium calculations and will be seldom be used in this book.

Unfortunately, there is no agreement among various authors about symbols to be used in thermodynamics or in equilibrium. All symbols used in this text are shown in the table of nomenclature. The general convention is to use uppercase letters for externally imposed conditions or conditions applying to whole systems, such as *P, T* and *V, U, H, S,* and to use lowercase letters for specific (or per unit mass or per mol) properties, such as *v, u, h, s.* For describing a property of one component in one of several phases in equilibrium, $x_i^{(1)}$ refers to the property *x* of component *i* in phase 1. If there is no possible confusion about which phase is meant, the phase superscript is dropped. This is done for mol fractions in vapor-liquid equilibrium, where we use y_i and x_i for the mol ftractions of component i in the gas and liquid phases, respectively, and drop the superscript.

1.6 REALITY AND EQUATIONS

Reality is complex, and our measurements of it are sparse and imperfect. When we want to know some piece of physical data, such as the density of water at some specified T and P, it is very unlikely that we will find in the literature a direct measurement of the density at that T and P. If the intended use of the data is crucial enough, we may be justified in making a direct measurement, but doing so to high precision is expensive. Instead, we normally look at tables and charts of such values. These do not represent direct measurements at all those values of T and P. Rather, they are values calculated from data-fitting equations, which are adjusted so that they reproduce the existing measurements to within its experimental uncertainty. These equations are then used to make up the useful tables, at even values of T and P, for example, the steam tables [5, 6], In the study of equilibrium, we could measure all the values we need, which are normally concentration values in phases in equilibrium (although that can be expensive for extreme values of T and P and/or for materials which are toxic or explosive). Instead, we normally try to find an equation that will reproduce the available experimental data and then use it to interpolate or extrapolate to the values we need. Much of this book is devoted to developing such equations. Normally, thermodynamics will show us that only some forms of such equations are possible and that others are not. Once we know the possible forms, we will then need many fewer expensive experimental data points to make a satisfactory predictive equation than we would if we did not know the possible forms and used simple, brute-force methods to find data-fitting equations.

In principle, we could program the original experimental data into our computers and let the computers decide how to interpret them. That would be very cumbersome. Instead, we almost always find some kind of equation to represent the data, and let our computers use that equation. Much of this book is devoted to showing the forms of the equations we use in our computers, and showing the reasons we choose those forms.

1.7 PHASES AND PHASE DIAGRAMS

Phase equilibrium deals with phases, so we need a working definition of a phase. A phase is a mass of matter, not necessarily continuous, in which there are no sharp discontinuities of any physical properties over short distances. An *equilibrium phase* is one that (in the absence of significant gravitational, electrostatic, or magnetic effects) has a completely uniform composition throughout. In this book we will deal almost exclusively with equilibrium phases.

All gases form one phase. All gases are miscible, so that there can be *only one gas phase* present in any equilibrium system at any time.

Liquids can form multiple phases. Figure 1.6 shows a graduate cylinder with layers of benzene, water, and mercury. These are all at equilibrium (because we have stirred them vigorously and then let them settle!). The right-hand part of the figure shows that within each layer the density is constant, but that there are sharp discontinuities in density at the borders of the layers. These are three separate phases. Hildebrand et al. [7] show an example of 10 liquid phases in equilibrium: hexane, analine, aqueous methyl cellulose, aqueous polyvinyl alcohol, aqueous mucilage, silicone oil, phosphorus, fluorocarbon, gallium, and mercury. That is probably the record. But examples with three separate liquid phases, as shown in Figure 1.6, can be constructed in any laboratory in minutes. (Four is not particularly difficult, but after that it gets harder.)

Homogeneous solids are single phases, for example diamond, pure metals, pure mineral crystals. Some apparently simple solids are not single phases, such as cast iron, steel, wood, bacon, grass. One can observe the "grain" in wood, showing that it consists of layers of at least two different compositions, and can similarly observe the fat and meat

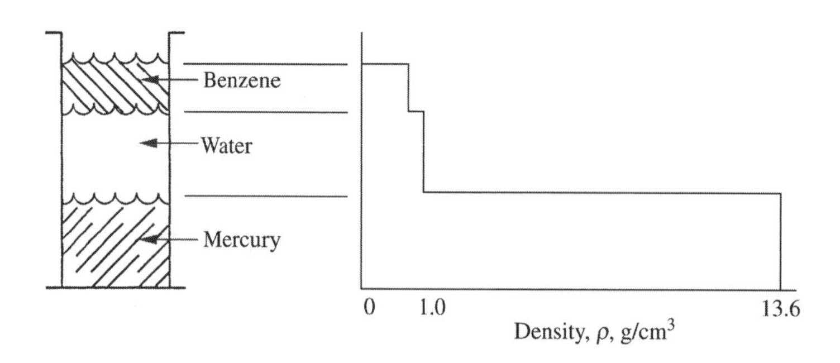

FIGURE 1.6 Appearance and elevation–density plot for three liquid phases at equilibrium.

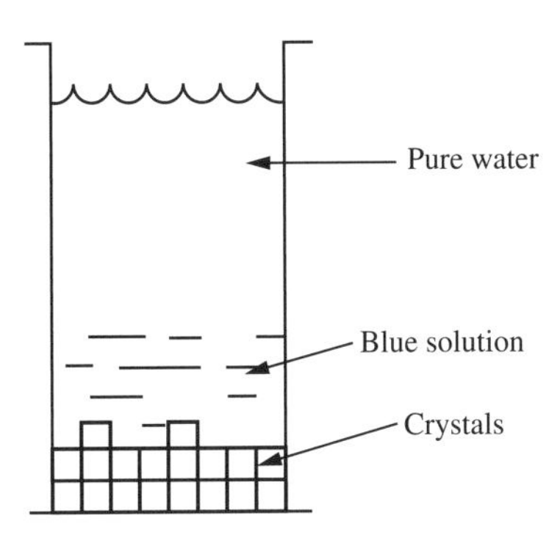

FIGURE 1.7 Copper sulfate crystals dissolving slowly in an unstirred graduate cylinder.

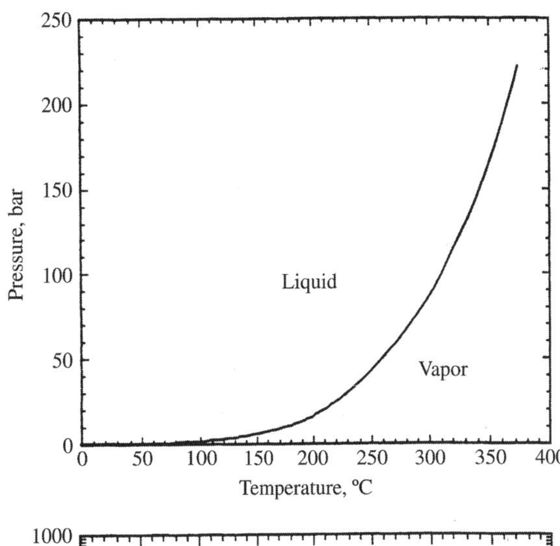

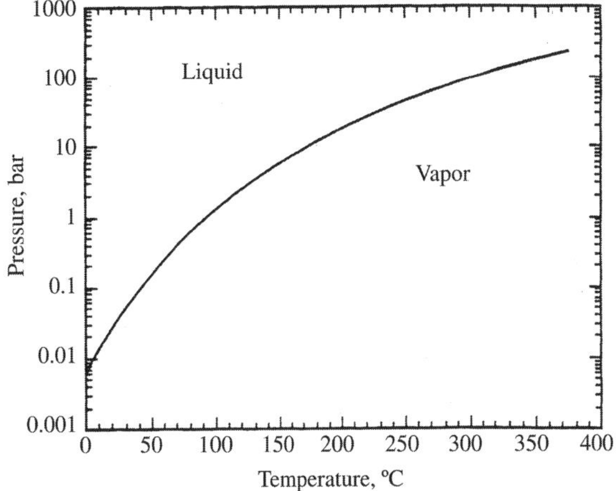

FIGURE 1.8 Vapor liquid equilibrium curve for water–steam, in arithmetic and logarithmic coordinates [6]. The range of values shown is so large that on arithmetic coordinates the low-temperature values disappear into the horizontal axis. On a logarithmic scale they are all visible.

layers in bacon. With a small microscope one can observe the same sort of thing in grass, and with a stronger microscope one can observe it in cast iron. Large numbers of pure solid phases can be in thermodynamic equilibrium with each other; they generally do not mix significantly. (Many metals are *solid solutions,* such as brass, which is a solution of copper and zinc, and bronze, which is a solution of copper and tin. These are formed by melting the metals together, in which state they dissolve each other, and then cooling. Most steels are mostly iron, with some dissolved carbon, and small amounts of other metals.)

Figure 1.7 shows a beaker of water with a layer of $CuSO_4$ crystals on the bottom. The crystals are slowly dissolving and diffusing through the water. The solution is one phase. It has no sharp discontinuities of properties. However, it is not of uniform chemical composition, or uniform density, color, and so forth. It is not an equilibrium phase. If we wait long enough (years!) for diffusion to make it uniform, it will become an equilibrium phase.

Here the $CuSO_4$ crystals are all one phase, although they are not continuous. They lie around in a pile at the bottom of the cylinder, but within any one crystal the properties are uniform and the properties of one crystal are the same as the properties of the next crystal.

Throughout this book we will present many forms of *phase diagram.* A phase diagram is a representation on some set of thermodynamic coordinates (many combinations of such variables are used in phase diagrams) showing which phase we would expect to find at a given set of values of the coordinates. The simplest phase diagram, and the one students are familiar with, is a vapor-pressure curve. Figure 1.8 shows the vapor-pressure curve for water.

In this figure we see that for pure water at combinations of temperature and pressure above and to the left of the vapor-pressure curve, only liquid water can exist. For combinations below and to the right of the curve, only gaseous water (steam or water vapor) can exist. Both can coexist at temperatures and pressures on the line (the vapor-pressure curve or equilibrium curve). Much more complex phase diagrams exist. Figure 1.9 shows the same diagram as Figure 1.8, extended to the left to $-15°C$, and showing only very low pressures. This takes us below the freezing temperature of water, and solid water (ice) appears on the diagram. Instead of two regions (two phases), we have three. But, again, we see that for pure H_2O at any temperature and pressure not on one of the curves, only one phase may occur, either solid (ice), liquid (water), or vapor (water vapor or steam).

If, instead of going to low pressures, we ask how Figure 1.8 looks at high pressures, we discover some surprises. Figure 1.10 shows the same data as in Figures 1.8 and 1.9,

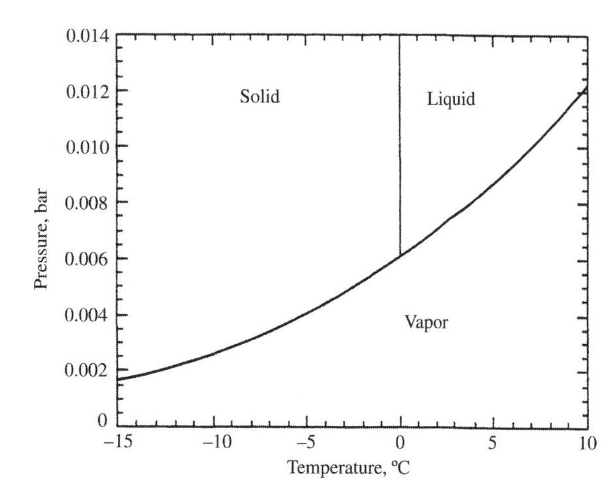

FIGURE 1.9 Extension of Figure 1.8 (arithmetic part only) to temperatures below the normal freezing point of water, showing the formation of ice [5, 6]. This is all at low pressures; the maximum pressure shown is ≈ 0.014 bar. The rightmost curve is the same as part of the curve in Figure 1.8, simply drawn on a much expanded pressure scale.

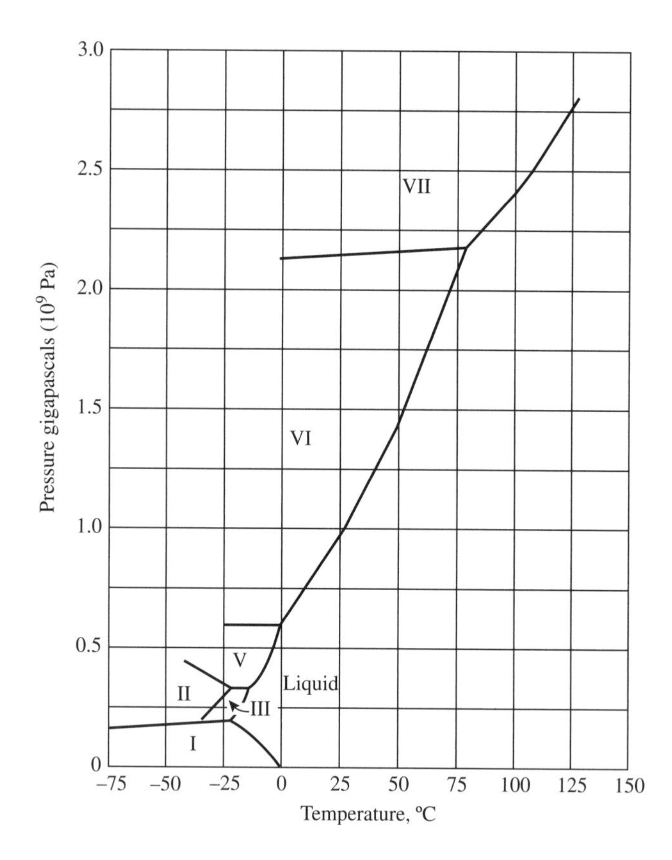

FIGURE 1.10 Phase diagram for water at high pressures, showing the five solid forms that do not exist at normal pressures. The pressures shown are so high that the normal vapor–liquid equilibrium curve (Figure 1.8) disappears into the horizontal axis. (The critical pressure, the highest value on that curve, is 3204 psia = 22.06 MPa = 0.022 GPa. It would barely be visible above that line, and it occurs at 374°C, far to the right of the figure. At 150°C, the vapor pressure, $p = 0.5$ MPa = 0.0005 GPa, indistinguishable from the horizontal axis.) (From Van Wylen, G. J., and R. E. Sonntag. *Fundamentals of Classical Thermodynamics*, ed. 3. © 1985, New York: Wiley, p. 40. Reprinted by permission of John Wiley and Sons.)

but also shows the five other forms of solid water (ice), which exist only at very high pressures. These temperature and pressure combinations are far beyond those near the surface of the earth, so these solid forms exist only in high-pressure research laboratories (at least on this planet).

We have examined the phase diagram for water in more detail than seems needed at this point. However, it is worth your while to study it to see that for even an apparently simple substance like H_2O the range of possible phase behavior is large. Observe that Figures 1.8 and 1.9 are merely expansions of part of the horizontal axis of Figure 1.10. Most of the time we will use Figure 1.8; we will refer to it occasionally in the rest of this book.

1.8 THE PLAN OF THIS BOOK

The first three chapters of the book are an introduction and review of basic thermodynamics and of very simple equilibrium. Chapters 4–7 set out the basic thermodynamics of equilibrium. Chapters 8–10 deal with the most common type of problem, vapor-liquid equilibrium. Chapter 11 deals with other kinds of phase equilibrium. Chapters 12–13 deal with chemical equilibrium, and Chapters 14,15 and 16 deal with a variety of related topics. Appendix A contains the data tables that are used for examples and homework problems. Appendixes B–G contain derivations and other material that supports the material in main text. It is placed there to keep the treatment in the texi as simple as possible. Appendix H contains answers to some of the problems.

1.9 SUMMARY

1. Equilibrium is one of the four basic tools of the chemical or environmental engineer. It is as important as the others, and is needed for a wide variety of engineering work.
2. As we will see later, nature minimizes Gibbs energy. A state of equilibrium is one at which the change of Gibbs energy for any infinitesimal change is zero because the Gibbs energy of the system is the lowest value possible, subject to the external constraints on the system.
3. Equilibriam states are stable, unstable, or neutral. On a molecular level all equilibria are dynamic; that is of

little concern at the level of most engineering problems.

4. We will work mostly with molar units of mass, and mol fractions as concentration units.
5. We will deal with a variety of phase diagrams, of which the vapor-pressure curve is the simplest.

PROBLEMS

See the Common Units and Values for Problems and Examples. An asterisk (*) on the problem number indicates that the answer is in Appendix H.

1.1 List the courses in your university work that correspond to the four basic tools of the chemical and environmental engineer.

1.2 Show the derivation of Eg. 1.D. Start with Eq. 1.C and multiply both numerator and denominator of the fraction by M_{solvent}. Then note the relative magnitudes of the two terms in the denominator.

1.3* Repeat Examples 1.1 and 1.2 for a solution made up of 5 g of sucrose and 995 g of water. The reported density of this solution at 20°C is 1.0178 g/cm^3.

1.4 The sucrose solution in Examples 1.1. and 1.2 is now heated enough that its volume expands to 105% of its volume at 20°C. At this higher temperature what are the values of all the concentration measures in Examples 1.1 and 1.2?

1.5 Sketch the equivalent of Figure 1.10, and on it sketch the average *P-T* curve for the earth. Does it intersect the region in which the high-pressure forms of ice occur? Take the temperature of the earth as 15°C at the surface, increasing with depth by about 30°C/km. The pressure inside the earth (near the surface) increases by about 30 MPa/km.

REFERENCES

1. Smil, V. Global population and the nitrogen cycle. *Sci. Am.* 227 (1):76–81 (July 1997).
2. Hooper, W. C. Ammonia synthesis: commercial practice. In *Catalytic Ammonia Synthesis, Fundamentals and Practice*, Jennings, J. R., ed. New York: Plenum, Chapter 7 (1991).
3. Levenspiel, O., and N. de Nevers. The osmotic pump. *Science* 183:157–160 (1974).
4. Bates, F., F. P. Phelps, and C. F. Snyder. Saccharimetry, the properties of commercial sugars and their solutions. In *International Critical Tables, 2*, Washburn, E. W. ed. New York: McGraw-Hill, p. 343 (1927).
5. Keenan, J. H., F. G. Keyes, P. G. Hill, and J. G. Moore. *Steam Tables: Thermodynamic Properties of Water Including Vapor, Liquid and Solid Phases*. New York: Wiley (1969).
6. Haar, L., J. S. Gallagher, and G. S. Kell. *NBS/NRC Steam Tables*. New York: Hemisphere (1984).
7. Hildebrand, J. H., J. M. Prausnitz, and R. L. Scott. *Regular and Related Solutions; The Solubility of Gases, Liquids and Solids*. New York: Van Nostrand Reinhold (1970).

2

BASIC THERMODYNAMICS

This chapter assumes that the reader has completed a course in basic mechanical engineering (ME) thermodynamics. It presents only a review and summary, to be referred to later in the text. A basic ME thermodynamics class is mostly about devices using pure substances, such as steam power plants, refrigerators, heating systems, and internal combustion engines (treated by the "air standard Otto cycle," which allows one to use pure-substance thermodynamics). Chemical engineering thermodynamics extends that approach to include devices treating mixtures (e.g., all separation processes like distillation or crystallization) and chemical reactors. The principles are the same, but the details and the viewpoints are often different.

In an elementary ME thermodynamics course the emphasis is on applying tables of thermodynamic properties, such as the steam tables, to a variety of processes. In this book the emphasis is on how we determine the values in those tables and how we produce the corresponding tables (or the parts of the tables we need) for mixtures. The subsequent courses in chemical engineering process design rely on the material in this book the same way that an elementary ME thermodynamics course relies on the steam tables.

2.1 CONSERVATION AND ACCOUNTING

Much of engineering is simply careful accounting of things other than money. The accountings are called mass balances, energy balances, component balances, momentum balances, and so on. Any balance begins by choosing some carefully specified region of space, called a *system* or a *control volume*.

The rest of the universe, outside the system is called the *surroundings*, (Figure 2.1). For the system we can list all the ways that the amount of some material, property, or set of individuals can be changed, add them with the proper algebraic signs, and thus have an accounting equation for the system of the form

$$\text{accumulation} = \text{creation} - \text{destruction} + \text{flow in} - \text{flow out} \tag{2.1}$$

This *general balance equation* and its variants form the basis of much of chemical engineering. If the creation and destruction terms are zero, then it is called a *conservation equation*. If they are not zero then Eq. 2.1 has no common name, but is widely used, for example, with chemical reactions, where it allows us to compute the changes in various chemicals as a chemical reaction destroys one species and creates another. Remember that it applies only to a system with properly defined boundaries. All balances can be changed to rate equations by dividing both sides by some time interval, dt:

$$(\text{accumulation rate}) = (\text{creation rate}) - (\text{destruction rate}) + (\text{flow rate in}) - (\text{flow rate out}) \tag{2.2}$$

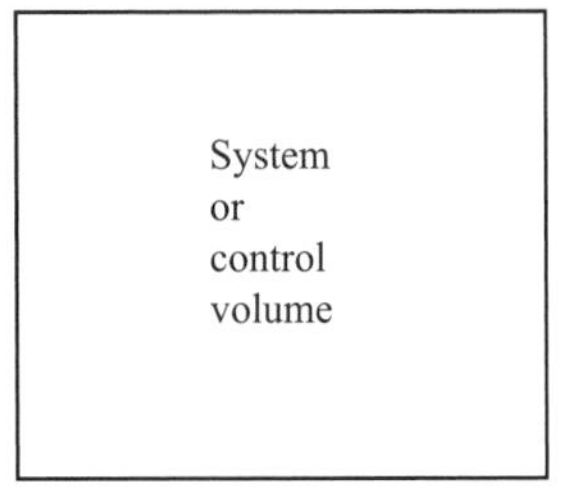

FIGURE 2.1 The system boundaries divide the whole universe into two parts: the system and the surroundings.

One also sees this equation and its later variants in the form

$$\text{(accumulation rate)} = \sum \text{(creation-destruction rates)} + \sum \text{(flow rates in and out)} \tag{2.3}$$

which is the same as Eq. 2.2, but in a mathematically more formal arrangement. In it we assign a positive value to creation and to flows in and negative values to destruction and flows out. This equation allows for multiple creations and destructions (e.g., multiple simultaneous chemical reactions) and multiple flows in and out. Almost every chemical engineering problem involves this equation either explicitly or implicitly. (We normally use the term *system* for some container with zero or a finite number of entrances, and the term *control volume* for some region in space that can have flow in or out at every point on its boundary. The balance equations are the same for either, with the sum of inflow flow terms in Eq. 2.3 replaced by a surface integral for a control volume.)

2.2 CONSERVATION OF MASS

One of the great human discoveries is that *mass* is conserved. According to Einstein's famous $E = mc^2$, there is a small conversion of mass to energy in all energy transformations (for example, your coffee cooling in its cup). This effect is small enough that, except for nuclear weapons or nuclear reactors, we can ignore it and slate as a general principle that mass obeys the general balance equation, with creation = destruction = 0. This is called the *law of conservation of mass,* the *principle of mass conservation,* or the *continuity equation.* There is no known way to derive it from any prior principle; it rests solely on its ability to predict the result of any experiment designed to test it.

Mass can exist in a variety of forms, for example, solid, liquid, gas, and some other bizarre forms, and can convert from one to the other. When liquid water evaporates we see the liquid disappear, but we have no visual evidence that the mass of the surrounding air increased by the mass of the water vapor thus produced. Lavoisier made the first clear statement of the law [1], and demonstrated that if processes similar to the evaporation of water were carried out in a closed glass jar resting on a balance, there was no loss of mass; the visible water had changed to invisible water vapor, but the mass of the contents of the jar did not change. The idea that mass is conserved seems quite obvious to us, but it was not known or believed by the human race before about 1780. The key discovery was that gases had mass, which was not intuitively obvious to scientists or the public before then. For some properly chosen system, we can restate Eq. 2.1 for mass as

$$\begin{pmatrix}\text{accumulation of}\\ \text{mass in the system}\end{pmatrix} = \begin{pmatrix}\text{flow of mass}\\ \text{into the system}\end{pmatrix} - \begin{pmatrix}\text{flow of mass}\\ \text{out of the system}\end{pmatrix} \tag{2.4}$$

In symbols

$$dm_{\text{system}} = dm_{\text{in}} - dm_{\text{out}} \tag{2.5}$$

or

$$\frac{dm_{\text{system}}}{dt} = \frac{dm_{\text{in}}}{dt} - \frac{dm_{\text{out}}}{dt} = \dot{m}_{\text{in}} - \dot{m}_{\text{out}} \tag{2.6}$$

The overdot indicates a flowrate. For systems with only one chemical species we usually use Eqs. 2.5 and 2.6 as written. However, in chemical engineering we very often deal with mixtures and with chemical reactions. For those we usually choose our unit of mass as one *mol* or one *pound mol* (lbmol) (= 454 mol). The relation between mass and mols, referred to often in this book, is given by Eq. 1.5. If we solve that equation for m_i and substitute everywhere in Eqs. 2.5 and 2.6, we find that all the Ms cancel, and we have the same equations for mols, with m_is replaced by n_i,s. However, mols are *not conserved.* For example, in the reaction

$$3H_2 + N_2 \Leftrightarrow 2NH_3 \tag{1.A}$$

the number of mols goes from 4 to 2. So if we write a general balance equation for mols, we must retain the creation and destruction terms, and the resulting equation is not a conservation equation.

We can now summarize the law of conservation of mass: It is an experimental law, not derivable from other laws, but thoroughly confirmed by experiment. It simply states that an abstract quantity called mass is conserved.

Mass obeys the general balance equation with neither creation nor destruction. From that statement we can write a very general mass balance, and then the more widely used simpler forms.

2.3 CONSERVATION OF ENERGY; THE FIRST LAW OF THERMODYNAMICS

Another of the great human discoveries is that *energy* is conserved. The restriction concerning $E = mc^2$ applies to this statement as well, but except for nuclear weapons or nuclear reactors, we can state with almost perfect accuracy as a general principle that energy obeys the general balance equation, with creation = destruction = 0. This is called the *first law of thermodynamics,* the *principle of conservation of energy,* the *energy principle,* or the *energy balance.* There is no known way to derive it from any prior principle; it rests solely on its ability to predict the result of any experiment designed to test it.

Like mass, energy can exist in a variety of forms, which we now call kinetic, potential, internal, electrostatic, magnetic, and surface. Before about 1800 the human race did not know that these were all the same thing in different forms. The principal discoverers of that fact were Mayer, Rumford, and Joule [2]. Like the law of conservation of mass, the law of conservation of energy seems intuitively obvious to us, but it was far from obvious to the scientists or the public before about 1800. Furthermore, there is no satisfactory simple, verbal definition of energy. The definitions can be simple or accurate, but not both. Simple definitions like "the ability to do work or warm things" are useful, but inaccurate or incomplete. The technically accurate definition is that energy is an abstract quantity, which can appear in various forms, which can be converted from one form to another subject to some restrictions, and which appears to be conserved in all energy transactions.

For some properly chosen system, we can restate Eq. 2.1 as

$$\begin{pmatrix}\text{accumulation of}\\ \text{energy in the}\\ \text{system}\end{pmatrix} = \begin{pmatrix}\text{flow of energy}\\ \text{into the system}\end{pmatrix} - \begin{pmatrix}\text{flow of energy}\\ \text{out of the system}\end{pmatrix} \tag{2.7}$$

If we let E stand for energy, then the energy balance, in symbols, it is the same as the mass balance, Eqs. 2.5 and 2.6, with all ms replaced by Es.

Mass can be transferred from one body to another by cutting a piece off of one and gluing it onto another (or pouring a liquid from one container to another), and that mass takes its energy of the above forms with it in such a transfer. In addition, bodies can exchange energy in the form of heat and work, which do not involve any transfer of mass.

In ME thermodynamics and in fluid mechanics, changes in kinetic energy and potential energy are often important, and we normally write

$$E = U + \text{kinetic energy} + \text{potential energy} \tag{2.A}$$

But in most equilibrium problems we can ignore all forms of energy except internal energy (but see Chapter 14!) and state

$$E = \begin{pmatrix}\text{energy of}\\ \text{some mass}\\ \text{of matter}\end{pmatrix} = U + \begin{pmatrix}\text{other forms}\\ \text{we will}\\ \text{ignore in}\\ \text{most}\\ \text{of this book}\end{pmatrix} \approx U = mu \tag{2.8}$$

where u is the *specific internal energy* or *internal energy per unit mass,* with dimensions Btu/lbm or J/kg, and U is the internal energy of some body, the product of the specific internal energy and the mass, with dimensions Btu or J. Intuitively, we may think of the internal energy as the energy due to being hot (relative to some arbitrary datum temperature) and the energy due to being able to cause a heat-releasing chemical reaction. If we ignite a mixture of gasoline and air in a constant-volume, adiabatic container, it will undergo a chemical reaction forming carbon dioxide and water. When the reaction is over (in a few milliseconds), the mixture will be much hotter (have a much higher temperature) than the starting mixture did. But its internal energy will not have changed; it will have converted "potential to undergo a heat-releasing chemical reaction" internal energy to "hotness" internal energy, without changing their algebraic sum. We may also think of this as changing from energy stored in chemical bonds within molecules to energy present as motion of the molecules; the former is "potential to undergo a heat-releasing chemical reaction" energy, the latter "hotness" internal energy. This is only an intuitive approximation, but it is useful.

Tables of thermodynamic properties are always in terms of specific properties (properties per lbm, or kg, or per mol). For systems that involve only one chemical species (e.g., steam power plants, refrigerators, the other systems in ME thermodynamics), the equations and tables are all per unit mass (lbm or kg). However, in chemical engineering thermodynamics, which deals with mixtures and with chemical reactions, we most often choose our unit of mass as one mol or one pound mol.

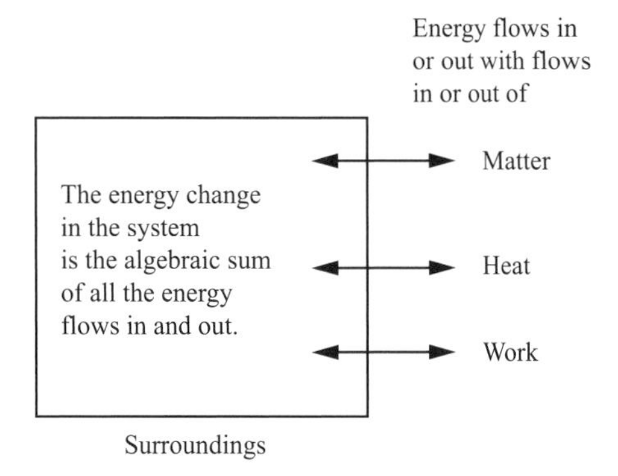

FIGURE 2.2 A pictorial representation of the energy balance.

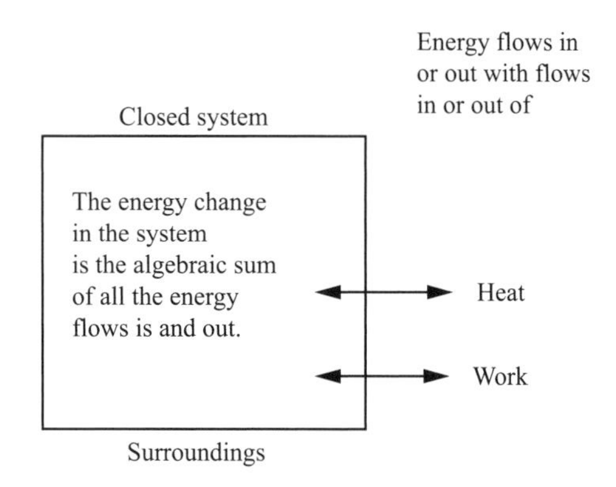

FIGURE 2.3 For a closed system, there is no flow of matter in or out, so the energy in the system can change only by the flow of heat and work.

Writing the balance for energy, using Eq. 2.8 we find

$$d(mu)_{\text{system}} = \sum u_{\text{in or out}}\, d(m)_{\text{in minus out}} + \sum dQ_{\text{in minus out}} + \sum dW_{\text{in minus out}} \tag{2.9}$$

where dQ stands for a heat flow into the system and dW stands for external work done on the system (see Figure 2.2).

Equation 2.9 would be a perfectly satisfactory form, except for a complication in the work term. When some amount of mass dm crosses the system boundary, it requires an amount of work of $Pv_{\text{in}}dm_{\text{in}}$ to force it across the system boundary. This is called the *injection work, flow work, injection energy,* and some other names. If we divide the work term into this injection work, and all other types of work, we can rearrange Eq. 2.9 into

$$d(mu)_{\text{system}} = \sum (u + Pv)dm_{\text{in minus out}} + \sum dQ_{\text{in minus out}} + \sum dW_{\text{in minus out excluding injection work}} \tag{2.10}$$

We normally see this equation with the "excluding injection work" deleted, because it is assumed that the reader knows that. Two special cases of Eq. 2.10 are widely used. If there is no flow of matter in or out, which means that we are considering some *closed system* containing a fixed mass of matter, then Eq. 2.10 becomes

$$dU_{\text{system}} = \sum dQ_{\text{in minus out}} + \sum dW_{\text{in minus out}} \tag{2.11}$$

This is most often written as (Figure 2.3)

$$dU = dQ \pm dW \tag{2.12}$$

The reason for the $\pm$ sign is that we regularly see this equation with a plus or a minus before the dW. In almost all uses of the general balance equation, all flows in arc positive and all flows out are negative. But thermodynamics was originally developed around the steam engine, whose net flows were of heat in and work out. The early thermodynamicists defined heat as positive flowing in and work as positive flowing out. That leads to a $-\ dW$ in Eq 2.12. Recently many thermodynamics authors have decided that this causes students more confusion than it is worth, and assert that we should follow the conventions in the general balance equation, not the historical convention, so they write Eq. 2.12 with a $+\ dW$. This book follows the latter usage, and takes work, heat flow, and mass flow *into* the system as positive, thus writing Eq. 2.12 with a $+\ dW$. The reader should remember that many books define work *leaving* the system as positive and write a minus dW in Eq. 2.12.

Equation 2.12 is often called the *chemist's version of the first law*. It is entirely appropriate for chemical reactions in closed systems, and will be used often in this book. The other widely used simplification of Eq. 2.10 is the steady-state, steady-flow form. We begin by dividing both sides by dt, thus converting it to the rate form. If we apply it to some system or device, for example, the turbine in a steam power plant, with a steady flow in and out, then the $d(mu)_{\text{system}}$ term must be zero, because there is no change with time of the energy content of the system. We remember that $dm/dt = \dot{m}$, so

$$0 = \sum (u + Pv)\dot{m}_{\text{in minus out}} + \sum \dot{Q}_{\text{in minus out}} + \sum \dot{W}_{\text{in minus out excluding injection work}} \tag{2.13}$$

Here $\dot{Q}$ is the heat flow rate, Btu/s or cal/s or (J/s = watt). Logically, $\dot{W}$ should be called the work flow rate, but instead we use its common name, the *power*, expressed in horsepower or watts (horsepower = 33,000 ft · lbf/minute = 0.746 kW). This is the most commonly used form in elementary ME thermodynamics books (Figure 2.4). In those

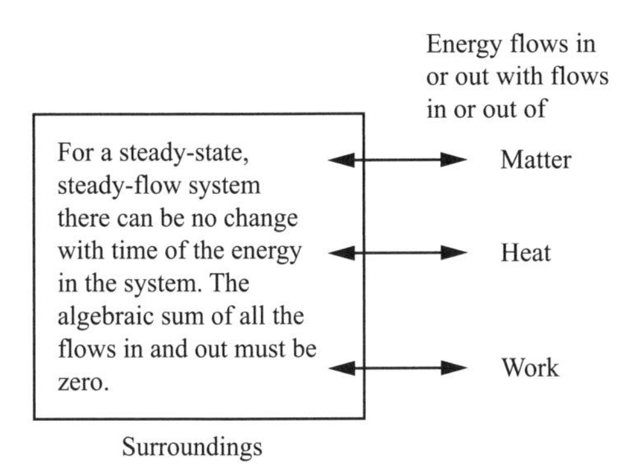

FIGURE 2.4 For a steady-state, steady-flow system we can have all three kinds of energy flow in and out, but their algebraic sum must be zero.

applications, kinetic and potential energies are often important, so that the $(u + Pv)$ term also includes terms for kinetic and potential energies.

We can now summarize the first law: It is an experimental law, not derivable from other laws, but thoroughly confirmed by experiment. It simply says that an abstract quantity called energy is conserved; energy obeys the general balance equation with neither creation nor destruction. Like mass, energy can exist in a variety of forms, and be converted from one to the other. We can write a very general energy balance, and then write the more-often-used restricted forms for closed systems or for steady-flow, steady-state systems, (and some other forms). It is used in some way in almost every chemical engineering problem.

2.4 THE SECOND LAW OF THERMODYNAMICS

The second law of thermodynamics is the most intuitively obvious of all the laws of nature. It states that processes that occur spontaneously in one direction do not occur spontaneously in the opposite direction.

2.4.1 Reversibility

Consider the following processes:

1. A drinking glass is dropped on a concrete floor and shatters.
2. A firecracker explodes.
3. An inflated toy balloon is released; the air rushes out while the balloon flies erratically around the room.
4. A baby is born, grows to be an adult, ages, and dies.

Will any of these processes ever run in the opposite direction without outside intervention? We all know that they won't; that is the universal result of human experience. They can be made to appear to run backward by making a movie of them and running it backward, but most adults will know that in these cases the film is being run backward. Most of us laugh at a movie run backward; we recognize it as "funny." This shows that our intuition supports the statement that in nature processes generally only occur in one direction.

There are other kinds of processes, in which we can run the movie backward or forward and not easily be able to tell whether it is running forward or backward; for example, a pendulum swinging, a ball rolling across a hard surface, a counterweighted garage door opening or closing, an elevator going up or down, or a gas expanding by driving down a piston and thereby raising a weight. In each of these cases, if we watched long enough, we could tell if the movie were running backward or forward because of friction. (Air resistance ultimately stops a pendulum unless it is driven by some outside agency; rolling friction eventually stops a ball, etc.) But, if there were no friction, we could never know whether these were going forward or backward. If a process will run either forward or backward without significant outside intervention, we call it a *reversible process.* There are no completely reversible processes in nature; but, for example, a pendulum swinging in a vacuum is practically reversible.

On the other hand, processes like the exploding firecracker or the shattering glass are highly irreversible. They cannot be made to go in the reverse direction without gigantic outside intervention. There exist in nature all kinds of intermediate behavior between the two extremes listed. So we normally speak of a degree of irreversibility, with processes like the swinging pendulum being slightly irreversible and processes like the exploding firecracker being very irreversible.

Why won't irreversible processes work equally well in the opposite direction? The "laws of nature" that we have previously discussed would be perfectly well satisfied if they did. For example, the exploding firecracker could go in the opposite direction without violating the principles of conservation of mass, conservation of energy, Newton's laws, the electromagnetic laws, and so on. Similar comments apply to the air going back into the balloon or to the shattered glass reassembling itself. Thus, these other laws do not tell us in which direction these processes occur.

Yet this observation of the one-way character of spontaneous natural processes is so universal that it is one of the most basic observations of nature. It has been given a name, *the second law of thermodynamics*, although, historically, it partially precedes the first. Those most important in formulating it were Carnot, Clausius, and Kelvin [2]. Like the first law, Newton's laws, or the law of conservation of matter, it cannot be derived from any more basic law; rather, it rests on its ability to explain all the observations ever made to test it. This law appears in many forms and has very far-reaching consequences. Many scientists believe that the second law is the most fundamental of all the laws of nature.

To quote A. S. Eddington [3]:

> The law lhat entropy always increases—the second law of thermodynamics—holds, I think, the supreme position among the laws of Nature. If someone points out to you that your pet theory of the universe is in disagreement with Maxwell's equations—then so much the worse for Maxwell's equations. If it is found to be contradicted by observation—well, these experimentalists do bungle things sometimes. But if your theory is found to be against the second law of thermodynamics, I can give you no hope; there is nothing for it but to collapse in deepest humiliation.

2.4.2 Entropy

While the above statements are intuitively satisfying, for calculational purposes we need a mathematical statement. Clausius provided that by defining a new quantity, which he named *entropy*. It is defined as a property of some mass of matter that, *for reversible processes in a closed system*, obeys Eq. 2.14:

$$(dS = m\,ds)_{\text{reversible processes in a closed system}} = \frac{dQ}{T} \quad (2.14)$$

Here the T is the absolute temperature (K or °R, never °F or °C) s is the specific entropy, and $ms = S$. Later workers added to Clausius' definition by showing that for irreversible processes there is always an entropy increase, which we may simply call $dS_{\text{irreversible}}$. In terms of the general balance equation, this is a creation term; entropy increases in the system without a corresponding entropy decrease in the surroundings. Using this idea, we can write the general balance equation for entropy as (Figure 2.5)

$$d(ms)_{\text{system}} = \sum (s\,dm)_{\text{in minus out}} + \sum \left(\frac{dQ}{T}\right)_{\text{in minus out}} + dS_{\text{irreversibile}} \quad (2.15)$$

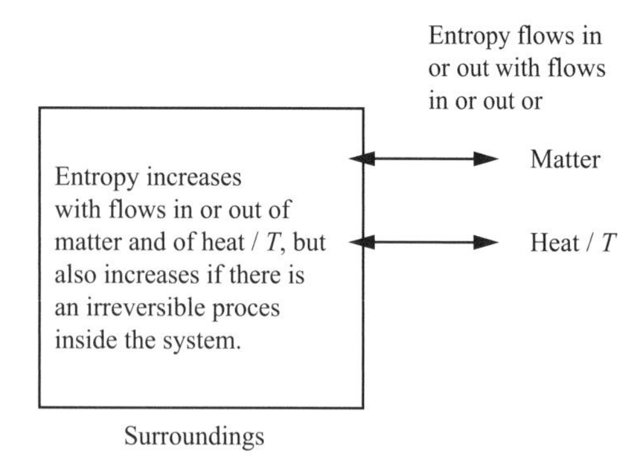

Entropy flows in or out with flows in or out or
Matter
Heat / T
Entropy increases with flows in or out of matter and of heat / T, but also increases if there is an irreversible proces inside the system.
Surroundings

FIGURE 2.5 The entropy balance has no term for work flow, but does have a term for entropy creation inside the system by any irreversible process.

Comparing Eq. 2.15 to our treatment of energy we see the following:

1. This is an accounting or balance equation, but not a conservation equation. Mass and energy are conserved, entropy is not. Clausius made that distinction clear in his famous formulation, "The energy of the universe is constant; the entropy of the universe increases toward a maximum," which irreverent students have rephrased as "You can't win; you can't even break even!"
2. If *we restrict our attention to reversible processes*, then entropy is conserved and this becomes a conservation equation. The increase of entropy in our system due to flow of matter in or out and heat flow in or out is exactly balanced by the decrease in entropy of the surroundings from which matter and heat flow into or out of our system.
3. There are no terms involving work, If we do work on the system or have it do work *in a reversible way*, then its entropy will not change. This leads to the most common use of the second law; for a reversible device (e.g., a steam turbine) consuming or producing work, the change in entropy (batch process) or the change of entropy of the streams flowing through (steady-state, steady-flow process) is zero. We may use Eq. 2.15 or its variants to find final or the outlet state of such a reversible device. The common definitions of the efficiencies of real devices are the ratios of the work produced or consumed by the real device to that of a reversible device with the same inputs.
4. We have *not* ignored other forms of entropy, as we did when we left kinetic, potential, electrostatic, electromagnetic, and surface energies out of Eq. 2.8. There is only one form of entropy.
5. By dropping the appropriate terms we can easily make up the chemist's (closed system) form of the second law,

$$dS_{\text{system}} = (m\,ds)_{\text{system}} = \sum \left(\frac{dQ}{T}\right)_{\text{in minus out}} + dS_{\text{irreversibile}} \quad (2.16)$$

and the steady-state, steady-flow form

$$0 = \sum \left(\frac{\dot{Q}}{T}\right)_{\text{in minus out}} + \sum (s\dot{m})_{\text{in minus out}} + \frac{dS_{\text{irreversible}}}{dt} \quad (2.17)$$

As with energy there is no satisfactory simple, verbal definition of entropy. The definitions can be simple or accurate, but not both. The simple definitions like "a measure of disorder, or of the uselessness of energy" are helpful but inaccurate or incomplete. The technically accurate definition is that entropy is an abstract quantity, equal to dQ/T for reversible heat transfer, which is conserved in all reversible energy transactions, and which increases in all irreversible energy transactions.

2.5 CONVENIENCE PROPERTIES

All of thermodynamics could be worked out in terms of P, T, m, v, u, and s; some purists do it that way. However, some combinations of these variables occur together so often that we can save writing and calculations by defining *convenience properties*, which are combinations of these basic variables. The most widely seen of these are

$$\text{enthalpy per unit mass} = h = u + Pv \tag{2.18}$$

$$\text{Gibbs energy per unit mass} = g = u + Pv - Ts = h - Ts \tag{2.19}$$

$$\text{Helmholz energy per unit mass} = a = u - Ts \tag{2.20}$$

In Chapter 7 we will encounter some more convenience properties. If we multiply each of these convenience properties by the mass we will find $H = mh$, and so on. Chemistry books often use f and F for the Gibbs energy, where chemical engineers use g and G.

One almost always sees Eq. 2.10 rewritten with the enthalpy substituted for $u + Pv$,

$$d(mu)_{\text{system}} = \sum (h_{\text{in}} dq)_{\text{in minus out}} + \sum dQ_{\text{in minus out}} + \sum dW_{\text{in minus out excluding injection work}} \tag{2.21}$$

and the common tables of thermodynamic properties all contain h, while only a few contain u, because in common engineering problems u appears most often only as part of the enthalpy.

2.6 USING THE FIRST AND SECOND LAWS

The typical problems addressed in an elementary ME thermodynamics course ask the following questions:

1. How much heat and/or work must be transferred into or out of some batch or flow system as the material in it goes from an initial to a final state?
2. If the amount of heat and/or work is specified, what is the final or outlet state of the matter?

If you have a table of properties of the material in the system, you can solve such problems using the equations in the previous section. Such tables are available (and included as appendices in all elementary thermodynamics books) for commonly used materials, such as steam, common refrigerants, and low molecular weight hydrocarbons. A typical table is made up of several parts, with the description shown in Table 2.1.

Figure 2.6 shows on a P-T diagram the regions described in each of the five subtables in a complete table of thermodynamic properties.

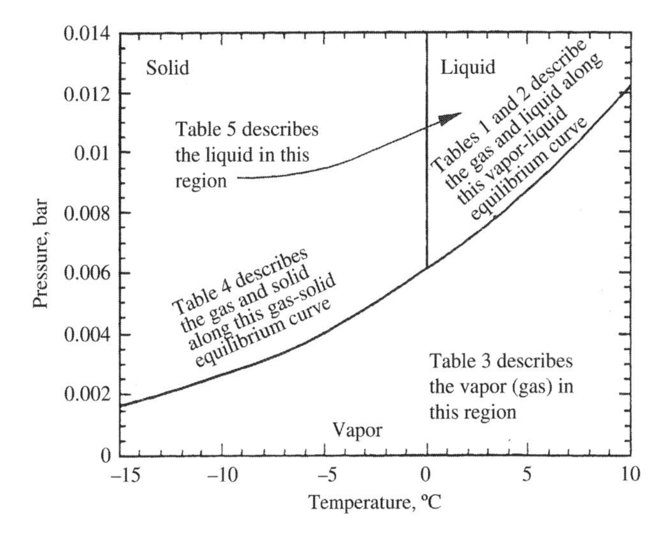

FIGURE 2.6 The regions covered by the five subtables in a set of thermodynamic tables. This is a reprint of Figure 1.9, which shows the actual data for water.

Table 2.1 Contents of a Typical Table of Thermodynamic Properties

Part of Table	Properties Shown	Found in
1. Vapor–liquid saturation at even values of T	P, v, h, s, of both vapor and liquid, sometimes u, sometimes f/P	All tables
2. Vapor–liquid saturation at even values of P	T, v, h. s, of both vapor and liquid, sometimes u, sometimes f/P	Some tables
3. Superheated vapor at even values of P and T	v, h, s, sometimes u, sometimes f/P	All tables
4. Gas–solid equilibrium at even values of T	P, v, h, s, of both vapor and solid, sometimes u, sometimes f/P	A few tables
5. Compressed liquid values at even values of P and T	v, h, s, sometimes f/P	A few tables

Because of its economic importance, we have studied water and steam more thoroughly than almost any other substance. The common steam tables [4] have all five of the parts shown in Table 2.1. Refrigerant tables normally show only the first and third; some show the first three. The common light hydrocarbon tables [5] shown only the first and third. These tables are all organized with T and P as the independent variables; one enters at a known T and P and looks up the other values (v, h, s, f/P) one needs. There is no thermodynamic reason why some other pair of variables, such as s and v, could not have been chosen as independent variables. The users find tables with T and P as the independent variables much more convenient; that choice is practically universal. (Before we had computers, the information in tables like the steam tables was also often presented in chart form. The most popular charts were T-s, h-s, P-h, and h-T. Those add some intuitive insight to problems, but are used much less often now that we all have computers.)

Example 2.1 One lbm of steam at 300°F and 14.7 psia is contained in the piston and cylinder device shown in Figure 2.7. The surrounding atmosphere is at 14.7 psia, and the piston is frictionless and weightless, so that the pressure in the container is always 14.7 psia.

We now add 50 Btu of energy as heat to the system. What are the increases in internal energy, enthalpy, and entropy of the steam? How much work is done by the piston, expanding against the atmosphere?

This is a closed system, for which Eqs. 2.11 and 2.16 are applicable. From the steam tables [4] we look up the properties of steam at 300°F and 14.7 psia, finding $u = 1109.6$ Btu/lbm, $h = 1192.6$ Btu/lbm, and $s = 1.8157$ Btu/(1bm · °R).

The work here is done by the system, equal to

$$-dW = PA_{\text{piston}}dx = P\,dV = Pm\,dv \qquad (2.B)$$

If we substitute this in Eq 2.11 and rearrange we find

$$m\,d(u + Pv) = m\,dh = dQ \qquad (2.C)$$

from which we can solve for the final specific enthalpy,

$$h_{\text{final}} = h_{\text{initial}} + dQ = 1192.6 + 50 = 1242.6\ \text{Btu/lbm} \qquad (2.D)$$

Next we look in the steam tables at 1 atm for the temperature at which $h = 1242.6$, finding that we must interpolate between 400 and 420°F. By linear interpolation we find at $h = 1242.6$ Btu/1bm and $P = 1$ atm, that $T = 405.7$°F, $u =$ 1147.7 Btu/1bm, and $s = 1.8772$ Btu/(1bm · °R). The increases in internal energy, enthalpy, and entropy are, respectively, 38.07 Btu/lbm, 50.00 Btu/lbm, and 0.0615 Btu/(1bm · °R). The work done on the atmosphere (which is generally useless) is the difference between the enthalpy increase and the internal energy increase, or 11.93 Btu/lbm. We can perform internal checks by looking up the values of v at the initial and final states, and computing the work by Eq. 2.B, finding the same result, and can compute the entropy change by dividing the heat inflow by the average of the (Rankine!) temperatures between start and finish, also finding the same value shown here. ■

Example 2.2 A reversible, adiabatic, steady-state, steady-flow steam turbine (see Figure 2.8) has input steam at 600°F and 200 psia. The exit pressure is 50 psia. What is the work output of the turbine in Btu/lbm of steam?

Because this is a steady-state, steady-flow process, we use Eqs. 2.21 and 2.17. Solving Eq. 2.21, we find

$$(\text{work per pound}) = \frac{\dot{w}}{\dot{m}} = -(h_{\text{in}} - h_{\text{out}}) \qquad (2.E)$$

(Remember the sign convention; here this is the work into the system. The work we are interested in is that leaving the system, which is negative according to this sign convention.)

From the steam table we can read the inlet enthalpy and entropy as 1322.1 Btu/1bm and 1.6767 Btu/(1bm · °R.) To solve Eq. 2.E we need the value of the outlet enthalpy. We know the outlet pressure (50 psia) but not the outlet

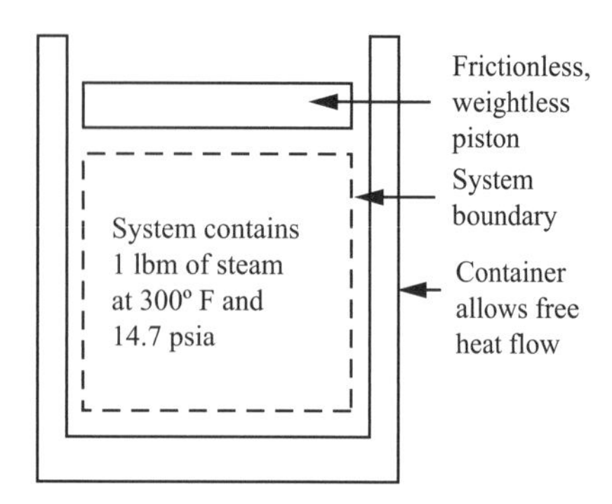

FIGURE 2.7 Piston and cylinder system for Example 2.1.

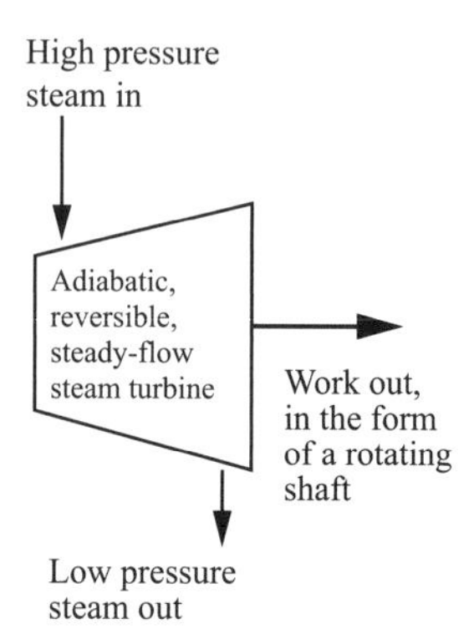

FIGURE 2.8 Flow diagram for Example 2.2.

temperature. However, from Eq. 2.17 we know that for a reversible adiabatic steady-state, steady-flow process

$$0 = \sum s\dot{m}_{\text{in minus out}} = (s_{\text{in}} - s_{\text{out}})\dot{m} \qquad (2.F)$$

which requires that the inlet and outlet entropies must be the same. Thus, we can find the outlet temperature by finding the value of the temperature in the steam table, for which the entropy at 50 psia is the same as the inlet entropy, 1.6767 Btu/(lbm·°R). By linear interpolation in the table we find $T_{\text{out}} = 307.1$ °R, and by linear interpolation in the same table we find that $h_{\text{out}} = 1188.1$ Btu/1bm. Substituting these values into Eq. 2.E, we find

$$\begin{aligned}(\text{work per pound}) &= -(h_{\text{in}} - h_{\text{out}}) = (1322.1 - 1188.1) \\ &= -134\ \text{Btu/1bm}\end{aligned} \qquad (2.G)$$

This is work leaving the system (which this system is designed to produce). In the traditional sign convention it would have a positive sign, but in the current sign convention it has a minus sign. The entropy does not appear in our final answer. We used it only to determine the outlet temperature, so that we could find the proper value to use in our first law statement. That is the normal situation with the second law; we don't really care what the entropy is, except to help us find the right value of the enthalpy or some other property to use in our energy balance. We use the energy balance to find the useful numbers, in this case how much work we can sell to our customers. ■

Both of these examples used h, not u. That is true for most but not all simple energy balance problems. It is not true, for example, for the combustion process in the piston and cylinder of your auto engine.

These two examples show that if we have the steam tables or their equivalent, then all first and second law problems for pure substances are straightforward. Industrial problems usually have more parts than these examples, but none of the individual steps is more complex than these two examples. In complex problems the hard part is choosing the right system boundaries and making sure that no terms in the balances have been omitted. If we had tables like the steam tables for all possible substances we could work all single-species thermodynamic problems from them. But we have such detailed tables for only a few substances, and chemical engineers work with a wide variety of substances. We often synthesize new substances whose thermodynamic properties are unknown. For these we must combine experimental data (if available) plus estimating methods to make up the equivalents of the parts of the steam tables we need. Naturally, we recognize that such estimated values are less reliable than the high-quality values in the steam tables.

Furthermore, this book is about equilibrium, mostly for mixtures. The steam tables and other such tables are for pure substances. Few if any such detailed tables exist for mixtures. We will see that much of the rest of this book is devoted to ways of measuring and/or estimating the thermodynamic properties of mixtures. Once we know those properties, we can carry out the material and energy balances on which to base our chemical plant designs. These estimating methods generally begin with the assumption that we have or can estimate the properties of the individual pure substances that make up the mixtures. The next few sections describe where pure substance tables like the steam tables come from, and how we would make up such a table (or the part we need of it) for some new substance.

2.7 DATUMS AND REFERENCE STATES

Equations 2.9 and 2.14 define u and s only in terms of their *changes*. To assign a numerical value to u or h and s in the steam tables or comparable tables for other substances, the general procedure is to assign an arbitrarily selected numerical value to u or h and to s in some specified state, called the *datum state* or *reference state*, and then to calculate all the other values in the table by calculating the change in u or h and s in going from the datum state to some other state in the table using Eqs. 2.9 and 2.14 (plus some other relations based on them). This may seem arbitrary, but in all the problems normally assigned in an elementary thermodynamics course we are concerned only with *changes* in energy or entropy, so this procedure (and the steam or other tables based on it) work perfectly well.

The *third law of thermodynamics* (see Appendix E) shows us that at 0.00 K all perfectly regular crystalline substances have the same value of the entropy, which we assign a datum value of $s = 0.00$. Entropies based on this value are called *absolute entropies*. They are used only in the study of chemical equilibrium (Chapter 12). The datum chosen for the steam tables is $u = s = 0.00$ for the saturated liquid at the triple point (solid, liquid, and gas in equilibrium at 32.018°F and 0.08866 psia). For refrigerants the common choice is $h = s = 0.00$ for the saturated liquid at −40°F = −40°C. The light hydrocarbon tables choose $h = s = 0.00$ for the elements at 0.00 K, which is one of the common reference states for chemical reaction calculations. There are a variety of choices of datums, all of which seem like the right choice for some class of substances or problems. These datum or reference state values can be used in Eqs. 2.9 and 2.14 to calculate those tables, by methods shown below. There is no logical or thermodynamic reason why the values in the datum state must be chosen as 0.00. They could just as well be chosen as 23.7 or $\sqrt{\pi}$ or the first three digits of your social security number. There is no thermodynamic reason why we cannot choose the datum for h at one condition and that for s at another. For most tables the choice of $h = s = 0.00$, both at the same datum state, is most convenient. Absolute entropies are all positive or zero. Steam table entropies can be positive or negative. The steam table entropy of ice at 32°F = −0.292 Btu/(lbm·°R).

In later chapters we will see other datum or reference states that are chosen for physical or chemical equilibrium calculations.

2.8 MEASURABLE AND IMMEASURABLE PROPERTIES

The quantities in thermodynamics that we can directly measure with simple instruments are pressure P, temperature T, mass m, and volume V. Although we can argue about whether these are direct (in a thermocouple we measure an emf and use a table to convert that value to a temperature), all of these values are measurable without recourse to thermodynamic calculations. We commonly combine m and V to get $v = V/m$ or its reciprocal, density $= \rho = 1/v = m/V$. Such measurements are normally called PvT measurements.

There is no known direct measurement of u or s (or the properties derived from them, h, g, and a, or the other convenience properties we will define in Chapter 7). These must be calculated from PvT (and heat capacity, see below) measurements. (In so doing we often rely on electrical measurements in which we measure heat flow rates as the product of voltage and current.) Thus, all the values you will ever see of s and u are based at least in part on calculations, based on the above four measurable variables.

2.9 WORK AND HEAT

The defining equation for changes in energy involves both work and heat, and that for entropy involves heat. We have no direct measurements for either of these quantities. However, from mechanics we can show that all work is equivalent to the product of a force and a distance. Forces and distances can be measured more or less directly. For the simple case of a system expanding by driving back a piston the force is the pressure times the piston area so that

$$-dW = F\,dx = (PA\,dx = PdV)_{\text{for a simple piston process}} \tag{2.22}$$

We can easily show the equivalent of this for rotating shafts or electrical or magnetic work. In the English engineering system the unit of work is the ft · lbf, and in SI the N · m = J. In the SI system the electrical units were chosen so that N · m = J = V · C = W · s = several other combinations. In the English engineering system there is no comparable simplification.

For heat there is no comparably simple, mechanical way of defining a unit quantity. The historic choice was to measure the amount of heat needed to raise the temperature of 1 unit mass of water by 1 degree and base the unit on that. The results of that are

$$(\text{British thermal unit}) = \text{Btu} = \begin{pmatrix}\text{heat requred to heat}\\ \text{1 lbm of water by 1°F}\end{pmatrix} \tag{2.23}$$

and

$$(\text{calorie}) = \text{cal} = (\text{heat required to heat 1 g of water by 1°C}) \tag{2.24}$$

The calorie is an unpractically small engineering unit; we normally use the kilocalorie (kcal) = 1000 cal. (The "calorie" in diet books is the kilocalorie. A normal adult doing moderate work needs to eat about 2500 kcal of food per day.) The Btu is also impracttcally small for industrial size equipment; we often use 10^6 Btu as the working unit. (In 2012 the world wholesale price of natural gas was about $\$4/10^6$ Btu, that of coal about $\$2/10^6$ Btu.)

One of the final steps in establishing the first taw of thermodynamics was Joule's measurements of the *mechanical equivalent of heat*. He used falling weights to drive paddles around in an insulated tank of water and measured the temperature increase (Figure 2.9). From the known mechanical work and mass and temperature rise of the water he found what was very close to the currently accepted value

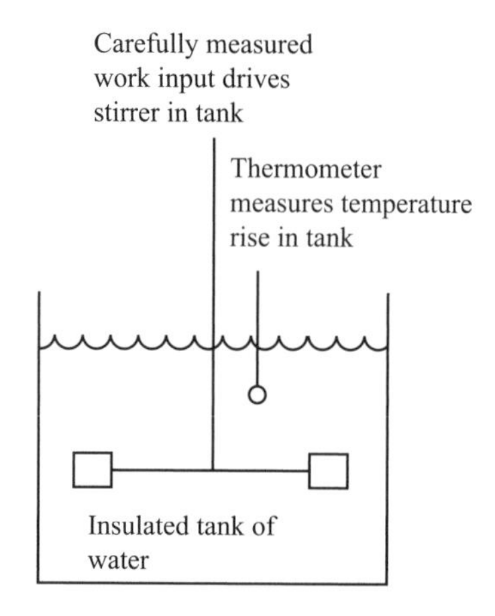

FIGURE 2.9 Schematic of Joule's calorimeter for measuring the mechanical equivalent of heat. All calorimeters are variations on this idea.

of the relation between heat and work,

$$1 \text{ Btu} = 778.17 \approx 778 \text{ ft} \cdot \text{lbf} \tag{2.25}$$

The corresponding SI statement is

$$1 \text{ cal} = 4.184 \text{ J} \quad 1 \text{ kcal} = 4.184 \text{ kJ} \tag{2.26}$$

In SI the use of the calories or kilocalories is discouraged; all heat quantities should be stated in joules or kilojoules. In current practice the calorie and kilocalorie are widely used because of their high intuitive content (and the reluctance of people to change).

Joule's falling-weight-paddlewheel–tank device was one of the first *calorimeters*, devices for measuring heat quantities by measuring the temperature increase of a known mass of some reference substance, almost often water. Most of our data for the changes of u and s with changes in temperature are based on measurements made in more refined versions of Joule's calorimeter. We most often report such information in terms of the heat capacity,

$$(\text{heat capacity}) = C = \begin{pmatrix} \text{energy required} \\ \text{to raise one unit mass} \\ \text{of the substance} \\ \text{by 1 degree} \end{pmatrix} = \frac{Q}{m\Delta T} \tag{2.27}$$

where the experimental measurements are Q, m, and ΔT. The units of heat capacity are Btu/(lbm·°F), kcal/(kg.°C) or J/(kg·°C) Because of its use as a reference substance

$$\begin{pmatrix} \text{heat capacity} \\ \text{of liquid} \\ \text{water is} \\ \text{defined as} \end{pmatrix} = C_{\text{water}} = 1.00 \frac{\text{Btu}}{\text{lbm} \cdot °\text{F}} = 1.00 \frac{\text{kcal}}{\text{kg} \cdot °\text{C}} = 1.00 \frac{\text{cal}}{\text{g} \cdot °\text{C}} \tag{2.28}$$

There are several different definitions of the heat capacity. For gases we regularly use the heat capacity at constant pressure C_P and also the heat capacity at constant volume C_V. These are significantly different from each other for gases, as discussed in any elementary thermodynamics book. For liquids and solids they are practically the same.

The same calorimeters, in somewhat modified form, are used to measure latent heats of phase transformations, heats of mixing, heats of chemical reactions, and some other thermochemical quantities.

The process of making up a table of thermodynamic properties is as shown conceptually in Figure 2.10. For some substance in some state (liquid, gas, or solid) the values of h and s in the datum state are known (normally = 0.00). We then compute along a constant pressure path to the desired final temperature, and then along a constant temperature path to the final pressure, thus computing the values of h and s at the desired T and P. Other paths are possible and are used for special circumstances, but the path shown is almost always the most convenient.

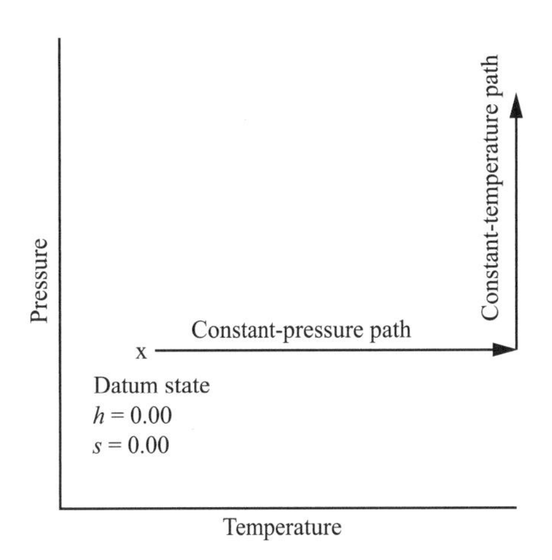

FIGURE 2.10 Conceptual procedure for making up a steam table.

For the constant pressure path we apply Eqs. 2.21 and 2.16 *at constant pressure* to find

$$h_{\substack{\text{at some } T \text{ and} \\ P = P_{\text{datum}}}} = h_{\substack{\text{at } T = T_{\text{datum}} \text{ and} \\ P = P_{\text{datum}}}} + \int_{T_{\text{datum}}}^{T} C_p \, dT \tag{2.29}$$

and

$$s_{\substack{\text{at some } T \text{ and} \\ P = P_{\text{datum}}}} = s_{\substack{\text{at } T = T_{\text{datum}} \text{ and} \\ P = P_{\text{datum}}}} + \int_{T_{\text{datum}}}^{T} \frac{C_P}{T} \, dT \tag{2.30}$$

For the normal choice of datum states, the first term on the right of these two equations is 0.00. The integrations are normally performed by fitting some simple algebraic relation, normally a power series, to the experimental C_p data, substituting that function in these equations, and integrating on a computer or spreadsheet. Table A.9 shows such relations for common gases in the ideal gas state, based on simply fitting the constants to experimental data.

2.10 THE PROPERTY EQUATION

To calculate the changes in h and s with changes in pressure along the vertical (constant temperature) path in Figure 2.10, we need the derivatives of h and s with respect to P at constant

T. To find these we first solve Eq. 2.14 for dQ, and substitute it into Eq. 2.12:

$$(du = T\,ds + d\,W)_{\text{reversible}} \tag{2.31}$$

This is true for any kind of work or the sum of several kinds of work. If we next restrict ourselves to work of expansion, then $dW = -PdV$. The minus sign appears because if the volume increases, then work is done on the surroundings and thus flows out of the system. (In Chapter 14 we will repeat the treatment here, taking more work forms into account.) Making this substitution, we find

$$(du = Tds - Pdv)_{\text{reversible}} \tag{2.32}$$

We show this as reversible, because we derived it that way. But various texts show that thermodynamic properties are *state functions*, because they do not depend on the route used to reach them. (The elevation on top of Mt. Everest is the same, no matter which of the three common routes are used to get there; the elevation above sea level, like thermodynamic properties, is a state function.) So we may drop this subscript, and recognize Eq. 2.32 as the *property equation*. We base most calculations of thermodynamic properties on it. (In Chapter 6 we will see an expanded form of Eq. 2.32.)

This equation shows ds and dv as independent variables. Many thermodynamicists choose s and v as independent variables for all calculations. However, there is no direct measurement of s, and v is more difficult to measure than are T and P, so for engineering work it is much more practical to take the readily measured P and T as independent variables. Some calculus and algebra, shown in most elementary thermodynamics books, allows us to make up Table 2.2 showing the derivatives of the main thermodynamic properties as functions of P and T.

From this table it is clear that we can work out all the needed derivatives if we have data on C_V and C_P and some way to evaluate the partial derivatives of v with respect to P and T. Values of C_V and C_P are measured in calorimeters. The derivatives of v are found from an equation of state. (Most modem equations of state are easy to solve for the derivatives of P and T with respect to v, but not for the v derivatives. For these equations we use some circuitous mathematical routes to provide the exact equivalents of the equations shown above.)

Table 2.2 Thermodynamic Properties in Terms of P and T

These relations can all be derived starting with the property equation (Eq. 2.32), and the definitions *of* h, g, a, C_P and C_V. The derivations are shown in many thermodynamics books and form a favorite exercise in calculus and algebra for graduate students. All 168 of the possible relations between the variables u, h, s, g, a, v, P and T can be worked out quickly and easily using a Bridgman Table, Appendix G, (thus missing out on all that fun calculus and algebra). The following five equations allow us to compute changes of u, h, s, g, and a with changes in T and P, the most common practical property calculations (but see Example G.2 and Problem G.3).

$$du = [C_P - P(\partial v/\partial T)_P]dT - [T(\partial v/\partial T)_P + P(\partial v/\partial P)_T]dP \tag{2.33}$$

$$dh = C_P dT + [v - T(\partial v/\partial T)_P]dP \tag{2.34}$$

$$ds = C_P/TdT - (\partial v/\partial T)_P dP \tag{2.35}$$

$$dg = -sdT + vdP \tag{2.36}$$

$$da = -[s + P(\partial v/\partial T)_P]dT - P(\partial v/\partial P)_T dP \tag{2.37}$$

2.11 EQUATIONS OF STATE (EOS)

An equation of state (EOS) is a mathematical relation between the absolute pressure P, the absolute temperature T and the specific or molar volume v of a pure substance or mixture. Mostly we use the term EOS to describe gases, liquids at high temperatures, or mixtures of the two. For solids or liquids well below their critical temperatures we use very different EOSs than those shown here (see Appendix D).

For a pure substance (like water) in the gas state we have perfectly satisfactory EOSs. For mixtures, the problem is harder, and we have less confidence in our EOSs. We use EOSs to calculate any one of the above variables when the other two are known, and to construct the partial derivatives of v with respect to P or T, which are needed for the calculations in Table 2.2. All EOSs are attempts to replace some table of experimental PvT data with an equation, which we can then use to interpolate and extrapolate between and beyond the values in the table, and which we can use to construct derivatives by simple mathematics. As a general proposition, simple EOSs can reproduce low-pressure gas PvT data with fair accuracy, but as the pressure becomes higher we require more and more complex EOSs to have the equation match the data.

A few of the simplest EOSs are based on theory (or had theory found for them after their utility was shown). The more complex EOSs start with the simple EOSs and add terms that have no theoretical basis at all, but with which they can match the experimental data to higher and higher pressures. We would all like one EOS that represented the liquid, the gas, the solid, and the two-phase or three-phase mixtures of gas, liquid, and solid. In principle, it should be possible to devise such an EOS, but none has been found so far. However, for making up tables like the steam tables, EOSs have been found that describe both the liquid and the gas to within the uncertainties of the best experimental PvT measurements. These EOSs also describe the two phase regions, but their values there do not correspond to reality (see Chapter 10). We will also see that simpler forms of these EOSs are widely used in vapor-liquid equilibrium calculations.

2.11.1 EOSs Based on Theory

The simplest and most widely used EOS is the ideal gas law:

$$\begin{aligned} Pv &= RT, \text{ or its exact equivalent} \\ PV &= nRT \end{aligned} \tag{2.38}$$

Here R is the universal gas constant, (see the end papers), and v is the volume per mol, or molar volume. This EOS can be derived from the kinetic theory of gases, in which one assumes that each gas molecule has zero volume (i.e., is a mathematical point) and that the individual molecules have no attraction for one another, but interact only by elastic collisions. Those assumptions are very close to reality for gases at low pressure and high temperatures (relative to the critical temperature of the gas), so this EOS would be expected to represent experimental PvT data very well under those conditions, and it does.

The first theoretical improvement on the ideal gas law is due to *van der Waals* (vdW). He suggested that the individual molecules do have some volume, b, and that there is some attraction between one molecule and another. Using those two ideas he wrote

$$P = \frac{RT}{(v-b)} - \frac{a}{v^2} \tag{2.39}$$

where the a/v^2 term represents attraction between the molecules. For large values of v, the vdW EOS becomes identical to the ideal gas law. For very low values of v, as $(v-b)$ approaches zero, it shows liquid-like behavior. The two constants, a and b, are to be determined from experimental data for each individual substance.

The vdW EOS is *not very good* at representing experimental PvT data, but it has had a profound influence on thermodynamics. Fairly simple, totally empirical modifications of it by Redlich and Kwong, Soave, and Peng and Robinson are very widely used in vapor-liquid equilibrium calculations, as discussed in Chapter 10 and Appendix F. Furthermore, it led to the principle of corresponding states, discussed below, which is very useful.

The other useful theoretical EOS, the *virial EOS*, begins by defining the *compressibility factor*

$$(\text{compressibility factor}) = (\text{“}z\text{ factor”}) = z = \frac{PV}{nRT} = \frac{Pv}{RT} \tag{2.40}$$

For an ideal gas the compressibility factor is 1.00 for all T and P. Thus, the compressibility factor is a simple, dimensionless measure of the real gas behavior, compared to ideal gas behavior. With this definition, we can write the basic virial EOS

$$z = 1 + \frac{B}{v} + \frac{C}{v^2} + \frac{D}{v^3} + \ldots \tag{2.41}$$

in which the coefficients, $B, C, D, \ldots$ are functions of T but not of P. Clearly, as v becomes very large (at high T and/or low P), this becomes the same as the ideal gas law. The reason this is listed as a theoretical EOS is that from statistical mechanics we can make some useful statements about the relations of the coefficients, B, C, $D, \ldots$ to various kinds of molecular interactions. This EOS is most often used for gas mixtures (see Chapters 9 and 10); with it we can make some fairly good property estimates for mixtures for which we have no experimental data. We most often see this EOS with only the B term, sometimes with the B and C terms, practically never with the higher terms. By convention, B is the *second virial coefficient*, C is the *third*, and so on.

2.11.2 EOSs Based on Pure Data Fitting

None of the theoretical EOSs can represent the experimental data for gases and liquids over a wide range of pressures with accuracy comparable to the accuracy of the best experimental data. For the most precise calculations, and for making up thermodynamic tables like the steam tables, we use EOSs that have little or no theoretical foundation (they all start with the ideal gas law and add terms) and that have many adjustable constants. (The idea of an "adjustable constant" may seem contradictory. But these are values that, once chosen for a particular pure substance, do not change with changes in T and P, so we call them constants.) Using computers we can select the values of those adjustable constants that minimize the differences between the EOS and the experimental data. One of the first of these was the Beattie–Bridgeman (BB) EOS, used in the 1920s and 1930s. It can be thought of as a virial EOS of state, Eq. 2.41, with the B, C, and D in that EOS written as

$$B_{\text{Eq. 2.41}} = B_0 - \frac{A_0}{RT} - \frac{c}{T^3} \tag{2.42}$$

$$C_{\text{Eq. 2.41}} = bB_0 + \frac{aA_0}{RT} - \frac{cB_0}{T^3} \tag{2.43}$$

$$D_{\text{Eq. 2.41}} = \frac{bcB_0}{T^3} \tag{2.44}$$

In these three equations, a, b, c, A_0, and B_0 are adjustable constants, with values determined from the experimental PvT data. Thus, the BB EOS has five adjustable constants.

The Benedict–Webb–Rubin (BWR) EOS developed in the 1940s, which has 11 adjustable constants, abandons the strict virial form, but looks somewhat like it:

$$z = 1 + \left(B_0 - \frac{A_0}{RT} - \frac{C_0}{RT^3} + \frac{D_0}{RT^4} - \frac{E_0}{RT^5}\right)\frac{1}{v} + \left(b - \frac{a}{RT} - \frac{d}{RT^2}\right)\frac{1}{v^2} + \frac{\alpha}{RT}\left(a + \frac{d}{T}\right)\frac{1}{v^5} + \frac{c}{RT^3}\left(1 + \frac{\gamma}{v^2}\right)\exp\left(-\frac{\gamma}{v^2}\right)\frac{1}{v^3} \tag{2.45}$$

It is widely used to prepare tables of thermodynamics properties, particularly for hydrocarbons [5].

The only substance for which we seem justified in using an even more complex EOS is water (and steam) for which we have much more detailed PvT data than for any other substance and for which we have great economic incentive to have a precise thermodynamic table. The current U.S. steam table [4] uses a special EOS with 58 adjustable constants. The more recent SI steam table [7] uses a similar EOS with 74 adjustable constants. These EOSs have little theoretical basis and little intuitive content. They would rarely be used for any substance other than water, and never for hand calculations.

With EOSs as complex (and with as many adjustable constants) as the BWR and those in the steam tables, one can compute the properties of both liquid and vapor from a single EOS. In the high-pressure, high-temperature region near the critical point this has some advantages; it is done in the steam tables and in some hydrocarbon tables.

2.12 CORRESPONDING STATES

One of the useful consequences of the vdW EOS is that, if it were indeed a correct representation of reality, then it could be written in the form

$$z = f\left(\frac{T}{T_c} \cdot \frac{P}{P_c}\right) = f(T_r, P_r) \tag{2.46}$$

where T_c and P_c are the critical temperature and pressure, and T_r and P_r are the *reduced temperature* and *reduced pressure*. The reduced properties and z are all dimensionless, so that Eq. 2.46 is a dimensionless EOS as is the virial EOS and others based on it. This says that on a plot of z vs. P_r the lines of constant T_r should be the same for every substance. This is known as the *principle of corresponding states*. It is *only approximately true*, but it is close enough to being correct to be very useful. Figure A.4 shows such a plot. It is, in effect, a graphical EOS. We can use it in all the ways we use an algebraic EOS. Equation 2.46 is a two-constant EOS, because T_c and P_c are determined from the experimental PvT data.

Plots like Figure A.4 are made up from the experimental data for some pure substance. Many are for propane, which seems to be an average substance and for which good PvT data have been available for a long time. The plots for different substances are similar but not identical. This led thermodynamicists to ask whether adding a third parameter to Eq. 2.46,

$$z = f(T_r, P_r, \text{some third dimensionless parameter}) \tag{2.47}$$

would make the corresponding states principle more accurate. It clearly does, suggesting that perhaps some fourth parameter should be added, and so on. However, the three-parameter versions seem satisfactory for most purposes, and the advantages of adding a fourth parameter do not seem to justify the extra effort. There have been two widely used choices for the third parameter in Eq. 2.47: the value of z at the critical point (the *critical compressibility factor* z_c) and the *accentric factor* ω.

If the simple corresponding states principle (Eq. 2.46) were true, then all substances would have the same value of z_c. the compressibility factor at the critical point (for which $P_r = T_r = 1.00$). Table A. 1 shows that the measured values of z_c for a variety of substances range from aboul 0.23 for water to 0.307 for ethyl amine; most are close to 0.27. Figure A.4 is for a substance with $z_c = 0.27$. Complete tables for using Eq. 2.47 with z_c as the third parameter are shown by Hougen et al. [6]. However, in recent years most workers have decided that the choice of the accentric factor ω as the third parameter is more satisfactory. Table A.1 also shows the values of this parameter for a variety of substances. It is defined to make it equal 0.00 for argon, 0.2 to 0.3 for most substances, and up to 0.7 for some. It is negative only for H_2 and He, which seem to be different from everything else because at their low normal boiling points quantum effects become important. The technical definition and physical meaning of ω are discussed in Section 5.5.

The most widely used corresponding states approach is that due to Pitzer and his co-workers [8], often called *Pitzer-type equations*. The common form of their approach is

$$z = z^0 + \omega z^1 \tag{2.48}$$

where z^0 and z^1 are both functions of T_r and P_r. The general idea is that there is some base function (z^0) that describes the behavior of a substance with $\omega = 0.00$ and there is also a correction function (z^1), which, when multiplied by ω and added to the base function, gives the best representation of the experimental data. Equation 2.48 is a three-constant EOS: T_c, P_c, and ω must be determined from the experimental PvT data. Pitzer and his co-workers presented tables of the functions z^0 and z^1 for various P_r and T_r. The z^0 table is very much like the values we would read from Figure A.4, except that the values are somewhat larger (because materials with $\omega = 0.00$ mostly have $z_c \approx 0.29$ and Figure A.4 is for materials with $z_c = 0.27$). The z^1 table shows small negative values for P_r and T_r less than 1.00, and small positive values for P_r and T_r greater than 1.00. Thus, for the most common uses the ωz^1 term in Eq. 2.48 is a small negative correction to the z^0 value.

Lee and Kessler [9] developed their own tables of those functions, which differ slightly from those of Pitzer et. al. The Lee and Kessler tables are probably the most widely used tables of this type. In addition to tables of z^0 and z^1 they present similar tables for other thermodynamic functions, all in the Pitzer-type format: a base function for $\omega = 0.00$ and a second function to be multiplied by ω and added to the base function.

While the tables, with proper interpolation are probably the best corresponding states estimates of fluid properties, they are inconvenient for computer use. Many attempts have been made to replace those tables with EOSs. The following simple, totally empirical estimating EOS called the *little EOS* [10, p. 89], is reasonably accurate at low pressures:

$$z^0 = 1 + \frac{P_r}{T_r}\left(0.083 - \frac{0.422}{T_r^{1.6}}\right) \tag{2.49}$$

and

$$z^1 = \frac{P_r}{T_r}\left(0.139 - \frac{0.172}{T_r^{4.2}}\right) \tag{2.50}$$

These two equations have no theoretical basis; they are simply data-correlating equations that do a surprisingly good job of reproducing pure species PvT data at low pressures.

Example 2.3 Estimate the compressibility factor of steam at 500°F and 680 psi, using the little EOS (Eqs. 2.48–50), and compare that estimate with the value from the steam table.

For water (see Table A.I), $T_c = 647.1\text{K} = 1164.78°\text{R}$. $P_c = 220.55$ bar, and $\omega = 0.345$, so that

$$T_r = \frac{(500 + 459.67)°\text{R}}{1164.78°\text{R}} = 0.8239 \tag{2.I}$$

$$P_r = \frac{680 \text{ psia}}{220.55 \text{ bar} \cdot \dfrac{14.51 \text{ psia}}{\text{bar}}} = 0.2125 \tag{2.J}$$

$$z^0 = 1 + \frac{0.2125}{0.8239}\left(0.083 - \frac{0.422}{0.8239^{1.6}} = 0.8730\right) \tag{2.K}$$

$$z^1 = \frac{0.2125}{0.8239}\left(0.139 - \frac{0.172}{0.8239^{4.2}}\right) = -0.0642 \tag{2.L}$$

$$z = 0.8730 + 0.345 \cdot (-0.0642) = 0.851 \tag{2.M}$$

Based on the steam table (which may be considered as reliable as the experimental data), the value of z is 0.804. ■

Figure 2.11 shows the results of similar comparisons over a range of pressures and for two temperatures, 500°F ($T_r = 0.82$) and 800°F ($T_r = 1.08$).

From this comparison we see the following:

1. For pressures less than 50 psia the experimental and calculated values of z are all ≈ 1.00, indicating that for this pressure and temperature range steam behaves practically as an ideal gas.
2. For the lower temperature the departure from ideal gas behavior is greater than for the higher temperature. This is the same behavior shown in Figure A.4, and is a general observation. At a given pressure, increasing the temperature makes the gas more like an ideal gas.
3. The little EOS does an excellent job of representing the data below pressures of 100 psia, and a good job at higher pressures. It is less reliable close to the condensation temperature. For the rest of this book we will generally use the little EOS for estimating the behavior of gases at modest pressures. We will see in Chapter 9 that its form makes it convenient for calculating the behavior or gas mixtures at low pressures.

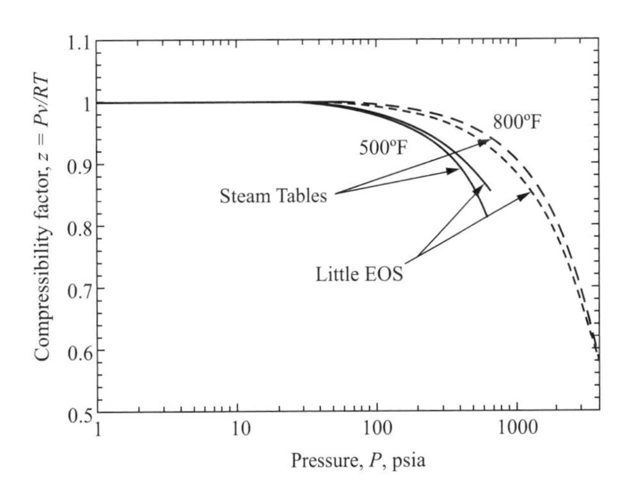

FIGURE 2.11 Comparison of compressibility factor from the steam tables (most reliable values) and computed from the little EOS. This has the same form as Figure A.4, but covers the P_r range from 0.0003 to 1.25. The 500°F curves end at 680 psia, at which pressure steam at 500°F condenses to water.

4. The ωz^1 terms contribute 14% of the answer. The common z charts like Figure A.4 are made up for an average substance, so that if we use them without the corrections for z_c not being equal to 0.27, we often make a small error. The same is not the case for the Pitzer-type equations, ω was defined to be zero for argon, which is not a typical substance, so if we used only the z^0 term for a typical substance we would make a serious error.

2.13 DEPARTURE FUNCTIONS

The compressibility factor z, defined by Eq. 2.40 and shown in Figures 2.11 and A.4, is the simplest of a family of *departure functions* that are widely used in chemical engineering. Most of these are based on selecting as a model the simplest possible behavior, and then correlating or predicting departures from this model. The compressibility factor shows departure from ideal gas behavior. In the limit (low pressure and/or high temperature) the departure becomes negligible, as shown in Figures 2.11 and A.4 as $P \to 0$, $z \to 1.00$.

For pure species gases, the most widely used departure functions are

$$(\text{enthalpy departure}) = \frac{h^* - h}{RT_c} \tag{2.51}$$

and

$$(\text{entropy departure}) = \frac{s^* - s}{R} \tag{2.52}$$

where h^* and s^* are the specific enthalpy and entropy that an ideal gas would have at the same T and P as the real gas. Obviously, these departure functions are both zero for ideal gases. The RT_c and R in the denominators are inserted to make the terms dimensionless. If the simple corresponding states statement were exactly true, then a plot of either of these functions for various T_r and P_r would be the same for all pure species. In practice these departure functions are mostly correlated in terms of T_r, P_r, and ω, using Pitzer-type equations. The forms shown in Eq. 2.51 and 2.52 fit logically with corresponding states estimates. For EOS-based estimates one often sees these with the RT_c and R dropped, so that, for example, the enthalpy departure becomes $(h^* - h)$.

We will see in Chapter 7 that an *ideal solution* will play the same role for solutions as an ideal gas does for real gases, and we will correlate and predict departures from ideal solution behavior much as the compressibility factor lets us deal with departures from ideal gas behavior.

2.14 THE PROPERTIES OF MIXTURES

This book is about equilibrium, mostly involving mixtures of pure species. We say much more about the properties of mixtures in subsequent chapters, but we introduce the subject here so we can use the results in the next section. The molar enthalpy and entropy of mixtures are described by

$$h_{\text{molar, mixture}} = \sum_{\text{all chemical pecies}} \left(x_i h_i^{\circ}\right) + \Delta h_{\text{isothermal mixing}} \tag{2.53}$$

and

$$s_{\text{molar, mixture}} = \sum_{\text{all chemical species}} \left(x_i s_i^{\circ}\right) + \Delta s_{\text{isothermal mixing}} \tag{2.54}$$

Where the $^{\circ}$ indicates pure species at the same temperature and pressure. We can find the enthalpy and entropy per pound or kg of mixture by replacing the molar pure species values by per unit mass values and the mol fractions by mass fractions.

The isothermal enthalpy of mixing is the amount of heat we must add or subtract when we mix the species adiabatically and then heat or cool as needed to bring the mixture to its starting temperature. There is no direct experimental way to measure the isothermal entropy change of mixing; we must infer it from other measurements.

In many cases, Δh_{mixing} is negligible so that the mixture enthalpy is $\approx$ the molar average of the enthalpies of the species present in the mix; however for the important mixture of sulfuric acid and water Δh_{mixing} is large and dangerous! Δs_{mixing} is never zero!

For ideal gases, (and ideal solutions of liquids, solids or nonideal gases, see Chapter 7) $\Delta h_{\text{mixing}} = 0$, so that

$$h_{\text{molar, mixture}} = \sum_{\text{all chemical species}} \left(x_i h_i^{\circ}\right) \quad \begin{bmatrix}\text{ideal gases} \\ \text{and ideal} \\ \text{solutions}\end{bmatrix} \tag{2.55}$$

However for ideal gases and other ideal solutions $\Delta s_{\text{isothermal mixing}} = -R \sum x_i \ln x_i$, so that

$$s_{\text{molar, mixture}} = \sum_{\text{all chemical species}} \left(x_i s_i^{\circ}\right) - R \sum x_i \ln x_i \quad \begin{bmatrix}\text{ideal gases} \\ \text{and ideal} \\ \text{solutions}\end{bmatrix} \tag{2.56}$$

2.15 THE COMBINED FIRST AND SECOND LAW STATEMENT; REVERSIBLE WORK

Figure 2.12 shows a very general process, not necessarily at steady state, with one or more material flows in and out, and with heat exchange at one or more temperatures. It is shown exchanging work with the surroundings, but it could be a zero-work process. What can we say about how much work it takes to make this process go? Or how much work it can produce if it is a work-producing process (e.g. a power plant)? We can say quite a bit!

We begin by multiplying Eq. 2.15 by minus T_o and adding it to Eq. 2.10, finding,

$$d(m(u-T_oS))_{\text{system}} = \sum(u+Pv-T_os)dm_{\text{in minus out}} + \sum\left(1-\frac{T_o}{T_{\text{in}}}\right)dQ_{\text{in minus out}} + \sum dW_{\text{in minus out}} - T_o ds_{\text{irreversible}} \quad (2.57)$$

Here T_o is the surroundings temperature, discussed below. If the process in the box is an oil refinery, Eq, 2.57 will have dozens of terms for flows of material, heat and work in and out. If it is a simple binary distillation column it will have one material flow in and two out, with heat exchange at two different temperatures, and (generally negligible) work flows for some pumps. Chemical engineers often use Eq. 2.57 for simple applications like the distillation column, although there is no reason (except for complexity) that it could not be applied to an oil refinery.

If we now restrict out attention to reversible processes, for which $ds_{\text{irreversible}}=0$, we can say that the dW must be the work that the process in the box would consume or produce if it had the same flows of mass and heat in and out, and it were reversible. We can solve for that reversible work amount as

$$\sum dW_{\text{reversible, in minus out}} = d(m(u-T_os))_{\text{system}} - \sum(u+Pv-T_oS)dm_{\text{in minus out}} - \sum\left(1-\frac{T_o}{T_{\text{in}}}\right)dQ_{\text{in minus out}} \quad (2.58)$$

The two Ts that appear here are T_{in}, the temperature at which heat flows in or out across the system boundary, and T_o the *reservoir temperature* – the lowest temperature of an unlimited reservoir of heat or cooling, normally the temperature of the nearest large body of water that can supply or accept heat, or of the atmosphere, or of an industrial plant's cooling water system. The reversible work in this equation is the *maximum* work that can be obtained from a work-producing device (some kind of motor or power plant) or the *minimum* work required by some work-consuming device (for example a compressor or air separation plant). Real devices are never this good; these minimum and maximum work values show the thermodynamic limits of real devices and processes.

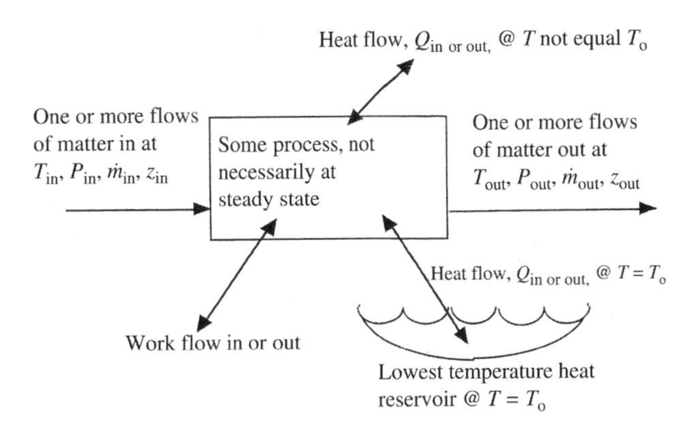

FIGURE 2.12 The flow diagram for a very general process, that exchanges multiple flows of mass, heat and work with its surroundings, and may change internally with time.

Example 2.4 Figure 2.13 shows the schematic steady-flow CO_2 separation device that will be used if we decide to separate and store the CO_2 in power-plant exhaust gases to prevent or minimize global climate change. Estimate the required work per lbmol of CO_2 captured for a reversible plant, using Eq. 2.58.

We begin by solving the general problem, in which a mixture of A and B (stream 1) is separated into two streams, one pure A (stream 2), the other pure B (stream 3). Since the plant in Figure 2.13 operates at steady state the $d(m(u-T_os))_{\text{system}}$ term is zero and since it exchanges heat with the surroundings only at T_o the $\sum(1-T_o/T_{\text{in}})dQ_{\text{in minus out}}$ term is also zero. Thus

$$\sum dW_{\text{reversible, in minus out}} = -\sum(u+Pv-T_os)dm_{\text{in minus out}} = (h-T_os)_2dm_2 + (h-T_os)_3dm_3 - (h-T_os)_1dm_1 \quad (2.59)$$

Switching now from mass to mols, dividing both sides by dn_1 and observing that Figure 2.13 shows only one work flow (which could be the algebraic sum of several) we find

$$\frac{dW_{\text{reversible}}}{dn_1} = (h-T_os)_2\frac{dn_2}{dn_1} + (h-T_os)_3\frac{dn_3}{dn_1} - (h-T_os)_1 \quad (2.60)$$

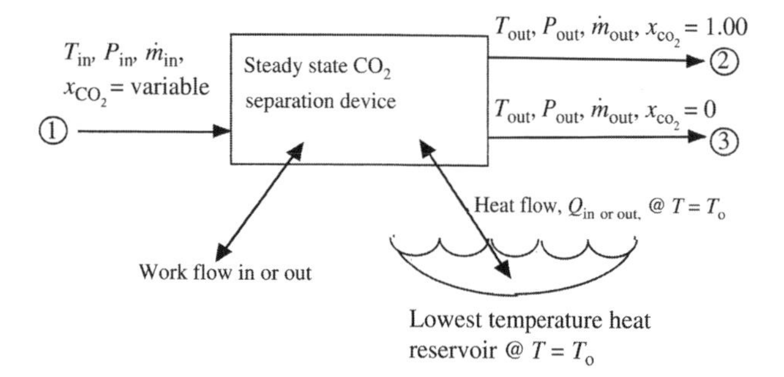

FIGURE 2.13 The flow diagram for a general steady-state process with one inlet stream (1) and two outlet streams (2 and 3) exchanging heat with the surroundings only at T_o, and also for the specific separation of CO_2 from power plant waste gas.

This is the reversible work per mol (lbmol or kgmol) of feed for *any* steady-flow processes separating one stream into two, and exchanging heat only at T_o (but not, for example, for a distillation device that exchanges heat at two temperatures). Next we observe that (n_2/n_1) is the mol fraction of A in the feed (x_a), and similarly $(n_3/n_1) = (x_b)$ in the feed so that

$$\frac{dW_{\text{reversible}}}{dn_1} = (h - T_o s)_2 x_a + (h - T_o s)_3 x_b - (h - T_o s)_1 \tag{2.61}$$

This equation is for any A and B; we now restrict it to ideal gases (or ideal solutions of liquids or solids), and replace the h and s terms with Eq. 2.55 and 2.56. The h terms cancel and that all the s terms cancel except the last, so that

$$\frac{dW_{\text{reversible}}}{dn_1} = -RT_o(x_a \ln x_a + x_b \ln x_b) \quad \begin{bmatrix}\text{ideal gases or}\\ \text{ideal solutions}\end{bmatrix} \tag{2.62}$$

This is the general equation for the reversible work of an isothermal, constant-pressure, steady-state process that exchanges heat only with the reservoir at T_o and separates an ideal gas stream (or ideal solution of liquids or solids) containing A and B into essentially pure A and B.

Returning to the CO_2 example, A is CO_2 so that $x_a = 0.12$, and $x_b = (1 - 0.12) = 0.88$. For $T_o = 70°\ \text{F} = 294.7\ \text{K}$,

$$\frac{dW_{\text{reversible}}}{dn_1} = -8.314 \frac{\text{J}}{\text{mol K}} \cdot 294.7\ \text{K} \cdot (0.12 \ln 0.12 + 0.88 \ln 0.88) = 0.897.7 \frac{\text{J}}{\text{mol}} \tag{2.N}$$

The work per mol of feed is positive (inflowing) so this is a work-consuming process. All real (nonreversible) work-consuming plants would require more than this reversible amount of work.

To find the reversible work per mol of substance A (in this case CO_2.) we divide the reversible work per mol of mixture by the x_a, finding

$$\frac{dW_{\text{reversible}}}{dn_a} = \frac{1}{x_a}\frac{dW_{\text{reversible}}}{dn_1} = \frac{1}{0.12} \cdot 897.7 \frac{\text{J}}{\text{mol}} = 7.48 \frac{\text{kJ}}{\text{mol}} = 3.2 \frac{\text{kBtu}}{\text{1bmol}} = 73 \frac{\text{Btu}}{\text{1b}} \tag{2.O}$$

To finish this example, we ask what would happen if instead of taking the CO_2 out of the power plant exhaust gas, we let it dilute to the concentration in the atmosphere (about 390 ppm) and then separated it? Substituting $x_a = 0.000390$ in the above equations we find that the required reversible work increases by a factor of 2.9. ■

This example shows some of the utility of the combined statement. Using it, we can quickly and simply estimate the reversible work requirement for this separation, and the reversible work ratio between starting with 12% CO_2 and 0.039% (390 ppm). This ratio shows that we should begin with the most concentrated a CO_2 stream as we can.

The combined statement (Eq. 2.58) leads naturally to the definition of several important convenience properties. If we consider a steady flow, isothermal process at the reservoir temperature, then $(T_o/T_{in}) = 1$, so that the dQ term is zero, and at steady state flow the $d(m(u - T_o s))_{\text{system}}$ term is also zero, so we have

$$\sum dW_{\text{reversible, in minus out}} = -\sum (u + Pv - T_o s) dm_{\text{in minus out}} \quad \begin{bmatrix}\text{steady flow,}\\ \text{isothermal}\end{bmatrix} \tag{2.63}$$

The term $(u + Pv - T_o s)$ becomes $(u + Pv - Ts)$ for this isothermal process. If we further restrict our attention to a steady flow device with only one flow in and out, this further simplifies to

$$\frac{dW}{dm} = -\Delta(u + Pv - Ts)_{\text{in minus out}} \begin{bmatrix}\text{steady flow,}\\ \text{isothermal at } T_o,\\ \text{reversible}\end{bmatrix} \tag{2.64}$$

This shows how the convenience function

$$(u + Pv - Ts) = (h - Ts) = g = (\text{Gibbs energy per mass or mol}) \tag{2.65}$$

naturally arises from a combined first and second law analysis of the minimum or maximum work. The Gibbs energy also arises naturally and plays a dominant role in the study of phase and chemical equilibrium in Chap. 4.

Returning to our piston and cylinder in Figure 2.7, we ask what is the minimum or maximum work for a reversible, isothermal, constant volume process at $(T_o/T_{in}) = 1$ in such a device. Dropping the unnecessary terms from Eq. 2.58, we find

$$\frac{dW}{dm_{\text{reversible}}} = \Delta(u - Ts)_{\text{system}} \quad \begin{bmatrix} \text{reversible isothermal} \\ \text{constant volume} \\ \text{process at } T = T_o \end{bmatrix} \tag{2.66}$$

This leads to another convenience function

$$(u - Ts) = a = (\text{Helmholz energy per mass or mol}) \tag{2.67}$$

which is not widely used in chemical engineering calculations (which are mostly made at constant T and P) but more often used in chemistry, (where many systems operate at constant T and V).

Eq. 2.58 naturally leads to two other convenience functions,

$$(u + Pv - T_o s) = (h - T_o s) = (\text{availability function}) \tag{2.68}$$

the favorite of the cryogenic engineers, and

$$(h - h_o) - T_o(s - s_o) = (\text{exergy}) \tag{2.69}$$

which the mechanical engineers like much more the chemical engineers seem to. Here h_0 and s_0 are the enthalpy and entropy per mass or mol in an appropriately chosen datum or reference state.

Much more details on the use of the combined statement are given in [11].

2.16 SUMMARY

1. The general balance equation is used in almost every chemical engineering problem. It can be applied only to some system or control volume with a precisely defined set of boundaries.
2. The law of conservation of mass states that *mass* obeys the general balance equation with creation = destruction = 0.
3. The first law of thermodynamics (law of conservation of energy) states that *energy* obeys the general balance equation with creation = destruction = 0.
4. The second law of thermodynamics states that all real processes are irreversible, and defines a reversible process as an ideal to which we may compare real processes.
5. The second law is most often used numerically by introducing the entropy, whose values increase by heat addition and by irreversible processes.
6. In elementary thermodynamics courses we generally solve problems for devices and systems, processing pure species, looking up values of v, h, and s in suitable tables. In this book we are mostly concerned with mixtures, and how we compute the equivalent of those tables of thermodynamic properties for mixtures.
7. The measurable properties in pure-species thermodynamics are P, T, m, and V. All other properties are calculated from measurements made of those three (including calorimetric measurements, where differences in T are used).
8. We make up pure-species thermodynamic property tables by defining a datum state to which we assign values of h and s, normally setting both to 0.00 in the same datum state. Then we calculate the values at other states using heat capacity data and an EOS.
9. The simplest EOS, the ideal gas law, is based in theory and quite reliable for low pressures and high temperatures. For higher pressures and lower temperatures we use more complex EOSs all of which begin with the ideal gas law, and then add other terms, generally with no theoretical basis, which allow us to fit the EOS to the experimental PvT data.
10. The corresponding states principle is only approximately correct, but it has proven very useful. It is, in effect, an EOS that applies to all substances and for which we need very little experimental data.
11. Departure functions show the deviation or departure of real substances from ideal substances. The compressibility factor z shows that departure for real gases.
12. The combined first and second law statement allows us to calculate the reversible work for any process, actual or hypothetical. It also leads to the definitions of several useful convenience functions.

PROBLEMS

See the Common Units and Values for Problems and Examples. An asterisk (*) on a problem number means that the answer is in Appendix H.

2.1 Write an energy balance for the earth, indicating which terms are probably important and which are probably negligible.

2.2* From an energy balance around the earth estimate the rate of energy liberated by radioactive decays in the earth. Assume that heat losses from the earth are at steady state. Use the following:
Earth is roughly a sphere 8000 mi in diameter.
Geothermal gradient of temperature, dT/dz is approximately 0.02°F/ft.
Thermal conductivity k of earth (near the surface), is about 1 Btu/(h · °F·ft.) Heat flow is estimated from $\dot{Q} = -kA(dT/dz)$, where A is area.

2.2 Also calculate the rate at which matter is being converted to energy in the earth.

2.3 In rating the energy release of nuclear explosives the Atomic Energy Commission uses the energy unit "kiloton," where 1 kton $= 10^{12}$ cal. This is roughly the energy release involved in detonating 10^3 tons of TNT. The Hiroshima bomb was reported to be about 14 kton. How much matter was converted to energy in it?

2.4 The groups $u + gz + V^2/2$ and $h + gz + V^2/2$ occur in most ME thermodynamics problems. To evaluate the relative magnitude of the individual terms, calculate gz and $V^2/2$ in Btu/lbm and J/kg for the following: $z = 10$, 100, 1000, 10,000 ft; $V = 10$, 100, 1000, 10,000 ft/s. Show these results on a log-log plot.

2.5* Typical high explosives liberate about 1800 Btu/lbm of thermal energy on exploding. It has been suggested that a high-velocity projectile might liberate as much thermal energy on being stopped, by conversion of its kinetic energy to thermal energy. How fast must such a projectile be going in order that its kinetic energy, if all turned to internal energy, would be the same as that of the typical explosive described above? Under what circumstances could a projectile have this kind of velocity?

2.6 List the sources of irreversibility in an ordinary household refrigerator. Show what methods we could use to overcome these if the cost were of no concern.

2.7 List two industrial processes that closely approach reversibility. List two industrial processes that are *very* irreversible.

2.8 Most investigators believe that the entropy of all perfect crystalline materials at 0 K is zero. Someone has suggested that the true situation is that the entropy of all perfect crystalline materials at 0 K is A, where A is the same for all materials (per gram atom). (This means that the entropy of a monatomic substance, such as He, would be A at 0 K, while that of a diatomic substance, such as H_2, would be $2A$, and so on. Thus, for any chemical reaction at 0 K, there would be no entropy change, because the number of atoms is conserved.) If $A = 0$, then the two views are the same. What experiments could be performed to determine whether A is equal to zero?

2.9 **a.** Give an example of a system undergoing a reversible change, during which the system's entropy (1) increases, (2) decreases, (3) remains constant,
b. Give an example of a system undergoing an irreversible change, during which the system's entropy (1) increases, (2) decreases, (3) remains constant.

2.10 Using values from a steam table, verify the statements in Example 2.1 that the difference between the internal energy and enthalpy changes is equal to the product of the pressure and the volume change.

2.11 Verify the statement in Example 2.1 that the change in entropy is approximately equal to the heat input divided by the average absolute temperature for the process.

2.12* One hundred kilograms of steam is contained in a cylinder with a frictionless, zero-mass piston, the other side of which is exposed to the atmosphere. The steam is initially 100% quality at 100°C. Enough heat is transferred through the cylinder walls to reduce the quality (weight fraction vapor) to 25%. How much heat was transferred? Which way?

2.13 Using the heat capacity constants in Table A.9, estimate the enthalpy and entropy changes when steam at 14.7 psia is heated from 250 to 500°F at constant pressure. Compare your answer to values from the steam table ($\Delta h = 118.5$ Btu/lbm, $\Delta s = 0.1431$ Btu/(lbm·°R)).

2.14 Show the forms of Eqs. 2.33, 2.34, and 2.35 for a gas that obeys the ideal gas law by substituting the appropriate values of the v derivatives calculated from the ideal gas law.

2.15 Figure 2.10 shows the constant-pressure (horizontal) integration at the same pressure as that in the datum state, which is normally not zero (0.08866 psia = 0.0061173 bar in both the US and SI steam tables). Often the available heat capacity equation is for an ideal gas at zero pressure.
a. If the pressure in the datum state is not zero, and we wish to use a zero- pressure heat capacity, sketch what the integration paths would look like on Figure 2.10.
b. Show the form of the integration for the vertical path on Figure 2.10, finding the appropriate derivative $(\partial h/\partial P)_T$ from Table 2.2.

c. Show the mathematical form of this derivative if the material is an ideal gas. Table A.9 indicates that the equations there for C_p are for the ideal gas state. Where would they be appropriate on Figure 2.10?

d. Show the value of this derivative for liquid water near its datum state (0.018°F, 0.08866 psia). The specific volume of water at this state is 0.016022 ft^3/lbm and $(\partial v/\partial T)_p \approx 2 \cdot 10^{-7}$ ft^3/(lbm·°F).

2.16 Estimate the z of steam at 500°F and 680 psia, using Figure A.4. Comment on the degree of agreement or disagreement with the experimental value reported in Example 2.3.

2.17 The steam table [4] shows that at 500 psia and 500°F the specific volume of steam is 0.9924 ft^3/lbm.

a. Calculate the value of z for steam at this temperature and pressure.

b. Estimate the same quantity from Figure A.4.

2.18 Show the detailed calculations going from Eq. 2.61 to 2.62.

2.19 Estimate the isothermal entropy change of mixing for binary mixtures of two species for $x_1 = 0.5$, 0.3, 0.1 and 0.01, assuming ideal solution.

2.20 Show the calculations leading to the statement in Ex. 2.4 that reducing the inlet concentration of CO_2 from 12 mol % to 0.039 mol % increases the required reversible work by a factor of 2.9.

2.21 For Ex. 2.3, plot $\frac{-1}{RT_o}\frac{dW_{\text{reversible}}}{dn_1}$ and $\frac{-1}{RT_o} \times \frac{dW_{\text{reversible}}}{dn_a}$ vs x_a for the range of x_a from 0 to 1.0.

2.22 The plot in Prob. 2.21 leads to the conclusion that as x_a goes to zero, the reversible work to separate per mol mixture goes to zero, while the reversible work to separate per mole of A goes to infinity!

a. Verify from the plot that this is correct.

b. Show that the equations demand it.

c. Explain physically what the equations are telling us.

d. This part of the problem requires some knowledge of distillation. Explain it in terms of a distillation separating A and B. If we have an existing column, and we cut it open and insert some additional trays, and then weld it together and run it at the same feed rate and reflux rate, will the heat input(s) to the reboiler and condenser increase? Will the quality of the separation increase?

2.23 The largest scale separation of ideal gas mixtures into practically pure gases is probably the separation of air into nitrogen and oxygen (and sometimes also argon). Air separation plants operate worldwide; there is probably at least one in your city.

a. Estimate the required reversible work (J/mol of air) for such a plant, using the same T_o as in Ex. 2.4.

b. Shreve [12] page 110 suggests that for a plant processing 15,000 lbmol/day of air delivering practically pure nitrogen and oxygen gases at practically atmospheric pressure the actual power input is 1 MW. What is the ratio of this power input to that required for a reversible plant?

REFERENCES

1. Poirier, J. P. *Lavoisier, Chemist, Biologist, Economist.* Philadelphia: University of Pennsylvania Press (1996).
2. Von Baeyer, H. C. *Maxwell's Demon: Why Warmth Disperses and Time Passes.* New York: Random House (1998).
3. Eddington, A. S. Quoted in S. W. Angust and L. G. Hepler, *Order and Chaos.* New York: Basic Books, p. 146 (1967). This semitechnical book gives many interesting insights into the second law.
4. Keenan, J. H., F. G. Keyes, P. G. Hill, and J. G. Moore. *Steam Tables: Thermodynamic Properties of Water Including Vapor, Liquid and Solid Phases*, New York: Wiley (1969).
5. Starling, K. E. *Fluid Thermodynamic Properties for Light Petroleum Systems.* Houston, TX: Gulf (1973).
6. Hougen, O. A., K. M. Watson, and R. A. Ragatz. *Chemical Process Principles, Part II: Thermodynamics*, ed. 2. New York: Wiley (1959).
7. Haar, L., J. S. Gallagher, and G. S. Kell. *NBS/NRC Steam Tables.* New York: Hemisphere (1984).
8. Pitzer, K. S., D. Z. Lippmann, R. F. Curl, Jr., C. M. Huggins, and D. E. Petersen. The volumetric and thermodynamic properties of fluids, II: compressibility factor, vapor pressure and entropy of vaporization. *J. Am. Chem. Soc.* 77:3433–3440 (1955).
9. Lee, B. I., and M. G. Kesler. A generalized thermodynamic correlation based on three-parameter corresponding states. *AIChE J.*, 21:510–527 (1975).
10. Smith, J. M., H. C. Van Ness, and M. M. Abbott. *Introduction to Chemical Engineering, Thermodynamics*, ed 5. New York: McGraw-Hill, p. 569 (1996). In the sixth edition of the same text, footnote at the bottom of page 102, the authors indicate that what I call "the little EOS" was developed by M. M. Abbot, and first published without attribution in the fifth edition, as cited above.
11. de Nevers, N. and J. D. Seader. Mechanical lost work, thermodynamic lost work, and thermodynamic efficiencies of processes. *Lt. Am. J. Heat Mass Transf.* 8:77–105 (1984).
12. Shreve, N. R. *Chemical Process Industries*, ed 3, New York: McGraw-Hill. p. 244 (1956).

3

THE SIMPLEST PHASE EQUILIBRIUM EXAMPLES AND SOME SIMPLE ESTIMATING RULES

In this chapter we discuss the simplest phase equilibrium examples. In various places the discussion is intuitive, not rigorous; that is pointed out where it occurs. Most of the material in this chapter is revisited in a rigorous way in Chapters 7–9, after we have developed the necessary thermodynamics in Chapters 4–6.

3.1 SOME GENERAL STATEMENTS ABOUT EQUILIBRIUM

We know that in a system at equilibrium there can be no spontaneous process occurring within the system. This leads to three simple, nonmathematical statements.

First, we know that if a system has temperature differences in it, then there will be spontaneous change as heat flows from hot to cold. Thus, for any system to be at equilibrium with its surroundings, it must be at the same temperature as its surroundings and must have the same uniform internal temperature. In many situations we have systems in which the temperature is not the same at every point in the system, but it is constant in time at every point; that is,

$$\left(\frac{\partial T}{\partial x}\right)_{y,z,t} \neq 0, \quad \text{but} \left(\frac{\partial T}{\partial t}\right)_{x,y,z} = 0 \qquad (3.1)$$

This is not an equilibrium situation. Rather it is the "steady state," which forms the basis of much of *irreversible thermodynamics.* If we could find a perfect heat insulator, then we could have a system at equilibrium with two parts at different temperatures. As yet, no one has any idea how such an insulator could be made, so that we are secure in making the sweeping statement that any *system at equilibrium is an isothermal system; and unless it is in an adiabatic container (which exists only in theory and in thermodynamics textbooks), it is at the same temperature as its surroundings* (Figure 3.1).

Second, we know from basic thermodynamics that any form of mechanical energy can be converted 100% to some other form of mechanical energy (if we have frictionless conversion devices, which, like adiabatic containers, exist only in theory and thermodynamics textbooks). Thus, conversion of mechanical, electrical, or kinetic energies from one form to another is possible in an equilibrium system. We also know from thermodynamics that we can convert any of these forms of energy 100% to heat or internal energy by a frictional process, but that the second law forbids 100% conversion in the opposite direction. Thus, we know that any frictional behavior is irreversible, and cannot be occurring in a system at equilibrium. Thus, in any equilibrium system there can be no processes involving friction. Under most circumstances this will mean that there are no moving parts of a system at equilibrium because in any real macroscopic system, motion leads to friction.

For most situations this means the system must be at a uniform pressure. If there is gravity, then there will be a pressure gradient that is opposed by gravity, and if one part of a system is in a piston restrained by a spring, or in a droplet or bubble restrained by surface forces, then that part may be at a pressure different from the other parts. But *in the absence of restraining gravity, spring, electrostatic, magnetic, osmotic,*

Physical and Chemical Equilibrium for Chemical Engineers, Second Edition. Noel de Nevers.

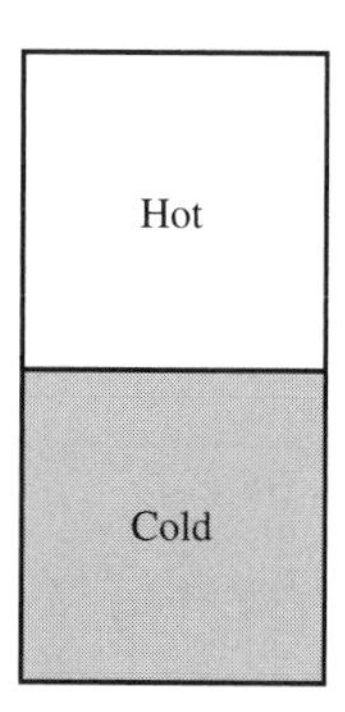

FIGURE 3.1 If two parts of a system are at different temperatures and are in thermal contact with each other, then there will be a spontaneous change as heat flows from hot to cold. So this is not an equilibrium system.

or surface forces, at equilibrium the system must be at a uniform pressure (Figure 3.2).

If there is a pressure gradient in a system, which is opposed by gravity (or a spring, etc.), the system is often said to be in *mechanical equilibrium.* Such systems can be in phase and chemical equilibrium with different pressures in different parts, as discussed in detail in Chapter 14. For most of the rest of this book we will assume that these pressure differences are small, and can be ignored, so that a system at phase and chemical equilibrium will have the same uniform pressure throughout.

This leaves open a question of oscillating systems. Is a frictionless pendulum in equilibrium with its surroundings? This is clearly a matter of definition and we can define it either way. If we define the "state of the system" to be one of regular oscillatory motion, then there is no spontaneous change from this state and, hence, equilibrium. If we define equilibrium to forbid spontaneous conversion of one form of energy to another, then we have defined the pendulum as a nonequilibrium system. This may be a good question for philosophers, but it seems to have no important engineering consequences. (We know that in any material not at 0 K there are internal vibrations of the atoms and molecules; these are present in all systems of matter at equilibrium, except at 0 K.)

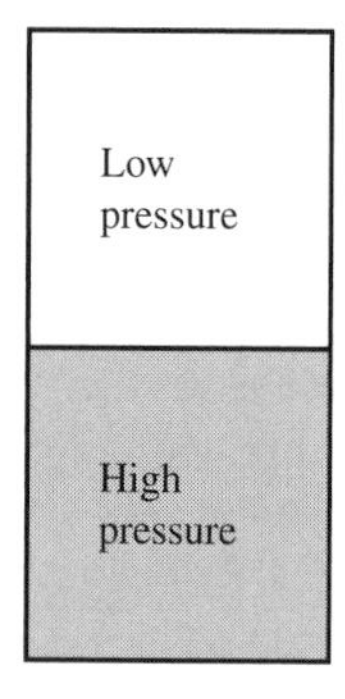

FIGURE 3.2 If two parts of system are in contact and do not have the same pressure, then a spontaneous process will occur unless the difference in pressure is opposed by gravity, a spring, or surface tension.

Third, there can be no flow of electric current through any resistor because this is a spontaneous change that causes the conversion of some other kind of energy into heat. This means that for an electric cell to be at equilibrium, it must be either disconnected (e.g., an auto storage battery with its terminals not connected) or must be opposed by an equal voltage in the opposite sense (e.g., a storage battery with its terminals connected through a balanced potentiometer) The situation in Figure 3.3 is metastable. As long as we do not close the switch, the two parts of the system can remain at different voltages (electric potentials) indefinitely. But when we close the switch, they will move toward a stable equilibrium state, in which they are both at the same voltage, by passing a current through the resistor. (Electrochemical equilibrium is discussed in Chapter 13.)

In effect this assumes that there exist perfect resistors. If there were no good resistors (i.e., electrical insulators), then it would be impossible to build a storage battery; charge would leak through the case of the battery and discharge the battery. As a practical matter, we all know that electrical insulators do exist, which are good enough that one can make excellent storage batteries whose current leakage rate is ≈ 0 (over a few months or years). Thus, on a human time scale, we can say that such resistors exist. The same is not the case with the flow of heat. Given the best practically available thermal insulation, it is not now possible to make a small portable icebox that will keep a block of ice from melting for a summer month in the Sahara desert.

The remaining, and interesting part of the study of equilibrium is determining the chemical compositions of phases at equilibrium and of chemicals within a phase that are in chemical-reaction equilibrium.

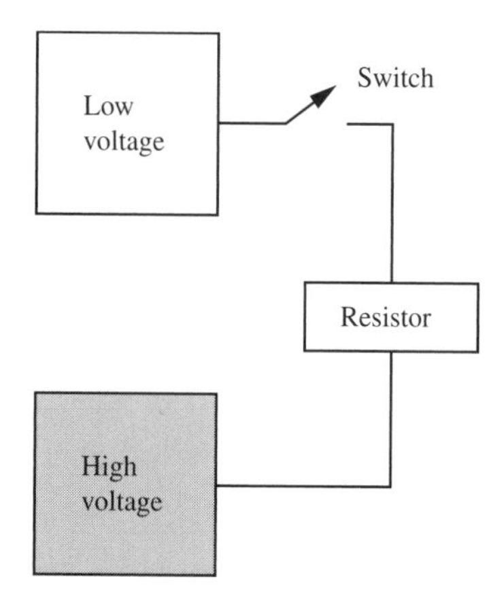

FIGURE 3.3 A system may be a metastable equilibrium system with two parts at different voltages, if they have no electrical connection. But if there is the possibility of a current flow through any resistance, then a spontaneous change will occur and this is not an equilibrium system.

3.2 THE SIMPLEST EXAMPLE OF PHASE EQUILIBRIUM

The simplest possible phase equilibrium is that between a pure liquid and the corresponding gas or vapor.

3.2.1 A Digression, the Distinction Between Vapor and Gas

It would be less confusing to students if the terms *vapor* and *gas* had not both been invented, because they are often used interchangeably. In many ways they have the same meaning, but in some situations we use only one of them, so the student must leam both, and learn the distinction between them. A *gas* is any substance in the gaseous state, which means that it will expand as needed to fill any container in which it is placed. Solids and liquids do not do that. A *vapor* is a gas that is at a temperature below its critical temperature. This means that if a vapor is compressed at a constant temperature it will turn into a liquid (Figure 1.8). A gas at a temperature above its critical temperature can be compressed at constant temperatures to very high pressures without changing to a liquid. Since a vapor is a gas, one could logically suggest that we dispense with the term vapor. But it is in very common usage. For example, the water that exists in the atmosphere is certainly in gaseous form, but it is below its critical temperature and is always referred to as *water vapor*; no other term is ever used. Similarly, Figure 1.8 and other figures like it show the relation between a liquid and a gas. But the gas is below its critical temperature, and such curves are always called *vapor-pressure curves.* Throughout this book I have tried to use the wording that the student will most often encounter in professional life, but confusion is sure to occur. Remember that a vapor is a gas, but that some gases are not vapors.

3.2.2 Back to the Simplest Equilibrium

We are all familiar with vapor–liquid equilibrium, in the case of water and steam. For pure water and steam to be in equilibrium (without air mixed in) the pressure of the gas must equal the vapor pressure of the liquid. If the pressure of the gas is less than the vapor pressure of the liquid, then the liquid will boil, expelling gas. If the pressure of the gas is greater than the vapor pressure of the liquid, then the gas will condense into the liquid. If these processes occur in a closed container (Figure 3.4), then they will continue until the two pressures (and temperatures) become the same, at which time we will have phase equilibrium. From our previous studies, we know that the vapor pressure of any pure liquid is a simple function of the temperature, called the vapor-pressure curve. Its values for many liquids are presented in textbooks and handbooks and are discussed in detail in Chapter 5.

We may profitably think of this as a situation in which the rate of escape of molecules from the liquid surface is

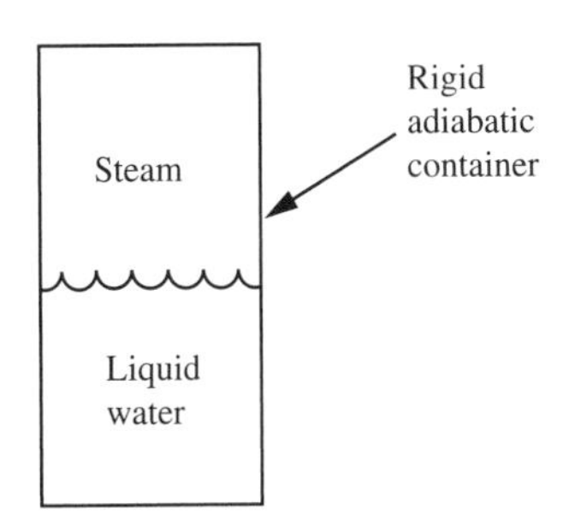

FIGURE 3.4 If we place a suitable amount of pure H_2O in a rigid adiabatic container and wait for equilibrium, we will eventually find steam and liquid water, both at the same temperature and pressure (except for the very small pressure differences caused by gravity and surface tension).

proportional to its vapor pressure, and the rate of condensation of molecules from the steam into the liquid is proportional to the pressure of the gas. These are equal and opposite (and we have phase equilibrium) when the gas pressure and the vapor pressure are equal. Figure 1.8 shows this relationship for steam and water. Equilibrium between liquid and gas occurs only along the curve shown. Above and to the left of the curve only liquid can exist; below and to the right of the curve only gas can exist. At temperatures below the critical temperature (374°C = 705.4°F for water) that gas would be called a vapor.

In Chapter 5 we will see that the Antoine equation (Eq. 5.12) represents experimental vapor-pressure data with reasonable accuracy in a simple algebraic form. It is shown, along with suitable values for a variety of chemicals, in Appendix A.2. We will use it in this chapter, while deferring the discussion of its origin until Chapter 5.

3.3 THE NEXT LEVEL OF COMPLEXITY IN PHASE EQUILIBRIUM

If we now consider the air–water system, we see that at temperatures near room temperature and 1 atm pressure there will be a liquid and a gas, as there was for pure water, but that there will be air dissolved in the water, and water vapor dissolved in the gas. Because this system is so important, we will describe it physically a bit before we begin the mathematics. Figure 3.5 shows a piston and cylinder arrangement containing air and water. It is assumed that heaters or refrigerators work to keep this system at some constant temperature and that the piston can move up and down to adjust the pressure. If we make up this system with very pure water and dry air, then at first the water will contain no dissolved air, and the air will contain no water vapor. If we wait a long enough time, air will dissolve into the water and water vapor will evaporate into the air, until finally an equilibrium state is reached in which none of the physical or chemical properties of the system is changing with time.

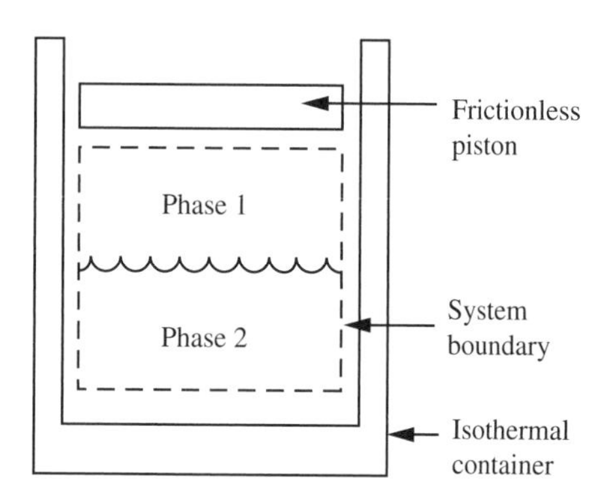

FIGURE 3.5 Two phases (mostly air and mostly water) in equilibrium in a piston and cylinder arrangement, at constant temperature.

If we wait for molecular diffusion to produce this result, we will have to wait a long time. If we stir or shake the system vigorously, we can reach the equilibrium state in a few seconds.

When equilibrium has been reached the composition of the two phases will be that shown in the Table 3.1. For the moment, please accept these values on faith; by the end of the chapter it will be clear how they were calculated.

Observe also that the two phases in equilibrium have very different chemical compositions. In Figure 3.4 both phases have the same chemical composition, because there is only one chemical species present. When we increase the complexity by adding a second or third species, the compositions of the two phases will generally not be the same (except for azeotropes; see Section 8.4.4). Most physical separation processes (distillation, crystallization, evaporation, drying) utilize this difference in composition between equilibrium phases to separate one chemical from another.

The amount of oxygen dissolved in the water is small, but it is needed for almost all life on this planet All living things conduct their biochemical business in dilute solutions of various materials in water; most need dissolved oxygen to conduct that biochemical business. We regularly oxygenate our fish bowls to provide the oxygen, dissolved in the water, that the fish must have to live. Our blood and that of most animals has chemicals in it (ours is hemoglobin in our red blood cells, which gives our blood its red color) that increase the equilibrium amount of oxygen dissolved in it making us much more efficient animals than we would be otherwise (see Problem 3.1). That dissolved oxygen makes iron and steel rust. Oxygen-free water will not rust iron or steel; the feed water for boilers is treated to remove the dissolved oxygen.

Table 3.1 Composition of Air and Water at Equilibrium at 20° C = 68°F, and 1.00 atm Pressure

	Gas Phase	Liquid Phase
Mol fraction water	0.023	0.999985
Mol fraction oxygen	0.205	5×10^{-6}
Mol fraction nitrogen	0.772	10×10^{-6}
Sum of mol fractions	1.00	1.00

Note: This treats air as 21% oxygen, 79% nitrogen, ignoring its other minor species.

The amount of water in the gas phase (air) is small, but plays a significant role in many processes. The water content of the atmosphere is responsible for many of the interesting, dramatic, and destructive things that our weather does. Without this water there would be no clouds, rain, lightning, hurricanes, and so on. Normally outdoor air contains less moisture than the amount shown in Table 3.1, except in very humid situations. For this reason, water evaporates from our skins and clothing into the air. For liquid water, the direction of equilibrium is to increase the water content of the atmosphere toward 0.023 mol fraction (at 20°C and 1 atm pressure). Until it does that, liquid water continues to evaporate at a rate that depends mostly on the air temperature and the wind velocity.

If we change the temperature and pressure, we will change the values in Table 3.1. Clothes dry faster in a clothes dryer than on a clothesline, because in the dryer we heat the clothes and the air around them, thus increasing the equilibrium concentration of water in air, and making the process go faster. Conversely, if we lower the temperature of the air, the equilibrium concentration of water will decline. If there is an available solid surface, the water will condense on it, forming dew. If there is no surface, but there are enough fine particles in the air, the water will condense on them, forming clouds or fog.

If we bring a cold drink into a warm room, water will condense on the glass (Figure 3.6), because the equilibrium water content of the air at the temperature of the drink is less than the typical water concentration in the room air. The cooled air next to the drink must reject water to reach equilibrium. In Figure 3.6 if the room is at 20°C = 68°F and 50% relative humidity, the water content of the air is

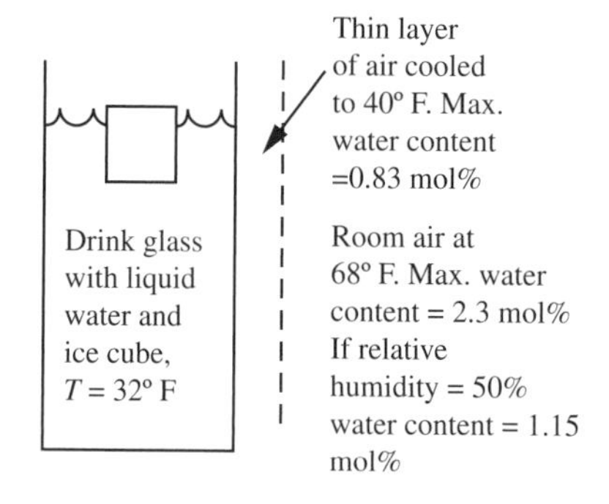

FIGURE 3.6 When a cold surface is brought into a warm room, the equilibrium moisture content of the cooled air layer near the surface will be lower than the moisture content of the air in the room, and moisture (dew) will form on the cold surface.

1.15 mol%. Air cooled to 40°F can hold only 0.83 mol% water, so the air next to the glass must reject water, normally by forming dew on the glass.

Raising the temperature increases the equilibrium concentration of water vapor in air, but it lowers the equilibrium concentration of oxygen and nitrogen in water. This is easily seen by filling a glass with very cold water from a tap and letting it sit on a kitchen counter. As the water warms, the amount of dissolved oxygen and nitrogen it contains becomes more than the equilibrium amount at the warmer temperature, so the water must reject oxygen and nitrogen to reach equilibrium. Bubbles of air form on the inside walls of the glass, where this oxygen and nitrogen comes out of solution. Raising the pressure has the opposite effect: It decreases the equilibrium amount of water water vapor in the air and increases the equilibrium amount of dissolved oxygen and nitrogen in water. The reader is certainly familiar with the effect of reducing the pressure on the solubility carbon dioxide in water, as shown by opening a carbonated beverage container and watching the carbon dioxide bubbles come out of solution. A more sinister example is the effect of air pressure on the solubility of nitrogen in our blood. Divers breathe air at pressures comparable to that of the water around them. At those pressures the equilibrium solubility of nitrogen in the blood is much larger than the value shown in Table 3.1. When the diver comes to the surface and reduces her body pressure, the amount of dissolved nitrogen in her blood is more than the equilibrium value at the new, much lower pressure, and the blood must emit nitrogen, normally through the lungs. If the pressure reduction is done too rapidly, the nitrogen comes out of solution in small bubbles in the blood veins, blocking the blood flow, causing a painful or fatal case of "the bends."

3.4 SOME SIMPLE ESTIMATING RULES: RAOULT'S AND HENRY'S "LAWS"

Before we begin to show the role of thermodynamics in understanding this simple kind of equilibrium, we consider two important and widely used estimating rules. The rest of this chapter sacrifices rigor for intuitive content. We will revisit this same material rigorously in Chapters 7–9. We start with our knowledge that for gas–liquid equilibrium of a single pure species (water and steam, with no air) the pressure of the gas must equal the liquid's vapor pressure. If we add another species (e.g., air), then it will dilute the water in both phases However, it is reasonable to assume that at equilibrium there will be something like the equality of the pressure of the vapor and the liquid's vapor pressure in the pure species case At equilibrium the flow of water molecules from liquid to gas, and from gas to liquid must be equal, just as it was for pure water and steam. To apply that equality for mixtures, we define two new quantities. The *partial pressure* of a species in a gas is the product of the total pressure P, of the gas, times the mol fraction of that species y_i, in the gas:

$$\begin{pmatrix}\text{partial pressure}\\ \text{of species } i\\ \text{in a gaseous mixture}\end{pmatrix} \equiv \begin{pmatrix}\text{mol fraction of}\\ i \text{ in the mixture}\end{pmatrix} \cdot \begin{pmatrix}\text{system}\\ \text{absolute}\\ \text{pressure}\end{pmatrix} = y_i P \tag{3.2}$$

For a pure gas the partial pressure is equal to the total pressure (because $y_i = 1.0$ for a pure gas). For any gas mixture the sum of the partial pressures equals the total pressure, $\sum y_i P = P \sum y_i = P$ because the sum of the mol fractions is unity. The partial pressure is defined only for gases; this term is not used for solids and liquids.

An equivalent term, not as widely used but perhaps equally useful, is the *partial vapor pressure* of one species in a liquid, which is defined as the *pure species* vapor pressure of that species p_i multiplied by its mol fraction in the liquid x_i

$$\begin{pmatrix}\text{partial vapor pressure}\\ \text{of species } i\\ \text{in a liquid mixture}\end{pmatrix} \equiv \begin{pmatrix}\text{mol fraction of}\\ i \text{ in the mixture}\end{pmatrix} \cdot \begin{pmatrix}\text{vapor pressure}\\ \text{of pure species } i\end{pmatrix} = x_i p_i \tag{3.3}$$

(We regularly use y_i, for mol fractions in the gas phase and x_i for mol fractions in the liquid phase, P for the total pressure of the gas and p_i for the pure-species vapor pressure of species i in the liquid.) For a pure liquid the partial vapor pressure is equal to the pure liquid's vapor pressure at that temperature (because $x_i = 1.0$ for a pure liquid). For one kind of ideal solution (defined in Chapter 7) the sum of the partial vapor pressures equals the total vapor pressure of the liquid. The partial vapor pressure is defined only for liquids; it is occasionally used for solids and never for gases.

Using these definitions we can state *Raoult's law* (which would be better named "Raoult's useful estimating approximation," because it is not a law like the laws of thermodynamics). It says that for gas–liquid equilibrium for *each chemical species* present, the partial pressure in the gas is equal to the partial vapor pressure in the liquid. Mathematically, Raoult's law is

$$y_i P = x_i p_i \tag{3.4}$$

and by simple extension

$$P = \frac{\sum x_i p_i}{\sum y_i} = \sum x_i p_i \tag{3.A}$$

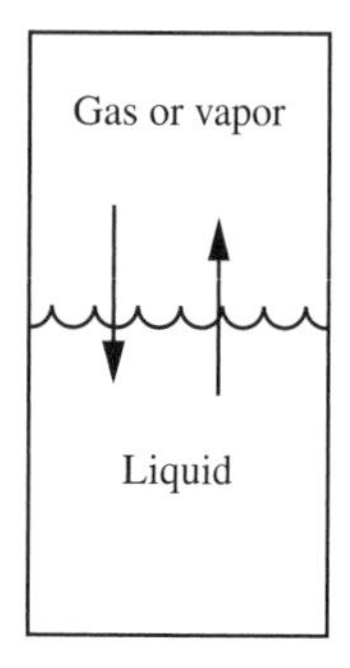

FIGURE 3.7 At equilibrium the net interchange of molecules of species i between gas and liquid is zero. This means that the two rates in opposite directions must be equal and opposite.

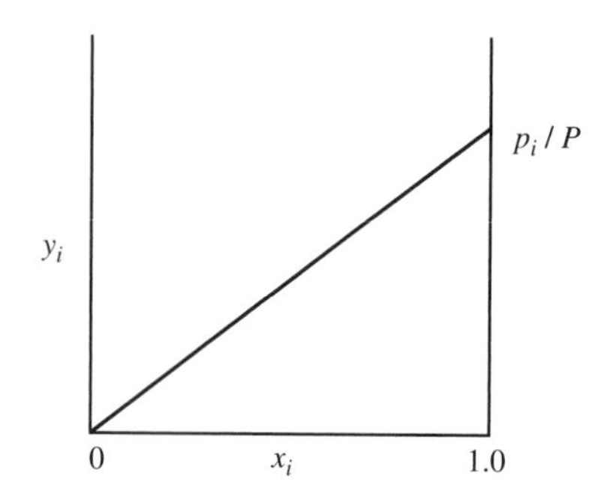

FIGURE 3.8 Raoult's law has a very simple geometrical interpretation. There is a corresponding plot for species j, species k, and so on.

Figure 3.7 shows a gas and liquid in equilibrium. If there is only one chemical species present, for example H_2O, then this is the same as Figure 3.4. In that case the number of molecules that pass from the gas to the liquid per s · m^2 is the number of gas molecules that strike the surface times the probability that they will stick. The number of molecules that pass from the liquid to the gas per s · m^2 is the number of molecules that strike the surface from the liquid side per s · m^2 times the probability that they have enough energy to pass through the surface and escape. The first of these is proportional to the gas pressure, the second to the liquid's vapor pressure. At equilibrium these are equal.

In the case of a mixture like air–water, the total number of gas molecules that hit the surface per s · m^2 is practically the same for a mixture as for a pure compound. But the number of water molecules that hit the surface per s · m^2 is that number times the fraction of the molecules in the mixture that are water, which is the mol fraction of water in the gas phase. Similarly, for the water molecules escaping from the liquid, the number of molecules hitting the liquid surface from below is more or less the same as for a pure liquid, but the percentage of those that are water is equal to the mol fraction of water in the liquid. This intuitive description is not rigorously correct. But it shows a simple way of looking at why the mol fractions play their crucial role in Raoult's law.

In all that follows on phase equilibrium, remember that for two or more phases to be in equilibrium, *for each chemical species* present in those phases the value of a quantity very much like the partial pressure or partial vapor pressure must be the same in all of the phases at equilibrium for the *net* rate of molecular movement from one phase to the other to be zero. The mathematical details may become complex, because nature seldom does things as simply as we would like, but ultimately all phase equilibrium calculations are based on finding the chemical compositions of the phases in equilibrium for which the values of this quantity, for each species, is the same in all the phases at equilibrium. For most common phase equilibrium problems that quantity will be a defined thermodynamic property called the *fugaciiy* (discussed in Chapter 7), which has the properties that for mixtures of ideal gases it is *identical* to the partial pressure and for liquid solutions of one common kind it is *identical* to the partial vapor pressure defined above. For real solutions that are not simple and for high pressures, we will still use the *equivalent* of Raoult's law, but using the fugacity instead of the simple partial pressure and partial vapor pressure we use here. Chapter 8 discusses this topic in more detail, with mathematical and thermodynamic rigor.

Figure 3.8 shows that Raoult's law has a very simple geometrical interpretation. At a constant value of p_i / P, y_i is linearly proportional to x_i. We can also see that if we increase the temperature while holding the total pressure constant, then p_i/P will increase because the vapor pressure p_i increases with increasing temperature. We also see that if $p_i / P = 1.0$, then $y_i = x_i$.

Example 3.1 Estimate the mol fraction of water vapor in air in equilibrium with water at 20°C = 68°F and one atmosphere pressure.

Rearranging Eq. 3.4, we have

$$y_i = \frac{x_i p_i}{P} \tag{3.5}$$

Here we have a ternary mixture of nitrogen, oxygen, and water. If we let the subscript i stand for water, we can say that

$$x_{\text{water}} = 1 - x_{N_2} - x_{O_2} \tag{3.B}$$

but we know from experience that the mol fractions of dissolved N_2 and O_2 in liquid water are quite small, so that we are safe in saying that $x_{\text{water}} \approx 1$. Later we must check to see that this approximation is satisfactory, which we do in Example 3.2. From any steam table we may look up the value of the vapor pressure of water at 20°C, finding $p_{\text{water}} = 0.023$ atm. We use this value in Eq. 3.5, with the total pressure, $P = 1.00$ atm, finding $y_{\text{water}} \approx 0.023$. ■

From previous parts of this course (see Figure 1.8) we know that for all liquids the vapor pressure increases with increasing temperature. From Eq. 3.5 we see that increasing the temperature, and hence the vapor pressure, increases the equilibrium water content of air in equilibrium with water. This explains, in a quantitative way, how dryers work, and how dew and clouds form. The value shown by Eq. 3.5 is the equilibrium concentration, in this case also called the *saturation concentration*, because it shows the maximum amount of water that air can hold at 20° C. If we start with air with this water concentration and cool it, P_{water} will decrease so that $(y_{water})_{equilibrium}$ will decrease; to come to the new position of equilibrium, the air must reject water. This is how dew occurs, how water condenses on the side of a cold drink container, and how clouds form.

If we want to use Eq. 3.5 to find the equilibrium concentration of oxygen and nitrogen in the water, we quickly find that we cannot look up the vapor pressures of pure liquid nitrogen and oxygen at 20°C = 298.15 K, because that is above the critical temperature of these materials (154.8 K for oxygen, 126.2 K for nitrogen, see Table A.1); they cannot exist as pure liquids at this temperature. The experimental measurements of the solubility of gases like oxygen and nitrogen in liquids like water show that we can still use the equivalent of Raoult's law, but in place of the liquid vapor pressure we must use a "pseudo vapor pressure," determined not from measuring the vapor pressure of the pure liquid as we did for water, but rather from the measured gas solubility data. This pseudo vapor pressure is called the *Henry's law constant*. Using it we may state *Henry's law* (better called "Henry's useful estimating approximation"), as follows: At modest pressures, gases that dissolve only to small amounts in liquids obey the equivalent of Raoult's law, with their pure species vapor pressure replaced by an empirical pseudo pressure called the "Henry's law constant." The mathematical form of Henry's law is the same as Eqs. 3.4 and 3.5, with the vapor pressure p_i, replaced by the Henry's law constant H_i:

$$y_i = \frac{x_i H_i}{P} \tag{3.6}$$

Table A.3 shows the reported values for the Henry's law constant for a variety of gases dissolved in water for several temperatures.

Example 3.2 Estimate the concentration of oxygen dissolved in water when air and water are at equilibrium at 20°C = 68°F and one atmosphere pressure. From Example 3.1 we know that $y_{water} = 0.023$, so that $Y_{N_2} + Y_{O_2} = 1 - 0.023 = 0.977$. The oxygen is 0.21 mol fraction of this mix, so that

$$y_{O_2} = 0.21 \cdot 0.977 = 0.205 \tag{3.C}$$

From Table A.3 we look up the Henry's law constant for oxygen in water at 20°C, finding $H = 40{,}100$ atm. Then, by direct substitution in Eq. 3.6,

$$x_{oxygen} = \frac{y_{oxygen} \cdot P}{H_{oxygen}} = \frac{0.205 \cdot 1\ \text{atm}}{40{,}100\ \text{atm}} = 5 \times 10^{-6} \tag{3.D}$$

This is the value shown in Table 3.1 as the mol fraction of oxygen dissolved in water at equilibrium at this temperature and pressure. By the same logic we find that

$$y_{N_2} = 0.79 \cdot 0.977 = 0.772 \tag{3.E}$$

and

$$x_{nitrogen} = \frac{y_{nitrogen} \cdot P}{H_{nitrogen}} = \frac{0.772 \cdot 1\ \text{atm}}{80{,}400\ \text{atm}} = 10 \times 10^{-6} \tag{3.F}$$

which is the value shown in Table 3.1. Often the concentrations of dissolved gases in water are shown as the equivalent mass, mols, or volume of the gas at standard temperature and pressure (stp), taken as 1 atm and 20°C in this book, for example,

$$\begin{pmatrix}\text{concentration of}\\ \text{dissolved oxygen}\\ \text{in equilibrium with}\\ \text{air at 1 atm and 20°C}\end{pmatrix} = 5 \times 10^{-6} \frac{\text{mol O}_2}{\text{mol solution}}$$

$$\cdot \frac{998.2 \dfrac{\text{g}}{\text{L solution}}}{18 \dfrac{\text{g}}{\text{mol solution}}} = 0.00028 \frac{\text{mol O}_2}{\text{L solution}} = 0.0090 \frac{\text{g O}_2}{\text{L solution}}$$

$$\frac{\begin{pmatrix}\text{volume of}\\ \text{O}_2\text{, stp}\end{pmatrix}}{\text{L solution}} = 0.00028 \frac{\text{mol O}_2}{\text{L solution}} \cdot 24.06 \frac{\text{L gas, stp}}{\text{mol}}$$

$$= 6.7 \frac{\text{mL O}_2\text{, stp}}{\text{L solution}} \quad \blacksquare \tag{3.G}$$

Figure 3.9 shows the relation of Henry's law to experimental data. The small circles represent the experimental data points; the curve is a simple parabolic fit of those points. The line is Henry's law, using the values from Table A.3 (linearly interpolated between 20 and 30°C).

This plot shows:

1. That the solubility of nitrogen in water is small; at a pressure of 1000 atm, the mol fraction of nitrogen in water (at 25°C) is just under 0.006. This is true for most common gases in liquids like water.

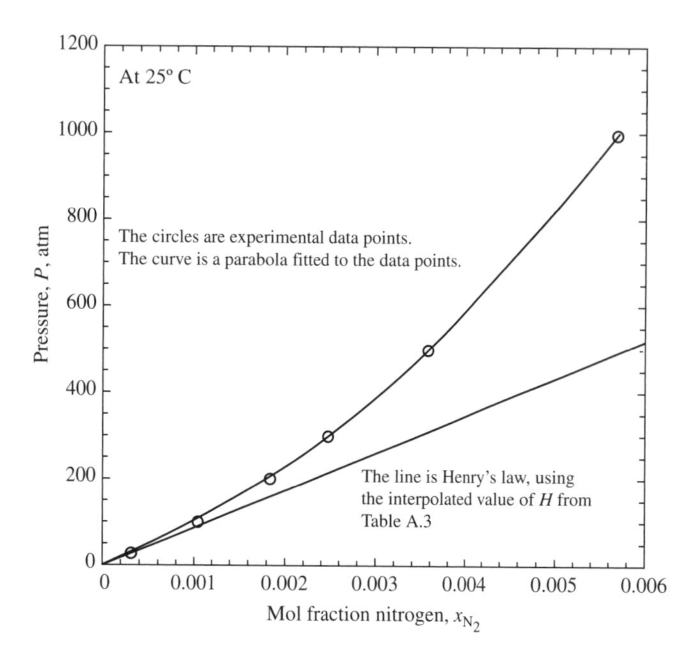

FIGURE 3.9 Experimental data [1] for the solubility of nitrogen gas in water at 25°C on *P*-*x* coordinates, compared to a Henry's law estimate of the same data, using an interpolated Henry's law constant from Appendix A.3.

2. For pressures of 25 atm or less, Henry's law reproduces the experimental data with sufficient accuracy for most (but not all!) purposes. For the 25 atm data point, Henry's law shows a nitrogen mol fraction of 103.5% of the measured value; for 100 atm it shows 106.9% of the measured value.
3. This agreement should not surprise us. The values in Table A.3 were found by plotting the experimental data as shown in this figure, drawing a straight line tangent to the curve as the mol fraction approaches zero, and extrapolating that line to $x_i = 1.0$. The intercept on that axis is the H shown in Table A.3.
4. This plot shows only the very left-hand side of a complete plot. If the horizontal axis went from 0 to 1.0, the data points would disappear into the left axis.
5. Using Henry's law for pressures greater than 100 atm (or nitrogen mol fractions greater than about 0.001) would lead to serious errors.
6. Figure 3.8 shows that Raoult's law covers the whole range of x_i from 0 to 1.0. The same is not true for Henry's law. If we were to continue to increase the nitrogen pressure in Figure 3.9 we would reach the pressure (thousands of atmospheres) at which the liquid would disappear.
7. At the low pressures and dissolved gas mol fractions for which we use Henry's law, its linear form is satisfactorily accurate. For higher pressures it is not.

We might try to estimate the Henry's law constant by extrapolating the vapor-pressure curve from the critical point, but the extrapolation is so large and thus uncertain that instead we normally use the experimental concentration values for dissolved gases, plotted as in Figure 3.9; this topic is discussed again in Chapter 9.

Of the gases shown in Table A.3, He, H_2, CH_4, N_2, O_2, and O_3 are above their critical temperature, so that they cannot exist as pure liquids at these temperatures. Acetylene, CO_2, ethane, and H_2S can exist as liquids at these temperatures, but only at high pressures. But most of these liquids do not dissolve to any substantial extent in water, so applying Raoult's law directly to them would lead to serious errors. However, we can use Henry's law, which applies at modest pressures to all gases that have only slight solubility in liquids. One may think about this situation as sketched in Figure 3.10. The mutual solubilities of liquid water and liquid ethane are very small. The ethane that dissolves in the water, and the water that dissolves in the ethane behave physically *as if* they had first evaporated, and then dissolved as gases in the liquid. CO_2 has a lower value of H than the other gases in Table A.3 because it enters into a chemical reaction with water, discussed in Chapter 13.

Raoult's law deals with vapor–liquid equilibrium; Henry's law deals with gas–liquid equilibrium. For normal atmospheric air, oxygen and nitrogen are above their critical temperatures and are gases, not vapors. The water vapor dissolved in air is also a gas, because it is part of a gaseous phase. But it is the kind of gas called a vapor, because it is below its critical temperature. Thus, this is a simultaneous gas-liquid equilibrium for N_2 and O_2 and a vapor–liquid equilibrium for H_2O. This mixed terminology does not seem to cause much confusion.

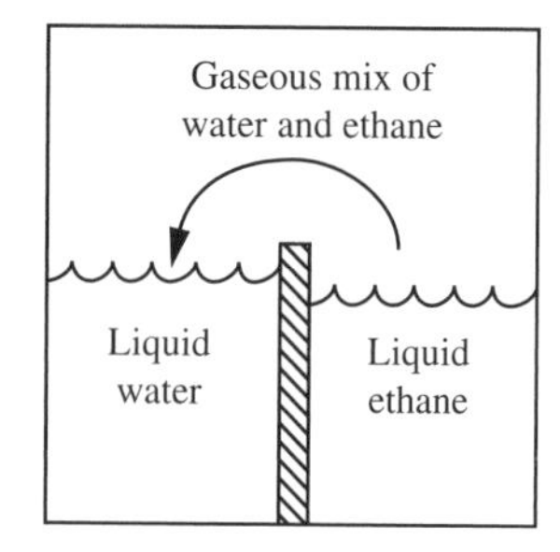

FIGURE 3.10 One may think of the application of Henry's law to the solubility of ethane in water by visualizing a closed container with an internal barrier, with water on one side and liquid ethane on the other. The ethane evaporates into the gas space and then dissolves in the water from the gas space, as shown by the arrow. Henry's law allows us to estimate the concentration of ethane in water in this situation. There is a corresponding equilibrium flow of water through the vapor into the liquid ethane, also estimable by Henry's law; both concentrations are small.

3.5 THE GENERAL TWO-PHASE EQUILIBRIUM CALCULATION

In Examples 3.1 and 3.2 the calculation was simple, because we knew from Table 3.1 that the liquid was practically pure water and that the concentration of water vapor in air was small. In general, for solutions involving more than two species we will not know this, and will have as many species equilibrium equations to solve simultaneously as there are species present, normally by trial and error. Before the age of computers this calculation was a giant pain, but our computers now do it for us quickly and easily. The three equilibrium equations from Examples 3.1 and 3.2 are

$$y_{\text{water}} P = x_{\text{water}} p_{\text{water}} \tag{3.H}$$

$$y_{\text{oxygen}} P = x_{\text{oxygen}} H_{\text{oxygen}} \tag{3.I}$$

$$y_{\text{nitrogen}} P = x_{\text{nitrogen}} H_{\text{nitrogen}} \tag{3.J}$$

a set of three equations with six unknowns. We have two additional equations that say that the mol fractions in each phase sum to 1.0, making this a system of five equations and six unknowns. The additional relation we need is supplied by the assumption that in the gas phase the molar ratio of nitrogen to oxygen $(y_{\text{nitrogen}}/y_{\text{oxygen}}) = 0.79/0.21$, independent of how much water vapor is dissolved in the air. In Examples 3.1 and 3.2 we were able to solve this set of six equations simply because we knew that three of the values were negligibly small. In general, we will not know that, so we (or our computers) will be solving sets of simultaneous equations of this type.

Example 3.3 Repeat the calculation of Table 3.1, using the above six equations, and not making the simplifications previously used.

Inserting numerical values, we have

$$\begin{aligned}
y_{\text{water}} \cdot 1 \text{ atm} &= x_{\text{water}} \cdot 0.023 \text{ atm} \\
y_{\text{oxygen}} \cdot 1 \text{ atm} &= x_{\text{oxygen}} \cdot 40{,}100 \text{ atm} \\
y_{\text{nitrogen}} \cdot 1 \text{ atm} &= x_{\text{nitrogen}} \cdot 80{,}400 \text{ atm} \\
y_{\text{water}} + y_{\text{oxygen}} + y_{\text{nitrogen}} &= 1 \\
x_{\text{water}} + x_{\text{oxygen}} + x_{\text{nitrogen}} &= 1 \\
\frac{y_{\text{oxygen}}}{y_{\text{nitrogen}}} &= \frac{0.21}{0.79} = 0.266
\end{aligned} \tag{3.K}$$

Solving this set of six linear equations in six unknowns by any of the standard simultaneous linear equation methods, we find the values shown in Table 3.A. ■

Table 3.A Values for Example 3.3

Variable	Value from Table 3.1, Based on Simplifications in Examples 3.1 and 3.2	Value Found by Solving Eqs. 3.K Simultaneously
y_{water}	0.023	0.0229997
y_{oxygen}	0.205	0.205273
y_{nitrogen}	0.772	0.771726
x_{water}	0.999985	0.9999853
x_{oxygen}	5×10^{-6}	5.119×10^{-6}
x_{nitrogen}	10×10^{-6}	9.598×10^{-6}

This calculation was done iteratively on a spreadsheet, although it could have been done analytically by any of several procedures. It shows that the shortcuts taken in Examples 3.1 and 3.2 lead to very small errors in Table 3.1. For more complex equilibria, treated in subsequent chapters, the equations will generally not be linear and often will be transcendental, so that analytic solutions will be impossible; the iterative computer solution works in those cases. (You should not believe that the values in Table 3.A are reliable to 7 significant figures; they are shown thus so that the small effect of the simplifying assumptions becomes clear!) Looking ahead to Chapters 7–9, we will see that they are largely devoted to understanding, computing, and correlating experimental values that replace the Raoult's law and Henry's law equations in the above examples. If we must compute the vapor–liquid equilibrium for a set of some three chemicals whose interactions are more complex than those for air and water, we will use the methods in Chapters 7–9 to find suitable replacements for the Raoult's and Henry's law statements in these examples. The material balance statements will be unchanged. As we deal with the complexities of Chapters 7–9, remember that there we are simply finding replacements for the simple equilibrium statements that took part in the six simultaneous equations in this example.

3.6 SOME SIMPLE APPLICATIONS OF RAOULT'S AND HENRY'S LAWS

If we know the system temperature we can use Raouh's law (and the assumption that the vapor is an ideal gas) to solve some simple vapor–liquid equilibrium problems.

Example 3.4 Estimate the vapor pressure and the composition of the vapor in equilibrium with a liquid that is 80 mol% benzene and 20 mol% toluene, at 20°C, assuming that benzene and toluene behave according to Raoult's law.

Here we calculate the vapor pressures of benzene and toluene at 20°C using the Antoine equation (see Chapter 5) and Table A.2. For benzene

$$\log \frac{p}{\text{torr}} = A - \frac{B}{T/°\text{C} + \text{C}}$$
$$= 6.90565 - \frac{1211.033}{20 + 220.79} = 1.87623 \quad \text{(3.L)}$$
$$p = 10\,\text{E}\,1.87623\ \text{torr} = 75.2\ \text{torr}$$

and similarly, for toluene, $p = 21.8$ torr. Then we can compute that

$$y_{\text{benzene}}\,P = x_{\text{benzene}} \cdot p_{\text{benzene}} = 0.8 \cdot 75.2\ \text{torr} = 60.2\ \text{torr} \quad \text{(3.M)}$$

and correspondingly for toluene, $y_{\text{toluene}}\,P = 0.2 \cdot 21.8 = 4.36$ torr. If we add these two values we will have

$$y_{\text{benzene}}\,P + y_{\text{toluene}}\,P = (y_{\text{benzene}} + y_{\text{toluene}})P = 64.6\ \text{torr} \quad \text{(3.N)}$$

But we know that $y_{\text{benzene}} + y_{\text{toluene}}$ must equal 1.00, so the total pressure must be 64.6 torr. To find either mol fraction in the gas phase we observe that the mol fraction of any species is equal to that species' partial pressure divided by the total pressure, so the two mol fractions are 60.2/64.6 = 0.932 and 4.36/64.6 = 0.068. ■

Example 3.5 Estimate the concentration of benzene in air that is saturated with benzene at 20°C.

This repeats Example 3.1, with the vapor pressure of benzene taken from Example 3.4. Thus,

$$y_i = \frac{x_i p_i}{P} = \frac{1.00 \cdot 75.2\ \text{torr}}{760\ \text{torr}} = 0.099 \approx 0.1 \approx 10\% \quad \text{(3.O)}$$

which is the value shown in Section 1.1. ■

We may repeat Example 3.4 for a variety of values of x_b and plot the results as shown in Figure 3.11. We see that the vapor pressure of the solution (at constant temperature) is linearly dependent on the mol fraction of benzene, and that the mol fraction of the species with the higher vapor pressure (benzene in this case) is higher in the vapor phase than in the liquid phase. Separation of miscible liquids like benzene and toluene by distillation is based on this fact.

If we know the system pressure, but not the temperature, in principle the problem is the same, but in practice it is more difficult, always leading to a numerical (trial-and-error) solution.

Example 3.6 At what temperature will the benzene–toluene mixture in Ex. 3.4 have a vapor pressure of one atmosphere? (That is, what is the normal boiling point temperature for this mixture?) Assume that both species obey Raoult's law.

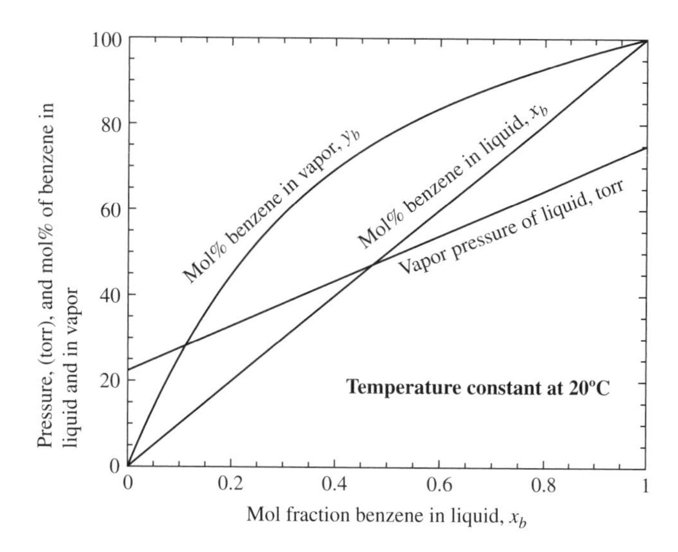

FIGURE 3.11 Computed liquid vapor pressure and vapor mol% benzene, according to Raoult's law, for a mixture of benzene and toluene at a constant temperature of 20°C. (Here we use mol percent instead of mol fraction on the ordinate so that both curves plot on the same scale.) The line for mol% benzene in the liquid is shown because it is traditionally shown in this type of diagram.

Writing the Raoult's law expression for each species, adding them, and grouping terms we have

$$(y_{\text{benzene}} + y_{\text{toluene}})P = x_{\text{benzene}} \cdot p_{\text{benzene}} + x_{\text{toluene}}\,p_{\text{toluene}} \quad \text{(3.P)}$$

The terms on the left of the equation are known, (1·1 atm) = 1 atm, and the two mol fractions on the right are known. The two pure species vapor pressures on the right depend on temperature alone (see Chapter 5, or Figure 1.8, which shows such a curve for water). In principle we can substitute the vapor pressure equations for benzene and toluene in the above equation and solve. But in practice the useful vapor pressure equations always involve logarithms, and equations containing two different logarithms have no analytic solutions, so this is inherently a trial-and-error problem. Using the Antoine equation and Table A.2 for each pure species vapor pressure and guessing values of T we may compute Table 3.B and we see that 84.377°C the vapor pressure is 760 torr = 1 atm. (We should not believe the computed temperature to more than 3 significant figures, but the value shown makes the calculated value of $P = 760.00$ torr.) From Raoult's law we see that the mol fractions in the vapor are

$$y_b = \frac{x_b p_b}{P} = \frac{0.8 \cdot 865.4\ \text{torr}}{760\ \text{torr}} = 0.911$$
$$y_t = \frac{x_t p_t}{P} = \frac{0.2 \cdot 338.3\ \text{torr}}{760\ \text{torr}} = 0.089 \quad \text{(3.Q)}$$

Table 3.B Solution to Example 3.6

Guessed T (°C)	p_{benzene} (torr) Calculated from the Antoine Equation	p_{toluene} (torr) Calculated from the Antoine Equation	P (torr) Calculated from Eq. 3.P
80	757.7	291.5	664.4
83	830.3	323.0	728.8
84.377	865.4	338.3	**760.0**
86	908.3	357.1	798.0
90	1021.0	407.1	898.2

This type of problem is a pain to calculate by hand (a giant pain for systems with a large number of species), but can be solved quickly and easily on a spreadsheet or other computer program (as was done here).

This solution was done intuitively. Formally, the set of equations we are solving is

$$
\begin{aligned}
x_b + x_t &= 1.00 \\
y_b + y_t &= 1.00 \\
y_b &= \frac{x_b p_b}{P} \\
y_t &= \frac{x_t p_t}{P} \\
\frac{p_b}{\text{torr}} &= 10\,\text{E}\left(6.90565 - \frac{1211.003}{T/°\text{C} + 220.79}\right) \qquad (3.\text{R}) \\
\frac{p_t}{\text{torr}} &= 10\,\text{E}\left(6.95334 - \frac{1343.943}{T/°\text{C} + 219.337}\right) \\
P &= 760 \text{ torr} \\
x_b &= 0.8
\end{aligned}
$$

■

The student may verify that the above intuitive solution satisfies all of Eqs. 3.R. Comparing this equation set to the equation set 3.K, we see that it is longer because instead of the vapor pressures being specified (which we can do for a known T), the vapor pressures are shown as functions of T. In Chapter 8 we will repeat this type of calculation, for more complex equilibria than those represented by Raoult's law, showing that the genera] procedure for this problem is to write Eq. 3.R in a spreadsheet and use the spreadsheet's numerical methods to find the value of T that solves this set of equations.

We may repeat this calculation for various values of x_b and plot the results as shown in Figure 3.12. This is a phase diagram, just as Figure 1.8 is, but a more complex one because we have two chemical species, compared to the one species there. To have a two-dimensional representation, we must hold some variable constant, in this case the pressure. (We may profitably think of this as a constant-pressure slice through a three-dimensional figure, whose axes are pressure, temperature, and mol fraction benzene; see Chapter 10.)

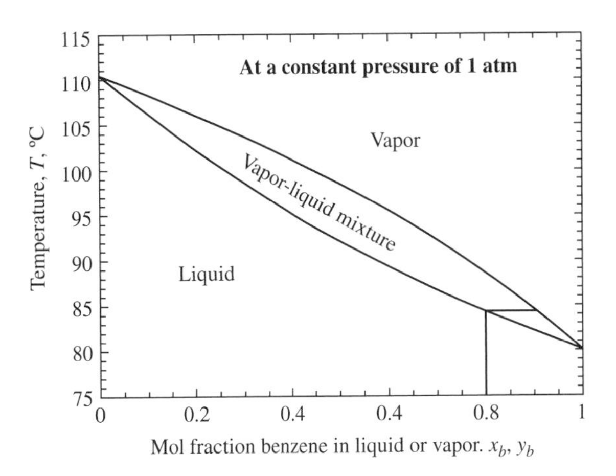

FIGURE 3.12 Computed temperature, liquid, and vapor compositions for benzene–toluene at a constant pressure of 1 atm. The results of Example 3.6 are shown at the right; for $x_b = 0.8$, $T = 84.4$°C, and $y_b = 0.911$.

Here, at a constant pressure of 1 atm, we see that at the left, the vapor–liquid equilibrium temperature is the normal boiling point (NBP) temperature of pure toluene, 110.6°C, and at the right the equilibrium temperature is the NBP of pure benzene, 80.1°C. (NBP is discussed in Chapter 5.) The two curves divide the space into three regions, just as the vapor-pressure curve divided Figure 1.8 into two regions. Above and to the right of the uppermost curve, only gas (or vapor) can exist. Below and to the left of the lowest line, only liquid can exist. In the region between the two lines vapor–liquid mixtures exist. In Chapter 8 we will return to this figure and discuss how to use it.

Example 3.7 An 0.25-L glass of water at 0°C is brought into a room at 20°C. Initially the water was in equilibrium with air at 0°C. How much oxygen and nitrogen must it reject to come to equilibrium at 20°C?

From Example 3.3 we know the mol fractions of oxygen and nitrogen in the water at 20°C. If we repeat the calculations in Example 3.3, using the Henry's law constants for oxygen and nitrogen at 0°C for water from Table A.3 (2.55 and 5.29×10^4 atm) and using the vapor pressure of water as 0.006 atm at 0°C, we will find the mol fractions of dissolved nitrogen and oxygen to be 8.19×10^{-6} and 14.84×10^{-6}. The amount of oxygen to be rejected is

$$
\begin{aligned}
\text{oxygen rejected} &\approx (0.25\,\text{L})(8.19 - 5.12) \\
&\quad \times 10^{-6} \frac{\text{mol}\,O_2}{\text{mol solution}} \frac{\text{mol solution}}{0.018\,\text{L}} \qquad (3.\text{S}) \\
&= 4.26 \times 10^{-5}\,\text{mol}\,O_2 = 1.36 \times 10^{-3}\,\text{g} \\
O_2 &= 1.03\,\text{mL}\,(\text{stp})\,O_2
\end{aligned}
$$

and correspondingly for nitrogen, 7.23×10^{-5} mol $= 2.03 \times 10^{-3}$ g $= 1.75$ mL (stp). The total volume of air rejected is $1.03 + 1.75 = 2.78$ mL (stp). Here the equation for oxygen rejected shows an approximately equal to sign ($\approx$) because it assumes that the mols of solution does not change. It actually decreases by $(4.26 + 7.23) \times 10^{-5}$ mols, making a negligible change in the answer (see Problem 3.20). ■

In cold weather, students may try this experimentally in their kitchens; they will see roughly the calculated volume of bubbles clinging to the side of the glass as it warms.

Example 3.8 A diver's blood and other bodily fluids are saturated with nitrogen at 5 atm pressure and 0.79 mol fraction nitrogen. The diver comes to the surface and her blood and other bodily fluids come to equilibrium with the atmosphere. How much nitrogen must she give up? Assume that her body mass is 55 kg, that 75% is water, that her body temperature is 37°C, and that the solubility of nitrogen in bodily fluids is the same as in pure water (only a fair assumption, [2]).

At 37°C (98.6°F) the Henry's law constant for N_2 interpolated from Table A.3 is 10.05×10^4 atm, so that

$$\begin{pmatrix}\text{mols } N_2\\ \text{rejected}\end{pmatrix} = \begin{pmatrix}\text{mols bodily}\\ \text{fluids}\end{pmatrix}(x_{N_2.5\text{ atm}} - x_{N_2.1\text{ atm}}) = \left(\frac{m_{\text{bodily fluids}}}{M_{\text{bodily fluids}}}\right) \times \left(\frac{y_i \cdot 5\text{ atm}}{10.05 \times 10^4\text{ atm}} - \frac{y_i \cdot 1\text{ atm}}{10.05 \times 10^4\text{ atm}}\right) \quad (3.T)$$

Here the two y_i, are slightly different from each other, because the water vapor content of the air is only 1/5 as large at 5 atm as at 1 atm. If we simplify by ignoring that and taking the $y_{\text{nitrogen}} \approx 0.79$ (but see Problem 3.8!), we can solve approximately:

$$\begin{pmatrix}\text{mols } N_2\\ \text{rejected}\end{pmatrix} \approx \left(\frac{55{,}000\text{ g} \cdot 0.75}{18\text{ g/mol}}\right) \cdot 0.79 \times \left(\frac{5\text{ atm}}{10.05 \times 10^4\text{ atm}} - \frac{1\text{ atm}}{10.05 \times 10^4\text{atm}}\right) = 0.072\text{ mol} = 2.02\text{ g} = 1.73\text{ L stp} \quad (3.U)$$

■

This surprisingly large value shows why "the bends" can be a serious or fatal problem. This treatment leaves out many interesting details about the bends, [2].

3.7 THE USES AND LIMITS OF RAOULT'S AND HENRY'S LAWS

Raoult's and Henry's laws are widely used because they are simple. However, as discussed above, these are strong simplifications of the real behavior of nature. Later in this book we discuss how thermodynamics lets us replace these laws with more accurate and reliable rules. However, we will see that these more reliable rules will enter into calculations like Example 3.3, and play exactly the same roles that Raoult's and Henry's laws did in that example. The following general guidelines may help you decide when you can use these simple rules, and when you cannot:

1. In a dilute solution of any kind, Raoult's law will apply satisfactory to the *solvent,* but probably not to the solute. For example, in beer ($\approx$3% ethanol, $\approx$96% water, $\approx$1% carbon dioxide, plus small amounts of other materials) the behavior of the water is well described by Raoult's law, but the behavior of the ethanol or carbon dioxide is not.
2. If solute and solvent are chemically similar, like benzene and toluene, then Raoult's law will apply satisfactorily for both solute and solvent, over the whole range of possible concentrations at modest pressures.
3. If the solute and solvent interact strongly chemically (for example, solutions of strong acids, such as H_2SO_4, and bases, such as NaOH in water), then Raoult's law gives poor estimates of the behavior.
4. Henry's law is useful for the solution of most gases in water, except for gases that interact chemically with water, such as HC1, NH_3, and SO_2.
5. Henry's law is widely used for liquids that are strongly immiscible with water, such as mercury and hydrocarbons, in which the minuscule amount of the other material dissolved in water behaves as if it had first vaporized and then dissolved *as a gas* in the water.
6. Henry's law can also be used for small amounts of gases dissolved in liquids other than water, as discussed in Chapter 9.
7. We will see in Chapter 8 that Raoult's law, modified by adding an activity coefficient to account for nonideal behavior, is applicable and widely used for vapor–liquid equilibrium calculations and some other kinds of equilibrium calculations.

3.8 SUMMARY

1. For any physical or chemical equilibrium, the temperature must be uniform throughout all the phases at equilibrium. Unless the pressure is balanced by gravity, spring, surface, or osmotic pressures (see Chapter 14), the pressure must also be uniform throughout all the

phases at equilibrium. Unless some parts of the system are separated from others with perfect electrical insulators, the electric potential must be the same throughout the system.

2. For gas–liquid equilibrium of more than one chemical species, the compositions of the two phases in equilibrium will be very different from each other. Many systems of great chemical and environmental engineering importance are of this type.
3. For these systems, at equilibrium, there is a phase equilibrium statement for each of the species. In addition, there is a statement for each phase that the mol fractions sum to 1.0. To complete the calculation we usually need more relations, normally the specification of some mol fractions or mol fraction ratios. This set of equations can be solved to find the compositions of both phases. Normally this set will not have an analytical solution; computers solve the set easily.
4. Raoult's and Henry's laws are widely used to supply the equilibrium relations in the above equation sets. These are *useful estimating approximations,* not laws like Newton's laws of mechanics or the laws of thermodynamics.
5. The more reliable estimating methods, based on thermodynamics and developed in Chapters 7–9, become the same as Raoult's and Henry's laws for some very simple (but very common) systems (e.g., ideal gases).

PROBLEMS

See the Common Units and Values for Problems and Examples. An asterisk (*) on the problem number indicates that the answers is in Appendix H.

3.1* Schmidt-Nielsen [3] reports that for almost all mammals the blood hemoglobin concentration is about the same ($\approx$ 130 g/L) and that the corresponding oxygen concentration of the blood leaving the lungs (for air at 1 atm and 21% oxygen) is 175 mL of oxygen (stp) per liter. For this problem assume that the blood leaving the lungs is in equilibrium with air at 1 atm and ignore the difference between the 20°C in the examples in this chapter and the 37°C temperature of the human body.

(a) How does the oxygen content in the blood (mL/L, stp) compare to the equilibrium volume of oxygen dissolved in pure water at 20°C, shown in Example 3.2?

(b) For this problem (but not in real life) you may assume that the hemoglobin in the red cells is a solid, so that a liter of blood consists of 130 g of solid red blood cells and 870 g of pure water (plus dissolved oxygen). What is the distribution of the oxygen between the pure water and the solid hemoglobin in the red blood cells?

(c) Using the results from (b) estimate the weight fraction of oxygen in the hemoglobin. Assume that the hemoglobin has the same density as water (only approximately true!)

3.2* Some safety officials are worried about exposing workers to mercury ($M = 200.6$ g/mol). If the air in a room is saturated with mercury (vapor pressure at 20°C = 0.0012 torr), assuming that the equilibrium between mercury and air obeys Raoult's law,

(a) What will the mol fraction of mercury in the air be?

(b) What will the concentration be, expressed in g/m^3?

(c) The TLV (TLV = threshold limit value = the permissible concentration to which workers may be exposed for an 8-h shift) for elemental and inorganic mercury was 25 $\mu g/m^3$ in the United States in 2012. What is the ratio of this value to the value you computed in part (b)?

(d) An average adult male inhales about 15 kg/day of air. How much mercury would such an adult inhale in an 8-h shift, if the air were saturated with mercury?

3.3 An underground soil layer is contaminated with benzene at 20°C (see Example 3.5). We wish to remove the benzene by blowing air through the soil, bringing the air to the surface, and incinerating it to destroy the benzene. We are able to pass 10 lbm/h of air through the soil. How many pounds per hour of benzene will we remove if we are able to bring the air to saturation, with respect to liquid benzene?

3.4 In Problem 3.3 it has been suggested that if we heated the air and the soil layer from the 20°C in that example to 40°C, the process would work much better. Estimate the benzene removal rate for this temperature and the same airflow rate as in Problem 3.3.

3.5 Estimate the mol fraction of toluene in air that is in equilibrium with practically pure toluene at 20°C,

(a) At a total pressure of 1 atm.

(b) At a total pressure of 5 atm.

3.6* Estimate the concentration of ethane dissolved in water, saturated with ethane gas at 20°C and 1 atm.

3.7 A piston and cylinder contains methane (CH_4) and water at 20°C and 1 atm pressure. There are one gas phase and one liquid phase present. Estimate the equilibrium concentration of methane in the liquid phase.

3.8 Repeat Example 3.8, taking the change in N_2 content due to the water vapor present in the equilibrium air at 5 atm and at 1 atm into account. The vapor pressure of water at 37°C ≈ 0.062 atm.

3.9 A kilogram of water is heated from 20 to 40°C. Initially, the water is in equilibrium with oxygen at 1 atm pressure and 20°C. Finally, the water is in equilibrium with oxygen also at 1 atm and 40°C. How much oxygen must flow into or out of the water, during this heating process, to maintain the equilibrium?

3.10 Repeat Example 3.8 for an initial pressure of 10 atm. What depth does this correspond to (see any fluid mechanics text)? Normal recreational divers do not go nearly this deep. The deepest recorded dive depth using ordinary (open-circuit) diving equipment is about 1000 ft.

3.11 The Henry's law version shown in this chapter is the most common version in chemical engineering. However, the same data can be presented as

$$y_i P = \begin{pmatrix} \text{some measure} \\ \text{of the concentration} \\ \text{of } i \end{pmatrix} \begin{pmatrix} \text{a Henry's law} \\ \text{constant in} \\ \text{appropriate units} \end{pmatrix} \tag{3.7}$$

For example, Clever et al. [4] show for the solubility of mercury in water at 298 K that if the measure of concentration is the mol fraction of mercury in the water, the Henry's law value is $H = 4.7 \times 10^4$ kPa. The same source also shows the values for the concentration being expressed in mol/kg. Another common way is to choose the concentration variable as mol/m^3.

Let H^* be the Henry's law constant for the concentration expressed in mol/kg, and H^{**} be that for the concentration expressed in mol/m^3. Show the numerical values and the dimensions of H^* and H^{**} for mercury in water.

3.12* Estimate the mol fraction of dissolved CO_2 in an ordinary soft drink at 20°C. Assume that the partial pressure of CO_2 in the gas phase = 2 atm. We will repeat this problem in Chapter 13, taking into account the chemical reaction of carbon dioxide with water, showing that the answer found here is close to correct.

3.13 Estimate the mass and volume (stp) of dissolved CO_2 in an ordinary soft drink at 20°C. Use the equilibrium mol fraction of CO_2 calculated in the preceding problem. An ordinary soft drink has a liquid mass of ≈ 0.75 lbm, and may be considered, for this problem, to be pure water with dissolved CO_2.

3.14 Rework Example 3.3 for the gas being 100% oxygen, instead of air. How many variables are there? How many equations? What is the resulting computed set of concentrations?

3.15 Show the equivalent of Table 3.A for 40°C. Include both the hand and the computer versions of the answer.

3.16 Rework Example 3.3 for air and water, including the carbon dioxide in the air. How many variables are there? How many equations? In atmospheric air the carbon dioxide concentration changes slightly from place to place and time to time, but on the average, for dry air, the concentration is about 390 ppm. Assume that the ratio of mole fractions of CO_2 to N_2 is always 0.000,390/0.79, independent of the water vapor content of the air.

3.17 Repeat Example 3.4 for 20 mol% benzene, 80 mol% toluene. Compare your results to Figure 3.11.

3.18 Repeat Example 3.6 for 20 mol% benzene, 80 mol% toluene. Compare your results to Figure 3.12.

3.19 A benzene–toluene vapor mixture has $y_{\text{benzene}} = 0.4$, at a pressure of 1 atm. Estimate

(a) The mol fraction benzene (x_{benzene}) in the liquid in equilibrium with this vapor.

(b) The temperature at which they are in equilibrium.

3.20 The calculation in Example 3.7 ignores the change in number of mols in the solution when the nitrogen and oxygen are rejected. Show the complete solution, taking that into account, and show how much difference this makes.

3.21 Show the calculation of the mol fractions at 0°C in Example 3.7.

3.22 Show that the trial-and-error solution to Example 3.6 is equal to the solution to equation set 3.R.

REFERENCES

1. Wiebe, R., V. L. Gaddy, and C. J. Heins. The solubility of nitrogen in water at 50, 75, and 100°C from 25 to 1000 atmospheres. *JACS* 55:947–953 (1933).
2. Phillips, J. L. *The Bends*. New Haven; Yale University Press, p. 214 (1998).
3. Schmidt-Nielsen, K. *Scaling, Why Is Animal Size So Important?* New York: Cambridge University Press, p. 115 (1984).
4. Clever, H. L., M. Iwamoto, S. H. Johnson, and H. Miyamoto. *Solubility Data Series, Vol. 29: Mercury in Liquids, Compressed Gases, Molten Salts and Other Elements*, Oxford, UK: Pergamon, International Union of Pure and Applied Chemistry, pp. 1–21 (1987).

4

MINIMIZATION OF GIBBS ENERGY

4.1 THE FUNDAMENTAL THERMODYNAMIC CRITERION OF PHASE AND CHEMICAL EQUILIBRIUM

After all the introductory descriptive material, we are finally ready to see precisely what thermodynamics tells us about phase and chemical equilibrium. Figure 4.1 shows a simple, isothermal piston and cylinder arrangement, containing some unspecified mass of some unspecified substance or mixture of substances, in one or more phases. We need not know the identity or state of the contents for the argument here, so we do not specify the contents, and we obtain a very general result.

The system is originally at equilibrium, and we now allow a small change in the system to occur, that is, a condensation or evaporation or crystallization of a small amount of material, or a small chemical reaction. Kinetic, potential, surface, electrostatic and electromagnetic energy effects are negligible, but the change will probably involve a small amount of heat given off or absorbed by the system, and a small volume change, which causes a small piston movement. We may write the energy balance for the contents of this system (a batch process) as

$$dU = dQ + dW = dQ - PdV \tag{4.1}$$

Here the only work term is that of driving back the piston. A differential change in any system at equilibrium is a reversible process; for a reversible batch process, the second law shows us that

$$dS = \frac{dQ}{T} \quad \begin{bmatrix}\text{only for reversible} \\ \text{batch processes}\end{bmatrix} \tag{4.2}$$

Eliminating dQ between Eqs. 4.1 and 4.2 we find

$$dU = TdS - PdV \tag{4.3}$$

Now we make the algebraic substitutions,

$$TdS = d(TS) - SdT \quad \text{and} \quad PdV = d(PV) - VdP \tag{4.4}$$

Which changes Eq. 4.3 to

$$d(U + PV - TS) = -SdT + VdP \tag{4.5}$$

But we have restricted the process to constant temperature and pressure, so the two terms on the right of Eq. 4.5 are zero, and we can say that for any reversible equilibrium change at constant T and P

$$d(U + PV - TS) = dG = 0 \tag{4.6}$$

We defined G in Section 2.5; here we see the most important application of G. For any small change of state at equilibrium with T and P held constant, there is zero change in G. This is the principal reason for bothering to define G. It is normally called the Gibbs energy after Josiah Willard Gibbs (1790–1861), who made the pioneering investigation of the thermodynamics of equilibrium. We use G for the Gibbs energy, but the older literature and present-day chemists sometimes use F. The Helmholz energy, defined as

$$A = U - TS = G - PV \tag{4.7}$$

is widely used for processes at constant temperature and volume, and will appear again later in this book.

Physical and Chemical Equilibrium for Chemical Engineers, Second Edition. Noel de Nevers.

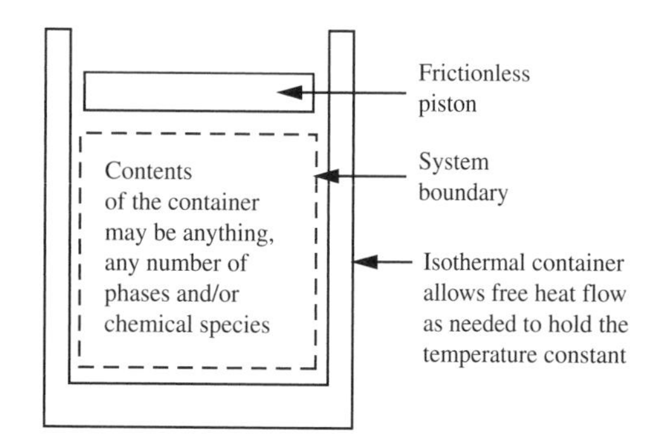

FIGURE 4.1 Very general system for illustrating the thermodynamics of a small, reversible, isothermal change.

Both the Gibbs and the Helmholz energies, G and A, are *convenience properties* as is the enthalpy H. We could have done all of thermodynamics and all of equilibrium without bothering to define them. We defined all of them because they are combinations that occur so often that their introduction allows a vast reduction in what we must write and speak to discuss thermodynamics. If that bothers you, go through the book, and wherever you see a G replace it with $U + PV - TS$.

From this simple example, we observe the following:

For any differential equilibrium change, chemical or physical or both, at constant T and P,

$$dG_{\text{sys}} = 0 \tag{4.8}$$

This result is placed in a box because it is so important. In Chapter 3 we saw that at equilibrium, T is uniform and that unless there are significant effects of gravity, springs, or surface tension, the phases and species at equilibrium are all at a constant pressure, so the restriction of constant T and P does not limit the applicability of this relation very much. The corresponding relations for other constraints (e.g., constant S and V) are shown in Appendix B.

Example 4.1 A large mass of steam and water are at equilibrium at 212°F = 671.7°R. We now allow 1 lbm of steam to condense to water at constant temperature and pressure. What is the Gibbs energy change? Using values from the steam table [1], we find

$$\Delta h_{\text{condensation}} = -970.3 \text{ Btu/lbm}$$
$$\Delta s_{\text{condensation}} = -1.4446 \text{ Btu/lbm} \cdot {}^\circ\text{R}$$

So

$$\Delta g_{\text{condensation}} = -970.3 - (671.7)(-1.4446) = 0.0 \text{ Btu/lbm} \tag{4.A}$$

■

This does not in any way prove that Eq. 4.8 is correct; the people who made up the steam tables knew about Eq. 4.8 and used it in their work, which guaranteed that Eq. 4.8 would be satisfied. But it does illustrate that the enthalpy change Δh for an isothermal vaporization, melting, or sublimation is equal to $T\,\Delta s$ for the same process in any correct table of thermodynamic properties.

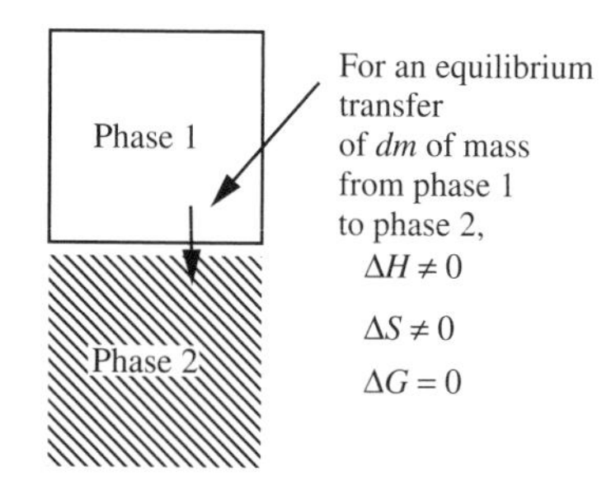

FIGURE 4.2 One more way to summarize the most important relation in phase equilibrium.

Figure 4.2 restates the conclusion of this section of the book. This is the most important idea in this book, and the basis for most of the others.

The fact that $dG_{\text{sys}} = 0$ for any differential change of a system at equilibrium at constant T and P shows that such a system must be a minimum or a maximum. But which? If we repeat the above derivation, and consider irreversible processes, then Eq. 4.2 will be replaced by

$$dS > \frac{dQ}{T} \quad \left[\begin{matrix}\text{for irreversible} \\ \text{batch processes}\end{matrix}\right] \tag{4.9}$$

and if we solve this for dQ, and substitute as we did before, we will find that for irreversible processes at constant T and P

$$d(U + PV - TS) = dG < 0 \quad [\text{for irreversible processes}] \tag{4.10}$$

Thus, the situation is as sketched in Figure 4.3. The equilibrium state is one at the bottom of a Gibbs energy basin. If the system can reduce its Gibbs energy, it will do so by an irreversible, spontaneous process. The equilibrium state is the state with the lowest Gibbs energy consistent with the constraints on the system, such as its temperature, pressure, and initial chemical composition.

Figure 4.4 shows the same plot as Figure 4.3, but includes a metastable state. In that state the system is at a local minimum of the Gibbs energy. Displacement from that local minimum (e.g., a small spark in a hydrocarbon–air mixture) can allow the system to make the transition to the true equilibrium state, with a lower Gibbs energy. (Many such metastable states are limited not by being in such a local Gibbs energy minimum, but by being limited by the rate of approach to equilibrium, i.e., a kinetic rather than a thermodynamic constraint.)

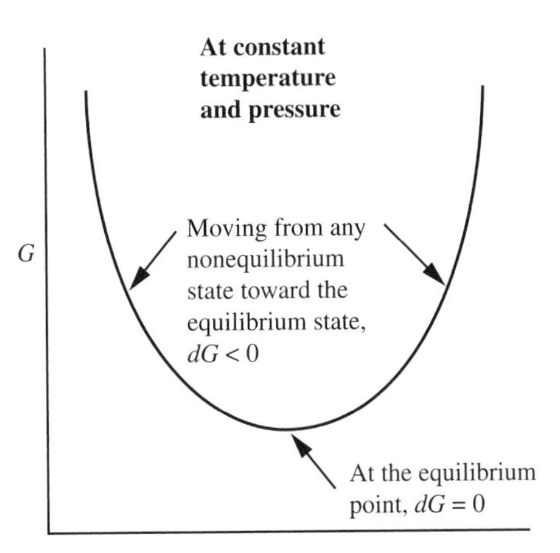
At constant temperature and pressure
G
Moving from any nonequilibrium state toward the equilibrium state, dG < 0
At the equilibrium point, dG = 0
Some property of the system, e.g., chemical composition, phase composition, or any other property

FIGURE 4.3 The equilibrium state is at a Gibbs energy minimum. All adjacent states have higher Gibbs energies. From these states any spontaneous processes takes the system toward the state of lowest Gibbs energy, which is the equilibrium state.

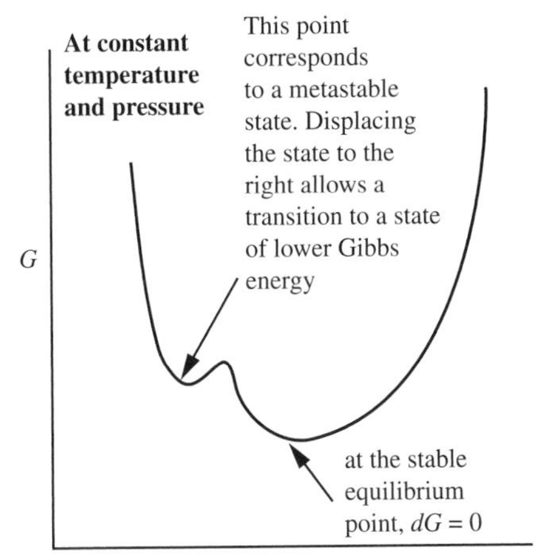
At constant temperature and pressure
This point corresponds to a metastable state. Displacing the state to the right allows a transition to a state of lower Gibbs energy
G
at the stable equilibrium point, dG = 0
Some property of the system, e.g., chemical composition, phase composition, or any other property

FIGURE 4.4 Same as Figure 4.3, but showing how a metastable state can have a local minimum Gibbs energy, but respond to a displacement by passing to the stable equilibrium, which has a lower Gibbs energy.

4.2 THE CRITERION OF EQUILIBRIUM APPLIED TO TWO NONREACTING EQUILIBRIUM PHASES

Figure 4.5 shows two phases at equilibrium inside a container in which the temperature and pressure are held constant. The phases could be gas–liquid, gas–solid, liquid–liquid, liquid–solid, or solid–solid without invalidating the mathematics that follows.

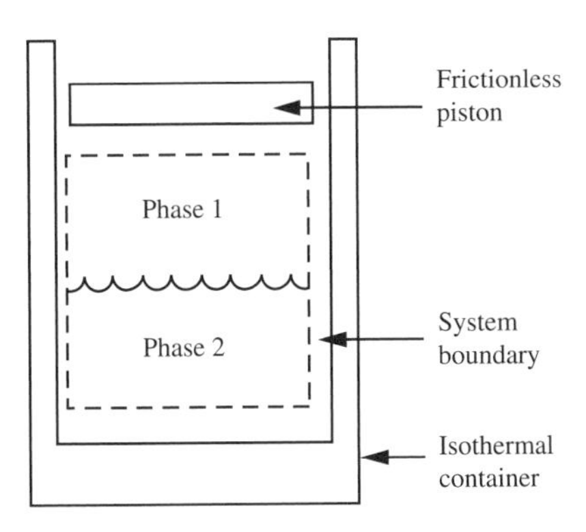
Frictionless piston
Phase 1
Phase 2
System boundary
Isothermal container

FIGURE 4.5 Simple, constant-pressure, two-phase system.

Here we have chosen the restraints as constant temperature and pressure. (The moveable piston allows volume change.) Other possible restraints are discussed in Appendix B. However, the constant-temperature, constant-pressure restriction is the one that proves useful for most industrial process calculations and for the calculation of tables of thermodynamic properties that we may then use for systems with some other kind of restraint, so we will use it here.

As shown in the figure, we have chosen as our system the entire contents of the container, including both phases. One might assume that at equilibrium, both phases would have the same chemical composition; that is not correct. Chapter 3 shows that for simple systems like air and water, the two phases in equilibrium have very different chemical compositions. Now, starting at equilibrium, let some small amount of mass (dm) move from phase 2 to phase 1. From Eq. 4.8 we know that there will be no change in the overall Gibbs energy of the system for this differential movement of mass.

The fact that the overall Gibbs energy of the system did not change with a small movement of mass from one phase to another is interesting, but we can find out much more detailed information by writing out the total differential of the Gibbs energy for each phase. To do so, we must decide what we consider the variables on which G depends. There are numerous choices, but the convenient one assumes that G of a phase depends on the temperature, pressure, and the number of mols of each species present. If we specify that only one phase is present and we specify the number of mols of each chemical species present and the temperature and pressure, we have completely defined the state of that phase. (To be convinced, try the "thought experiment" of considering how you could change such a single phase in any way without altering T, P, or one of the numbers of mols of some chemical species present. Here we omit the effects of electrostatic, electromagnetic, tensile, or surface energies.) In mathematical terms

$$G = G(T, P, n_a, n_b, \ldots) \tag{4.11}$$

where n_a is the number of mols of species a, and so on. We could also have done all of this in terms of a unit of mass, $dm_i = M_i \cdot dn_i$, where M_i is the molecular weight of i. The resulting equations would look the same, and the logic would be the same, until we came to chemical reactions, where the equations would become much more complex, because they would contain multiple values of M_i. Some of the following relations are shown in both dn_i and dm_i forms in Chapter 6. We could also have done it all in terms of mol fractions; the equations would be much more complex.

Performing the total differentiation for each phase we obtain

$$dG^{(1)} = \left(\frac{\partial G}{\partial P}\right)^{(1)} dP + \left(\frac{\partial G}{\partial T}\right)^{(1)} dT + \left(\frac{\partial G}{\partial n_a}\right)^{(1)} dn_a^{(1)} + \left(\frac{\partial G}{\partial n_b}\right)^{(1)} dn_b^{(1)} + \cdots \quad \text{[phase 1]} \tag{4.12}$$

$$dG^{(2)} = \left(\frac{\partial G}{\partial P}\right)^{(2)} dP + \left(\frac{\partial G}{\partial T}\right)^{(2)} dT + \left(\frac{\partial G}{\partial n_a}\right)^{(2)} dn_a^{(2)} + \left(\frac{\partial G}{\partial n_b}\right)^{(2)} dn_b^{(2)} + \cdots \quad \text{[phase 2]} \tag{4.13}$$

Here, the (number) superscripts refer to the phases, so that $n_b^{(2)}$ is the number of mols of species b in phase 2, and so on. In each of the partial derivatives, all of the other variables on the right-hand side of Eq. 4.11 are held constant. By material balance we know that the combined number of mols of species a *in both phases* cannot change (excluding a chemical reaction), so

$$dn_a^{(1)} + dn_a^{(2)} = 0 \quad \text{or} \quad dn_a^{(1)} = -dn_a^{(2)} \tag{4.14}$$

and similarly for the dn_bs. Thus, we can substitute and replace the values of $dn_a^{(2)}$ and $dn_b^{(2)}$ in Eq. 4.13 with minus $dn_a^{(1)}$ and $dn_b^{(1)}$. If we then add the two equations, we will find

$$\begin{aligned} dG_{\text{sys}} &= dG^{(1)} + dG^{(2)} \\ &= \left[\left(\frac{\partial G}{\partial P}\right)^{(1)} + \left(\frac{\partial G}{\partial P}\right)^{(2)}\right] dP + \left[\left(\frac{\partial G}{\partial T}\right)^{(1)} + \left(\frac{\partial G}{\partial T}\right)^{(2)}\right] dT \\ &\quad + \left[\left(\frac{\partial G}{\partial n_a}\right)^{(1)} - \left(\frac{\partial G}{\partial n_a}\right)^{(2)}\right] dn_a^{(1)} \\ &\quad + \left[\left(\frac{\partial G}{\partial n_b}\right)^{(1)} - \left(\frac{\partial G}{\partial n_b}\right)^{(2)}\right] dn_a^{(1)} + \cdots \end{aligned} \tag{4.15}$$

But we know from Eq. 4.8 that $dG_{\text{sys}} = 0$ and that we have held T and P constant so that $dP = dT = 0$, from which it follows that

$$\begin{aligned} 0 &= \left[\left(\frac{\partial G}{\partial n_a}\right)^{(1)} - \left(\frac{\partial G}{\partial n_a}\right)^{(2)}\right] dn_a^{(1)} \\ &\quad + \left[\left(\frac{\partial G}{\partial n_b}\right)^{(1)} - \left(\frac{\partial G}{\partial n_b}\right)^{(2)}\right] dn_b^{(1)} + \cdots \end{aligned} \tag{4.16}$$

This must be true for any choice we make of $dn_a^{(1)}$, $dn_b^{(1)}$, and for any number of species. That can be true only if all the terms in brackets are zero. Thus, we can say that there will be phase equilibrium in the all system if and only if

$$\left(\frac{\partial G}{\partial n_a}\right)^{(1)} = \left(\frac{\partial G}{\partial n_a}\right)^{(2)} \tag{4.17}$$

$$\left(\frac{\partial G}{\partial n_b}\right)^{(1)} = \left(\frac{\partial G}{\partial n_b}\right)^{(2)} \tag{4.18}$$

and so on for species $c, d, \ldots$. This result is so important that we spend some time here and all of Chapter 6 discussing the nature of the partial derivatives in Eqs. 4.17 and 4.18. $(\partial G/\partial n_a)^{(1)}$ is the increase in Gibbs energy in phase 1 that occurs if we add 1 mol of species a while holding constant the values of T, P and the number of mols of all other species $(b, c, d, \ldots)$ in that phase. In the most formal terms, we would write it as $[(\partial G/\partial n_a)_{T,P,n_b \cdots}]^{(1)}$ to remind ourselves what is being held constant. This derivative at constant T, P, n_b, and so on has been given a name and a symbol in thermodynamics; it is called *a partial molar derivative* or *partial molar property*. [For much of its life this kind of derivative was called a partial *molal* derivative, but that name is now rarely used. The terms partial *molar* and partial *molal* mean the same thing.]

The symbol for a partial molar property, widely used in thermodynamics, is

$$\left[\left(\frac{\partial(PROPERTY)}{\partial n_a}\right)_{T,P,nb,\text{ etc.}}\right]^{(1)} = \left(\overline{property}\right)_a^{(1)} \tag{4.19}$$

so that

$$\left[\left(\frac{\partial G}{\partial n_a}\right)_{T,P,n_b,\text{etc.}}\right]^{(1)} = \bar{g}_a^{(1)} \tag{4.20}$$

However, the most interesting partial molar property, $\bar{g}_a^{(1)}$, has a name and symbol of its own. J. Willard Gibbs, who was the first to show its importance, gave it a name and symbol before modern terminology was adopted. He called it the "chemical potential" and used the symbol μ_i. Thus, we have

$$\begin{pmatrix}\text{partial molar}\\ \text{Gibbs energy}\\ \text{of } a \text{ in phase 1}\end{pmatrix} = \left[\left(\frac{\partial G}{\partial n_a}\right)_{T,P,n_b,\text{etc.}}\right]^{(1)} = \bar{g}_a^{(1)} = \mu_a^{(1)}$$

$$= \begin{pmatrix}\text{chemical potential}\\ \text{of species } a \text{ in}\\ \text{phase 1}\end{pmatrix} \quad (4.21)$$

The later symbol is widely used.

We are all used to the idea of an electrical potential, often called voltage. Electrons flow as a result a potential difference or potential gradient (voltage difference or voltage gradient). For heat flow by conduction, the temperature plays the same role as does the electrical potential for flow of electrons, so that temperature is the "potential" for conductive flow of heat, and the same equations used for flow of electrons can be used for conductive heat flow with the symbols renamed. Gibbs recognized that the partial molar Gibbs energy plays the same role for chemical diffusion of one species through another and also for chemical reactions, so he called it the "chemical potential."

Substituting this definition in Eqs. 4.17 and 4.18, we get the working form of the equilibrium relationship:

$$\mu_a^{(1)} = \mu_a^{(2)} \quad (4.22)$$

$$\mu_b^{(1)} = \mu_b^{(2)} \quad (4.23)$$

and the same for c, d, ... This says that the chemical potential of a is the same in phase 1 as in phase 2, and that the chemical potential of b is the same in phase 1 as in phase 2. It does not say that the chemical potential of a is the same as that of b. That may or may not be so, but is not necessary for phase equilibrium. This does not say that the Gibbs energy per mole of phase 1, $g^{(1)}$, is the same as that of phase 2, $g^{(2)}$. For a pure species, such as in Example 4.1, they are the same, but for mixtures they generally are not.

As we proceed into the study of phase equilibrium, we will see that the chemical potential plays the key role. If we knew the chemical potential for every species of every possible chemical mixture in every phase, we would know all there possibly is to know about phase equilibrium and chemical equilibrium.

This whole discussion has been in terms of a small amount of matter moving from phase 1 to phase 2 for example by molecular diffusion. If we made up the system in Figure 3.5 by first putting pure liquid water in the container and then pure gaseous oxygen, some oxygen would dissolve in the water and some water would evaporate into the oxygen, both by diffusion. When the process ended, the system would be at equilibrium. If we repeat the whole derivation, allowing the small amount of matter to move by diffusion, we find that we will not change anything. Our result will still be Eqs. 4.22, 4.23, and so on. Thus, we can see that those are the equilibrium relations for diffusion. When we make up the system with pure oxygen and pure water, the value of the chemical potential of water in the pure liquid water $\mu_{\text{water}}^{(\text{liquid})}$ is greater than the chemical potential of water in the pure gaseous oxygen $\mu_{\text{water}}^{(\text{gas})}$. Water will evaporate into (diffuse into) the oxygen until the chemical potential of the water vapor in the oxygen gas is the same as the chemical potential of the water in the (practically pure) liquid water. When the chemical potential of water in both phases is the same and that of oxygen in both phases is the same, then the process stops and we have equilibrium. On a molecular level the process continues at equal rates in both directions; water and oxygen move into and out of the gas phase at equal rates so that the net rate is zero.

4.3 THE CRITERION OF EQUILIBRIUM APPLIED TO CHEMICAL REACTIONS

Figure 4.6 shows another of our isothermal, constant pressure containers, in this case containing a single phase in which a differential chemical reaction occurs. From Eq. 4.8 we know that if the system is at equilibrium, then $dG_{\text{sys}} = 0$ for a differential change of any kind, including a chemical reaction. Suppose the chemical reaction is of the form A $\rightarrow$ B; then the total differential of the Gibbs energy of the entire contents of the container is

$$dG = \left(\frac{\partial G}{\partial P}\right)dP + \left(\frac{\partial G}{\partial T}\right)dT + \mu_a dn_a + \mu_b dn_b \quad (4.24)$$

(Here we drop the superscript, because there is only one phase.) We know from the restraints of the problem that dP

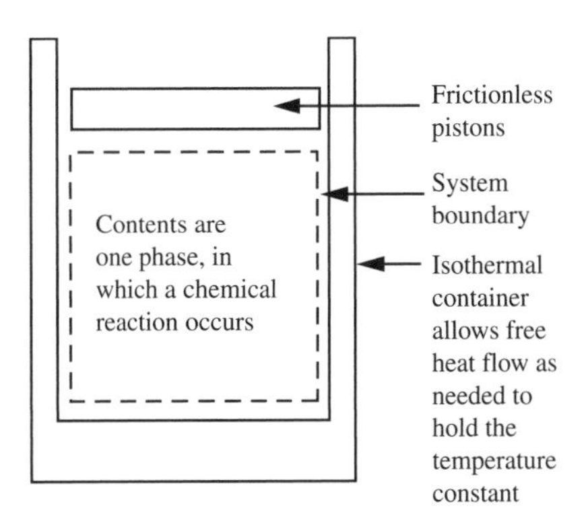

FIGURE 4.6 An isothermal, constant-pressure container containing only one phase, in which a chemical reaction occurs.

and dT are zero, and as shown above, dG is zero. Furthermore, from stoichiometry we know that

$$dn_a + dn_b = 0 \tag{4.25}$$

Taking all of these into account, we see that the restraints can all be satisfied only if

$$\mu_a = \mu_b \tag{4.26}$$

By entirely similar arguments, we can show that for a reaction of the form A + B → C, the condition for chemical equilibrium is

$$\mu_a + \mu_b = \mu_c \tag{4.27}$$

and similarly for any other balanced chemical equation we can write.

Thus, we see that the chemical potential, in addition to playing the key role in the description of equilibrium between two phases, also plays the key role in the description of equilibrium in chemical reactions. Chapter 12 shows the detailed consequences of this statement for chemical equilibrium.

4.4 SIMPLE GIBBS ENERGY DIAGRAMS

The Gibbs energy is the dominant property for equilibrium. This section shows graphical examples of how the Gibbs energy behaves, to help the reader develop an intuitive picture of this all-important function. Let us first consider a system consisting of only one chemical species in the form of two coexisting phases, gas (phase 1) and liquid water (phase 2). From the criterion listed above, we know that both phases are at the same temperature, pressure, and electrical potential, and that $\mu_a^{(1)} = \mu_a^{(2)}$. For any single *pure* species

$$\mu_a^{(1)} = \bar{g}_a^{(1)} = \left(\frac{\partial G}{\partial n_a}\right)_{T,P,n_b,\ldots} = g_a^{(1)} \tag{4.28}$$

We see that for a single pure species, the partial molar Gibbs energy is the same as the pure species Gibbs energy per mol. Furthermore, we see that for this equilibrium, the Gibbs energy per mol (or per g, kg, or lbm) is the same in each phase. Precisely the same results were obtained in Example 4.1 by writing the enthalpy and entropy changes for the transition from one phase to the other and solving for the Gibbs energy change, which was found to be zero.

To make up Gibbs energy diagrams we need the derivatives of G with respect to T and P. These are shown in Table 2.2. Their derivation is shown here because the values are so important for equilibrium calculations. First we write

$$G = H - TS = U + PV - TS \tag{4.29}$$

and take the total derivative; we obtain

$$dG = dU + PdV + VdP - TdS - SdT \tag{4.30}$$

But we know that for any uniform mass of matter, from the property equation (Chapter 2)

$$dU = TdS - PdV \tag{2.32}$$

so we may subtract Eq. 2.32 from Eq. 4.30 obtaining

$$dG = VdP - SdT \tag{4.31}$$

This result was shown as derived from a Bridgman table in Section 2.10. Here we see a somewhat more intuitive route to it. From Eq. 4.31 the two necessary derivatives follow by inspection:

$$\left(\frac{\partial G}{\partial P}\right)_T = V \tag{4.32}$$

and

$$\left(\frac{\partial G}{\partial T}\right)_P = -S \tag{4.33}$$

We can obviously divide each of these by the number of mols or the mass to obtain derivatives on a per-mol, per-pound, or per-kilogram basis,

$$\left(\frac{\partial g}{\partial P}\right)_T = v \quad \text{and} \quad \left(\frac{\partial g}{\partial T}\right)_P = -s \tag{4.34}$$

The S and s in Eqs. 4.33 and 4.34 are absolute entropies, not entropies relative to some arbitrary datum like steam table entropies (see Appendix E).

Now we construct a three-dimensional g-T-P plot for some substance, such as water. For one phase it will look like Figure 4.7.

Here we show the surface sloping upward in the positive P direction; it must do this because, as Eq. 4.34 shows us, the slope (at constant T) is equal to v; and all real substances have positive values of the specific volume. Similarly, it shows the surface curving downward in the positive T direction because, as Eq. 4.34 shows, the slope is minus s; all real substances have positive values of the absolute entropy. Furthermore, we can see that the downward slope of the surface must increase as we increase T because raising the temperature at constant pressure will always increase the

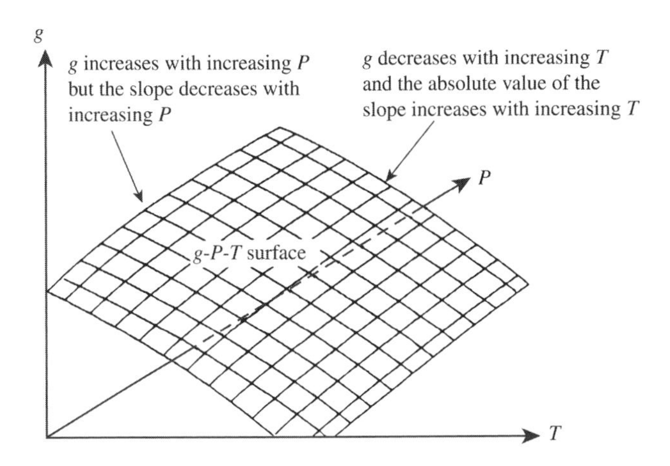

FIGURE 4.7 *g-T-P* plot for one phase of any pure substance, such as water.

value of s and hence increase the absolute value of the slope. The slope in the P direction must decrease slightly with increasing P because the specific volume must decrease slightly as the pressure increases.

Figure 4.7 shows the g-T-P relation for one phase. Now suppose we plot the same relation for two phases, for example, liquid and gaseous water, on a single plot, as shown in Figure 4.8. The figure shows two surfaces that intersect along a curve labeled Equilibrium Curve. Everywhere else there are two surfaces, one above the other. We know that for a temperature and pressure not on the equilibrium curve, only one phase can exist so at each P-T point not on the equilibrium curve, one of these two surfaces must represent an unstable (or metstable) state, which cannot exist at equilibrium. How are we to decide which of the two is the stable one?

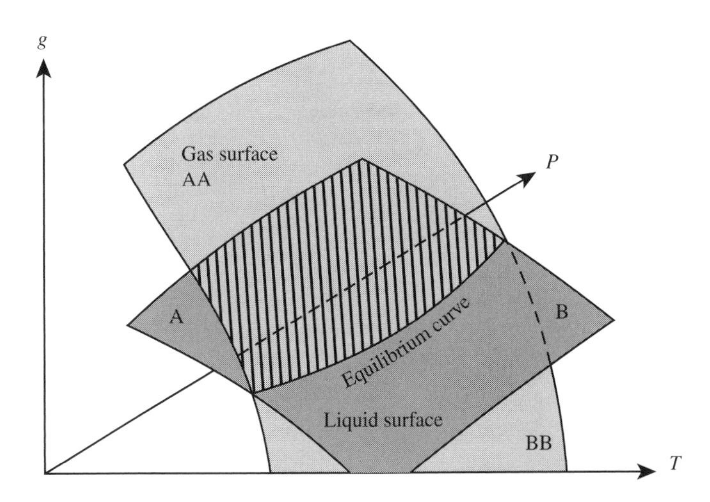

FIGURE 4.8 *g-T-P* plot for a pure substance, such as, H_2O, showing both gas and liquid surfaces and their intersection. Points A and B are on the liquid surface; points AA and BB are on the gas surface.

Consider points A and AA. These are on the two surfaces, each at the same temperature and pressure. They have different Gibbs energies per pound. As sketched in Figure 4.3, natural systems move toward the lowest Gibbs energy consistent with the restrictions on the system. Thus, the state with the higher Gibbs energy is the unstable one. A spontaneous change can occur from higher to lower Gibbs energy, but never the reverse. Therefore, we can conclude that for the pressures and temperatures on Figure 4.8 at which the two surfaces do not coincide (not on the equilibrium curve), the *lower* Gibbs energy surface is the stable one; and the higher one represents an unstable or metastable state. Thus, from Figure 4.8 we see that at low temperatures and high pressures (to the left in the figure), the liquid is the stable phase; and at high temperatures and low pressures (to the right in the figure), the gas is the stable phase. We can also see that the surfaces must cross because the two phases have different values of v and s. If we start from the temperature and pressure corresponding to A and AA and raise T at constant P, the gas surface must come down more rapidly than the liquid surface because s_{gas} is larger than s_{liquid}. Similarly, if we start from the temperature and pressure corresponding to B and BB and increase the pressure at constant temperature, the gas surface must rise more rapidly because v_{gas} is larger than v_{liquid}. Finally, we may consider the projection of the equilibrium curve (the intersection of the two surfaces) on the P-T plane; this is merely the vapor-pressure curve with which we are all familiar (Figure 1.8). If we redrew Figure 4.8 with all six crystalline varieties of ice, then there would be eight g surfaces, one above the other. We would trace on the P-T plane the intersection between the *two lowest surfaces* at any T and P, forming Figure 1.10.

4.4.1 Comparison with Enthalpy and Entropy

Figure 4.9 compares the behavior of the enthalpy, entropy, and Gibbs energy for water at 1 bar (0.987 atm) being heated through the boiling point. We see that the enthalpy and entropy have large changes at a constant temperature as the liquid boils to form a vapor, but the Gibbs energy has no such change in value at the boiling point. There is no "latent Gibbs energy ??" of boiling or freezing, corresponding to the latent enthalpy and entropy changes corresponding to boiling. Referring to Figure 4.2 we see that this is obvious, after we think about it. In this sense, a g-T plot has more similarity to a P-T plot (Figure 1.10) than to the h-T and s-T plots, which form the top part of Figure 4.9.

4.4.2 Gibbs Energy Diagrams for Pressure-Driven Phase Changes

It is instructive to make up a g-P diagram for a pressure-driven phase change, as shown below.

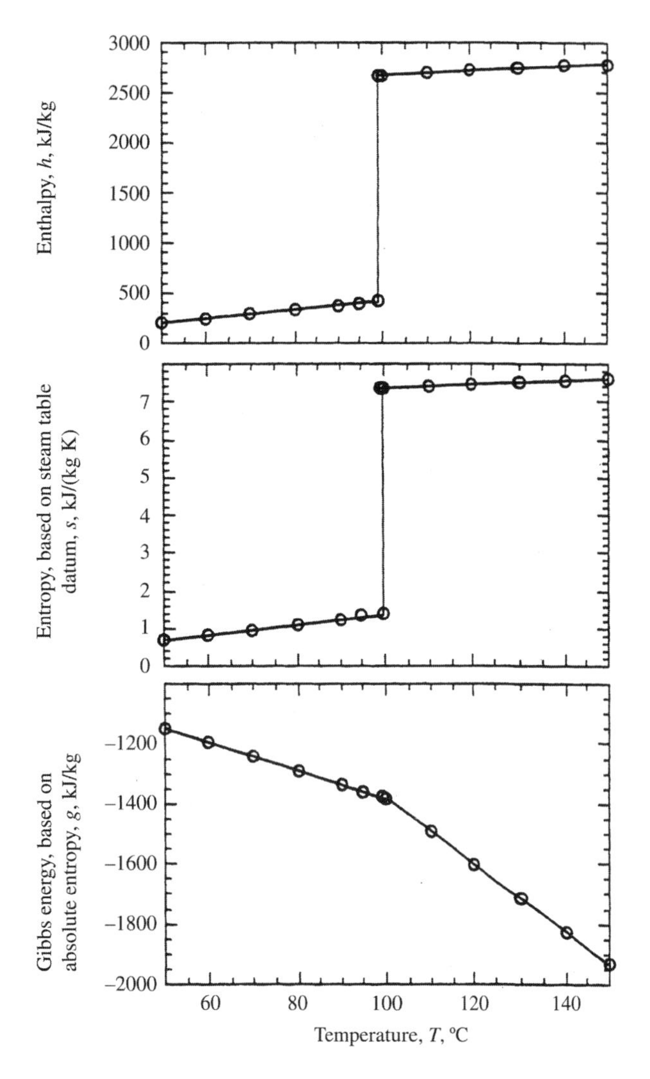

FIGURE 4.9 Comparison of the calculated enthalpy, entropy, and Gibbs energy, as a function of temperature, for water at temperatures near 100°C, all at a pressure of 1 bar (0.987 atm). The points are data from [2]. The entropy plot uses values directly from the table. The Gibbs energy plot uses absolute entropies, where $s_{\text{absolute}} = s_{\text{steam table}} + 3.515$ kJ/(kg K) and steam-table enthalpies, relative to $h = 0$ for liquid water at the triple point.

Example 4.2 Construct a Gibbs energy–pressure diagram for the graphite–diamond system at 25°C, using the data in Table 4.A.

From Eq. 4.33 we know that the slopes of the curves $(\partial g/\partial P)_T$ are equal to the specific volumes v. So, for example, for graphite at 25°C

$$\left(\frac{\partial g}{\partial P}\right)_T = v = 5.31\,\frac{\text{cm}^3}{\text{mol}}\cdot\frac{\text{m}^3}{10^6\ \text{cm}^3}\cdot\frac{\text{J}}{\text{Nm}}\cdot\frac{\text{N}}{\text{m}^2\text{Pa}}$$
$$= 5.31\times10^{-6}\,\frac{\text{J/mol}}{\text{Pa}} \tag{4.B}$$

Table 4.A Data for Graphite and Diamonds at 25°C and 1 bar

	Gibbs Energy (kJ/mol, see Table A.8)	Specific Volume (mL/mol)
Graphite	0.00	5.31
Diamond	2.90	3.42

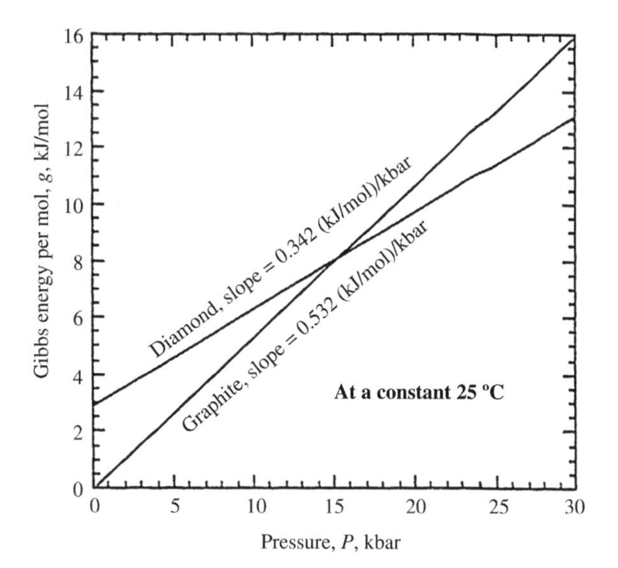

FIGURE 4.10 Gibbs energy–pressure diagram for the graphite–diamond system at 25°C. The two lines cross at 15,300 bar = 15,100 atm = 2.2×10^5 psi. At pressures less than that, graphite has the lower g, so it is the stable phase. At pressures higher than that, diamond has the lower g, so it is the stable phase. See Problem 4.5.

and similarly for diamond, the slope is 3.42×10^{-6} (J/mol)/Pa. If we assume that the specific volumes do not change with changes in pressure, a fair but not excellent assumption (see Problem 4.6), then we can see that the above derivatives must be constant. Each substance must be represented by a straight line on a g-P plot. We can locate the points on the $P = 1$ bar ≈ 0 axis from the values of the Gibbs energy, and thus make up Figure 4.10. ■

We see that at 25°C for all pressures below 15,100 atm the Gibbs energy of diamond is more than that of graphite, while at pressures higher than this, graphite has the higher Gibbs energy. This means that at the lower pressures graphite is the stable phase, and diamond is metastable. If we had a catalyst that would cause this phase change to occur at room temperature, and touched it to any diamond, the diamond would change to graphite. No such catalyst is known, so your diamonds are quite safe. Figure 4.10 is a constant temperature slice through a figure like Figure 4.8. If we wish to show the pressure–temperature curve for the equilibrium, we must be able to express the Gibbs energy of each phase as a function of T and P. Rossini and Jessup did just that in 1938 [3] (Problem 4.7) Based on their work, we can construct

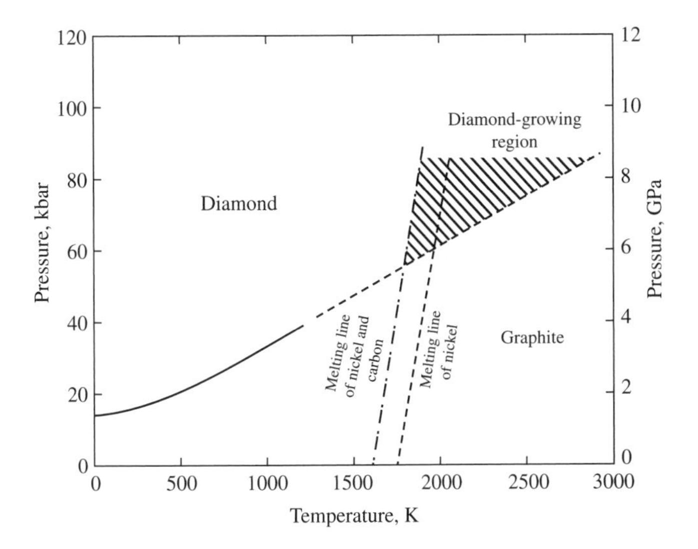

FIGURE 4.11 Diamond–graphite equilibrium, showing the calculated equilibrium curve and (shaded) the region of commercial diamond production. (From Wentorf, R. H. J. Diamond synthetic. In *Kirk-Othmer Encyclopedia of Chemical Technology,* Vol. 4, ed. 3. Grayson, M., ed. © 1978, New York: Wiley. Reprinted by permission of John Wiley and Sons.)

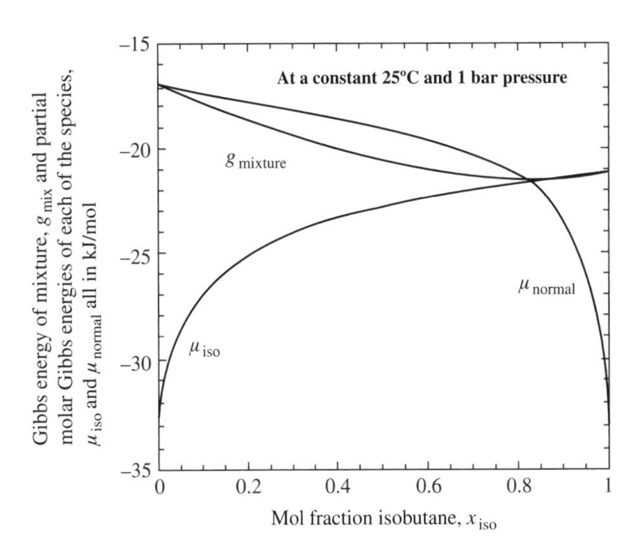

FIGURE 4.12 Calculated Gibbs energy and chemical potentials for an equilibrium mixture of normal and isobutane. (This plot assumes ideal solution behavior as discussed in Chapter 7, and pure species Gibbs energies as discussed in Chapter 12 and Appendix F.)

Figure 4.11 [4]. This shows that the equilibrium pressure calculated in Example 4.2 forms one point (at 25°C = 298.15 K) on the *P-T* equilibrium curve.

Their published data (and other similar data) spurred inventors to try to make synthetic diamonds. The problems were

1. Devising high-pressure cells that could work at industrially useful sizes at pressures of up to 100,000 atm.
2. Finding a catalyst that would cause the two phases to come to equilibrium at commercially useful rates.

As Figure 4.11 shows, the catalyst (molten nickel) was found. (The molten nickel is often referred to as a catalyst-solvent. Apparently the thermodynamically unstable graphite dissolves in the molten nickel, and then precipitates as stable diamond.) At temperatures above about 2000 K = 3140°F the conversion rate in the presence of molten nickel is fast enough. Ingenious mechanical designs of high-pressure presses were devised. Using them, we have facilities that make diamonds from graphite on an industrial scale [5] (initially only industrial quality, later of gem quality.)

4.4.3 Gibbs Energy Diagrams for Chemical Reactions

Figure 4.12 shows the Gibbs energy-composition diagram for mixtures of isobutane and normal butane, two species that are completely miscible. With an appropriate catalyst, they will convert from one to the other at moderate temperatures until they reach an equilibrium concentration. (This is done on a large scale in modern petroleum refineries, because isobutane is more valuable than normal butane.)

In this figure there are three curves, one for the g (Gibbs energy per mol) of the mixture and one for each of the two values of μ. At the two extremes where we have pure substances, the curves must meet because for any *pure substance*, $\mu = g$. The curve for g of the mixture lies below the line joining the pure species values because in any mixture there is an entropy increase on mixing; an entropy increase lowers the Gibbs energy. The curves for the individual μs are shown continuing downward; they each approach negative infinity (see Chapter 6) and thus are tangent to the vertical axes.

The equilibrium situation is the one where the two curves for the individual μs cross (i.e., $\mu_{n\text{-butane}} = \mu_{\text{isobutane}}$, see Eq. 4.26. This point is the minimum point on the g curve (see Figure 4.3). Thus, if we start with either pure substance and introduce a catalyst that allows that substance to convert to the other, the system will lower its Gibbs energy spontaneously. Moving in whichever direction will take it to the minimum g, which in this case corresponds to the lowest point on the g curve.

Example 4.3 In elementary chemistry we learn "the law of mass action," which says that for a reaction like the isobutane–normal-butane isomerization the concentrations at equilibrium can be represented by an equation of the form

$$K = \frac{\text{mol fraction of isobutane}}{\text{mol fraction of } n\text{-butane}} = \frac{x_{\text{iso}}}{x_{\text{normal}}} \tag{4.35}$$

where K is the chemical equilibrium constant. (This is a simplified form, which assumes some types of ideal behavior; we will see the more complex forms in Chapter 12.) For this reaction, at 25°C, $K = 4.52$, so that, at equilibrium

$$4.52 = \frac{x_{\text{iso}}}{x_{\text{normal}}} = \frac{x_{\text{iso}}}{1 - x_{\text{iso}}}$$

which we may solve, finding that $x_{\text{iso}} = 0.82$. ■

Looking back at Figure 4.12, we see that the minimum in the g curve occurs (as closely as we can read it) at $x_{\text{iso}} = 0.82$. This is also the value of x_{iso} at which the two partial molar Gibbs energies (chemical potentials) are equal. Thus, we see that the well-known law of mass action or chemical equilibrium relation is a shorthand form of the statement that at a state of chemical equilibrium the system has taken up those values of the chemical composition at which the Gibbs energy is a minimum. This is the same as saying that Eq. 4.26 is obeyed (for a reaction of the form A $\rightarrow$ B), or that its more complex equivalent for a more complex reaction is obeyed. In Chapter 12 we will return to this topic, and see how the statement in Eq. 4.8 leads directly to the law of mass action and the common expressions for calculating chemical equilibrium.

None of the Gibbs energy diagrams shown in this section are the normal way of treating these problems in process calculations. There are other, more convenient ways. But seeing these diagrams should help you connect the other, more common ways of dealing with these problems with the basic statement of Eq. 4.8. Nature minimizes Gibbs energy!

4.5 LE CHATELIER'S PRINCIPLE

Le Chatelier's principle, presented in most physical chemistry books, says (among other things) that if a two-phase system is compressed, the equilibrium will shift in the direction of the phase with the lower specific volume, and that if a system that can undergo a chemical reaction is heated, it will shift its equilibrium in that reaction in the heat-absorbing direction. Both of these statements can be restated that natural systems respond to changes in their external environment by internally readjusting to minimize their Gibbs energy. Thus, from Eq. 4.32 we see that if we increase P, the rate of increase of g will be less the smaller v is. Thus, systems that can respond to increases in P by going to a lower v will do so. That is what happens in the graphite–diamond equilibrium. As we squeeze the system, it shrinks to make it harder for us to raise the pressure. At a high enough pressure g_{graphite} becomes greater than g_{diamond}, so the system minimizes its Gibbs energy by converting to the lower specific-volume phase, diamond. The two statements are equivalent.

If we heat a system that can undergo an equilibrium reaction, it will move in the endothermic direction, making it harder for us to raise the temperature. We see from Eq. 4.34 that if we increase T, the rate of increase of g will be less (the rate of decrease will be greater) the larger the value of s. Thus, systems that can respond to an increase in T by going to a larger value of s (by undergoing an endothermic reaction) will do so. Many chemical reactions are of this form. Thus, Le Chatelier's principle, like most of this book, is simply the working out of the detailed consequences of Eq. 4.8. *Nature minimizes Gibbs energy*!

In this chapter the values of G and K appear without a source being shown. The basis for computing these values is shown in Chapter 12 and Appendix F.

4.6 SUMMARY

1. The fundamental criterion for all kinds of physical and chemical equilibrium is that at constant T and P, $dG_{\text{sys}} = 0$ for any infinitesimal change. The constant T and P restriction does not prevent this from being almost universally applicable.
2. For finite changes away from the equilibrium state, in any direction, $dG_{\text{sys}} > 0$. Thus, the equilibrium state is one of *minimum* Gibbs energy, subject to the external constraints.
3. For two or more phases in equilibrium there is a separate equilibrium relationship for each of the chemical species present, and that relationship is that the partial molar Gibbs energy, called the chemical potential, for species i is the same in all of the phases (1, 2, ...). The same statement applies to species j, k, and so on.
4. A system is at chemical equilibrium when it has adjusted its chemical compositions so that the overall Gibbs energy of the system is the minimum possible, subject to the external constraints (T, P, initial chemical composition, etc.). The shorthand way of showing this is "the law of mass action" or "chemical equilibrium constant." The relation between the two is explored in Chapter 12.
5. Le Chatelier's principle is a detailed restatement of the fact that natural systems minimize their Gibbs energy. They respond to changes in T and P by moving in the direction that minimizes Gibbs energy.

PROBLEMS

See the Common Units and Values for Problems and Examples. An asterisk (*) on a problem number means that the answer is in Appendix H.

4.1 Repeat the derivation in Section 4.1 leading to Eq. 4.7, holding T and V constant instead of T and P. Show that the result is that, for this constraint, nature minimizes the Helmholz energy (Eq. 4.7). See the further discussion of this problem in Appendix B.

4.2* Repeat Example 4.1 for saturated steam at 100 psia.

4.3 **a.** Sketch the equivalent of Figure 4.9 for heating water at 1 bar from $-10°C$ to $+10°C$. Show no numerical values; show only the right relationship between the various curves.

b. On Figure 4.9 write the equation for the slopes of the various curves.

c. Do the curves have upward, zero, or downward curvature for enthalpy? entropy? Gibbs energy? Assume that C_P is practically independent of temperature.

4.4 The type of phase transitions shown in Figure 4.9 and Problem 4.2 are called first-order phase transitions to distinguish them from the group called second-order phase transitions. Most second-order phase transitions are order–disorder transitions in solids. The one interesting second-order transition known in a liquid is the "lambda point" in liquid helium. This marks the transition from helium I to helium II. At this point, there is no entropy or energy change of transition, but there is a jump discontinuity in the heat capacity at constant pressure. The heat capacity curve looks like a Greek lambda, as sketched in Figure 4.13. Assuming that this heat capacity does not reach infinite values, make a plot for helium similar to Figure 4.9. Indicate why such a transition is called a second-order transition in contrast to the first-order transition in Figure 4.9.

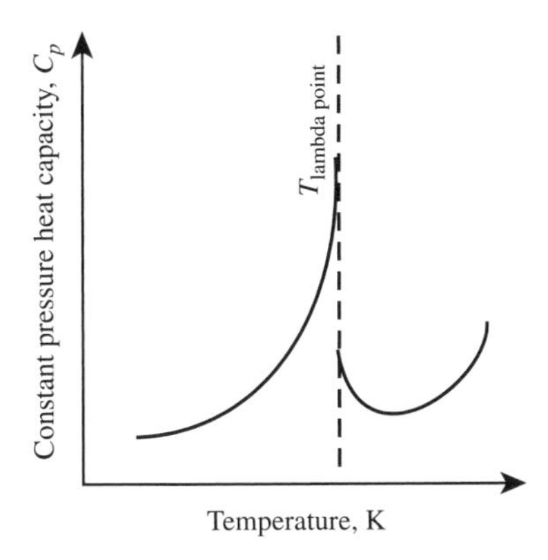

FIGURE 4.13 The heat capacity of liquid helium at the lambda point temperature. The lambda point temperature depends on pressure, with values from 1.763 K at 3.01 MPa to 2.172 K at 5.04 kPa.

4.5 The graphical procedure for finding the equilibrium pressure on Figure 4.10 can be reduced to a simple equation involving P, and the gs and vs of graphite and diamonds. Show that equation, solved for $P_{\text{equilibrium}}$. The calculation is simplified if one assumes that 1 bar pressure $\approx$ zero pressure.

4.6 In Example 4.2 we assumed that the specific volumes v of diamond and graphite were independent of pressure. Actually, they compress slightly with increasing pressure. For a constant temperature of 25°C the specific volumes of diamond and graphite are given [3] by equations of the form

$$v = v_{\text{zero pressure}} + \alpha P \tag{4.36}$$

and the appropriate values are shown in Table 4.B.

a. Show the integrated form of the equation for the pressure at which the Gibbs energy change is zero, at this temperature.

b. Compute the numerical value of the equilibrium pressure, which is about 7% higher than that computed in Example 4.2 because the graphite is more compressible than the diamond.

c. Sketch the equivalent of Figure 4.10, with the straight lines replaced by the appropriate curves. If drawn to scale the curves still appear practically straight; exaggerate the curvature to show its effect.

d. Observe that of the quadratic equation in part (b) gives 2 solutions. Show on the sketch in part (c) what these represent. The higher pressure solution is unreal physically, because it requires extrapolating Eq. 4.36 beyond the range of its applicability.

e. In Appendix D, the equation for the change of density with pressure is given as

$$\rho = \rho_0[1 + \beta(P - P_0)]_{T\ \text{constant}} \tag{D.2}$$

Show that this is approximately the same as Eq. 4.36 for small values of $\beta(P - P_0)$. *Hint:* Write Eq. (D.2) in volume form as

$$v = \frac{v_0}{1 + \beta(P - P_0)} = v_0[1 + \beta(P - P_0)]^{-1} \tag{4.C}$$

Expand the $v_0[1 + \beta(P - P_0)]^{-1}$ term by binomial expansion, and drop the higher-order terms.

Table 4.B More Graphite and Diamond Data, at 25°C

	v_0 (mL/mol)	α (mL/(mol atm))
Diamond	3.42	-0.55×10^{-6}
Graphite	5.31	-16.1×10^{-6}

4.7 Rossini and Jessup [3] show the complete development of the Gibbs energy values for graphite and diamond. Their final equation is

$$\begin{aligned} g_{\text{diamond}} - g_{\text{graphite}} = {} & 541.82 + \frac{6700}{T} + 1.17662T \log T \\ & -2.43723T - 0.000221T^2 \\ & -(0.045660 + 0.91236 \times 10^{-6}T \\ & -0.7830 \times 10^{-10}T^2 - 0.3623 \\ & \times 10^{-12}T^3)P + 0.19 \times 10^{-6}P^2 \end{aligned} \tag{4.D}$$

with g in cal/mol, T in K, and P in atm.

a. Show that for 25°C = 298.15 K and 1 atm, Eq. 4.D gives the same value for $g_{\text{diamond}} - g_{\text{graphite}}$ as the data used in Example 4.2. (Observe the different units!)

b. Show that if one sets $g_{\text{diamond}} - g_{\text{graphite}}$ equal to zero, for $T = 25°\text{C} = 298.15$ K, Eq, 4.D gives approximately the same equilibrium pressure as is shown in Example 4.2. See the preceding problem for the reason for the small difference.

c. Compute the equilibrium pressure from Eq. 4.D for 1400 K (the highest temperature for which Eq. 4.D is claimed to be reliable), and compare it with the value in Figure 4.11.

4.8 We have brought our sample of graphite to the diamond-growing region in Figure 4.11 and held it there long enough to convert all the graphite to diamonds. We now want to bring our diamonds back to room temperature and pressure.

a. Should we reduce the pressure first, and the temperature second, or the temperature first and the pressure second?

b. Why?

The following 3 problems are suitable for graduate students, probably not for undergraduates.

4.9 Sketch a plot analogous to Figure 4.8, showing solid, liquid, and gas.

4.10 **a.** For a substance like water, which expands on freezing, sketch a Gibbs energy-pressure diagram showing solid, liquid, and gas states, two-phase regions, and the critical and triple points. Show no numerical values, only the correct relations of the areas.

b. Sketch on the figure a curve of constant temperature for some pressure between the critical and triple-point temperatures.

4.11 Do the same as in the preceding problem, but for a substance like benzene, which contracts on freezing.

REFERENCES

1. Keenan, J. H., F. G. Keyes, P. G. Hill, and J. G. Moore. *Steam Tables: Thermodynamic Properties of Water Including Vapor, Liquid and Solid Phases.* New York: Wiley (1969).
2. Haar, L., J. S. Gallagher, and G. S. Kell. *NBS/NRC Steam Tables.* New York: Hemisphere (1984).
3. Rossini, F. D., and R. S. Jessup. Heat and free energy of formation of carbon dioxide, and the transition between graphite and diamond. *J. Res. NBS* 21:491–513 (1938).
4. Wentorf, R. H. J. Diamond, synthetic. In *Kirk-Othmer Encyclopedia of Chemical Technology*, Vol. 4, ed. 3. Grayson, M., ed. New York: Wiley, pp. 676–688 (1978).
5. Wentorf, R. H. J. Diamond, synthetic. In *Kirk-Othmer Encyclopedia of Chemical Technology*, Vol. 4, ed. 4. Howe-Grant, M., ed. New York: Wiley, pp. 1082–1096 (1992). (In 2010 the Wikipedia article on Synthetic Diamonds was quite good.)

5

VAPOR PRESSURE, THE CLAPEYRON EQUATION, AND SINGLE PURE CHEMICAL SPECIES PHASE EQUILIBRIUM

The vapor pressure is crucially important in a wide variety of physical and chemical equilibrium situations, such as those in Chapter 3. This short chapter discusses vapor pressure and related topics.

5.1 MEASUREMENT OF VAPOR PRESSURE

The vapor pressure for a pure species is that pressure exerted by a pure sample of the liquid at a fixed temperature. The experimental procedure to measure vapor pressure is sketched in Figure 5.1. A sample of the material to be tested is placed in a closed container, with the amount chosen so that there will be both vapor and liquid present. Then the temperature is made constant over the whole container, normally by placing the whole container in a constant-temperature bath, with circulating water or some other heat transfer fluid. When the temperature and pressure readings have reached constant values, these are recorded, and the constant temperature bath is set for a new temperature. When the temperature and pressure readings are again unchanging, the values are again recorded, and a new temperature is chosen, continuing until the desired range of temperatures has been tested.

Real vapor-pressure measuring devices are more refined versions of that shown in Figure 5.1, with special attention directed to getting very accurate measurements of the temperature and pressure, and to making sure that all the air and other possible contaminants are removed from the container, so that the sample being measured is as pure as possible. In principle, the measurement of the vapor pressure of solids is conducted exactly the same way as that of liquids; in practice, the pressure measurements are difficult because the pressures are very low. The vapor pressure of liquid mixtures is not as often measured as that of pure liquids, but the procedure is the same.

5.2 REPORTING VAPOR-PRESSURE DATA

Figure 1.8 shows the simplest representation of vapor-pressure data for one pure substance, H_2O, a plot of vapor pressure or its ln or log vs. T. This is easy to understand, but not very convenient because of the huge range of values. A more common and useful plot is shown in Figure 5.2. There we see, for a wide variety of substances, a plot of the log of the vapor pressure vs. an unusual temperature scale, which makes the curves close to being straight lines (see Problem 5.6). This plot and others like it are useful for quick lookup of values, but they are hard to read to more than one significant figure.

Many handbooks have tables of vapor pressures, as do more detailed tables like the steam tables [1–4]. The values shown there are not direct experimental measurements. Instead, they are computed from equations that have been fitted to the experimental data. In the computer age we wish to represent the experimental data by equations that our computers can manipulate. There are some strong limits on the forms those equations can take, which we will see in Section 5.4.

5.2.1 Normal Boiling Point (NBP)

The temperature at which the vapor pressure is 1.00 atm is called the *normal boiling point, NBP*. The most widely known NBP is that of water, $100°C = 212°F$. In Figure 5.2

Physical and Chemical Equilibrium for Chemical Engineers, Second Edition. Noel de Nevers.

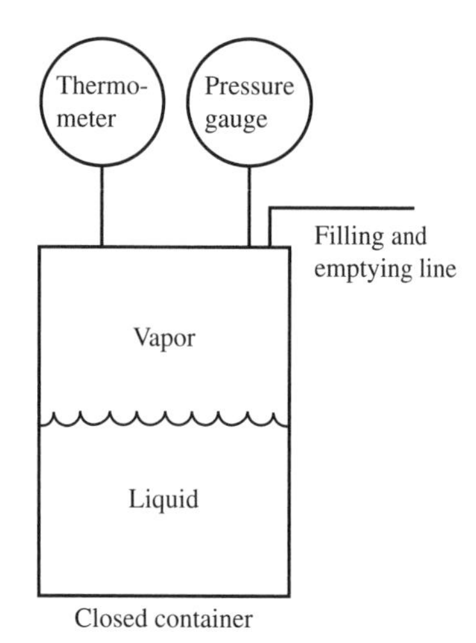

FIGURE 5.1 Simplified schematic of the device for measuring the vapor pressure.

the NBP is the temperature at which an individual compound's vapor pressure curve crosses 1 atm (14.696 psia) with values ranging from −44°F for propane to 675°F for mercury. Handbooks show the NBP for many substances. Normal boiling points increase more or less regularly with increasing molecular weight. This increase is quite regular for members of one chemical family (e.g., alkanes) but not between families. For high molecular weight materials such as sugar (sucrose, $C_{12}H_{22}O_{11}$, $M = 342.3$ g/mol), the NBP is higher than the chemical decomposition temperature, so the NBP cannot be measured. Cooks know that melted sugar will "caramelize" before it boils, and use this fact in cooking.

5.3 THE CLAPEYRON EQUATION

If two phases (1 and 2) of a pure substance are in equilibrium (gas–liquid, gas–solid, liquid–liquid, liquid–solid, or solid–solid) at T_1 and P_1, then the Gibbs energy per mol (or per lbm or kg) must be the same in each of the equilibrium phases.

$$g^{(1)} = g^{(2)} \tag{5.1}$$

Now, we raise the temperature from T_1 to $(T_1 + dT)$. At this new temperature we will have a new equilibrium (Figure 5.3) at which

$$g^{(1)} + dg^{(1)} = g^{(2)} + dg^{(2)} \tag{5.2}$$

So for this change

$$dg^{(1)} = dg^{(2)} \tag{5.3}$$

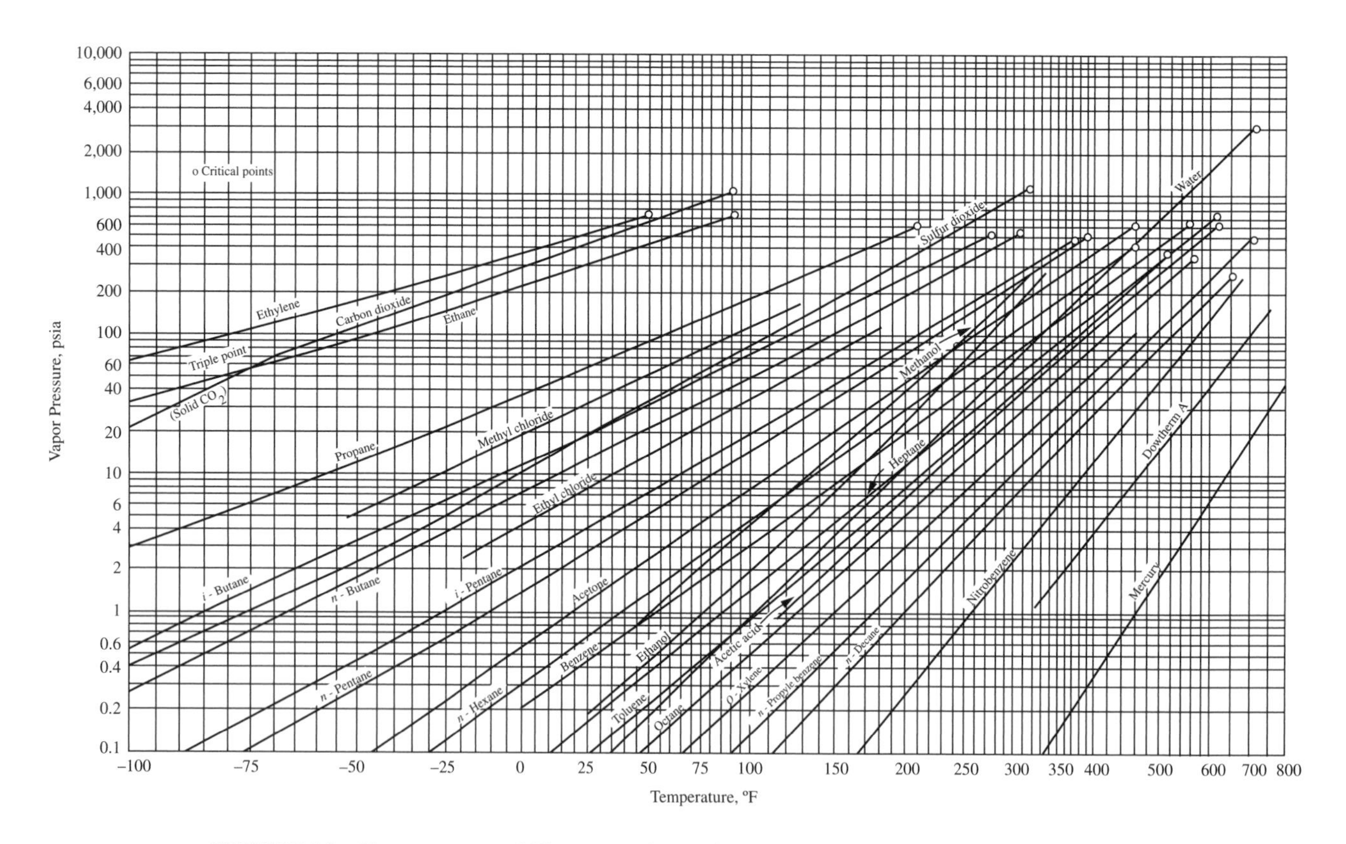

FIGURE 5.2 Vapor pressure of 27 compounds as a function of temperature. (From Brown, G. G. et al. *Unit Operations*. © 1951, Wiley: New York. Reprinted by permission of John Wiley and Sons.)

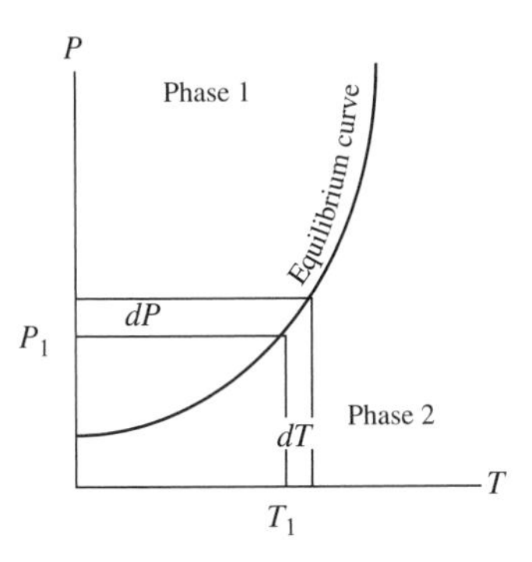

FIGURE 5.3 On a *P-T* diagram for a single pure chemical, we increase the temperature by *dT*, which causes the pressure to increase by *dP*. On the equilibrium curve $g^{(1)} = g^{(2)}$.

Using the derivatives in Eq. 4.33 we can write

$$dg^{(1)} = v^{(1)}dP - s^{(1)}dT = dg^{(2)} = v^{(2)}dP - s^{(2)}dT \quad (5.4)$$

Factoring gives

$$(v^{(1)} - v^{(2)})dP = (s^{(1)} - s^{(2)})dT$$

$$\frac{dP}{dT} = \frac{(s^{(1)} - s^{(2)})}{(v^{(1)} - v^{(2)})} = \frac{-\Delta s}{-\Delta v} = \frac{\Delta s}{\Delta v} = \frac{\Delta h}{T\Delta v} \quad (5.5)$$

This is the Clapeyron equation, which is *rigorously correct* for *any* phase change of a single pure chemical species. All of the curves in Figures 1.7, 1.8, 1.9, and 5.2 agree with the Clapeyron equation. If your experimental data do not agree with it, then you have made an experimental error!

Example 5.1 Compute the value of *dP/dT* for the steam–water equilibrium at 212°F using the Clapeyron equation, and compare it with the value in the steam table [5].

From the steam table, we have $\Delta h = 970.3$ Btu/lbm, $\Delta v = 26.78$ ft^3/lbm, and $T = 671.7$°R. Thus,

$$\frac{dP}{dT} = \frac{(970.3\ \text{Btu/lbm})(778\ \text{ft lbf/Btu})}{(671.7°\text{R})(26.78\ \text{ft}^3/\text{lbm})(144\ \text{in}^2/\text{ft}^2)}$$

$$= 0.2914\ \text{psi/°R} \quad (5.A)$$

Using the nearest adjacent steam table entries for vapor pressure, we have

$$\frac{dP}{dT} \approx \frac{\Delta P}{\Delta T} = \frac{15.291 - 14.125}{214 - 210} = 0.2914\ \text{psi/°R} \quad ■ \quad (5.B)$$

As in Example 4.1, this agreement does not demonstrate the correctness of the Clapeyron equation; the authors of the steam tables used the Clapeyron equation in making up that table. (If the values did not agree, it would show an error in the steam table!) In the Clapeyron equation we normally use *P* for pressure, because the Clapeyron equation applies to any equilibrium between any two phases of the same pure substance. If one of the phases is a gas or vapor, then the (*dP/dT*) relation is a vapor-pressure curve, and we normally use *p* for vapor pressure, so the above derivatives would be written as *dp/dT*. This distinction is not very important, but the use of *P* for pressure in general and *p* for vapor pressure is common.

5.4 THE CLAUSIUS–CLAPEYRON EQUATION

The Clapeyron equation is *rigorous and exact*. It applies not only to gas–liquid equilibrium, but to any two-phase equilibrium of a pure species (e.g., liquid–solid, gas–solid) or change between two different crystal forms (e.g., graphite to diamonds, see Example 4.2). By adding some simplifications, we find the Clausius–Clapeyron (C-C) equation, which is only *approximate* but is a *surprisingly good* approximation of observed behavior for gas–liquid and gas–solid equilibria (vapor pressures) of single pure species.

For most gases at low and moderate pressures, $v^{(\text{gas})}$ is reasonably represented by the ideal gas law, and $v^{(\text{gas})}$ is so much larger than $v^{(\text{liquid})}$ that we may write

$$\Delta v = v^{(\text{gas})} - v^{(\text{liquid})} \approx v^{(\text{gas})} \approx \frac{RT}{p} \quad (5.6)$$

Making this substitution in Eq. 5.5 and rearranging, we find

$$\frac{dp}{dT} = \frac{\Delta h}{T(RT/p)} \quad (5.7)$$

$$\frac{dp}{p} = \left(\frac{\Delta h}{R}\right)\frac{dT}{T^2} \quad (5.8)$$

For small ranges of temperature we may assume that Δh is practically a constant and integrate, finding the C-C equation

$$\ln p = \left(\frac{\Delta h}{R}\right)\left(\frac{-1}{T}\right) + \text{constant} \quad (5.9)$$

or its alternative form

$$\ln\left(\frac{p_2}{p_1}\right) = \left(\frac{\Delta h}{R}\right)\left(\frac{1}{T_1} - \frac{1}{T_2}\right) \quad (5.10)$$

This equation works fairly well for low-pressure gas–liquid and gas–solid equilibria.

Example 5.2 Estimate the temperature at which the vapor pressure of ice (solid H_2O) is 0.005 psia. From the steam tables [5] at the triple point, we find $T = 32.018°F$, $p = 0.0887$ psia, $\Delta h_{\text{solid-to-gas}} = \Delta h_{\text{sublimation}} = 1218.7$ Btu/lbm. Assuming that the enthalpy change of vaporization is independent of temperature (a fairly good approximation in this case) we start with Eq. 5.10 and rearrange:

$$\frac{1}{T_2} = \frac{1}{T_1} - \ln\left(\frac{p_2}{p_1}\right) \cdot \frac{R}{\Delta H}$$

$$= \frac{1}{492.018°\text{R}} - \ln\left(\frac{0.005}{0.0887}\right) \cdot \frac{\left(\dfrac{1.987\,\text{Btu}}{\text{lbmol}\cdot°\text{R}}\right)}{\left(\dfrac{1218.7\,\text{Btu}}{\text{lbm}}\right)} \cdot \frac{\text{lbmol}}{18\,\text{lbm}}$$

$$= 2.293 \times 10^{-3} \frac{1}{°\text{R}} \tag{5.C}$$

$$T_2 = \frac{1}{\left(\dfrac{2.293 \times 10^{-3}}{°\text{R}}\right)} = 436.09°\text{R} = -23.5°\text{F} \tag{5.D}$$

By interpolation in the steam tables, one finds $-23.8°F$; these values are practically equal. ■

Again, this close agreement does not prove correctness. The authors of the steam table used equations practically identical to the C-C equation to make up their table of the vapor pressure of solid water (ice).

Returning to Eq. 5.9, we see that the C-C equation implies that if we plot the ln (or log) of p (the vapor pressure) vs. (1/absolute T), the experimental data should fall on a straight line. Since we derived the C-C equation by making the ideal gas assumption, we would expect that the data plotted this way would form a straight line only over the low-pressure region. To our surprise, we observe (see Figure 5.4, [6]) that the experimental vapor-pressure data for most substances form practically straight lines on this kind of plot over their entire temperature ranges. Why?

In deriving the C-C equation, we assumed that Δh was constant and that Δv was proportional to (T/p). As the pressure rises, both of these assumptions become incorrect, but since we are taking the ratio of Δh and $(p\Delta v)/T$, both assumptions can be wrong and still have the ratio $(T\Delta h)/(p\Delta v)$ remain constant. Table 5.1 shows that for saturated steam this ratio varies by only 13% over the pressure range from 1 to 3000 psia. Thus, the inaccuracies of the two assumptions tend to cancel each other, and the resulting equation is surprisingly good.

Example 5.3 Using the C-C equation, estimate the vapor pressure of water at 1155.2°R, based on the vapor pressures at 652.9 and 787.5°R, which are 10 and 100 psia. Compare the result with that shown in Table 5.1.

Here we can write Eq. 5.9 in the form most often seen,

$$\ln p = A - \frac{B}{T} \tag{5.11}$$

where A and B are constants to be determined from the pair of T and p values above. We simply write

$$\ln 10 = A - \frac{B}{652.9°\text{R}} \quad \text{and} \quad \ln 100 = A - \frac{B}{787.5°\text{R}} \tag{5.E}$$

By straightforward algebra we find that $A = 15.715$ and $B = 8755°R$. Thus, for 1155.2°R we have

$$\ln p = 15.715 - \frac{8755}{1155.2} = 8.136 \tag{5.F}$$

and $p = 3416$ psia. From Table 5.1 we see the correct value is 3000 psi. Thus, there is an error of 11 % in the predicted pressure. ■

This example illustrates the advantages and disadvantages of the C-C equation. Using it, one can very easily make useful vapor-pressure estimates. This example shows a 30-fold extrapolation from the input pressure (100 psia) with only an 11% error. For many applications, this is adequate.

5.5 THE ACCENTRIC FACTOR

In Section 2.12 we introduced the accentric factor ω as a dimensionless correlating parameter in the theorem of corresponding states, and promised to explain it in this chapter. It is most easily explained in terms of Figure 5.4 and the Clausius–Clapyron equation. If the vapor-pressure data for some pure species form a straight line in Figure 5.4, then the equation of that line is

$$\ln P_r = 1 - \frac{B}{T_r} \tag{5.12}$$

By rewriting Eq. 5.11 in terms of P_r and T_r we have eliminated A (and changed the value and dimensions but not the meaning of B). If the simple corresponding states theorem were correct, then all the curves in Figure 5.4 would be identical (but not necessarily straight), and if they were also all straight then all substances would be represented by Eq. 5.12 with the same value of B. In Figure 5.4 the curves for various species are close to straight, but not exactly straight, and they have different slopes and thus different values of B.

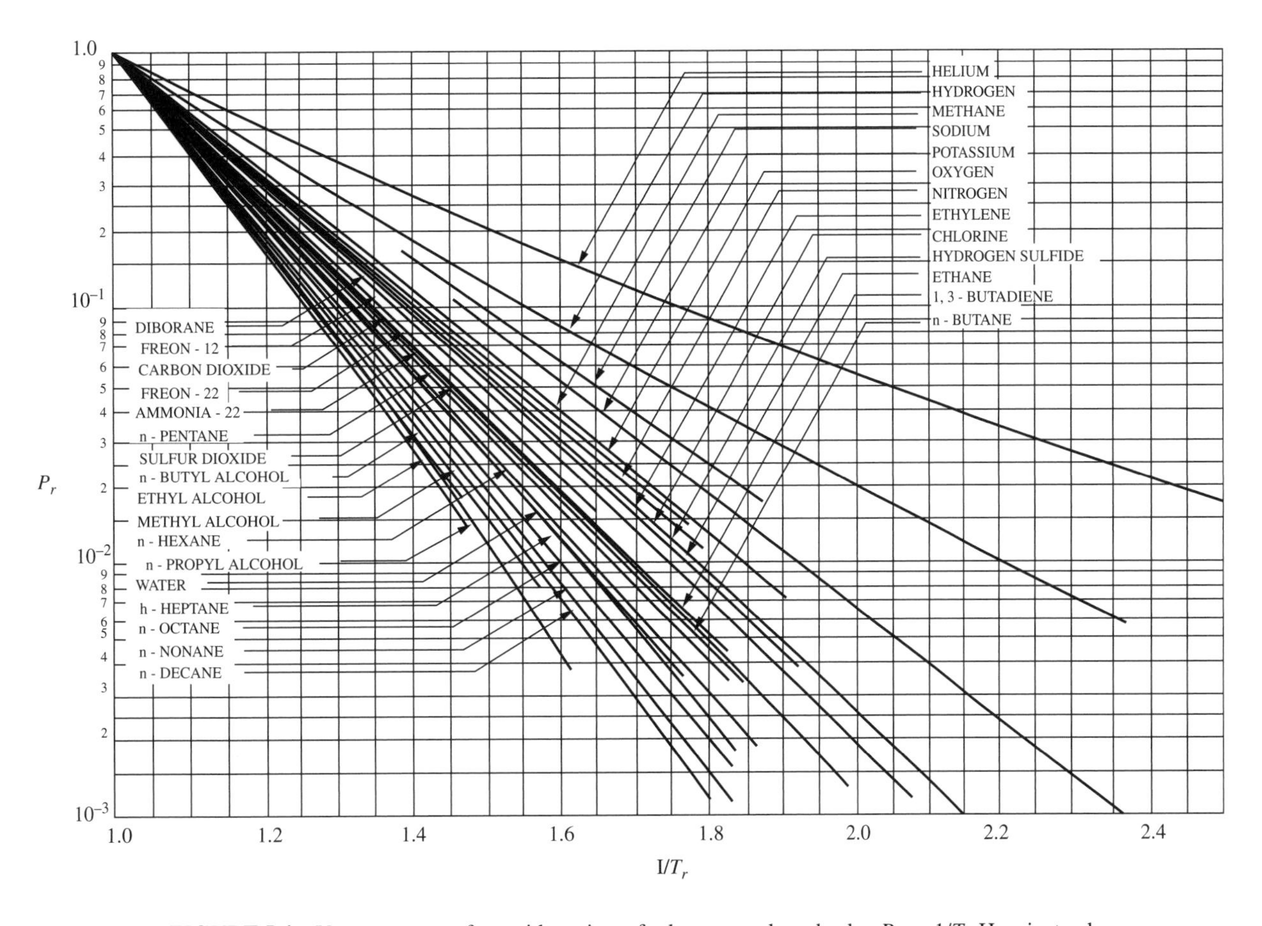

FIGURE 5.4 Vapor pressures for a wide variety of substances, plotted as log P_r vs. $1/T_r$. Here instead of using the vapor pressure and temperature directly, which would have required a much larger plot because of the wide range of pressures and temperatures involved, the plot uses the reduced vapor pressure (pressure/critical pressure) and the reduced temperature (absolute temperature/critical temperature). This makes all the curves come together at (1, 1), which corresponds to the critical point. We see that most of the curves are close to being straight lines, except for helium and hydrogen, which are always different from everything else. (If the primitive form of the theorem of corresponding states were true we would expect there to be only one curve on this plot; that is clearly not the case, and the mathematical equivalent of this plot is widely used as a correction to the primitive form of that theorem, through the *accentric factor* (Sections 2.12 and 5.4). (From Martin, J. J., and J. B. Edwards. Correlation of latent heats of vaporization. *AIChEJ* 11: 331–333 (1965). Reproduced with permission of the American Institute of Chemical Engineers.)

This suggests that if we replaced each of the curves in Figure 5.4 by a best-fit straight line and determined its corresponding value of B in Eq. 5.12, then B would be a good third parameter for corresponding states. It would. The accentric factor is the practical equivalent of that B. Pitzer et al. [7] observed that the curves in Figure 5.4 (and other plots of the same type with many more species on them) fall into three categories: (1) He and H_2, whose behavior is

Table 5.1 Values of Parameters Used in the Derivation of the C-C Equation Taken from the Steam Tables [5]

P (psia)	T (°R)	Δh (Btu/lbm)	Δv (ft^3/lbm)	$(T\,\Delta h)/(p\,\Delta v)$ (Btu·°R/ft^3·psi)
1	561.4	1036.0	333.6	1743
10	652.9	982.1	38.38	1670
100	787.5	889.2	4.416	1586
500	926.8	755.8	0.9083	1542
1000	1004.4	650.0	0.4249	1537
2000	1095.7	464.4	0.16213	1569
3000	1155.2	213.0	0.04974	1649
3200	1164.9	29.3	0.00639	1669

different than everything else because at the very low temperatures at which they are liquid, quantum effects are important; (2) spherical, nonpolar molecules like CH_4, Ar, and Kr (of which only CH_4 is shown in Figure 5.4), which all have practically identical curves on this type of figure; and (3) other substances, whose values depart more and more from those of the spherical, nonpolar materials as their shape departs from spherical and their polarity increases. He then observed that for the spherical, nonpolar molecules (which he called *simple fluid*) at $T_r \approx 0.7(1/T_r = 1.428)$ the experimental $P_r \approx 0.1$. He defined

$$\begin{pmatrix}\text{accentric}\\ \text{factor}\end{pmatrix} = \omega = -\log(P_r)_{at\ T_r=0.7} - 1 \tag{5.13}$$

which makes the accentric factor $=$ 0 for the spherical, nonpolar molecules, positive for nonspherical and/or polar molecules, and negative for He and H_2. Please review the values of ω in Table A.1 to see that this is true!

Example 5.4 Estimate the accentric factor for ethanol from Figure 5.4. (Ethanol is chosen because its curve is the lowest one in Figure 5.4.)

At $T_r \approx 0.7(1/T_r = 1.428)$, we read $P_r \approx 0.023$, so that

$$\omega = -\log 0.023 - 1 = -(-1.64) - 1 = 0.64 \tag{5.J}$$

Table A.1 shows that the value based on the best data is 0.645. ■

The accentric factor is used exclusively in making corresponding-states estimates of PvT, EOSs that estimate PvT, and other thermodynamic properties derived from PvT. It is almost never used in making vapor-pressure estimates, and will not be used again in this chapter.

5.6 THE ANTOINE EQUATION AND OTHER DATA-FITTING EQUATIONS

The Clapeyron equation is rigorous. The C-C equation is derived from it, using some easily understood approximations. The C-C equation has only two arbitrary constants (A and B), which can be determined by using two experimental data points. It is the *best two-constant* vapor-pressure estimating equation. For any substance whose vapor pressure curve in Figure 5.4 is a straight line, the C-C equation gives a perfect representation of the data. But most substances produce a slightly curved line in Figure 5.4. To represent the experimental data for substances whose plots in Figure 5.4 are curved (most substances), we use equations with more adjustable constants, which are neither rigorous nor based on any theory.

The Antoine equation

$$\ln p \text{ or } \log p = A - \frac{B}{T + C} \tag{5.14}$$

introduces a third arbitrary constant, C, to represent that curvature. Here A, B, and C are arbitrary constants to be obtained from the experimental data (e.g., by a least-squares fit of the experimental vapor-pressure data). There is no theoretical basis for introducing the third constant; it is simply the most successful simple way to modify the C-C equation, improving its ability to fit experimental data. Normally this equation is applied with T in °C, not K. If $C = 273.15$, then Eq. 5.14 becomes the same as Eq. 5.11. Table A.2 shows a sampling of published constants for the Antoine equation. They show that most values of C are less than 273.15, so that for most substances, a plot of ln or log p_r vs. $1/T_r$ will show a slight curvature, as do most of the curves in Figure 5.4.

To be mathematically precise, we would write of the logarithm as log (p/torr) and write B and C as B, °C and C, °C. That would make it unnecessary to state with the equation what units of p and T are used. While this is mathematically precise, it is almost never done. The constants are almost always presented as in Table A.2, and it is assumed that the user is not terribly upset about taking the dimensionless log of a quantity that is not itself dimensionless.

Example 5.5 Estimate the NBP of water, using the Antoine equation and Table A.2.

Equation 5.14, solved for T, is

$$T = \frac{B}{A - \log p} - C \tag{5.H}$$

Inserting the values of the constants for water from Table A.2, and the value of 1.00 atm expressed in torr we find

$$T = \frac{1668.21}{7.96681 - \log 760} - 228.0 = 100.001\,°\text{C} \tag{5.I}$$

This does not prove the overall accuracy of the Antoine equation, but does show that whoever fitted the constants to the experimental data for water made them represent the NBP (100°C) very well. ■

There seems to be widespread agreement that the Antoine equation is the *best three- constant* vapor-pressure equation; it is very widely used. But now that our computers have grown so large and powerful, we can easily use more complex equations.

Poling et al. [8] use a five-constant vapor-pressure equation

$$\ln p = A + \frac{B}{T} + C \ln T + DT^E \quad (5.J)$$

where $A \ldots E$ are data-fitting constants (called C1 ... C5 in [8]). They present values of the constants for 345 pure compounds (see Problem 5.13). An even larger compilation, with both equations and plots, is in [9]. Poling et al. [10] review the field of vapor-pressure equations, recommend which is best for various applications and present a table of constants for their recommended equations.

Comparing Eq. 5.J to the C-C equation we see that the first two terms on the right are those of the C-C equation. The subsequent terms are always smaller in absolute value than the sum of the first two terms. They correct for the approximate nature of the C-C equation. If we have a large amount of high-quality vapor-pressure data on some substance, then we are justified in using more complex equations to represent it. The steam table [5] uses a purely empirical data-fitting vapor-pressure equation that has 12 data-fitting constants; with it the authors claim they can represent the best available experimental vapor-pressure data with an average disagreement of 0.008% and maximum disagreement of 0.017%. In all of these equations the first two constants (C-C) have some theoretical basis; the remaining constants have none.

5.6.1 Choosing a Vapor-Pressure Equation

In choosing among vapor-pressure equations, we trade accuracy for simplicity. The complex equations are more accurate than the simple ones. This is illustrated by Figure 5.5. We see that, for the whole range of the vapor pressure of water from the triple point to the critical point, the C-C equation makes errors up to 14%, and that the sign of the errors changes from plus, to minus, and then back to plus as we go from low temperatures to high. The Antoine equation, using the constants from Table A.2, has a maximum error of 5% over the same range, and its error pattern is the reverse of that of the C-C equation. Equation 5.J using the constants from [8], reproduces the steam table values over this entire range with a maximum error of 0.5%; it is not shown in Figure 5.5 because it would plot as nearly a straight line with value 1.00 ± 0.005.

In most of the rest of this book we will use the Antoine equation and the constants in Table A.2. If we are writing a computer program for important calculations, the extra effort to program in Eq. 5.J and the correspondingly larger table of constants is probably worthwhile. In a few examples we will use the C-C equation, for reasons that will be clear in those examples.

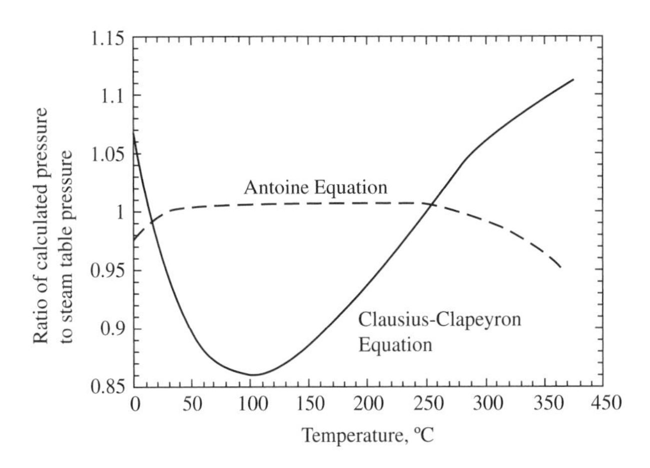

FIGURE 5.5 Comparison of the C-C and Antoine equations to the steam table vapor-pressure values over the range between the triple point and the critical point.

5.7 APPLYING THE CLAPEYRON EQUATION TO OTHER KINDS OF EQUILIBRIUM

As discussed in Section 5.2, the Clapeyron equation (but *not* the C-C approximation) applies rigorously to liquid–solid and solid–solid equilibria, as well as to the more common vapor pressure (vapor–liquid or vapor–solid) equilibria. We can also use it to compute the behavior of a phase change involving only solids and liquids.

Example 5.6 Estimate the freezing pressure of water at $-22°C = -7.6°F$. Here, if we consider the phase change to be going from ice to water at 32°F, then [5]

$$\Delta h = \Delta h_{\text{fusion}} = 143.35\,\text{Btu/lbm}$$

$$\Delta v = v^{(\text{water})} - v^{(\text{ice})} = 0.01602 - 0.01747 = -0.00145\,\text{ft}^3/\text{lbm}$$

and

$$\left(\frac{dP}{dT}\right)_{32°F} = \frac{\left(143.35\frac{\text{Btu}}{\text{lbm}} \cdot \frac{778\,\text{ft lbf}}{\text{Btu}}\right)}{(492°\text{R})\left(-0.00145\frac{\text{ft}^3}{\text{lbm}}\right)\left(\frac{144\,\text{in}^2}{\text{ft}^2}\right)}$$

$$= -1085.6\frac{\text{psi}}{°\text{R}} \quad (5.K)$$

This gives the rigorously correct slope of the liquid–solid curve at 32°F on a P-T diagram. Here we use P instead of p because neither phase is a gas, so this is not a vapor pressure. If we further assume that the solid-liquid curve is a straight line, which is equivalent to assuming that $\Delta h/(T\,\Delta v)$ is a constant over the region of interest, then we can estimate the pressure at $-22°C = -7.6°F$ by

$$\Delta P = \int \left(\frac{dP}{dT}\right) dT \approx \left(\frac{dP}{dT}\right)_{32^\circ\mathrm{F}} \Delta T$$

$$= -1085.6 \frac{\mathrm{psi}}{^\circ\mathrm{R}}(-7.6-32)^\circ\mathrm{F} = 43{,}000\,\mathrm{psi} \quad (5.L)$$

From this we would estimate the final pressure as 43,000 + 0.09 ≈ 43,000 psia. In this case, the experimental pressure is well known, because this temperature corresponds to the triple point between liquid water, ice I (the common variety), and ice III, a variety that does not exist at pressures below about 30,000 psia (see Figure 1.10). The measured value is 30,000 psia, which shows that our assumption of a straight line on a *P-T* plot ($\Delta h/(T\Delta v)$ = constant) is only approximately correct. ■

If we do not make the linear assumption (which the shape of Figure 1.10 shows is a mediocre assumption), we would have to write

$$\Delta P = \int_{T_1}^{T_2} \frac{\Delta h}{T\Delta v} dT \quad (5.15)$$

and use measured or estimated values of Δh and Δv to perform the integration. At these high pressures it is much easier to measure pressures, temperatures, and volume changes than enthalpy changes, so we are more likely to use Eq. 5.15 to estimate enthalpy changes from the measured *PvT* values than the reverse.

Instead of assuming that $\Delta h/(T\Delta v)$ is a constant as we did in Example 5.5, we could assume that $\Delta h/\Delta v$ is constant, and integrate Eq. 5.15, finding

$$\Delta P = \frac{\Delta h}{\Delta v}\int_{T_1}^{T_2} \frac{dT}{T} = \frac{\Delta h}{\Delta v}\ln\frac{T_2}{T_1} \quad (??) \quad (5.M??)$$

which seems plausible and which regularly appears on students' homework and exam papers. However, if we substitute the above values into Eq. 5.M??, we find a calculated ΔP of 45,000 psia, which is a poorer approximation of the experimental value than the one shown in Example 5.5. Figure 5.6 compares the observed behavior to the two approximate applications of the Clapeyron equation discussed above. For all of the cases I know of, the linear extrapolation of the tangent (Eq. 5.L) gives a better estimate of the experimental findings than does Eq. (5.M??) which, as shown in Figure 5.6, curves in the wrong direction! Neither the linear extrapolation or Eq. 5.M?? is very good. Both give an order-of magnitude estimate of the right value. Remember that Eq. 5.15 is rigorously correct, so that if we knew the values of Δv and Δh as a function of *T*, we could use them in Eq. 5.15 to compute the *P-T* curve with complete accuracy. Observe that both Eq. 5.L and 5.M?? are tangent to the experimental curve and give good estimates at low values of ΔT.

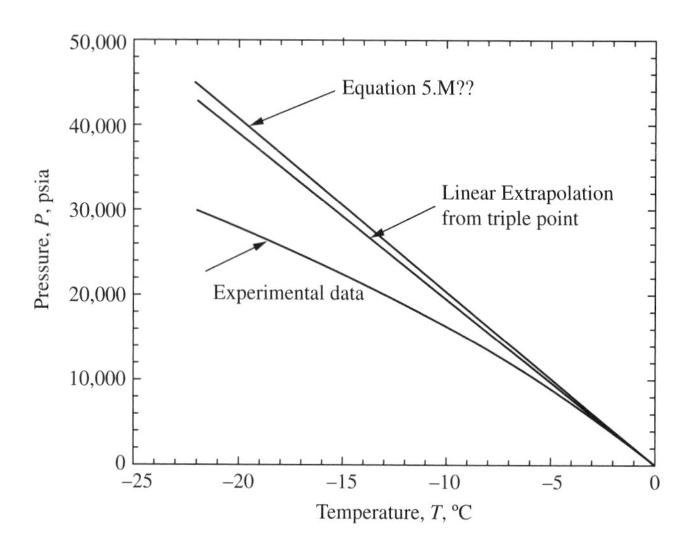

FIGURE 5.6 Comparison of the experimental results, the linear extrapolation of the equilibrium curve and of Eq. 5.M?? for estimating the pressure at the liquid–ice I–ice III triple point of water. Both the linear extrapolation and Eq. 5.M?? are tangent to the experimental curve at 0°C.

5.8 EXTRAPOLATING VAPOR-PRESSURE CURVES

A true equilibrium vapor-pressure curve extends only from the triple point (the lowest temperature at which vapor and liquid can be in equilibrium) to the critical point (the highest temperature at which vapor and liquid have separate existence). What would happen if we extrapolated the vapor-pressure curves? The logical coordinates on which to do this are ln *p* vs. 1/*T*, which correspond to the C-C equation, as shown in Figure 5.7.

On these coordinates the heavy line represents the true vapor pressure. The extrapolation to temperatures above the critical temperature (to the left) represents physically unreal states, which surprisingly have some meaning, described in Chapter 9. The extrapolation to temperatures below the triple point (to the right) represents unstable states, which can actually be produced and measured in the laboratory! Some values are shown in Figure 5.8. We see that between 0 and 10°C the vapor pressure of water plots as practically a straight line on these coordinates, as we would expect for these low pressures at which the C-C equation is quite reliable.

Below 0°C there are two curves, one for subcooled water and one for ice. The subcooled water curve (practically a straight line) is almost a linear extension of the liquid line, as logic suggests it should be. The ice curve (also practically a straight line) falls below the subcooled water line and has

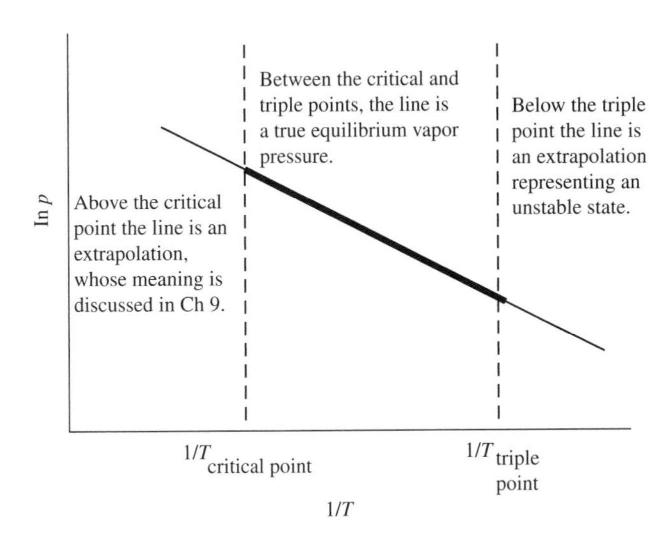

FIGURE 5.7 Extrapolation of the vapor-pressure curve on $\ln p$ vs. $1/T$ coordinates, to temperatures above the critical and below the triple points.

a different slope. At equilibrium, the liquid will turn to ice. We may think of this by imagining a piece of ice separated from some mass of subcooled water, but both sharing the same vapor space. The vapor pressure of the ice is less than that of the water, so that if we select some system pressure between the vapor pressures of ice and water then the water will evaporate because its pressure is above the system pressure, while vapor will deposit on the ice, because the system pressure is above the ice's vapor pressure. If a suitable crystallization nucleus is brought in contact with the water (almost any solid surface will serve as this nucleus) then the liquid water will rapidly convert to its stable form at these temperatures, ice.

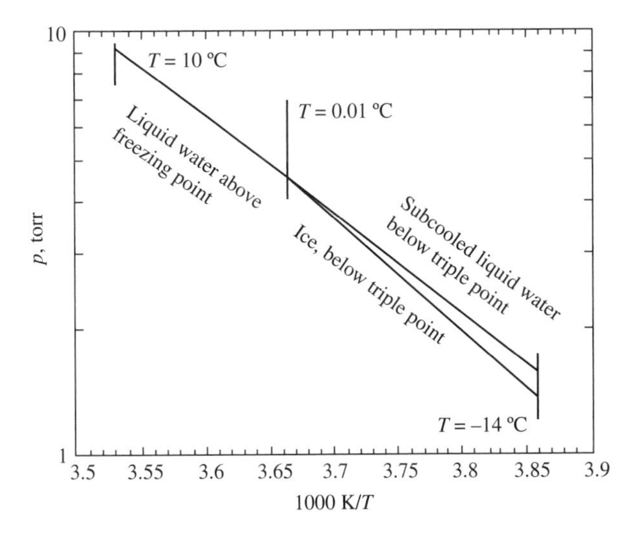

FIGURE 5.8 Experimental vapor pressures of water, subcooled water, and ice [1, pp. 2–48 and 2–49], plotted on $\ln p$ vs. 1000 K/T coordinates. Actual temperatures run from right to left; the values of −14°C, 0°C, and +10°C are shown.

The most important examples of this subcooling of liquid water are weather phenomena. Most cloud seeding takes place at temperatures below 0°C, in conditions where water drops have been cooled below their equilibrium freezing temperature, but for want of a solid nucleus have not frozen. Providing the nuclei causes them to freeze, grow in size, and fall as ice or (upon passing through a higher temperature air layer) as rain. The most spectacular example of this subcooling is the ice storm that regularly paralyzes the eastern United States. Rain forms in a warm air layer. The raindrops fall through an air layer that is much colder than 0°C, and turn to subcooled water. When they reach the ground the solid ground provides the crystallization nucleus they need to turn to ice. This covers streets with sheet ice, and covers trees with ice, adding enough weight to break limbs, which tear down power lines. The result is a major power outage, with impassable streets. Similar unstable or metastable liquids exist in other substances, but we have studied the water–ice–subcooled water system shown in Figure 5.8 more than any other, because of its great practical significance.

5.9 VAPOR PRESSURE OF SOLIDS

As Figure 5.8 makes clear, solids have vapor pressures, just as liquids do. The same is obvious from Figure 1.9 for water and Figure 5.2 for carbon dioxide. Vapor pressures of solids behave the same as the vapor pressures of liquids. We have fewer data for them, and most of those data are at temperatures at which their vapor pressure is modest, and well represented by the C-C equation. You may not think of wood, steel, or concrete as having vapor pressures. But they do. The values, at normal temperatures, are so low that the evaporation rate is nearly zero, even on a geologic time scale, (Problem 5.25).

5.10 VAPOR PRESSURES OF MIXTURES

This whole chapter is devoted to the vapor pressure (and related phenomena) of single pure species. We use that information, plus other information to estimate the vapor pressures of mixtures, discussed in detail in Chapters 8–10.

5.11 SUMMARY

1. Vapor pressure is fairly easy to measure and vapor-pressure data are widely used and widely available, as plots, tables, or tables of coefficients to be used in vapor-pressure equations.
2. The Clapeyron equation follows directly from the basic statement that if any two phases of one pure substance are in equilibrium, their molar or specific Gibbs energies are the same. It is a fundamental,

rigorous thermodynamic relationship, which applies to any two-phase equilibrium of a single chemical species, either gas–liquid, gas–solid, liquid–liquid, liquid–solid or solid–solid.

3. If one of the phases is a low-pressure gas, then we can use the ideal-gas approximation to convert the Clapeyron equation to the C-C equation, which is very accurate for low-pressure vapor–liquid and vapor–solid equilibrium.
4. Surprisingly, the C-C equation is fairly accurate for high-pressure vapor-liquid equilibrium, where its ideal gas assumption is clearly inaccurate. The reason is that the other assumption, constant heat of vaporization, is also inaccurate, and the two inaccuracies work in opposite senses and tend to cancel one another.
5. More complex vapor-pressure equations are empirical, with little if any theoretical basis. They mostly start with the C-C equation and add or modify terms to make their equation fit the experimental data more accurately than the C-C equation. The most widely used of these is the Antoine equation, which adds one arbitrary correction constant to the C-C equation, thus modestly increasing its complexity and vastly improving its ability to fit experimental data.
6. Vapor-pressure data are often needed in equilibrium calculations; the material in this chapter will be used in subsequent ones.

PROBLEMS

See the Common Units and Values for Problems and Examples. An asterisk (*) on a problem number indicates that the answer is in Appendix H

5.1 Figure 5.2 shows only one line for every substance except CO_2, for which two lines are shown.
- **a.** Explain.
- **b.** How low in pressure would the plot have to go to show a similar situation for water?

5.2
- **a.** Estimate the vapor pressure of acetic acid at $150°F = 65.56°C$.
- **b.** Estimate the temperature at which the vapor pressure of acetic acid is 100 psia.

5.3
- **a.** Estimate the vapor pressure of propane at $100°F = 37.78°C$.
- **b.** Estimate the temperature at which the vapor pressure of propane is 1.00 atm.

5.4* Estimate the vapor pressure of solid carbon dioxide ("dry ice") at −75°F.

5.5 Estimate the temperature at which the vapor pressure of n-hexane is $0.5\ \text{atm} = 0.5065\ \text{bar}$.

5.6 Figure 5.2 is a Cox chart. On it the curve for a reference substance, water, in this case, is drawn as a perfectly straight line with a logarithmic ordinate. Then the temperature scale on the abscissa is made up to match the tabulated values of the vapor pressure of the reference substance. Then the vapor-pressure curves of other substances are drawn in on this plot, whose ordinate is the vapor pressure and whose abscissa is a special, nonlinear scale based on the vapor-pressure line of the reference substance. The temperature scale is approximately $(1\ \text{K})/(T - 43.15)$, plotted from right to left. This is equivalent to using the Antoine equation with $C = 230°C$. Show that this is so by observing that the water curve on Figure 5.2 is perfectly straight, and then that a plot of the $\ln p$ values for water vs $1/(T - 43.15)$ is also perfectly straight.

5.7 Estimate the heat of vaporization ΔH_{vap} in Btu/lbm of ethane at 0°F from the data shown in Table 5.A.

5.8* Estimate the vapor pressure of ice at 100 K, using the values in Example 5.2.

5.9
- **a.** Estimate the temperature at which the vapor pressure of ice is 0.001 psia, using the values in Example 5.2.
- **b.** Estimate the vapor pressure of ice at $-40°F = -40°C$, using the values from Example 5.2.

5.10 Figure 5.2 shows a curve (close to a straight line) for n-decane. As best I can read that line, it shows that the vapor pressure of n-deqane is 0.1 psia at $T \approx 111°F$ and that the vapor pressure is 100 psia at $T \approx 570°F$. Based on these values estimate the values of the constants A and B in the C-C equation in its most commonly used form (Eq. 5.11).

5.11 In Example 5.3, we used the C-C equation to estimate the vapor pressure of water at 1155.2°R, and compared the calculated value to the steam table value. Repeat that example using the Antoine equation (Eq. 5.14) and the constants in Table A.2.

5.12 Estimate the accentric factor ω for water
- **a.** Based on Figure 5.4.
- **b.** Based on steam table values.
- **c.** Compare your result to the value in Table A.1. Comment?

Table 5.A Data for Ethane

	$T = 0°F$	$T = 5°F$
p (psia)	219.7	237.0
$v^{(liq)}$ (ft^3/lbm)	0.0357	
$v^{(vap)}$ (ft^3/lbm)	0.5754	

5.13 Estimate the vapor pressure of propane at 100°F.

a. Using the Antoine equation (Eq. 5.14) and the constants in Table A.2.

b. Using Eq. 5.J, with the following constants from [8]: A, B, C, D, $E = 59.078$, -3492.6, -6.0669, 1.0919×10^{-5}, 2. Here the vapor pressure is in Pa, and the temperature in K.

c. Compare these values with the reported [11] value of 188.320 psia.

5.14 From the data shown in Example 5.6, estimate the freezing temperature of water at 10,000 psia.

5.15* Estimate the pressure at which the freezing temperature of water is 31°F.

5.16 Two laboratories have been studying the vapor–liquid equilibria of monochloro-difluoromethane ($CHClF_2$) in connection with its use as a refrigerant. They report conflicting data, shown in Table 5.B. Which set of data appears to be the more reliable? (Never let your boss see any data of yours that can be checked by the Clapeyron equation before you have made that check yourself!)

5.17 In the preceding problem we concluded that there was at least one error in the data from the April Fool Laboratory. We have studied their values and conclude that most likely the error is in the reported values of the enthalpy of the vapor. If all of the other values in their table are correct, what value would the enthalpy of the vapor at 30°F have to be, to make their data agree with the Clapeyron equation?

5.18 Percy Bridgman, the famous high-pressure physicist, measured the properties of water at very high pressures.

Table 5.B Two Sets of Thermodynamic Data

	Trick or Treat Laboratory			April Fool Laboratory		
Temperature (°F)	20	30	40	20	30	40
Vapor pressure (psia)	58.0	70.0	83.5	57.9	69.9	83.7
Liquid density (lb/ft^3)	81.6	80.4	79.2	81.5	80.5	79.3
Vapor, sp vol (ft^3/lb)	0.937	0.782	0.656	0.937	0.781	0.65
Liquid enthalpy (Btu/lb)	16.0[a]	18.7[a]	21.7[a]	0[b]	2.6[b]	5.5[b]
Vapor enthalpy (Btu/lb)	107.1[a]	108.1[a]	109.1[a]	81.0[b]	82.0[b]	83.0[b]

[a] Datum, −40°F, liquid.
[b] Datum, +20°F, liquid.

Table 5.C Some High-Pressure Equilibrium Data for Water

Pressure (atm)	Temperature (°C)	$v^{(liquid)} - v^{(solid)}$ (mL/g)
25,162	110.3	0.0847
30,969	149.5	0.0763
36,744	182.5	0.0694
38,710	192.3	0.0674

Table 5.C shows some of his data on the equilibrium between liquid water and one of the crystal modifications of ice, which he calls ice VII (see Figure 1.10).

a. Using these data, estimate the heat of fusion of ice VII at 149.5°C.

b. Sketch what kind of apparatus you would use to make these measurements.

5.19 The melting temperature of bismuth at 1 atm is 968°F. At this condition the volume change on melting is -5.52×10^{-5} ft^3/lbm and the enthalpy change is + 18.33 Btu/lbm.

a. Estimate the melting temperature at a pressure of 4000 psig.

b. Bismuth is one of very few substances (including water) that expand on freezing. What industrial use is made of this property?

5.20 If you are very careful to use very pure water, and to exclude all crystallization nuclei, it is possible to subcool liquid water to temperatures far below its equilibrium freezing point (see Figure 5.8).

a. Estimate the vapor pressure of subcooled liquid water at −10°C, using the Clausius–Clapeyron equation, starting at the triple point, at which point $\Delta h_{vaporization} = 1075.4$ Btu/lbm.

b. Compare your answer to the experimental value of 2.149 torr [8 p. 2–48]

c. Estimate the vapor pressure of subcooled water at this temperature using the Antoine equation and the constants in Table A.2.

d. The vapor pressure of ice at −10°C is 1.950 torr. Using this information, sketch the equivalent of Figure 1.9, adding the subcooled water curve. (This is the same plot as Figure 5.8, but on *P-T* coordinates.)

5.21* In the previous problem you compared the vapor pressure of ice at −10°C (=14°F) to that of subcooled liquid water at the same temperature. Using values from that problem, your thermodynamics textbook, or any other source you like, estimate by how much subcooled liquid water at −10°C will increase or decrease its Gibbs energy when it changes to ice at the same temperature.

5.22 **a.** Refer to Figure 1.9. At the triple point, is *dP/dT* for the gas–solid equilibrium greater than, equal to, or less than *dP/dT* for the gas–liquid equilibrium?

b. Refer to Figure 5.2. Is *dP/dT* for the gas–solid equilibrium of carbon dioxide greater than, equal to, or less than *dP/dT* for the gas–liquid equilibrium of carbon dioxide?

c. Can the results of parts (a) and (b) be explained by the Clapeyron equation?

5.23 The constants in Table A.2 are for the Antoine equation with pressure expressed in torr and temperature in °C. Those are not the only possible units. Using the values from Table A.2 for acetaldehyde, show the revised values of *A*, *B*, and *C* if we:

a. Express pressures in torr, but express temperatures in K instead of °C.

b. Express temperatures in °C, but express pressures in psia instead of torr.

c. Express pressures and temperatures in torr and °C, but use ln *p* instead of log *p*.

5.24* Estimate the slope of the *P-T* curve for the graphite–-diamond equilibrium at 25°C. At this temperature and at 1 atm, $\Delta h_{\text{graphite to diamond at 1 atm}} = +1.9$ kJ/mol (see Table A.8). For graphite and diamonds, $M = 12$ g/mol. See Example 4.2 for volume data.

a. Show the value of *dP/dT* you would estimate if you assume that the $\Delta h_{\text{graphite to diamond at 1 atm}} = +1.9$ kJ/mol (shown above for 1 atm) is applicable for all pressures.

b. Show the value of *dP/dT* you would estimate if you do not make the assumption in (a), but instead assume that the enthalpy of diamond and graphite both change with pressure according to Eq. 2.34. We may show that for solids $T(\partial v/\partial T)_P \ll v$.

c. Compare the results of parts (a) and (b) with Figure 4.11. Are both plausible? Is only one plausible? Which one?

5.25 The melting temperature of pure iron at 1 atm is 1535°C ≈ 1808 K. At that temperature the vapor pressure of liquid and of solid iron are the same, 0.00037 atm. The enthalpy change of sublimation of solid iron at that temperature is 418.3 kJ/mol. Using these data

a. Estimate the vapor pressure of solid iron at 20°C.

b. The evaporation rate of solids and liquids is more or less proportional to the vapor pressure. Comment on the rate at which steel (which is mostly iron) is likely to evaporate at 20°C [12].

REFERENCES

1. Poling, B. E., G. H. Thompson, D. G. Friend, R. L. Rowley, and W. V. Wilding. Physical and chemical data. In *Perry's Chemical Engineers' Handbook*, ed. 8, D. W. Green, ed. New York: McGraw-Hill, pp. 2–412 (2007).
2. Chu, J. C., S. L. Wang, S. L. Levy, and R. Paul. *Vapor–Liquid Equilibrium Data*, Ann Arbor, MI: J. W. Edwards (1956).
3. Speight, J. G. *Lange's Handbook of Chemistry*, ed. 16. New York: McGraw-Hill, pp. 1–201 (2005).
4. Lide, D. R. *CRC Handbook of Chemistry and Physics*, ed. 71. Boca Raton, FL: CRC Press, pp. 6–118 (1990).
5. Keenan J. H., F. G. Keyes, P. G. Hill, and J. G. Moore. *Steam Tables: Thermodynamic Properties of Water Including Vapor, Liquid and Solid Phases*. New York: Wiley (1969).
6. Martin, J. J., and J. B. Edwards. Correlation of latent heats of vaporization. *AIChE J.* 11:331–333 (1965).
7. Pitzer, K. S., D. Z. Lippmann, R. F. Jr., Curl, C. M. Huggins, and D. E. Petersen. The volumetric and thermodynamic properties of fluids, II: compressibility factor, vapor pressure and entropy of vaporization. *J. Am. Chem. Soc.* 77:3433–3440 (1955).
8. Poling, B.E., et. al., *op.cit.* p. 2–55.
9. Yaws, C. L. *Handbook of Vapor Pressure*, vols. 1–4. Houston, TX: Gulf (1994).
10. Poling, B. E., J. M. Prausnitz, and J. P. O'Connell. *The Properties of Gases and Liquids*, ed. 5. New York: McGraw-Hill, pp. 2–412 (2001).
11. Starling, K. E. *Fluid Thermodynamic Properties for Light Petroleum Systems*. Houston, TX: Gulf (1973).
12. The data for this problem were generated, working backwards from the vapor pressure data in *CRC Handbook of Chemistry and Physics*, ed. 71. Boca Raton, FL: CRC Press, pp. 5–71 (1990).

6

PARTIAL MOLAR PROPERTIES

In Chapter 4 we saw that for physical equilibrium, the key thermodynamic relations were equality between phases of the partial molar Gibbs energy or chemical potential, $\mu_i = \bar{g}_i = (\partial G/\partial n_i)_{T,P,n_j}$, and that for a chemical reaction the key thermodynamic relation was the equality of the algebraic sum of these same quantities on either side the reaction. The chemical potential is the only really important partial molar property, although a few others are useful. As mentioned in Chapter 4, partial molar properties (also called partial molar derivatives) have a strange and counterintuitive property, discussed in Section 6.9. This chapter illustrates some of the behavior of partial molar derivatives, and then explores that counterintuitive property, showing why it leads naturally to the definition of the fugacity, which we explore in the next chapter.

6.1 PARTIAL MOLAR PROPERTIES

It is difficult to visualize the partial molar Gibbs energy or chemical potential, $\mu_i = \bar{g}_i = (\partial G/\partial n_i)_{T,P,n_j}$, and there is no direct way of measuring it experimentally. So we first study the partial molar volume and enthalpy, which are easier to visualize and which can be measured experimentally. Then we use the insight gained with them to consider the partial molar Gibbs energy, or chemical potential. Furthermore, there are some practical applications of partial molar volume and/or partial molar enthalpies. We could also do all of this chapter in terms of a unit of mass instead of mol $dm_i = M_i dn_i$, where M_i is the molecular weight of i. The resulting equation would look practically the same, and the logic would be the same, until we came to chemical reactions; there the equations would become much more complex, because they would contain multiple values of M_i. This possibility is explored in detail in Section 6.7.

The basic definition of the partial molar property (in phase 1) is

$$\left[\left(\frac{\partial(PROPERTY)}{\partial n_a}\right)_{T,P,n_b,\text{etc}}\right]^{(1)} = \left(\overline{property}\right)_a^{(1)} \quad (4.18)$$

Here we show *PROPERTY* in capitals and its partial molar derivative, $\left(\overline{property}\right)_a^{(1)}$, in lowercase letters to emphasize that the derivative is normally taken of an extensive property, such as the enthalpy of a system, but the resulting $\left(\overline{property}\right)_a^{(1)}$ is intensive, for example, enthalpy per mol Because a partial molar property is the derivative of an extensive property with respect to number of mols it is an intensive property itself. Partial molar values normally exist only for *extensive* properties (*V, U, H, S, A, G*). They do not exist for *intensive* properties (*T, P*, viscosity, density, refractive index, all "specific" or "per unit mass" properties). There is no meaning to the terms "partial molar temperature" (degrees per mol at constant *T*??) or "partial molar specific volume" (cubic feet per mol per mol??).

In Chapters 7 and 9 we will use the partial molar derivative of the compressibility factor *z*, which is an intensive, dimensionless quantity. This usage seems to contradict the previous paragraph. However, if we define an extensive property $Z = nz$ and insert its value in Eq. 4.18 we will find that $\bar{z}_i$, is perfectly well behaved and has the right dimensions. This procedure is also sometimes used for other intensive

Physical and Chemical Equilibrium for Chemical Engineers, Second Edition. Noel de Nevers.

properties, like the coefficients in some equations of state for mixtures. Multiplying these intensive coefficients by the total number of mols present converts them to extensive properties, which we can then insert in Eq. 4.18 (see Example 6.5).

6.2 THE PARTIAL MOLAR EQUATION

Now let us consider some extensive property Y (here Y may stand for $V,S,U,H,A,G,$ etc.). Then $Y = Y(T.P,n_a,n_b \ldots)$ and

$$dY_{T,P} = \left(\frac{\partial Y}{\partial n_a}\right)_{T,P,n_b,n_c,\ldots} dn_a + \left(\frac{\partial Y}{\partial n_b}\right)_{T,P,n_a,n_e,\ldots} dn_b \ldots$$

$$= \bar{y}_a dn_a + \bar{y}_b dn_b \ldots \quad (6.1)$$

This equation is simply a quite general first-order Taylor series. It states that the differential change of any variable is the sum of the product of its partial derivatives times the differential changes in the independent variables. It is slightly modified from the Taylor series because we have held T and P constant, thus eliminating the terms in dT and dP. However, we see from it that the derivatives that appear on the right are the partial molar derivatives of Y. For example, if we let Y be volume, then

$$dV_{T,P} = \bar{v}_a dn_a + \bar{v}_b dn_b + \cdots \quad (6.2)$$

Here at constant T and P the $\bar{y}_i$ (or $\bar{v}_i$) are functions of composition only.

Now in (Figure 6.1), we take an empty vessel ($V = 0$ for the contents) and add species a and b simultaneously, at constant T and P, mixing all the time so that the composition in the vessel is always constant. Thus, the $\bar{v}_i$ are all constant during this process of addition. Integrating Eq. 6.2 yields

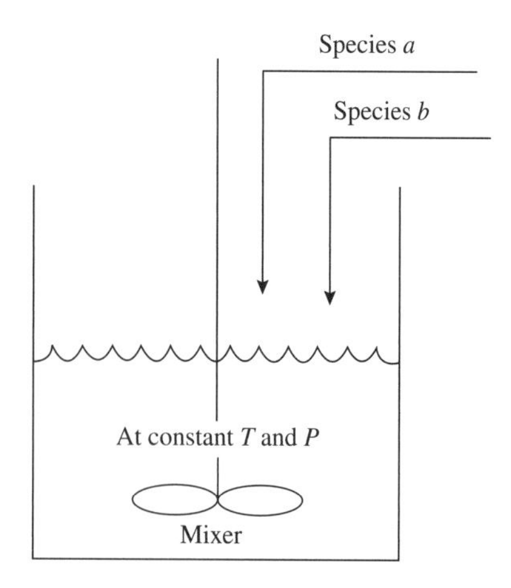

FIGURE 6.1 If we add species a and b at constant rates and run the mixer vigorously, then the composition of the fluid in the tank will remain constant, while its volume increases.

$$\begin{aligned} \int dV &= \int \bar{v}_a dn_a + \int \bar{v}_b dn_b \cdots = \bar{v}_a \int dn_a + \bar{v}_b \int dn_b \cdots \\ V &= \bar{v}_a n_a + \bar{v}_b n_b + \cdots \end{aligned} \quad (6.3)$$

Here we did the additions by a special path, which allowed us to perform the integrations. However, V is a state function, dependent only on P, T, and the various n_i. Thus, this relation is true, no matter what path we follow. For any extensive property Y follows that

$$Y = \bar{y}_a n_a + \bar{y}_b x_b + \cdots \quad (6.4)$$

We can divide both sides of Eq. 6.4 by the total number of moles n_T, which changes the Y to a y and changes the number of mols of each species to the mol fraction of that species; the result is

$$y = \frac{Y}{n_T} = \bar{y}_a x_a + \bar{y}_b x_b + \cdots \quad (6.5)$$

which we will use in the section 6.4, and then again in Section 6.9 and subsequent chapters.

Equation 6.4 and its alternative form Eq. 6.5 have no common names. A good name for them is the *partial molar equation.*

6.3 TANGENT SLOPES

To visualize this kind of property, suppose we have a graduate cylinder containing 1000 g of pure water at 20°C (Figure 6.2) We add some material, such as ethanol, a bit at a time, all at a constant T and P. After each addition, we stir and wait for equilibrium, holding temperature and pressure constant. Then we measure the volume and plot it as shown in Figure 6.3. The slope of the curve at any point is

$$\text{slope} = \left(\frac{\partial V}{\partial n_i}\right)_{T,P,n_j} = \bar{y}_i. \quad (6.6)$$

so the partial molar volume can be found directly from this plot, at any value of the number of mols added. Calculating

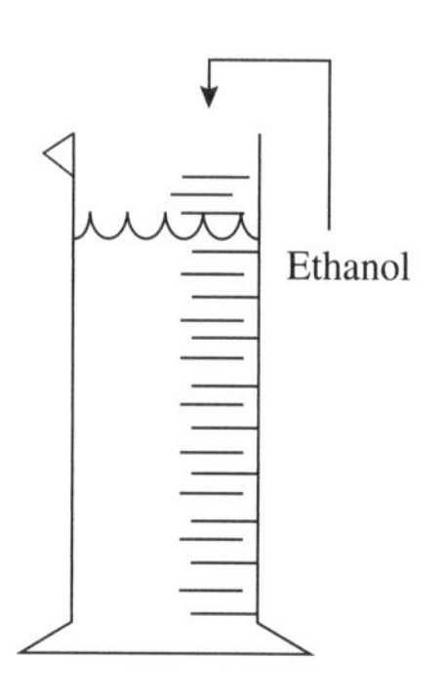

FIGURE 6.2 Measuring the partial molar volume by dissolving ethanol in 1000 g of water at 20°C and 1 atm.

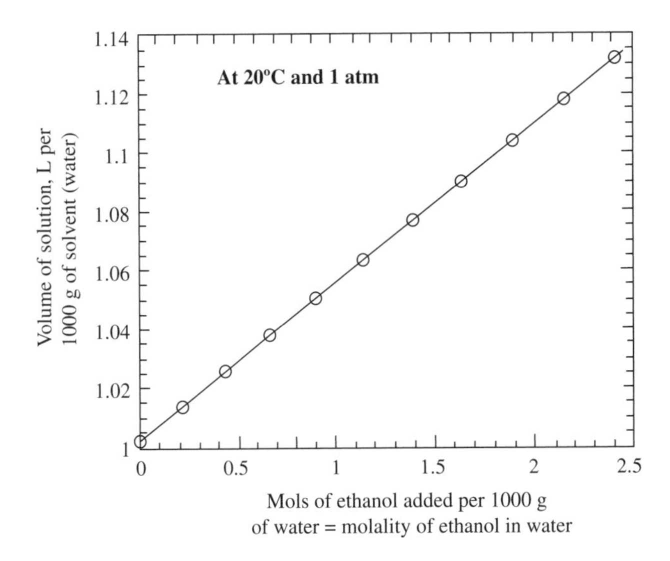

FIGURE 6.3 Volume of solution plotted vs mols of ethanol added, for a constant 1000 g of water. The points represent data from Gillespie et al. [1]. See Problem 6.1.

the partial molar volume this way is called the *method of tangent slopes.* (Logically, it should be called simply the method of slopes, but before we had computers, we drew a tangent line to the curve by hand and eye and measured the slope of that line geometrically.)

Generally, we would like the plot like Figure 6.3 to have some concentration variable as an abscissa rather than simply mols added. The logical one for the above procedure is

$$\text{molality} = \frac{\text{mols of solute}}{1000\text{ g of solvent}} \tag{6.7}$$

The corresponding ordinate must be volume of solution per 1000 g of solvent. Thus, on a plot of (V per 1000 g solvent) vs. molality of the solute, the slope is partial molar volume of the solute, $\bar{v}_{\text{solute}}$.

The molality is *not* the same as molarity (mols of solute per 1000 mL of solution). Molarity is the convenient concentration unit for aqueous analytical chemistry using burettes and standard solutions. Molality has no practical use other than in making up of plots like Figure 6.3. Most tables of chemical thermodynamic properties (discussed in Chapter 12) use molality as the standard unit of concentration, because it is independent of temperature, which molarity is not. For very dilute solutions molality and molarity are practically the same and are often used interchangeably. For dilute solutions molality is nearly proportional to mol fraction.

Example 6.1 From the data shown in Figure 6.3 estimate the partial molar volume of ethanol in water solution at ethanol molalities of 0 and 1 molar, at 20°C.

The data shown in Figure 6.3 appear to form a straight line, but careful examination shows a modest curvature. These data can be represented with excellent accuracy by Eq. 6.A, a simple data-fitting equation, which is applicable only for the range of molalities shown in Figure 6.3.

$$\begin{pmatrix}\text{solution volume,}\\ \text{liters per 1000 g}\\ \text{of water}\end{pmatrix} = 1.0019 + 0.054668m - 0.000418m^2 \tag{6.A}$$

where m is the molality of ethanol in water. The partial molar volume is

$$\bar{v}_{\text{ethanol}} = \frac{d\begin{pmatrix}\text{solution volume,}\\ \text{liters per}\\ \text{1000 g of water}\end{pmatrix}}{dm}$$

$$= 0.054668 - 2\cdot 0.000418m \tag{6.B}$$

So that at zero molality ($m = 0$),

$$\bar{v}_{\text{ethanol}} = 0.054668\text{ L/mol} = 54.7\text{ cm}^3/\text{mol} \tag{6.C}$$

and at $m = 1$

$$\bar{v}_{\text{ethanol}} = 53.8\text{ cm}^3/\text{mol}$$ ■

Example 6.2 Estimate the volume change on mixing for 1 mol of ethanol with 1000 g of water at 20°C.

For pure ethanol at 20°C, $v^{\text{o}}_{\text{ethanol}} = 58.4\text{ cm}^3/\text{mol}$, and $v^{\text{o}}_{\text{water}} = 1.0019\text{ L}/1000\text{ g}$. From Eq. 6.A we compute that the mixed solution volume is 1.05615 L. The volume expansion on mixing is

$$\begin{pmatrix}\text{volume expansion}\\ \text{on mixing}\end{pmatrix}$$

$$= V_{\text{final, mix}} - V_{\text{initial, water}} - V_{\text{initial, ethanol}}$$

$$= 1.05615 - 1.0019 - 0.0584 = -0.00415\text{ L} = -4.15\text{ cm}^3 \tag{6.D}$$

We see that there is a net contraction on mixing of (4.15/58.4 = 7%) of the volume of the ethanol added; the solution has less volume than did the parts that were mixed to make it. ■

We can also express this in terms of the partial molar volume, by writing

$$V_{\text{solution, final}} = V_{\text{solution, initial}} + \int \bar{v}_i\, dn_i \tag{6.8}$$

(If $\bar{v}_i$ is independent of composition, then this integral becomes simply $n\bar{v}_i$.) Before we mixed the (n_i mols of i into the solution

$$V_{\substack{\text{solution, and material}\\ \text{to be mixed in}}} = V_{\text{solution, initial}} + \left(nv_i^{\text{o}}\right) \qquad (6.9)$$

where v_i^{o} is the pure species volume per mol (often called the *molar volume*), so the volume change on mixing is

$$\Delta V_{\text{mixing}} = V_{\text{solution, final}} - V_{\substack{\text{solution, and material}\\ \text{to be mixed in}}} = \int \left(\bar{v}_i - v_i^{\text{o}}\right) dn \qquad (6.10)$$

Example 6.3 Estimate the volume change on mixing in Example 6.2 by Eq. 6.10.

Here the integrated average value of $\bar{v}_i$ over the molality range from 0 to 1 is 0.05425 L/mol, so that

$$\Delta V_{\text{mixing}} = (v_{i,\text{ average}}^{\text{o}} - v_i^{\text{o}})\Delta n = (0.05425 - 0.0584) \cdot 1$$

$$= -0.004175\ \text{L} = -4.18\ \text{cm}^3 \qquad (6.\text{E})$$

which is within round-off error of the value in Example 6.2. ■

For a pure substance, any partial molar property is the same as the molar property; for example, the pure species partial molar volume is the same as the pure species molar volume. If we add one mol of pure water to a large number mols of pure water, the volume of the original sample of water will increase by the exact amount of the volume of the water added, because there is no volume change on mixing pure anything with itself. However, in mixtures of more than one species, the two are generally not the same, as shown by Examples 6.1, 6.2, and 6.3.

For chemically similar materials like benzene and toluene, the volume change on mixing is negligible, while for less similar materials like the ethanol and water shown in Figure 6.3 and these three examples, the volume change on mixing is significant. The volume shrinks by ≈7% of the volume of ethanol added, which greatly complicates the measurement of solutions of ethanol in water for liquor tax purposes. If there is zero volume change on mixing, then

$$\Delta V_{\text{mixing}} = 0 = n(\bar{v}_i - v_i^{\text{o}}) \quad \text{[for zero volume change on mixing]} \qquad (6.11)$$

so that $\bar{v}_i$ must equal v_i^{o} (the pure species molar volume of the solute), which is a constant independent of the concentration. The plot must be a straight line, and its slope must equal the pure species molar volume of the solute (Figure 6.4). In Figure 6.3 the curve is close to, but not exactly straight, but its slope is only ≈ 93% of the pure species molar volume of the solute.

For the experiment shown in Figure 6.2, we could also have measured the heat added to or subtracted from the

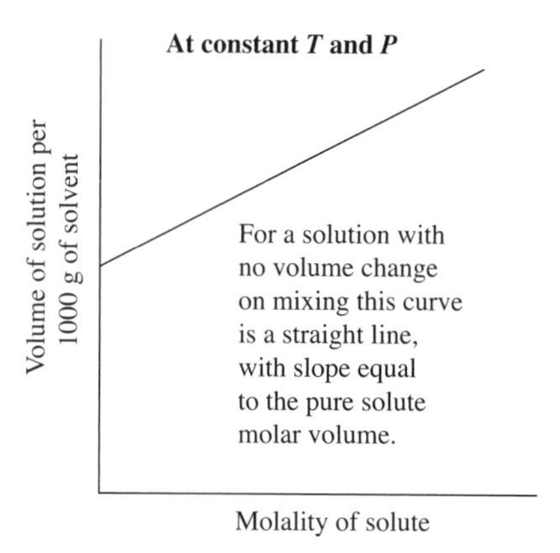
At constant T and P
Volume of solution per 1000 g of solvent
For a solution with no volume change on mixing this curve is a straight line, with slope equal to the pure solute molar volume.
Molality of solute

FIGURE 6.4 A plot of volume per 1000 g of solvent for a solution with no volume change of mixing.

system to hold the temperature constant. If we take as our system the solution (i.e., an open system with mass flow in), then the energy balance for the process of adding dn moles of ethanol is

$$d(nu)_{\text{sys}} = h_{\text{in}} dn_{\text{in}} + dQ + dW \qquad (6.12)$$

The system pressure is constant, so the only work is the work of driving back the surroundings

$$-dW = PdV_{\text{sys}} = Pd(nv)_{\text{sys}} = d(nPv)_{\text{sys}} \qquad (6.13)$$

Substituting and rearranging, we have

$$\begin{aligned} d(nu)_{\text{sys}} + d(nPv)_{\text{sys}} &= d(nh)_{\text{sys}} = dH_{\text{sys}} \\ &= h_{\text{in}} dn_{\text{in}} + dQ \end{aligned} \qquad (6.14)$$

If we now divide by dn_{in} and note that we have held T, P, and n_{water} constant, then we have

$$\left(\frac{dH_{\text{sys}}}{dn_{\text{in}}}\right)_{T,P,n_j} = \left(\frac{\partial H}{\partial n_i}\right)_{T,P,n_j} = \bar{h}_i = h_{\text{in}}^{\text{o}} + \frac{dQ}{dn_{\text{in}}} \qquad (6.15)$$

Here $\bar{h}_i$ is the partial molar enthalpy and h_{in}^{o} is the pure species molar enthalpy.

As with the volume, we can represent $\bar{h}_i$ as the slope of a plot of the enthalpy of the solution vs. the mols of i added. The enthalpy, unlike the volume, is known only to plus or minus a constant which is equal to the value of the enthalpy in the datum state. Since that datum is normally chosen arbitrarily, we can make any convenient choice, as long as we do not use a different datum for the same species somewhere else in the problem.

From Eq. 6.15 we see that the partial molar enthalpy is equal to the pure species molar enthalpy of the material added, plus the heat added per mol of material added. We may illustrate this by rewriting Eq. 6.15 as

$$\frac{dQ}{dn_{\text{in}}} = \bar{h}_i - h^{\circ}_{\text{in}} \tag{6.16}$$

If the partial molar enthalpy is the same as the pure species molar enthalpy, then $(dQ/dn)_{\text{in}}$ is zero and there is zero *heat of mixing*. Most solutions of chemically similar materials, such as benzene and toluene, have negligible heats of mixing, so for them the partial molar enthalpy is practically the same as the pure species molar enthalpy. For a mixture with zero heat of mixing, a plot of enthalpy per 1000 g of solvent vs. molality would have to be a straight line with slope $\overset{\circ}{h}_i$, the same as is shown for volume in Figure 6.4.

6.4 TANGENT INTERCEPTS

Molality is a satisfactory unit of concentration for dilute solutions. However, if we wish to cover the entire range of concentrations from pure solute to pure solvent it becomes unsatisfactory. (For pure solute, the solute molality is infinite!) A more convenient plot to work with is one of (property) per mol vs. mol fraction. For (property) = volume this is shown for the ethanol–water mixture in Figure 6.5.

The curve on this plot shows the molar volume (volume per mol) as a function of mol fraction of species *a*, ethanol. For $x_a = 0$ the molar volume is that of pure species *b*, water, $v^{\circ}_b \approx 0.018$ L/mol, while for $x_a = 1$ it is the molar volume of pure species *a*, ethanol, $v^{\circ}_a \approx 0.058$ L/mol. The curve appears to be straight, but careful examination shows that is it slightly S-shaped.

Figure 6.6 shows the same kind of plot as Figure 6.5, but with the curvature greatly exaggerated, to allow us to see the

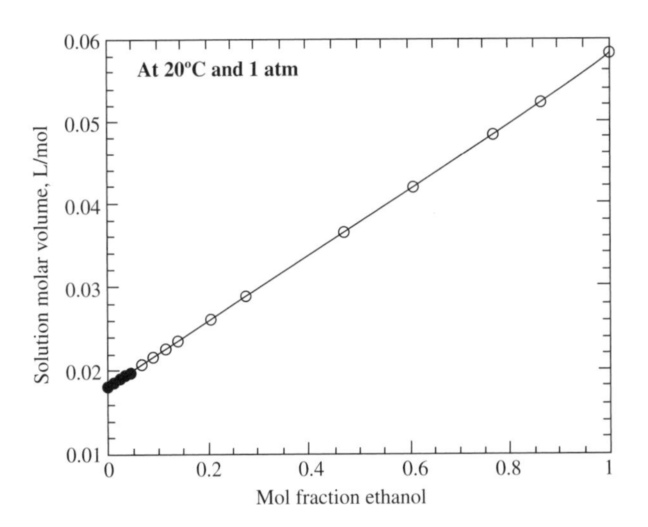

FIGURE 6.5 Molar volume as a function of mol fraction of ethanol, x_{ethanol}, for ethanol and water at 20°C. Data from [1].

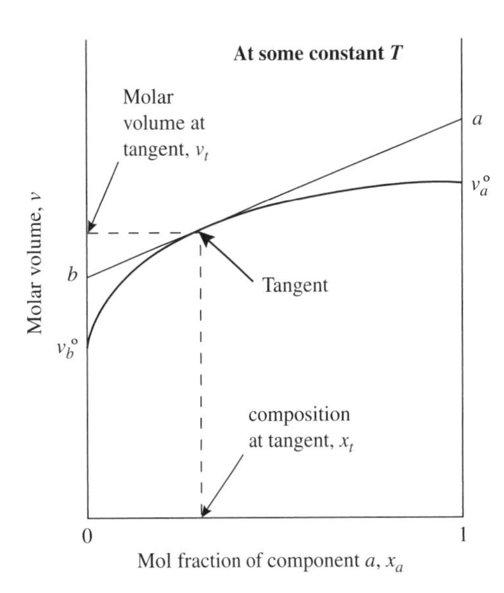

FIGURE 6.6 Redraft of Figure 6.5, with curvature exaggerated, to be used in demonstrating the idea of tangent intercepts.

difference between the curve and a tangent drawn to it, at an arbitrarily chosen value of x_a. The tangent line intersects the two vertical axes at *a* and *b*. As we choose different values of x_a at which we draw the tangent, the tangent line rotates, causing the values of *a* and *b* to change. For $x_a = 0$, $b = v^{\circ}_b$, and *a* is larger than the value shown in Figure 6.6. For $x_a = 1$, $a = v^{\circ}_a$, and *b* is larger than the value shown on Figure 6.6.

The equation of the tangent line is

$$v = b + (a-b)x_a = ax_a + b(1-x_a) = ax_a + bx_b \tag{6.17}$$

with, discussed above, the values of *a* and *b* changing with changes in x_a. Comparing Eqs. 6.5 and 6.17 we see that they are the same if *a* and *b* are identical to the partial molar volumes of species *a* and *b*. This same result can be shown by geometrical arguments (Problem 6.3). Thus we see that by the simple geometrical construction in Figure 6.6 we can find the needed partial molar volumes by simply reading the intercepts of the tangent line. This result is general, for all partial molar properties. It is not restricted to partial molar volume.

Example 6.4 Using Figure 6.5, estimate the partial molar volumes of ethanol and of water in a solution that is 1 molar in ethanol, by Eq. 6.17.

First we convert from 1.00 molal to mol fraction

$$\begin{pmatrix}\text{mol fraction}\\ \text{ethanol}\end{pmatrix} = x_{\text{ethanol}}$$

$$= \frac{1\ \text{mol}}{1\ \text{mol} + 1000\ \text{g water} \cdot \dfrac{\text{mol}}{18\ \text{g water}}} = 0.01768 \tag{6.F}$$

Next we fit an equation to the low range of data points on Figure 6.5, finding

$$\begin{pmatrix}\text{specific volume,}\\ \text{L/mol}\end{pmatrix}$$
$$= 0.018032 + 0.037002\, x_{\text{ethanol}} - 0.039593(x_{\text{ethanol}})^2 + 0.21787(x_{\text{ethanol}})^3 \quad (6.G)$$

which is applicable only for $0 < x_{\text{ethanol}} < 0.04$ (roughly the same range as Figure (6.3). Then from Eq. 6.G we can compute that at this ethanol mol fraction, $v = 0.018675$ L/mol and $dv/dx_{\text{ethanol}} = 0.035806$ L/mol. Then by simple geometry from Figure 6.6 we find

$$a = v_{\tan} + (1 - x_{\tan})\left(\frac{dv}{dx_1}\right)_{\tan} \qquad b = v_{\tan} - x_{\tan}\left(\frac{dv}{dx_1}\right)_{\tan} \quad (6.18)$$

where the *tan* subscript means the values at the point of tangency. Inserting values gives

$$a = \bar{v}_{\text{ethanol}} = 0.018675\frac{\text{L}}{\text{mol}} + (1 - 0.01768)\cdot 0.035806\frac{\text{L}}{\text{mol}}$$
$$= 0.0538\frac{\text{L}}{\text{mol}}$$
$$b = \bar{v}_{\text{water}} = 0.018675\frac{\text{L}}{\text{mol}} - 0.01768\cdot 0.035806\frac{\text{L}}{\text{mol}}$$
$$= 0.01804\frac{\text{L}}{\text{mol}} \quad (6.H)$$

The partial molar volumes for ethanol computed here and in Example 6.1 are the same. The method used in Example 6.1 does not give a value for the partial molar volume of water, which this method does. See also Problem 6.4. ■

This method of calculating partial molar quantities is called the *method of tangent intercepts.* (The required partial molar property is the *intercept* of the *tangent line* to the curve on the pure-species axis.) We can also readily make up plots of enthalpy of solution vs. mol fraction and read from them the partial molar enthalpies by the method of tangent intercepts. We will see in the rest of this course that while the method of tangent slopes is more intuitively obvious, the method of tangent intercepts is more widely used, because plots like Figures 6.5 and 6.6 are more widely available than those like Figure 6.3, and because the mol fraction is much more widely used as a concentration variable than the molality.

6.5 THE TWO EQUATIONS FOR PARTIAL MOLAR PROPERTIES

If we have experimental data, or an equation that is believed to represent such experimental data for some extensive property as a function of concentration, we can compute the partial molar values by the method of tangent slopes or the method of tangent intercepts. In either case we could plot the data and make the geometric constructions. However, the mathematical procedure, which our computers can do for us is much more useful.

If we have the data in the form of an equation for the molar value of some property (e.g., the molar volume) as a function of mol fraction, then from the geometry of Figure 6.6, and from Eq. 6.18, we can see that, for any value of x_a

$$\bar{v}_a = v_{\text{at } x_a} + (1 - x_a)\left(\frac{dv}{dx_a}\right)_{\text{at } x_a} \qquad \bar{v}_b = v_{\text{at } x_a} - x_a\left(\frac{dv}{dx_a}\right)_{\text{at } x_a} \quad (6.19)$$

This equation has no common name; a good name for it would be the *tangent intercepts calculating equation.* It is shown here for volume, but equally applicable for any property whose molar values can be expressed as a function of mol fraction. It is not restricted to binary mixtures; it can be applied to species a (or b or c) in mixtures of any number of species. Example 6.4 shows the application of this equation.

The corresponding equation for the method of tangent slopes (Eq. 4.19) requires that we have the property equation in the form of an extensive property, stated as a function of the number of mols present of each species. If we have, for example, an equation for v as a function of x_i we can multiply both sides of that equation by the total number of mols present, n_T. This makes the following changes

$$v n_T = V \quad \text{and} \quad x_a n_T = n_a \quad (6.I)$$

Example 6.5 shows its application.

Example 6.5 Repeat Example 6.4, for ethanol only, using the method of tangent slopes. Multiplying both sides of Eq. 6.G by n_T and inserting those in Eq. 4.19 we find

$$\left[\frac{\partial(n_T v)}{\partial n_a}\right]_{T,P,n_b} = \left(\frac{\partial}{\partial n_a}\right)_{T,P,n_b}\left(a n_T + b n_a + c\frac{n_a^2}{n_T} + d\frac{n_a^3}{n_T^2}\right) \quad (6.J)$$

where a, b, c, and d are the numerical constants from Eq. 6.G. Then we observe that when we increase n_a at constant n_b, then n_T is a variable and $dn_T/dn_a = 1$. Performing the differentiation in Eq. 6.1J we find

$$\bar{v}a = \left(a + b + c\frac{2n_T n_a - n_a^2}{n_T^2} + d\frac{3n_T^2 n_a^2 - 2n_T n_a^3}{n_T^4}\right) \quad (6.K)$$

If we then divide out the fractions on the right, thus changing the numbers of mols back to mol fractions, we find

$$\bar{v}_a = (a + b + c(2x_a - x_a^2) + d(3x_a^2 - 2x_a^3)) \quad (6.L)$$

A little algebra (Problem 6.7) shows that Eq. 6.L is identical to the solution to Example 6.4. ■

Example 6.5 is clearly a longer and messier way to do by tangent slopes what we did more easily by tangent intercepts in Example 6.4. So why bother? Because some functions are easier to do by tangent slopes, and this method appears in the historical literature, so you must understand it to understand that literature. Most often we use tangent intercepts and Eqs. 6.4 and 6.5

6.6 USING THE IDEA OF TANGENT INTERCEPTS

Although we most often make equilibrium calculations using the fugacity (as described in the next chapter), we can gain some further insight into the partial molar Gibbs energy by one more example, which also shows the utility of the idea of tangent intercepts.

Example 6.6 For liquid mixtures of α and β (two imaginary chemicals) at some fixed T and P the Gibbs energy per mol of the mixture is given by

$$g = Ax_\alpha + B(1 - x_\alpha) - C \sin 3\pi x_\alpha \quad (6.M)$$

Here x_α is the mol fraction of a; A, B, and C are constants, all positive. At this temperature and pressure, at equilibrium, this mixture forms two liquid phases. What are their compositions?

First, observe that no real mixtures have this simple an equation for Gibbs energy as a function of composition; however, this equation leads to simple mathematics, so please bear with it. Here we know that we have two relations to satisfy, namely that the partial molar Gibbs energy of α is the same in both phases and that the partial molar Gibbs energy of β is the same in both phases,

$$\mu_\alpha^{(1)} = \mu_\alpha^{(2)} \quad \text{and} \quad \mu_\beta^{(1)} = \mu_\beta^{(2)} \quad (6.N)$$

where $\mu_\alpha^{(1)}$ is the partial molar Gibbs energy of α in phase 1, and so on.

The easiest way to proceed is to sketch the function as shown in Figure 6.7. We see that g has two minima and one maximum. Because this is a plot of g vs. x, we can determine both of the species partial molar Gibbs energies at any value of x_α by drawing a tangent line to the curve at that value of x_a, and reading the intercepts on the axes at $x_\alpha = 1$ and $x_\beta = 1$.

We can draw a tangent line at any point on the curve, but if the conditions of equilibrium are to be satisfied then the only satisfactory tangent line is the one that is tangent to the curve twice. For it, the intercepts on the pure-species axes are the values of the μs that are the solutions to the two equilibrium equations. Then by simple algebra we can compute that the equilibrium concentrations are those for which $\sin 3\pi x_\alpha = 1$ (i.e., $x_\alpha = \frac{1}{6}$ and $x_\alpha = \frac{5}{6}$). (The plot was drawn for arbitrarily selected values of A, B, C. The answer is independent of those values.) ■

Although the equation used in this example is fanciful, the curve is similar in shape to that which would be observed in nature. If, for example, we attempted to sketch such a curve for a system like benzene-water, for which the mutual solubilities are only a few hundred parts per million, we would find that the curve would have two minima, and the center part would approach positive infinity. The center part of Figure 6.7 (between the two minima) represents an unstable solution, which can lower its Gibbs energy by splitting into two phases. If we could produce such a solution (e.g., 50% benzene, 50% water) we would expect it to have a very high Gibbs energy per mol and to split almost immediately into two phases, water with a few hundred ppm of benzene, and benzene with a few hundred ppm of water (see

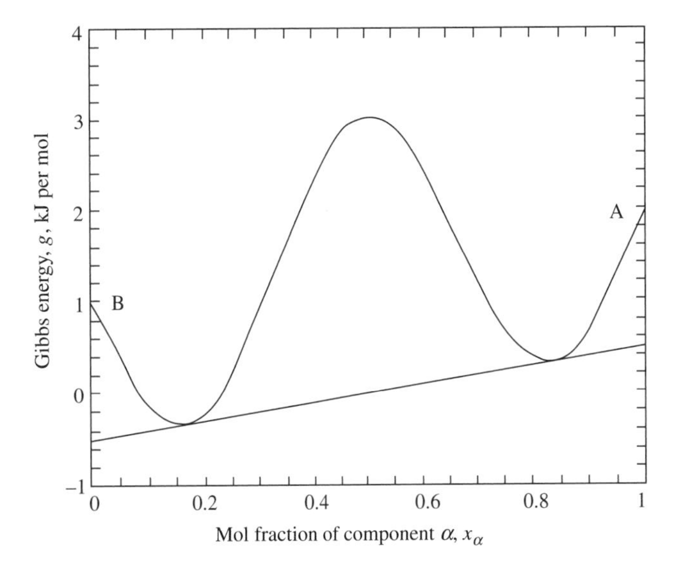

FIGURE 6.7 Gibbs energy composition plot for a very simple binary that forms two liquid phases. (This example is drawn for Eq. 6.M, with A, B, $C = 2$, 1, 1.5, all kJ/mol.)

Chapter 11), thereby greatly lowering its Gibbs energy. For the imaginary solution sketched in Figure 6.7, if we could produce a single phase with 50 mol% α, it would be expected to split into two liquid phases, with $x_\alpha = \frac{1}{6}$ and $\frac{5}{6}$.

For this example, a solution with 50 mol% α has $g = 3$ kJ/mol. If one mol of such a solution splits into 0.5 mol each of liquid phases with $x_\alpha = \frac{1}{6}$ and $\frac{5}{6}$, they will have $g = -0.333$ kJ/mol and $+0.333$ kJ/mol, so that the change will result in a Gibbs energy change of

$$\Delta G = (0.5\ \text{mol})\left(-0.333\frac{\text{kJ}}{\text{mol}}\right) + (0.5\ \text{mol})\left(0.333\frac{\text{kJ}}{\text{mol}}\right) - (1.00\ \text{mol})\left(3.0\frac{\text{kJ}}{\text{mol}}\right) = -3\ \text{kJ} \tag{6.O}$$

Any system that can reduce its Gibbs energy will do so; this system would. In Chapter 11 we will see the application of the calculation in Ex. 6.6 to real liquid-liquid equilibrium problems.

Example 6.6 is for two species that do not react chemically with each other, the common situation in phase equilibrium (Chapters 7–11), but not for those that can react chemically (Chapters 12 and 13). If the species do not react, then the values of A and B in Example 6.6 do not influence the mol fractions at equilibrium; the student can verify that algebraically. For chemical equilibrium (Chapter 12 and 13), the values of A and B (the pure component molar Gibbs energies) play a dominant role.

6.7 PARTIAL MASS PROPERTIES

For equilibrium calculations we almost always use the partial molar Gibbs energy or chemical potential. But the partial volumes and enthalpies appear most often in the form of partial mass properties, and are used for calculations other than equilibrium. Wherever a dn_i appears in this chapter (or elsewhere in this book) it could be replaced by dm_i/M_i

$$dn_i = \frac{dm_i}{M_i} \tag{6.20}$$

If we do that (Problem 6.9) we find analogs to most of the equations and procedures shown in this chapter, with partial molar properties replaced by those same properties divided by the molecular weight. Such properties have no common name; perhaps they are best called *partial mass properties*. The most commonly seen application is with plots of enthalpy per unit mass vs. mass fraction, as shown for water and sulfuric acid in Figure 6.8. As shown in Problem 6.9, if we apply the method of tangent intercepts to this figure, the intercept values are the partial mass enthalpies.

Example 6.7 Using Figure 6.8 and the method of tangent intercepts, estimate the partial molar enthalpies of both water and H_2SO_4 in a mixture of 60 wt% H_2SO_4, balance water, at 200°F.

Drawing the tangent to the 200°F curve at 60 wt% H_2SO_4, we find that it intersects the 0% (pure water) axis at ≈25 Btu/lbm, and the 100% H_2SO_4 axis at ≈−100 Btu/lbm. Using Eq. 6.20 we find

$$\bar{h}_{\text{water}} = \bar{h}_{\text{water, per pound}} \cdot M_{\text{water}} = 25\frac{\text{Btu}}{\text{lbm}} \cdot 18\frac{\text{lbm}}{\text{lbmol}} = 450\frac{\text{Btu}}{\text{lbmol}} \tag{6.P}$$

$$\bar{h}_{H_2SO_4} = \bar{h}_{H_2SO_4,\text{per pound}} \cdot M_{H_2SO_4} = -100\frac{\text{Btu}}{\text{lbm}} \cdot 98\frac{\text{lbm}}{\text{lbmol}} = -9800\frac{\text{Btu}}{\text{lbmol}} \tag{6.Q}$$

■

6.8 HEATS OF MIXING AND PARTIAL MOLAR ENTHALPIES

Figures like Figure 6.8 are widely used for *heat of mixing* calculations. These have practicaly nothing to do with equilibrium, the subject of this book, but are related to partial molar properties, the subject of this chapter.

6.8.1 Differential Heat of Mixing

Example 6.8 One lbm of water at 200°F is added to a large mass of H_2SO_4-water solution at 200°F and 60 wt% H_2SO_4. How much heat must be added or subtracted to keep the temperature constant at 200°F?

From Eq. 6.16 we have

$$\frac{dQ}{dm_{\text{in}}} = \bar{h}_i - h^{\text{o}}_{\text{in}} = 25\frac{\text{Btu}}{\text{lbm}} - 168\frac{\text{Btu}}{\text{lbm}} = -143\frac{\text{Btu}}{\text{lbm}} \tag{6.R}$$

$$\frac{dQ}{dn_{\text{in}}} = M_i\frac{dQ}{dm_{\text{in}}} = 18\frac{\text{lbm}}{\text{lbmol}}\left(-143\frac{\text{Btu}}{\text{lbm}}\right) = -2574\frac{\text{Btu}}{\text{lbmol}} \tag{6.S}$$

Here the $\bar{h}_i$ value is taken from Example 6.7, and the h^{o}_{in} is read from the intersection of the 200°F curve with the left axis (pure water). The enthalpy datum for Figure 6.8 is practically the same as that of the common steam tables [2], so we could have looked up h^{o}_{in} from the steam tables and found the same answer. The negative value shows that this mixing is exothermic; we must remove 143 Btu/lbm of water added. ■

The quantity computed here is called the *differential heat of mixing* because it refers to adding a differential amount of pure material into a large amount of solution. If the amount

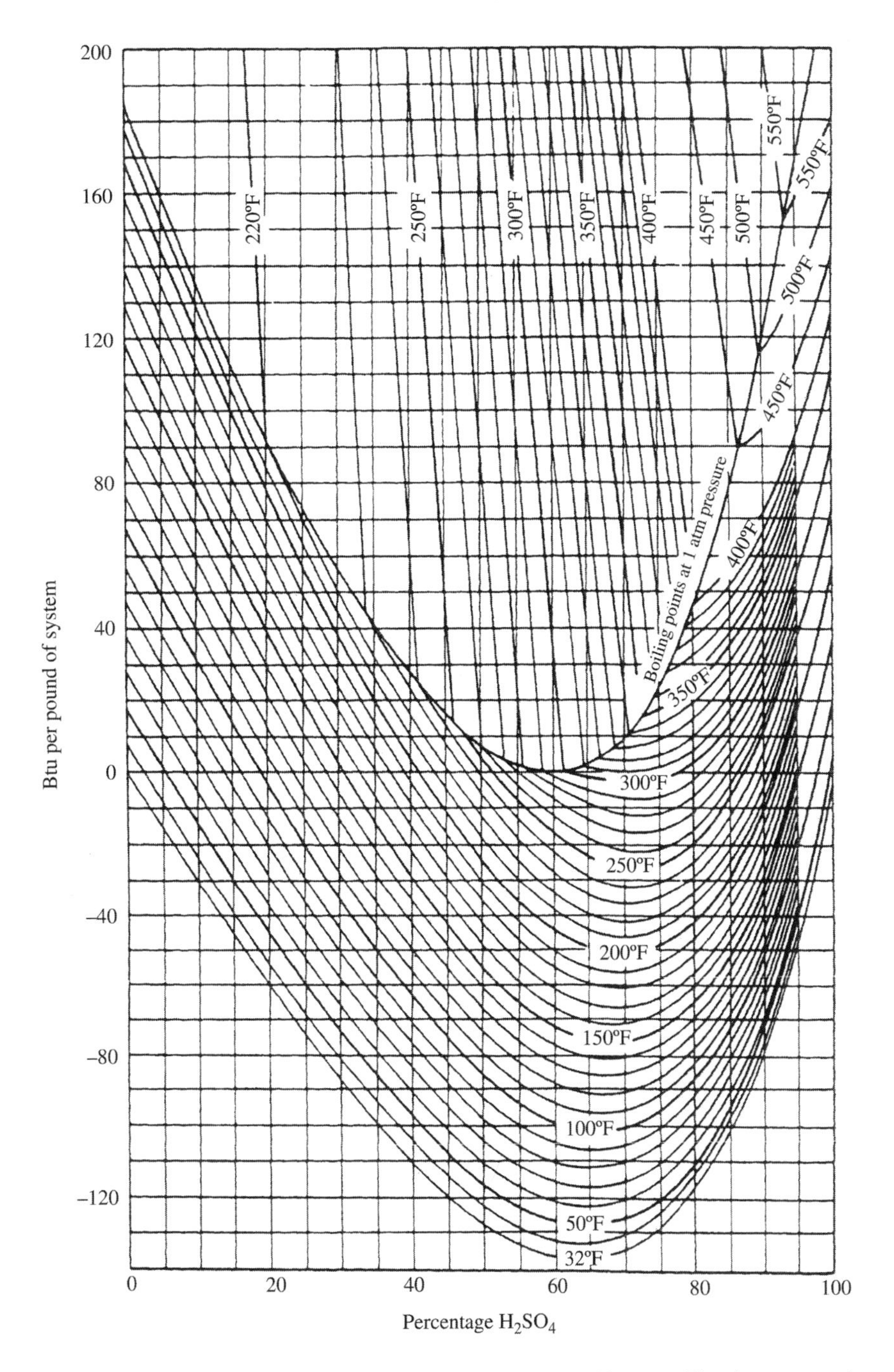

FIGURE 6.8 Enthalpy–concentration diagram for sulfuric acid-water. The datum state is zero enthalpy for pure water and for pure sulfuric acid as liquids at 32°F = 0°C, and the pure species vapor pressure. The "percentage" is weight percent. (The data are from the International Critical Tables, as plotted in Hougen, O. A., K. M. Watson, and R. A Ragatz, *Chemical Process Principles Charts.* © 1960, New York: Wiley. Reprinted by permission of the estate of O. A. Hougen.)

added were large enough to substantially change the composition of the mixture, then the value of $\bar{h}_i$ used would have to be the average value over that composition range. We see that the differential heat of mixing per mol is the partial molar enthalpy minus the pure species molar enthalpy at the same temperature and that the differential heating per unit mass is simply that per mol divided by the molecular weight.

6.8.2 Integral Heat of Mixing

Example 6.9 Four-tenths of a pound of water is mixed with 0.6 lbm of H_2SO_4 to form 1.00 lbm of 60 wt% H_2SO_4 solution. If the starting materials and the mixed solution are all to be at 200°F, how much heat must be added or removed?

Here at 200°F we can read the solution enthalpy from Figure 6.8 as a ≈−43 Btu/lbm, and the pure H_2SO_4 enthalpy from its rightmost axis as ≈53 Btu/lbm. Then by energy balance, using h^{o}_{water} from Example 6.8 we find

$$\Delta Q = (mh)_{\text{solution}} - \sum (mh)_{\text{inlet streams}}$$
$$= 1 \cdot (-43) - (0.4 \cdot 168 + 0.6 \cdot 53) = -243.2\text{Btu} \tag{6.T}$$

We must remove 243.2 Btu per pound of solution to hold the temperature constant. ■

This quantity is called the *integral heat of mixing*. It is the total amount of heat that must be removed to hold the temperature constant while making up the solution, starting with pure species, both at the solution temperature. Again we may convert between integral heat of mixing per mass and per mol by multiplying or dividing by the molecular weight.

Example 6.10 Estimate the enthalpy of a solution of 60 wt% H_2SO_4 balance water at 200°F, from the partial mass enthalpies computed in Example 6.7.

From Eq. 6.5, rewritten for masses instead of mols (Problem 6.9) we have

$$h_{\text{solution}} = \bar{h}_{\text{water, mass}} \cdot x_{\text{water, mass}} + \bar{h}_{H_2SO_4,\text{ mass}} \cdot x_{H_2SO_4,\text{ mass}}$$
$$= 25\frac{\text{Btu}}{\text{lbm}} \cdot 0.4 + \left(-100\frac{\text{Btu}}{\text{lbm}}\right) \cdot 0.6 = -50\frac{\text{Btu}}{\text{lbm}} \quad ■ \tag{6.U}$$

The fact that Eq. 6.5, rewritten for masses gives the same value that we read from Figure 6.8 in Example 6.9 should not surprise us; the equations and the geometry on which they are based demand that this must occur.

6.9 THE GIBBS–DUHEM EQUATION AND THE COUNTERINTUITIVE BEHAVIOR OF THE CHEMICAL POTENTIAL

So far nothing this chapter has been very counterintuitive. But so far we have not applied the partial molar idea to Gibbs energy (except in Example 6.6, where we did it only in a graphical way, and not for very dilute solutions). The previous parts of this chapter are preparation for this part, where the counterintuitive behavior appears.

Now we return to the partial molar equation (Eq. 6.5) and differentiate it at constant T and P:

$$dV_{T,P} = \bar{v}_a dn_a + n_a d\bar{v}_a + \bar{v}_b dn_b + n_b d\bar{v}_b + \ldots \tag{6.21}$$

Subtracting Eq. 6.2, we find

$$0 = (n_a d\bar{v}_a + n_b d\bar{v}_b + \cdots)_{T,P} \tag{6.22}$$

Equation 6.22 is true for any partial molar property, $\bar{h}_i$, $\bar{s}_i$, and so on. Equation 6.22 has no common name; when we write it for G, we find the Gibbs–Duhem equation:

$$0 = (n_a d\bar{g}_a + n_b d\bar{g}_b + \cdots)_{T,P} \tag{6.23}$$

When we divide Eq. 6.23 by n_T we find its mol-fraction equivalent, also called the Gibbs–Duhem equation:

$$0 = (x_a d\bar{g}_a + x_b d\bar{g}_b + \cdots)_{T,P} \tag{6.24}$$

which we will see in Chapter 9 is widely used in vapor–liquid equilibrium calculations.

Now we restrict our attention to binary (two species) mixtures and divide Eq. 6.24 by $x_a dx_a$ at constant T and P, finding

$$\left[\left(\frac{d\bar{g}_a}{dx_a}\right) + \left(\frac{x_b}{x_a}\right)\left(\frac{d\bar{g}_b}{dx_a}\right) = 0\right]_{T,P} \tag{6.25}$$

which is also sometimes also called the Gibbs–Duhem equation. Equation 6.25 is also true if we substitute $\bar{v}_a$ or $\bar{h}_a$ for $\bar{g}_a$, because it is a general consequence of the partial molar equation. This says that for a binary mixture if we have a plot of $\bar{v}_a$ vs. x_a over the whole range of x_a, (Figure 6.9), then, using the Eq. 6.25, we can construct the curve of $\bar{v}_b$ over the whole range of x_a, except for a constant of integration. Normally, the curve must pass through the pure species value of the volume of species b, (v^{o}_b), which supplies the necessary constant of integration.

Now what happens to $d\bar{v}_b/dx_a$ as x_b goes to zero? From Eq. 6.25,

$$\frac{d\bar{v}_b}{dx_a} = -\frac{(d\bar{v}_a/dx_a)}{(x_b/x_a)} \tag{6.26}$$

$$\text{as } x_b \to 0, \quad \frac{d\bar{v}_b}{dx_a} \to -\frac{(d\bar{v}_a/dx_a)}{0} \tag{6.27}$$

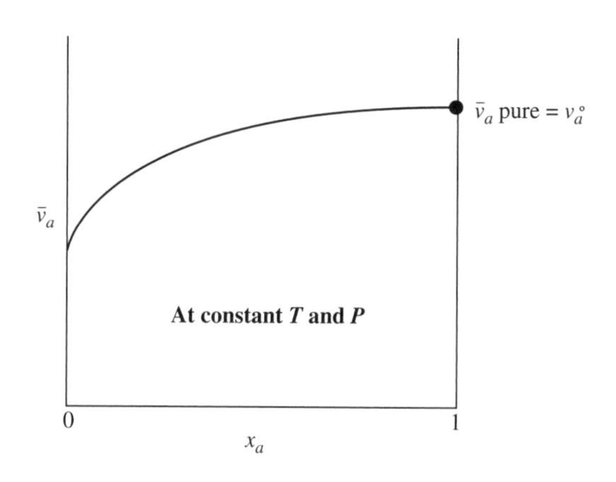

FIGURE 6.9 Partial molar volume of species a, as a function of composition.

Case 1 If the derivative $(d\bar{v}_a/dx_a)$ goes to zero as x_b goes to zero, then this leads to an indeterminate form, which may be zero, finite, or infinite.

Example 6.11 If

$$\frac{d\bar{v}_a}{dx_a} = 3x_b^2 + 2x_b \tag{6.V}$$

what is the value of $d\bar{v}_b/dx_a$ at $x_b = 0$?

$$\frac{d\bar{v}_b}{dx_a} = -\left(\frac{x_a}{x_b}\right)(3x_b^2 - 2x_b) = x_a(-3x_b - 2) \tag{6.W}$$

$$\lim_{x_b \to 0} \frac{d\bar{v}_b}{dx_a} = -2x_a = -2 \quad \blacksquare \tag{6.X}$$

See Figure 6.10.

Case 2 If the derivative $(d\bar{v}_b/dx_a)$ does not go to zero as x_b goes to zero, then Eq. 6.26 indicates that $(d\bar{v}_b/dx_a)$ must go to plus or minus infinity as x_b goes to zero.

Example 6.12 If

$$\frac{d\bar{v}_a}{dx_a} = 3x_b^2 + 2x_b + 1 \tag{6.Y}$$

what is the value of $d\bar{v}_b/dx_a$ at $x_b - 0$? Starting with Eq. 6.26

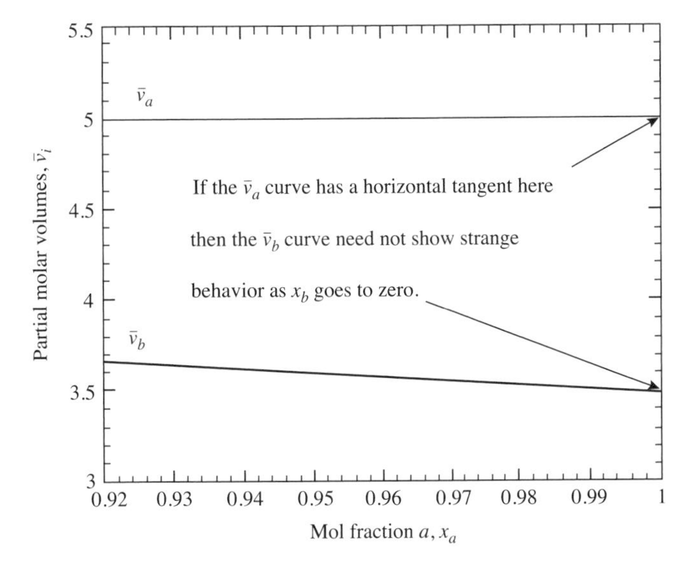

FIGURE 6.10 Case 1: The partial molar volume of species a has a horizontal tangent as $x_a \to 1.00$, so that the partial molar volume of species b does not show strange behavior as the mol fraction of species b goes to zero. This plot is drawn to scale for the equations in Example 6.11, with both pure species value of $v_i^{\circ} = 5.00$. Only the rightmost part of the whole figure is shown.

$$\frac{d\bar{v}_b}{dx_a} = \left(\frac{-x_a}{x_b}\right)(3x_b^2 + 2x_b + 1) = -x_a\left(\frac{3x_b^2 + 2x_b + 1}{x_b}\right) \tag{6.Z}$$

$$\lim_{x_b \to 0} \frac{d\bar{v}_b}{dx_a} = -\infty \quad \blacksquare \tag{6.AA}$$

See Figure 6.11.

For U, H, V, C_P, and so on $(d\bar{v}_a/dx_a)$ or $(d\bar{h}_a/dx_a)$, and so on, all approach zero as $x_a \to 1.00$. All of these are case 1 above, in which there is no unusual behavior at the dilute end of the curve. But as will be shown in Chapter 7, for a mixture, as $x_a \to 1.00$, $(d\bar{s}_a/dx_a)$ does not go to zero. In fact, for a binary mixture of ideal gases, $(d\bar{s}_a/dx_a) = -R/x_a$, so that as $x_a \to 1.00$ this derivative approaches $-R$. Thus, the derivative of the other partial molar entropy $(d\bar{s}_b/dx_a)$ must go to plus infinity as $x_b \to 0$! This does not mean that the entropy of all the species i in the solution becomes infinite as the mol fraction of i goes to zero. As the mol fraction of i goes to zero the entropy of all the i present must also go to zero. But that entropy goes to zero more slowly than the mol fraction does, so that their ratio, $\bar{s}_i$, becomes infinite as the mol fraction goes to zero.

Our main reason for studying partial molar properties was to gain understanding of and facility in computing $\mu_i = \bar{g}_i$. However, by definition

$$\mu_i = \bar{g}_i = \bar{h}_i - T\bar{s}_i \tag{6.28}$$

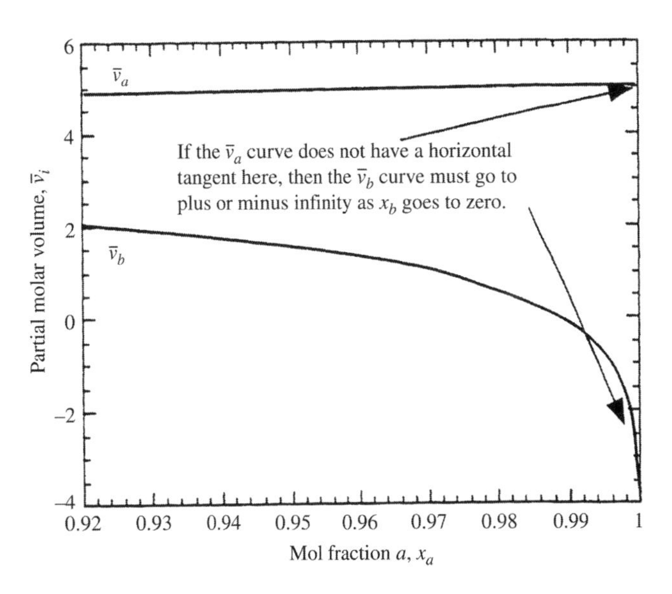

FIGURE 6.11 Case II: The $\bar{v}_a$ curve does not have a horizontal tangent as $x_a \to 1.00$, so that the, $\bar{v}_b$ curve must go to plus or minus infinity. This plot is drawn to scale for the equations in Example 6.12, with both pure species value of $v_i^{\circ} = 5.00$. Only the rightmost part of the mol fraction range is shown. The $\bar{v}_a$ curve appears close to horizontal, but careful examination will show that it does indeed have a slope of $+1.00$.

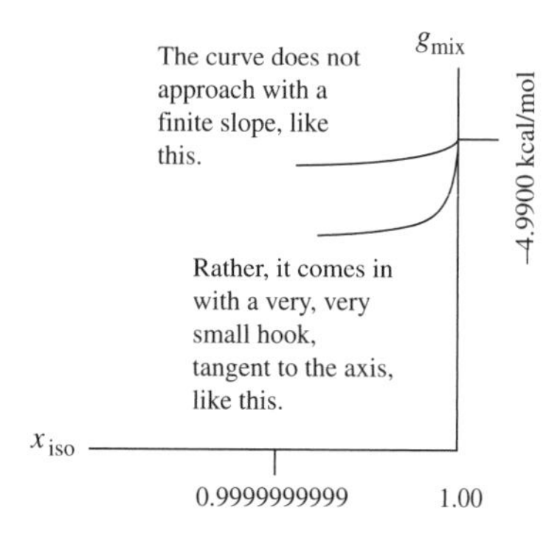

FIGURE 6.12 If we carry out the computation of the extreme ends of the g_{iso} curve in Figure 4.12, we find that the slope of the curve increases continually, thus forming a microscope hook, tangent to the vertical axis.

and we have shown above that as x_i goes to zero, $\bar{s}_i$ becomes infinite. Thus, we conclude that as the concentration of *i* goes to zero, its chemical potential in the solution must become minus infinity. If we refer back to Figure 4.12, a plot of *g* and of the two chemical potentials vs. x_a, we see that the *g* curve does not have horizontal tangents at the borders of the plot, and that the μ_i curves do indeed become asymptotic to the borders of the plot, thus approaching minus infinity as the concentrations approach zero.

It is not obvious from Figure 4.12 that the two ends of the g_{mixture} curve not only do not have horizontal tangents, but that they do not have finite slopes at the extremes. Instead, they meet the axis as tangents, as shown in Figure 6.12. To see this one must use a computer that carries many significant figures, and continue the calculation to values closer and closer to 1.00, or to 0.00 at the other end.

Seeing the difficulty of working numerically with the chemical potential, whose value goes to minus infinity when the concentration goes to zero, G. N. Lewis invented a new property, which he named the fugacity f_i, defined by

$$\mu_i = \bar{g}_i = RT \ln f_i + g_i^{\text{o}}(T) \tag{6.29}$$

Here $g_i^{\text{o}}(T)$ is some function of temperature alone. Thus, as $\left(\bar{g}_i - g_i^{\text{o}}(T)\right)$ goes to minus infinity, f_i, which is proportional to exp $\left[\bar{g}_i - g_i^{\text{o}}(T)\right]$, goes to zero, indicating that the fugacity of the solute goes to zero as its concentration in the solution goes to zero. This makes it a convenient property for equilibrium calculations. It also makes it intuitively satisfying; we all prefer a property that becomes zero as the concentration becomes zero to one that becomes minus infinity when the concentration becomes zero. The fugacity is the working form of the partial molar Gibbs energy for most equilibrium calculations. We see from this section that its form is practically forced upon us by the counterintuitive property of the partial molar equation, which shows that the partial molar Gibbs energy approaches minus infinity as the concentration approaches zero.

6.10 SUMMARY

1. The only important partial molar property is the partial molar Gibbs energy, or chemical potential $\mu_i = \bar{g}_i = (\partial G/\partial n_i)_{T,P,n_j}$. Our real reason for studying partial molar properties is to gain understanding of this function.
2. The partial molar volume and enthalpy are easier to visualize than the chemical potential, and can be measured directly, which the chemical potential cannot. They have some uses, mostly in heat-of-mixing or volume-change-on-mixing calculations.
3. The partial molar equation shows a unique and important relation between the partial molar properties in a mixture. When the differential of the partial molar equation is applied to the Gibbs energy, the result is the Gibbs–Duhem equation, which we will use in Chapter 9.
4. The Gibbs–Duhem equation shows that as the cocenntration of one species of a mixture approaches zero, its chemical potential approaches minus infinity. This makes the chemical potential (partial molar Gibbs energy) an inconvenient working property for equilibrium calculations. For this reason we use the fugacity (Chapter 7) instead.

PROBLEMS

See the Common Units and Values for Problems and Examples. An asterisk (*) on a problem number indicates that the answer is in Appendix H.

6.1 The actual data for Figure 6.3 are presented in [1] as a table of densities and wt% ethanol. For 10 wt% ethanol at 20°C the density is reported as 0.98187 g/cm^3. Using these values, calculate the molality and the volume of solution per 1000 g of water. Compare your result to the value plotted in Figure 6.3.

6.2* Solutions of $MgSO_4$ in water at 18°C have the properties shown in Table 6.A.

a. Using the method of tangent slopes, calculate $\bar{v}_{MgSO_4}$ at molality = 0.02, 0.1, and 0.2.

b. Comment on the physical–chemical reasons for these surprising values.

6.3 The method of tangent intercepts is shown in the main text, based on the partial molar equation, Eq. 6.5. It may also be shown purely geometrically as follows. Figure 6.13 is the same as Figure 6.6, but the point of tangency has been moved to move points *b* and *d* further apart, points *c*, *d*, and *e* have been added, and unnecessary text has been deleted.

Table 6.A Volumetric Properties of Solutions of $MgSO_4$ in Water at 18°C

Molality = mols $MgSO_4$/1000 g H_2O	Volume (mL per 1000 g H_2O)
0.00	1001.33
0.02	1001.27
0.04	1001.23
0.06	1001.22
0.08	1001.22
0.10	1001.23
0.12	1001.25
0.14	1001.28
0.16	1001.32
0.18	1001.37
0.20	1001.44
0.22	1001.51

By simple geometry we see that

$$eb = ed - bd = v - ec\frac{dv}{dx_a} \tag{6.AB}$$

where eb is the distance from e to b in Figure 6.13, and so on. From the definitions of the molar volume and the mol fraction

$$v = \frac{V}{n_a + n_b} \quad \text{and} \quad x_a = \frac{n_a}{n_a + n_b} \tag{6.AC}$$

we have the derivatives with respect to n_b, at constant n_a:

$$\frac{dV}{dn_b} = \frac{-V}{(n_a+n_b)^2} + \frac{dV/dn_b}{(n_a+n_b)} \quad \text{and} \quad \frac{dx_a}{dn_b} = \frac{-n_a}{(n_a+n_b)^2} \tag{6.AD}$$

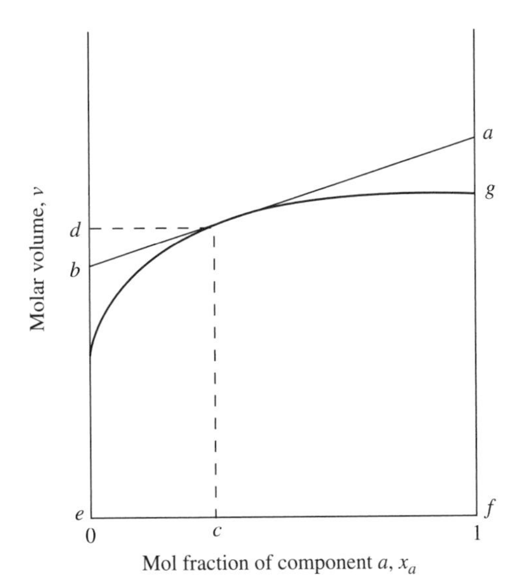

FIGURE 6.13 Same as Figure 6.6, but the point of tangency has been moved to the right to allow more room for symbols, and unnecessary text has been deleted.

Putting the first over a common denominator and dividing one by the other, we have

$$\frac{dV}{dx_a} = \frac{V}{n_a} - \frac{(n_a+n_b)(dV/dn_b)}{n_a} = \frac{1}{x_a}\left(v - \frac{dV}{dn_b}\right) \tag{6.AE}$$

Substituting this value in Eq. 6.AB and observing that $ec = x_a$, we find

$$eb = v - x_a\frac{dv}{dx_a} = v - x_a\left[\frac{1}{x_a}\left(v - \frac{dV}{dn_b}\right)\right] = \frac{dV}{dn_b} \tag{6.AF}$$

which shows on purely geometrical grounds that the tangent intercept at the left is the partial molar volume of species b. Show the corresponding derivation for dV/dn_a. This argument is independent of which partial molar property is involved; it is not restricted to the partial molar volume.

6.4 In Example 6.4 we used a curve fit of Figure 6.5 that covered only mol fraction from 0 to 0.04. If we curve-fit the mol fraction range from 0 to 1.00, we find

$$\left(\begin{array}{l}\text{specific volume,}\\ \text{L/mol}\end{array}\right) = 0.018056 + 0.034223x_{\text{ethanol}} + 0.0088468(x_{\text{ethanol}})^2 - 0.0028901 \cdot (x_{\text{ethanol}})^3 \tag{6.AG}$$

Show that by using this equation in Example 6.4 we compute a partial molar volume for ethanol of 0.0554 L/mol. Explain why the answer is different from that in Example 6.4.

6.5* The specific volumes of ethanol-water mixtures at 10°C are shown in Table 6.B [3].

Table 6.B Volumetric Properties of Ethanol-Water Solutions at 10°C

Wt% Ethanol	Specific Volume (mL/g)
0	1.00027
2	1.00399
4	1.00746
6	1.01065
8	1.01358
10	1.01633
12	1.01890
14	1.02133
16	1.02362
18	1.02592
20	1.02826

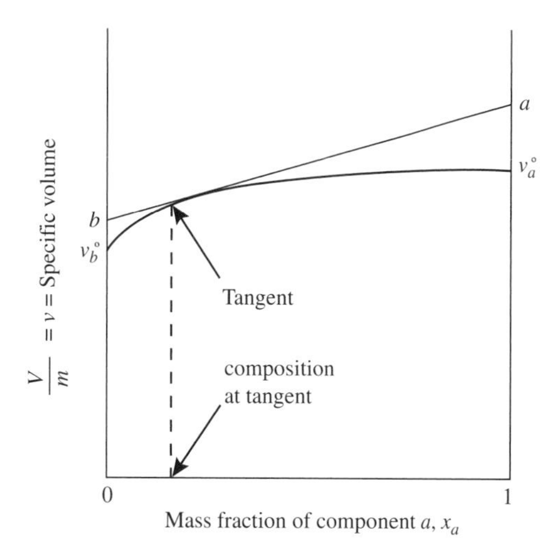

FIGURE 6.14 Same as Figure 6.6, but both the composition variable changed to mass fraction and the volume variable are changed from liters per mol to liters per unit mass.

a. From these data, calculate the partial mass volume of ethanol at 0, 2, 4, 6, 8, 10, and 12 wt% ethanol, at 10°C, using the method of tangent intercepts.

b. Convert the partial mass volume of ethanol for 10 wt% alcohol to partial molar volume at 10°C.

6.6 Figure 6.14 is the same as Figure 6.6, but the mol fraction has been replaced by the mass fraction, and the molar volume has been replaced by the specific volume (volume per unit mass). Show that, for the construction shown, the two tangent intercepts, a and b, are the partial mass volumes. Start with Eq. 6.5 and observe that each mol fraction is the number of mols that species divided by the total number of mols, n_T. Cancel the n_Ts in the denominators. Then replace each of the n_i values with the mass of that species divided by its molecular weight. Then divide both sides by the total mass of the system.

6.7 Show that Eq. 6.L is the same as the corresponding result by tangent intercepts in Example 6.4. *Hint:* Write the second term as $bx_a + b(1-x_a)$, and then factor it to be the equivalent of the b part of Eq. 6.19. Then use an analogous procedure on the c and d parts.

6.8* In a mixture of water and ethanol in which the mol fraction of alcohol is 0.4, $\bar{v}_{\text{ethanol}}$ is 57.5 cm^3/mol, and the density of the mixture is 0.8494 g/cm^3, what is $\bar{v}_{\text{water}}$?

6.9 Repeat Example 6.7 for the following H_2SO_4 concentrations: 20, 40, and 80 wt%.

6.10* Repeat Example 6.8 for the following H_2SO_4 concentrations: 20, 40, and 80 wt%.

Table 6.C Enthalpy of MeOH-Water Solutions at 50°C

Mol% MeOH	Enthalpy (Btu/lbmol of solution)
0	1620
30	1537
40	1543
50	1557
60	1577
70	1600
100	1733

6.11 Repeat Example 6.9 for the following H_2SO_4 concentrations: 20, 40, and 80 wt%.

6.12 **a.** One lbmol of methanol (MeOH) is added isothermally to a large mixture of MeOH-water at 50°C. The initial composition of the mixture is 50 mol% MeOH. The addition of 1 mol does not change the concentration measurably. How much heat must be added to or subtracted from the system to hold the temperature constant? The enthalpies MeOH–water solutions at 50°C are shown in Table 6.C.

b. Same as part (a) except that instead of 1 Ibmol of MeOH, we add 1 Ibmol of water.

c. One Ibmol of water is mixed isothermally with 1 lbmol of MeOH, all at 50°C. How much heat must be added or subtracted? Make clear whether this is heat added or heat subtracted.

d. The following is an *incorrect* solution to part (a) of this problem. Where is the error in this solution?

> Consider the system to consist originally of two parts: the lbmol of MeOH and the large mass of solution. The two are added and mixed isothermally, for which the energy balance is
>
> $$d(mu) = dQ - P\,dV$$
> $$H_f - H_i = dQ$$
> $$H_i = nh_{\text{system}} + h_i^{\text{o}}$$
> $$H_f = (n+1)h_{\text{system}}$$
> $$dQ = (n+1)h_{\text{system}} - (nh_{\text{system}} + h_i^{\text{o}}) = h_{\text{system}} - h_i^{\text{o}} \tag{6.AH}$$

6.13 For a mythical binary at constant T and P, the partial molar volume of species a is

$$\bar{v}_a = v_a^{\text{o}} - 0.2\frac{\text{cm}^3}{\text{mol}} \cdot x_b^2 \tag{6.AI}$$

where v_a^{o} is the molar volume of pure species a, and x_b is the mol fraction of species b. If the molar volume of pure species b is v_b^{o} write the equation for $\bar{v}_b$. Draw a sketch showing $\bar{v}_a$ and $\bar{v}_b$ vs. x_b.

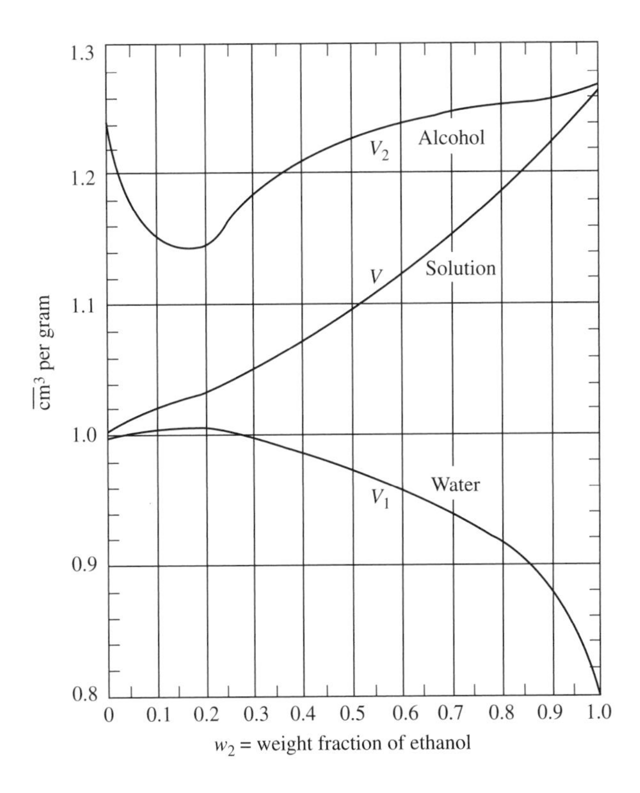

FIGURE 6.15 The solution volume per unit mass and partial mass volumes for ethanol–water at 20°C. (From Hougen, O. A., K. M. Watson and R. A. Ragatz, *Chemical Process Principles,* Part I: *Material and Energy Balances,* ed. 2. © 1954, New York: Wiley, p. 333. Reprinted by permission of the estate of O.A. Hougen.) This figure uses V_2 for $\bar{v}_{\text{ethanol, mass}}$, and so forth.

6.14 One Ibmol of a and 2 lbmol of b are mixed in an isothermal constant-pressure mixer. Estimate the heat added to or subtracted from the mixture from the following data:

$$h_a^{\text{o}} = 10.0 \text{ Btu/lbmol} \qquad h_b^{o} = 20.0 \text{ Btu/lbmol}$$

$$\bar{h}a = h_a^{o} + 10 \frac{\text{Btu}}{\text{lbmol}} \cdot x_b^2 - 5 \frac{\text{Btu}}{\text{lbmol}} \cdot x_b^3 \tag{6.AJ}$$

6.15 Figure 6.15 shows the measured specific volume and calculated partial mass volumes for mixtures of ethanol and water as a function of weight fraction. Using it,

a. Compute the partial *molar* volumes of water and ethanol at an ethanol concentration of 1.00 molar, and compare them to the values found in Example 6.4.

b. Estimate the increase in solution volume when 2 g of ethanol is added to 8 g of water at 20°C.

6.16 In Example 6.6, is the shape of the curves as x approaches zero and x approaches unity plausible? If not, why not?

6.17 For phase equilibrium between two partly miscible liquids, called phases 1 and 2, consisting of two chemical species, A and B, we know that $\mu_A^{(1)} = \mu_A^{(2)}$ and $\mu_B^{(1)} = \mu_B^{(2)}$ (or in the other notation $\bar{g}_A^{(1)} = \bar{g}_A^{(2)}$ and $\bar{g}_B^{(1)} = \bar{g}_B^{(2)}$). Does this mean that the Gibbs energies per mol of the two phases must necessarily be equal, $g^{(1)} = g^{(2)}$? Or are these quantities not equal (or not necessarily equal)?

REFERENCES

1. Gillespie, L. J., et al. Density (specific gravity) and thermal expansion (under atmospheric pressure) of aqueous solutions of inorganic substances and strong electrolytes. In *International Critical Tables of Numerical Data, Physics*, Vol. 3: *Chemistry and Technology*, Washburn, W. E. et al., ed. New York: McGraw-Hill, pp. 116–119 (1928).
2. Keenan, J. H., F. G. Keyes, P. G. Hill, and J. G. Moore. *Steam Tables: Thermodynamic Properties of Water Including Vapor, Liquid and Solid Phases.* New York: Wiley (1969).
3. Haynes, W. M. *Handbook of Chemistry and Physics*, ed. 91. Boca Raton, FL: CRC Press, pp. 8–56 (2010)

7

FUGACITY, IDEAL SOLUTIONS, ACTIVITY, ACTIVITY COEFFICIENT

7.1 WHY FUGACITY?

Equality of the chemical potentials (partial molar Gibbs energies) is the *fundamental criterion* for phase and chemical equilibrium, but we seldom use it directly because its form makes it difficult to use. The partial molar derivative (see Chapter 6) of the Gibbs energy of species *i* approaches minus infinity as the concentration of species *i* in the mixture approaches zero. This is highly inconvenient and counterintuitive; we expect the properties of interest of any species to approach zero as the concentration of that species in the mixture approaches zero. Seeing this inconvenience, G. N. Lewis [1] invented a new quantity called the *fugacity* f_i, which remedies this defect and forms a much more convenient working criterion for equilibrium.

7.2 FUGACITY DEFINED

Lewis defined the fugacity by

$$\mu_i = \bar{g}_i = RT \ln f_i + g_i^\circ(T) \tag{7.1}$$

He chose $\ln f_i$ to have the right property that as $\mu_i \to -\infty$, the natural log (ln) of f_i also approaches minus infinity, which makes $f_i \to 0$. The RT was needed because the natural log of f_i is dimensionless and the other terms in the equation have dimension of (energy/mol).

The fugacity (as we will see later) has the dimension of pressure, so mathematical purists would have us write the natural log as $\ln(f_i/1$ unit of pressure), for example, $\ln(f_i/\text{psia})$, to make the argument of the logarithm dimensionless. Most of us ignore that minor mathematical nicety. To see the meaning of $g_i^\circ(T)$, consider what happens when $f_i \to 1.0$. That makes $\ln f_i = 0$, so Eq. 7.1 becomes

$$\mu_i = \bar{g}_i = 0 + g_i^\circ(T) \tag{7.2}$$

showing that $g_i^\circ(T)$ is the value of $\mu_i = \bar{g}_i$ in the state for which $f_i = 1$. Unfortunately, f_i has the dimension of pressure; we regularly see it expressed in psia, bar, atm or torr. This means that the numerical value of $g_i^\circ(T)$ is different for different units of pressure. In *chemical reaction* equilibrium (Chapter 12) this causes no problem, and we regularly use $g_i^\circ(T)$. But in *phase* equilibrium the variable value of $g_i^\circ(T)$ would cause real problems. For that reason we practically never use $g_i^\circ(T)$ in phase equilibrium. Instead we work out all the necessary relations beginning with the derivative of Eq. 7.1 at constant temperature,

$$(d\bar{g}_i)_T = d(RT \ln f_i)_T \tag{7.3}$$

The fugacity is a convenience function, like the enthalpy. We use the enthalpy to replace a more complex set of symbols, $h = u + Pv$. If we solve Eq. 7.1 for f_i and replace G in terms of its basic components, we find

$$f_i = \exp\left[\frac{\left(\dfrac{\partial(U + PV - TS)}{\partial n_i}\right)_{T,P,n_j} - g_i^\circ(T)}{RT}\right] \tag{7.A}$$

Physical and Chemical Equilibrium for Chemical Engineers, Second Edition. Noel de Nevers.

Students who do not like to learn new convenience properties can simplify this book by substituting for the fugacity from Eq. 7.A, wherever the fugacity appears in this book.

7.3 THE USE OF THE FUGACITY

If phases 1 and 2 are in equilibrium, then we know that for each species i

$$\mu_i^{(1)} = \mu_i^{(2)} \quad \text{or} \quad \bar{g}_i^{(1)} = \bar{g}_i^{(2)} \tag{7.4}$$

If we substitute Eq. 7.1 twice in Eq. 7.4 and simplify, we find that Eq. 7.4 is equivalent to

$$f_i^{(1)} = f_i^{(2)} \tag{7.5}$$

and thus we may use equality of the individual species' fugacities between phases as a working criterion of physical equilibrium. This is the convenient criterion for computing phase equilibrium concentrations. We will see that the fugacities are functions of the mol fractions. Looking back at Example 3.3 (*please* look at that example now!), we can state that Raoult's and Henry's laws (which appear there) are both special cases of Eq. 7.5, which is the general case. We will explore this further in Chapter 8. This is the principal use of the fugacity. We will also see in Chapter 12 that fugacities play an important role in chemical equilibrium. The rest of this chapter is devoted to the calculation or estimation of the fugacities of individual species in various mixtures, as a preliminary step to calculating the compositions of phases in equilibrium, or of mixtures at chemical equilibrium.

7.4 PURE SUBSTANCE FUGACITIES

Our principal reason to be interested in pure substance fugacities is that we use them in estimating the fugacities of individual species in mixtures. If we never dealt with mixtures, we would not have bothered to define fugacity at all. In addition, computing the fugacities of pure substances illustrates some ideas more simply than the same computation for mixtures. So we begin here by studying the fugacity of pure substances. Remember that almost the only use of pure-substance fugacities is as one of the input data in the computation of individual species fugacities in mixtures. One exception to this statement is shown in Chapter 10.

Appendix C shows the mathematics of the fugacity of pure substances and of mixtures. We may summarize the findings for pure substances from Appendix C as follows. *For a pure substance,*

$$g = RT \ln f + g_i^\circ(T) \tag{7.6}$$

$$\left(\frac{\partial \ln f}{\partial P}\right)_T = \frac{v}{RT} \tag{7.7}$$

$$\lim_{P \to 0} \frac{f}{P} = 1 \tag{7.8}$$

$$\left(\frac{\partial \ln f}{\partial T}\right)_P = \frac{(h^* - h)}{RT^2} \tag{7.9}$$

The asterisk (*) indicates the property of an ideal gas at this temperature and pressure.

$$\frac{f}{P} = \phi = \exp\left(\frac{-1}{RT}\int_0^P \alpha dP\right) = \exp \int_0^P \frac{(z-1)}{P} dP \tag{7.10}$$

where ϕ is the *fugacity coefficient* (f/P) which we will use later,[1] and α is another convenience property, the *volume residual:*

$$\alpha = \text{volume residual} = \left(\frac{RT}{P} - v\right) = v\left(\frac{1}{z} - 1\right) = \frac{RT}{P}(1 - z) \tag{7.11}$$

where z is the compressibility factor (Pv/RT). For an ideal gas the volume residual is identically zero. Thus, from Eq. 7.10 we can see that for an ideal gas $f/P = \exp 0 = 1$, or $f = P$ and $\phi = 1.00$.

At this point we may give a simple answer to the obvious question, What *is* the fugacity? For a pure ideal gas the fugacity is identical to the pressure, and has the same dimensions as the pressure. Thus, we may think of it as a "corrected pressure" that enters many equilibrium calculation in place of the real pressure. Equations 7.8 and 7.10 show that we commonly show pure species fugacities as the dimensionless ratio of f/P. The mathematics of Appendix C show why that occurs. For pure species, the plot of the fugacity most often seen is Figure 7.1, which is somewhat similar to the common compressibility factor chart (z chart). This plot is shown in Appendix A.5 along with the same information in an alternative format. In both f/P and z charts, an ideal gas is represented by a horizontal line with value

[1] In the older literature, the fugacity coefficient ϕ had the symbol ν (Greek nu).

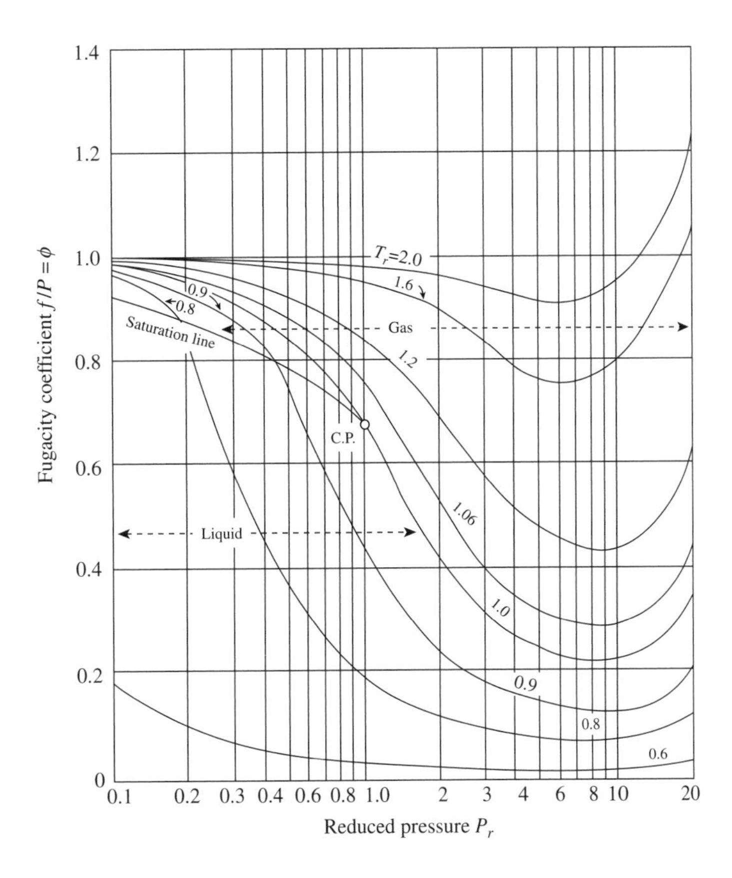

FIGURE 7.1 Pure species f/P as a function of P_r with curves of constant T_r. This plot, like Appendix A.4, is based on the simple, two-parameter version of the theorem of corresponding states, which is *only approximately correct.* The figure gives a visual insight into the behavior, but is reliable only to perhaps ±5%. See also Appendix A.5. (From Hougen, O. A., K. M. Watson, and R. A. Ragatz, *Chemical Process Principles,* Part II: *Thermodynamics,* ed. 2. © 1959, New York: Wiley, p. 600. Reprinted by permission of the estate of O. A. Hougen.)

1.00, and for real gases the curves of constant T_r (for T_r large enough for the material to be a gas) all become ≈ 1.00 as P_r goes to zero. The two plots are similar, but not the same. If we have one plot, it is possible to construct the other plot from it, using the relations in Appendix C.

From Figure 7.1 we see that for low-pressure gases (e.g., $P_r < 0.1$), and for gases up to medium pressures at high temperatures (e.g., $T_r > 1.4$), the pure species fugacity and the pressure are within a few percent of one another. For liquids the same is not the case; the fugacity of liquids is normally much less than the pressure. We will see below that the ways of estimating the fugacities of gases, liquids, and solids are the same in principle (they all obey Eqs. 7.6 to 7.10), but quite different in practice.

7.4.1 The Fugacity of Pure Gases

Example 7.1 shows the calculation of the fugacity of pure gases.

Example 7.1 Estimate the fugacity of propane gas at 220°F and 500 psia. We will proceed in several ways, showing several possibilities.

a. If we have direct and reliable PvT measurements for the pure substance, we can compute the fugacity directly from them. Figure 7.2 shows the measured compressibility. factor data for propane [2]. To use these data in Eq. 7.10, we read the z values at various pressures (from the table in [2]) and compute the volume residual as shown in Table 7.A. The value of α for 0 psia is extrapolated, because the definition of the volume residual (Eq. 7.11) shows that the volume residual becomes indeterminate (0/0), as $P \to 0$ (see Problem 7.1). Observe that while z changes considerably over this pressure range, α changes much less. This is one of the reasons we define and use α.

We then perform the integration by trapezoid rule, finding

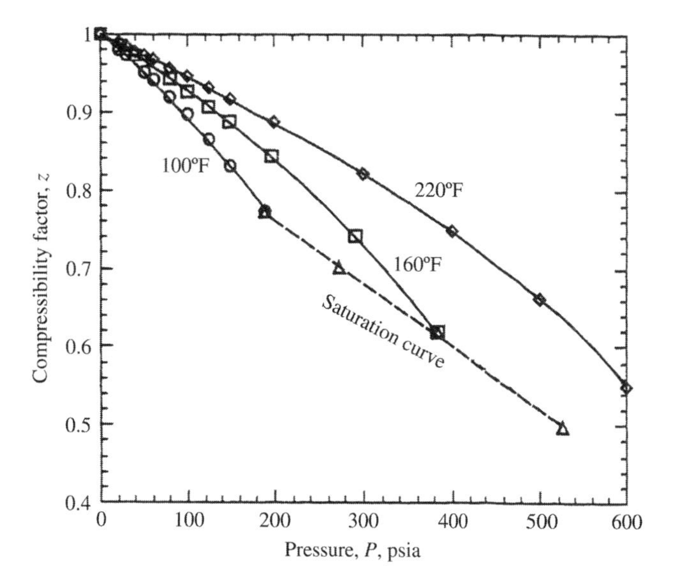

FIGURE 7.2 Compressibility factor z for propane gas at three temperatures and along the saturation curve. Data from [2].

$$\int_{0\text{ psia}}^{500\text{ psi}} \alpha\, dP \approx 2128 \frac{\text{psi}\cdot\text{ft}^3}{\text{lbmol}} \tag{7.B}$$

(This corresponds to an average value of $\alpha \approx 4.25$ ft^3/lbmol, which the reader may verify is approximately the average value in Table 7.A.) Then, inserting this value in Eq. 7.10 we find

$$\frac{f}{P} = \exp\left(\frac{-1}{RT}\int_0^P \alpha dP\right)$$

$$= \exp\frac{-2128\dfrac{\text{psi}\cdot\text{ft}^3}{\text{lbmol}}}{10.73\dfrac{\text{psi}\cdot\text{ft}^3}{\text{lbmol}\cdot{}^\circ\text{R}}.679.67^\circ\text{R}} = 0.747 \tag{7.C}$$

and $f = 0.747.\ 500$ psia $= 374$ psia.

This is the most direct and rigorous method of calculating a pure species fugacity. Its reliability is equal to the reliability of the experimental PvT data on which it is based. However, we will most likely not find z data specific for all the chemicals for which we wish to estimate the fugacity. So instead of using this calculation, we normally rely on EOSs, which are the mathematical equivalent of z charts like Figure 7.2.

Table 7.A Volume Residuals of Propane at 220°F

Pressure (psia)	z (dimensionless)	α (ft^3/lbmol)
0	1.0	3.8
100	0.9462	3.923
200	0.8873	4.110
300	0.8221	4.324
400	0.7787	4.582
500	0.6621	4.929

b. The simplest EOS is the ideal gas law. If we assume that propane behaves as an ideal gas, or practically so at all temperatures and pressures, then we can see from Eq. 7.11 that $\alpha = 0$ and $f/P = 1$, so $f = 500$ psia. This is *not* a very good estimate of f for this high a pressure. (If all the gases of practical interest were ideal, we might not have bothered to define the fugacity.)

c. If we can represent the nonideal gas behavior of propane by an EOS, we can estimate the fugacity from that EOS. This is the most widely used method, using a variety of EOSs. Here we will use the little EOS (Eqs. 2.48–2.50). Substituting these in Eq 7.10 we find

$$\frac{f}{P} = \exp\int_0^P \frac{(z-1)}{P} dP = \exp\int_0^P \frac{(P_r/T_r)f(T_r)}{P} dP \tag{7.D}$$

Then we factor P_r into P/P_c, cancel the Ps, and see that the integration is simple indeed, finding

$$\frac{f}{P} = \exp\left(\frac{P_r}{Tr}\cdot f(T_r)\right) \tag{7.E}$$

From Chapter 2 we know that in this formulation

$$f(T_r) = \left(0.083 - \frac{0.422}{T_r^{1.6}}\right) + \omega\cdot\left(0.139 - \frac{0.172}{T_r^{4.2}}\right) \tag{7.F}$$

from Table A.1 we find that for propane, $T_c = 369.8 =$ K $= 665.6$°R, $P_c = 42.48$ bar $= 41.9$ atm, and, $\omega = 0.152$ so that

$$T_r = \frac{680^\circ\text{R}}{665.6^\circ\text{R}} = 1.022$$
$$P_r = \frac{500\text{ psia}}{41.9\cdot 14.7\text{ psia}} = 0.812 \tag{7.G}$$

$$f(T_r) = \left(0.083 - \frac{0.422}{1.022^{1.6}}\right) + 0.152\cdot\left(0.139 - \frac{0.172}{1.022^{4.2}}\right)$$
$$= -0.3248 + 0.152\cdot(-0.018) = -0.3276 \tag{7.H}$$

and

$$\frac{f}{P} = \exp\frac{-0.3276\cdot 0.812}{1.022} = \exp(-0.2603) = 0.771 \tag{7.I}$$

Thus, based on this *very simple* EOS, we would estimate a fugacity of $500 \cdot 0.771 = 385$ psia.

d. From Figure 7.1 for $T_r = 1.022$ and $P_r = 0.812$, we may estimate $f/P \approx 0.76$, from which we would estimate a fugacity of ≈ 380 psia. Figure 7.1 should be used only for rough estimates.

e. The hydrocarbon thermodynamics tables of Starling [3] use an extended version of the Benedict–Webb–Rubin (BWR) EOS (Eq. 2.45) to match the experimental data. Using that equation, the table shows a computed value of $f/P = 0.7493$ for this T and P. This is probably our *best current estimate*, $f = 500 \cdot 0.7493 = 375$ psia. (Starling reports that the constants in his EOS were chosen to match the data in Figure 7.2, so the excellent agreement here should not surprise us!)

f. Most pure species thermodynamic tables (like the steam or refrigerant tables) do not show f/P. But we can easily calculate it from the values shown in those tables. If we write Eq. 7.6 twice, for states 1 and 2 at the same temperature, and subtract the equation for state 1 from that for state 2, we will find

$$\left(g_2 - g_1 = RT \ln\frac{f_2}{f_1}\right)_T \tag{7.12}$$

We now take the exponential, divide both sides of this equation by $(P_1 P_2)$ and rearrange, to find

$$\left[\frac{f_2}{P_2} = \frac{f_1}{P_1}\frac{P_1}{P_2}\exp\left(\frac{g_2 - g_1}{RT}\right)\right]_T \tag{7.13}$$

which is correct for *any* states 1 and 2 *at the same temperature,* for *any* pure substance.

If state 1 is at a low enough pressure that it is practically an ideal gas, then $f_1/P_1 = 1.00$ and we can write

$$\left[\frac{f_2}{P_2} = \frac{P_1}{P_2}\exp\left(\frac{g_2 - g_1}{RT}\right)\right]_T \quad \text{(correct only if } f_1/P_1 = 1.00\text{)} \tag{7.14}$$

Starling's tables show that at 220°F and 1 psia the calculated value of $f/P = 1.0057 \approx 1.00$, so that 1 psia is practically the ideal gas limit shown in Eq. 7.8. If we now decide that state 1 is at the $P \to 0$ limit in Eq. 7.8, for which $f_1 = P_1$, then we compute the values in Table 7.B, from which

Table 7.B Values for Propane at 220°F [3]

P (psia)	1	500
h (Btu/lbm)	−588.73	−624.97
s (Btu/lbm·°R)	1.6873	1.3673
$g = h - Ts$ (Btu/lbm)	−1735.54	−1554.28

Table 7.C Summary of Example 7.1

Calculation Method	Calculated *f/P*
(a) Directly from measured *PvT* data	0.747
(b) Ideal gas law	1.00
(c) Little EOS	0.771
(d) Figure 7.1	0.76
(e) Starling's table	0.7493
(f) Starling's table *h* and *s* values	0.744

$$\frac{f}{P} = \frac{1\text{ psia}}{500\text{ psia}}\exp \times \left(\frac{-1554.28 - (-1735.54)\dfrac{\text{Btu}}{\text{lbm}}}{1.987\dfrac{\text{Btu}}{\text{lbmol}\cdot{}^\circ\text{R}}\cdot 679.67^\circ\text{R}}\cdot\frac{44.062\text{ lbm}}{\text{lbmol}}\right)$$

$$= 0.002\exp(5.9139) = 0.740 \tag{7.J}$$

If we wish to take the 1.0057 above into account we would compute an f/P value of $0.740 \cdot 1.0057 = 0.744$. The difference between the 0.744 here and the 0.749 in part (e) is most likely round-off error in the calculations. The equations used to compute h, s, and f/P in any proper table of thermodynamic properties guarantee that the values should be the same (see Problem 7.6). These values are summarized in Table 7.C. ■

What are we to make of these values? Clearly, the ideal gas law is highly unreliable in this case. The remaining values are all in the range 0.74–0.77. The values based directly or indirectly on the experimental data for propane are all in the range 0.744–0.749, with the differences due to round-off error. Figure 7.1 and the little EOS, which are based on two versions of the (approximate!) theorem of corresponding states, are not as reliable as the values based on the experimental data for propane, but are close enough that they could be used with negligible errors in this case. Although the s in the free energy definition is an absolute entropy, the above calculation uses only Δs, which is the same for absolute entropies, or for those in common steam and refrigerant tables, which are based on arbitrarily chosen datums. Thus, this procedure works just as well with common steam and refrigerant tables as it does here. Concluding this example, we see that the calculations by an industrial-strength EOS give practically the same values as those based directly on the experimental data, while those based on the (approximate!) theorem of correspond states give values within a few percent of those values in this case. This good agreement is somewhat misleading because propane is a compound whose properties are often used in making up corresponding-states equations and plots; the agreement would not be as good for some highly polar substance.

In current industrial practice, we almost always estimate the fugacity of pure gases from an EOS. For simple estimates, we use a simple EOS; the computer programs that compute and use such fugacity data normally use complex EOSs, which are more accurate than simple EOSs like the little EOS.

7.4.2 The Fugacity of Pure Liquids and Solids

In principle, we could compute the fugacity of pure liquids and solids the same way we computed that of gases. This is impractical, however, so we use other methods. The reason is that the molar volume v of liquids and solids is so small that the volume residual α is practically the same as the ideal gas volume.

Example 7.2 Estimate the compressibility factor z and the volume residual α for liquid water at 100°F and 1 psi.

From the steam table, the specific volume of water at $101.7°F \approx 100°F$ and 1 psia is $0.016136\,ft^3/lbm = 0.290\,ft^3/lbmol$ and we compute that

$$z = \frac{Pv}{RT} = \frac{1\,\text{psia} \cdot 0.290\dfrac{ft^3}{lbmol}}{10.73\dfrac{psia \cdot ft^3}{lbmol \cdot °R} \cdot 560°R} = 0.000048 \approx 0.00005 \tag{7.K}$$

and

$$\alpha = \frac{RT}{P}(1-z) = \frac{10.73\dfrac{psia \cdot ft^3}{lbmol \cdot °R} \cdot 560°R}{1\,psia}(1-0.00005)$$
$$= 6008\frac{ft^3}{lbmol}(0.99995) = 6007.7\frac{ft^3}{lbmol} = 0.99995\frac{RT}{P} \quad ■ \tag{7.L}$$

This value of α is large enough that if we were to use it in Eq. 7.10 we would have a very large exponential, which would be very sensitive to round-off errors and hence very unreliable. (If we had *perfect* liquid PvT data, we could use it. We never have *perfect* data.) So we proceed a different way.

Example 7.3 Estimate the fugacity of pure liquid water at 100°F and 1000 psia. Here it is easiest to see what we are doing by drawing the 100°F isotherm on a P-v diagram (Figure 7.3).

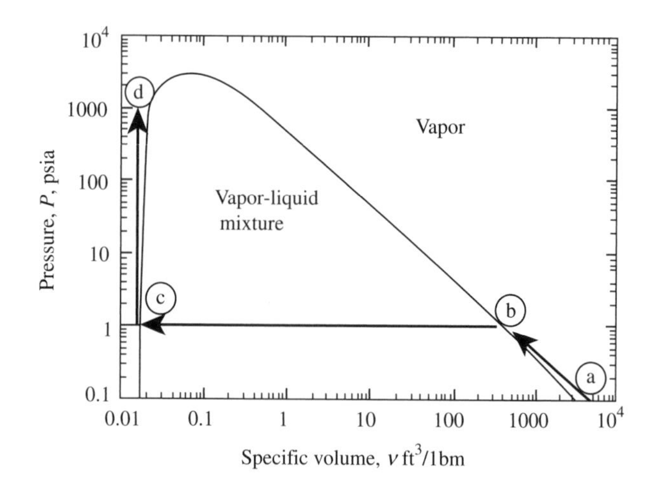

FIGURE 7.3 Integration path (from a to d) to find the fugacity of pure liquid water at 100°F and 1000 psia. Both axes are on logarithmic scales. The position of point a has been shifted slightly to the right, to make line a-b distinguishable from the vapor specific volume curve. Line c-d lies in the liquid region, which is not labeled because there isn't room for a label.

We know that as $P \to 0$, $(f/P) \to 1$, so $f \to 0$. Therefore, on the right in Figure 7.3, $f_a \to 0$. We may find f_b by

$$\left(\frac{f}{P}\right)_b = \exp\left(\frac{-1}{RT}\int_0^{P=0.9503\,\text{psia}} \alpha\, dP_T\right)$$

However, it is easy to show that at (b) $P_r = 0.9503\,\text{psia}/3203.6\,\text{psia} = 0.00030$, $T_r = 560°R/1165°R = 0.481$ for which, from the tabular equivalents of Figure 7.1 in the same source, $f/P = 0.9997 \approx 1.00$, so we are safe in calling the material at (b) an ideal gas with $f_b = P_b$. See also Problem 7.10, where it is shown that estimating this f/P by the other methods in Example 7.5 gives the same result.

We also know from Eq. 7.5 that $f_c = f_b = 0.95$ psia. To find f_d we use the integrated form of Eq. 7.7

$$\int (d \ln f)_T = \int \left(\frac{v}{RT} dP\right) T \tag{7.M}$$

Here v is practically constant (for a liquid), so we can take it out of the integral sign and have

$$\ln\frac{f_d}{f_c} \approx \frac{v}{RT}(P_d - P_c) = \frac{0.016\dfrac{ft^3}{lbm} \cdot 18\dfrac{lbm}{lbmol} \cdot 999\,psi}{10.73\dfrac{psi \cdot ft^3}{lbmol°R} \cdot (560°R)} = 0.048 \tag{7.N}$$

$f_d = f_c \exp(0.048) = 0.95\ \text{psia} \cdot (1.049) = 1.0$ psia. (See Problem 7.9) ■

This example shows why most discussion of pure species fugacities is about gases, rather than liquids. Compressing liquid water from 1 psia to 1000 psia changes the fugacity by only 4.9%; normally we ignore this and assume that for pure liquids and solids the fugacity is equal to the vapor pressure. (Compressing an ideal gas by this same ratio raises its fugacity by a factor of 1000!) This factor (1.049 in the above example) is called the *Poynting factor* or *Poynting correction factor* in many physics texts and in some chemical engineering equilibrium publications [4] (see also Section 7.11). In Chapter 14 we will see that we cannot ignore the effect of pressure on the fugacity of liquids when we compute osmotic equilibrium or small drop equilibrium. In those situations, the Poynting factor is the same size as the other terms in the equilibrium expressions.

7.5 FUGACITIES OF SPECIES IN MIXTURES

Now let us return to the more interesting case of mixtures. (After all, the fugacity is used almost exclusively for mixtures; we discussed pure species fugacities mostly to find out what the fugacity "feels like" and because we often use pure species fugacities to help us compute the fugacities of individual species in mixtures.) Appendix C also shows the mathematics of the fugacities of species in mixtures. From it we can summarize the findings as

$$\bar{g}_i = RT \ln f_i + g_i^{\circ}(T) \tag{7.1}$$

$$\left(\frac{\partial \ln f_i}{\partial P}\right)_T = \frac{\bar{v}_i}{RT} \tag{7.15}$$

$$\lim_{P \to 0} \frac{f_i}{P x_i} = 1 \tag{7.16}$$

$$\left(\frac{\partial \ln f_i}{\partial T}\right)_P = \frac{(h_i^* - \bar{h}_i)}{RT^2} \tag{7.17}$$

$$\frac{f_i}{P x_i} = \hat{\phi}_i = \exp\left(\frac{-1}{RT}\int_0^P \bar{\alpha}_i dP\right) = \exp\int_0^P \frac{(\bar{z}_i - 1)}{P} dP \tag{7.18}$$

Comparing these equations with the corresponding ones for pure species, we see that the pressure P has been replaced by the partial pressure Px_i and the volume, enthalpy, compressibility factor, and volume residual have been replaced by their partial molar equivalents. Here we introduce $\hat{\phi}_i$, the fugacity coefficient for species i in a mixture, sometimes called the partial fugacity coefficient.

We will discuss the estimation of the fugacities of species in mixtures in Section 7.12 after we have introduced some more useful definitions.

7.6 MIXTURES OF IDEAL GASES

For a mixture of ideal gases

$$V = \frac{n_T RT}{P} \tag{7.19}$$

$$\bar{v}_i = \left(\frac{\partial}{\partial n_i}\right)_{T,P,n_j} \frac{n_T RT}{P} = \frac{RT}{P} \tag{7.20}$$

and

$$\bar{\alpha}_i = \frac{RT}{P} - \bar{v}_i = 0 \tag{7.21}$$

It follows from Eq. 7.21 that the integral in Eq. 7.18. is zero, or that

$$f_i = P y_i \text{ or its exact equivalent } \hat{\phi}_i = 1.00 \qquad (7.22)^{(*)}$$

for ideal gases. Thus, we see that *for mixtures of ideal gases, the fugacity of each species is equal to that species' partial pressure.* The asterisk (*) on an equation number indicates that this equation is applicable only to ideal gases.

7.7 WHY IDEAL SOLUTIONS?

If, as shown above, for ideal gas mixtures the fugacity of one species in the mixture is equal to its partial pressure, then we would like to extend that simple idea to nonideal gas mixtures, and to solutions of liquids and solids. We can, using the definition of an *ideal solution.* An ideal solution is like an ideal gas in the following respects:

1. Neither exists *exactly* in nature. There are no gases that show *exactly* ideal gas behavior over a wide range of pressures, and no solutions that show *exactly* ideal solution behavior over a wide range of compositions.
2. There are many examples of gases and solutions that have *practically* ideal behavior. All gases at very low pressures and most gases at moderate pressures and temperatures well above their critical temperatures are *practically* ideal gases. Most mixtures of gases and

most liquid mixtures made of members of homologous series are *practically* ideal solutions.

3. It is often convenient to work with deviations from ideal behavior, as a mathematical artifice, rather than to work directly with the properties of the same material. This is the approach used in the compressibility factor z, which is a measure of departure from ideal gas behavior. We will see that the activity coefficient γ plays a similar role for departure from ideal solution behavior.
4. An ideal solution, like a ideal gas, is the simplest kind of behavior imaginable. Real gases have more complex EOSs than ideal gases; real solutions have more complex behavior than ideal solutions.

7.8 IDEAL SOLUTIONS DEFINED

An ideal solution is one that obeys Eq. 7.23:

$$f_i = x_i f_i^{\text{o}} \qquad (7.23)^{(\dagger)}$$

The symbol † on an equation indicates that it is applicable only to ideal solutions. Here f_i is the fugacity of species i in the solution, x_i is the mol fraction of species i in the solution, and f_i^{o} is some constant that will generally depend on T and P, but not on composition. (Frequently this constant is called the "standard state fugacity" or "reference state fugacity.") This can be any constant, so long as it has the dimensions of f_i and does not depend on x_i. Very often it is taken as the fugacity of pure i at the temperature and pressure of the solution, which is why it has the symbol f_i^{o}. But this is not the only possible choice; the real definition of an ideal solution is that for all possible mixtures the fugacity of a species in the solution is some constant (which may depend on P and T, but not on composition) times the mol fraction of that species in the solution. We will say more about the various choices that are regularly used for f_i^{o} in Chapters 8, 9, 12, and 13.

7.8.1 The Consequences of the Ideal Solution Definition

From mathematics shown in Appendix C, we may show that any ideal solution (of gases, liquids, or solids) has the following properties:

$$f_i = x_i f_i^{\text{o}} \quad \text{[definition of ideal solution!]} \qquad (7.23)^{(\dagger)}$$

$$\bar{g}_i - g_i^{\text{o}} = RT \ln x_i \qquad (7.24)^{(\dagger)}$$

$$\bar{v}_i - v_i^{\text{o}} = 0 \qquad (7.25)^{(\dagger)}$$

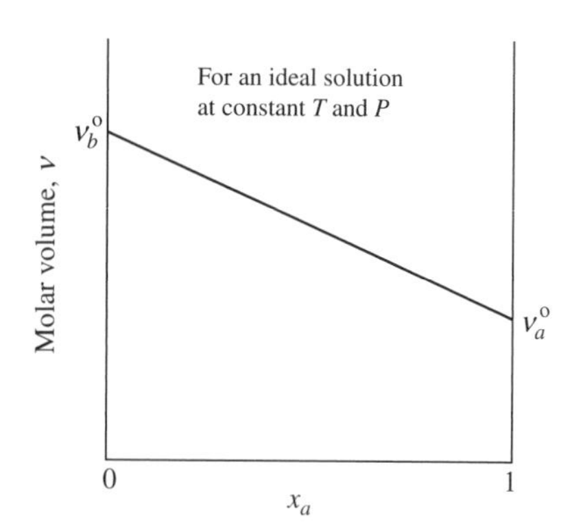

FIGURE 7.4 Molar volume-mol fraction plot for an ideal solution.

$$\bar{s}_i - s_i^{\text{o}} = -R \ln x_i \qquad (7.26)^{(\dagger)}$$

$$\bar{h}_i - h_i^{\text{o}} = 0 \qquad (7.27)^{(\dagger)}$$

Equation 7.25 says that the partial molar volume of one species in an ideal solution is equal to its pure-species molar volume; hence, if we plot molar volume of solution versus mol fraction of species a as shown in Figure 7.4, we will have a straight line. Stated another way, there can be no volume change on mixing to make up an ideal solution.

Equation 7.27 says that the partial molar enthalpy of a species in an ideal solution is equal to the pure species molar enthalpy, or that a molar enthalpy-mol fraction diagram must have the form sketched in Figure 7.5. This may also be stated that there is no heat effect of mixing for any ideal solution.

From the observations that there is no volume change or heat effect on mixing for an ideal solution, we may infer an intuitive model of such a solution, which may help us compare it with real solutions. Since there is no volume change on mixing, it follows that the average intermolecular distance in the solution must be the same as the average

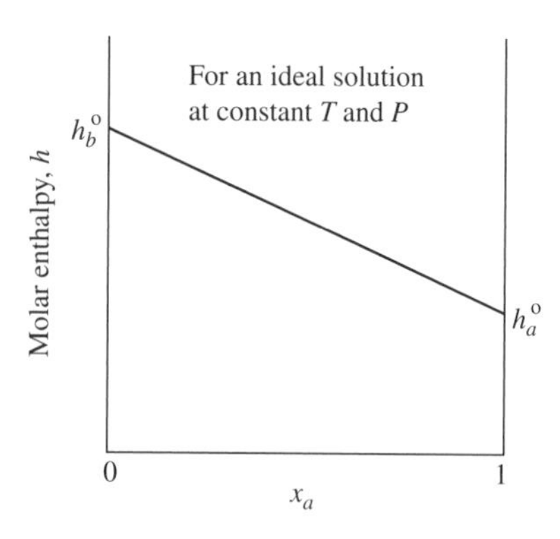

FIGURE 7.5 Molar enthalpy-mol fraction plot for an ideal solution.

intermolecular distance in the pure species. If the two pure species have different average intermolecular distances, then each molecule must, on the average, take up the same distance from its neighbors in the solution that it would take up in the pure species. Similarly, the fact that there is no heat effect (neither heat absorbed or heat evolved) on mixing for an ideal solution suggests that the strength of the intermolecular attractive or repulsive forces is the same between the various kinds of molecules present in an ideal solution as it would be between those same molecules present in their pure state. If we think of the molecules as people, we would say that in an ideal solution the molecules are neither more attracted to nor more repelled by molecules of the other kind than they are by molecules of their own kind, and that they form intermolecular bonds of equal strength with their own kind and the other kind. This useful, intuitive model is only approximately right; do not use it as substitute for detailed calculations.

The fact that there is no volume change or heat effect on mixing for an ideal solution does not mean that there is no entropy change on mixing. Mixing is always irreversible (an increase in disorder), so ideal solutions have greater entropies than the same species would have if they existed in the pure state, unmixed. If we substitute Eq. 7.26 twice in the partial molar equation (Eq. 6.5) and simplify, we find that for an ideal solution

$$s = \left(x_a s_a^{\mathrm{o}} + x_b s_b^{\mathrm{o}}\right) - R\sum x_i \ln x_i \qquad (7.28)^{(\dagger)}$$

This function is sketched in Figure 7.6.

We see that if there were no entropy change on mixing, then the first term in Eq. 7.28 would represent the entropy of the mixture. But all mixing produces an entropy increase. The second term in Eq. 7.28 represents this entropy increase.

The solid curve in Figure 7.6 represents the sum of the two terms in Eq. 7.28; the difference between these curves shows the molar entropy of mixing for ideal solutions. The maximum entropy increase on mixing, for an ideal solution, corresponds to $x_a = x_b, = 0.5$, for which

$$-R\sum x_i \ln x_i = -8.314\frac{\mathrm{J}}{\mathrm{mol\,K}} \cdot 2(0.5 \ln 0.5) = 5.76\frac{\mathrm{J}}{\mathrm{mol\,K}} \qquad (7.\mathrm{O})^{(\dagger)}$$

independent of the values of T and P.

If the entropy changes on mixing, then the Gibbs energy must also change. We may substitute Eq (7.24) twice in Eq. (6.5) and rearrange, finding that for any ideal solution

$$g = \left(x_a g_a^{\mathrm{o}} + x_b g_b^{\mathrm{o}}\right) + RT\sum x_i \ln x_i \qquad (7.29)^{(\dagger)}$$

We use this equation to make up the molar Gibbs energy–mol fraction plot shown in Figure 7.7. The maximum Gibbs energy change on mixing, for an ideal solution, corresponds to $x_a = x_b = 0.5$, for which at 25°C = 298.15 K,

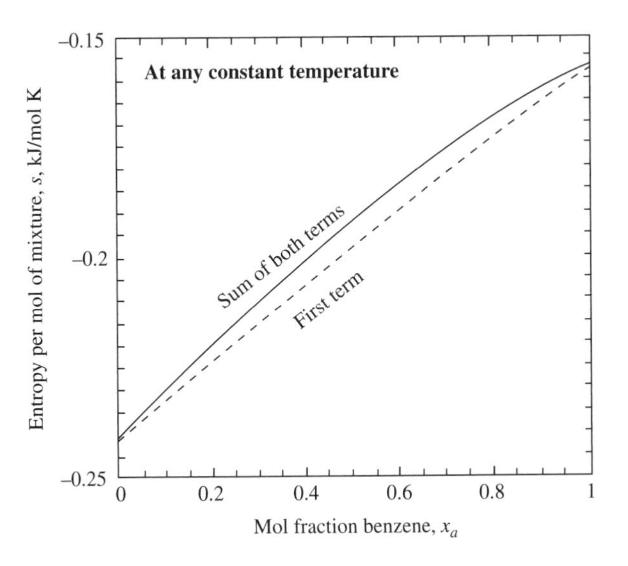

FIGURE 7.6 Molar entropy-mol fraction plot for an ideal solution at any constant temperature. Here the values are for benzene–toluene, with the pure species entropy values being those of formation from the elements (not absolute entropies) calculated from the values in Table A.8. The straight (dotted) line represents the first term in Eq. 7.28, the curved (solid) line the sum of the two terms.

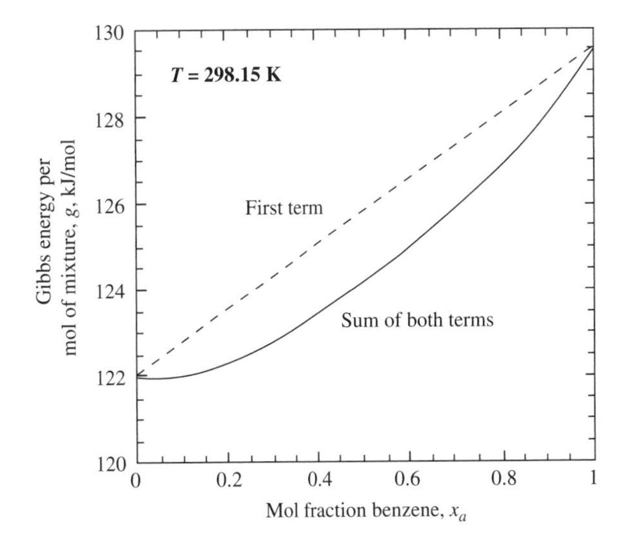

FIGURE 7.7 Molar Gibbs energy–mol fraction plot for an ideal solution. Here the values are for benzene–toluene, with the pure species Gibbs energy values being those of formation from the elements calculated from the values in Table A.8. The straight (dotted) line represents the first term in Eq 7.29 the curved (solid) line the sum of the two terms.

$$RT\sum x_i \ln x_i = 8.314\frac{\text{J}}{\text{mol K}} \cdot 298.15\,\text{K} \cdot 2(0.5 \ln 0.5) = -1.718\frac{\text{kJ}}{\text{mol}} \tag{7.P}$$

This negative increase (decrease) corresponds to the difference between the two curves in Figure 7.7.

7.9 WHY ACTIVITY AND ACTIVITY COEFFICIENTS?

The fugacity has the dimension of pressure. Often we want a nondimensional representation of the fugacity, for example, in mass-action (chemical equilibrium) calculations. We will see in Chapter 12 that this requirement leads naturally to the definition of the activity. Furthermore, when we apply the ideal solution idea to nonideal solutions, we will need a measure of "departure from ideality," just as the compressibility factor z is a measure of departure from ideal gas behavior. The logical choice for that measure is the activity coefficient, defined below. We will see that the activity and activity coefficient are dimensionless, and that for ideal solutions and many practical solutions the activity is equal to the mol fraction.

7.10 ACTIVITY AND ACTIVITY COEFFICIENTS DEFINED

Now we define two new quantities:

$$\text{activity} = \frac{f_i}{f_i^{\text{o}}} = a_i \tag{7.30}$$

$$\text{activity coefficient} = \gamma_i = \frac{a_i}{x_i} = \frac{f_i}{x_i f_i^{\text{o}}} \quad \text{or} \quad f_i = \gamma_i x_i f_i^{\text{o}} \tag{7.31}$$

These two new quantities have the merit that they are dimensionless (which the fugacity is not) and, as we will see later, they lead to very useful correlations of liquid-phase fugacities. We will also see in Chapter 12 that the normal chemical equilibrium statement, the law of mass action, is given in terms of activities.

The activity is almost never used in discussing *phase* equilibrium, and will almost never appear in the rest of this chapter or Chapters 8–11. However, it is widely used in the common formulation of *chemical* equilibrium, and will appear very often in Chapters 12 and 13. Furthermore, we needed to define it before we could introduce the activity coefficient, which we will see plays a major role in this chapter and the next few. (If we did not define the activity for use in chemical equilibrium, we might have chosen some other name for what we call the activity coefficient!) From the definition we see that the activity is equal to the mol fraction, multiplied by a coefficient that shows how much more or less "chemically active" the species is than it would be in an ideal solution.

From these definitions it is clear that for a pure species or for a species in an ideal solution the activity coefficient is identically 1.00 and the activity is equal to the mol fraction. We may then redefine an ideal solution as a solution with activity coefficient = 1.00; this definition is completely equivalent to Eq. 7.23. (This is like defining an ideal gas as a gas whose compressibility factor $z = 1.00$.) Similarly, we may see that the activity coefficient is the ratio of the fugacity to the fugacity the same species would have in the same solution, if that solution were an ideal solution. Thus, we see that the activity coefficient is a simple, dimensionless measure of the departure of the species fugacity from ideal solution behavior. Activity coefficients can be greater than one ("positive deviations from ideality") or between zero and one ("negative deviations from ideality"), but never negative; most have values between 0.1 and 10. Correspondingly, activities can have any positive value; most are in the range (0.1 to 10) times the mol fraction.

From mathematics shown in Appendix C,

$$\left(\frac{\partial \ln \gamma_i}{\partial P}\right)_{T,x_i} = \frac{\bar{v}_i - v_i^{\text{o}}}{RT} \tag{7.32}$$

$$\left(\frac{\partial \ln \gamma_i}{\partial T}\right)_{P,x_i} = \frac{h_i^{\text{o}} - \bar{h}i}{RT^2} \tag{7.33}$$

We previously noted that for an ideal solution $\gamma = 1$, independent of pressure, temperature, or composition. This means that for an ideal solution the derivatives shown in Eqs. 7.32 and 7.33 must be identically zero. From Eqs. 7.25 and 7.26 we can see that this is indeed the case.

There is no direct way to measure fugacity, activity, or an activity coefficient. (If you can invent an instrument to do it, you will become rich and famous!) All values you will ever see have been calculated, either by the estimating methods shown in this and later chapters or computed from the things we can measure experimentally, such as temperature, pressure, density, and the concentrations (normally mol fractions) of the various species in the coexisting phases at equilibrium. The following example (which will be referred to and

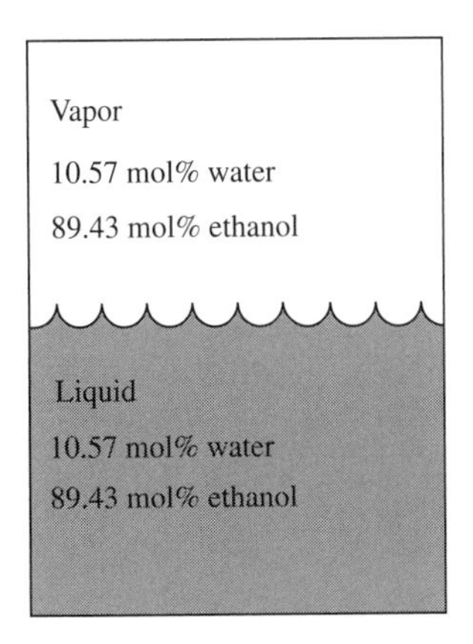

FIGURE 7.8 At any azeotrope, the chemical compositions of the vapor and the liquid are identical. These are the values for the ethanol–water azeotrope, at 1 atm and 78.15°C [5].

expanded upon in the next chapter) shows that approach, and gives some idea of the expected values.

Example 7.4 At 1 atm pressure, the ethanol–water azeotrope (discussed more in the next chapter) has the same composition of 10.57 mol% water, (and thus 89.43 mol% ethanol) in both vapor and liquid phases at a temperature of 78.15°C [5] (Figure 7.8). At this temperature the pure species vapor pressures are water 0.434 atm and ethanol 0.993 atm. Estimate the fugacity and activity coefficient for each species in each phase.

Here we name ethanol as species *a*, and water as species *b*, and name the vapor as phase 1 and the liquid as phase 2. Thus, from the above values

$$y_a = x_a = 0.8943 \qquad y_b = x_b = 0.1057 \tag{7.Q}$$

Again, observe the common convention of using *y* for mol fractions in gases and vapors, and *x* for mol fractions in liquids and solid; with this convention we drop the (superscript), which tells us which phase we are describing.

To find the liquid-phase fugacities from this type of data, we begin with some way of estimating the gas-phase fugacities. At a pressure of 1 atm all gases are practically ideal gases, so that we may safely assume that the vapor is a mixture of ideal gases, for which Eq. 7.22 tells us that the fugacity is equal to the partial pressure. Thus, the vapor-phase fugacities of ethanol and water are approximately equal to their partial pressures, mol fractions times the total pressure, $y_i P$, 0.8943 atm, and 0.1057 atm. From Eq. 7.5 we know that these are the same as the fugacities of the two species in the liquid, so now we know the liquid-phase fugacities. Observe that there is no way to know the liquid-phase fugacities in mixtures directly from experimental measurements; instead, we infer them from the gas and liquid compositions and the very good assumption that the gas phase is practically an ideal gas mixture. For pressures high enough that we cannot safely assume ideal gas behavior we need to use an EOS for the vapor. We will discuss this further later in this chapter, and again in Chapters 8 and 9.

To compute the activity coefficients, we need to assign values of f_i^o. We might think that one choice would do for both phases, but, alas, that doesn't work. If we look back to Section 7.4, we see that we use very different methods to compute the fugacities of gases than those used for liquids (because their specific volumes are so different). Whatever choice of f_i^o we make for either phase must have the property that it gives the right value of f_i as x_i or y_i approaches unity. For the gas (assumed ideal) that means that f_i must approach P, so that the logical choice for ideal gases and vapors is $f_i^o = P$. For pure liquids (see Example 7.3) the fugacity, practically independent of total pressure, is approximately equal to the vapor pressure p_i. Thus, we choose

$$f_i^o = P \text{ (the system pressure) for the gas} \tag{7.34}$$

and

$$f_i^o = p_i \begin{pmatrix} \text{the pure species vapor} \\ \text{pressure for this species} \\ \text{at this temperature} \end{pmatrix} \text{for the liquid} \tag{7.35}$$

We now write Eq. (7.5 for each species, inserting the definition of the activity coefficient (Eq. 7.31) and the definitions of f_i^o in Eq. 7.34 and 7.35, finding

$$f_{\text{ethanol}}^{\text{(liquid phase)}} = f_{\text{ethanol}}^{\text{(gas phase)}} = (y\gamma P)_{\text{ethanol}}^{\text{(gas phase)}} = (x\gamma p)_{\text{ethanol}}^{\text{(liquid phase)}} \tag{7.36}$$

and

$$f_{\text{water}}^{\text{(liquid phase)}} = f_{\text{water}}^{\text{(gas phase)}} = (y\gamma P)_{\text{water}}^{\text{(gas phase)}} = (x\gamma p)_{\text{water}}^{\text{(liquid phase)}} \tag{7.37}$$

These equations are the correct description, in terms of the quantities defined so far, for any vapor–liquid equilibrium of two species (if we substitute their names for those of ethanol and water). Observe that this equation has four values of γ, one for each species in each of two phases. For any ideal solution, $\gamma = 1.00$. Mixtures of ideal gases are all ideal solutions, so if the gas (or vapor) is practically an ideal gas, then both of the γs for the gas phase in Eqs. 7.36 and 7.37 are ≈ 1.00. In Chapters 8 and 9 we will see that this is an excellent approximation, so for most pressures we normally drop those two γs out of Eqs. 7.36 and 7.37 and assume that any γ we encounter is for one species in a liquid. Then we can solve Eqs. 7.36 and 7.37 for these γs, finding

Table 7.D Summary of Example 7.4

Phase	Ethanol, $i=a$	Water, $i=b$
VAPOR, PHASE 1		
y_i	0.8943	0.1057
$f_1^{(1)}$, atm	0.8943	0.1057
$f_1^{(1)o}$, atm	$=P=1$	$=P=1$
$\gamma_i^{(1)}$ (assumed)	1.0000	1.0000
LIQUID, PHASE 2		
x_i	0.8943	0.1057
$f_1^{(2)}$, atm	0.8943	0.1057
$f_1^{(2)o}$, atm	$=p_a=0.993$	$=p_b=0.434$
$\gamma_i^{(2)}$	1.007	2.31

$$\gamma_{\text{ethanol}} = \left(\frac{yP}{xp}\right)_{\text{ethanol}} = \left(\frac{0.8943 \cdot 1.00\text{ atm}}{0.8943 \cdot 0.993\text{ atm}}\right) = 1.007 \tag{7.R}$$

and

$$\gamma_{\text{water}} = \left(\frac{yP}{xp}\right)_{\text{water}} = \left(\frac{0.1057 \cdot 1.00\text{ atm}}{0.1057 \cdot 0.434\text{ atm}}\right) = 2.31 \tag{7.S}$$

These results are summarized in Table 7.D. ■

This example shows the interrelations between fugacity, total pressure, vapor pressure, mol fraction, and activity coefficient. If we dealt only with ideal gas mixtures and ideal liquid solutions, we would scarcely have bothered to define fugacity, activity, or activity coefficient, because for ideal gases the fugacity is equal to the partial pressure $(y_i \cdot P)$ and for ideal solutions of liquids and solids the fugacity is equal to the mol fraction times the vapor pressure $(x_i \cdot p_i)$ making $\gamma = 1.00$ for both. However, Table 7.D (and the experimental data on which it is based) show that this liquid is not an ideal solution, because the activity coefficients are not unity. (The activity coefficient of ethanol $= 1.007 \approx 1.00$, but that of water is 2.31!) This is an important industrial system, which we will speak about more in the next chapter.

7.11 FUGACITY COEFFICIENT FOR PURE GASES AND GAS MIXTURES

In Figure 7.1 we showed and in example 7.1 we estimated the fugacity coefficient $f/P = \phi$ for pure species. Equation 7.10 shows that this quantity could logically be used for both liquids and gases, but normally is only used for gases, where ϕ represents the ratio of the pure gas fugacity to the fugacity of a pure ideal gas at the same T and P, computed by Eq. 7.10 or its equivalent.

The reason that we don't use ϕ for liquids is shown in Example 7.3 where we computed the effect of compressing pure liquid water at 100°F from 1 psia to 1000 psia, showing that the fugacity increased by a factor of 1.049. However that factor is equal to f/p, the fugacity divided by the pure species vapor pressure, not f/P, the fugacity divided by the system pressure. One could use the results of that example to define

$$\phi_{\text{Liquid in Exmple 7.3}} = \frac{f_{\text{liquid water}}}{P_{\text{system}}} = \frac{1\text{ psia}}{1000\text{ psia}} = 0.001 \tag{7.T}$$

which is formally correct, but not very useful and seldom seen.

Turning now to the fugacity of individual species in *gas* mixtures, we normally use Eq. 7.18 or its equivalent. We find the values of $\bar{\alpha}_i$ or $\bar{z}_i$, using an EOS for the mixture (Appendix F). The resulting $\hat{\phi}_i$ shows both the effects of the nonideal behavior of the individual gas (i), and also the nonideal mixing of the various gases. The EOSs normally used in this computation do not separate these effects, but simply show the combined effect. We might formulate this as

$$\hat{\phi}_i = \left(\text{pure species nonideal } \frac{f}{P}\right) \cdot \left(\begin{matrix}\text{nonideality}\\ \text{of mixing}\end{matrix}\right) = \phi_i \cdot \gamma_{i \text{ in the gas phase}} \tag{7.37}$$

This is helpful in understanding what the computations are doing, but not used in actual calculations. The γ in the vapor phase that we show in Table 7.D is logically consistent, but almost never seen. Instead it is combined with ϕ_i to form $\hat{\phi}_i$ as shown in Eq. 7.37.

7.12 ESTIMATING FUGACITIES OF INDIVIDUAL SPECIES IN GAS MIXTURES

Our real uses of fugacity involve the fugacities of individual species in mixtures. These cannot be measured by any direct-reading instrument. They can be computed (or estimated) from PvT data, EOSs, or vapor–liquid equilibrium measurements.

7.12.1 Fugacities from Gas PvT Data

If we have reliable PvT data for gas mixtures, we can compute individual species fugacities from it; the calculated values are as reliable as the original PvT data.

Example 7.5 Hougen et al. [7, p. 865] present Table 7.E of the volume residuals for gaseous mixtures of methane and n-butane at 220°F, based on the experimental PvT data

Table 7.E Volume Residuals for a Mixture of Methane with *n*-Butane (all at 22°F)

Mol% Methane ↓	Volume Residual α (ft^3/lbmol) at Pressure, psia → 100	200	400	600	800	1000
28.7	5.55	5.55				
47.5	4.12	4.00	3.88			
60.8	2.99	2.91	2.68	2.44	2.26	2.12
70.7	2.30	2.22	2.01	1.81	1.645	1.52
78.4	1.86	1.745	1.55	1.383	1.252	1.147
84.5	1.42	1.348	1.202	1.062	0.950	0.860
89.4	1.065	1.020	0.921	0.826	0.745	0.679
93.5	0.780	0.746	0.695	0.637	0.579	0.529
97.0	0.546	0.528	0.500	0.461	0.426	0.394

of [8]. Using these values, estimate the fugacity of methane and of *n*-butane in a gaseous mixture of 78.4 mol% methane (50 wt% methane), balance butane, at 220°F and 1000 psia.

To use these values in Eq. 7.18 we need the partial molar volume residual $\bar{\alpha}_{\text{methane}}$. We find its value at 100 psia by plotting the volume residuals at 100 psia as a function of mol fraction, as shown in Figure 7.9, drawing the tangent to the data points at $x_{\text{methane}} = 0.784$ and reading its intercept on the 100 mol% methane axis as $\approx 0.6\,\text{ft}^3/\text{lbmol}$.

In the same way we find the value of $\bar{\alpha}_{\text{methane}}$ for all of the other pressures, and plot them vs. pressure, as shown in Figure 7.10. From this plot we find the integral we need by numerical integration (trapezoid rule) as 290 psi · ft^3/lbmol. Thus, for methane

$$\frac{f_i}{P_{y_i}} = \hat{\phi}_i = \exp\left(\frac{-1}{RT}\int_{P=0}^{P=P} \bar{\alpha}_i dP\right) = \exp\frac{-290\dfrac{\text{psia}\cdot\text{ft}^3}{\text{lbmol}}}{10.73\dfrac{\text{psia}\cdot\text{ft}^3}{\text{lbmol}\cdot{}^\circ\text{R}}\cdot 680\cdot{}^\circ\text{R}} = 0.961 \tag{7.U}$$

$$f_i = 0.961\,Py_i = 0.961\cdot 1000\,\text{psia}\cdot 0.784 = 753\,\text{psia} \tag{7.V}$$

From Figure 7.9 we can also read the other intercept, finding that $\bar{\alpha}_{\text{n-butane}} \approx 6.6\,\text{ft}^3/\text{lbmol}$. We can do the same for other pressures, and make up the equivalent of Figure 7.10 for *n*-butane (Problem 7.15), finding that the integral is 5859 psi · ft^3/lbmol, so that for *n*-butane

$$\frac{f_i}{Py_i} = \hat{\phi}_i = 0.448$$

$$f_i = 0.448\cdot 1000\,\text{psia}\cdot(1-0.784) = 96.8\,\text{psia} \tag{7.W}$$ ■

The values of $\bar{\alpha}_{\text{methane}}$ actually used to make up Figure 7.10 were found on a spreadsheet, using Eq. 6.5 and numerical differentiation of the data, which is not as obvious as the graphical procedure shown in Figure 7.9, but is much more reliable and much quicker for a group of pressures. The original authors [8], using graphical differentiation (on huge graph paper) and then graphical integration, found $f_i/Py_i = 0.968$ for methane, instead of the 0.961 found here. For *n*-butane they found $f_i/Py_i = 0.436$, instead of the 0.448 found here. The differences are small, but the values from [8] are probably the more reliable.

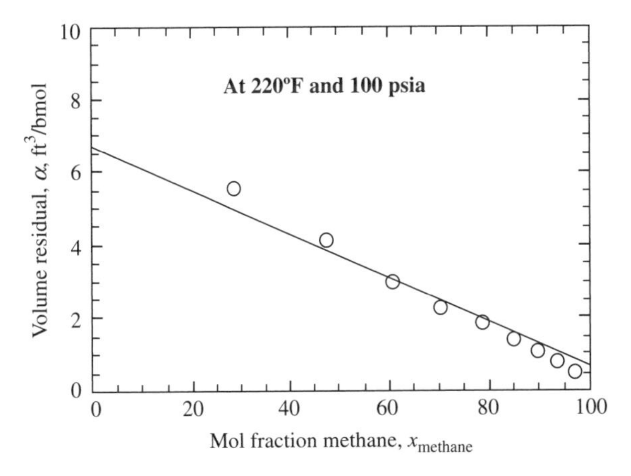

FIGURE 7.9 Finding the partial molar volume residual $\bar{\alpha}_{\text{methane}}$ by the method of tangent intercepts. The line is tangent to the curve at $y_{\text{methane}} = 0.784$. Its intercept on the right hand axis is $\approx 0.6\,\text{ft}^3$/lbmol.

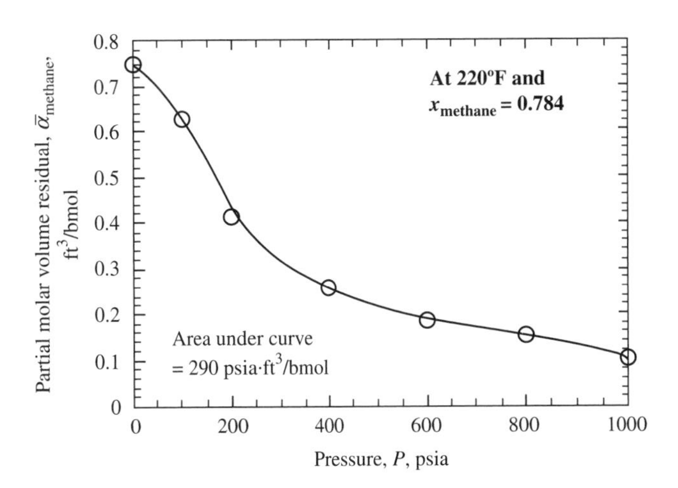

FIGURE 7.10 Partial molar volume residual of methane at 220°F and $y_{\text{methane}} = 0.784$ for various pressures. The area under the curve = 290 psia · ft^3/lbmol.

In principle, we could use this same procedure for liquids, but, as discussed in Section 7.4.2, the values of α for liquids are huge, so small uncertainties in them cause huge uncertainties in the calculated fugacities. This procedure is rarely used for liquids.

7.12.2 Fugacities from an EOS for Gas Mixtures

We have experimental PvT data of this quality for only a few mixtures, so the procedure in Example 7.5, while theoretically the most reliable, is seldom used. However, if we have an EOS that we believe accurately reproduces or estimates the PvT data of a mixture, we can repeat the above calculation using it. When working with experimental PvT data as in Example 7.5 it is generally most convenient to work with the first form of Eq. 7.18:

$$\frac{f_i}{Px_i} = \hat{\phi}_i = \exp\left(\frac{-1}{RT}\int_{P=0}^{P=P} \bar{\alpha}_i dP\right) \tag{7.18}$$

while with an EOS the mathematics are generally simpler if we use the second form:

$$\frac{f_i}{Px_i} = \hat{\phi}_i = \exp\int_{P=0}^{P=P} \frac{(\bar{z}-1)}{P} dP \tag{7.18}$$

All the EOSs in Chapter 2 are for one single pure species and can be written in the form $z = f(T, P,$ and various constants). In the corresponding-states formulation, they are mostly written as $z = f(T_r, P_r, \omega)$. Most EOSs for mixtures begin with the EOSs for the individual pure species, and then use empirical or semitheoretical *mixing rules* to fill in the region between the pure species.

If we have a mixture of a and b, at some T and P then we will have two pure species zs, z_a, and z_b, Any mixing rule must be of the form

$$z_{\text{mix}} = f(z_a, z_b, y_a, y_b)_{T,P} \tag{7.38}$$

The mixing rules must give back the pure species values as the concentration approaches either pure species, so only some mathematical forms are possible. Many of the empirical, data-fitting rules are of the form

$$z_{\text{mix}} = \left(x_a z_a^{1/n} + x_b z_b^{1/n}\right)^n \tag{7.39}$$

where n is an arbitrarily chosen constant. Equation 7.39 has the correct property that for pure a or b it gives the right value of z, for any choice of n. If we choose $n = 1$ in Eq. 7.39, we find the simplest possible mixing rule, linear molar mixing:

$$z_{\text{mix}} = y_a z_a + y_b z_b \tag{7.X}$$

A plot of z_{mix} vs. y_a according to Eq. 7.X is a straight line, like Figures 7.4 and 7.5. One may show (Problem 7.17) that Eq 7.X is equivalent to an ideal solution of *nonideal* gases. (For ideal gases, $z = 1.00$ for all P and T.) For this mixing rule

$$\bar{z}_a = z_a \text{ and } \bar{z}_b = z_b \tag{7.Y}$$

for all possible values of y_a. For most simple EOSs like the little EOS, this mixing rule leads (Problem 7.19) to equations of the form

$$\frac{f_a}{Py_a} = \hat{\phi}_a = \exp\left(\frac{P_{r,a}}{T_{r,a}} f\left(T_{r,a}, \omega_a\right)\right) \tag{7.Z}$$

in which f_a/Py_a in a mixture depends only on the pressure, temperature, mol fraction of a, and *the properties of pure a.* The calculated f_a/Py_a would be the same if we replaced all the b in the mixture with the same number of mols of some other gas c, independent of what b and c are. This cannot be rigorously correct, but, as we will see, it is a very useful approximation. A separate equation gives f_b/Py_b for b, as a function of P, T, and the properties of pure b. This is called the *Lewis* and *Randall fugacity rule* which may also be stated as $\hat{\phi}_i = \phi_i$, or that the $\gamma_{\text{in the gas phase}}$ in Eq. 7.37 is $= 1.00$.

7.12.3 The Lewis and Randall (L-R) Fugacity Rule

At pressures up to a few atmospheres, we normally use this simplest possible mixing rule, which assumes that the gas mixture is an ideal solution of nonideal gases, even though the corresponding liquid in equilibrium (Chapter 8) may be quite nonideal. For many gas mixtures at moderate or even high pressures it appears that there is no measurable volume change or heat effect on mixing of gases, so that the gas mixture may be considered an ideal solution, even though the individual species are at pressures at which they may not be considered ideal gases (i.e., their z does not equal 1.00).

Example 7.6 In Example 7.5 we estimated the fugacity of methane in a mixture of 78.4 mol% methane, balance n-butane at 1000 psia and 220°F, directly from the measured PvT data. For that example, estimate the values of ϕ_i and $\hat{\phi}_i$. Use these values to evaluate the L-R rule for this example, and to evaluate the relative effects of pure-gas nonideality and of nonideality of mixing.

In that example, we found directly from the PvT data that for methane

$$\frac{f_i}{Px_i} = 0.961 = \hat{\phi}_i \tag{7.AA}$$

From Starling's tables of hydrocarbon properties [3, p. 14] we read that for pure methane at this T and P, $f_i/P = \phi_i = 0.954$, from which it follows that

$$\frac{\hat{\phi}_i}{\phi_i} = \frac{0.961}{0.954} = 1.007 \quad \blacksquare \tag{7.AB}$$

If we had assumed ideal gas behavior, we would have computed (for methane) $\hat{\phi}_i = f_i/Px_i = 1.00$ which is 4% higher than the value calculated from the PvT data. Using the L-R rule we would have computed $\hat{\phi}_i = \phi_i = f_i/Px_i = 0.954$ which is 0.7% less than the value from the experimental data. The 0.7% difference between 0.961 and 0.954 is probably less than the uncertainty in the measured PvT data and in our use of it in Figures 7.9 and 7.10, so that we may say that within experimental accuracy the estimate based on the L-R rule and that based directly on the PvT data are the same.

The L-R rule is widely used because it is simple and is the next step in complexity (and reliability) over the ideal gas law. However, frequently gases exist in states for which we cannot compute ϕ_i. Consider the n-butane in Example 7.5. At 220°F the vapor pressure of n-butane is 241.6 psia, far less than 1000 psia, so that pure n-butane cannot exist as a gas under these conditions. For n-butane at these conditions, $T_r = 0.889$ and $P_r = 1.815$. If we attempt to find a suitable value of ϕ_i from Figure 7.1 for these values we find that the pure substance is a liquid for which we would estimate $\phi_i \approx 0.2$. How can the n-butane exist as a gas? We may form an intuitive model of this behavior by considering that in a pure n-butane vapor each n-butane molecule collides only with other n-butane molecules, with which it forms attractive bonds. The molecules of n-butane in this mixture mostly collide with methane molecules (which make up 78.4% of the molecules present). The n-butane molecules do not form comparably strong attractive bonds with methane molecules, so the n-butane molecules in this mixture will not condense at a T and P at which pure n-butane would condense. (The water vapor in air at 20°C is in the same situation.)

If we wish to use the L-R rule for the n-butane, we must estimate the ϕ_i for it, not using the observed value of the nonexistent pure vapor at this T and P. The only practical way is to use some EOS.

Example 7.7 Estimate the value of ϕ_i that n-butane would have *if it could exist* as a gas at $T_r = 0.889$ and $P_r = 1.815$, using the little EOS.

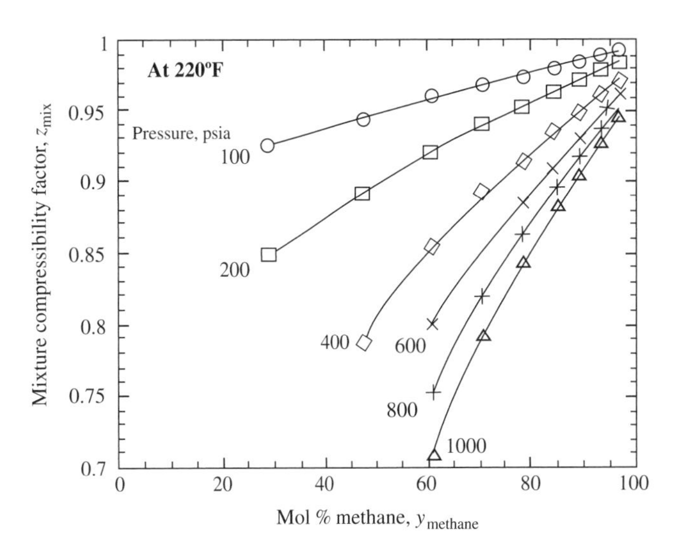

FIGURE 7.11 Compressibility factor z as a function of pressure and methane %, $y_{methane}$, based on the data in Table 7.E.

Using the properties of n-butane from Appendix A.1 and Eq. 7.F we find that

$$\frac{f_i}{P} = \phi_i = \exp\left(\frac{P_r}{T_r} \cdot f(T_r, \omega)\right) = \exp\left[\frac{1.815}{0.889} \cdot (-0.48553)\right]$$
$$= 0.397 \quad \blacksquare \tag{7.AC}$$

This is 89% of the value found directly from PvT data. If the L-R rule were absolutely correct, and if the extrapolation of the little EOS—which only applies to gases—from pressures at which butane would be a gas to a pressures at which butane could not exist as a pure gas were absolutely correct, then the two values would agree completely. Thus in this case the L-R rule and the little EOS are only a fair approximation of the experimental value. We will discuss the L-R rule a little more in Chapter 8.

7.12.4 Other Mixing Rules

Figure 7.11 shows the data from Table 7.E in the form of a z–$y_{methane}$ plot. If the L-R rule were absolutely correct (that is, if Eq. 7.39 with $n = 1.00$ were absolutely correct), then each of the constant-pressure lines would be absolutely straight. Figure 7.11 shows that while the data form practically a straight line for 100 psia, for the higher pressures a straight line is not a very good representation of these data. (The ideal solution approximation is fairly good for 100 psia, but not for higher pressures. However, as the previous example shows, the resulting fugacity calculations are not very sensitive to this nonideality.) If we wish our mixture EOS to represent nonideal solutions (which are not straight lines on Figure 7.11), then we must use the more complex mixing rules shown in Chapters 9 and 10.

7.13 LIQUID FUGACITIES FROM VAPOR-LIQUID EQUILIBRIUM

If we have any way of estimating the fugacity of a species in the gas phase, by ideal gas law, the L-R rule, or any EOS with appropriate mixing rules, we can use that way to compute the fugacity of any species in a gas phase. If that gas phase is in equilibrium with a liquid, the fugacity of that species must be the same in the liquid as in the gas, so that if we have VLE phase composition data and a way of estimating the fugacity in the gas, we use that to estimate the fugacity in the liquid. This is shown in Example 7.4 and used many times in Chapter 8.

7.14 SUMMARY

1. Fugacity was invented to remedy the counterintuitive behavior of the chemical potential, which makes it approach minus infinity as the concentration approaches zero. For pure ideal gases the fugacity is the same as the pressure, and for ideal gas mixtures the fugacity of one species is equal to that species' partial pressure.
2. For pure gases we normally correlate and compute $f/P = \phi$ based on either measured PvT data or an EOS.
3. For pure liquids and solids we normally compute the fugacity from the vapor pressure. The effect of increases in pressure above the vapor pressure on the fugacity of liquids and solids (the Poynting factor) is generally negligible (because their molar volume is so small).
4. Ideal solutions are like ideal gases—an approximation, but a very useful one.
5. Activity and activity coefficient are nondimensional ways of representing fugacities, based on comparing the behavior of the solution to that of an ideal solution.
6. Fugacity, activity, and activity coefficient are computed quantities; none can be measured directly.
7. For mixtures of gases, we can determine the fugacity of each species directly from the measured PvT data if they are available. If not, we can use an EOS as a substitute for the experimental PvT data, We normally begin with the EOSs for the pure gases, and use mixing rules to estimate the EOS for the mixture.
8. The simplest mixing rule leads to the L-R rule, an ideal solution of nonideal gases, which is very widely used and fairly reliable for modest pressures.
9. For mixtures of liquids we normally estimate fugacities of species from measured VLE data, as described in Chapter 8.

PROBLEMS

See the Common Units and Values for Problems and Examples. An asterisk (*) on the problem number indicates that the answer is in Appendix H.

7.1* What is the value of α at $P = 0$, in terms of R, T, and z?

7.2 See Problem 7.1. The Boyle-point temperature is defined as that temperature at which $\alpha = 0$ for $P = 0$. From Figure A.4 estimate the Boyle-point temperature for gases that follow that chart. What is the practical significance of the Boyle point?

7.3 For small pressures α is practically a constant. Show that if α = constant and $\alpha \ll 1$, $(f/P) \cong z$.

7.4 **a.** Show the computation of the volume residuals in Table 7.A.

b. Show the trapezoid rule integration in Example 7.1(a)

7.5 Repeat Example 7.1, parts (b), (c), (d), and (f), for water at 2000 psia and 700°F, using values from any steam table.

7.6 The calculation in Example 7.1(f) is quite sensitive to small errors or changes, because it leads to the exponential of a large number.

a. Verify this by repeating that calculation, with the following common simplifications: $(f/P)_{1\text{psia}} = 1.0057 \approx 1.00$: $M = 44.062$ g/mol ≈ 44 g/mol, and $T = 679.67°\text{R} \approx 680°\text{R}$. How much do these changes change the calculated value of $(f/P)_{500\text{psia}}$ compared to the values in Example 7.1(f)?

b. The value of f/P at 220°F and 1 psia in Starling's table (1.0057) is a misprint. A value greater than 1.00 for this P_r and T_r is impossible. Estimate the correct value, by repeating Example 7.1(a) for this pressure, taking $\alpha = 4.23$ ft^3/lbmol. Compare it to the value now suggested by Starling [9] of 0.9995.

7.7 **a.** The two-term, pressure-explicit virial EOS is

$$z = \frac{Pv}{RT} = 1 + \frac{BP}{RT} \tag{7.AD}$$

where B is a function of temperature alone. For a gas that obeys this EOS, derive the equation for $\ln(f/P)$.

b. Show that the little EOS (Eqs. 2.48–2.50) is equivalent to the two-term, pressure-explicit virial

EOS. Show the value of B in Eq. 7.AD that makes it the same as the little EOS.

c. Show the equation for $\ln(f/P)$ that corresponds to the little EOS.

7.8* Estimate the fugacity of pure liquid water at 100°F and 10,000 psia. For this problem only, you may assume that the specific volume of liquid water at 100°F is 0.01613 ft^3/lbm, independent of the pressure.

7.9 In Example 7.3 we assumed that the specific volume of water at 100°F was a constant, independent of pressure. The steam table [10, p. 104] shows that at 100°F and 1 psia, 500 psia and 1000 psia, the specific volumes are 0.01613, 0.016106, and 0.016082 ft^3/lbm. Rework Example 7.3 taking this change in water specific volume into account. How much difference does it make?

7.10 In Example 7.3, we estimated the value of f/P for pure water vapor at 100°F and 0.9503 psia as 0.9997. Repeat that estimate using the method used in part (c) of Example 7.1.

7.11 In Figure 7.3, sketch the area that corresponds to $\int \alpha dP$. A simple free-hand sketch will be satisfactory.

7.12 Estimate $(\partial \ln f/\partial T)_P$ for Freon 12 at 400°F and 1000 psia from the Freon 12 chart or table which is presented in many introductory thermodynamics books. The molecular weight of Freon 12 is 121 lbm/lbmol.

7.13 Assuming that n–pentane and n-hexane form an ideal solution, sketch plots of p–x_{pentane}, volume per mol–x_{pentane}, enthalpy per mol–x_{pentane}, entropy per mol–x_{pentane}, and Gibbs energy per mol–x_{pentane} for liquid mixtures of n-pentane and n-hexane at 25°C (see Table 7.F). Simple sketches with a few numerical values will be satisfactory.

7.14 Show the calculation of $\bar{\alpha}_{\text{methane}}$ at 100 psia in Example 7.5 by numerical differentiation and application of Eq. 6.5. To do this, approximate $d\alpha/dx_1$ by

$$\frac{\Delta\alpha}{\Delta x_a} = \frac{1.86 - 2.30}{0.784 - 0.707} \tag{7.AE}$$

which is the *backward difference* approximation. Then set this function and Eq. 6.5 up on a spreadsheet to compute the values of $\bar{\alpha}_{\text{methane}}$ for all the pressures in Table 7.E. Compute the values and compare them to those shown in Figure 7.10.

7.15 Using the method shown in the previous problem, estimate the fugacity of n-butane in the same mixture examined in Example 7.5. Compare your results to the values shown in that example.

7.16 If a substance has the v-x behavior shown in Figure 7.4 what does its z-x plot look like?

7.17 Show that Eq. 7.X (Eq. 7.39 with $n = 1.0$) is an ideal solution of nonideal gases.

a. Sketch a plot z_{mix} vs. y_a.

b. Show by tangent intercepts that $\bar{z}_a = z_a$ and $\bar{z}_b = z_b$.

c. Show that this is equivalent to an ideal solution of nonideal gases.

7.18 Sketch the equivalent of Figure 7.1 and show the basis for the statement in Example 7.6 that for the T_r and P_r of n-butane it would be a liquid with $\phi_i \approx 0.2$.

7.19 Show that Eq. 7.Y follows from Eq. 7.X.

7.20 For the two-term, pressure-explicit virial EOS, Eq. 7.AD,

a. Show the general form it takes for f_i/Py_i of a binary mixture.

b. Then show the form that takes for the L-R rule, which is equivalent to Eq. 7.39 with $n = 1$.

c. Then show the form that takes for the mixing rule produced by substituting $n = 0.5$ in Eq. 7.39.

7.21 We occasionally [11] see Eq. 7.10 rewritten and integrated to

$$\ln\frac{f}{f_0} = \int_{P_0}^{P} z\, d(\ln P) \tag{7.AF}$$

where f_0 is the fugacity at P_0. Show the derivation of this equation.

Table 7.F Data for Problem 7.13

Properties at 25°C	n-Pentane	n-Hexane
Molar volume (mL/mol)	116.4	130.5
Vapor pressure (torr)	572.5	151.26
Molar enthalpy (kJ/mol)	−173.1	−166.9
Molar entropy (kJ/mol K)	−0.549	− 0.560

REFERENCES

1. Lewis, G. N. The law of physico-chemical change. *Proc. Am. Acad. Arts Sci.* 37:49–69 (1901).
2. Reamer, H. H., B. H. Sage, and W. N. Lacey. Phase equilibrium in hydrocarbon systems: volumetric behavior of propane. *IEC* 41:482–484 (1949).
3. Starling, K. E. *Fluid Thermodynamic Properties for Light Petroleum Systems*. Houston, TX: Gulf (1973).
4. Wilson, L. C., W. V. Wilding, and G. M. Wilson. Vapor–liquid equilibrium measurements on four binary mixtures. *AIChE Symp Ser.* 85(271):25–43 (1989).

5. Haynes, W. M. *Handbook of Chemistry and Physics*, ed. 91. Boca Raton, FL: CRC Press, pp. 6–209 (2010).
6. Walas, S. M. *Phase Equilibria in Chemical Engineering*. Boston: Butterworth, p. 535 (1985).
7. Hougen, O. A., K. M. Watson, and R. A. Ragatz. *Chemical Process Principles*, Part II: *Thermodynamics,* ed. 2. New York: Wiley (1959).
8. Sage, B. H., R. A. Budenholzer, and W. N. Lacey. Phase equilibria in hydrocarbon systems: methane–*n*-butane system in the gaseous and liquid regions. *IEC* 32:1262–1277 (1940).
9. Starling, K. E. Personal communication (August 1999).
10. Keenan, J. H., F. G. Keyes, P. G. Hill, and J. G. Moore. *Steam Tables: Thermodynamic Properties of Water Including Vapor, Liquid and Solid Phases*. New York: Wiley (1969).
11. Sage, B. H., J. G. Schaafsma, and W. M. Lacey. Phase equilibria in hydrocarbon systems, V: pressure–volume–temperature relations and thermal properties of propane. *IEC* 26:1218–1224 (1934).

8

VAPOR–LIQUID EQUILIBRIUM (VLE) AT LOW PRESSURES

Vapor–liquid equilibrium (VLE) is at the heart of many chemical and environmental engineering processes and activities. Distillation, drying, and evaporation are all based in VLE. In Figure 1.1, the ammonia synthesis process, VLE determines the behavior of the separator. The liquid ammonia product leaving it contains some dissolved nitrogen and hydrogen and also some impurities from the feed, which we wish to keep out of the reactor; the recycle gas leaving it contains about 2% ammonia. The pressure and temperature of the separation are chosen to have this separation be as complete as possible (we would like zero H_2 and N_2 in the liquid and zero NH_3 in the vapor, but VLE requires that there be some). In Figure 1.2 the removal of the benzene contaminant depends on VLE. All of Chapter 3 and the air–water example discussed there are about VLE. Raoult's and Henry's laws are VLE laws (really estimating approximations). Although the applications of VLE are very broad, the principal chemical engineering application, and the one for which VLE has been studied most thoroughly, is *distillation.* Distillation is separation by boiling point. For ideal solutions it is straightforward; we can separate any mixture of species with different boiling points. For nonideal solutions, the process is more complex, as discussed in this chapter.

If miscible liquids can be separated by distillation, then that is probably the least expensive way to separate them. The chemical and petroleum industries are full of distillation; the visual appearance of a chemical plant or petroleum refinery is a forest of distillation columns. Figure 8.1 shows the mercaptan production unit at the Borger, Oklahoma facility of the Chevron Phillips Chemical Company. In it we see at least 10 distillation columns.

This chapter discusses low-pressure VLE. At high pressures, VLE is different, mostly because we approach the critical pressures of the vapor–liquid mixtures. Chapter 10 starts from what we see in this chapter and shows how the experimental behavior and our mathematical approaches to it change at high pressures.

8.1 MEASUREMENT OF VLE

Figure 8.2 shows, schematically, the simplest possible VLE experiment. A liquid sample of the mixture of interest is placed in an Erlenmeyer flask and heated to a boil. The boiling continues until the vapor has displaced all the air from the flask. This means that the liquid composition will no longer be equal to that originally prepared, because the vapor leaving the system does not have the same composition as the liquid. When we are sure that all the air is gone, we measure the temperature and take samples of liquid and vapor, which we analyze (by any of several laboratory techniques, e.g., chromatography).

In practice, the simple device in Figure 8.2 is not used because it is very difficult to collect a large enough vapor sample to analyse without contaminating it with drops thrown up from the boiling liquid, and because precise temperature measurements require special care. The most common device for measuring the equilibrium temperature and compositions of the vapor and liquid in equilibrium (a refined version of Figure 8.2) is sketched and described in Figure 8.3. It is most often used for binary (two-species) mixtures, but can equally well be used for mixtures with any

Physical and Chemical Equilibrium for Chemical Engineers, Second Edition. Noel de Nevers.

FIGURE 8.1 The mercaptan manufacturing facility at the Borger (Texas) facility of the Chevron Phillips Chemical Company. At least 10 distillation columns are visible. (Courtesy of the Chevron Phillips Chemical Company.)

number of chemical species. It is most often used at or near atmospheric pressure or less, although equivalent devices have been built for high pressures. Table 8.1 shows the results of such experiments for one binary mixture.

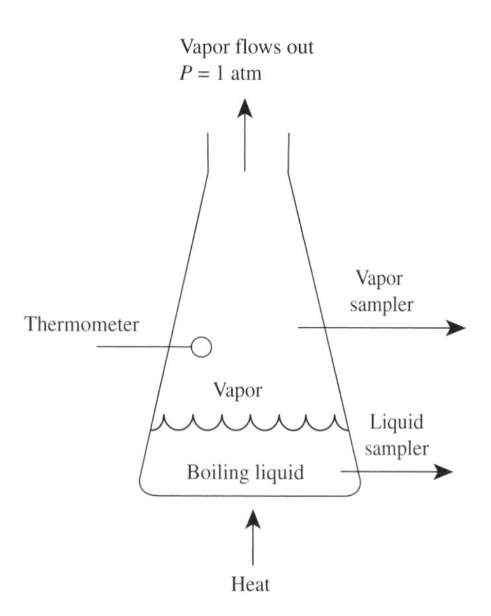

FIGURE 8.2 Schematic view of the simplest possible VLE measurement.

In this table and throughout VLE it is the custom that the lowest-boiling (most volatile) species in the mixture is species *a*, and that the species are listed in order of increasing normal boiling point (NBP). Thus, a mixture of methane, *n*-butane, and *n*-pentane would almost always be listed in that order. In a table like Table 8.1 the vapor and liquid mol fractions of species *a* (in this case acetone) are shown, with the mol fractions of species *b* (water) to be computed from the values in the table. Similarly, when properties are plotted vs. mol fraction (e.g., Figure 8.4), the mol fraction is almost always that of species *a*, running from left to right.

If such data were available for all the possible mixtures of industrial interest, at the pressures of industrial interest, then we would not need the estimation and correlation procedures described in Sections 8.6 and 8.7 and Chapter 9. Although such tables are available for several thousand binary mixtures, they are rarely available for ternary and more complex mixtures. Most of the available tables of this kind are for atmospheric pressure; many of the industrial applications are at pressures far from atmospheric.

There are other ways of measuring VLE that are less direct but quicker and cheaper [5]; their goal is the same as the methods described above, to produce the equivalent of Table 8.1 for the mixture of interest at the pressure of interest.

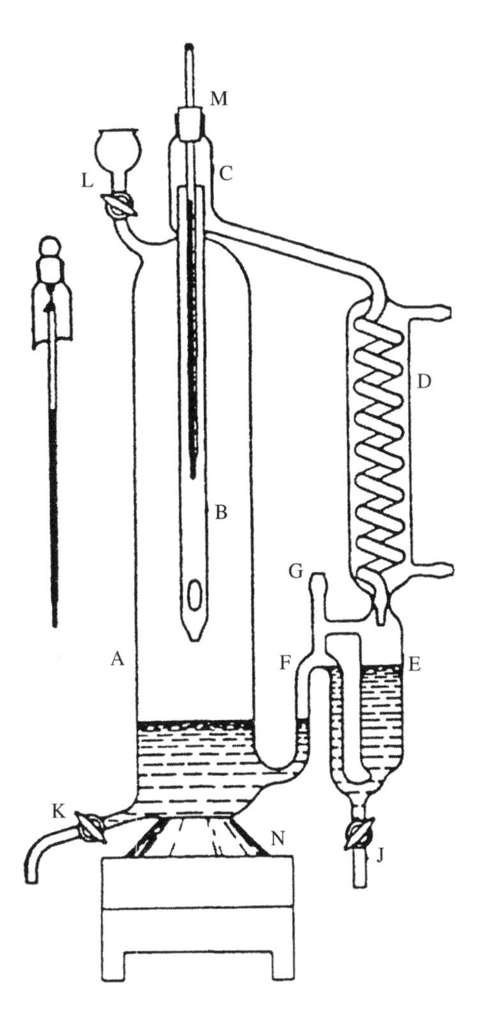

FIGURE 8.3 Sketch of the standard way of measuring low-pressure VLE, called an Othmer still. The device is first evacuated through connection G, and the liquid mixture admitted through opening L. The mixture is then boiled by the heater N, and the vapor condensed in condenser D. After enough time elapses for steady state to occur, the liquid in reservoir E will have the same composition as the vapor flowing up from the pool in A. The boiling must be vigorous enough to produce compete mixing in the liquid pool in A, but not so vigorous as to throw up drops of liquid to be carried up through the aperture in the bottom of tube B to pass through the condenser and contaminate the condensed-vapor sample in E. The liquid sample taken from K is in equilibrium with the vapor sample from J. The volume of reservoir E is much less than the total liquid volume, so that the composition of the liquid in the main reservoir will be close to—but not identical to—the composition of the liquid originally introduced into the apparatus. The apparatus can operate under modest pressures, but it normally operates at atmospheric pressure with the valve L open and the boiling rate high enough to have a continual small flow out that opening, thus excluding air. Many modified versions of this device are in common usage. (Reprinted with permission from D. F. Othmer, Composition of vapors from boiling binary solutions. *Ind. Eng. Chem.* 20: 743–746. Copyright (1928), American Chemical Society.)

Table 8.1 Vapor–liquid Equilibrium Data for Acetone and Water at 1.00 Atm

Boiling Temperature, T(°C)	Mol Fraction Acetone in Liquid, $x_{acetone}$	Mol Fraction Acetone in Vapor, $y_{acetone}$
100	0	0
74.8	0.05	0.6381
68.53	0.1	0.7301
65.26	0.15	0.7716
63.59	0.2	0.7916
61.87	0.3	0.8124
60.75	0.4	0.8269
59.95	0.5	0.8387
59.12	0.6	0.8532
58.29	0.7	0.8712
57.49	0.8	0.895
56.68	0.9	0.9335
56.3	0.95	0.9627
56.15	1	1

These data are from [1], which shows such tables for 21 mixtures. Data on 466 such mixtures are given in [2]. Very extensive compilations are given in [3, 4]. For binary mixtures (mixtures with only two species) the table shows only one mol fraction in each phase; the other is found by subtracting this value from 1.0. These values and the values shown in most such tables are not direct experimental values. Rather, the experimental data points are plotted as in Figure 8.4, and then the appropriate temperatures and vapor mol fractions are read from those curves (by eye or by computer curve fit and computer look up) at the even values of the liquid mol fractions as shown in this table.

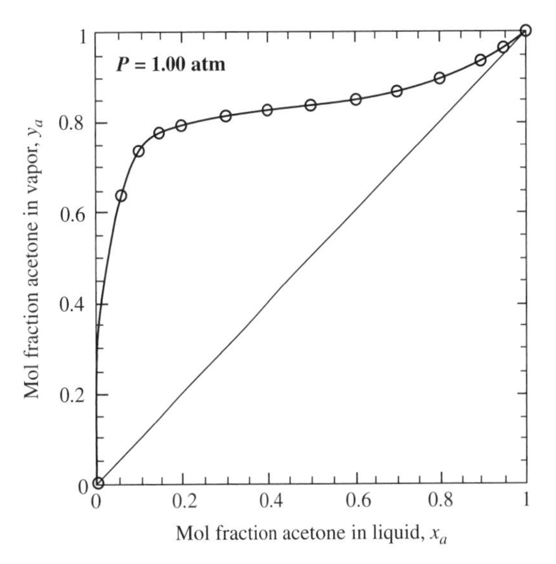

FIGURE 8.4 Data from Table 8.1 plotted as an y-x plot for the more volatile species (which is in this case is acetone, which boils at 56.15°C, compared to 100.0°C for water). The data points from the table are shown, and a simple interpolation curve is added. The 45° line represents the condition $y_a = x_a$. This line is customarily shown because the equilibrium curve and this line interact in the McCabe–Thiele method of calculation of equilibrium stages in distillation. The plot is all for 1 atm pressure (as are most such plots). The temperatures from Table 8.1 are not shown.

8.2 PRESENTING EXPERIMENTAL VLE DATA

The most common way to present data of this type is as a table, like Table 8.1. The next most common way to present them is a y-x plot for the more volatile species (Figure 8.4). Such plots are widely used for simple binary distillation calculations, but for multispecies mixtures there is no correspondingly simple plot. The general approach for calculating distillation columns for any number of species defines and uses

$$K_i = \frac{y_i}{x_i} = \text{“}K\text{ factor”} = \text{“}K\text{ value”}$$

$$= \begin{pmatrix}\text{equilibrium constant}\\ \text{for distillation}\end{pmatrix} = \begin{pmatrix}\text{vaporization}\\ \text{equilibrium}\\ \text{ratio, VER}\end{pmatrix} \quad (8.1)$$

A separate value of K_i exists for each species in a mixture being distilled, and its value changes with changes in temperature, pressure, and composition. One also frequently sees

$$\alpha = \text{relative volatility} = \frac{K_{\text{more volatile species}}}{K_{\text{less volatile species}}} = \frac{y_a \cdot x_b}{y_b \cdot x_a} \quad (8.2)$$

which also changes with changes in temperature, pressure, or composition.

Example 8.1 Compute the two "K factors" and the relative volatility for a liquid with 0.05 mol fraction (5 mol%) acetone, balance water.

Using the values from Table 8.1,

$$K_{\text{acetone}} = \frac{0.6381}{0.05} = 12.76 \quad (8.A)$$

$$K_{\text{water}} = \frac{1-0.6381}{1-0.05} = 0.381 \quad (8.B)$$

$$\alpha = \frac{12.76}{0.381} = 33.5 \quad (8.C)$$

Figure 8.5 shows these values, and the corresponding values for the other points in Table 8.1. ■

We see that the individual Ks can take on values greater or less than one, but never negative values. As $x_i \to 1.0$, $K_i \to 1.0$ for each species. The relative volatility is always positive and greater than 1 if there is no azeotrope present (Section 8.4.4). For quick estimates of the difficulty of a separation by distillation, the relative volatility α is the chemical engineer's favorite. If α is greater than 1.5 to 2 over the whole range of composition values, then distillation will almost always be the cheapest separation method. If α is less than 1.1, then we seriously consider other separation methods. The lowest α mixtures that are separated industrially by distillation have $\alpha \approx 1.05$; those separations are difficult and expensive. Figure 8.5 shows α values from 33.5 to 1.4 for this mixture, which is easily separated by distillation, except near 100% pure acetone, where the curve for α drops toward 1.00 and the separation becomes difficult.

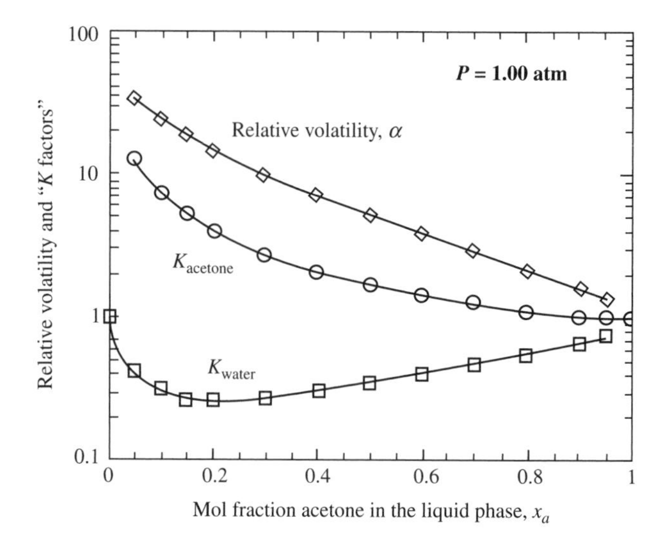

FIGURE 8.5 K_{water}, K_{acetone}, and the relative volatility α plotted as a function of acetone mol fraction in the liquid at 1 atm pressure, using data from Table 8.1. The curves are simple interpolations.

Most distillation and equilibrium parts of process-design computer programs (e.g., Aspen, ChemCad, ProMax) report K_i values for each species in the VLE data for mixtures, and some distillation programs report α. It must be clear from their definitions that they are made up from values of the y_i and x_i. The computer programs mostly first compute the y_i, and x_i, by the methods shown in the rest of this chapter and the next, and then compute and report the K_i and α values for the convenience of users who prefer that formulation; the, x_i and y_i, are the basic values. The common hand-calculation methods of estimating VLE are presented in terms of the K values in Section 8.9.

Section 8.4 shows and discusses several other common ways of plotting experimental VLE data like that in Table 8.1.

8.3 THE MATHEMATICAL TREATMENT OF LOW-PRESSURE VLE DATA

The tables and plots shown in the previous section are clear and have some intuitive content. But in the age of computers we wish to reduce all such information to equations, which can then be used by computers to interpolate and extrapolate the data and to apply it to practical problems. Furthermore, these equations will allow us to estimate VLE for systems for which we have little or no data, often with considerable success.

The normal way of representing and correlating such data is via equations for the fugacity coefficient in the gas and the activity coefficient in the liquid. We know from chapter 7 that for each species in a VLE, $f_i^{\text{vapor}} = f_i^{\text{liquid}}$ for each species i. Substituting the definitions of the fugacity coefficient, $\hat{\phi}_i$, in the vapor and the activity coefficient, γ_i, in the liquid we find

$$f_i^{\text{vapor}} = \hat{\phi}_i y_i P = f_i^{\text{liquid}} = \gamma_i x_i p_i \quad \text{or} \quad \hat{\phi}_i y_i P = \gamma_i x_i p_i \tag{8.3}$$

The right hand side of Eq. 8.3 is called the "phi-gamma" representation of VLE, and is used almost universally in chemical engineering (and in the rest of this book). For most of low-pressure VLE the vapor phase is close enough to being an ideal gas that we set $\hat{\phi}_i = 1.00$, which makes our working equation

$$y_i P = \gamma_i x_i p_i \quad \left(\begin{array}{l}\text{working equation for low-pressure VLE,} \\ \text{with vapor phase assumed practically ideal gas}\end{array}\right) \tag{8.4}$$

Example 8.2 Estimate the liquid-phase activity coefficients for acetone and water from the data in Table 8.1.

Here we need to know the vapor pressures p_i corresponding to the temperatures of each of the values in the table. We estimate them from the Antoine equation using the values in Table A.2. For the data point at 74.8°C, the computed vapor pressures are 1.812 and 0.377 atm, so

$$\gamma_{\text{acetone}} = \frac{y_{\text{acetone}} P}{x_{\text{acetone}} p_{\text{acetone}}} = \frac{0.6381 \cdot 1 \text{ atm}}{0.05 \cdot 1.812 \text{ atm}} = 7.04 \tag{8.D}$$

and

$$\gamma_{\text{water}} = \frac{y_{\text{water}} P}{x_{\text{water}} p_{\text{water}}} = \frac{(1-0.6381) \cdot 1 \text{ atm}}{(1-0.05) \cdot 0.377 \text{ atm}} = 1.01 \tag{8.E}$$

We may repeat this calculation for each point in the table (which is tedious by hand but very easy on a spreadsheet) and plot the results as shown in Figure 8.6 ■

From the definition of the activity coefficient, it is clear that it must be unity for a pure species. In Figure 8.6 we can see that each activity coefficient curve does indeed become 1.00 and also becomes tangent to the $\gamma_i = 1.00$ line as the mol fraction of that species becomes unity. We also see that both activity coefficients increase as the concentration of that species decreases. Intuitively, we can say that as we approach $x_i = 1.00$, most of the i molecules are surrounded by other i molecules, so their interaction with each other is practically the same as in a pure solution, and $\gamma_i = 1.00$. Conversely, as we approach $x_i = 0.00$, each lonely i molecule is surrounded entirely by the other kind of molecule, so that any difference in intermolecular

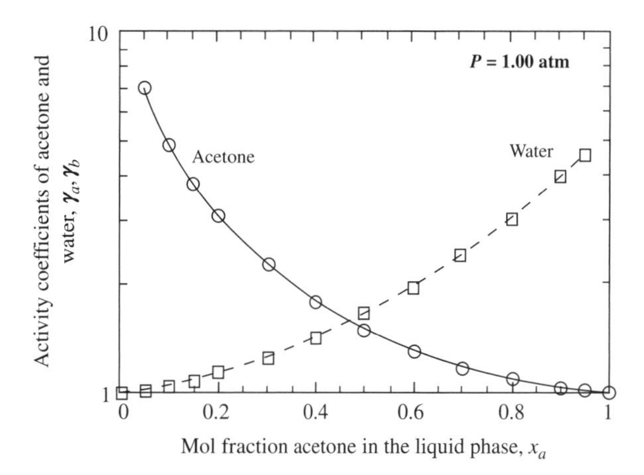

FIGURE 8.6 Calculated activity coefficients for acetone and water at 1 atm based on the experimental values in Table 8.1. The values in Example 8.2 are the first points from the left in this plot, corresponding to $x_{\text{acetone}} = 0.05$. The curves are simple smoothed interpolations. Activity coefficient plots are almost always presented on semi-logarithmic coordinates, as is this one. In the text we often refer to plotting ln (γ_i); that is equivalent to what we show here, plotting γ_i on a logarithmic scale.

behavior between the two kinds of molecule will be at its maximum for the few i molecules that remain. That is also very common behavior. We will see in Chapter 9 that there are strong restrictions on the possible shapes the curves on Figure 8.6.

Figure 8.6 contains, implicitly, all the data in Table 8.1. If we were told, for example, that at 1 atm pressure a liquid with $x_{\text{acetone}} = 0.05$ had $\gamma_{\text{acetone}} = 7.04$ and $\gamma_{\text{water}} = 1.01$ (and that the gas phase was practically an ideal gas), then that information plus the Antoine equation constants from Table A.2 would be enough information to compute both the equilibrium temperature and the value of y_{acetone}.

8.3.1 Raoult's Law Again

We can rewrite Eq. 8.4 as

$$y_i = \frac{\gamma_i x_i p_i}{P} \tag{8.5}$$

Comparing this to Eq. 3.5, Raoult's law, we see that they are the same, except that Eq. 8.5 has a liquid-phase activity coefficient, while Raoult's law has set that equal to 1.0. Since we defined an ideal solution as one in which $\gamma_i = 1.00$ for all values of x_i, we can see that Raoult's law is the ideal-solution simplification of the more general form shown in Eq. 8.5. From the fact that the calculated activity coefficients for acetone-water (Example 8.2) are greater than 1.00, we can see that this mixture is *not an ideal solution and does not obey Raoult's law.*

Table 8.A Trial and Error Solution in Example 8.3

Assumed $T(°C)$	Calculated p_a (atm)	Calculated p_b (atm)	$y_a=x_ap_a/P$	$y_b=x_bp_b/P$	$\sum y_i$
80.0000	2.1114	0.4674	0.1056	0.4441	0.5496
90.0000	2.7939	0.6919	0.1397	0.6573	0.7970
95.0000	3.1927	0.8342	0.1596	0.7925	0.9521
96.4060	3.3123	0.8783	0.1656	0.8344	**1.0000**
97.0000	3.3638	0.8975	0.1682	0.8526	1.0208

Example 8.3 How much difference does nonideal solution behavior make in the acetone-water VLE? To answer this, compute the boiling temperature and vapor composition that would correspond to a liquid with $x_{\text{acetone}} = 0.05$, if this were an ideal solution ($\gamma_i = 1.00$),—Raoult's law—and compare them to the experimental values.

This is a repeat of Example 3.5. We must find, by trial and error, the temperature at which the sum of the computed ideal solution vapor-phase mol fractions is 1.00. For our first try, we guess $T = 80°C$. Using the Antoine equation constants in Table A.2, we compute that at 80°C the two pure species vapor pressures for acetone and for water are 2.11 and 0.47 atm. Then multiplying each of these by the corresponding liquid mol fractions and dividing by 1 atm, we find that the computed vapor mol fractions are 0.106 and 0.444, and that their sum is 0.55. This is less than 1.00, so our assumed temperature is too low. These values are shown as the first data row in Table 8.A. The calculation was done on a spreadsheet, with which one can quickly repeat the calculation for various assumed temperatures and display the results in subsequent rows of Table 8.A. The assumed temperature that makes the sum of the vapor-phase mol fractions equal 1.00 is $T = 96.406$ °C. (We should not believe that we know any boiling temperature ±0.001°C, we should report the calculated boiling temperature as 96.4°C). ■

The results of this calculation are compared with the experimental values in Table 8.B. The ideal solution assumption leads to *very poor* estimates of the vapor composition and boiling temperature. Looking back at the values of the activity coefficients in Figure 8.6, we see that at this x_a the activity coefficient of acetone is ≈7. The ideal solution assumption, that these activity coefficients = 1.00, is a very poor assumption here. (If all solutions were ideal solutions, industry would need far fewer chemical engineers than it currently employs!)

Table 8.B Comparison of Experimental Values to Those Computed by the Ideal Solution (Raoult's Law) Assumption, for $x_{\text{acetone}} = 0.05$ and $P = 1.00$ atm

	Experimental Values from Table 8.1	Values Calculated in Example 8.3. Assuming Ideal Solution
Equilibrium (boiling) temperature T°C	74.8	96.4
Mol fraction acetone in the vapor phase (y_a)	0.6381	0.1656

The choices of values for f_i^{o} made here are often called "Raoult's law type" choices and the γ_i thus found are called "Raoult's law-type activity coefficients." We can see why with Table 8.2. This formulation of simple, low-pressure VLE in terms of Raoult's law-type activity coefficients is the most commonly used formulation. When we see an activity coefficient for low-pressure VLE without a description of which type it is (i.e., what choices have been made for $f_i^{\text{o, vapor phase}}$ and $f_i^{\text{o, liquid phase}}$), then it is most likely of this type.

8.4 THE FOUR MOST COMMON TYPES OF LOW-PRESSURE VLE

Before we continue to the correlation and prediction of low-pressure VLE, which mostly means the correlation, prediction, and use of liquid-phase activity coefficients of the Raoult's law type, we first consider the four most common outcomes of experimental VLE measurements. These four differ from each other mostly because of the magnitude and direction of the intermolecular forces between the two types of molecules in the liquid phase, as is discussed when we discuss each of the various types. The vast majority of experimental VLE data can be represented as one of these four types of behavior. This section follows closely the treatment in the classic text of Hougen, Watson and Ragatz [6].

Figures 8.7, 8.8, 8.9, and 8.12 show examples of the four most common types of outcomes of the simple low-pressure, binary VLE experiments described in Section 8.1, called types I, II, III, and IV. In each of these figures the vapor is assumed to be an ideal gas, as in Table 8.2. In each of these figures the same information is presented four ways. In part (a), the partial pressure of each species is plotted against mol fraction of the more volatile (lower boiling) species in the liquid, and the total equilibrium pressure (sum of the partial pressures) is shown, all at a constant temperature (not the same temperature in all four plots, but one chosen for each mixture to have the total pressure near 1 atm). Part (b) shows

Table 8.2 Comparison of Raoult's Law, and Raoult's Law-type Activity Coefficients

Variable in Eq. 8.3	Replacement in Raoult's Law	Replacement with Raoult's Law-type Activity Coefficient
$\hat{\phi}_i$	1.00	1.00
y_i	y_i	y_i
$f_i^{o,\text{vapor phase}}$	P	P
γ_i	1.00	γ_i
$f_i^{o,\text{liquid phase}}$	p_i	p_i
Working equation for y_i	$y_i = \dfrac{x_i p_i}{P}$	$y_i = \dfrac{\gamma_i x_i p_i}{P}$

the activity coefficients (Raoult's law type) plotted the same way as Figure 8.6. As is customary, these show the activity coefficient on semi-logarithmic coordinates. Part (c) shows the vapor composition as a function of the liquid composition at 1 atm, exactly as is done in Figure 8.4. Finally, part (d) shows both vapor and liquid compositions and boiling temperatures at 1.00 atm, just as was done in Figure 3.12.

Parts (d) of these figures are phase diagrams. On them we can enter at a specific temperature and liquid or vapor phase composition, and determine whether the material is present as a vapor, a liquid, or a two-phase mixture and if so, the composition of the other phase (at the pressure for which the plot was made, 1.00 atm in this case.)

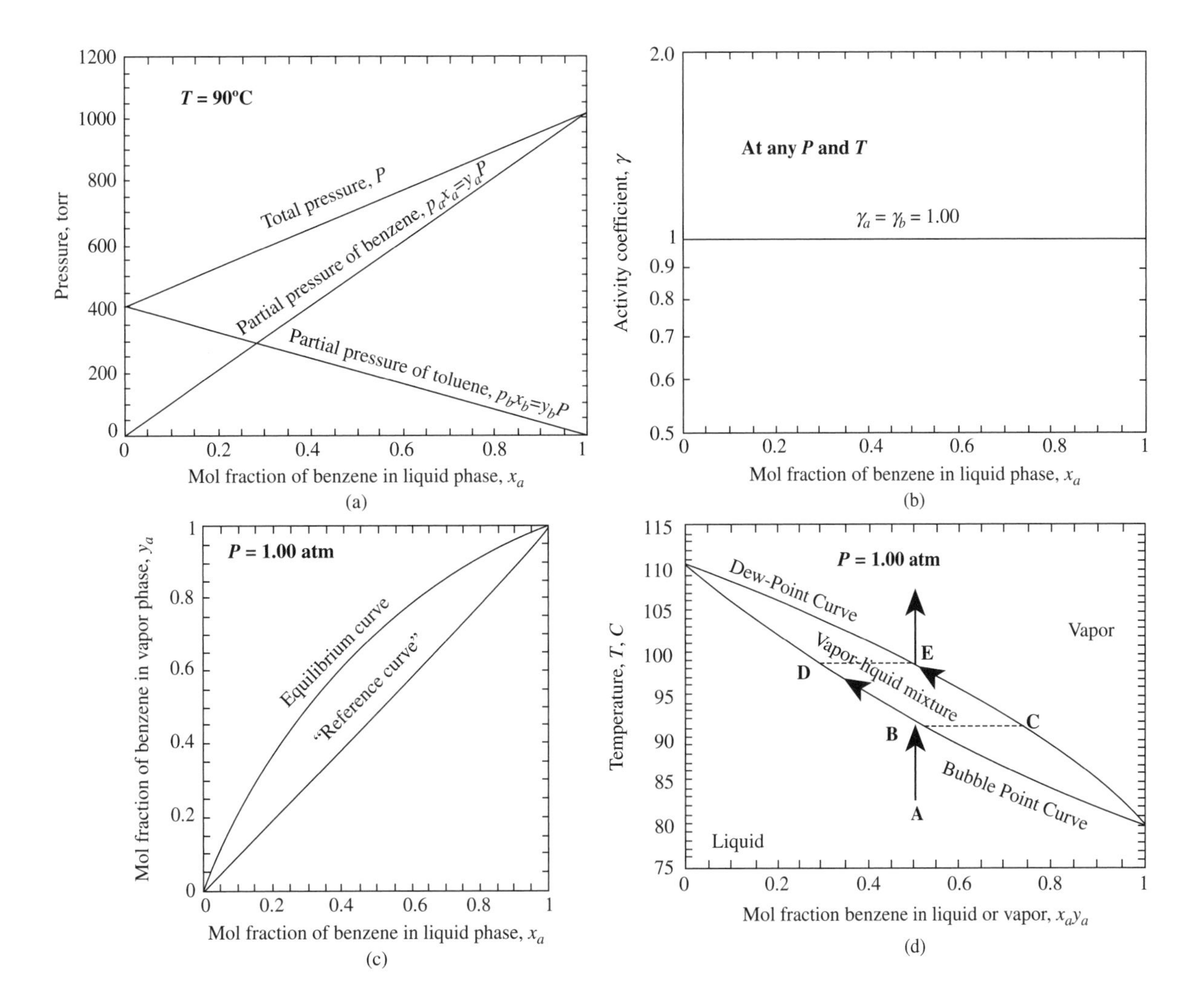

FIGURE 8.7 (a) Partial and total pressure of benzene–toluene solutions at 90°C. (b) Activity coefficients for benzene–toluene solutions, independent of pressure and temperature. (c) Vapor–liquid equilibrium diagram for benzene–toluene solutions at 1.00 atm. The "reference curve" is simply a 45° line on these coordinates. It is traditionally shown because it and the equilibrium curve are used in the McCabe–Thiele calculation method for distillation columns, (d) Vapor–liquid equilibrium phase diagram for benzene–toluene solutions at 1 atm. This is the same as Figure 3.12, with the behavior of a liquid turning to a vapor sketched on it, as described in the text.

8.4.1 Ideal Solution Behavior (Type I)

Figure 8.7 is for the benzene-toluene mixture, which is regularly cited as an example of ideal solution behavior; its measured behavior is very close to ideal solution behavior (see Problem 8.7). Figure 8.7a shows that for an ideal solution at a constant temperature of 90°C, the total pressure and the two partial pressures are linear functions of liquid-phase mol fraction of the more volatile species. The total pressure goes from the pure species vapor pressure of toluene, 407 torr, to the pure species vapor pressure of benzene, 1021 torr. The partial pressures range from zero to the pure species vapor pressures. Increasing the temperature would increase the individual species' values of p_i, thus increasing the endpoints of the lines, but not the overall shape of the figure.

Figure 8.7b shows that for an ideal solution all activity coefficients are 1.00, forming a remarkably dull plot. Its analogs for nonideal solutions (types II, III, and IV) are much more interesting. Figure 8.7c shows that for ideal solutions of species with different boiling points, at any liquid composition the equilibrium vapor contains more of the lower-boiling species than does the liquid. This makes type I mixtures ideally suited for separation by distillation; we may prepare a and b (e.g., benzene and toluene) from such a mixture by simple distillation to whatever purity we are willing to pay for.

Figure 8.7d is an exact copy of Figure 3.12 with additional information. The two-phase boundary curves are labeled as the *bubble-point* and *dew-point curves*, and the course of a simple equilibrium vaporization (from **A** to **E**) is sketched. If we heat a liquid mixture of 50 mol% benzene, 50 mol% toluene at 1 atm, starting at point **A** at temperatures below 92°C only liquid will be present. At 92°C and 50 mol% benzene, point **B**, the *bubble point,* the liquid begins to boil and the first vapor bubble appears. This vapor must be at the same temperature as the liquid, 92°C, but, as shown on the figure, its composition is that at point **C**, about 72 mol% benzene. As we continue to heat the liquid, it boils, so that the mass and volume of liquid decrease and the mass and volume of vapor increase. The temperature rises, and the compositions of the two phases change. The vapor has more benzene than the liquid, so as we produce vapor we must reduce the benzene content of the remaining liquid. As we do so, the vapor subsequently produced is poorer in benzene than that first produced, so the benzene content of the vapor also falls (at equilibrium all phases are internally completely mixed!). The path followed by the liquid is from **B** to **D**, while that of the vapor is from **C** to **E**. At 98°C the last droplet of liquid, with composition **D**, evaporates, so we have all vapor. This vapor must have the same composition as our starting liquid, because we have vaporized all the starting liquid. Point **E** has the same composition as points **A** and **B**.

If we ran the process backward, starting with vapor at a temperature greater than 98°C and cooling, the path followed would appear the same in Figure 8.7d, with the direction of the arrows reversed. When the vapor reached 98°C, the *dew point*, the first droplet of liquid would appear, with composition at point **D**. From this we see that the upper curve, which is labeled *dew-point curve* on Figure 8.7d represents those states at which the vapor is in equilibrium with a liquid at the same temperature, with the liquid composition on the *bubble-point* curve. *If any liquid and any vapor are in equilibrium, then the liquid is at its bubble point and the vapor is at its dew point*. If we add a little heat to the mixture, an additional bubble will form. If we remove a little heat, an additional drop of dew will form. Both phases must be at the same temperature and pressure, but, as this figure shows, their compositions will generally be different.

8.4.2 Positive Deviations from Ideal Solution Behavior (Type II)

Figure 8.6 shows that for the acetone–water system the liquid-phase activity coefficients are $\geq$1.00 for all possible mixtures. This is type II, *positive deviation from ideal solution behavior* (the logarithms of the activity coefficients are *positive*). Figure 8.8 has the same format as Figure 8.7, and shows that type of deviation for mixtures of isopropanol and water.

Figure 8.8a shows that for positive deviation from ideal solution behavior the partial pressure curves and the total pressure curve all bow upward, relative to the straight line connecting their endpoints, as Eq. 8.5 says they must. For this set of activity coefficients, the total pressure curve (at constant temperature) has a maximum. This corresponds in Figure 8.8d to a minimum in the boiling-point T-x_a curve. This is produces a *minimum-boiling azeotrope.*

Figure 8.8b shows the same kind of ln γ_a, vs. x_a plot as Figure 8.6. As expected, the activity coefficient curves approach 1.00 asymptotically as the individual mol fractions approach 1.00, and have values greater than 1.00 for all other concentrations, becoming larger as the concentration of the species becomes smaller.

Figure 8.8c shows that for liquid isopropanol concentrations less than the azeotrope (which corresponds to $x_a = y_b \approx 0.685$), the vapor contains a higher percentage of isopropanol than the liquid, while at liquid isopropanol concentrations greater than the azeotrope the vapor contains less isopropanol than the liquid. At the azeotrope (where the equilibrium curve crosses the reference curve) the vapor and liquid have the same composition, and the boiling-point temperature of this mixture, 80.4°C, is less than the boiling point of either pure species or of any mixture of them with a different composition (at 1.00 atm). This type of azeotrope is common, and makes separation by distillation difficult. If we start with a liquid mixture of, say, 10 mol% isopropanol and attempt to separate it into pure isopropanol and pure water by distillation, we find that it is easy to get practically

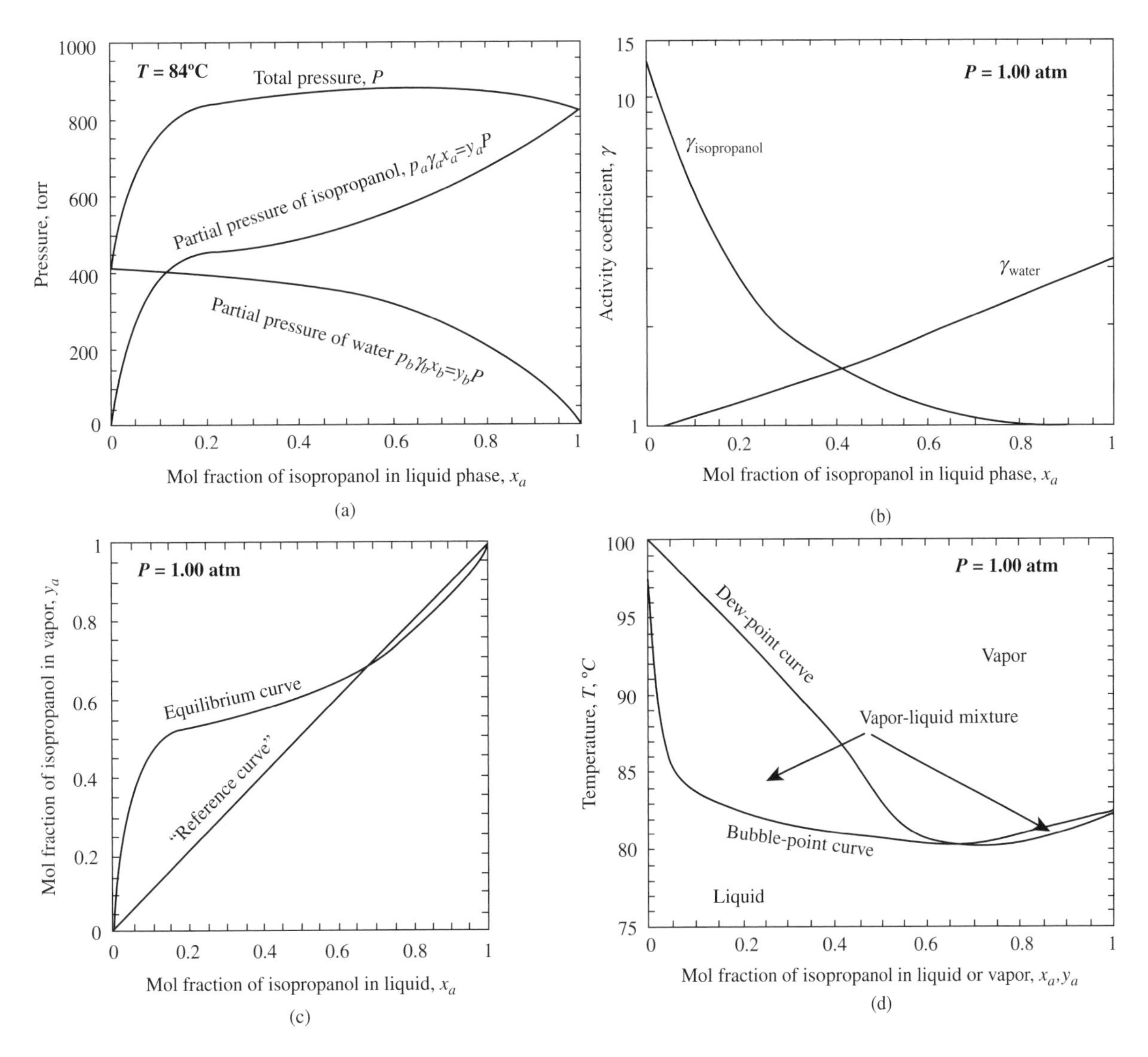

FIGURE 8.8 (a) Partial and total pressure of isopropanol–water solutions at 84°C. (b) Activity coefficients for isopropanol and water at 1 atm. This is a more interesting plot than Figure 8.7(a). (c) Vapor–liquid equilibrium diagram for isopropanol–water solutions at 1.00 atm. (d) Vapor–liquid equilibrium phase diagram for isopropanol–water at 1 aim.

pure water from the mixture; the low isopropanol part of Figure 8.8c is similar to that in Figure 8.7c. But it is not possible to produce pure isopropanol by distillation because, as Figure 8.8c shows, at high concentrations of isopropanol, the vapor contains a lower concentration of isopropanol than the liquid.

Figure 8.8d shows that this system has a minimum boiling point, and two regions with different behaviors on either side of the composition at which that minimum occurs. For liquid solutions to the left of the azeotrope in Figure 8.8d ($x_a < 0.685$) the figure is of the same general type as Figure 8.7d. If we attempted to separate a liquid in this composition range by simple distillation we would produce practically pure water and the azeotrope. If we started with a solution to the right of the azeotrope in Figure 8.8d, simple distillation would produce pure isopropanol and the azeotrope. But simple distillation will not produce practically pure isopropanol and practically pure water from any liquid solution shown on this diagram.

8.4.3 Negative Deviations from Ideal Solution Behavior (Type III)

Figure 8.9 shows type III, the opposite of the type II behavior in Figure 8.8, for mixtures of acetone and chloroform. The activity coefficients (Figure 8.9b) for all possible mixtures are ≤ 1.00. This is called *negative deviation from ideal solution behavior* (the logarithms of the activity coefficients

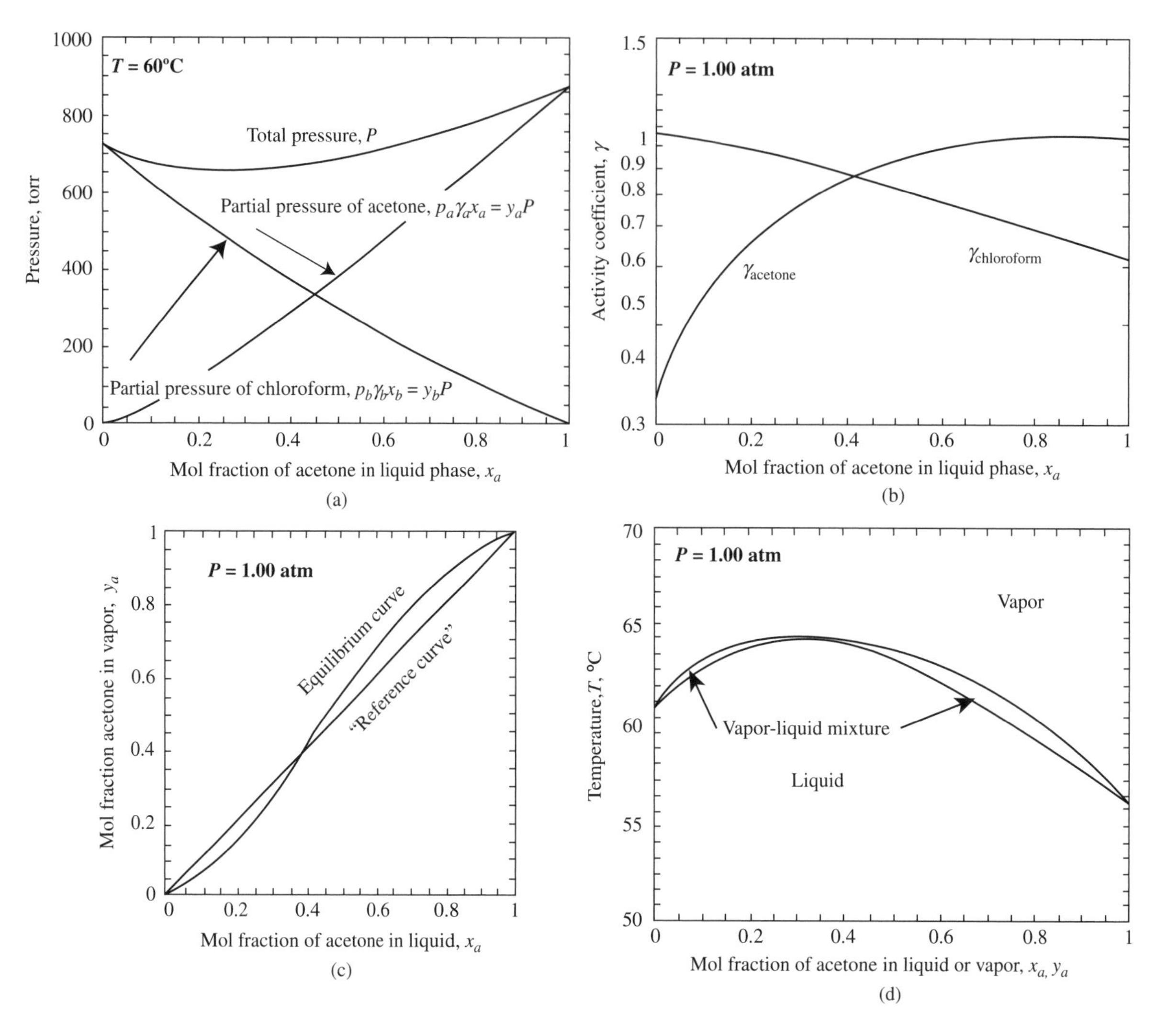

FIGURE 8.9 (a) Partial and total pressure of acetone–chloroform at 60°C. (b) Activity coefficients for acetone and chloroform at 1.00 atm. (c) Vapor–liquid equilibrium diagram for acetone–chloroform solutions at 1.00 atm. (d) Vapor–liquid equilibrium phase diagram for acetone–chloroform at 1.00 atm.

are *negative*). Almost all the statements made in Section 8.4.2 apply here, with the words minimum and maximum (and higher and lower and positive and negative) interchanged.

Figure 8.9a shows that for negative deviation from ideal solution behavior the partial pressure curves and the total pressure curve all bow downward, relative to the straight line connecting their endpoints, as Eq. 8.5 shows they must. For this set of activity coefficients, the total pressure curve (at constant temperature) has a minimum. This corresponds in Figure 8.8d to a maximum in the boiling point T-x_a curve, producing a *maximum-boiling azeotrope.*

Figure 8.9b shows a $\ln \gamma_a$ vs. x_a plot, which is conceptually the opposite of Figure 8.8b. As expected, the activity coefficient curves approach 1.00 asymptotically as the individual mol factions approach 1.00, and have values less than 1.00 for all other concentrations, becoming smaller as the concentration of the species becomes smaller.

Figure 8.9c shows that for liquid acetone concentrations greater than the azeotrope (which corresponds to $x_a = y_a \approx 0.345$), the vapor contains a greater percentage of acetone than the liquid, while at liquid acetone concentrations less than the azeotrope the vapor contains less acetone than the liquid. At the azeotrope the vapor and liquid have the same composition, and the boiling-point temperature of this mixture, 64.5°C, is greater than the boiling point of either pure species or of any mixture of them with a different composition (at 1.00 atm).

Figure 8.9d is practically the vertical mirror image of Figure 8.8d. It shows that this system has a maximum boiling point, and two regions with different behaviors on either side of the liquid composition at which that maximum occurs. This type of azeotrope is less common than minimum boiling azeotropes, but it also makes separation by distillation difficult. If we start with a mixture to the left of the azeotrope in

Figure 8.9d ($x_a < 0.345$), simple distillation will produce practically pure chloroform and the azeotrope, while for $x_a > 0.345$, simple distillation produces practically pure acetone and the azeotrope. But simple distillation will not produce practically pure acetone and practically pure chloroform from any liquid solution shown on this diagram.

Almost all examples of type III behavior (activity coefficients <1.00) are for mixtures in which the individual species form some kind of attractive bond with each other, typically hydrogen bonds. Normally, these are not as strong as, for example, the covalent bonds that bind hydrogen to oxygen in a water molecule, but they are strong enough to produce the results shown in Figure 8.9. Most often these bonds affect the behavior of the liquid significantly, but have little effect on the behavior of the vapor (where the average intermolecular distances are much greater). Thus the assumption of ideal gas behavior here is a good one. An interesting exception is discussed in Chapter 13. The evidence for the formation of such bonding in liquid acetone–chloroform mixtures is given in [7, p. 86]. A more extreme example of type III is the water–sulfuric acid system. Figure 8.10 shows that in dilute solutions of sulfuric acid in water the sulfuric acid activity coefficient is approximately 10^{-5} to 10^{-6}.

8.4.4 Azeotropes

A minimum-boiling azeotrope occurs when the system pressure curve (at constant temperature) has a maximum; a maximum-boiling azeotrope occurs when the system pressure curve (at constant temperature) has a minimum. A minimum-boiling azeotrope can occur only for activity coefficients greater than 1.00; a maximum-boiling azeotrope only for activity coefficients less than 1.00. Activity coefficients greater than 1.00 indicate that the molecules of the two species repel one another or at least are less attracted to each other than to their own kind, for example, acetone and water. Activity coefficients less than 1.00 indicate that the molecules of the two species are more attracted to the other kind than to their own kind, for example, H_2SO_4–H_2O.

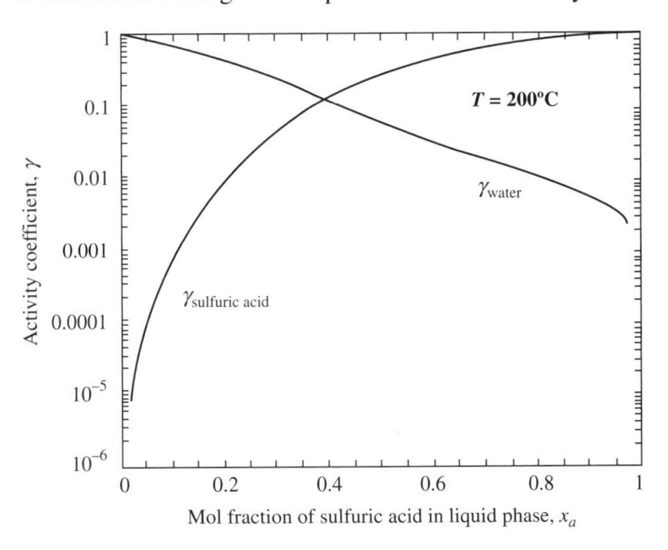

FIGURE 8.10 Activity coefficients for water–sulfuric acid at 200°C [8, p. 2–83]. Observe that the least volatile species, sulfuric acid, is chosen as species *a*. This is contrary to the standard convention, but the normal way of showing water–sulfuric acid diagrams. This equilibrium is complicated by the formation of several weak intermolecular compounds and by the presence of free SO_3 in the equilibrium vapor. Nonetheless, it shows that for species that form such weakly–bonded quasi-compounds, the equilibrium activity coefficients can be quite small.

For a given degree of mutual repulsion or attraction, the probability of forming an azeotrope becomes greater as the difference in boiling points of the two species becomes smaller. On a plot like parts (a) of Figures 8.7, 8.8, and 8.12 (below), nonideal behavior bends the curve upward (types II and IV) from the ideal solution line for positive deviation or downward (type III) for negative deviation. If the ideal solution line is steep (the boiling points are far apart), then a large deviation from ideal solution is needed to produce a minimum or a maximum. If the line is close to flat (the boiling points are close together), then a modest deviation will produce a minimum or a maximum. In Figure 8.8 the activity coefficients are less than shown in Figure 8.6, but the difference in NBP is small (100 and 82°C). The mixture in Figure 8.6 is more strongly nonideal than that in Figure 8.8 (larger activity coefficients), but because of its wide difference in NBP (100°C and 56.1°C) acetone–water does not form an azeotrope at 1.00 atm.

Figure 8.11 shows the *T*-*x* diagram for acetone–water based directly on Table 8.1. Comparing this to Figure 8.8d we see that although the dew-point and bubble-point lines deviate strongly from an ideal-solution bubble-point and dew-point lines (Figure 8.7d), the large difference in NBPs

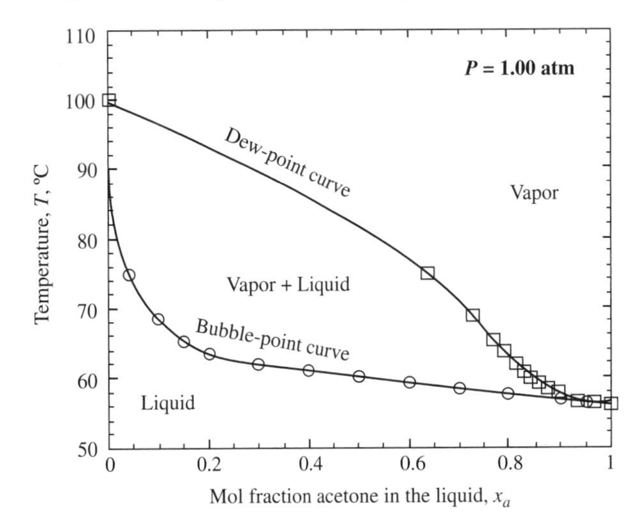

FIGURE 8.11 Temperature–composition diagram for acetone–water at 1 atm. The points are from Table 8.1; the curves are simple interpolations.

prevents the curves from showing a minimum, and hence there is no azeotrope.

Binary azeotropes between compounds with widely different boiling points are rare. Table A.6 shows a sample table of azeotropes. The *Handbook of Chemistry and Physics* [9] lists over a thousand binary azeotropes. For water (NBP – 100°C) the highest boiling is the azeotrope with hydrogen iodide at 127°C; the vast majority fall in the boiling range 70–99°C. The few that have lower temperatures have very little water in the azeotrope; for example, the water-isoprene azeotrope boils at 32.4°C and is 99.86 wt% isoprene. About 90% of the known azeotropes are of the minimum-boiling variety. It is worth the student's while to spend a little time looking at Table A.6, to form an intuitive idea of what kind of binary mixtures form azeotropes.

8.4.5 Two-Liquid Phase or Heteroazeotropes (Type IV)

Figure 8.12 shows the same four plots for type IV, a *two-liquid phase* system or *hetewazeotrope.* This may be considered an extreme example of positive deviation from ideality. In Figure 8.8, type II (isopropanol–water), the

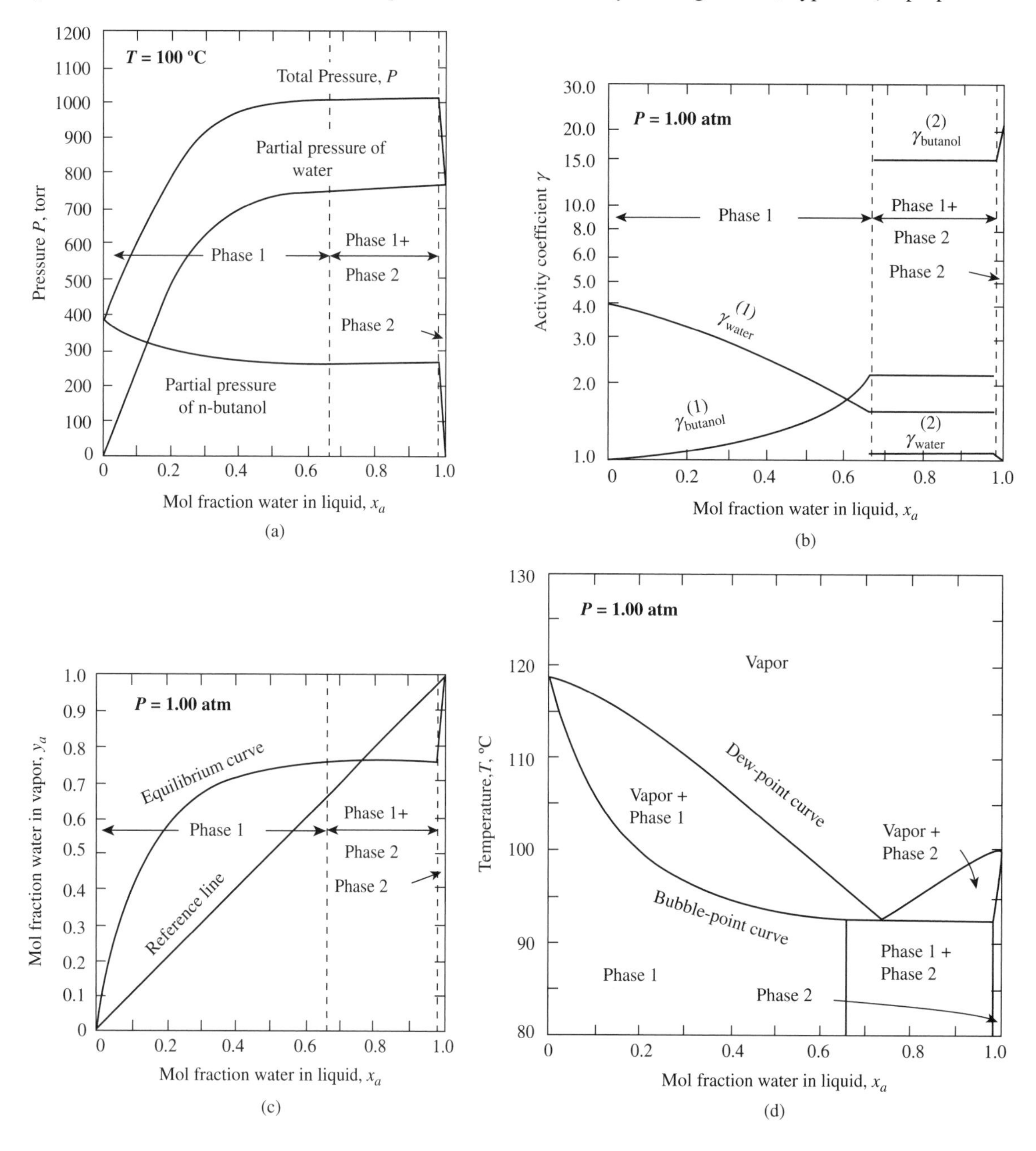

FIGURE 8.12 (a) Partial and total pressure of water–*n*–butanol solutions at 100°C. (b) Activity coefficients for water–*n*–butanol solutions at 1.00 atm. (c) Vapor–liquid equilibrium diagram for water–*n*–butanol at 1.00 atm. (d) Vapor–liquid equilibrium phase diagram for water–*n*–butanol at 1 atm.

positive deviation from ideality was strong enough to produce an azeotrope, but not so strong that the two liquids were not miscible in all proportions. In Type IV the mutual repulsion between the two kinds of molecules is strong enough that over the concentration range from $\approx$ 65 mol% water to $\approx$ 98 mol% water no single liquid phase exists (at 1 atm and the boiling point). So, for example, if we were to make up a mixture of 80 mol of water and 20 mol of *n*-butanol, at equilibrium we would find two liquids, one with 65 mol% water and one with 98 mol% water, but no liquid with 80 mol% water. We discussed this type of situation in Section 6.5 and Figure 6.9. There the hypothetical binary had symmetric properties, and the two liquids were $\frac{1}{6}$ and $\frac{5}{6}$ mol fraction of one species. Real situations are never as simple as that, as this example shows.

Figure 8.12a shows that for positive deviation from ideal solution behavior the partial pressure curves and the total pressure curve all bow upward, relative to the straight line connecting their endpoints, as Eq. 8.5 shows they must. Comparing this figure to Figure 8.8a we see that the total pressure and partial pressure of *n*-butanol rise very steeply in the water mole fraction range 0.98 to 1.00, and that all three pressure curves are totally horizontal in the two-liquid-phase region. The maximum in the vapor-pressure curve (at constant *T*) produces a *minimum-boiling azeotrope* like type II, which is called a *heteroazeotrope* to make clear that two liquid phases are present.

Figure 8.12b shows the same kind of ln γ_a vs. x_a plot as Figure 8.6 or 8.8. In the two regions with only one liquid phase there are only two activity coefficients whose curves approach 1.00 asymptotically as the individual mol fractions approach 1.00 and have values greater than 1.00 for all other concentrations, becoming larger as the concentration of the species becomes smaller. In the two-phase region there are two liquids, each with two activity coefficients, so there are four straight lines across the two-phase region, as discussed in Example 8.6.

Figure 8.12c has the same general shape as Figure 8.8c, but the departure from ideal solution is much stronger. In the two-liquid-phase region the mol fraction of water in the vapor is constant, as discussed below. This type of azeotrope is also common, and makes separation by distillation difficult. If we start with a mixture to the left of the azeotrope, for example, 10 mol% water, and attempt to separate it into pure water and pure *n*–butanol by distillation we find that it is easy to get practically pure *n*–butanol from the mixture; the low-water part of Figure 8.12c is similar to that in Figure 8.7c. But it is not possible to produce pure water because, as Figure 8.12c shows, at high concentrations of water the vapor contains a lower concentration of water than the liquid.

Figure 8.12d shows that this system has a minimum boiling point, and three regions with different behaviors, one on either side of the composition at which two liquid phases are present, and a third type of behavior for the region in which two liquid phases are present.

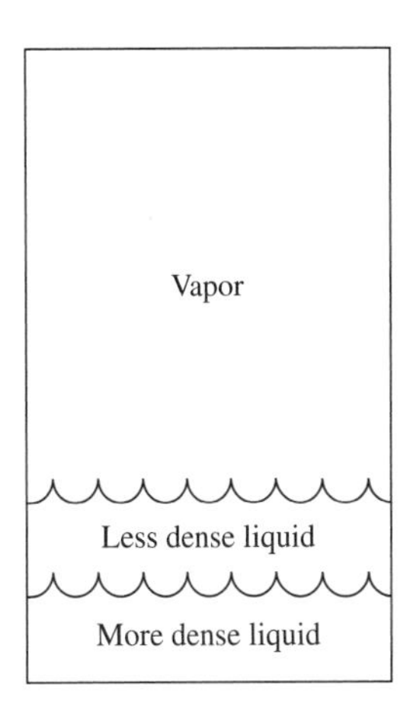

FIGURE 8.13 Two liquid phases and one vapor phase in equilibrium.

This type of equilibrium is not seen in everyday experience, but is very common in petroleum refining, in which steam and water are often mixed with hydrocarbons, and the mixture is then treated by distillation; liquid water and liquid hydrocarbons repel each other strongly; "water and oil don't mix" (but see Chapter 11). We may visualize the situation as shown in Figure 8.13. The two liquid phases will generally not have equal densities, so the less-dense liquid will float upon the more dense liquid (low-molecular weight hydrocarbons like gasoline float on water). For the three phases to be in equilibrium, the liquids generally will have to be boiling, so that vapor bubble formation provides vigorous agitation. If the liquids are not agitated, and the pressure or temperature are changed, the vapor will quickly come to equilibrium with the less dense liquid, which it contacts, but will not quickly come to equilibrium with the more dense liquid, which it can exchange matter with only by slow diffusion through the less dense liquid. The entire discussion here is of equilibrium with good agitation, normally by boiling, which brings all three phases to equilibrium.

Example 8.4 Eighty mols of water and 20 mols of *n*-butanol are mixed at 92°C, coming to equilibrium. Two liquid phases form. Based on Figure 8.12 estimate how many mols of each of the liquids are present.

Here we know the mol fraction of water in each phase. Let n_T, $n^{(1)}$, and $n^{(2)}$ be the total number of mols and the number of mols in each of the two phases. Let z_{feed} be the mols of water (species *a*) fed/total mols fed. Then by material balance on water

$$\begin{aligned} n_T z_{\text{feed}} &= n^{(1)} x_a^{(1)} + n^{(2)} x_a^{(2)} = n^{(1)} x_a^{(1)} + (n_T - n^{(1)}) x_a^{(2)} \\ \frac{n^{(1)}}{n_T} &= \frac{z_{\text{feed}} - x_a^{(2)}}{x_a^{(1)} - x_a^{(2)}} = \frac{0.8 - 0.98}{0.65 - 0.98} = 0.545 \end{aligned} \tag{8.F}$$

We would expect to find 54.5 mols of the 65 mol% water liquid and (100 – 54.5 = 45.5) (mols of the 98 mol% water liquid. ■

Outside the two-liquid phase region (i.e., for x_{water} less than 65 mol% or greater than 98 mol%) the four parts of Figure 8.12 have the same shapes as Figure 8.8, but with more extreme deviation from ideal solution behavior (larger activity coefficients than Figure 8.8). Please make that comparison to see that this is true. If we repeat Example 8.4, always for 100 total mols, for any value of the mols of water in the feed between 65 and 98, we will find some value of the above ratio between 0.0 and 1.0. As long as both liquid phases are present, all liquid–phase intensive properties in both phases are constant, independent of the ratio of mass or mols one phase to another. The ratio of the mass or mols of one phase to the other, which is not represented in these figures, does change as this feed ratio changes. Thus, all the curves representing the liquids are exactly horizontal in this two-liquid-phase region. However, only one vapor composition can be in equilibrium with these two liquids, regardless of their mol ratios. Figure 8.12d shows this; that vapor has about 73 mol% water.

The previous examples have all been of two phases in equilibrium. Here we have three. The basic relations must be the same, but with more terms in the equations. For three phases (1, 2, 3) with two chemical species (a, b),

$$f_a^{(1)} = f_a^{(2)} = f_a^{(3)} \quad \text{and} \quad f_b^{(1)} = f_b^{(2)} = f_b^{(3)} \tag{8.G}$$

The fugacity of water is the same in all three phases; the fugacity of n-butanol is the same in all three phases.

Example 8.5 An equilibrium (dew-point) vapor at 1.00 atm has $y_{water} = 60$ mol%, balance n-butanol. What is the composition of the equilibrium (bubble-point) liquid? Repeat the example for vapors containing 90 mol% water and 73 mol% water.

From Figure 8.12d, if we start at 130°C and 60 mol% water and cool, we will have only vapor until we meet the dew-point line at 99°C, and at the same temperature the bubble-point curve shows $x_{water} \approx 0.22$. For $y_{water} = 90$ mol%, the same procedure brings us to the rightmost dew-point curve at 98°C. In this case, the corresponding bubble-point line is not the one at the left, but the steeply sloping one at the right, from which we read $x_{water} \approx 0.99$. For $y_{water} = 73$ mol% we see that the two dew-point lines meet at 92°C. Vapor of this composition is in equilibrium with both liquid phases, as sketched in Figure 8.12d. Vapor with any other composition is in equilibrium with only one liquid, with $x_{water} < 0.65$ if $y_{water} < 0.73$, and $x_{water} > 0.98$ if $y_{water} > 0.73$. ■

Table 8.C Summary of Example 8.6

Phase	x_{water}	f_{water} (atm)	γ_{water}	$x_{n\text{-butanol}}$	$f_{n\text{-butanol}}$ (atm)	$\gamma_{n\text{-butanol}}$
1	0.65	0.73	1.51	0.35	0.27	2.10
2	0.98	0.73	1.005	0.02	0.27	36.9

Example 8.6 Estimate the activity coefficients for water and for n-butanol for the situation sketched in Figure 8.13, in which two liquid phases and the gas phase are all in equilibrium. From Figure 8.12d we know the mol fractions of water in all three phases, and the temperature, 92°C. From the Antoine equation and Table A.2 we estimate the vapor pressures of pure n-butanol and water at this temperature as 5.38 and 10.9 psia.

We make the same assumptions as in Section 8.2 (ideal gas), from which we see that the fugacity of water in the gas phase is the same as its partial pressure, 0.73 atm, and that of n-butanol is 0.27 atm. Then, from the criterion of equilibrium, we know that the fugacity of water in each of the two liquid phases must be 0.73 atm, and the fugacity of n-butanol in each of the two liquid phases must be 0.27 atm. We see that we need to compute four liquid-phase activity coefficients, instead of the two in Section 8.3, because there are two liquid phases. For water in the phase with $x_{water} \approx 0.65$,

$$\gamma_{water} = \frac{y_{water}P}{x_{water}p_{water}} = \frac{0.73 \cdot 14.7\,\text{psia}}{0.65 \cdot 10.9\,\text{psia}} = 1.51 \tag{8.H}$$

We then make up Table 8.C, showing all four liquid-phase activity coefficients.

These are the four activity coefficients shown in Figure 8.12d. (The experimental data for this system [10, p. 328] show considerable data scatter and disagreement between data sets. The value for $\gamma_{n\text{-butanol}}$ calculated here is roughly twice the value shown in that figure. The experimental data differ at least that much between data sets.) ■

From Figure 8.12a we see that the partial pressures form a maximum, similar to Figure 8.8a, but that over the whole range where two liquid phases exist that pressure is constant. From Figure 8.12c we see that this is the same type of azeotrope as Figure 8.8c, except that the vapor composition is constant over the range of liquid variables for which two liquid phases are present.

8.4.6 Zero Solubility and Steam Distillation

A logical extension of Figure 8.12 (type IV) is a pair of liquids that have practically zero solubility, such as mercury and water. There is no known case of absolute zero solubility, either theoretically or experimentally (see Chapter 11) but at room temperature the measured solubility of mercury in water is only 4 ppb mol fraction [11]. The solubility of water in mercury is not easily found in the literature but is believed to be comparable to the solubility of mercury in water, a few parts per billion by mol. The analog of Figure 8.12 would have the two-phase region extending from $x_{water} = 4 \times 10^{-9}$ to about $x_{water} = (1.00$ minus a few parts per billion). This makes the two miscible regions so thin that they cannot be distinguished from the vertical axes, and makes the mercury-in-water activity coefficient $\approx 2.5 \times 10^8$ and that of water-in-mercury about the same. These high activity coefficients indicate the strong molecular incompatibility between water and mercury. This is an extreme case of type IV (Figure 8.12).

Example 8.7 Suppose that the solubilities of water in *n*-butanol and of *n*-butanol in water were zero, instead of the actual solubilities shown in Figure 8.12d. Then, at 1 atm, what would be the boiling-point temperature of the system shown in Figure 8.12, and what would be the composition of the vapor?

In this case the two liquids shown in Figure 8.13 would be pure water and pure *n*-butanol. We assume that there is vigorous boiling, so that both liquids are in contact with the vapor, and thus in equilibrium with it. Raoult's law should apply to each of the two phases separately. If we write Eq. 8.5 for either one of the pure phases we see that for each $y_i = p_i/P$, because in any pure phase $x_i = \gamma_i = 1.00$. Writing this equation for each species, adding the equations, and solving for P, we find

$$P = \sum y_i P = \sum \frac{p_i}{P} P = \sum p_i \qquad (8.1)$$

The total pressure is the sum of the individual pure species vapor pressures. To find the boiling-point temperature we perform a trial and error like that in Example 8.3 (see Table 8.D), with the difference that each of the liquid mol fractions is taken as ≈ 1.00. The result is compared to the experimental values (from Figure 8.12d) in Table 8.D. We see fair agreement, indicating that for mixtures of limited mutual solubility, the zero-solubility assumption and Raoult's law lead to fair estimates of the boiling point and vapor composition. This result is the basis of *steam distillation.* If we had a sample of *n*-butanol contaminated with some high-boiling material, we could boil off the *n*-butanol, with steam, at a temperature of 92°C. The boiling point of pure *n*-butanol is 117.5°C. By using the water we lower the combined boiling temperature by 25°C. For heat-sensitive or high-boiling materials this lowering of the boiling point is important, and steam distillation is widely used. (We could accomplish the same result by vacuum distillation; for some cases it is more economical than steam distillation, for others not.) If we have measured equilibrium data for the mixture, like Figure 8.12, we use that data to estimate the boiling point and vapor composition. If we do not have such data, we use the method in this example, and, as shown, find a fair estimate of the observed behavior. ■

Table 8.D Solution to Example 8.7

	Experimental Value from Figure 8.12.d	Value from Example 8.7, Assuming Zero Mutual Solubility
Boiling temperature T (°C)	92	89
Mol fraction water in vapor phase, y_{water}	0.73	0.67

It may seem strange to have practically pure water boiling at 89°C and 1 atm pressure. But because the vapor is only 65 mol% water this is really water boiling at a partial pressure of 0.65 atm, for which the pure-water boiling temperature really is 89°C! The values shown above are independent of how much of each liquid phase is present, as long as both are present. The same is true for Figure 8.12, as is confirmed by the phase rule (Chapter 15).

8.4.7 Distillation of the Four Types of Behavior

In Figure 8.7d (type I) we saw that if we began with a mixture that was 50 mol% benzene and heated it until it began to boil, the liquid and vapor would have different compositions. If we vaporized about half of that mixture (halfway between **B** and **D** in that figure) the vapor would have roughly 65 mol% benzene, and the liquid roughly 35 mol% benzene. If we separated the vapor and condensed it, we would have performed a simple distillation, changing one liquid with 50 mol % benzene into two liquids with roughly 65 and 35 mol% benzene. If we then repeated the process with each of these, we would have four liquids, etc., with a range of benzene concentrations. Modern *fractional distillation* columns do this in an efficient way, producing one *overhead product*, which in this case would be practically pure benzene, and one *bottom product*, which in this case would be practically pure toluene. In any simple fractional distillation, the lower-boiling material concentrates in the overhead, and the higher-boiling material in the bottoms. For type I the purity of the two products would be limited only by how much we were willing to pay.

For types II, III, and IV, the situation is more complex. If we follows the same logic shown above, we find the results in Table 8.3. The reader is encouraged to work these relations out, using the same logic shown above, for types II, III, and IV.

Table 8.3 Outcomes of Simple Fractional Distillation for Types II, III, and IV

	Type II	Type III	Type IV
For $x_a < x_{\text{azeotrope}}$			
Overhead product	Azeotrope that forms one liquid on condensation	b	Azeotrope that separates into two liquids on condensation
Bottoms product	b	Azeotrope	b
For $x_a > x_{\text{azeotrope}}$			
Overhead product	Azeotrope that forms one liquid on condensation	a	Azeotrope that separates into two liquids on condensation
Bottoms product	a	Azeotrope	a

8.5 GAS–LIQUID EQUILIBRIUM, HENRY'S LAW AGAIN

In the previous sections of this chapter, VLE has mostly concerned equilibrium of vapors and liquids (see Section 3.2.1 for the distinction between gases and vapors). In all of the plots previously presented in this chapter, both of the species can exist as a pure liquid at the temperature of interest. In all of the figures shown so far, the vapor–liquid equilibrium region extends over the whole range of x_a. Do the same ideas and equations apply to gas–liquid equilibrium, such as the air–water system we spent so much time on in Chapter 3? In principle, yes; in practice, yes and no.

First, reconsider Henry's law. We saw in Chapter 3, that Henry's law was Raoult's law, rewritten to replace the pure liquid's vapor pressure p_i with an experimentally determined Henry's law constant H_i, which for common gases at room temperature has values of 10,000 – 40,000 atm (see Table A.3). If we apply the same logic as we did in Section 8.3.1, we see that we can simply expand Table 8.2 into Table 8.4.

We see that most of the statements made about Raoult's law apply to Henry's law. It is normally shown as an ideal solution law, $\gamma_i = 100$, with the pure species vapor pressure replaced by the Henry's law constant. Table 8.4 also shows the equation for x_i, in addition to the working equation for y_i because Henry's law is most often used to estimate the concentration of a gas dissolved in the liquid. We occasionally see these equations written with a Henry's law-type $f_i^{\text{o, liquid phase}}$ and a liquid-phase activity coefficient, $y_i = \gamma_i x_i H_i / P$, but that is uncommon in ordinary VLE. Henry's law is most often used for gases dissolved in water, but can also be used for gases dissolved in other solvents. This topic is discussed further in Chapter 9, and an example of Henry's law plus simultaneous chemical reaction is explored in Chapter 13.

Thus, we see that Henry's law fits into our computational scheme for VLE, as an ideal solution law, with the choice of $f_i^{\text{o,liquid phase}} = H_i$. Often it is applied in examples like the air–water example in Chapter 3, in which one species in the gas (e.g., water) exists as a vapor, while one or more other species in the gas (e.g., nitrogen and oxygen) are present as gases above their critical temperatures. Purists would describe that as part VLE (for the water) and part gas–liquid equilibrium (for the nitrogen and oxygen). The gaseous phase would be called a gas, not a vapor, but we see that this is a matter of arbitrary definitions.

8.6 THE EFFECT OF MODEST PRESSURES ON VLE

Most experimental VLE data are for pressures at or near 1 atm. For pressures up to perhaps half of the critical pressure

Table 8.4 Comparison of Henry's Law, Raoult's Law, and Raoult's Law-type Activity Coefficients

Variable in Eq. 8.3	Replacement in Henry's Law	Replacement in Raoult's Law	Replacement with Raoult's Law-type Activity Coefficient
$\hat{\phi}_i$	1.00	1.00	1.00
y_i	y_i	y_i	y_i
$f_i^{\text{o,vapor phase}}$	P	P	P
γ_i	1.00	1.00	γ_i
$f_i^{\text{o,liquid phase}}$	H	p_i	p_i
Working equation for y_i	$y_i = \dfrac{x_i H_i}{P}$; $x_i = \dfrac{P y_i}{H}$	$y_i = \dfrac{x_i p_i}{P}$	$y_i = \dfrac{\gamma_i x_i p_i}{P}$

we can estimate the VLE from data at or near 1 atm by the methods shown below, if we make suitable corrections for the effect of pressure. The corrections used for the liquid and the vapor are not the same.

8.6.1 Liquids

For liquids and solids the effect of pressure on fugacity is small, and is most often ignored (but see Chapters 10 and 14!). Example 7.3 shows, for instance, that at 100°F, raising the pressure of liquid water from 1 to 1000 psia increases its fugacity by only 4.9%. For the most careful VLE work, the procedure in Example 7.3 is used. For most routine, low-pressure VLE calculations this small change is ignored, and the effect of pressure on the fugacity of the liquid is considered to be nearly zero. The process-design computer programs make that small correction, because it costs them practically nothing to do so.

8.6.2 Gases, the L-R Rule

For an ideal gas mixture, increasing the pressure increases the fugacity of each species in the mixture; $f_i = y_i P$. If there is no corresponding increase f_i in the liquid, then the gas will condense into the liquid. (The liquid can practically increase its fugacity only by increasing its temperature and thus its vapor pressure. So, the vapor pressure curve shows that the temperature must increase as the pressure is increased to maintain equilibrium.) However, for real gases Figure 7.1 shows that as the pressure increases, $f_i/P = \phi_i$ decreases for gases, with the decrease being faster for lower temperatures than for higher ones. Thus, we need some way to estimate the change in $f_i/P = \phi_i$ of individual species in gas mixtures as the pressure increases, to estimate the VLE behavior.

The Lewis–Randall (L-R) fugacity rule, introduced in Section 7.12.3, is widely used to estimate the nonideal behavior of gases because it is simple and is the next step in complexity (and reliability) over the ideal gas law. It gives reasonably good estimates of gas-phase fugacities up to pressures about half the critical pressure (typically ≈ 300 psia for common systems). However, frequently gases exist in states for which we cannot compute ϕ_i. Consider the water vapor that is present in the air at 68°F and 1 atm (see Chapter 3). If we look in a steam table [12], we see that at this temperature and pressure pure water can exist only as a liquid, not as a vapor. We may form an intuitive model of this behavior by considering that in a pure water vapor each water molecule collides only with other water molecules, with which it forms weak attractive bonds. The molecules of water vapor in air mostly collide with air molecules (which make up more than 95% of the molecules present). The water molecules do not form comparably strong attractive bonds with air molecules, so the water molecules will not condense, at a T and P at which pure water would condense.

If we wished to calculate ϕ_i for the water vapor in the air at 68°F, we would compute $P_r = 1$ atm/218.3 atm $= 0.00458$ $T_r = 528°\text{R}/1165°\text{R} = 0.453$. Referring to Figure 7.1 we see that this condition is at too low a T_r to be represented as a gas. It is clearly in the liquid region where, if we guess its f/P, we will find a value like 0.02. From the steam table we know that for liquid water at 68°F and 1.00 atm, $f/P \approx 0.023$. However, experimentally we can show that the water vapor dissolved in air behaves as if its value of f/P were practically 1.00. We can understand the problem in terms of Figure 8.14, where we see the line for $P_r = 0.1$ on an f/P versus T_r plot. At the right, in the gas phase we have a continuous curve and at the lower left in the liquid phase we have a continuous curve. These have an extremely sharp change in slope at the saturation curve. When we remember that

$$\left(\frac{\partial \ln f_i}{\partial T}\right)_P = \frac{h_i^* - h_i}{RT^2} \tag{7.9}$$

we see that the enormous change in enthalpy due to the latent heat of vaporization must cause an extremely sharp change in slope at the saturation curve. However, the water vapor in the air does not exist as a pure liquid. It has not given up the latent heat of condensation, and thus the term on the right of Eq. 7.9 is practically zero.

If we wish to use the L-R rule for this water vapor, we must find some hypothetical state that would represent it. The most sensible approach to this problem seems to be to extrapolate the gas curve into the liquid region. This is shown as a dotted curve in Figure 8.14. In performing this extrapolation we may best fit an EOS to the behavior of the gas from some significant temperature down to the saturation temperature and simply continue extrapolating with that EOS. The resulting extrapolated values are called "hypothetical standard states" [13].

Example 8.8 Estimate the value $f/P = \phi_i$ for water vapor in the hypothetical state of 1 atm pressure and 68°F, using the little EOS.

Using the above values of T_r and P_r, the value of ω from Table A.1, and applying the result for the little EOS from Example 7.1, we find

$$\frac{f}{P} = \phi_i = \exp\left(\frac{P_r}{T_r} \cdot f(T_r)\right) \tag{7.E}$$

$$\begin{aligned} f(T_r) &= \left(0.083 - \frac{0.422}{0.453^{1.6}}\right) + 0.345 \cdot \left(0.139 - \frac{0.172}{0.453^{4.2}}\right) \\ &= -1.425 + 0.345 \cdot (-4.645) = -3.018 \end{aligned} \tag{8.J}$$

$$\begin{aligned} \frac{f}{P} &= \phi_i = \exp\frac{-3.018 \cdot 0.00458}{0.453} = \exp(-0.0305) \\ &= 0.970 \approx 1.0 \end{aligned} \quad \blacksquare \tag{8.K}$$

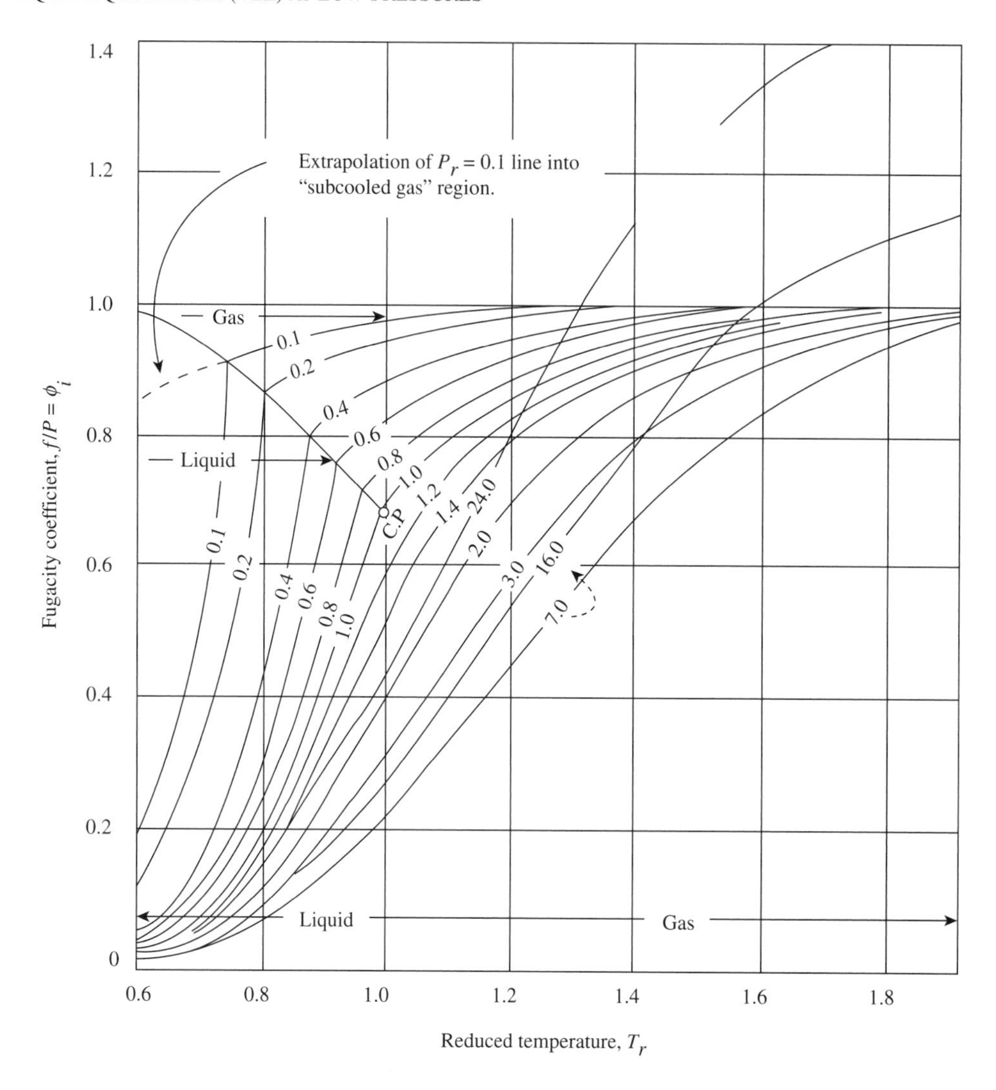

FIGURE 8.14 Pure species $f/P = \phi_i$ as a function of T_r with lines of constant P_r showing extrapolation of the $P_r = 0.1$ curve into the "subcooled gas" region. This is drawn onto an alternative version of Figure 7.1, see Appendix A.5.

For this low a pressure the little EOS is fairly reliable although the extrapolation to this low a T_r is speculative. We may safely conclude that although water cannot exist at a pure gas at 1 atm and 68°F, its vapor, dissolved in air will behave practically as an ideal gas. A similar problem occurs at substantial pressures in gas mixtures in which the temperature is below the boiling point of some of the species at the system pressure, such as the *n*-butane in Examples 7.5 and 7.7. In those cases, the procedure shown in Figure 8.14 or its EOS equivalent is needed. There is a kind of symmetry between Henry's law and the region where we must use hypothetical standard states in the L-R rule, as illustrated in Figure 8.15.

For hand calculation the L-R rule is normally satisfactory. If we are writing a computer program to do many of these calculations, then the additional cost of additional complexity is small, so it is common to consider the possible nonideal behavior of gas mixtures, that is, not to make the assumption that $\hat{\phi}_i = \phi_i$. Instead we compute $\hat{\phi}_i$ from an EOS for the mixture, using some set of mixing rules believed to be more accurate than the ideal solution mixing rule that leads to the L-R rule. In most cases the results are almost the same as those produced by the L-R rule.

8.7 STANDARD STATES AGAIN

In most of this chapter we have expressed equilibrium in terms of Eq. 8.3, which involves a standard state fugacity f_i^{o}

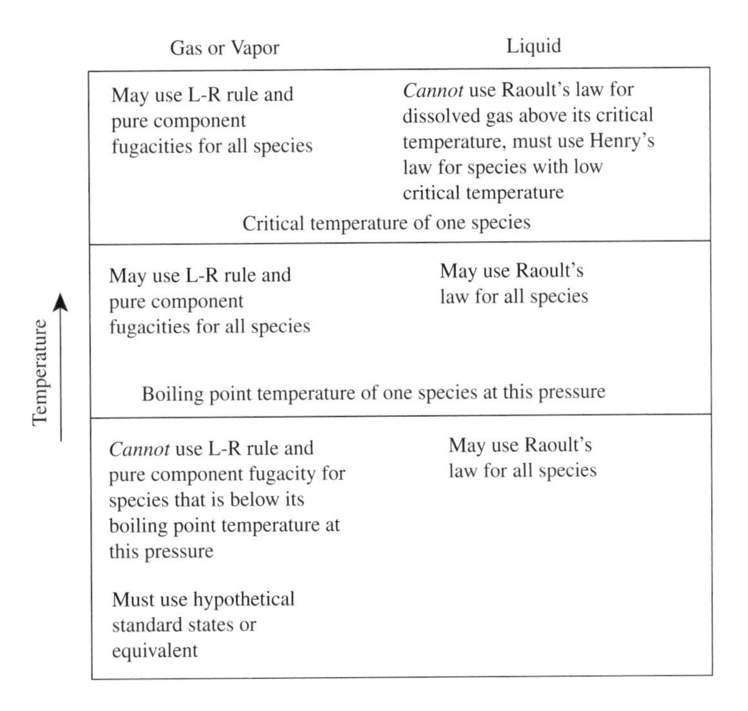

FIGURE 8.15 Symmetry between region where one cannot use Raoult's law and must use Henry's law, and region where one cannot use the L-R rule with pure species fugacity, because the pure species cannot exist as a vapor.

for each phase. It would be nice if we could choose the same standard state, but in the most common approach—the one you will see most often in the literature—we chose different standard states for gas ($f_i^o = P$) and liquid ($f_i^o = p_i$). You may not have noticed that two other standard states appeared in this chapter; $f_i^o = H_i$ for Henry's law (gases dissolved in liquids), and $f_i^o = \phi_i P$ for the L-R rule (ideal solution of nonideal gases). These four standard states are the most common in VLE calculations and publications. There are others, used in other places as well (see Chapter 12). Always check which standard state is being used in publications you read; good publications will make that very clear.

Henry's law makes a strange standard state, because its $f_i^o = H_i$ does not correspond to the behavior of pure i but rather to the behavior of i in the solution as $x_i \to 0.00$. This is one of the reasons it says in Chapter 7 that f_i^o is sometimes chosen to correspond to the fugacity of pure i at this temperature and pressure and sometimes not. For dissolved oxygen in liquid water at 68°F there is no "pure state" of pure liquid oxygen, because oxygen cannot exist as a liquid at this temperature. The fact that Henry's law leads to this strange value of f_i° seems to have no practical consequences, and is pointed out here only to remind the reader that we make several choices for f_i^o, some of them not very intuitive. (Some authors make much of this unimportant distinction, under the name *unsymmetrical standard states.*)

8.8 LOW-PRESSURE VLE CALCULATIONS

We made several VLE calculations in Chapter 3, one in Chapter 7, and several in this chapter. Most of those were simple enough that we used shortcut, manual methods. In this section we consider the six standard types of VLE calculation, more formally than in previous sections.

For the most *low-pressure* VLE (up to several hundred psia) we don't worry much about nonideality in the gas phase; if we use the L-R rule we will have reasonable confidence in our estimates in the gas phase. All of the examples before Example 8.8 assumed that the gas phase was practically an ideal gas. However, those examples showed that the liquid phase was often quite nonideal. Our attempts to correlate and predict the VLE have mostly been attempts to correlate and predict liquid-phase activity coefficients. Many mixtures with widely different chemical structures and widely different vapor pressures can be represented reasonably well by Eq. 8.5 and fairly simple equations or prediction methods for liquid-phase activity coefficients. The next chapter shows how that is done. Before we begin that, we will borrow a result from that chapter, and show how the results are used.

Of the many equations used to correlate and predict liquid-phase activity coefficients showing the type and extent of

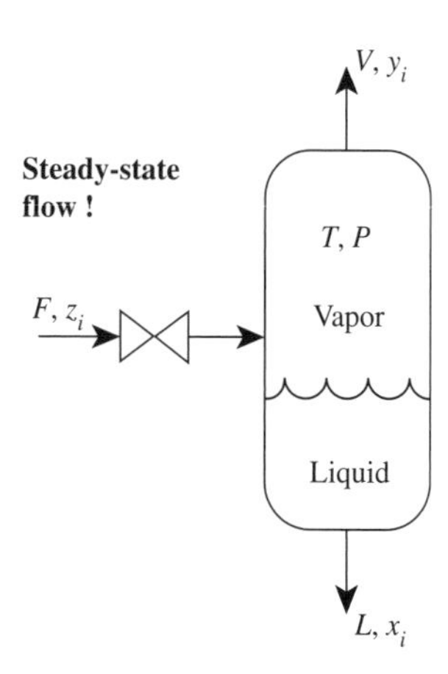

FIGURE 8.16 Figure and notation for *flash* calculations. F is the feed entering the equilibrium vessel, V the vapor leaving it, L the liquid leaving it, and T and P the temperature and pressure at which the contents of the vessel are at equilibrium. Mol fractions in the feed are z_i, those in the vapor are y_i and those in the liquid x_i.

nonideal behavior, one of the simplest and most useful is the van Laar equation, which says

$$\log \gamma_a = \frac{Ax_b^2}{\left(\dfrac{A}{B}x_a + x_b\right)^2} \qquad \log \gamma_b = \frac{Bx_a^2}{\left(x_a + \dfrac{B}{A}x_b\right)^2} \tag{8.6}$$

These two equations are most often presented this way. For programming them into our computers it is often easier to write their equivalent form

$$\log \gamma_a = \frac{B^2Ax_b^2}{(Ax_a + Bx_b)^2} \qquad \log \gamma_a = \frac{A^2Bx_a^2}{(Ax_a + Bx_b)^2} \tag{8.L}$$

We will use this equation in numerous examples in this section, *not* because it is the most reliable equation for this purpose, but because it is simple, easily understood, and easily programmed into computers. We will use the ethanol–water system as the example nonideal solution, both because of its industrial significance and because experimental data are available for it, to be compared with our calculations. For the ethanol–water system at 1 atm pressure, taking ethanol as species a and water as species b, the values of the constants in Eq. 8.6 (based on fitting the experimental VLE data) are $A = 0.7292$, $B = 0.4104$ (Table A.7, with A and B interchanged as described in that table.). In Chapter 9 we will consider other, more complex equations that play the same role as the van Laar equation, which are superior for some circumstances and are more widely used. Table A.7 shows a sampling of van Laar equation constants for binary mixtures.

The six most common VLE problems can all be formulated in terms of *equilibrium flashes.* This name (or simply *flashes* or sometimes *equilibrium flash vaporizations*, EFV) is used for the computation modules that do this type of calculation in process-design computer programs. The schematic and the basic notation are shown in Figure 8.16. There, a feed F (which may be a gas, a liquid, or a gas–liquid mixture) is reduced in pressure through a throttling valve into a pressure vessel, which acts as a gas-liquid gravity separator. The vapor V exits the top of the vessel and the liquid L exits the bottom of the vessel. This is called a flash, because if the pressure is reduced significantly, some or all of the liquid will change to vapor so rapidly that it occurs "in a flash." The six basic types of VLE calculations are summarized in Table 8.5. There are others, for example, specified S or V, which many computer programs also supply, but the six shown here are the most commonly seen.

Table 8.6 summarizes the equations that are used to solve low-pressure VLE problems. It appears formidable, but we will see that in practice these computations are not difficult. Most of the equations have been used and discussed previously. Here we add two more equations. By material balance in Figure 8.16 we may say that

$$F = V + L \tag{8.7}$$

where F, V, and L are most often expressed in mol/h or some equivalent molar flow rate. A material balance on species i leads to

$$Fz_i = Vy_i + Lx_i \tag{8.8}$$

where z_i is the mol fraction of i in the feed F.

The examples in this chapter are all for binary (two-species) mixtures. If we have more than two species, then

Table 8.5 The Six Basic Types of VLE Calculations ($K_i = y_i/x_i$)

Calculation Type	Given	To Find	Computation
Bubble point, T known	T, all x_i	P, all y_i	Find P for which $\sum y_i = \sum K_i x_i = 1.00$
Bubble point, P known	P, all x_i	T, all y_i	Find T for which $\sum y_i = \sum K_i x_i = 1.00$
Dew point, T known	T, all y_i	P, all x_i	Find P for which $\sum x_i = \sum y_i/K_i = 1.00$
Dew point, P known	P, all y_i	T, all x_i,	Find T for which $\sum x_i = \sum y_i/K_i = 1.00$
Isothermal flash	T, P, all z_i	L/F, V/F, all x_i all y_i	$\sum x_i - \sum y_i = 0$
Adiabatic flash	P, all z_i, H_F	T, L/F, V/F, all x_i, all y_i	$\sum x_i - \sum y_i = 0$, and $H_V + H_L = H_F$

Table 8.6 Equations to Be Solved in a General Equilibrium Flash Calculation

Type of Equation, Number of Equations	Form
Mol fractions sum to 1.00 in vapor	$\sum y_i = 1.00$
Mol fractions sum to 1.00 in liquid	$\sum x_i = 1.00$
Overall molar material balance	$F = V + L$
Material balance on a species, one equation for each species	$Fz_i = Vy_i + Lx_i$
Equilibrium statement, one equation for each species	$K_i = \frac{y_i}{x_i} = \frac{\gamma_i p_i}{\hat{\phi}_i P}$ most often simplified to $K_i = \frac{y_i}{x_i} = \frac{\gamma_i p_i}{\hat{\phi}_i P} = \frac{\gamma_i p_i}{P}$
Feed specification, as many equations as the number of species minus one.	z_i = specified value
Vapor-pressure equation, one for each species	$p_i = f(T)$ most often the Antoine equation
Liquid-phase activity coefficient, one for each species	$\gamma_i = f(x_i, x_j, \ldots)$ using the van Laar equation in this chapter
Energy balance, **for adiabatic flashes only**	$H_F = H_V + H_L$

for each additional species we add two variables, x_i and y_i and for each additional species we add one equilibrium statement and one feed specification, z_i. Thus, the number of variables and of equations both increase by 2 for each additional species; the solution procedures shown below are the same, independent of the number of species. If we do not make the ideal gas assumption, then we leave the $\hat{\phi}_i$ in our equation for K_i and add one equation for $\hat{\phi}_i$ for each species in the mixture, using the L-R rule or some more complex mixing rule that allows for nonideal mixing.

8.8.1 Bubble-Point Calculations

In Figure 8.16 we first consider a case in which $L \approx F$, $V \approx 0$. This is the bubble point calculation, in which we are asking the composition of the vapor that is in equilibrium with a liquid of specified composition. For this condition Eq. 8.8 shows that $x_i = z_i$ for all i; all except for an infinitesimal amount of the material in the feed goes to the liquid. Bubble points need not be thought of as flashes; we can think of them as "given the composition of the liquid phase and either P or T, find the composition of the equilibrium vapor phase and T or P." But the process-design programs normally include them in the flash module, and they have parallels to the other kinds of flashes in Table 8.5 so we include them here.

8.8.1.1 Temperature-Specified Bubble Point

Example 8.9 Estimate the boiling pressure and the y_is in equilibrium with a liquid that is 0.1238 mol fraction ethanol, balance water, at 85.3°C.

The equations to be solved are shown in Table 8.E. The Antoine equation constants are from Table A.2. We divided the two Antoine equations by 760 torr/atm, to have the vapor pressure in atm.

This and the following examples are solved on a spreadsheet, which is a fast, reliable, intuitive way of doing the computer equivalent of hand calculations. The solution of this problem, and the prototype for the rest of the solutions in this chapter is summarized in Table 8.F. The first column shows all the variables. The second column shows that T and liquid mol fractions are given. The next four variables are intermediate values, calculated directly from the given values. The pressure is the trial variable; for any value of P we can compute all the variables below it in the table. The check value is the sum of the vapor mol fractions, which must equal 1.00. There is only one value of the trial variable, P, which makes the check value = 1.00. The third column shows the values that result from an initial guess of $P = 0.8$ atm. The check variable is 1.25, indicating that a larger value of the trial variable is needed. The fourth column shows the solution, found using "goal seek" on an Excel spreadsheet.

Table 8.E Equations in Example 8.9

$$x_a + x_b = 1.00; \; y_a + y_b = 1.00$$

$$\frac{y_a}{x_a} = \frac{\gamma_a p_a}{P}; \quad \frac{y_b}{x_b} = \frac{\gamma_b p_b}{P}$$

$$p_a = 10\text{E}[8.04494 - 1554.3/(222.65 + T)]/760$$

$$p_b = 10\text{E}[7.96681 - 1668.21/(228.0 + T)]/760$$

$$\gamma_a = 10\text{E}\left[\frac{B^2 A x_b^2}{(Ax_a + Bx_b)^2}\right] = 10\text{E}\left[\frac{(0.4104)^2 \cdot 0.7292 x_b^2}{(0.7292 x_a + 0.4104 x_b)^2}\right]$$

$$\gamma_b = 10\text{E}\left[\frac{A^2 B x_a^2}{(Ax_a + Bx_b)^2}\right] = 10\text{E}\left[\frac{(0.7292)^2 \cdot 0.4104 x_a^2}{(0.7292 x_a + 0.4014 x_b)^2}\right]$$

$$z_a = x_a = 0.1238$$

$$T = 85.3°\text{C}.$$

Table 8.F Trial-and-Error (spreadsheet) Solution to Example 8.9

Variable	Type	Initial Guess	Solution
T(°C)	Given	85.3000	85.3000
x_a	Given $= z_a$	0.1238	0.1238
x_b	Given $= 1 - z_a$	0.8762	0.8762
p_a (atm)	Intermediate	1.3088	1.3088
p_b (atm)	Intermediate	0.5772	0.5772
γ_a	Intermediate	2.9235	2.9235
γ_b	Intermediate	1.0388	1.0388
$\boldsymbol{P}$ **(atm)**	**Trial variable**	**0.8000**	**0.9991**
y_a	Result	0.5921	0.4741
y_b	Result	0.6567	0.5259
$y_a + y_b$	**Check value**	1.2489	**1.0000**

We find $P = 0.9991$ atm and $y_a = 0.474$. The experimental values are 1.00 atm and 0.470 [1]. The small differences between experimental and calculated values remind us that the Antoine equation, used for the vapor pressures and the van Laar equation, used for the activity coefficients, are both approximate representations of experimental data. In this example (and the subsequent ones) the approximation is good in some cases, excellent in others. ■

If we were solving this equation set by hand, we would compute the two pure species vapor pressures and the two liquid-phase activity coefficients (the 4 intermediate values in Table 8.F) first, using the known values of T and the two x_i, thus eliminating four equations and four unknowns. After we did that, we could solve the problem analytically, without using the spreadsheet; that solution is shown for benzene–toluene in Example 3.4. If the liquid is an ideal solution, then the two activity coefficients become 1.00, which simplifies the calculation.

This type of bubble-point calculation has no simple graphical solution on the four types of plots shown in the previous examples because the pressure is unknown. Figure 8.17 shows a T-x diagram for ethanol–water at 1 atm, similar to part d in Figures 8.7, 8.8, 8.9, and 8.12 for a pressure of 1.00 atm. In principle, we could have a separate plot of this type for each possible pressure. (If we wanted the equivalent of Figure 8.17 for some other pressure we could make it up by repeating Example 8.10 (below) at that pressure, for a variety of liquid compositions and plotting the results.) With a set of such plots we would try to find the plot on which the specified T and x_a lie exactly on the liquid composition curve. Then we would read the pressure at which that plot was made, and the y_a at that temperature. In this example T and x_a were chosen to make $P = 1.00$ atm, to match reported experimental value! So in Figure 8.17 we see that we can enter the plot at $z_a = x_a = 0.1238$, read vertically to the liquid line, finding $T = 85.3$°C, and then read horizontally to the vapor line, finding $y_a = 0.474$. But if

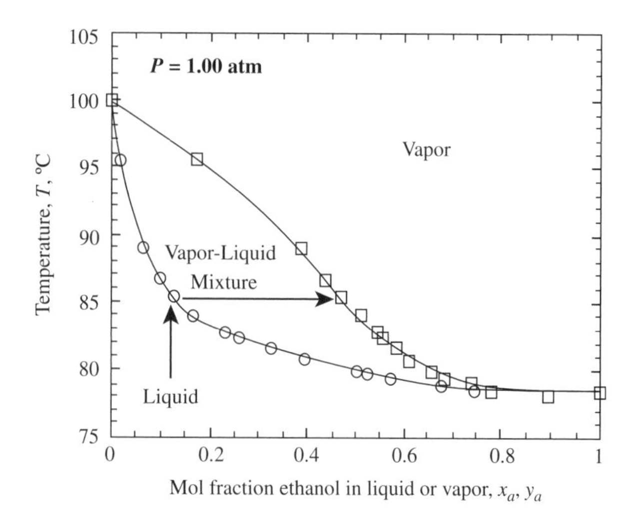

FIGURE 8.17 Temperature–composition diagram for ethanol–water at 1 atm, using data points from [1]. The curves are simple smooth interpolations. The arrows show the graphical solution for the bubble point (temperature-specified) and vapor composition.

the problem statement had asked for the same $z_a = x_a = 0.1238$ but $T = 100$°C instead of 85.3°C, then the pressure we found in the numerical calculation would have been greater than 1.00 atm, and we could not solve the problem on this 1.00-atm graph.

*8.8.1.2 **Pressure-Specified Bubble Point*** If the pressure is specified instead of the temperature, we will have the same equations as above, but will not be able to calculate the vapor pressures in advance. Instead, we must assume a temperature and compute the corresponding pure species vapor pressures, varying the assumed temperature until the computed vapor mol fractions sum to 1.00.

Example 8.10 Estimate the boiling temperature and the y_is in equilibrium with a liquid that is 0.2608 mol fraction ethanol, balance water, at $P = 1.00$ atm.

The equation set to be solved is the same as in Table 8.E, with the $T = 85.3$°C replaced by $P = 1.00$ atm and $z_a = x_a = 0.1238$ replaced by $z_a = x_a = 0.2608$. The solution is shown in Table 8.G. We find $T = 82.0$°C, $y_a = 0.568$. The experimental values are 82.3°C and 0.558 [1]. ■

This problem is inherently a trial and error problem, because the Antoine equations are inside the trial-and-error loop and are transcendental; more than one transcendental equation cannot be simultaneously solved analytically. The hand trial-and-error solution (for a benzene-toluene solution) is shown as Example 3.6.

This type of bubble–point problem is easily solved graphically on a figure like Figure 8.17, which is for the specified pressure. We enter at the bottom at $z_a = x_a = 0.2608$, and read upward to the liquid line, finding a temperature of 82.3°C. We

Table 8.G Trial-and-Error (spreadsheet) Solution to Example 8.10

Variable	Type	Initial Guess	Solution
P(atm)	Given	1.0000	1.0000
x_a	Given $= z_a$	0.2608	0.2608
x_b	Given $= 1 - z_a$	0.7392	0.7392
γ_a	Intermediate	1.8859	1.8859
γ_b	Intermediate	1.1506	1.1506
T(°C)	**Trial variable**	**80.0000**	**82.0400**
p_a (atm)	Intermediate-trial	1.0678	1.1558
p_b (atm)	Intermediate-trial	0.4674	0.5074
y_a	Result	0.5252	0.5684
y_b	Result	0.3976	0.4316
$y_a + y_b$	**Check value**	0.9228	**1.0000**

Table 8.H Trial-and-Error (spreadsheet) Solution to Example 8.11

Variable	Type	Initial Guess	Solution
T (°C)	Given	80.7000	80.7000
y_a	Given $= z_a$	0.6122	0.6122
y_b	Given $= 1 - z_a$	0.3878	0.3878
p_a (atm)	Intermediate	1.0973	1.0973
p_b (atm)	Intermediate	0.4809	0.4809
P (atm)	**Trial variable**	**0.8000**	**0.9965**
x_a	Result	0.6000	0.3754
x_b	Result	0.5000	0.6246
γ_a	Result-intermediate	1.1867	1.4809
γ_b	Result-intermediate	1.5495	1.2866
$x_a + x_b$	**Check Value**	1.1000	**1.0000**

then read horizontally from that point to the vapor curve, finding $y_a = 0.558$.

Summarizing bubble-point calculations we see that by hand they are easy if T is specified, and harder if P is specified. Both are easy in computers. We will see below that dew points are more difficult, because the nonideality is in the phase we are looking for, not in the specified phase.

8.8.2 Dew-Point Calculations

Dew-point calculations are the mirror image of bubble-point calculations. Referring to Figure 8.16 we see that if $F \approx V$ and $L \approx 0$, then according to Eq. 8.8, $y_i = z_a$ for all i, because all but an infinitesimal amount of the material in the feed goes to the vapor. Dew-points need not be thought of as flashes; we can think of them as follows: "given the composition of the vapor phase and either P or T, find the composition of the equilibrium liquid phase and T or P." But the process-design programs normally include them in the flash module, and they have parallels to the other kinds of flashes in Table 8.5, so we include them here.

8.8.2.1 ***Temperature-Specified Dew Point*** Of the two dew-point calculation types, the temperature-specified is the easiest, because we do not need a trial-and-error-procedure on the temperature.

Example 8.11 Estimate the condensing pressure and the x_is in equilibrium with a vapor that is 0.6122 mol fraction ethanol, balance water, at 80.7°C.

The equations to be solved are the same as in Table 8.E, with the $T = 85.3$°C replaced with $T = 80.7$°C, and the $z_a = x_a = 0.1238$ replaced with $z_a = y_a = 0.6122$. The solution is shown in Table 8.H.

Here the trial variable is P; the check variable is the sum of the *liquid* mol fractions = 1.00. (For a bubble point it is the sum of the *vapor* mole fractions = 1.00.) However, the activity coefficients are now a mixture of results and intermediates. We need them to compute the liquid mol fractions, but we need the values of the liquid mol fractions to compute the activity coefficients. In the Initial Guess column the liquid mol fractions are guessed, not calculated. Normally, we would guess x_i values that sum to 1.00, but that makes the check value be 1.00, so the unusual values shown are chosen. In the solution column the liquid-phase activity coefficients are computed using the van Laar equation and the current values of the liquid mol fractions; then those activity coefficients are used to update the estimates of the liquid mol fractions. This leads to "circular reference" problems, which can be managed iteratively in most spreadsheets. The calculated values are $P = 0.997$ atm and $x_a = 0.375$. The experimental values [1] are 1.000 atm and 0.397. We see that our calculated pressure is quite accurate, but the calculated composition is a good but not excellent approximation of the experimental value. ■

This problem is tedious by hand, but if we must do it, we solve for the vapor pressures, then guess a pressure and a set of activity coefficients. From those guesses we compute the liquid mol fractions. Using these, we revise the guess of the activity coefficients and then again the pressures until the check value = 1.00. For ideal solutions the problem is much easier, because the activity coefficients are 1.00, and we have a simple trial and error on P.

As with the temperatare-specified bubble point, to solve this equation set graphically we would look for the constant-pressure plot on which the specified T and y_a met exactly on the vapor line. Figure 8.18 is the same plot as Figure 8.17, showing that the specified y_a and T do meet on the vapor line, and reading horizontally (i.e., at the same temperature) we find $x_a = 0.397$.

8.8.2.2 ***Pressure-Specified Dew Point*** This is the more difficult dew point because both the temperature and the liquid mol fractions (and hence liquid-phase activity coefficients) are unknown.

Table 8.I Trial-and-Error (spreadsheet) Solution to Example 8.12

Variable	Type	Initial Guess	Solution
P (atm)	Given	1.0000	1.0000
y_a	Given $= z_a$	0.1700	0.1700
y_b	Given $= 1 - z_a$	0.8300	0.8300
T (°C)	**Trial variable**	**80.0000**	**95.3500**
p_a (atm)	Intermediate-trial	1.0678	1.8897
p_b (atm)	Intermediate-trial	0.4674	0.8450
x_a	Result	0.1327	0.0187
x_b	Result	1.4797	0.9813
γ_a	Result-intermediate	1.2000	4.8108
γ_b	Result-intermediate	1.2000	1.0010
$x_a + x_b$	**Check value**	1.6123	**1.0000**

Example 8.12 Estimate the equilibrium T and the liquid mol fractions that are in equilibrium with a vapor mixture of 0.1700 mol fraction ethanol, balance water, at 1.00 atm pressure.

The set of equations to be solved is the same as in Table 8.E, with $T = 85.3°C$ replaced by $P = 1.00$ atm and $z_a = x_a = 0.1238$ replaced with $z_a = y_a = 0.1700$. The solution is shown in Table 8.I

The procedure, on a spreadsheet is similar to that in Example 8.11, but here the trial variable is T, and the vapor pressures are inside the trial-and-error loop, along with the liquid mol fractions, as are the liquid-phase activity coefficients that depend on the liquid mol fractions. The solution is $T = 95.3°C$, $x_a = 0.0187$. The experimental values [1] are 95°C and 0.0190. ■

The graphical solution of this problem can be carried out in Figure 8.18, by entering from above at $y_a = 0.1700$.

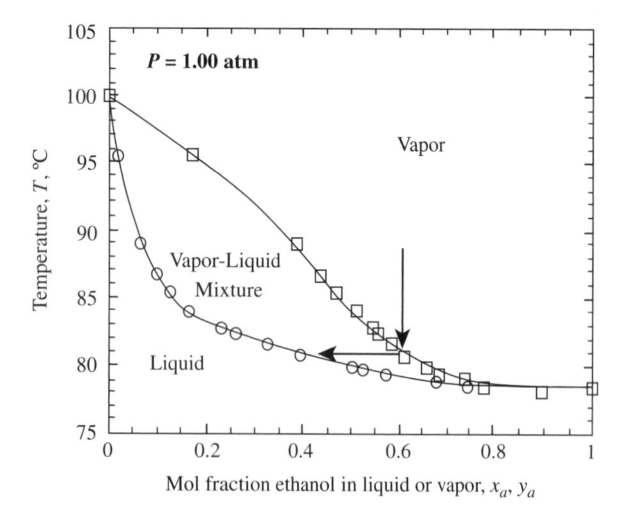

FIGURE 8.18 Temperature-composition diagram for ethanol–water at 1 atm, using data points from Seader et al. [1]. The curves are simple smooth interpolations. The arrows show the graphical solution for the dew-point temperature and liquid composition.

reading down to the vapor curve, finding $T = 95.5°C$, and then reading horizontally (isothermally) to the liquid line, finding $x_a = 0.0190$.

8.8.3 Isothermal Flashes (T- and P-Specified Flashes)

Returning to Figure 8.16, we now consider the case in which T and P at equilibrium are both specified. This is commonly called an *isothermal flash,* although a much better name would be a *T- and P-specified flash.* In the bubble-point calculations, $V \approx 0.00$, and in the dew-point calculations $L \approx 0.00$. In T- and P-specified flashes both L and V have nonzero values. This means that we have gained two variables, but we have also gained two equations, Eqs. 8.7 and 8.8. (It might appear that we have two Eq. 8.8s, but one of them is derivable from Eq. 8.7 and the other from Eq. 8.8, so only one of the two, Eq. 8.9 below, is independent and useful.)

$$\frac{y_a}{x_a} = K_a; \quad \frac{y_b}{x_b} = K_b$$

$$z_a = \frac{V}{F}y_a + \left(1 - \frac{V}{F}\right)x_a = \frac{V}{F}y_a + \left(1 - \frac{V}{F}\right)\frac{y_a}{K_a}$$

$$y_a = \frac{z_a}{\dfrac{V}{F} + \dfrac{1}{K_a}\left(1 - \dfrac{V}{F}\right)}; \quad y_b = \frac{z_b}{\dfrac{V}{F} + \dfrac{1}{K_b}\left(1 - \dfrac{V}{F}\right)} \tag{8.9}$$

Example 8.13 An ethanol–water mixture with $z_a = 0.126$ is brought to equilibrium (flashed) to a pressure of 1 atm and a temperature of 91.8°C. Estimate the vapor fraction, and the mol fractions in both phases.

The applicable equation set is shown in Table 8.J. The trial variable is V/F the check variable is $\sum x_i - \sum y_i = 0.00$. For

Table 8.J Equations to Be Solved in Example 8.13

$$\sum x_i - \sum y_i = 0.00$$

$$y_a = \frac{z_a}{\dfrac{V}{F} + \dfrac{1}{K_a}\left(1 - \dfrac{V}{F}\right)}; \quad y_b = \frac{z_b}{\dfrac{V}{F} + \dfrac{1}{K_b}\left(1 - \dfrac{V}{F}\right)}$$

$$\frac{y_a}{x_a} = K_a = \frac{\gamma_a p_a}{P}; \quad \frac{y_b}{x_b} = K_b = \frac{\gamma_b p_b}{P}$$

$$\gamma_a = 10\text{E}\left[\frac{B^2 A x_b^2}{(Ax_a + Bx_b)^2}\right] = 10\text{E}\left[\frac{(0.4104)^2 \cdot 0.7292 x_b^2}{(0.7292 x_a + 0.4104 x_b)^2}\right]$$

$$\gamma_b = 10\text{E}\left[\frac{A^2 B x_a^2}{(Ax_a + Bx_b)^2}\right] = 10\text{E}\left[\frac{(0.7292)^2 \cdot 0.4104 x_a^2}{(0.7292 x_a + 0.4014 x_b)^2}\right]$$

$$p_a = 10\text{E}[8.04494 - 1554.3/(222.65 + T)]/760$$

$$p_b = 10\text{E}[7.96681 - 1668.21/(228.0 + T)]/760$$

$$z_a = 0.126$$

$$T = 91.8°\text{C}$$

$$P = 1.00 \text{ atm}$$

Table 8.K Trial-and-Error (spreadsheet) Solution to Example 8.13

Variable	Type	Initial Guess	Solution
T(°C)	Given	91.8000	91.8000
P (atm)	Given	1.0000	1.0000
z_a	Given	0.1260	0.1260
z_b	Given = 1 z_a	0.8740	0.8740
p_a (atm)	Intermediate	1.6642	1.6642
p_b (atm)	Intermediate	0.7406	0.7406
V/F	**Trial variable**	**0.5000**	**0.3495**
$y_a/x_a = K_a$	Intermediate-result	1.9971	7.1308
$y_b/x_b = K_b$	Intermediate-result	0.8887	0.7439
y_a	Result	0.1679	0.2859
y_b	Result	0.8225	0.7141
x_a	Intermediate-result	0.0841	0.0401
x_b	Intermediate-result	0.9255	0.9599
γ_a	Intermediate-result	1.2000	4.2848
γ_b	Intermediate-result	1.2000	1.0045
$y_a + y_b$	Intermediate-result	0.9904	1.0000
$x_a + x_b$	Intermediate-result	1.0096	1.0000
$\Sigma x_i - \Sigma y_i$	**Check value**	0.0192	**0.0000**

any assumed value of V/F, we can solve for all the values shown in Table 8.J.

The calculated values are shown in Table 8.K. In the initial guess column we not only guess a value of ***V/F*(0.5000)** but also the activity coefficients (both 1.2000). In the solution column the program computes the values of all three of these, needed to make $\sum x_i - \sum y_i = 0.0000$. Table 8.L compares the values calculated here with those read from the chart in [14] (chart reading accuracy only!, see Problem 8.45). We see good, but not excellent agreement. However, remember that the values in the rightmost column in Table 8.L are read from a chart, which involves both chart-reading errors and any errors made in constructing the chart from the experimental data. The check value is not very sensitive to changes in the trial variable; we may think we are close enough to the solution, and be mistaken. ■

This problem can be solved graphically if we have a T-x_a diagram for the specified pressure. Figure 8.19 shows such a diagram with horizontal and vertical lines corresponding to the specified T and z_a. We see that they cross in the two-phase region, as they must for a T- and P-specified flash if both liquid and vapor are present. The vapor and liquid compositions in equilibrium (at 1.00 atm and 91.8°C) can be read

Table 8.L Comparison of Example 8.13 and Published Values

Variable	Value Calculated in this Example	Value/Read from the Chart in [14] (see Problem 8.45)
V/F	0.3495	0.326
y_a	0.2859	0.2977
x_a	0.0401	0.0416

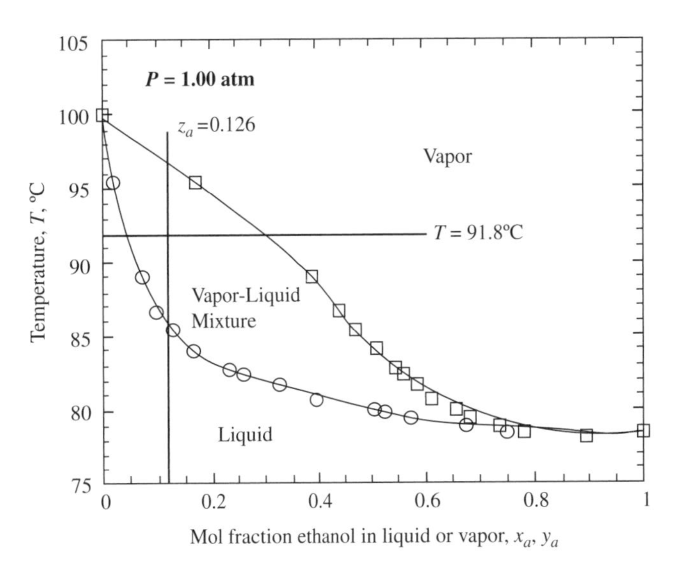

FIGURE 8.19 Temperature–composition diagram for ethanol–water at 1 atm, using data points from Seader et al. [1]. The curves are simple smooth interpolations. The horizontal and vertical lines show that the specified z_a and T correspond to a vapor–liquid mixture.

directly from the figure. Using those and the specified feed composition, we can compute V/F from Eq. 8.8.

8.8.4 Adiabatic Flashes

A more complex set of VLE problems involve *adiabatic flashes*. In these, some liquid, vapor, or vapor–liquid mixture is reduced in pressure at constant enthalpy, as, for example, through an insulated throttling valve, into an insulated pressure vessel. These problems involve all the information in Example 8.13, plus an energy balance on the system. In them we normally guess a final temperature, solve for V/F and the vapor and liquid mol fractions exactly as in Example 8.13, and then use the energy balance to ask if the outlet energy equals the inlet energy. If not, a new guess is made of the outlet temperature, and the process repeated until the unique T is found for which the energy balance is satisfied. By hand this is a giant pain, but our computers do it fairly quickly and easily.

If we must perform an S or V specified flash, the procedure is the same as for an adiabatic (H specified) flash. We guess a final T and P, solve for V/F and then the mol fractions, and compare the computed S or V with the specification. If the specification is not met, we adjust the guessed T and P and repeat the calculation until agreement is reached. This is seldom done by hand, but is offered as an option in some process-design computer programs.

The computer methods shown in these five examples can all be easily extended to any number of species; we simply add variables and equations. There is no comparably easy way to solve multispecies VLE graphically, because the compositions are not easily represented in two dimensions. Thus, the graphical solutions shown here are of little current industrial use, but are included because of their high intuitive content.

8.9 TRADITIONAL *K*-FACTOR METHODS

Before we had computers, flash calculations were done by hand. The most widely used methods, for mixtures of hydrocarbons and occasionally for other mixtures, were based on K factors, defined in Eq. 8.1. These are mostly of historical interest, but also are of some use for simple, noncomputer, multispecies VLE. In all the examples in Section 8.8, the ratio $K_i = y_i/x_i$ was assumed to depend on P, T, and the liquid composition. If the liquid is an ideal solution, K depends on P and T, but not on liquid composition. For light hydrocarbon systems over a range of industrial conditions the liquid phases are close to ideal solutions or have activity coefficients that do not depend strongly on composition, so that it is possible to correlate K_i as a function of P and T alone. The most widely used correlation, by DePreister [15], consists of two charts, one of which is reproduced as Figure 8.20. DePreister made very clear that this chart was to be used for making *initial VLE estimates*, to be used as starting points for the more complex trial-and-error method developed in his paper, which does take into account the effect of composition on K values. Others have used Figure 8.20 widely because it is simple and its predictions are in fair agreement with experiments.

Example 8.14 Estimate the bubble-point temperature at 200 psia and the dew-point composition of a liquid mixture of 1.29 mol% methane, 26.50 mol% ethane and 72.21 mol% propane, using Figure 8.20.

The equations to be solved are shown above in Table 8.M. We begin by guessing that $T = 30°F$. We draw a straight line across Figure 8.20 from 200 psia to 30°F, and read the three K values as 8.7, 1.43 and 0.400. Then the calculated y_a is $0.0129 \times 8.7 = 0.1118$. The other two calculated ys are 0.3790 and 0.2888. These sum to 0.7796, which is less than the required 1.00, indicating that a higher T is needed. Guessing $T = 60°F$ leads to a sum of 1.0713, but for $T = 50°F$ the Ks (as best we can read them), are 9.5, 1.78 and 0.54, leading to ys of 0.1221, 0.4718 and 0.3899 and a sum of 0.9838, as close to 1.00 as we are justified in seeking, given the two-figure accuracy with which we can read Figure 8.20.

The corresponding experimental values [17] are 50°F and ys of 0.1279, 0.4691 and 0.3979. The remarkably good performance of Figure 8.20 in this example (and in Problem 8.52) explains its 60-year-long popularity. ■

Although this was devised as a hand method of calculation, the trial-and-error solution is faster and easier using a spreadsheet. We read the values from Figure 8.20 by hand, and enter them into a spreadsheet trial-and-error solution. Problems 8.53 and 8.54 show the application of Figure 8.20 to estimating dew points.

Plots like Figure 8.20 are only available for the aliphatic hydrocarbons, because of their practically ideal solution behavior in the liquid phase. No such plots seem to be available for seriously nonideal solutions, like ethanol-water.

8.10 MORE USES FOR RAOULT'S LAW

In this chapter we have mostly used Raoult's law and the modified form with an activity coefficient of the Raoult's law type. Most low-pressure VLE problems can be satisfactorily solved that way. In addition, there are some problems that are not traditional VLE, but for which we can use Raoult's law to good effect.

The four types of behavior in Figures 8.7, 8.8, 8.9, and 8.12 all show that as we approach either edge of the γ_a-x_a plot, the activity coefficient of the species that is becoming practically pure approaches 1.00, and does so asymptotically. This is true for even strongly nonideal solutions like sulfuric acid–water (Figure 8.10). Logically, this must be so, because as we approach practically pure anything, the molecules of that species encounter almost entirely molecules of their own kind, with which they have ideal-solution interactions. Thus, even though the *solute* may have very nonideal behavior, in any dilute solution the behavior of the *solvent* is practically ideal solution and can be described with good accuracy by Raoult's law. This does not mean that Raoult's law applies to the solute, but often we are satisfied lo know the behavior of the practically pure solvent. The following two examples illustrate this.

8.10.1 Nonvolatile Solutes, Boiling-Point Elevation

A logical extension of type I behavior (Figure 8.7) assumes that we dissolve some material with a very high boiling point in a solvent, for example, sugar in water. At temperatures near the boiling point of water the vapor pressure of pure sugar is ≈ 0. As long as the solution is dilute, the behavior of the water is practically that predicted by Raoult's law. We may represent this as being practically the same as Figure 8.7a, with the vapor pressure line for the solute being replaced by a horizontal line $x_b p_b, = 0$, and the total pressure being the same as the partial pressure of the water.

Example 8.15 One mol of sugar (sucrose, $C_{12}H_{22}O_{11}$, $M = 342.3$ g/mol whose vapor pressure $p_{\text{surcrose}} \approx 0$) is dissolved in 1000 g (1000/18 = 55.6 mol) of water. What is the vapor pressure of this solution at 100°C = 212°F? At what temperature will this solution boil at 1 atm?

The mol fraction of water in the solution is 55.6/(55.6 + 1) = 0.982, and at 100°C, $p_{\text{water}} = 1.00$ atm, so that

$$P = x_{\text{water}} p_{\text{water}} + x_{\text{sugar}} p_{\text{sugar}}$$
$$= 0.982 \cdot 1.00 \text{ atm} + 0.018 \cdot 0 \text{ atm} = 0.982 \text{ atm} \quad (8.M)$$

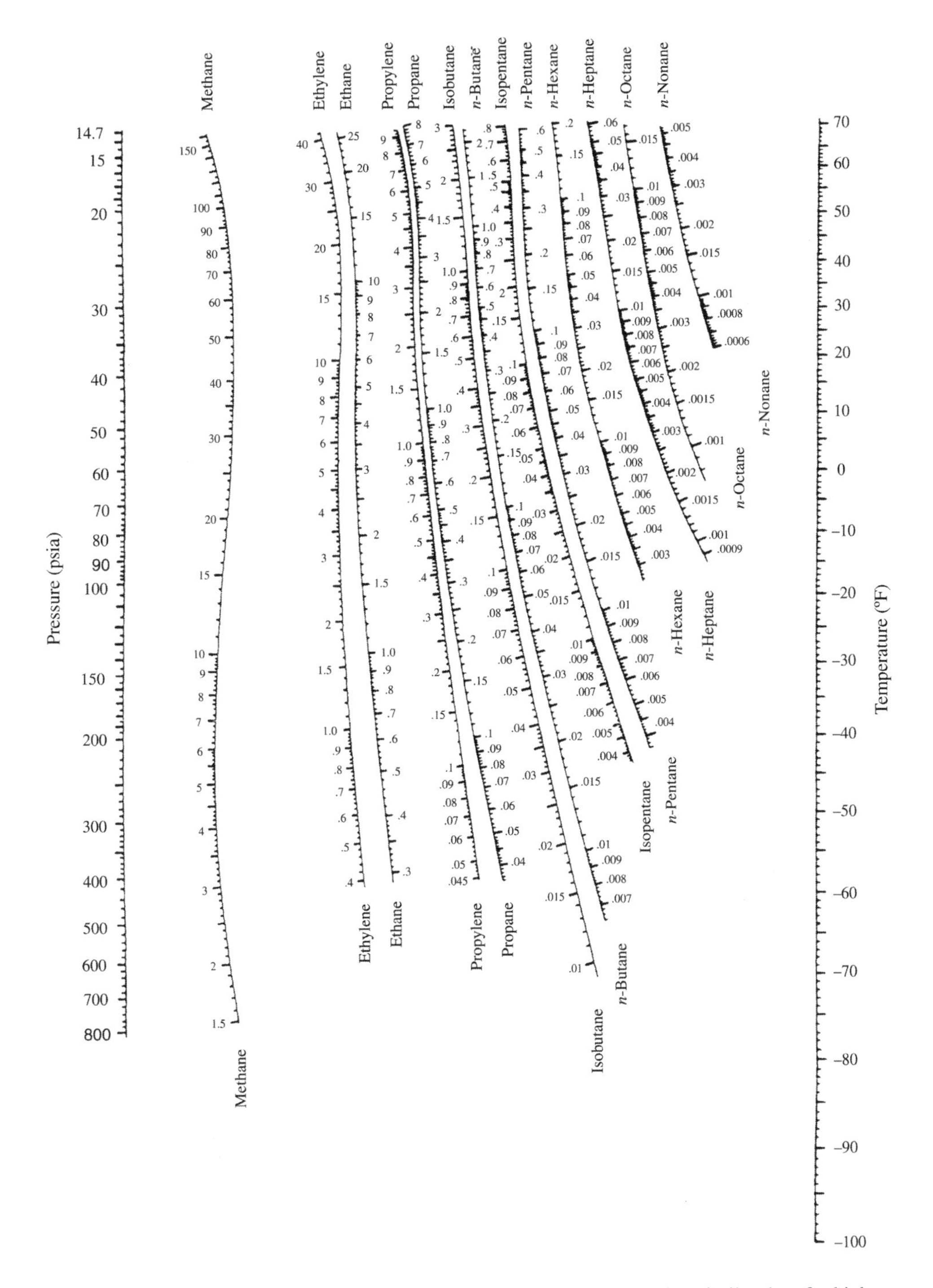

FIGURE 8.20 DePreister's K-factor chart, for low temperatures. There is a similar chart for higher temperatures. An SI equivalent is in [16]. This nomograph is intended only for preliminary estimates, but is widely used because it is simple. (From DePreister, C. L. Light hydrocarbon vapor–liquid distribution coefficients. *Applied Thermodynamics, CEP Symp Ser.* 7–49: 1–43 (1953). Reproduced with permission of the American Institute of Chemical Engineers.)

The situation is as sketched in Figure 8.21. The total pressure for equilibrium is the mol fraction of water in the solution times its pure-solvent vapor pressure.

To find the temperature at which the solution will boil, we see on the figure that we must raise the temperature to increase p_{water} to a value high enough that the total pressure $P = 1.00$ atm, with $x_{\text{water}} = 0.982$.

$$p_{\text{water}} = \frac{P}{x_{\text{water}}} = \frac{1 \text{ atm}}{0.982} = 1.018 \text{ atm} = 14.97 \text{ psia} \quad (8.\text{N})$$

Table 8.M Equations to Be Solved in Example 8.14

$x_a + x_b + x_c = 1.00$; $y_a + y_b + y_c = 1.00$
$\frac{y_a}{x_a} = K_a$; $\frac{y_b}{x_b} = K_b$; $\frac{y_c}{x_c} = K_c$
$z_a = x_a = 0.0129$, $z_b = x_b = 0.2650$, $z_c = x_c = 0.7221$
$P = 200$ psia

Interpolating in the steam table [12] we find $T = 212.92°F = 100.51°C$. We may restate this that the *boiling-point elevation* caused by this dissolved, nonvolatile solute is $0.92°F = 0.51°C$. ■

The amounts of species a and b in this example (1 mol of solute, 1000 g of solvent, which produce a solute concentration of 1.00 molal) were chosen to match the common presentation of boiling point elevation in chemistry textbooks. They normally report the boiling-point elevation coefficient for various solvents. *For water* at one atmosphere the *molal boiling-point elevation constant* $K_b \approx 0.51°C$; it has different values for other solvents. It is common in chemistry books to write that as

$$\begin{pmatrix}\text{boiling-point}\\ \text{elevation}\end{pmatrix} = T_{\substack{\text{boiling}\\ \text{solution}}} - T_{\substack{\text{boiling}\\ \text{pure solvent}}} = K_b \begin{pmatrix}\text{molality}\\ \text{of solute}\end{pmatrix} \tag{8.10}$$

where the molality (see Chapter 6) is the moles of solute per 1000 g of solvent, = 1.00 in this example. This is only approximately correct; the statement in terms of Raoult's law and Example 8.15 is theoretically better. However, for dilute solutions the results are practically identical (see Problem 8.56). As always seems to be the case, nature is more complex than our simple models. Figure 8.22 compares the experimental values of the boiling-point elevation of sucrose [18] in water with the predictions of Raoult's law, and of the simple linear relation of Eq. 8.10.

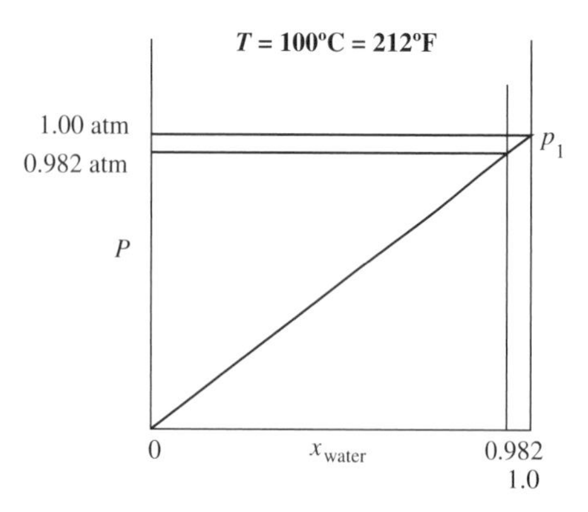

FIGURE 8.21 Vapor pressure of a dilute solution of a zero-vapor pressure solute. (This figure is not to scale; if it were, the 0.982 line would disappear into the rightmost axis.)

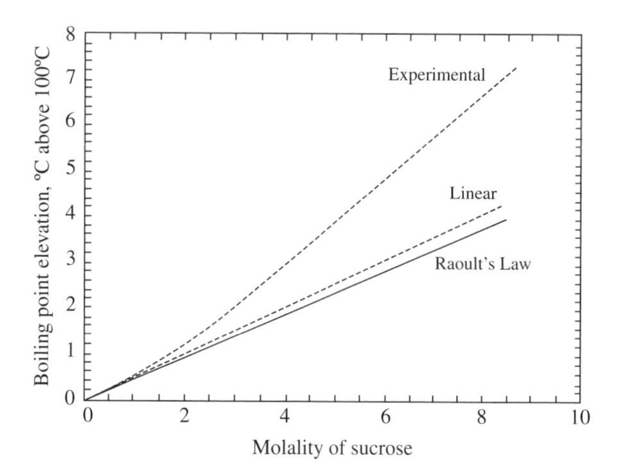

FIGURE 8.22 Comparison of experimental [18] and two computed boiling-point elevations curves for sucrose solutions in water at 1 atm. Below about 1 molal the three curves are nearly identical.

At the low concentrations at which boiling-point elevations are most often observed, the experimental results and the two computed values are indistinguishable. For higher concentrations, the simple linear values and the Raoult's law values are practically the same, while the experimental boiling-point elevations are substantially larger, indicating mild type III behavior, in this case most likely caused by weak association of the water and sugar molecules (see Problem 8.31).

This type of behavior is regularly observed for solutions of other organic materials in water, or for solutions of very low volatility organic solutes in organic solvents. Boiling-point elevation tests are regularly used to estimate the molecular weight of some unknown substance; a known weight of the substance is dissolved in a solvent, the boiling point of the solution is measured, and the solute molality estimated from Eq. 8.10. From that molality and the known weight, the molecular weight is calculable. This type of behavior is not observed for solutions of electrolytes (salts, acids, bases) in water, because they ionize. If we repeat Example 8.15, dissolving 1.00 mol of table salt (NaCl, $M = 58.5$ g/mol), in 55.6 mol (1000 g) of water, we would expect the same boiling-point elevation as for 1 mol of sugar, $0.92°F = 0.51°C$. The experimental value is roughly twice this value. The reason is that electrolytes ionize. If all the NaCl converted to Na^+ and Cl^- ions, then the mol fraction of water in this solution is

$$x_{\text{water}} = \frac{n_{\text{water}}}{n_{\text{water}} + n_{Cl^-} + n_{Na^+}} = \frac{55.6}{55.6 + 1 + 1} = 0.949 \tag{8.O}$$

instead of the $x_{water} = 0.982$ we would have if each mol of NaCl did not produce 2 mols of ions (see Problem 8.62). This ionization is nearly complete for strong acids and bases and their salts (like NaCl), but is only partial for weak acids and bases and their salts (e.g., carbonic or acetic acid), so that we need experimental data or theoretical estimates of the degree of ionization to carry out this kind of calculation for them. The experimental data for NaCl solutions in water (Problem 8.63) are similar to those shown for sucrose in Figure 8.22. For dilute solutions the experimental data, Raoult's law for total ionization and Eq. 8.10 for total ionization are practically identical. For high concentrations the observed boiling-point elevation is greater than is predicted by Raoult's law or Eq 8.10, indicating mild type III behavior.

8.10.2 Freezing-Point Depression

Adding a nonvolatile solute raises the boiling point of a solvent, but it lowers the freezing point. Again, this is all explicable in terms of Raoult's law.

Example 8.16 One mol of sugar (sucrose, $C_{12}H_{22}O_{11}$, $M = 342.3$ g/mol) is dissolved in 1000 g ($1000/18 = 55.6$ mol) of water. What is the freezing-point temperature of this solution at 1 atm pressure?

The sugar concentration in this example and the previous one is about twice that in common soft drinks. Figure 8.23 shows the situation: an ice cube floating in a glass of the liquid. We are used to seeing this with the surroundings warmer than the glass and its contents, so the ice slowly melts and the liquid never gets as cold as the ice – not an equilibrium situation. For equilibrium, the temperature of ice, solution and surroundings must all be the same. In addition, we normally see this situation with the surrounding gas being air at 1 atm, but for this example we consider the surrounding gas to consist only of water vapor at a pressure far below 1 atm.

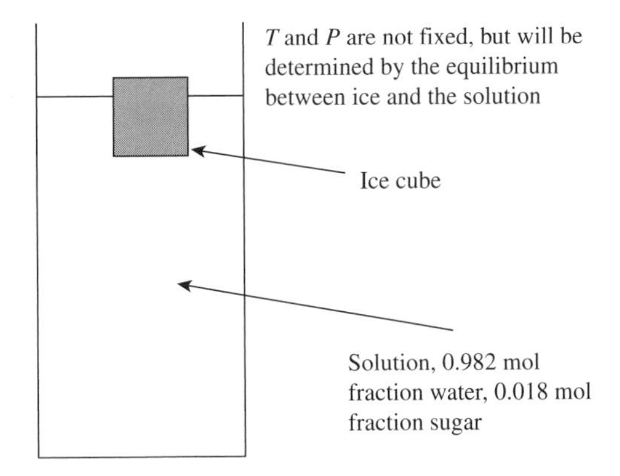

FIGURE 8.23 An ice cube floating in a dilute sugar solution both at the same *T* and *P*.

Next we observe that this is the same sugar solution as in example 8.15, so we can rewrite the pressure form of Raoult's law, Eq. 8.M, as

$$P = x_{water}p_{water} + x_{sugar}p_{sugar} = 0.982 \cdot p_{water} + 0.018 \cdot p_{sugar} \tag{8.P}$$

and because $p_{sugar} \approx 0$, the pressure of the system at equilibrium must be

$$P = 0.982 \cdot p_{water} + 0.018 \cdot 0 = 0.982 \cdot p_{water} \tag{8.Q}$$

where p_{water} is the vapor pressure of pure water at the equilibrium temperature. Because the vapor is also in equilibrium with the solid ice, this must also be the vapor pressure of the ice, so that

$$P = 0.982 \cdot p_{water} = p_{ice}; \quad \frac{p_{ice}}{p_{water}} = 0.982 \tag{8.R}$$

If the sugar dissolved significantly in the ice then the p_{ice} in this equation would have to reflect that fact. But it doesn't, so we can use the vapor pressure of pure ice.

Figure 5.8 shows the vapor pressure of subcooled water and of ice. The values in the table from which that figure was made can be represented by the following totally empirical data-fitting equation

$$\frac{p_{ice}}{p_{water}} \approx 1 + \frac{0.0096686}{°C}T + \frac{4.0176 \times 10^{-5}}{(°C)^2}T^2 \tag{8.S}$$

If we set this equal to 0.982 and solve for T we find $-1.84°C$; the commonly reported value is $-1.86°C$. If we now repeat the calculation assuming that the system shown in Figure 8.23 is open to the air at 1 atm we will see that the amount of air dissolved in the solution and in the ice is small enough to ignore, and that the above value is correct. ■

Almost all the statements made about boiling-point elevation can be repeated for freezing-point depression. It is commonly shown in chemistry books as

$$\begin{pmatrix}\text{freezing-point}\\\text{depression}\end{pmatrix} = T_{\substack{\text{freezing}\\\text{solution}}} - T_{\substack{\text{freezing}\\\text{pure solvent}}}$$

$$= K_f \begin{pmatrix}\text{molality}\\\text{of solute}\end{pmatrix} \tag{8.11}$$

where K_f is the *molal freezing-point depression constant,* which equals – 1.86°C for water and has different values for other solvents. If the solute ionizes, then we must compute the mol fraction taking that into account just as we did for boiling-point elevation. If we plot the experimental data on freezing-point depression for sucrose in the same form as Figure 8.22, we find the same result; at concentrations of 1 molal or less all three curves are substantially identical. For higher concentrations we see the same curvature, indicating again mild type III behavior between water and sucrose.

Before the invention of compression-expansion refrigerators (about 1850) the coldest readily available refrigerant, at $-21.12°C = -6.016°F$, (See Figure 11.14), was the practically equilibrium mixture of ice, solid salt and the saturated solution of salt in water. This mixture was used in laboratories and in hand-crank ice cream makers until the 1940s. It seems counter-intuitive that one can take crushed ice and solid salt pieces, both at 0°C, mix them vigorously and practically adiabatically, and produce a three-phase, slush mixture at −21.12°C. However to get the saturated solution one must supply the heat to melt the ice and the heat to dissolve the salt making the saturated salt-water solution, and this can only be supplied by a decrease in temperature of the unmelted solid ice and undissolved solid salt (both in fairly small pieces). When Daniel Fahrenheit was devising the temperature scale that bears is name, he used a similar water-ice-salt slush using NH_4Cl instead of NaCl as the point for 0°F (and human body temperature for 100°F). Later workers have modified his values slightly.

The use of road salt is based on this effect. When ice forms on a street or sidewalk it adheres well and is dangerously slippery. Salt crystals (typically a few millimeters in size) are scattered on the ice. Each crystal forms a small pool of concentrated salt solution on the ice surface. The crystal and its pool dissolve its way down through the ice to the ice-pavement interface. The pool then spreads laterally, greatly weakening the bond between ice and pavement, allowing the ice to break up and be swept away.

8.10.3 Colligative Properties of Solutions

The boiling-point elevation, together with the freezing-point depression and the change in osmotic pressure caused by dissolving a solute in a solvent are called the *colligative properties of solutions.* We will discuss osmotic pressure change in Chapter 14. In each of colligative properties, the change is proportional to the change in mol fraction of the *solvent*, (solvent behaves almost as an ideal solution), with the assumption the solute is practically inert (has a negligible fugacity for boiling point elevation or freezing point depression, or is totally excluded from passing through an osmotic membrane). The simple Raoult's law estimates of colligative properties are only accurate for very low concentrations of solute, but these properties themselves are often of considerable technical importance in cases where the solute concentrations are significant (e.g., reverse osmosis desalination, or boiling point elevation in multiple-effect evaporators). For these cases the Raoult's law approximations give useful first estimates, but the nonideality of the more concentrated solutions must be taken into account. Study of the colligative properties of solutions was one of the cornerstones of the new field of physical chemistry in the 1890s.

8.11 SUMMARY

1. VLE forms the basis of distillation, one of the most important chemical engineering processes. It is also important for drying, humidification, and some chemical reactions.
2. Experimental measurement of VLE at modest temperatures and pressures is fairly easy. Several thousand binary mixtures have been tested, and the results catalogued. High pressure, high temperature, or mixtures with more than two species make the measurements more expensive and difficult. There are many fewer measurements of this type in the literature than measurements of binary systems at modest temperatures and pressures.
3. In most low-pressure VLE the vapor phase is assumed to be a mixture of ideal gases. For higher pressures its nonideal behavior must be taken into account. High-pressure VLE is discussed in Chapter 10.
4. If the liquid forms an ideal solution (or close to it), then Raoult's law (and/or Henry's law) is used to estimate VLE behavior.
5. If the liquid does not form an ideal solution, then Raoult's law-type activity coefficients are added. About 90% of nonideal liquid mixtures show positive deviation from Raoult's law: activity coefficients greater than 1.00: about 10% show negative deviation: activity coefficients less than 1.00. Large positive deviations and close boiling points lead to minimum-boiling azeotropes. Large negative deviations and close boiling points lead to maximum-boiling azeotropes. If the activity coefficients are large compared to 1.00, the liquid may separate into two liquid phases, leading to a heteroazeotrope (see Chapter 11).
6. With correlations of liquid-phase activity coefficients (Chapter 9) we can make very accurate estimates of the low-pressure VLE of many systems. These estimates normally require an equation for the sum of the mol fractions in each phase and a statement of equality of the fugacities for each species present. The latter may involve subsequent equations for pure species vapor pressures, liquid-phase activity coefficients, and vapor-phase fugacity coefficients. For hand calcula-

tions we often ignore some of these, but computer programs normally evaluate them all.

7. The six basic types of VLE calculations are illustrated in Section 8.8 for one particular two-species system (ethanol–water) at modest pressures. The same calculation methods are used for systems with more species, simply adding variables and equations. If vapor-phase nonideality must be taken into account, the examples in Section 8.8 are easily modified to do that.
8. The colligative properties of solutions—boiling-point elevation, freezing–point depression, and osmotic pressure (Chapter 14)—are all easily understood in terms of Raoult's law.

PROBLEMS

See common units and values for problems and examples. An asterisk (*) on a problem number indicates that the answer is in Appendix H.

8.1* Repeat Example 8.1 for $x_{\text{acetone}} = 0.95$.

8.2* **a.** Repeat Example 8.2 for $x_{\text{acetone}} = 0.10$.
b. Repeat Example 8.3 for $x_{\text{acetone}} = 0.10$.

8.3* At 1 atm (760 torr) pressure, the normal boiling points of benzene and toluene are 80.1 and 110.6°C. Benzene and toluene may be assumed to form ideal liquid solutions. The vapor pressures of benzene and toluene at these temperatures are as follows:

T (°C)	p_{benzene} (torr)	p_{toluene} (torr)
80.1	760.0	292.5
110.6	1783.4	760.0

a. If we have a boiling liquid mixture of 99.99 mol% benzene, balance toluene at 760 torr, what are the approximate values of K_{benzene}, K_{toluene}, and α? What is the mol fraction of toluene in the equilibrium vapor phase?
b. If we have a liquid boiling mixture of 0.01 mol% benzene, balance toluene at 760 torr, what are the approximate values of K_{benzene}, K_{toluene}, and α? What is the mol fraction of benzene in the equilibrium vapor phase?

8.4 In Figures 8.8b, 8.9b and 8.10b there is some liquid-phase mol fraction, near the center, at which the two individual activity coefficient curves cross. At that composition the two activity coefficients are equal.
a. Show that at that liquid-phase mol fraction, the calculated vapor-phase composition is the same as we would calculate from the ideal solution assumption,
b. Show that the equilibrium temperature is not what we would calculate from that assumption.
c. Is the actual boiling temperature greater or less than that we would calculate from the ideal solution assumption for Figure 8.8? For Figure 8.9?
d. Is it a general proposition that for all binary solutions that do not form two liquid phases there is some value of x_a for which y_a is the same as the value for ideal solution behavior? Is it true that a clock that is stopped shows the correct time twice a day?

8.5 **a.** Show the equations for K_i and α that correspond to the assumptions that both vapor and liquid phases are ideal solutions.
b. For the binary mixture benzene–toluene at 1 atm (practically an ideal solution) estimate K_{benzene}, K_{toluene}, and α for liquid mixtures of 1, 50, and 99 mol% benzene. Use the Antoine equation values in Table A.2.

8.6* For the mixture of a and b at 0°C, the vapor pressures of the pure species are $p_a = 200$ torr; $p_b, = 500$ torr. For a 50-mol% a mixture, the liquid-phase activity coefficients are $\gamma_a = 1.2$, $\gamma_b = 1.3$. The vapor phase may be considered an ideal solution.
a. What is the composition of the equilibrium vapor phase?
b. What is the vapor pressure?

8.7 In this text and most others, benzene–toluene is normally given as the example of an ideal solution. No solution is totally ideal, including benzene-toluene. How much does it deviate from ideal solution behavior? Table 8.N shows some of the experimental VLE data at 120°C for benzene and toluene [19]. Using it,
a. Estimate the activity coefficients for benzene and toluene that correspond to each of these data points.
b. Comment on how good the ideal solution assumption is for this data set.

8.8 Benzene and o-xylene are assumed to form an ideal solution with each other. The normal boiling points of benzene and o-xylene are 80.1 and 144.°C. Sketch a T-x diagram for this mixture at 1 atm, showing the

Table 8.N Experimental VLE for Benzene-Toluene at 120°C

Pressure P (psia)	Mol Fraction Benzene in Liquid, x_a	Mol Fraction Benzene in Vapor, y_a
21.9	0.117	0.220
25.6	0.264	0.453
25.7	0.258	0.425
29.3	0.440	0.639
36.1	0.682	0.842

liquid, vapor, and vapor-liquid regions. Include simple numerical values where appropriate.

8.9* From measurements in a low-pressure equilibrium cell, we have the following data on the equilibrium of the binary mixture of a and b:

$$T = 80°\text{C} \quad P = 1 \text{ atm}$$
$$y_a = 0.8 \quad x_a = 0.6$$

At 80°C the vapor pressures of the pure species are $p_a = 1.5$ atm, $p_b = 0.7$ atm. Assuming that at this pressure the gas phase is an ideal solution of ideal gases, calculate the activity coefficients of species a and b in the liquid phase.

8.10 A binary vapor mixture of 50 mol% e, 50 mol% f at 100°F has the following properties:

$$p_e = 2.0 \text{ atm} \quad p_f = 0.4 \text{ atm} \quad \gamma_e = 1.5 \quad \gamma_f = 1.8$$

The vapor is an ideal solution of ideal gases. What is the composition of the equilibrium liquid? What is the vapor pressure of this mixture at 100°F?

8.11* **a.** Estimate the mol fraction of a in the liquid that is in equilibrium with a vapor (assumed an ideal solution of ideal gases) of 50 mol% a and 50 mol% b at 20°C, using the following data, all at 20°C:

$$p_a = 15 \text{ atm} \quad \gamma_a = 1.8 \quad p_b = 7 \text{ atm} \quad \gamma_b = 1.2$$

b. Estimate the equilibrium pressure.

8.12 **a.** Estimate the mol fraction of a in the vapor (assumed an ideal solution of ideal gases) in equilibrium with a liquid solution of 50 mol% a and 50 mol% b at 20°C, using the following data, all at 20°C.

$$p_a = 15 \text{ atm} \quad \gamma_a = 1.2 \quad p_b = 7 \text{ atm} \quad \gamma_b = 1.8$$

b. Estimate the equilibrium pressure.

8.13 At temperature T, a liquid mixture is 40 mol% a, balance b, for which $p_a = 0.8$ atm and $p_b = 1.3$ atm. The vapor in equilibrium with this liquid is an ideal solution of ideal gases. Estimate the mol fraction of a in the vapor, if

a. The liquid is an ideal solution.

b. The liquid is not an ideal solution, but $\gamma_a = 1.4$ and $\gamma_b = 1.2$.

8.14 Repeat Example 8.4 for 30 mols of n-butanol, balance water.

8.15* Repeat Example 8.5 for $y_{\text{water}} = 0.10$.

8.16 For a vapor–liquid equilibrium mixture at of water and n-butanol at 1.00 atm and 110°C (see Figure 8.12d), estimate the composition of the vapor and the liquid in equilibrium.

8.17 A vapor is 40 mol% water, 60 mol% n-butanol, at 130°C and 1 atm absolute pressure. We now cool this vapor at constant pressure.

a. At what temperature will the first drop of liquid appear?

b. What will be the mol% water in this first drop?

c. Problem 8.29 is the same as this problem, with the n-butanol replaced with n-butane. Why are the answers so different?

8.18 Eight mols of water and 2 mols of n-butanol are placed in a container and boiled. The resulting vapor is removed, condensed, and sent to storage.

a. What is the mol fraction of water in the vapor initially leaving the container?

b. What is the mol fraction of water in the vapor leaving the container when the container just runs dry.

c. Sketch a plot of mol fraction water vs. fraction evaporated for this process (as shown in Figure 8.24 showing the right shape of the curve and as many numerical values as you can.

8.19 Show the calculation of the three other activity coefficients in Example 8.6.

8.20 Show the trial and error computation in Example 8.7.

8.21 For the n-butanol–water binary shown in Figure 8.12, sketch a g-x_a diagram, like Figure 6.7. Show no numerical values, only the correct shapes of the various parts of the curve.

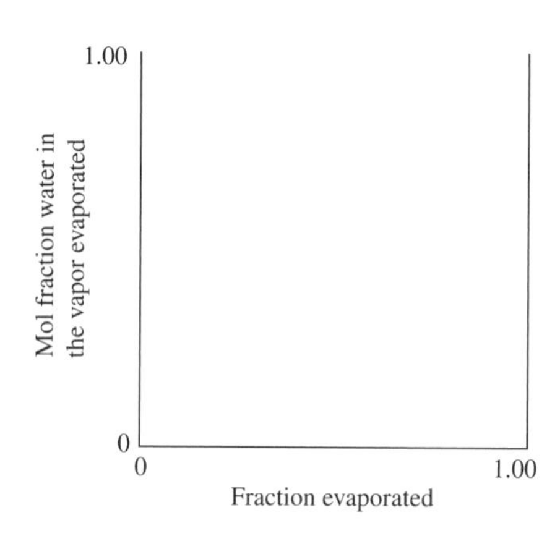

FIGURE 8.24 Coordinates for Problem 8.18 (c).

8.22 Figure 8.7d shows the behavior when a liquid mixture of 50% benzene, balance toluene, is heated until it is all vaporized. Prepare the equivalent of that plot for water-*n*-butanol, sketching it on a copy of Figure 8.12d. The liquid mixture originally placed in the container was

a. 3 mols of butanol and 7 mols of water.

b. 5 mols of butanol and 5 mols of water

c. 1 mols of butanol and 9 mols of water.

8.23 A liquid mixture of 3 mols of water and 7 mols of *n*-butanol is placed in a constant pressure (1 atm) container and heated.

a. At what temperature does the liquid begin to boil?

b. At what temperature does the last liquid disappear, leaving only vapor?

8.24 **a.** On one plot of ln γ_i vs. x_a sketch in the values from Figures 8.7, 8.8, and 8.9, all to the same scale.

b. Comment on the relation of the numerical values of the activity coefficient and the existence of azeotropes and heteroazeotropes.

c. Is there an analog of the heteroazeotrope for very small values of γ_i (very large negative deviations from ideal solution behavior)? If so, what is it?

8.25 Sketch the equivalent of Figure 8.7 (all four parts) for a binary system consisting of two chemicals, *a* and *b*, which form an ideal solution with each other,

a. If they have identical NBPs, both 0°C.

b. The NBP's are *a:* 0°C, *b*: 300°C. Draw the P-x_a diagram for 0°C, all the other parts for 1 atm.

8.26 Same as Problem 8.25, except that the system of interest is nitrogen–water. Cover only the nitrogen mol fraction range from 0 to 0.01. Make the P-x diagram for 20°C for which the Henry's law constants are shown in Table A.3. Make the T-x diagram for 14.7 psia. Here take x_a as the mol fraction of nitrogen, so that pure water is at the left. (See Figure 8.9.)

8.27 Same as Problem 8.25, except that the system of interest is sodium chloride–water. For the purpose of this problem assume that the vapor pressure of pure sodium chloride is zero at temperatures near room temperature and that the NaCl is 100% ionized. Omit any solid phases. Assume that both $\gamma_i = 1.00$. Draw the P-x_a diagram for 212°F and the T-x_a diagram for 14.7 psia. (Sodium chloride melts at 801°C and boils at 1413°C.)

8.28 Same as Problem 8.25, for the system water–mercury. Assume that the solubilities of water in mercury and mercury in water are exactly zero. Show the resulting plots. The NBP of mercury is 357°C. The vapor pressure of mercury at 100°C is 0.28 torr. Be prepared to discuss how different the plots would look if we took the minuscule solubilities of water in mercury and mercury in water into account.

8.29 A vapor is 40 mol% water, 60 mol% *n*-butane, at 130°C and 1 atm absolute pressure. We now cool this vapor at constant pressure. From Chapter 11 we know that liquid water and liquid *n*-butane are practically insoluble in each other (much less than 1 mol% solubility) and can be considered for this problem to be completely insoluble.

a. At what temperature will the first drop of liquid appear?

b. What will be the mol% water in this first drop?

8.30* The Henry's law constant for ethane in water at 0°C is 1.26×10^4 atm (Table A.3).

a. If the partial pressure of ethane over water is 1 atm, what is the mol fraction ethane in the water?

b. To what maximum pressure should this "law" be used for ethane and water?

8.31 Table 3.A shows the calculated mol fractions for the air–water equilibrium.

a. What are the values of K_i for all three species in that example?

b. What is the value of α between oxygen and water? (It is not common to use K_i and α notation for Henry's law problems, but it is perfectly logical to do so.)

8.32 The ethane-water system forms two immiscible liquid phases over most of the composition range. At 0°C in the water-rich liquid (i.e., the liquid that is almost pure water), if we take the standard-state fugacities to be Raoult's law type (0.09 psia for water and 23 atm for ethane), then the activity coefficients are practically independent of composition and are equal to 1.0 for water and 550 for ethane. Based on these data, estimate the mol fraction of ethane in the water-rich phase for pressures high enough that two liquid phases are present.

8.33* A vessel contains liquid water, and a gas which is mostly N_2, with some contained water vapor, all equilibrium at 20°C and 10 atm (absolute). At this temperature the vapor pressure of water is 0.023 atm. What is the mol fraction of water in the vapor phase, if:

a. The mol fraction of N_2 in the liquid is negligible, and the poynting factor (Eq. 7.N) = 1.00?

b. Same as (a) but we take into account the solubility of N_2 in the water?

c. Same as (b) but we take into account the effect of the poynting factor on the vapor pressure of water?

8.34* A liquid with 50 mol% a and 50 mol% b is fed to a continuous distillation column. The column is a very high efficiency column, which makes a very good separation. What are the approximate compositions of the overhead and bottoms products

a. If a and b are benzene–toluene?

b. If a and b are isopropanol–water?

c. If a and b are acetone–chloroform?

d. If a and b are water–n-butanol?

8.35 For the acetone–water system, shown in the first few examples of this chapter, the activity coefficients are >1.00. If we feed an acetone–water mixture to the distillation column in the previous problem, do the activity coefficients >1.00 help us

a. Get high-purity acetone out the top of the column?

b. Get high-purity water out the bottom of the column?

8.36 Estimate the composition of the vapor in equilibrium with a 50 mol% liquid solution of a and b and the equilibrium pressure from the following data at 20°C:

$$p_a - 15\text{ atm} \quad p_b = 7\text{ atm} \quad \gamma_a = 1.2 \quad \gamma_b = 1.8$$

The Lewis–Randall fugacity rule may be assumed to hold for the gas phase, with

$$\phi_a = 0.8 \quad \phi_b = 0.6$$

8.37 A binary liquid solution of 50 mol% a, 50 mol% b at 100°F has the following properties:

$$p_a = 20\text{ atm} \quad p_b = 0.4\text{ atm} \quad \gamma_a = 1.5 \quad \gamma_b = 1.8$$

What is the minimum reversible power needed to separate this solution into pure species in an isothermal steady-flow device? Express your answer in Btu/Ibmol. $T = T_o = 100$°F.

8.38 A new type of steady-flow electrochemical cell has been proposed that will generate electricity by mixing liquid a and b. The cell will operate at 100°F, in contact with a heat reservoir at 100°F. At this temperature, the vapor pressures of a and b are 5 and 8 atm, respectively. The mixing ratio will be 1 mol to 1 mol, and in the final mixture the activity coefficients are

$$\gamma_a = 0.5 \quad \gamma_b = 0.25$$

What is the maximum amount of electrical work that this cell can generate per lbmol of a?

8.39 Sketch a figure like Figure 8.8b for a substance that obeys the van Laar equation (Eq. 8.6):

a. For the case in which $A = B = 0.0$.

b. For the case in which $A = B = 1.0$.

c. For the case in which $A = B = -1.00$.

d. For the case in which $A = 0.5$, $B = 1.0$.

8.40 The van Laar equation (Eq. 8.6) is most often seen with $\log \gamma_i$, (see Table A.7). But it can also be shown with $\ln \gamma_i$, which is the most commonly seen form for most other activity coefficient equations. Using Eq. 8.6, show the values that replace A and B to change the $\log \gamma_i$ form to the $\ln \gamma_i$ form.

8.41 Set up Example 8.9 on a spreadsheet.

a. Show that the answers in that example are correctly calculated.

b. Repeat the calculation for ethanol-water, with $x_a = 0.0966$ and $T = 86.7$°C. The experimental values are $P = 1.00$ atm, $y_a = 0.4375$.

c. Repeat the calculation for benzene and toluene, which may be assumed to form an ideal solution, with $x_a = 0.5$, $T = 80$°C.

8.42 Set up Example 8.10 on a spreadsheet.

a. Show that the answers in that example are correctly calculated.

b. Repeat the calculation for ethanol–water, with $x_a = 0.3273$ and $P = 1.00$ atm. The experimental values are $T = 81.5$°C, $y_a = 0.5826$.

c. Repeat the calculation for benzene and toluene, which may be assumed to form an ideal solution, with $x_a = 0.3$, $P = 1$ atm.

8.43 Set up Example 8.11 on a spreadsheet.

a. Show that the answers in that example are correctly calculated.

b. Repeat the calculation for ethanol–water, with $y_a = 0.6564$ and $T = 79.8$°C. The experimental values are $P = 1.00$ atm, $x_a = 0.5079$.

c. Repeat the calculation for benzene and toluene, which may be assumed to form an ideal solution, with $y_a = 0.5$, $T = 70$°C.

8.44 Set up Example 8.12 on a spreadsheet.

a. Show that the answers in that example are correctly calculated.

b. Repeat the calculation for ethanol–water, with $y_a = 0.3891$ and $P = 1.00$ atm. The experimental values are $T = 89.0$°C, $x_a = 0.0721$.

c. Repeat the calculation for benzene and toluene, which may be assumed to form an ideal solution, with $y_a = 0.3$, $P = 1$ atm.

8.45 Set up Example 8.13 on a spreadsheet.

a. Show that the answers in that example are correctly calculated.

b. Repeat the calculation for ethanol–water with $z_a = 0.228$, $T = 87.3°C$, and $P = 1.00$ atm. The values of *V/F*, x_a and y_a (based on chart reading in [14]) are 0.405, 0.089, and 0.432.

c. Repeat the calculation for benzene and toluene, which may be assumed to form an ideal solution, with $z_a = 0.3$, $P = 2$ atm, and $T = 100°C$.

8.46 Show that instead of using $\sum x_i - \sum y_i = 0$ as the equilibrium criterion in Example 8.13, one could use $\sum x_i = 1.00$ or $\sum y_i = 1.00$. Suggest why this might be computationally less satisfactory.

8.47* At 1 atm pressure, the ethanol–water azeotrope has a composition of 10.57 mol% water, at a temperature of 78.15°C [9]. The liquid-phase activity coefficients are computed in Example 7.4. Calculate the corresponding values of the liquid-phase activity coefficients from Eq. 8.6. Are they the same? similar? The constants with Eq. 8.6 are chosen to give the best average representation of the VLE data over the whole data range; the answers to Example 7.4 are best for the azeotrope.

8.48* For a binary mixture of a and b at temperature T and pressure P, the liquid phase activity coefficients (Raoult's law type) are given by

$$\ln \gamma_a = Cx_b^2 \quad \ln \gamma_b = Cx_a^2 \tag{8.T}$$

where C is an experimental constant. (This is the *symmetrical equation*, see Chapter 9.) The pure species Gibbs energies per mole of a and b are identical: $g_a^o = g_b^o$. If C is a small number (e.g., 10^{-6}), then a and b will form one miscible liquid solution. If it is a large number (e.g., 10^6), then they will form an immiscible liquid pair (as, for example, mercury–water). What is the maximum value of C for which they form one liquid? (This is the same as the minimum value of C for which they form two phases; for that value of C, they are just beginning to form two phases.)

8.49 Liquid mixtures of a and b obey Eq. 8.T, with $C = 0.5$. Does this mixture form an azeotrope at a constant temperature of T, for

a. $p_a/p_b = 1.0$.

b. $p_a/p_b = 10$.

c. What is the value of p_a/p_b for which it forms an azeotrope with $x_a = 0.9900$?

d. What is the value of C for which a system with $p_a/p_b = 10$ forms an azeotrope with $x_a = 0.9900$?

8.50 The values in Table 8.L were calculated from values read from Figure 330 of [14], which has values shown in °F and weight fractions. The actual readings from the chart were $T = 197.2°F$, *L/F* by weight = 0.600, z_a, y_a, and x_a, all by weight, not mol, 0.27, 0.52, 0.10. Show the conversion of these values to the values shown in Table 8.L.

8.51 **a.** Estimate the K value for n-pentane at 100 psi and 0°F from Figure 8.20 (see Example 8.14).

b. Estimate the same K value from Raoult's law using the values in Table A.2. Compare the value to that in part (a).

c. Repeat part (b) using the L-R rule for the vapor. (You must extrapolate to a hypothetical standard state, as shown in Figure 8.14.)

8.52 Repeat Example 8.14 for 400 psia and liquid mol fractions: 0.110, 0.175 and 0.715. Compare them to the experimental values, 50°F, and $y_i = 0.557$, 0.193 and 0.250.

8.53 Find the dew point temperature and the liquid (bubble-point) composition for a vapor that is 5 mol% methane, 85 mol% n-butane, and 10 mol% n-pentane at 20 psia, using Figure 8.20. The procedure is the same as in Example 8.14, but we solve for the value of T at which the x_i, calculated from the y_i and the K_i, sum to 1.00. In this problem and the next two, although the chart reading must be done by hand and repeated for each stage of the trial and error, the solution is easiest if the computations are done on a spreadsheet.

8.54 Same as the preceding problem, but instead of a known pressure of 20 psia we have a known temperature of 70°F, and must estimate the dew-point pressure.

8.55* Estimate the 1-atm boiling temperature of an 0.5 molal solution of sugar (sucrose) in water.

8.56 **a.** Show the equation for mol fraction of solute as a function of molality of solute and the molecular weight of the solvent.

b. Then solve that relation for the molality of solute as a function of mol fraction of solute and the molecular weight of the solvent.

c. Then show the limiting forms of these two equations for very small values of the solute molality.

8.57 In Figure 8.22, the experimental data differ by 10% from that calculated by the two equations at $m_{\text{sucrose}} \approx 1.0$. At this molality, what is the wt% sugar in the solution? Is this a dilute solution?

8.58 Repeat Example 8.15 for sugar molalities of 0.1, 0.5 and 2. Compare the boiling-point elevations from the Raoult's law expression in Example 8.15 and from the equation normally shown in elementary chemistry books, Eq. 8.10.

8.59 Repeat Example 8.15 where the solvent is benzene (C_6H_6, $M = 78$ g/mol). Compare your result to the commonly reported value, $K_b = 2.53°C/(\text{molal})$.

8.60 In Example 8.15 we computed the boiling-point elevation from the mol fraction of the solvent and the vapor pressure curve of the solvent. In general, we must have $P = p_b$

$$\frac{dT}{dm_a} = \frac{dT}{dp_b}\frac{dP}{dx_a}\frac{dx_a}{dm_a} \tag{8.U}$$

where a is the solute and b is the solvent. For pressure near 1 atm we may replace dT/dp_b in terms of the C-C equation (Eq. 5.8).

a. Show that for dilute solutions ($x_a \ll 1.00$), $dP/dx_a \approx p_b$ and $dx_a/dm_a \approx M_b/1000\text{ g}$.

b. Make all three substitutions in Eq. 8.U, and show that the result is

$$\frac{dT}{dm_a} = \frac{RT_0^2}{\Delta h_{b,\text{molar}}/M_b} \cdot \frac{1}{1000\,g} \tag{8.V}$$

c. Use Eq. 8.V to estimate the value of the boiling-point elevation constant K_b for water at its NBP, using the value $(\Delta h_{\text{water, molar}}/M_{\text{water}})_{\text{NBP}} = 539.4$ cal/g. Compare it to the value shown in Example 8.15.

d. Based on Eq. 8.V list the thermodynamic properties we should choose to have a solvent with the highest possible boiling-point elevation constant K_b.

8.61* At 1 atm, a saturated solution of NaCl has a boiling temperature of 108.7°C and an NaCl molality of 6.87. Assuming that the NaCl is 100% ionized to Na^+ and Cl^- ions, estimate the activity coefficient of water in this mixture. (At this temperature the vapor pressure of pure water is 1.3533 atm.)

8.62 At 1 atm, a saturated solution of NaCl in water has a salt content of ≈ 390 g/1000 g of water. Estimate its boiling point by Raoult's law (see Example 8.15) and also by Eq. 8.10,

a. For the assumption that the salt does not ionize.

b. For the assumption that each mol of NaCl ionizes completely to form 2 mole of ions.

c. Compare the results to the observed boiling point, 108.7°C [19], and comment.

8.63 Table 8.O shows the experimental data for the boiling-point elevation of sodium chloride in water at 1 atm [17, p. 326]. Using this table (which is based on NaCl, not Na^+ and Cl^-),

a. Prepare the equivalent of Figure 8.22, assuming full ionization of the NaCl.

b. Estimate the Raoult's law-type activity coefficients for water in these solutions, again assuming full ionization of the NaCl.

Table 8.O Reported Boiling-Point Elevation for NaCl

Molality of NaCl in Water, m_{solute}	$\frac{\Delta T_B}{x_{\text{solute}}}$ (°C), where x_{solute} is the Mol Fraction Based on Nonionized NaCl
0.5	53.3
1	54.7
2	58.7
3	63
4	67
5	72
6.78 (saturation)	79.6

8.64 Table 8.P shows the experimental data for the freezing-point depression of sucrose in water at 1 atm [20]. Using this table,

a. Prepare the equivalent of Figure 8.22.

b. Estimate the Raoult's law-type activity coefficients for water in these solutions.

8.65 Table 8.Q shows the experimental/data for the freezing-point depression of sodium chloride in water at 1 atm [20]. Using this table,

a. Prepare the equivalent of Figure 8.22.

b. Estimate the Raoult's law-type activity coefficients for water in these solutions.

Table 8.P Reported Freezing-Point Depression for Sucrose

Molality of Sucrose in Water, m_{solute}	$\frac{-\Delta T_F}{m_{\text{solute}}}$ (°C/molal)
0.005	1.86
0.05	1.87
0.5	1.96
1.00	2.06
2.00	2.3
4.00	2.7

Table 8.Q Reported Freezing-Point Depression for NaCl

Molality of NaCl in Water, m_{solute}	$\frac{-\Delta T_F}{m_{\text{solute}}}$(°C/molal), where m_{solute} is Based on Nonionized NaCl
0.001	3.66
0.01	3.604
0.1	3.478
1	3.37
2	3.45
4	3.78
5.2 (eutectic, see Figure 11.14)	4.061

8.66 How big an error did we make in Example 8.16 by ignoring the air dissolved in the water at 1 atm? To simplify the calculation assume that the equilibrium concentrations at 20°C shown in Table 3.1 apply in this case.

REFERENCES

1. Seader, J. D., J. J. Siirola, and S. D. Barnicki. Distillation, In *Perry's Chemical Engineers' Handbook* ed. 7, D. W. Green, and J. O. Maloney, eds. New York: McGraw-Hill, pp. 13–11 (1997). This edition of *Perry's* shows binary data for 21 mixtures. The 8th edition (2007) shows such data for only 4 mixtures, probably reflecting the widespread availability of such data elsewhere.
2. Chu, J. C., S. L. Wang, S. L. Levy, and R. Paul. *Vapor-Liquid Equilibriun Data*. Ann Arbor, MI: J. W. Edwards (1956).
3. Oe, S. *Vapor-Liquid Equilibrium Data*. New York: Elsevier (1989).
4. Gmehling, J., et al. *Vapor–Liquid Equilibrium Data Collection*. Frankfurt am Main: Dechema (additional volumes appear regularly, starting in 1977).
5. Wilson, L. C., W. V. Wilding, and G. M Wilson. Vapor–liquid equilibrium measurements on four binary mixtures. *AIChE Symp Ser*. 85(271):25–43 (1989).
6. Hougen, O. A., K. M. Watson, and R. A. Ragatz. *Chemical Process Principles*, Part II: *Thermodynamics,* ed. 2. New York: Wiley (1959).
7. Prausnitz, J. M., R. N. Lichtenthaler, and E. Gomez de Azevedo. *Molecular Thermodynamics of Fluid-Phase Equilibria*, ed. 3. Upper Saddle River, NJ: Prentice Hall, Chapter 12 (1999).
8. Liley, P. E., G. H. Thompson, D. G. Friend, T. E. Daubert, and E. Buck. Physical and chemical data. In *Perry's Chemical Engineer's Handbook*, ed. 8, D.W. Green, ed. New York: McGraw-Hill, pp. 2–48 (2008).
9. Haynes, W. M. *CRC Handbook of Chemistry and Physics* ed. 91. Boca Raton, FL: CRC Press, pp. 6–209 (2010).
10. Sorensen, J. M., and W. Arlt. *Liquid–Liquid Equilibrium Data Collection,* Vol. 1: *Binary Systems,* 1. Frankfurt/Main: Dechema, p. 341 (1979).
11. Clever, H. L., M. Iwamoto, S. H. Johnson, and H. Miyamoto. *Mercury in Liquids, Compressed Gases, Molten Salts and Other Elements, Solubility Data Series*, Vol. 29. Oxford, UK: Pergamon, International Union of Pure and Applied Chemistry, pp. 1–21 (1987).
12. Keenan, J. H., F. G. Keyes, P. G. Hill, and J. G. Moore. *Steam Tables: Thermodynamic Properties of Water Including Vapor, Liquid and Solid Phases*. New York: Wiley (1969).
13. Prausnitz, J. M. Hypothetical standard states and the thermodynamics of high pressure phase equilibria. *AIChEJ*. 6:78–82 (1960).
14. Brown, G. G., et. al. *Unit Operations*. New York: Wiley, p. 327 (1950).
15. DePreister, C. L. Light hydrocarbon vapor–liquid distribution coefficients. *Applied Thermodynamics, CEP Symp Ser*. 7–49:1–43 (1953).
16. Dadyburjor, B. D. SI units for distribution coefficients. *CEP* 74:85–86 (1978).
17. Price, A. R., and R. Kobayashi, Low temperature vapor-liquid equilibrium in light hydrocarbon mixtures: methane-ethane-propane system. *J. Chem. Eng. Data,* 4:40–53 (1959).
18. Austin, J. B., H. G. Dietrich, F. Fenwick, A. Fleischer, G. L. Frear, E. J. Roberts, R. P. Smith, M. Solomon, and H. M. Spurlin. Boiling-point elevations, non-volatile solutes. In *International Critical Tables*, 3, Washbum, E. W., ed. New York: McGraw–Hill, pp. 324–350 (1928).
19. Griswold, J., D. Andres, and V. A. Klein. Determination of high pressure vapor–liquid equilibria: the vapor–liquid equilibrium of benzene-toluene. *Trans. AIChE*. 39:223–240 (1943).
20. Kobe, K. A. *Inorganic Process Industries*. New York: Macmillan, p. 28 (1948).
21. Hall, R. E., and M. S. Sherrill. Freezing point lowerings of aqueous solutions. In *International Critical Tables*, 4, Washburn, E. W., ed. New York: McGraw-Hill, pp. 254–263 (1928).

9

CORRELATING AND PREDICTING NONIDEAL VLE

In the Section 8.9 we borrowed the van Laar equation for liquid-phase activity coefficients from this chapter without showing its logical basis. If we do not make the ideal solution (Raoult's law) assumption for the liquid, then we need some equation of that kind to perform *any* vapor–liquid equilibrium (VLE) calculations. The hardest and most interesting part of VLE is seeking mathematical relations for the liquid-phase activity coefficients, to use in this type of calculation. (Many consider all the other parts trivial, and take VLE to mean the study of liquid-phase activity coefficients.) Our ultimate goal is to write out completely reliable predictive equations based on molecular theory alone. We are far from that goal. Our most plausible goal is to write out reasonably reliable predictive equations, either based on simplified molecular or empirical theories, or based on small amounts of experimental data. We are much closer to that goal.

The most reliable predictive equations are so complex that they are used only in computer programs. In this chapter we consider several predictive equations that are simple enough for hand (or spreadsheet) calculations, showing their logical basis and application, and indicate their relation to the more complex equations used in large computer programs. We also consider briefly how such programs estimate the effect of nonideal behavior in the vapor phase and discuss solubility parameter and gas–liquid equilibrium. This chapter is all about nonideal behavior in one phase, mostly in the liquid, occasionally in the gas. For that reason the superscripts that identify phases will not appear in this chapter.

9.1 THE MOST COMMON OBSERVATIONS OF LIQUID-PHASE ACTIVITY COEFFICIENTS

As discussed in Chapter 8, we have no direct measurements of liquid- or gas-phase activity coefficients. We compute them from observed VLE (or other kinds of physical measurements [1, p. 173]) or estimate them from molecular interaction models. To compute them from VLE observations we must make some kind of assumption about the behavior of the gas. At low pressures we normally assume ideal gas behavior, at modest pressures we use the methods shown in Section 9.7, and at high pressures we use the methods in Chapter 10.

Figures 8.7, 8.8 and 8.9 show the most common types of activity coefficient behaviors (types I, II, and III), which are summarized in Figure 9.1. Type IV (heteroazeotrope) behavior (Figure 8.12) is an extreme example of type II. The activity coefficients are larger than for type II, but have the same general shape. In all three patterns shown in Figure 9.1 the activity coefficients are both $= 1.00$ (type I), both >1.00 (types II and IV), or both <1.00 (type III). We are suspicious of any reported activity coefficients that do not fall into one of these three patterns. However, there is another pattern, sketched in Figure 9.2, that is uncommon, but that exists for the important ammonia–water system [2].

9.1.1 Why Nonideal Behavior?

What kinds of molecules would be expected to show the above four types of behavior? We can offer the following

Physical and Chemical Equilibrium for Chemical Engineers, Second Edition. Noel de Nevers.

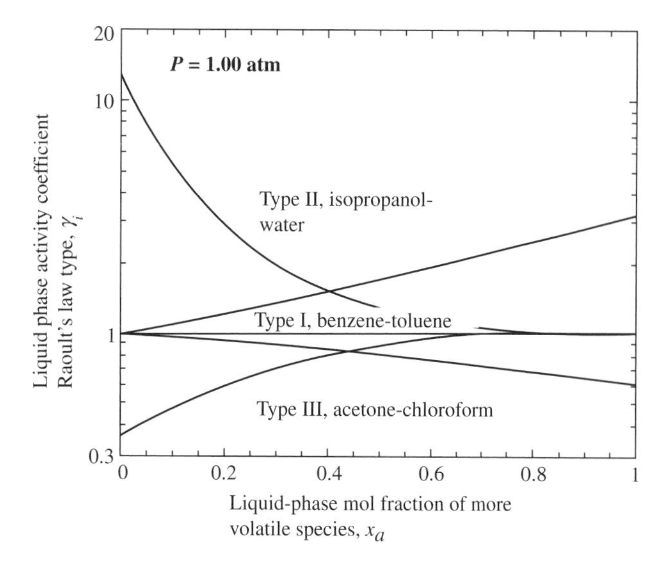

FIGURE 9.1 The three most common patterns of activity coefficients, copied from Figures 8.7, 8.8, and 8.9.

approximate rules, which help us guess what an activity coefficient plot would look like, without any data:

1. Molecules of the same homologous series generally have close to ideal behavior, type I, but as their difference in size becomes large they show weak type II behavior. The hydrocarbon *K*-value plot in Figure 8.20 generally corresponds to activity coefficients between 1 and 1.5, reflecting the fact that in the hydrocarbon systems for which it applies there is generally a range of molecular sizes, and hence they have activity coefficients in this range.

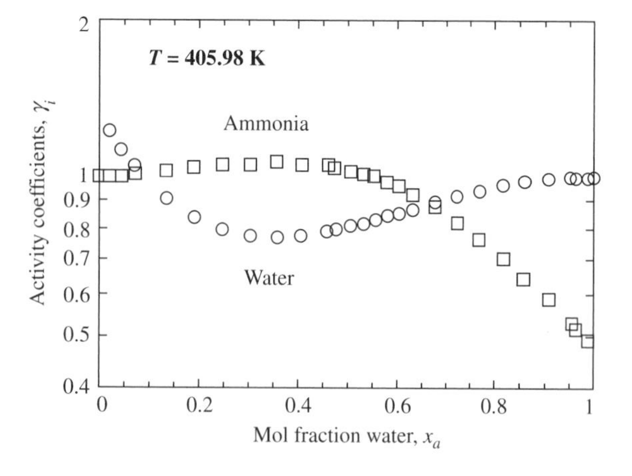

FIGURE 9.2 Activity coefficients for ammonia–water, which do not fit any of the types shown in Figures 8.7, 8.8, 8.9 and 8.12. The original article shows such plots for 8 different temperatures [2]. Observe that here species *a* is water, the higher-boiling species. This is contrary to the common usage, but follows the usage in the original article.

2. Molecules with differing polarity tend to have strong type II or type IV behavior. Water is polar and most hydrocarbons are not. This leads to strong type IV behavior for most water–hydrocarbon systems, with limited liquid-phase solubility and very high activity coefficients. Adding a polar group to a hydrocarbon, for example, a hydroxyl group, makes its water repulsion weaker, leading to complete water solubility for the lower molecular weight alcohols and decreasing solubility with increasing molecular weight with the higher ones. The corresponding activity coefficients for alcohol–water solutions increase as the number of carbon atoms in the alcohol increases (see Figures 8.9 and 8.12!). (This topic is explored further in Chapter 11.)
3. Chemical or quasi-chemical reactions between species lead to type III behavior. Solvation of molecules and ions is a weak quasi-chemical reaction, shown for sucrose and Na^+ and Cl^- in Chapter 8. Strong quasi-chemical reactions like that between water and sulfuric acid lead to strong type III behavior.
4. Some chemicals reversibly form dimers or trimers in both liquid and vapor phases, for example, carboxylic acids. This leads to type III behavior (see Chapter 13).
5. Water–organic solutions are almost always complex, because of the variety of ways in which water can interact with organic molecules.

We will discuss this topic a bit more in Section 9.9.

9.1.2 The Shapes of ln $\gamma - x$ Curves

As discussed in Section 8.3, the γ_i for all i must become 1.00 and the approach must be asymptotic to the $\gamma_i = 1.00$ axis as x_i approaches 1.00 because as x_i approaches 1.00 each i molecule becomes surrounded by only other molecules of i, with which it has ideal-solution interactions. As x_i approaches 0.00 (and γ_i approaches γ_i^∞) we expect the most strongly nonideal behavior (γ_i moving most rapidly away from 1.00, either up in type II or down in type III) because each lonely i molecule is surrounded by molecules of the other kind with which it has its strongest nonideal interactions. The pattern in Figure 9.2 suggests that something other than simple VLE must be occurring. In this case we know that water and ammonia react by

$$NH_3 + H_2O \Leftrightarrow NH_4OH \Leftrightarrow NH_4^+ + OH^- \tag{9.A}$$

On the right side of Figure 9.2 we have a small amount of NH_3 dissolved in water, so we would expect Reaction 9.A to move to the right, thus removing dissolved NH_3 from the solution and producing type III behavior (activity coefficients less than 1.00). On the left side of the figure we have almost pure ammonia, near its critical temperature (405.5 K) in which the small number of water molecules do not react and ionize and are apparently weakly repelled by the ammonia molecules. This leads to simple type II behavior (activity coefficients increase with increasing dilution). In the middle of Figure 9.2 we have the anomalous situation that one activity coefficient is greater than 1.00 and the other less. This is very uncommon behavior; in Figure 9.2 the deviations of the γ_i from 1.00, positive and negative, are both small.

For liquid mixtures for which we do not have experimental VLE data, or for which we have only partial data, we mostly proceed by extrapolating and interpolating the existing data and making estimates where the data are missing, using activity coefficient correlations. We might logically try to represent those curves with various empirical data-fitting equations, but there is a strong restriction on what forms of such data-fitting equations we can use.

9.2 LIMITS ON ACTIVITY COEFFICIENT CORRELATIONS, THE GIBBS–DUHEM EQUATION

It is shown in Chapter 6 that for any partial molar property (for example, $\bar{q}$ where Q is any extensive property) of any mixture with any number of species in one phase the partial molar equation requires that

$$(n_a d\bar{q}_a + n_b d\bar{q}_b + n_c d\bar{q}_c + \cdots)_{T,P} = 0 \tag{6.22}$$

If we let $\bar{q}$ be the partial molar Gibbs energy, then Eq. 6.22 becomes the Gibbs–Duhem equation:

$$\begin{aligned}(n_a d\bar{g}_a + n_b d\bar{g}_b + n_c d\bar{g}_c + \cdots)_{T,P} \\ = (n_a d\mu_a + n_b d\mu_b + n_c d\mu_c + \cdots)_{T,P} = 0\end{aligned} \tag{6.23}$$

If we consider a binary mixture for which $(n_c = n_d = \cdots = 0)$, then a little algebra converts this to

$$\left(\frac{d\bar{g}_a}{dx_a} + \frac{x_b}{x_a} \cdot \frac{d\bar{g}_b}{dx_a}\right)_{T,P} = \left(\frac{d\mu_a}{dx_a} + \frac{x_b}{x_a} \cdot \frac{d\mu_b}{dx_a}\right)_{T,P} = 0 \tag{9.1}$$

which is the most commonly used form of the Gibbs–Duhem equation. In principle, it applies only at constant T and P, but in practice that seems to be a restriction we can often ignore with negligible error (see Section 9.5). A little more algebra shows that we can substitute fugacities and then activity coefficients in this equation and find

$$\left(\frac{d\ln\gamma_a}{dx_a} + \frac{x_b}{x_a} \cdot \frac{d\ln\gamma_b}{dx_a}\right)_{T,P} = 0 \tag{9.2}$$

which is often also called the Gibbs–Duhem equation, because it is equivalent to Eq. 9.1. This equation says that if we have some experimental values of the activity coefficients for a binary mixture, as in Figure 8.6 or 9.1, and we attempt to represent this set of values by some empirical curve-fitting equations, there is a severe restriction on the form our curve-fitting equations can take.

For example, if we tried

$$\begin{aligned}\ln\gamma_a &= Ax_a + B \\ \ln\gamma_b &= Cx_b + D\end{aligned} \quad ?? \tag{9.B}$$

where A, B, C, and D are curve-fitting constants, and then took the derivatives (noting that $dx_b = -dx_a$) and inserted them in Eq. 9.2, we would find

$$\left(\frac{d\ln\gamma_a}{dx_a} + \frac{x_b}{x_a} \cdot \frac{d\ln\gamma_b}{dx_a}\right)_{T,P} = \left[A - \frac{x_b}{x_a}C\right] \quad ?? \tag{9.C}$$

which is equal to zero when $(x_b/x_a) = A/C$, but not for any other value of the mol fractions. Equation 9.2 must be obeyed for *all* values of x_a, so the set of data-fitting equations for the activity coefficients proposed in Eq. 9.B *cannot be correct.*

Intuitively, the reason for the Gibbs–Duhem equation is that for each species γ_i depends on the fraction of the molecules with which a molecule interacts that are of the same kind and the fraction that are of the other kind. Thus, as we change the ratio of one kind of molecule to the other (change the mol fractions), we would expect both γ_is to change; the Gibbs–Duhem equation shows how those changes relate to each other.

In Figures 8.7, 8.8, 8.9, 8.12, 9.1, and 9.2 the mol fractions, pressures, and temperatures are shown on arithmetic scales, while the activity coefficients are shown on log scales. They are almost always shown in that form, to correspond to Eq. 9.2, which says that in that representation, for any value of x_i, the slopes of the two ln γ_i curves must have opposite signs (one positive, one negative, or both zero) and that the magnitude of the slopes must be equal and opposite at $x_a = x_b = 0.5$. As best we can read the slopes on those figures, Eq. 9.2 is obeyed. In Eq. 9.2 as $x_b \to 0$, $(d\ln\gamma_a)/dx_a \to 0$ so that the ln γ_a curve must become horizontal, and thus tangent to the $\gamma_a = 1.00$ line.

Several data-fitting equations for activity coefficients have been proposed and used. All of these have forms that guarantee that Eq. 9.2 will be obeyed for all x_i. One of the simplest of these, one of few likely to be used in noncomputer calculations, is the van Laar equation, used in the examples in Chapter 8:

$$\log\text{ or }\ln\gamma_a = \frac{Ax_b^2}{\left(\dfrac{A}{B}x_a + x_b\right)^2} \qquad \log\text{ or }\ln\gamma_b = \frac{Bx_a^2}{\left(x_a + \dfrac{A}{B}x_b\right)^2} \tag{8.6}$$

Example 9.1 From Table A.7 for the isopropanol–water system at 1 atm pressure the van Laar constants (log form) are

$A = 1.0728$, $B = 0.4750$ (in that table water is shown as a, but here we use isopropanol as a, which interchanges the values of A and B). Using these values, estimate the two liquid-phase activity coefficients for $x_a = x_{\text{isopropanol}} = 0.4720$, and $x_b = x_{\text{water}} = 0.5280$.

$$\log \gamma_a = \frac{1.0728 \cdot 0.5280^2}{\left(\dfrac{1.0728}{0.4750} \cdot 0.4720 + 0.5280\right)^2} = 0.1177 \quad \gamma_a = 1.31$$

$$\log \gamma_b = \frac{0.4750 \cdot 0.4720^2}{\left(0.4720 + \dfrac{0.4750}{1.0728} 0.5280\right)^2} = 0.2124 \quad \gamma_b = 1.63 \tag{9.D}$$

The values computed directly from the measured VLE data are 1.32 and 1.63. Thus, for this particular data point, the van Laar equation does a very good *but not perfect* job of estimating these activity coefficients. ■

9.3 EXCESS GIBBS ENERGY AND ACTIVITY COEFFICIENT EQUATIONS

The van Laar equation is simple and useful, but more complex equations do a better job of correlating, interpolating, and extrapolating activity coefficients (and thus VLE data and calculations). Most of these methods begin by defining and then correlating a quantity called the *excess Gibbs energy*, G^E. If we write the equation for the Gibbs energy of some arbitrary solution, using the property from Chapter 6 that for any phase $G = \sum n_i \bar{g}_i$ and then subtract from it the Gibbs energy that the same phase would have if it were an ideal solution, we will have

$$\begin{aligned} G^E &= G_{\text{actual}} - G_{\substack{\text{solution of the same}\\ \text{composition, but ideal}}} \\ &= \sum n_i \bar{g}_i - \sum n_i (\bar{g}_i)_{\text{ideal solution}} \\ &= \sum n_i \left[RT \ln(x_i \gamma_i f_i^{\text{o}})\right] - \sum n_i \left[RT \ln(x_i f_i^{\text{o}})\right] \\ &= \sum n_i (RT \ln \gamma_i) \end{aligned} \tag{9.3}$$

from which it follows (see Problem 9.8) that

$$\left(\frac{dG^E}{dn_i}\right)_{n_j,\,T,\,P} = \left[\frac{d(g^E n_T)}{dn_i}\right]_{n_j,\,T,\,P} = \bar{g}_i^E = \mu_i^E = RT \ln \gamma_i \tag{9.4}$$

To find the above relations in terms of mol fraction and excess Gibbs energy g^E per mol of solution, we divide each part of Eq. 9.3 by n_T, finding

$$\begin{aligned} g^E &= g_{\text{actual}} - g_{\substack{\text{solution of the same}\\ \text{composition, but ideal}}} \\ &= \frac{G^E}{n_T} = \sum x_i \bar{g}_i - \sum x_i (\bar{g}_i)_{\text{ideal solution}} \\ &= \sum x_i \left[RT \ln(x_i \gamma_i f_i^{\text{o}})\right] - \sum x_i \left[RT \ln(x_i f_i^{\text{o}})\right] \\ &= \sum x_i (RT \ln \gamma_i) \end{aligned} \tag{9.5}$$

For a binary solution, $x_{c,\ldots} = 0$ and $(dx_b/dx_a) = -1$, so that

$$\left(\frac{dg^E}{dx_a} = RT \ln \frac{\gamma_a}{\gamma_b}\right)_{T,\,P} \tag{9.6}$$

and also (see Problem 9.12), for binary solutions only,

$$RT \ln \gamma_a = g^E + (1 - x_a) \frac{dg^E}{dx_a} \tag{9.7}$$

In most activity coefficient calculations we use g^E/RT as the dimensionless correlating variable, which simplifies the computations; an additional correlating parameter, $g^E/(RTx_a x_b)$, has advantages that will become clear after the next example.

Example 9.2 Calculate and plot g^E/RT and $(g^E/RTx_a x_b)$ for the acetone–water solution at 1 atm pressure, using the VLE data from Example 8.2.

Continuing Example 8.2, we have, for $x_{\text{acetone}} = 0.05$,

$$\begin{aligned} \frac{g^E}{RT} &= \sum x_i \ln \gamma_i \\ &= (0.05 \cdot \ln 7.04 + 0.95 \cdot \ln 1.01) \\ &= (0.0976 + 0.0094) = 0.107 \end{aligned} \tag{9.E}$$

$$\frac{g^E/RT}{x_a x_b} = \frac{0.107}{0.05 \cdot 0.95} = 2.25 \tag{9.F}$$

We then use a spreadsheet to make similar computations for the other data points, and plot the results as shown in Figure 9.3. ■

From Figure 9.3 we see that g^E/RT is a dome-shaped function that goes to zero at the two extremes. If we wish to represent it by some data-fitting equation we will need at least a quadratic equation, more likely a cubic or higher-power equation. However, $g^E/(RTx_a x_b)$ is close to being a

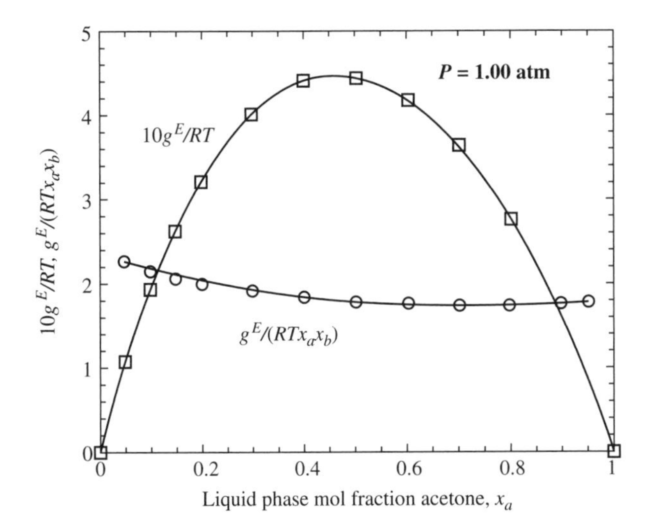

FIGURE 9.3 Calculated values of $10g^E/RT$ and (g^E/RTx_ax_b) plotted as a function of the mol fraction of acetone in the liquid phase for acetone and water. The g^E/RT values are multiplied by 10 to make the values more or less the same size.

straight line; we can often fit it with a linear equation, or even approximately by a constant (see Example 9.3). This kind of behavior is common; that is the reason for bothering with $g^E/(RTx_ax_b)$.

Example 9.3 Show the equations for the activity coefficients that result from the following two simplifications of the $g^E/(RTx_ax_b)$ curve in Figure 9.3:

$$\frac{g^E}{(RTx_ax_b)} = A = 1.886 \tag{9.G}$$

$$\frac{g^E}{(RTx_ax_b)} = ax_a + b = -0.51145x_a + 2.127 \tag{9.H}$$

The first of these is simply an average of the values, corresponding to a horizontal line through the middle of the data points in Figure 9.3. The second is the result of a least-squares straight-line fit of the data, corresponding to a line with a mild negative slope on Figure 9.3, with a and b as linear equation coefficients.

For Eq. 9.G, some algebra (Problem 9.17) leads to

$$\ln\gamma_a = Ax_b^2 = 1.886x_b^2 \quad \ln\gamma_b = Ax_a^2 = 1.886x_a^2 \tag{9.I}$$

For Eq. 9.H, we first rewrite the linear equation as

$$ax_a + b = (a+b)x_a + bx_b = cx_a + bx_b \quad \text{where}$$
$$c = (a+b) = 1.6156 \tag{9.J}$$

Then some more complex algebra (Problem 9.18) leads to

$$\begin{aligned}\ln\gamma_a &= x_b^2[b + 2(c-b)x_a] = x_b^2[2.127 + 2(1.6156 - 2.127)x_a]\\ &= x_b^2(2.127 - 1.023x_a)\\ \ln\gamma_b &= x_a^2[c + 2(b-c)x_b] = x_a^2[1.6156 + 2(2.127 - 1.6156)x_b]\\ &= x_a^2(1.6156 + 1.023x_b)\end{aligned} \tag{9.K}$$

Equation 9.I, the *symmetrical equation*, is the *simplest* mathematical representation of binary solution activity coefficient data that does not violate the Gibbs–Duhem equation. Equation 9.K is the *Margules equation.* (Equation 9.I is sometimes called the *two-suffix Margules* equation, and Eq. 9.K the *three-suffix Margules* equation. If we represent the $g^E/RTx_a x_b$ curve on Figure 9.3 by a quadratic equation in x_i, the result is sometimes called the *four-suffix Margules* equation. Here the number of the *suffix* is the number of the highest power of x_i or $x_i^n x_j^m$ in the resulting equations for the activity coefficients.) To see how well these equations represent the experimental data they are plotted in Figure 9.4, along with the experimental data from Figure 8.6.

From Figure 9.4 it is clear that the two curves predicted by the symmetrical equation are each other's mirror images, and that the two predicted by the Margules equation are similar in shape, but with different slopes. The symmetrical curves pass almost exactly through

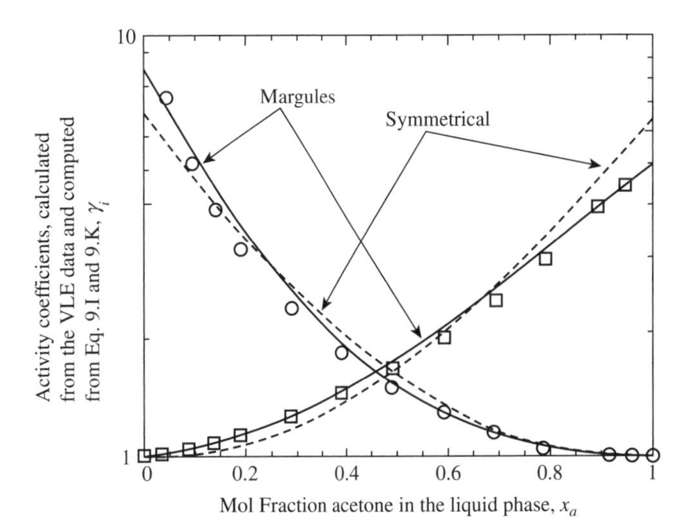

FIGURE 9.4 Comparison of the acetone–water liquid-phase activity coefficients computed from the VLE data (copied from Figure 8.6) and their representation by the symmetrical and Margules equations.

Table 9.1 Comparison of Various Equations for Representing and Estimating Liquid-Phase Activity Coefficients

Equation Name	$\dfrac{g^E}{RTx_ax_b}$	$\ln \gamma_a$ and γ_b
Symmetrical	A	$\ln \gamma_a = Ax_b^2$ $\ln \gamma_b = Ax_a^2$
Margules	$cx_a + bx_b$	$\ln \gamma_a = x_b^2[b + 2(c - b)x_a]$ $\ln \gamma_b = x_a^2[c + 2(b - c)x_b]$
van Laar	$\dfrac{AB}{Bx_b + Ax_a}$	$\ln \gamma_a = \dfrac{Ax_b^2}{\left(\dfrac{A}{B}x_a + x_b\right)^2}$ $\ln \gamma_b = \dfrac{Bx_a^2}{\left(x_a + \dfrac{B}{A}x_b\right)^2}$
Wilson	$\dfrac{-\ln(x_a + \Lambda_{ab}x_b)}{x_b} + \dfrac{-\ln(\Lambda_{ba}x_a + x_b)}{x_a}$	$\ln \gamma_a = -\ln(x_a + \Lambda_{ab}x_b) + x_b\left(\dfrac{\Lambda_{ab}}{x_a + \Lambda_{ab}x_b} - \dfrac{\Lambda_{ba}}{\Lambda_{ba}x_a + x_b}\right)$ $\ln \gamma_b =$ same as $\ln \gamma_a$, with subscripts interchanged
NRTL	$\left[\dfrac{\tau_{ba}G_{ba}}{x_a + G_{ba}x_b} + \dfrac{\tau_{ab}G_{ab}}{G_{ab}x_a + x_b}\right]$	$\ln \gamma_a = x_b^2\left\{\tau_{ba}\left(\dfrac{G_{ba}}{x_a + G_{ba}x_b}\right)^2 + \left[\dfrac{\tau_{ab}G_{ab}}{(G_{ab}x_a + x_b)^2}\right]\right\}$ $\ln \gamma_b =$ same as $\ln \gamma_a$, with subscripts interchanged
Scatchard–Hildebrand	$\dfrac{(\delta_a - \delta_b)^2}{RT\left(\dfrac{x_b}{v_a} + \dfrac{x_a}{v_b}\right)}$	Same as van Laar, with right-hand side multiplied by $\dfrac{(\delta_a - \delta_b)^2}{RT}$ and $A = v_a$; $B = v_b$

Note: A more complete table of this type, showing more details is in Walas [1, Chapter 4]. Most of the coefficients in these tables, A, B, etc. are data-fitting values, obtained from experimental VLE measurements. Some have semi-theoretical bases. The constants in the Scatchard–Hildebrand equation are based on the "Regular Solution" theory, and are calculable from pure species properties, without any data for tha mixture. Many authors replace all the symbols for the coefficients with a universal set, A_{ab} and A_{ba}, etc. Here the originally used symbols are shown.

the points nearest $x_{\text{aeetone}} = 0.5$, while the Margules curves pass just below those values. At the extremes ($x_{\text{acetone}} \rightarrow 1.00$ and $\rightarrow 0.00$) the Margules equation reproduces the experimental values substantially better than does the symmetrical equation, which should not surprise us, when we consider that in Figure 9.3, the symmetrical equation is a horizontal line through the average of the data points, which is sure to represent the middle of the data well, while the Margules equation is a sloping line, which compromises between fitting the middle and the extreme values well.

Most of the common activity coefficient correlation equations can be shown to be relatively simple algebraic representations of a plot of g^E/RTx_ax_b vs. x_a, (see Problem 9.20). Most of them were not originally derived that way, but were based on some type of theoretical model. But they all have that mathematical property. Table 9.1 shows this relationship for an assortment of widely used (or historically interesting) activity coefficient correlations. ■

9.4 ACTIVITY COEFFICIENTS AT INFINITE DILUTION

The values of the activity coefficients as concentrations approach zero, called infinite dilution values, ln γ_a^∞ and ln γ_b^∞ are often of considerable interest. The reasons [1, p. 215] are as follows:

1. It is often easier and cheaper to measure these than to measure the activity coefficients near the middle of plots like Figure 8.6.
2. For the design of distillation columns this is often the most important value, because the difficulty of the separation is often controlled by the difficulty of producing a high-purity product, and that is related to the activity coefficient near infinite dilution of the species other than the one we want at high purity.
3. The constants in many common activity coefficient equations can be estimated easily from infinite-dilution activity coefficients.

Example 9.4 Show the relations between ln γ_a^∞ and ln γ_b^∞ and the constants in the Margules equation.

ln γ_a^∞ is the value of ln γ_a as x_a approaches zero and x_b approaches 1.00. Substituting those values in Eq. 9.K leads to

$$\ln \gamma_a^\infty = b \tag{9.L}$$

and correspondingly

$$\ln \gamma_b^\infty = c \quad ■ \tag{9.M}$$

Example 9.5 By simple extrapolation of the data points in Figure 8.6, it appears that $\gamma_{\text{acetone}}^\infty \approx 10$ and $\gamma_{\text{water}}^\infty \approx 5$. Based on these values, estimate the values of the constants in the Margules equation for acetone–water at 1 atm.

From Eqs. 9.L and 9.M, we estimate

$$\begin{aligned} b &= \ln \gamma_{\text{acetone}}^\infty \approx \ln 10 = 2.303 \\ c &= \ln \gamma_{\text{water}}^\infty \approx \ln 5 = 1.609 \end{aligned} \tag{9.N}$$

From Example 9.3 we know that the estimates of these coefficients, based on all the data and not just the infinite dilution data, are 2.127 and 1.616. ■

If we had estimated the Margules equation constants this way, they would have represent the experimental activity coefficient values at the two extremes better than is shown in Figure 9.4, but represented the values in the middle of the plot more poorly. For systems like acetone–water these large values of ln γ_i^∞ make it harder to produce pure acetone by distillation (because the water's effective vapor pressure in practically pure acetone is increased) and make it easier to produce pure water (because the acetone's effective vapor pressure in practically pure water is increased).

9.5 EFFECTS OF PRESSURE AND TEMPERATURE ON LIQUID-PHASE ACTIVITY COEFFICIENTS

The VLE data for acetone and water in Table 8.1, reproduced several ways in Chapters 8 and 9, are all for 1 atm pressure. The temperatures range from the normal boiling point (NBP) of acetone, 56.15°C, to the NBP of water, 100°C. Most of the older experimental VLE data were taken at constant pressure in devices like that shown in Figure 8.2. Most distillation processes operate at practically constant pressure, but not at a constant temperature; the older data match that condition. In recent years new techniques (which would have been impractical before the computer age) have made it quicker and cheaper to collect VLE data at constant temperature, with the pressure varying from the boiling pressure of one species to that of the other at this temperature [2]. Much of the current data is in that form. To apply these to distillation calculations, we need a way of estimating $[\gamma_i = f(x_i)]_{\text{at constant } P}$ from $[\gamma_i = f(x_i)]_{\text{at constant } T}$. The general problem is to estimate the effects of changes in P and T on activity coefficients. Changes in pressure most often have little effect, but changes in temperature may have significant effects. The equations we need (see Appendix C) are

$$\left(\frac{\partial \ln \gamma_i}{\partial P}\right)_{T,\, x_i} = \frac{\bar{v} - v_i^{\text{o}}}{RT} \tag{7.31}$$

$$\left(\frac{\partial \ln \gamma_i}{\partial T}\right)_{p,\, x_i} = \frac{(h_i^{\text{o}} - \bar{h}_i)}{RT^2} \tag{7.32}$$

9.5.1 Effect of Pressure Changes on Liquid-Phase Activity Coefficients

Example 9.6 Estimate the activity coefficient of ethanol in a solution of 0.1238 mol fraction ethanol, balance water, at 85.3°C and 10 atm.

From Example 8.9 we know that at this T and x_a and 1 atm, $\gamma_{\text{ethanol}} = 2.9235$ (as estimated by the van Laar equation). We can compute that 0.1238 mol fraction ethanol corresponds to 0.265 mass fraction ethanol, and then read in Figure 6.15 that at 20°C and 0.265 mass fraction ethanol,

$$\bar{v}_{\text{ethanol}} \text{ by mass at } 0.265 \text{ wt fraction ethanol} \approx 1.16 \frac{\text{cm}^3}{\text{g}} \tag{9.O}$$

$$v_{\text{ethanol}}^{\text{o}} \text{ by mass} \approx 1.27 \frac{\text{cm}^3}{\text{g}} \tag{9.P}$$

$$(\bar{v} - v^{\text{o}})_{\text{ethanol}} \text{ by mass} \approx (1.16 - 1.27)\frac{\text{cm}^3}{\text{g}} = -0.11 \frac{\text{cm}^3}{\text{g}} \tag{9.Q}$$

$$\begin{aligned} (\bar{v} - v^{\text{o}})_{\text{ethanol}} &= -0.11 \frac{\text{cm}^3}{\text{g}} \cdot 46 \frac{\text{g}}{\text{mol}} = -5.06 \frac{\text{cm}^3}{\text{mol}} \\ &= -0.00506 \frac{\text{L}}{\text{mol}} \end{aligned} \tag{9.R}$$

If we assume that this value is more or less independent of temperature, we can use it as the corresponding value at 85.3°C, and compute

$$\left(\frac{\partial \ln \gamma_i}{\partial P}\right)_{T,x_i} = \frac{-0.00506\,\text{L/mol}}{0.08206\dfrac{\text{L}\cdot\text{atm}}{\text{mol}\cdot\text{K}}\cdot(85.3+273.15)\text{K}}$$
$$= -1.7278\times 10^{-4}\frac{1}{\text{atm}} \tag{9.S}$$

$$\Delta \ln \gamma_i \approx \left(\frac{\partial \ln \gamma_i}{\partial P}\right)_{T,x_i}\cdot \Delta P = -1.7278\times 10^{-4}\frac{1}{\text{atm}}\cdot 9\,\text{atm}$$
$$= -1.548\times 10^{-3} \tag{9.T}$$

$$\ln \gamma_{i,\,10\,\text{atm}} = \ln \gamma_{i,\,1\,\text{atm}} + (-1.548\times 10^{-3})$$
$$= 1.07278 - 0.001548 = 1.0712 \tag{9.U}$$
$$\gamma_{i,10\,\text{atm}} = \exp(1.0712) = 2.9189$$

$$\gamma_{i,10\,\text{atm}} - \gamma_{i,1\,\text{atm}} = 2.9189 - 2.9235 = -0.0045 \quad ■ \tag{9.V}$$

If there were no volume change on mixing, then the effect of pressure changes on activity coefficients would be zero. Ethanol–water is an extreme case. Its volume change on mixing, in this example is −8.7% of the specific volume of pure ethanol. Hildebrand and Scott [3, p. 142] show a table of volume changes on mixing. Most are less than ± 2% of the volume of the substance added. Even with ethanol–water's extreme volume change on mixing, the effect of raising the pressure from 1 to 10 atm is negligible (≈0.15%). The reason is that the specific volume of liquids is so small that even for large percentage values of the volume change on mixing, the value of $(\bar{v}_i - v_i^{\circ})$ must be small and thus the effect of pressure on the activity coefficient is normally negligible for liquids (and solids). See also Example 7.3.

9.5.2 Effect of Temperature Changes on Liquid-Phase Activity Coefficients

Example 9.7 Estimate the activity coefficient of ethanol in a liquid of 0.1238 mol fraction ethanol, balance water, at 70°C and 1.00 atm.

From Example 8.9 we know that at 1 atm this solution boils at 85.3°C and has a calculated $\gamma_{\text{ethanol}} = 2.9235$. Figure 9.5 [4] shows the values of $(\bar{h}_i - h_i^{\circ})$ for both ethanol (species a) and water (species b) at a variety of temperatures, calculated from measured heats of mixing.

For ethanol at $x_a = 0.1238$ and temperatures of 90 and 70°C (363.15 and 343.15 K) we read $(\bar{h}_i - h_i^{\circ}) \approx 0.2$ and 1.0 kJ/mol. The average of these is 0.6 kJ/mol = 600 J/mol. Using this average value as a constant in Eq. 7.32 (observe the sign change!) and rearranging to make the integration, we find

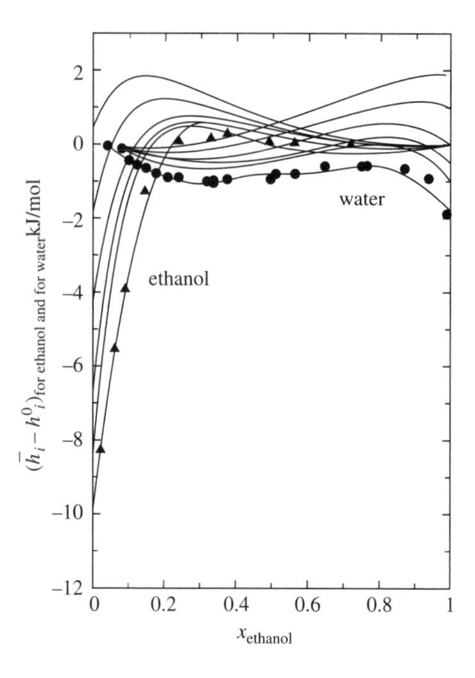

FIGURE 9.5 Values of $(\bar{h}_i - h_i^{\circ})$ for ethanol (species a) and water (species b) calculated from measured heats of mixing. The two sets of curves, from the lowest to the highest, correspond to 298.15, 323.15, 331.15, 343.15, 363.15, and 383.15 K. The triangles and circles are comparisons to data from another author. (From Larkin, J. A. Thermodynamic properties of aqueous nonelectrolyte mixtures, I: excess enthalpy for water + ethanol at 298.15 to 383.15 K. *J. Chem. Thermodyn.* 7: 137–148 (1975). Reproduced by permission of the publisher.)

$$\int \partial \ln \gamma_a = \int \frac{(h_a^{\circ} - \bar{h}_a)}{RT^2}\partial T \approx \frac{(h_a^{\circ} - \bar{h}_a)_{\text{average}}}{R}\int \frac{\partial T}{T^2} \tag{9.8}$$

$$\ln \frac{\gamma_{a\text{ at }T_2}}{\gamma_{a\text{ at }T_1}} \approx \frac{(h_a^{\circ} - \bar{h}_a)_{\text{average}}}{R}\cdot\left(\frac{1}{T_1} - \frac{1}{T_2}\right) \tag{9.9}$$

$$= \frac{-600\dfrac{\text{J}}{\text{mol}}}{8.314\dfrac{\text{J}}{\text{mol}\cdot\text{K}}}\cdot\left(\frac{1}{358.45\,K} - \frac{1}{343.15\,K}\right) = 8.98\times 10^{-3} \tag{9.W}$$

$$\gamma_{a,T_2} = \gamma_{a,T_1}\exp(8.98\times 10^{-3})$$
$$= 2.9235\cdot 1.009 \approx 2.950 \quad ■ \tag{9.X}$$

We find that a temperature change of ≈15°C changes the ethanol activity coefficient by 0.9%. The previous example showed that a pressure change of 9 atm changed the activity coefficient by 0.15%. Thus, we see that for small changes in temperature and pressure we can normally ignore the changes in liquid-phase activity coefficients. However, if

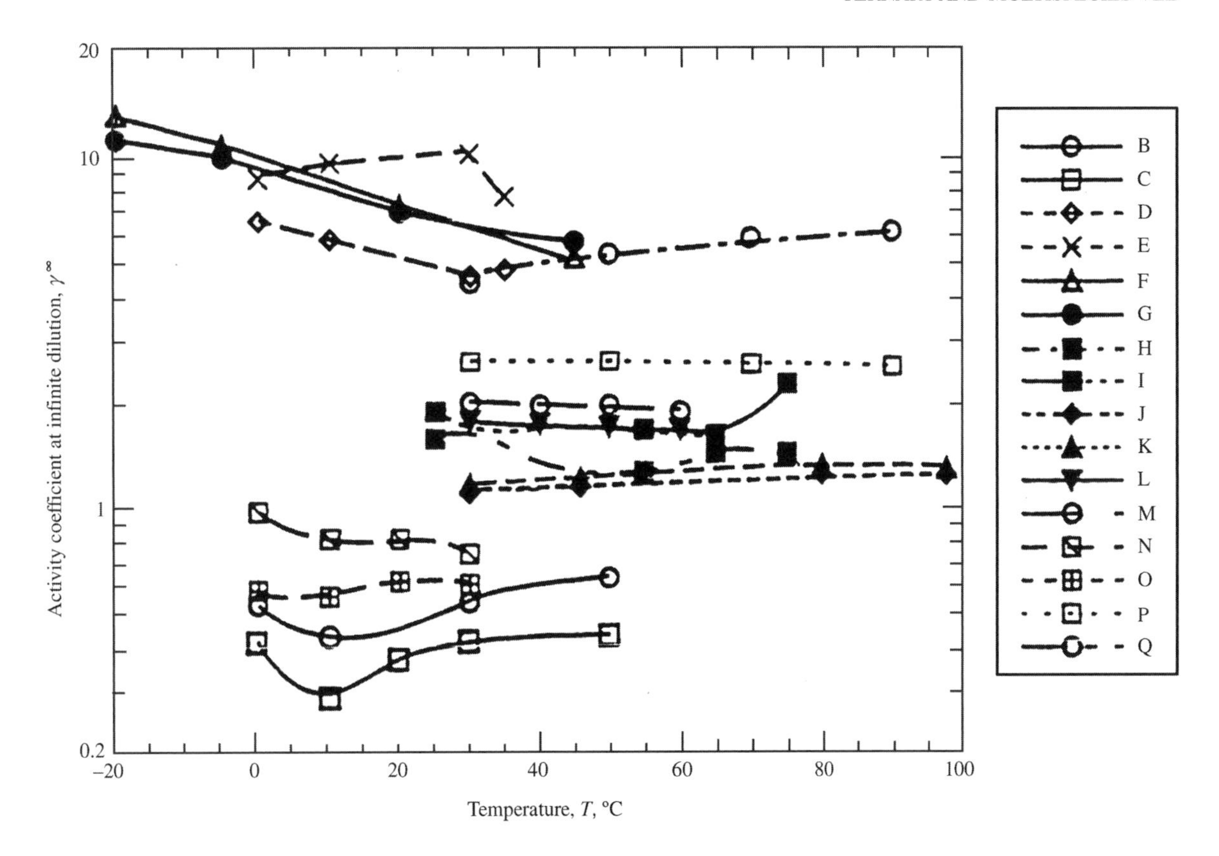

FIGURE 9.6 Measured activity coefficients at infinite dilution γ_i^{∞} for a variety of solutions [1, p. 233–235]. The systems corresponding to the legend are B, acetone in chloroform; C, chloroform in acetone; D, carbon disulfide in acetone; E, acetone in carbon disulfide; F, acetone in hexane; G, hexane in acetone; H, methyl ethyl ketone in ethyl benzene; I, ethyl benzene in methyl ethyl ketone; J, 1,4-dioxane in nitromethane; K, nitromethane in 1,4-dioxane; L, 1,4-dioxane in *N*,*N*-dimethyl formamide; M, *N*,*N*-dimethyl formamide in 1,4-dioxane; N, diethyl ether in chloroform; O, chloroform in diethyl ether; P, water in ethanol; and Q, ethanol in water.

the values of $(h_i^{\text{o}} - \bar{h}_i)$ are larger, which they often are near infinite dilution, then the changes in activity coefficient with temperature can also be significant.

Figure 9.6 shows the measured activity coefficients at infinite dilution for a variety of solutions based on [1, pp. 233–235]. The data in Figure 9.6 show that for a variety of systems the infinite dilution activity coefficients change little with changes in temperature. (The log scale minimizes such changes!) From Figure 9.5 we see that as the ethanol concentration approaches zero, the values of $(h_i^{\text{o}} - \bar{h}_i)_{\text{ethanol}}$ become large and negative. The same does not happen for water; as its concentration approaches zero, the values of $(\bar{h}_i - h_i^{\text{o}})_{\text{water}}$ remain small. Based on this observation we would assume that the activity coefficient at infinite dilution for ethanol should change substantially with temperature, while that for water would not. Curves P and Q in Figure 9.6 show that this is found experimentally. The scatter in the data in Figure 9.6 reminds us that activity coefficient measurements are not easy to make; most likely the true values form smoother curves than the experimental values shown here. We conclude that changes in pressure have little effect on liquid-phase activity coefficients, and that changes in temperature can have greater but still modest effects. If we must estimate an activity coefficient for some T and P from experimental data at some other T and P we try to find data as close to the desired T as possible and make the P correction as shown in Example 9.6. Unfortunately, data on $(\bar{h}_i - h_i^{\text{o}})$ and $(\bar{v}_i - v_i^{\text{o}})$ are scarce. For that reason we often assume that they are negligible, which is equivalent to assuming that the activity coefficients depend on concentration, but not on T and P. As Figure 9.6 shows, this is a fair, but not excellent assumption.

9.6 TERNARY AND MULTISPECIES VLE

The general low-pressure VLE calculation procedure for mixtures with more than two species is the same as for two species; we attempt to find equations for the fugacities in gas and liquid phases that present f_i as a function of T, P, and the mol fractions in that phase. As the number of species goes up, the mathematical complexity increases and the intuitive content of the equations and the calculations declines. For most low-pressure VLE we continue to assume that the vapor is either an ideal gas or is an ideal solution of nonideal gases, using the Lewis–Randall (L–R) rule, written for each individual species in the mixture.

9.6.1 Liquid-Phase Activity Coefficients for Ternary Mixtures

For liquids, we use the same equations listed in Table 9.1, in forms that extend to any number of species. Holmes and Van Winkle [5] show the ternary (three species) forms for the common activity coefficient equations. Here we consider only the Margules equation, which seems more intuitive than the others (not necessarily the most accurate, but perhaps with the best product of accuracy times intuitive content). The form shown is

$$\begin{aligned}\log \gamma_a = {} & x_b^2[A_{ab} + 2x_a(A_{ba} - A_{ab})] \\ & + x_c^2[A_{ac} + 2x_a(A_{ca} - A_{ac})] \\ & + x_b x_c[0.5(A_{ba} + A_{ab} + A_{ac} + A_{ca} - A_{bc} - A_{cb}) \\ & + x_a(A_{ba} - A_{ab} + A_{ca} - A_{ac}) \\ & + (x_b - x_c)(A_{bc} - A_{cb}) - (1 - 2x_a)C^*]\end{aligned} \tag{9.10}$$

If we examine this term by term, we will see that it has more intuitive content than appears at first glance. Comparing it to Eq. 9.K we see that it is in log form, while Eq. 9.K is in ln form. Then we see that the first term on the right, $x_b^2[A_{ab} + 2x_a(A_{ba} - A_{ab})]$, is the same as the right-hand side of Eq. 9.K, if

$$A_{ab} = b = \log \gamma^{\infty}_{b,\text{ in a mixture with } a}$$

and

$$A_{ba} = c = \log \gamma^{\infty}_{a,\text{ in a mixture with } b} \tag{9.11}$$

Thus, this term represents the behavior of species a and b, in the absence of c. (If we set $x_c = 0$, all except the first term in Eq. 9.10 disappears, and we have the simple binary form of the Margules equation.) Similarly, the second term on the right represents the behavior of species a and c, in the absence of b, with

$$A_{ac} = b = \log \gamma^{\infty}_{c,\text{ in a mixture with } a}$$

and

$$A_{ca} = c = \log \gamma^{\infty}_{a,\text{ in a mixture with } c} \tag{9.12}$$

These two first terms are binary-interaction terms, representing two of the three possible binary combinations of species a, b, and c. The final term contains all three mol fractions and the four As shown above, plus the two additional As (A_{bc} and A_{cb}) for the third possible combination of these three species. It represents interactions between three molecules at a time. It also contains C^*, an adjustable parameter that can be used if ternary VLE data are available, to make the equation fit that data. If we have no such data, we set $C^* = 0.00$.

If we set $C^* = 0.00$, then Eq. 9.10 allows us to estimate the activity coefficient for species a in mixtures of a, b, and c, using only the mol fractions and the Margules equation constants for the three binary combinations of a, b, and c. To find the activity coefficients for species b and c we rotate the subscripts, $a \rightarrow b$; $b \rightarrow c$; $c \rightarrow a$ or simply renumber the species, making the one we are now interested in species a.

Example 9.8 Estimate the activity coefficient of acetone in a mixture at 1 atm and 66.70°C with

Species Number	Identity	Mol Fraction in Liquid, x_i
a	Acetone	0.1200
b	Methanol	0.1280
c	Water	0.7520

From [5] we read the following values (see Problem 9.28):

acetone–methanol (a-b)	$A_{ab} = 0.2634$	$A_{ba} = 0.2798$
acetone–water (a-c)	$A_{ac} = 0.9709$	$A_{ca} = 0.5579$
methanol–water (b-c)	$A_{bc} = 0.3794$	$A_{cb} = 0.2211$

The first term on the right of Eq. 9.10 is

$$\begin{aligned} & x_b^2[A_{ab} + 2x_a(A_{ba} - A_{ab})] \\ & = (0.128)^2 \cdot [0.2634 + 2 \cdot 0.12 \cdot (0.2798 - 0.2634)] \\ & = 0.00438 \end{aligned} \tag{9.Y}$$

By similar calculations (not very difficult or time-consuming on a spreadsheet!) the second and third terms are 0.4930 and 0.0550, so that the whole term on the right is 0.5523, and $\gamma_a = 10^{0.5523} = 0.357$. For this temmperature we may estimate the vapor pressure of acetone as 1.417 atm, so that we estimate

$$y_{\text{acetone}} = (0.12 \cdot 0.357 \cdot 1.417 \text{ atm}/1 \text{ atm}) = 0.6068 \tag{9.Z}$$

The experimental value is $y_{\text{acetone}} = 0.698$ [6]. ■

What are we to make of this result?

1. The calculated y_{acetone} is $(0.6068/0.698) = 87\%$ of the experimental y_{acetone}. If we had assumed ideal solution

we would have calculated $y_{acetone} = 0.170$, 24% of the experimental $y_{acetone}$. Using only the published data for the three possible binary mixtures we made a much, much better estimate than that. If we had one data point for a mixture of these three species we could have used it to estimate a value of C^* other than zero, and possibly have gotten a somewhat better estimate (see Problem 9.34).

2. From the three values we see that the dominant term is the one for the acetone–water interaction. If we had assumed that the mixture was 0.12 mol fraction acetone, balance water, then we would have estimated that $\gamma_a = 10^{0.6751} = 4.73$, 133% of the value shown above. (Looking back at the table of As we see that those for acetone–water are much larger than those for the other two combinations, so perhaps we should have known this result simply from looking at that table!) Thus, this calculation not only gives us an excellent estimate of the experimental behavior, but also shows us which of the three binary interactions provides most of the nonideality.
3. The three-molecule interaction term contributed $(-0.0988/0.5523 = -18\%)$ of the calculated exponent.
4. This mixture would be expected to be fairly nonideal, because of a variety of molecule types (water, an alcohol, and a ketone) and still the prediction was fairly close to the experimental value.
5. The chosen liquid concentration was for a low value of $x_{acetone}$, where we would expect the prediction, based on constants for acetone at infinite dilution, to be fairly reliable. Problem 9.32 shows that for other compositions the prediction is not as good, but still fairly good.

Summarizing ternary and multispecies liquid activity coefficient estimation, we see that the mathematical and computational complexity increases, but that using only the readily available binary data we can make a fair estimate of the observed behavior. More complex schemes are more accurate and more time-consuming, but are easily done in computer programs.

9.7 VAPOR-PHASE NONIDEALITY

In all of Chapter 8, the preceding parts of this chapter, and in almost all hand calculations of VLE, we assumed that the vapor was a mixture of ideal gases. Before we had computers we used the L-R rule for hand calculations when the pressures were high enough that the ideal gas assumption seemed unreasonable (see Section 8.6.2). Now that we all have computers and mostly have our VLE calculations done by process-design programs, the extra cost of estimating vapor phase nonideality and taking it into account seems small, so most of these programs do that. For modest pressures the resulting changes in the calculations are small, but for high pressures (Chapter 10) and unusual situations they can be significant.

The calculation is based on Eq. 7.18

$$\frac{f_i}{Py_i} = \hat{\phi}_i = \exp\left(\frac{-1}{RT}\int_{P=0}^{P=P} \bar{\alpha}_i dP\right) = \exp\int_{P=0}^{P=P} \frac{(\bar{z}_i - 1)}{P} dP \tag{7.18}$$

In Example 7.5 we showed how this calculation is made from the experimental PvT data. In Section 7.12.2 we showed that we can base the calculation on the EOS values for the pure species, using mixing rules. There we showed that the simplest possible mixing rule led to the L-R rule, an ideal solution of nonideal gases. Here we examine a very simple case of a nonideal solution of nonideal gases. In principle, we could use the same kind of equations we use for PvT calculations and for making up the steam tables or their equivalent. But those equations are designed to reproduce very accurately the measured PvT data for a pure substance, and are not very good for constructing the derivatives with respect to concentration, which are needed for Eq. 7.18 Instead, we use different equations, which are not as good at the steam-table problem, but are better at the composition-derivative problem.

There are many such EOSs, of varying complexity. In Chapter 10 and Appendix F we show examples of the same type as shown here, with more complex EOSs. Here we consider only one very simple EOS, the two-term, pressure-explicit form of the *virial EOS*,

$$z = \frac{PV}{RT} = 1 + \frac{BP}{RT} \tag{7.AD}$$

(see Problem 7.7), where B for the mixture is a function of temperature and composition, but not of pressure. If we solve this equation for $(z - 1)$, take the partial molar derivative, and insert the value in Eq. 7.18, we can integrate to

$$\frac{f_i}{Py_i} = \hat{\phi}_i = \exp\int_{P=0}^{P=P} \frac{(\bar{b}_i P/RT)}{P} dP = \exp\left(\frac{\bar{b}_i P}{RT}\right) \tag{9.13}$$

where $\bar{b}_i$ is the partial molar derivative of B. We can take $\bar{b}_i$ out of the integral in Eq. 9.13 because $\bar{b}_i$ depends on T and composition, but not on pressure.

This very convenient result means that if we have reasonable ways of estimating $\bar{b}_i$, then we can easily compute $f_i/Py_i = \hat{\phi}_i$ (see Table 7.D). For pure species at high pressures this EOS is not very good; we would never use it to make up the equivalent of the steam tables. But it has the great merit that if we know the values of B for each of the pure species in a mixture, then molecular interaction theory allows a reasonable estimate of the value of B of the mixture and thus of $\bar{b}_i$. The more complex equations regularly used in place of this simple one are also based to some extent on molecular interaction theory.

From molecular interaction theory we know that for the two-term pressure-explicit virial EOS, the value of B for any mixture is given by the following mixing rule:

$$B = \sum_i \sum_j y_i y_j B_{ij} \tag{9.14}$$

This equation can be extended to mixtures with more than two species, by increasing the number of summation signs and adding y_k, and so on. Here we will restrict the treatment to binary mixtures. If we expand the sums in Eq. 9.14 we find

$$B = y_a y_a B_{aa} + y_b y_b B_{bb} + y_a y_b B_{ab} + y_b y_a B_{ba} \tag{9.15}$$

Here B_{aa} and B_{bb} are the two pure species values of B. Like Eq. 7.39, Eq. 9.14 has the property that it gives the correct values for B for the two pure species. B_{aa} and B_{bb} account for the interactions of molecules of species a with other molecules of species a and of molecules of species b with other molecules of species b. The B_{ab} and B_{ba} account for the interactions of molecules of species a with those of species b, and of species b with those of species a. Since the latter two must be the same, the *cross-coefficient* B_{ab} is equal to B_{ba} and Eq. 9.14 reduces to

$$B = y_a y_a B_{aa} + y_b y_b B_{bb} + 2 y_a y_b B_{ab} \tag{9.16}$$

In principle, B_{aa}, B_{bb}, and B_{ab} can be calculated from molecular theories. In practice, that has proven difficult, so that while molecular interaction theory provides some information, we currently use correlations like the little EOS (Eqs. 2.48, 2.49 and 2.50) to estimate B_{aa} and B_{bb}, and use the semitheoretical or empirical mixing rules described below to estimate B_{ab}. Equation 9.14 assumes that while B of the mixture depends on vapor composition (so that we can find its needed partial molar derivative), B_{aa}, B_{bb}, and B_{ab} depend only on the properties of the pure species and the temperature.

Although we could work directly with Eq. 9.16, it is conventional to rearrange it as follows:

$$\begin{aligned} B &= y_a(1 - y_b)B_{aa} + y_b(1 - y_a)B_{bb} + 2y_a y_b B_{ab} \\ &= y_a B_{aa} + y_b B_{bb} + y_a y_b (2B_{ab} - B_{aa} - B_{bb}) \end{aligned} \tag{9.17}$$

and then the rightmost expression in parentheses is given its own symbol:

$$\begin{aligned} \delta &= (2B_{ab} - B_{aa} - B_{bb}) \\ B &= y_a B_{aa} + y_b B_{bb} + y_a y_b \delta \end{aligned} \tag{9.18}$$

Figure 9.7 shows the form of Eq. 9.18. From this figure (and tangent intercepts, Chapter 6) it is clear that if $\delta = 0$,

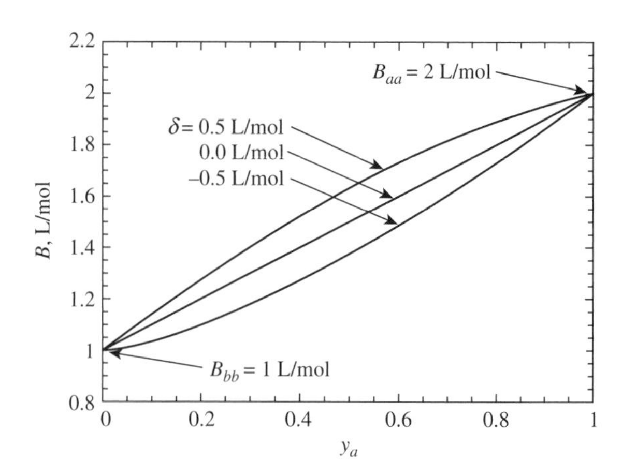

FIGURE 9.7 Equation 9.18 for arbitrarily selected values of B_{aa} = 2 L/mol, B_{bh} = 1 L/mol, and three choices of δ.

then $\bar{b}_a = B_{aa}$ and $\bar{b}_b = B_{bb}$ for all y_i. That corresponds to the L-R rule (an ideal solution of nonideal gases) (Section 8.6.2). So it is scarcely worth bothering with this formulation (and this whole section of this chapter) unless δ is known or believed to be significantly different from zero, and unless we have some way of estimating its value for mixtures of interest. If B_{ab} equals the average of the two pure species values, B_{aa} and B_{bb}, then from Eq. 9.18 we see that $\delta = 0$. Thus, what we are really looking for here is the difference between the cross-coefficient B_{ab} and the average of the pure species coefficients B_{aa} and B_{bb}. In principle, we can calculate B_{ab} from molecular theories, but in practice it is most often estimated from semitheoretical correlations.

The widely used estimating procedure for B_{ab} due to Prausnitz and his co-workers [7] is illustrated below. In it, instead of using one mixing rule for z, as we did in Chapter 7, we choose an EOS and then use separate mixing rules for each of the parameters that enter into it. These rules are usable in any EOS that takes as its input the corresponding-states variables T_r, P_r, z_c, and ω. Here we will use them in the little EOS, which is one of the simplest EOSs of that type. The computed mixing rule values are shown with the subscript *ij*, while the pure species values are shown with subscripts *ii* and *jj*, to match common convention.

The semitheoretical estimating rules begin with

$$T_{cij} = (T_{cii} \cdot T_{cjj})^{1/2}(1 - k_{ij}) \tag{9.AA}$$

where k_{ij} is an empirical factor whose value is determined by testing predictions against experimental data for the mixture. It is the only place where experimental mixture properties enter into this set of mixing rules. Of the several quantities that are called the *binary interaction parameter*, the k_{ij} in Eq. 9.AA is the most widely used. If you encounter a reference to a binary interaction parameter without a definition of which one is intended, your best guess is that it is the

k_{ij} in Eq. 9.AA. If no suitable data are available, k_{ij} is set equal to some small, positive quantity or to zero. Its values for a variety of mixtures are shown in [1, p. 596]. The remaining mixing rules are

$$\omega_{ij} = \frac{\omega_{ii} + \omega_{jj}}{2} \tag{9.AB}$$

$$z_{cij} = \frac{z_{cii} + z_{cjj}}{2} \tag{9.AC}$$

$$V_{cii} = \frac{z_{cii} R T_{cii}}{P_{cii}} \tag{9.AD}$$

$$V_{cij} = \left(\frac{V_{cii}^{1/3} + V_{cjj}^{1/3}}{2} \right)^3 \tag{9.AE}$$

$$P_{cij} = \frac{z_{cij} R T_{cij}}{P_{cij}} \tag{9.AF}$$

These mixing rules are based entirely on the pure species values, except for T_{cij}, which depends on k_{ij}, and P_{cij}, which depends on T_{cij}. The mixture composition enters via Eq. 9.18. We will illustrate the use of these rules in the little EOS for two examples previously considered.

Before we plunge into the mathematics, we may summarize where we are going. The procedure is: first estimate T_{cij}, P_{cij} and ω_{cij} of the mixture from the pure component values and k_{ij} by the mixing rules (9.AA to 9.AF) or by some other set of mixing rules which seem more appropriate; second compute the corresponding value of the mixing terms (also called *cross* terms) in whatever EOS we are using (B_{ab} in Eqs. 9.14 to 9.18 and in the little EOS) by inserting the appropriate values of T_{cij}, P_{cij} and ω_{cij} in the equations that define each such term in the EOS; then take the appropriate partial molar derivative of z from the combined EOS, and use it in Eq. 7.18 to estimate $\hat{\phi}_i$.

Example 9.9 Using the above mixing rules and the little EOS, estimate ϕ_i, $\hat{\phi}_i$ and $(\hat{\phi}_i/\phi_i)$ for ethanol and for water in a vapor at 85.3°C, 1 atm, and $y_{\text{ethanol}} = 0.4741$ (see Example 8.9). In Chapter 8 we worked numerous examples of the VLE of ethanol and water at 1 atm, all with the assumption that the vapor was practically an ideal solution of ideal gases. The results of this example will help us see how good (or bad) that assumption was.

The calculation is summarized in Table 9.A. We read the values of T_c, P_c, ω, and z_c for each pure species from Table A.1. For pure ethanol and pure water we use the procedure shown Exammple 7.1(c), leading to ϕ_{ethanol} and ϕ_{water} values of 0.9749 and 0.9863. The value of B for pure ethanol (see Problem 7.7) is

Table 9.A Estimation of ϕ_i, $\hat{\phi}_i$ and $(\hat{\phi}_i/\phi_i)$ for Ethanol and for Water in a Vapor at 85.3°C, 1 atm, and $y_{\text{ethanol}} = 0.4741$, using the Little EOS and Mixing Rules 9.AA Through 9.AF

Property	Ethanol, ii	Water, jj	Mix, ij, Assuming $k_{ij} = 0.00$	Mix, ij, Assuming $k_{ij} = 0.01$
T_c (K)	513.9	647.1	576.67	570.90
P_c (bar)	61.48	220.55	109.91	108.62
ω	0.645	0.345	0.495	0.495
z_c	0.24	0.229	0.2345	0.2345
V_c from Eq. 9.AE (mL/mol)	166.79	55.86	101.14	101.14
T_r	0.697	0.554	0.6216	0.6279
P_r	0.01647	0.004593	0.00911	0.009206
$f(T_r)$ in little EOS	−1.082	−1.664	−1.378	−1.338
$\frac{P_r \cdot f(T_r)}{T_r}$ in little EOS	−0.0256	−0.1380		
ϕ	0.9749	0.9863		
B (L/mol)	−0.7518	−0.4055	−0.5945	−0.5771
δ (L/mol)			−0.0313	0.00351
$\bar{b}_{\text{ethanol}}$ (L/mol)			−0.7605	−0.7509
$\bar{b}_{\text{water}}$ (L/mol)			−0.4125	−0.4047
$\hat{\phi}_{\text{ethanol}}$			0.9745	0.9750
$\hat{\phi}_{\text{water}}$			0.9861	0.9863
$(\hat{\phi}_i/\phi_i)_{\text{ethanol}}$			0.9995	0.9999
$(\hat{\phi}_i/\phi_i)_{\text{water}}$			0.9997	0.9999

$$B = f(T_r)\frac{RT_c}{P_c}$$

$$= -1.082 \cdot \frac{0.08206 \dfrac{\text{L} \cdot \text{atm}}{\text{mol} \cdot \text{K}} \cdot 513.9\ \text{K}}{61.48\ \text{bar}} \cdot \frac{1.013\ \text{bar}}{\text{atm}}$$

$$= -0.7518 \frac{\text{L}}{\text{mol}} \tag{9.AG}$$

and similarly $B = -0.4055$ L/mol for water.

We compute the mixture value of B_{ij} using Eq. 9.18, and Eq. 9.AA to AF. Here we have no information on k_{ij} so we will compute for assumed values of 0.00 and 0.01, which are suggested by various authors as plausible. Thus, to find T_{cij} for $k_{ij} = 0.00$, we compute

$$T_{cij} = (513.9\ \text{K} \cdot 647.1\ \text{K})^{0.5} = 576.67\ \text{K} \tag{9.AH}$$

and for $k_{ij} = 0.01$ we find 99% of that value. The other values are computed from the equations; for example, ω_{ij} is the arithmetic average of the two pure species values, independent of k_{ij}. B_{ij} is computed the same way as B in Eq. 9.18, but using the mixture values of P_r, T_r, and ω.

For $k_{ij} = 0.00$, we have

$$\delta = 2(-0.5955) - (-0.7518) - (-0.4059) = -0.0313 \frac{\text{L}}{\text{mol}} \quad (9.\text{AI})$$

By the normal methods of forming partial molar derivatives (see Problem 9.36) we find that that

$$\bar{b}_a = B_{aa} + y_b^2\delta \quad \text{and} \quad \bar{b}_b = B_{bb} + y_a^2\delta \quad (9.19)$$

so

$$\bar{b}_{\text{ethanol}} = -0.7518 + (1 - 0.4741)^2 \cdot (-0.0313) = -0.7605 \frac{\text{L}}{\text{mol}} \quad (9.\text{AJ})$$

Returning to Eq. 7.18 we see that

$$\frac{f_i}{Py_i} = \hat{\phi}_i = \exp\left(\frac{\bar{b}_i P}{RT}\right) \quad (9.20)$$

Substituting the value from Eq. 9.AJ we find

$$\hat{\phi}_{\text{ethanol}} = \exp\left(\frac{\bar{b}_i P}{RT}\right) = \exp\left(\frac{-0.7605 \frac{\text{L}}{\text{mol}} \cdot 1\ \text{atm}}{0.08206 \frac{\text{L} \cdot \text{atm}}{\text{mol} \cdot \text{K}} \cdot 358.45\ \text{K}}\right) = 0.9745 \quad (9.\text{AK})$$

In the same way we compute the other three values $\hat{\phi}$ shown in Table 9.A.

Finally, we may evaluate the relative importance of the nonideality of the pure species and the nonideality of mixing by computing

$$(\hat{\phi}_i/\phi_i)_{\text{ethanol}} = \frac{0.9745}{0.9749} = 0.9995 \quad (9.\text{AL})$$

and the corresponding other three values shown in Table 9.A. ■

In this 1-atm example, the ϕ_i, which represent the departure of the pure species from ideal gas behavior, are 0.975 and 0.986, independent of concentration and of our assumption for the binary interaction parameter k_{ij}. The $\hat{\phi}_i/\phi_i$, which represent the departure from ideal solution behavior and which depend both on the mol fractions and on our assumed values of the binary interaction parameter k_{ij}, are all practically unity. This EOS with this set of mixing rules suggests that under these conditions the vapor mixture is practically an ideal solution ($\hat{\phi}_i/\phi_i \approx 1.00$) of nonideal gases ($\phi_i \neq 1.00$). This is the description of the L-R rule (Section 8.6.2).

Looking back at Figure 9.7, we see that for the small values of δ in this example (-0.03 and $+0.003$) the predicted behavior is practically the same as the straight line for $\delta = 0$, the L-R rule; that makes the values of $(\hat{\phi}_i/\phi_i)$ practically 1.00. This whole calculation reinforces the statements made earlier that at modest pressures the nonideality of gas mixing is so small that we can ignore it with negligible errors. At pressures near the critical pressures of any of the species involved, this is not the case, as we will see in Chapter 10.

9.8 VLE FROM EOS

In all of Chapter 8 and the previous parts of this chapter we have used the Raoult's law type of liquid phase activity coefficients, as defined by $f_i = x_i\gamma_i p_i$. This is the most commonly used approach for most low-pressure VLE. However, for higher pressures (see Chapter 10) and for mixtures in which some species are above their critical temperatures, an alternative approach is widely used. This expresses the fugacity of species i in the *liquid* in terms of the system pressure P, instead of in terms of the vapor pressure p_i, by

$$f_i = x_i\hat{\phi}_i P \quad (9.21)$$

This has the merit that we do not need to know the pure species vapor pressure p_i, which is not defined for substances above their critical temperatures, such as the hydrogen that is dissolved in many industrially important liquids during hydroprocessing. Several EOSs have been designed to simultaneously calculate $\hat{\phi}_i$ for the various i in both the liquid and the vapor from the same equation. The most widely known equations designed for this purpose are the *Soave–Redlich–Kwong* (SRK) and *Peng–Robinson* (PR) equations often used for hydrocarbon mixtures. When we use a process-design computer package for VLE calculations, it will often offer the SRK or PR options for estimating the VLE. These compute the liquid-phase fugacities by Eq. 9.21; this approach is explored in Chapter 10 and Appendix F.

9.9 SOLUBILITY PARAMETER

This section and the next do not fit very logically here, but they fit as well here as any other place in the book and do not deserve a chapter of their own. So far this book has treated equilibrium as an experimental science, in which we use thermodynamics to help us correlate, interpolate, and extrapolate experimental data, without much effort to explain what is going on at the level of the molecules. The molecular explanations advanced so far are very simple intuitive pictures, chosen because they have high intuitive content, and do not do much violence to molecular theory. There is a vast literature on applying molecular, theory to phase and chemical equilibrium problems [8]. Most of it is too complex for an undergraduate textbook.

One theoretical idea, the *solubility parameter*, widely used for understanding and predicting VLE and also liquid–liquid solubility (LLE), is very simple and has a very high intuitive content. It is not as powerful or accurate as some of the more advanced theoretical methods, but its ratio of (intuitive content)/(complexity) is high enough that it makes sense to introduce it here. In Section 8.4 we discussed the various kinds of experimental VLE behavior and showed some simple intuitive ideas about how the interactions between the molecules influenced liquid behavior. The solubility parameter idea for describing those interactions, due to Scatchard and Hildebrand [3], is that in any pure-species liquid there is a certain force of attraction between the molecules. If two pure-species liquids have equal values of this force of attraction, then they should form ideal solutions. But if this force of attraction is quite different between two liquids, then the more strongly attracting liquid should "squeeze out" the less strongly attracting liquid, and the activity coefficients should be >1.00. The larger the difference between this force of attraction for the two liquids, the larger the activity coefficients should be.

Scatchard and Hildebrand suggested that the measure of this force of attraction is

$$\begin{pmatrix}\text{cohesive energy}\\ \text{density}\end{pmatrix} = \frac{\begin{pmatrix}\text{internal energy change}\\ \text{of vaporization}\end{pmatrix}}{(\text{liquid volume})} = \frac{\Delta u_{\text{vap}}}{v_L} = \frac{\Delta h_{\text{vap}} - RT}{v_L} \tag{9.22}$$

This is an intuitively satisfying idea. Liquids that have a high latent heat of vaporization require a large energy input to overcome the strong intermolecular attractions to change to a vapor. A more dense liquid (one with a lower value of v_L) must have a higher intermolecular attractive force per unit mass than one with a lower density. They then defined

$$\begin{pmatrix}\text{solubility}\\ \text{parameter}\end{pmatrix} = \delta = \sqrt{\begin{pmatrix}\text{cohesive energy}\\ \text{density}\end{pmatrix}} = \sqrt{\frac{\Delta u_{\text{vap}}}{v_L}} \tag{9.23}$$

They chose this form because in the *regular solution theory* [9], which they devised to go with the solubility parameter,

$$RT \ln \gamma_a = v_a \Phi_b^2 (\delta_a - \delta_b)^2 \tag{9.24}$$

where

$$\Phi_b = \begin{pmatrix}\text{volume}\\ \text{fraction of } b\end{pmatrix} = \frac{x_b v_b}{x_a v_a + x_b v_b} \tag{9.25}$$

The internal energy change of vaporization and the liquid volume may be expressed per mol or per gram; the molecular weight cancels. The traditional unit of the solubility parameter (sometimes called the Hildebrand parameter) used in most older publications is $(\text{cal/cm}^3)^{0.5}$, which some authors refer to as a *Hildebrand.* In SI, used in most current publications, the unit is $(\text{J/cm}^3)^{0.5} = (\text{MPa})^{0.5}$:

$$\left(\frac{\text{cal}}{\text{cm}^3}\right)^{0.5} = \sqrt{\frac{4.184\ \text{J}}{\text{cal}}} \cdot \left(\frac{\text{cal}}{\text{cm}^3}\right)^{0.5} = 2.045\left(\frac{\text{J}}{\text{cm}^3}\right)^{0.5}$$
$$= 2.045\ (\text{MPa})^{0.5} \tag{9.26}$$

If the solubility parameter idea and regular solution theory were exactly true, then we could compute all activity coefficients, based only on measurement of pure species heats of vaporization and liquid densities. This is not the case, as shown below. But the solubility parameter idea has been productive enough that we see it cited widely in the literature, and tables of solubility parameters are widely available [10, 11].

Example 9.10 Estimate the infinite dilution activity coefficients of n-hexane and diethyl ketone at 65°C. Table 9.B shows the molar volumes and solubility parameters at 25°C. Here we use these, with the assumption that $\gamma_{i,65°\text{C}} \approx \gamma_{i,\,25°\text{C}}$.

At infinite dilution, the volume fraction of the other species is 1.00, so, for n-hexane

$$\ln \gamma_a = \frac{v_a \Phi_b^2 (\delta_a - \delta_b)^2}{RT}$$
$$= \frac{131.6\dfrac{\text{mL}}{\text{mol}} \cdot 1^2 \cdot (14.9 - 18.1)^2 \text{MPa}}{8.314\dfrac{\text{m}^3\text{Pa}}{\text{mol}\cdot\text{K}} \cdot 338.15\ \text{K}} \cdot \frac{\text{m}^3}{\text{M(mL)}}$$
$$= 0.479 \tag{9.AM}$$

Table 9.B Molar Volumes and Solubility Parameters at 25°C

Substance	Molar Volume (mL/mol)	Solubility Parameter (MPa)$^{0.5}$
n-Hexane	131.6	14.9
Diethylketone	106.4	18.1

and

$$\gamma^{\infty}_{n\text{-hexane}} = \exp 0.479 = 1.62 \tag{9.AN}$$

Proceeding the same way for diethyl ketone, we find

$$\gamma^{\infty}_{\text{diethyl ketone}} = \exp 0.388 = 1.46 \tag{9.AO}$$

The experimental values at 65°C [1, p. 196] are 2.25 and 3.67. ■

This is only fair agreement but is about as good as we ever find with this approach. Regular solution theory excludes systems that associate or form hydrogen bonds, which lead to type III behavior (negative deviation from ideality). Equation 9.24 can only predict activity coefficients $\geq$1.00. For polar molecules its predictions are poor, especially for water–organic systems (see Problems 9.41 and 9.42). So why bother with it? Extensions of it to polar and hydrogen bonding systems have been made [11] and found very useful, particularly in the solvents and coatings industry. We regularly see it applied to systems where we do not have any experimental data. While it often does not give quantitatively useful information, it has been widely used to correlate solubilities of solids, liquids, and gases in liquids, often with very good success. Here and in Problems 9.41 and 9.42 we calculate the infinite dilution activity coefficients because they are easy ($\Phi_b = 1.00$). If we know the infinite dilution values we can use them in Eq. 9.24 or in many of the other common activity coefficient equations (see Section 9.4).

9.10 THE SOLUBILITY OF GASES IN LIQUIDS, HENRY'S LAW AGAIN

The solubility parameter helps us understand and correlate the solubility ot gases in liquids. Table 9.2 summarizes the experimental solubilities of various common gases in a range of solvents, at 1.00 atm and 25°C. The values are times 10^4, so that, for example, the mol fraction of He in perfluoro-n-hexane under these conditions is $x_{\text{helium}} = 0.00092$.

In this table the gases are arranged with increasing critical temperature from left to right, and the solvents arranged with increasing solubility parameter from top to bottom. There are some exceptions, but mostly the solubilities increase from left to right and from bottom to top. The values in Table 9.2 (excluding the water values) are plotted vs. the solubility parameter in Figure 9.8.

From Table 9.2 and Figure 9.8 we see the following:

1. The lines on Figure 9.8 are simple least-squares fits of the equation

$$\begin{pmatrix}\text{gas}\\ \text{solubity}\end{pmatrix} = x_{\text{solute}} = a\exp(b\delta) \tag{9.AP}$$

in which a and b are data-fitting constants (b is a small negative number), see Problem 9.46.

2. The data scatter about these lines, indicating that solubility parameter, while a good correlating parameter for gas solubility, is far from a perfect one. (For the 8 linear curve fits on Figure 9.8 the average value of R^2 is 0.82 – good correlation, but not perfect.)

Table 9.2 Mol Fraction of Dissolved Gas, $x_{\text{solute}} \times 10^4$ in Various Solvents, at 25°C and a Gas Pressure of 1.00 atm [1, p 329; 14, 15]

Gas →		He	H_2	Ne	N_2	Ar	O_2	CH_4	CO_2
Gas T_c (K) →		5.2	33.19	44.4	126.2	150.9	154.6	190.6	304.2
Solvent ↓	δ_{solvent}, $\left(\frac{\text{cal}}{\text{cm}^3}\right)^{0.5}$↓	Gas Solubility in Solvent, $x_{\text{solute}} \times 10^4$ ↓ →							
Pefluoro-n-hexane	5.9	9.2	14		36	54	54	78	203
n-Hexane	7.27	2.604	6.315	3.699	14.02	25.12	19.3	50.37	
n-Octane	7.54	2.397	6.845	3.626	13.04	24.26	20.83	29.27	
Cyclohexane	8.19	1.217	4.142	1.792	7.61	14.8	12.48	32.76	76.0
CCl_4	8.55		3.349		6.48	13.51	12.01	28.7	105.3
Toluene	8.93	0.974	3.171	1.402	5.74	10.86	9.09	24.14	101.3
Benzene	9.16	0.771	2.58	1.118	4.461	8.815	8.165	20.77	97.3
Acetone	9.62	1.081	2.996	1.577	5.395	9.068	8.383	18.35	185.3
Chlorobenzene	9.67	0.691	2.609	0.979	4.377	8.609	7.91	20.47	98.06
Nitrobenzene	10.8	0.35		0.436		4.448	4.95		99.8
Dimethyl sulfoxide	12.0	0.284	0.761	0.368	0.833	1.54	1.57	3.86	90.8
Ethanol	12.78	0.769	2.067	1.018	3.593	6.231	5.481	12.8	63.66
Methanol	14.5	0.595		0.814	2.747	4.491	4.147	8.695	55.78
Water	23.53	0.068	0.142	0.082	0.119	0.254	0.231	0.248	

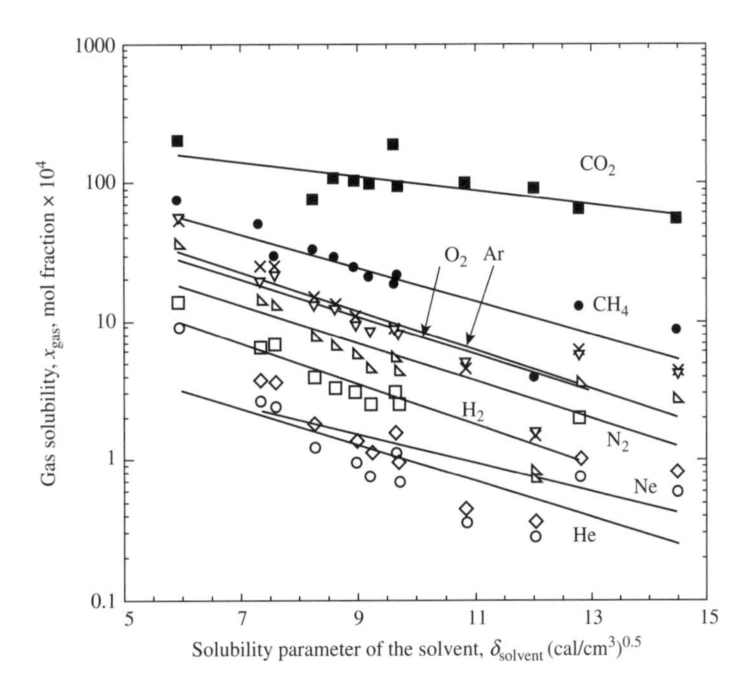

FIGURE 9.8 The solubility of a variety of gases in a variety of solvents, at 1 atm gas pressure and 25°C. The solubility is shown multiplied by 10^4, so that a 1 on the left axis corresponds to $x_{\text{gas-in-solution}} = 10^{-4}$ mol fraction. The liquid solvents are those in Table 9.2, in the order from the top of the table, omitting the values for water.

3. The solubility parameter for water is almost twice the highest of the others, so that including it in the plot would have put its points in the lower right corner of the figure, and bunched all the other points on the left side of the figure. Leaving water out makes a better figure.
4. For all of the gases the solubility decreases as the solubility parameter of the solvent increases. The solubility of O_2 in perfluoro-*n*-hexane is 234 times the solubility of O_2 in water. For all the gases in Table 9.2 the solubility in water is about one-tenth or less that gas's solubility in any of the other solvents. Water is a poor solvent for gases, because of its high solubility parameter (because of its very strong internal attractive forces – very high cohesive energy density) that makes it harder for gas molecules to enter than it is for gas molecules to enter perfluoro-*n*-hexane.
5. The family of perfluorocarbons has the lowest solubility parameters of any known set of compounds and correspondingly the highest gas solubilities.
6. If one remakes the plot and the linear correlations including water, whose solubility parameter is almost twice that of the highest value on Figure 9.8, one finds that the slopes and R^2s are very little changed; the lines on Figure 9.8, extrapolated to the solubility parameter of water pass very close to the solubility data points for water.
7. The lower critical temperature gases have lower solubilities than the higher critical temperature gases, with some exceptions. All of the gases shown are above their critical temperatures at 25°C, except for CO_2 (T_c = 31°C); its solubility is by far the highest shown.

This last observation suggests that we should be able to correlate the solubilities as a function of temperature. Returning now to Henry's law (see Sections 3.4 and 8.5) we recall that we found Henry's law by substituting a data-fitting constant, H_i, for the vapor pressure p_i in Raoult's law, because the temperatures of interest were above the critical temperatures of the gases, so there can be no pure liquid and hence no true vapor pressure. If we wished to estimate the Henry's law constant of the gases in Figure 9.8 from vapor pressure data, we would presumably use the linear extrapolation (on ln p vs. $1/T$ coordinates) shown on the left side of Figure 5.7. The efficient way to do this is simply to substitute $T = 25°C$ in the Antoine equation, using the values in Table A.2.

Example 9.11 Estimate the Henry's law constant for O_2 by extrapolation of the vapor-pressure curve as shown in Figure

5.7, and compare the resulting solubility of pure O_2 at 1.00 atm and 25°C with the values shown in Table 9.2.

Using the constants for O_2 in Table A.2 in the Antoine equation, we find an extrapolated vapor pressure at 25°C of 521.15 atm. If we take that value as equal to the Henry's law constant we find

$$x_i = \frac{y_i P}{p_i} = \frac{y_i P}{H_i} = \frac{1.00 \cdot 1 \text{ atm}}{521.15 \text{ atm}} = 19.2 \times 10^{-4} \quad ■ \tag{9.AQ}$$

This wild extrapolation ($P_r = 10.5!$) is independent of which solvent the O_2 was dissolved in. The quantity calculated is called the *ideal solubility of a gas.* Table 9.2 and Figure 9.8 makes clear that the solubility is highest in solvents with the lowest values of the solubility parameter. The lowest solubility parameters known are for perfluorocarbons of which perfluoro-*n*-hexane has the lowest value shown on that table and figure. The reported solubility, $x_{\text{oxygen in perfluoro-}n\text{-hexane}} = 54 \times 10^{-4}$. If instead of comparing with perfluoro-*n*-hexane we compare with *n*-hexane, we find the experimental solubility agrees very closely to the ideal solubility; if we make the same comparison for all the gases shown in Figure 9.8 in *n*-hexane we find the results shown in Table 9.3.

We see that for the three lowest-boiling gases, helium, hydrogen, and neon, the extrapolated values are not very close to the observed values. But for the other four gases the extrapolated values are between 0.8 and 1.31 times the observed values. This suggests that if we needed to know the solubility (i.e., the Henry's law constant) of some unknown gas in some unknown liquid, if we knew the Antoine equation constants for the gas and the solubility parameter of the liquid, we could calculate the ideal solubility as shown above, and use Figure 9.8 or its equivalents to *estimate* the fraction of the ideal solubility to be expected in that liquid. There is no theoretical basis tor this extrapolation to find the ideal solubility of gases at temperatures above their critical temperatures. The only reason for doing so is that using the resulting values in Raoult's law-type equilibrium calculations gives values that are often quite close to experimental values, and thus this approach gives useful estimates. The shorter the extrapolation, the more likely the result is to be useful. An additional example of this extrapolation is given in Chapter 10. This type of calculation is obviously speculative, but it does show that gas solubility is related to the extrapolated vapor-pressure curve and to the solubility parameter of the liquid.

Table 9.3 Comparison of Ideal Solubility of Several Gases to the Observed Solubility in *n*-Hexane, at 1.00 atm and 25°C

	Calculated, Ideal Solubility, $x_{gas} \times 10^4$ at 25°C and 1.00 atm	Observed Solubility in *n*-Hexane, $x_{gas} \times 10^4$ at 25°C and 1.00 atm	Ratio (calculated/observed)
Helium	40.63	2.604	15.60
Hydrogen	15.76	6.315	2.50
Neon	11.52	3.699	3.11
Argon	20.23	25.12	0.81
Nitrogen	18.33	14.02	1.31
Oxygen	19.19	19.3	0.99
Methane	40.64	50.37	0.81

Example 9.12 The above estimate of the solubility of O_2 in *n*-hexane was, in effect, the application of Raoult's law to a Henry's law situation, with the liquid-phase activity coefficient of the O_2, $\gamma_{O_2,\text{ in hexane}} = 1.00$. If we apply the same approach to O_2 in water at 25°C, what will the calculated activity coefficient be?

Rewriting Eq. 8.4, and using the above values we find

$$\gamma_a = \gamma_{O_2,\text{ in water}} = \frac{y_a P}{x_a p_a} = \frac{1.00 \cdot 1.00 \text{ atm}}{0.231 \times 10^{-4} \cdot 521.15 \text{ atm}} = 83.1 \quad ■ \tag{9.AR}$$

We see that we can look upon the solubility of O_2 in water at 25°C as a Raoult's law problem with the extrapolated vapor pressure and the very high activity coefficient shown, or we can look at it as a Henry's law problem. We can see that the result is the same, by writing that

$$H_{O_2} = p_{O_2,\text{ extrapolated}} \cdot \gamma_{O_2,\text{ in water}} = 521.15 \text{ atm} \cdot 8.31 = 43{,}300 \text{ atm} \tag{9.AS}$$

which is very close to the value we would interpolate in Table A.3.

If we look at the solubility of gases in liquids from the Raoult's law viewpoint we would expect a plot of the $\gamma_{O_2,\text{ in water}}$ vs. x_{oxygen} to be a type II plot, like Figure 8.8b, with a very large value of $\gamma^{\infty}_{O_2,\text{ in water}}$. This suggests that as the pressure increases, and the mol fraction of dissolved gas increases, the activity coefficient should decrease, so that (solubility/pressure) should increase with increasing pressure. However, to find out how large that effect is, we continue Example 9.12. According to Henry's law, at 25°C we would expect the mol fraction of dissolved O_2 in water at 1 atm to be 0.231×10^{-4}, and at 10 atm to be 0.231×10^{-3}. If we next assume that the activity coefficient for O_2 in water can be tolerably represented by the symmetrical equation (Eq. 9.G), then we would expect that

$$\frac{\gamma_{O_2 \text{ in water, 10 atm}}}{\gamma_{O_2 \text{ in water, 1 atm}}} = \frac{(x_{\text{water, 10 atm}})}{(x_{\text{water, 10 atm}})} = \frac{(1 - x_{O_2,\text{ 10 atm}})^2}{(1 - x_{O_2,\text{ 1 atm}})^2} = \frac{(1 - 0.000231)^2}{(1 - 0.0000231)^2} = 0.9996 \tag{9.AT}$$

We see that for the symmetrical equation (and most other useful activity coefficient equations) the activity coefficient of the *solute* is proportional to the square of the concentration of the *solvent*. For most gases dissolved in liquids, the change in solute concentrations is so small that the concentration of the solvent is practically constant, $\approx$1.00. Thus, over the range of practical interest the activity coefficient—Raoult's law type—of the dissolved solute gas is practically constant, which is the same as saying that Henry's law is obeyed within experimental accuracy, with $H_i \approx p_{i,\text{extrapolated}}\,\gamma_i^{\infty}$.

Extensive reviews present the experimental data and the theory of solubility of gases in liquids [8,14]. (Mostly such data compilations do not report Henry's law constants, but rather mol fractions of dissolved gas at 1.00 atm and various temperatures. Calculating Henry's law constants from these data is easy.) These compilations show that if the gas does not interact chemically with the solvent, and the mol fraction of the gas in the solution is small (as in Figure 9.8 and Table 9.2) then Henry's law is practically obeyed up to fairly large pressures, with the Henry's law constant being of the order of magnitude of the (1.00/ideal solubility) for solvents with low values of the solubility parameter and being substantially larger for solvents like water with high values of that parameter. Experimental data also show that Henry's law works much better at temperatures well below the critical temperature of the solvent than close to that temperature. Close to the critical temperature of the solvent we normally abandon Henry's law, and use the methods shown in Chapter 10 and Appendix F.

We often hear that the solubility of gases in liquids decreases with increasing temperature. If the activity coefficient calculated in Example 9.12 for O_2 were independent of temperature, then that would certainly be true because the extrapolated vapor pressure increases with increasing temperature, at all temperatures. However, as the temperature increases, the $\gamma_{O_2,\text{ in water}}$, in Eq. 9.AS increases, then goes through a maximum, and then decreases, so that eventually the solubility of oxygen begins to increase with increasing temperature, with the minimum solubility at about 380 K $\approx$ 100°C [14, p. 294]. This behavior is observed in most common gases. However, the temperature at which the solubility starts to increase with increasing temperature is above room temperature for most gases, so that the common statement that gas solubility decreases with increasing temperature is true for our most common experiences. The theory of why dissolved gases behave this way is in [8].

We will consider the combination of Henry's law and chemical reactions in the liquid in Chapter 13.

9.11 SUMMARY

1. The most common observation of VLE liquid-phase activity coefficients are types I, II, and III, shown in Figure 9.1. Type IV (heteroazeotrope) is an extreme version of type II. The type of activity coefficient pattern shown in Figure 9.2 is uncommon, but does occur.
2. The Gibbs–Duhem equation limits the possible mathematical forms we may choose to represent liquid-phase (or vapor- or solid-phase) activity coefficients.
3. Most widely used liquid-phase activity coefficient equations represent the group $g^E/(RTx_a x_b)$ as some relatively simple algebraic function of the liquid mol fractions. If we choose $g^E/(RTx_a x_b) = $ some constant, we find the symmetrical equation, which is the simplest activity coefficient equation which is consistent with the Gibbs–Duhem equation. More complex functions are more successful at fitting experimental VLE data.
4. Changes in pressure have very little effect on activity coefficients. Changes in temperature sometimes have little effect, and sometimes have significant effects.
5. We can estimate ternary liquid-phase activity coefficients from the measured (or calculated) activity coefficients of the three binary pairs with fair accuracy. The extension to larger number of species in the solution is more complex mathematically.
6. In most low-pressure VLE hand calculations, we assume that the vapor phase is an ideal solution of ideal gases. Most computer VLE programs include vapor-phase nonideality in their calculations. One easily understood approach uses the two-term, pressure-explicit form of the virial EOS, for which molecular theory provides the basis for estimating the values of the constants for various molecular interactions. For low pressures the result is practically the same as the L-R rule, because the computed deviations from ideal solution behavior of the mixture are much smaller than the calculated deviations from ideal gas behavior of the pure species.
7. Most low-pressure VLE calculations use the Raoult's law type of formulation, which uses the pure-species vapor pressures. A widely used alternative approach that works much better at high pressures computes both liquid- and gas-phase fugacities from the same EOS, most often using the SRK or PR EOSs, discussed in Chapter 10 and Appendix F.
8. The solubility parameter is widely used to correlate and predict solid–liquid, liquid–liquid, and gas–liquid solubilities. It has fair accuracy for nonpolar liquids, but poor accuracy for polar liquids.
9. Henry's law is satisfactory for most solutions of sparingly soluble gases at temperatures well below the critical temperatures of the solvent. The ideal solubility, calculated by extrapolating the vapor-pressure equation above the critical temperature, often leads to fair

estimates of the solubility of gases in nonpolar liquids. The difference between the ideal solubility and that estimated by Henry's law can be estimated from the solubility parameter of the solvent. Thus, the same problem can be seen as a Henry's law problem, or a Raoult's law problem with an extrapolated vapor pressure and an experimentally-determined γ_i^∞.

PROBLEMS

See common units and values for problems and examples. An asterisk (*) on a problem number indicates that the answer is in Appendix H.

9.1 Show the algebra leading from the statement of the Gibbs–Duhem equation in terms of chemical potentials to the form in terms of activity coefficients (Eq. 9.1 to Eq. 9.2).

9.2 Sketch what Eq. 9.B would look like on ln γ_a vs. x_a coordinates. Assume that A, B, C, and D are all ≥ 0. Does it agree with the shapes shown on Figure 9.1?

9.3 The Gibbs–Duhem equation (Eq. 9.2) requires that on ln γ_i vs. x_i coordinates the two curves must each become tangent to $\gamma_i = 1.00$ as x_i, approaches 1.00 and that at $x_i = 0.5$ the two curves must have equal and opposite slopes. Check Figures 8.7, 8.8, 8.9, 8.12, and 9.2 to see whether these statements appear to be obeyed there.

9.4 If we have $\ln \gamma_a = Ax_b^2$, show that the Gibbs–Duhem equation requires that $\ln \gamma_b = Ax_a^2$.

9.5 Sketch the equivalent of Figure 9.1 for the case in which both ln γ_i had positive slopes for all x_i. Show that pattern does not match any of the patterns shown in Figures 9.1 and 9.2 and that it contradicts the Gibbs–Duhem equation.

9.6 Show that the van Laar equation is consistent with the Gibbs–Duhem equation. (This is easiest if you start with Eq. 8.L.)

9.7* **a.** Repeat Example 9.1 for $x_a = x_{\text{isopropanol}} = 0.8020$.

b. The measured equilibrium temperature at 1.00 atm for this mixture is 80.55°C, and the measured $y_{\text{isopropanol}} = 0.768$. Estimate the activity coefficients for water and isopropanol based on these experimental values and compare them to the values computed in part (a).

9.8 Show the derivation of Eq. 9.4 from Eq. 9.3. *Hint*: Write Eq. 6.4, letting Y be G^E, and equate it to Eq. 9.3 finding

$$G^E = \bar{g}_a^E n_a + \bar{g}_b^E n_b + \cdots = n_a(RT \ln \gamma_a) + n_b(RT \ln \gamma_b) + \cdots$$

Take the derivative with respect to n_i.

9.9* If $g^E/x_a x_b RT = A$ for all values of x_a, what is the value of γ_a for $x_a = 0.4$:

a. For $A = 0$?

b. For $A = 0.4$?

9.10* The van Laar equation constants (log form, i.e., log γ_a = etc.) for acetone (species a) and water (species b) at 1 atm are shown in Table A.7. Based on these constants estimate

a. The activity coefficient for acetone in a solution that is 99.9999 mol% water.

b. The activity coefficient for water in a solution that is 99.9999 mol% acetone.

9.11* **a.** Estimate K_{ethanol} at infinite dilution (i.e., $x_{\text{ethanol}} = 10^{-9}$) in water, in a solution at its boiling point at 1 atm. Use the van Laar equation, with the constants for ethanol–water used in Chapter 8 (and Table A.7).

b. Repeat part (a) for K_{water} at infinite dilution (i.e., $x_{\text{water}} = 10^{-9}$) in ethanol, in a solution at its boiling point at 1 atm (≈78.4°C, at which $p_{\text{water}} \approx 0.438$ atm).

9.12 Show the derivation of Eq. 9.6 from Eq. 9.5.

9.13 Show the derivation of Eq. 9.7. *Hint.* This is the application of the method of tangent intercepts (Section 6.3) to Eq. 9.5.

9.14 If we separate variables in Eq. 9.6 and integrate, we find

$$\int dg^E = g_{\text{final}}^E - g_{\text{initial}}^E = RT \int \ln \frac{\gamma_a}{\gamma_b} dx \qquad (9.27)$$

If we perform the integration from $x_i = 0.00$ to 1.00, then both of the values of g^E are zero (see Figure 9.3) so that the value of the integral on the right must be zero.

This forms the basis for a widely used method of testing experimental VLE data for thermodynamic consistency. Using the data in Table 8.1 and following the procedure in Example 8.2, compute the value of ln (γ_a/γ_b) for each data point in that table, and plot the values vs. x_a. If the consistency test is met, then the areas above and below the $\ln(\gamma_a/\gamma_b) = 0$ line should be equal. Test visually or mathematically to see if they are.

9.15 As x_a goes to zero both the numerator and denominator of $g^E/(RTx_ax_b)$ approach zero. What is the value of $g^E/(RTx_ax_b)$ for $x_a = 0$? for $x_b = 0$?

9.16 Sketch the equivalent of the g^E/RT curve on Figure 9.3 for the data shown in Figures 8.7, 8.8 and 8.9. Rough sketches will do.

9.17 Show the derivation of Eq. 9.1 from Eq. 9.G in two ways (see Examples 6.4 and 6.5):

a. Show the derivation using Eq. 9.4 which is the equivalent of the method of tangent slopes (Section 6.3). This type of derivation is done in terms of the partial molar derivative, which involves total mols not mole fractions. Start with Eq. 9.G, rewritten as $g^E = aRTx_ax_b$. Replace each of the x_i with (n_i/n_T) and substitute in Eq. 9.4, leading to

$$\begin{aligned} RT \ln \gamma_a &= \frac{\partial}{\partial n_a}(G^E)_{T,\,P,\,n_b} \\ &= \frac{\partial}{\partial n_a}(n_T g^E)_{T,\,P,\,n_b} \\ &= \frac{\partial}{\partial n_a}\left(\frac{aRTn_an_b}{n_T}\right)_{T,\,P,\,n_b} \end{aligned} \tag{9.AU}$$

Perform the partial molar differentiation, then convert (n_i/n_T) back to x_i, using $x_a = (1 - x_b)$ and simplify.

b. Show the derivation using Eq. 9.7, which is the equivalent of the method of tangent intercepts (Section 6.4). Replace the x_b, in Eq. 9.G with $(1 - x_a)$, carry out the differentiation, simplify, and then replace $(1 - x_a)$ with x_b.

9.18 Repeat the preceding problem parts (a) and (b), starting with Eq. 9.H and finding Eq. 9.K.

9.19 Show both algebraically and graphically on Figure 9.3 that if $b = c$, then the Margules equation (Eq. 9.K) becomes the symmetrical equation.

9.20 Show that the van Laar equation (Eq. 8.11) is equivalent to

$$\frac{g^E}{RTx_ax_b} = \frac{AB}{Ax_a + Bx_b} \tag{9.AV}$$

Hint: Start with the Eq. 8.L form of the van Laar equation.

9.21 Show the values of the infinite dilution activity coefficients for

a. The van Laar equation.

b. The Scatchard–Hildebrand equation.

The infinite dilution activity coefficient relations for other common activity coefficient equations are shown in [1, p. 182].

9.22 In Figure 9.2, one of the activity coefficients at infinite dilution is >1.00, while the other is <1.00. This means that one of the $\ln \gamma_i^\infty$ is positive, the other negative. These are believed to be very reliable experimental data.

a. Can this situation be represented by the van Laar equation? Show the form that the equation would take and what difficulties it would encounter.

b. Can this situation be represented by the Margules equation? Show the form that the equation would take. Then, estimating the values of the γ_i^∞ from Figure 9.2, make a table of the γ_i values computed from the Margules equation and sketch it. Does it have the same general form as Figure 9.2?

c. In Figure 9.2 each of the activity coefficients has a minimum or maximum (horizontal part of the curve) inside the figure. Can that same behavior arise in the Margules equation with both coefficients >1.00? To test this compute the values and sketch the equivalent plot for the water–*n*–butanol binary, using the following published values of the Margules constants (ln form): $b = 0.8608$, $c = 3.2051$.

9.23 One may view the uncommon aspect of Figure 9.2 that the individual activity coefficient curves have local maxima or minima inside the figure.

a. Is it possible for one of the curves to have a maximum or minimum $(d \ln \gamma_i)/dx_i = 0$ without the other curve showing a minimum or maximum at the same value of x_i?

b. Show that for the Margules equation (Eq. 9.K) we expect $(d \ln \gamma_i)/dx_i = 0$ for $x_i = 1.00$, for any choice of b and c, and that we expect $(d \ln \gamma_i)/dx_i = 0$ for some value of x_i, inside the figure if $(b/(c - b)) \leq 1.00$. *Hint:* Differentiate Eq. 9.K with respect to x_i set the derivative $= 0.00$, divide out $(c - b)$, and consider the limiting case for which the local maximum or minimum occurs at $x_i = 0.00$.

9.24 Repeat Example 9.6 for the water in the same mixture.

9.25 Repeat Example 9.7 for the water in the same mixture.

9.26 In Example 9.7 we averaged $(h_a^\circ - \bar{h}_a)$ over the temperature range 70–90°C, and integrated $\partial T/T^2$. One could argue that we should have averaged $(h_a^\circ - \bar{h}_a)/T^2$ over the same temperature range and then integrated ∂T. How much difference would it make? Repeat Example 9.7 doing the averaging this way, and compare the computed value of $\gamma_{a \text{ at } T_2}/\gamma_{a \text{ at } T_1}$ to the value of the same ratio shown in Example 9.7.

9.27 Show the calculations for the second and third parts of the right-hand side of Eq. 9.10 in Example 9.8, then show the sum of all three terms, the activity coefficient, and $y_{acetone}$.

9.28 The Margules equation constants in Example 9.8 were taken from [5]. The van Laar constants for those three binary systems are shown in Table A.7. Prepare a table comparing the Margules values in Example 9.8 and the van Laar values from Table A.7. Should they be the same? Are they the same? Practically the same? Or very different? Explain.

9.29 **a.** Show that if we write the excess Gibbs energy as the ternary equivalent of the symmetrical equation

$$\frac{g^E}{RT} = Ax_a x_b + Bx_a x_c + Cx_b x_c \tag{9.AW}$$

then

$$\ln \gamma_a = Ax_b^2 + Bx_c^2 + x_b x_c(A + B - C) \tag{9.AX}$$

b. Show that if in Eq. 9.10 we let $A_{ba} = A_{ab}$ and the same for the other two sets of coefficients, then Eq. 9.10 becomes the same as Eq. 9.AX (without the added C term which we normally ignore), showing that this is the (approximate) ternary equivalent of the symmetrical equation (Eq. 9.I).

c. Using the spreadsheet for Example 9.8, replace the values of the six individual coefficients by using the averages, so for example we replace A_{ab} = 0.2634 and A_{ab} = 0.2798 with $A_{ba} = A_{ab} = 0.5*(0.2634 + 0.2798) = 0.2716$. Show how much this changes the computed value of $y_{acetone}$.

9.30 Show the ideal-solution calculations corresponding to Example 9.8.

9.31 Show the calculations supporting the statement below Example 9.8 that if we had assumed that the mixture were equivalent to 12 mol% acetone, balance water, we would have computed $\gamma_a = 10^{0.6751} = 4.73$.

9.32 Repeat Example 9.8 for the values shown in Table 9.C.

9.33 Using your spreadsheet for Example 9.8, compute the values of $y_{methanol}$ and y_{water} in that example. Do the ys sum to 1.00? The reported experimental values [6] are 0.082 and 0.2218. The 0.082 value is almost certainly an experimental error; the value you compute here is likely to be closer to correct than that value.

Table 9.C Values for Problem 9.32

Part of Problem	x_a	x_b	$T_{equilibrium}$ (°C)	$y_{a,experimental}$
(a)	0.1700	0.7050	64.90	0.3260
(b)	0.5570	0.0630	59.10	0.885
(c)	0.0730	0.6550	69.40	0.196

9.34 Using the spreadsheet program you developed for the preceding three problems,

a. Find the value of C^* that makes the calculated $y_{acetone}$ in Example 9.8 equal to the experimental value.

b. Using this value of C^*, repeat the above problem.

c. Discuss the result. Does using one experimental point to adjust the estimate of the three-molecule interactions improve the predictive power of the equation for other compositions? Make it worse? Make no change?

9.35 Sketch a three-dimensional figure showing the ln $\gamma_{acetone}$ as a function of the mol fractions in Example 9.8. Make the base of your figure a triangular mol fraction diagram, with pure acetone, water, and methanol as the vertices, and plot ln $\gamma_{acetone}$ vertically upward. A simple sketch of the three-dimensional ln $\gamma_{acetone}$ surface will suffice.

9.36 Show the derivation of Eq. 9.19 from Eq. 9.18.

9.37 Show that combining Eqs. 9.19 and 9.20 produces

$$\ln \frac{f_a}{Py_a} = \ln \hat{\phi}_a = \left(\frac{\bar{b}_a P}{RT}\right) = \left(\frac{P}{RT}\right)(B_{aa} + y_b^2 \delta) \tag{9.AY}$$

Then show that since $\hat{\phi}_a = \phi_a \cdot \dfrac{\hat{\phi}_a}{\phi_a}$ we can write

$$\frac{RT}{P} \ln \hat{\phi}_a = \frac{RT}{P} \ln \phi_a + \frac{RT}{P} \ln \frac{\hat{\phi}_a}{\phi_a} = (B_{aa} + y_b^2 \delta) \tag{9.AZ}$$

For pure a this becomes $\phi_a = \exp(PB_{aa}/RT)$ from which a little algebra shows that $\hat{\phi}_a/\phi_a = \exp(Py_b^2\delta/RT)$. Thus in this formulation, B_{aa} accounts for the pure-species departure from ideality and $y_b^2\delta$ accounts for the nonideality of mixing.

9.38* Repeat Example 9.9 for the same composition, $P = 5$ atm, $T = 300$°C, and $k_{ij} = 0.00$ (Here we use 300°C because at 85.3°C and 5 atm this mixture is all liquid.)

9.39 Repeat Example 8.9, taking into account the vapor-phase nonideality computed in Example 9.9. How much difference does it make?

9.40 Show that the Scatchard–Hildebrand equation (Eq. 9.24) has the same form as the van Laar equation by showing how one would compute A and B in the van Laar equation from the variables in Eq. 9.24.

9.41 Repeat Example 9.10 for isopropanol–water at 90°C using the values from Table 9.D and compare the values to those in Figure 8.8b. This should convince

Table 9.D Molar Volumes and Solubility Parameters at 25°C

Substance	Molar Volume (mL/mol)	Solubility Parameter ((MPa)$^{0.5}$)
Isopropanol	76.8	23.52
Water	18	47.83
Benzene	89	18.8

you that regular solution theory is not a reliable predictor for polar solvents like water.

9.42* Repeat Example 9.10 for benzene–water at 90°C, using the values from Table 9.D. The experimental values, calculated from liquid–liquid solubility data (see Example 11.4) are $\gamma_{\text{benzene}}^{\text{(water-rich phase)}} = 2500$ and $\gamma_{\text{water}}^{\text{(benzene-rich phase)}} = 333$. Comment on the applicability of regular solution theory to solutions involving water and organics.

9.43 Show that the extrapolation to values of $T > T_c$ in Figure 5.7 and Example 9.11 is practically the same as applying the Antoine equation at $T > T_c$.

9.44 Do the values for the solubility of various gases in water in Table 9.2 agree with the (interpolated) values of the Henry's law constants in Table A.3?

9.45* Equation 9.AT shows that for oxygen dissolved in water, for pressures of 1 and 10 atm, we would expect the ratio of the solute activity coefficients to be 0.9996. What would the value of the solubility ($x_{\text{solute gas}}$) at 1 atm have to be for this ratio to be 0.99? 0.95?

9.46 The straight lines on Figure 9.8 are calculated from Eq. 9.AP. The constants in the equations are shown in Table 9.E. Show the extrapolated values for the solubility in water ($\delta = 23.53$ (cal/cm^3)$^{0.5}$). Compare them to the experimental values in the last line of Table 9.2. Are they identical? Close to one another? Wildly different? Comment on the results of this extrapolation.

9.47 Propane and nitrogen are in equilibrium at 25°C and 200 psia. The vapor is 76.1 mol% propane, balance nitrogen. Estimate the mol fraction of nitrogen in the liquid. The solubility parameter of propane is 6.4 (cal/cm^3)$^{0.5}$. At this temperature and pressure the vapor is not quite an ideal gas, but may be assumed to be a mixture of ideal gases for the purposes of this problem. The experimental value [15] is 0.9 mol% which is similar to but not the same as the answer based on Section 9.10.

Table 9.E Constants for Linear Fits on Figure 9.8

Gas	a	b
Ne	12.614	−0.23391
He	19.145	−0.29947
N_2	111.2	−0.30965
H_2	73.063	−0.33658
O_2	177.06	−0.30921
Ar	217	−0.32102
CH_4	294.62	−0.27635
CO_2	327.33	−0.11907

REFERENCES

1. Walas, S. M. *Phase Equilibria in Chemical Engineering*. Boston: Butterworth, p. 535 (1985).
2. Gillespie, P. C., W. V. Wilding, and G. M. Wilson. Vapor–liquid equilibrium measurements on the ammonia–water system from 313 K to 589 K. *AIChE Symp. Ser.* 83(256): 97–127 (1987).
3. Hildebrand, J. H., and R. L. Scott. *The Solubility of Non-Electrolytes*, ed. 3. New York: Wiley (1950).
4. Larkin, J. A. Thermodynamic properties of aqueous non-electrolyte mixtures, I: excess enthalpy for water + ethanol at 298.15 to 383.15 K. *J. Chem. Thermodyn.* 7:137–148 (1975).
5. Holmes, M. J., and M. Van Winkle. Prediction of ternary vapor–liquid equilibria from binary data. *Ind. Eng. Chem.* 62:21–31(1970).
6. Bunch, D. W., W. J. James, and R. S. Ramalho, A rapid method for obtaining vapor-liquid equilibrium data. *I&EC Proc. Des. Dev.* 2(4):282–284(1963).
7. Smith, J. M., H. C. Van Ness, and M. M. Abbot, *Introduction to Chemical Engineering, Thermodynamics*, ed. 7. New York: McGraw-Hill, p. 409 (2005).
8. Prausnitz, J. M., R. N. Lichtenthaler, and E. Gomez de Azevedo. *Molecular Thermodynamics of Fluid-Phase Equilibria*, ed. 3. Upper Saddle River, NJ: Prentice Hall, Chapter 12 (1999).
9. Hildebrand, J. H., J. M. Prausnitz, and R. L. Scott. *Regular and Related Solutions: The Solubility of Gases, Liquids and Solids*, New York: Van Nostrand Reinhold (1970).
10. Barton, A. F., *CRC Handbook of Solubility Parameters and Other Cohesion Parameters*, Boca Raton, FL: CRC Press (1991).
11. Hansen, C., and A. Beerbower. Solubility parameters. In *Kirk–Othmer Encyclopedia of Chemical Technology*, ed. 2, *Supplement Volume*, Mark, H. F., J. J. McKetta, Jr., and D. Othmer, eds. New York: Wiley–Interscience, pp. 889–910 (1971).
12. Clever, L., and R. Battino. The solubility of gases in liquids. In *Techniques of Chemistry, Vol. 8: Solutions and Solubilities*, Dack, M. R. J., ed. New York: Wiley, p. 386 (1975).
13. Hougham, G. (ed.) *Fluoropolymers, Volume 2: Properties*, Hingham, MA: Kluwer Academic Publishers, p. 118 (1999).
14. Fogg, P. G. T., and W. Gerrard. *Solubility of Gases in Liquids: A Critical Evaluation of Gas/Liquid Systems in Theory and Practice*. Chichester, UK: Wiley (1991).
15. Schindler, D. L., G. W. Swift, and F. Kurata. More low temperature V–L design data. *Hydrocarbon Processing* 45(11): 205–210 (1966).

10

VAPOR–LIQUID EQUILIBRIUM (VLE) AT HIGH PRESSURES

Chapters 8 and 9 showed the common observations and calculation methods for low-pressure VLE. However, the observations at high pressures are different, and those calculation methods run into trouble at high pressures. Figure 10.1 shows why.

The species shown are all straight-chain aliphatic hydrocarbons, which would be expected to have very close to ideal-solution behavior with each other. The presentation in terms of K_i is the standard petroleum engineering practice (see Section 8.2). If all possible mixtures of these species formed ideal solutions with each other and the vapor is an ideal gas (Raoult's law), then the K_is would be independent of composition, and be given by $K_i = p_i/P$. For any value of T the value of p_i is fixed for each species, so we would expect the data to form a series of parallel straight lines with slope (−1) on a plot of log K_i vs. log P. Such lines are shown, dotted, in Figure 10.1. We see that below 100 psia all the data come close to the calculated Raoult's law lines. The temperature, 150°F = 339 K, is above the critical temperatures of methane and ethane (190.6 and 305.3 K), so the values of p_i for methane and ethane used in Raoult's law are Antoine equation extrapolations above the critical temperature (see Figure 5.7 and Example 9.12). Even with this long extrapolation, the experimental data are close to Raoult's law for methane and ethane. (However, the extrapolated-value $K_{\text{methane and ethane}}$ are larger than the experimental ones, which influences the calculated results in Examples 10.1 and 10.2.) This figure shows that at 150°F, for pressures below 100 psia, we could estimate the VLE for all mixtures of straight-chain aliphatic hydrocarbons (up to $C_{10}H_{22}$) from Raoult's law with fair accuracy. This partly explains the long-term popularity of Figure 8.20; it is practically this set of assumptions, with minor modifications.

Above about 100 psia (sooner for the higher molecular weight species than for the lower molecular weight ones), the data in Figure 10.1 begin to deviate significantly from Raoult's law. As the pressures approaches 3000 psia, the K_i all approach 1.00, indicating that the composition of the liquid and the vapor are approaching one another, eventually becoming identical. Obviously, something is happening here that is quite different from what we observe in low-pressure VLE. At the right side of Figure 10.1 we are entering the region near the mixture critical state, in which liquid and vapor are becoming more and more alike. The thermodynamic fundamentals are still the same; for each species in each VLE the fugacity of that species is the same in each of the coexisting phases. But the ways of estimating those fugacities that worked well at low pressures work poorly here, so we will need different ones.

10.1 CRITICAL PHENOMENA OF PURE SPECIES

Figure 10.2 shows the pressure-volume behavior of pure water near its critical point. We see that the vapor specific volume falls rapidly with increasing pressure (and temperature; the pressure temperature relation is shown in Figure 1.8). Table 10.1 shows the changes from 2000 psia to the critical pressure (3203.5 psia).

Physical and Chemical Equilibrium for Chemical Engineers, Second Edition. Noel de Nevers.

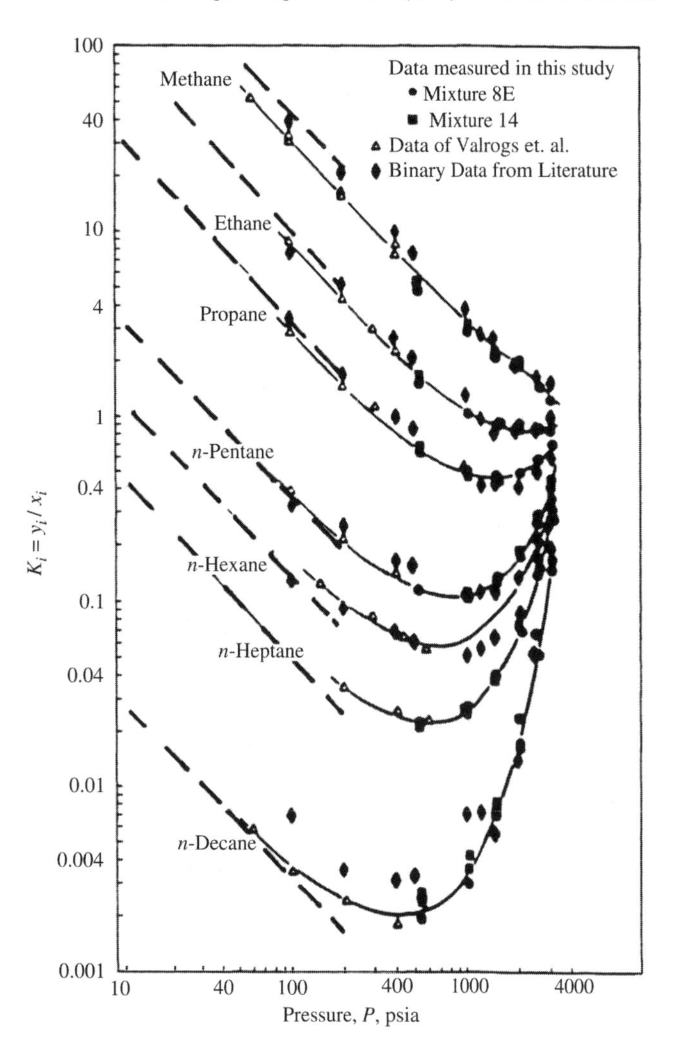

FIGURE 10.1 Observed K values for a series of aliphatic hydrocarbons at 150°F. The data are from a variety of sources and represent a variety of compositions. The solid lines are empirical curve fits for the individual species. The dotted lines are calculated by Raoult's law. (Reprinted with permission from Yarborough, L. Vapor–liquid equilibrium data for multicomponent mixtures containing hydrocarbon and nonhydrocarbon components. *J. Chem. Eng. Data* 17: 129–133. Copyright (1972) American Chemical Society.)

Over this 68.4°F temperature range the pressure increases by 60%. The compressibility factor of the liquid increases by almost a factor of ≈ 3. It is becoming more and more like an ideal gas (even though the pressure is rising!). Simultaneously, the vapor's compressibility factor falls by a factor of ≈ 2. It is becoming less like an ideal gas and more like a liquid. At the critical point the liquid and the vapor become the same. If we are watching this happen in a transparent high-pressure equilibrium cell, we see the interface between liquid and gas become fainter and fainter, and finally simply disappear. In Figure 1.8, the vapor-pressure curve for water simply ends at the critical point; vapor and liquid do not have separate existence above that temperature. (In Figure 10.2 a fluid

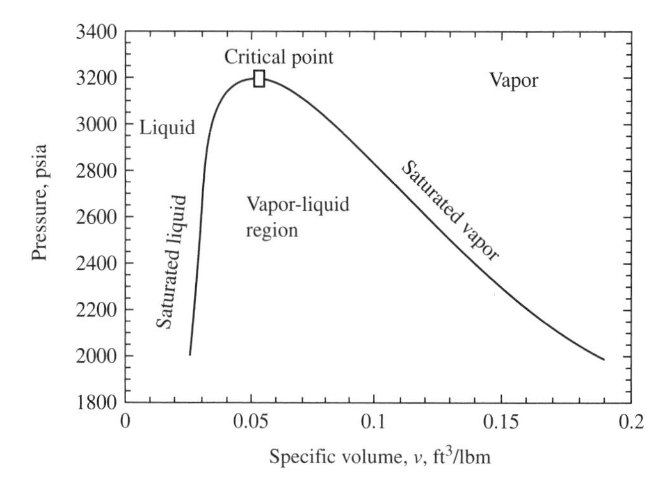

FIGURE 10.2 Pressure-specific volume plot for water near the critical point [1].

with $v \approx 0.05$ ft^3/lbm and $P \approx 3203.5$ psia would commonly be called a *dense fluid* because its behavior is not what we would expect of either a liquid or a gas.)

10.2 CRITICAL PHENOMENA OF MIXTURES

Naturally, the critical phenomena of mixtures are more complex and varied than those of single pure species. Here we consider only the simplest (and most common) type. The experimental data are mostly obtained in devices like that shown in Figure 10.3.

If we place a mixture of known composition in such a device at a low pressure, set the temperature, and slowly introduce mercury to reduce the volume and thus increase the pressure, we find that initially the mixture is all vapor. At the dew-point pressure the first drop of liquid appears. As the pressure is further increased the liquid volume increases and the vapor volume decreases, until at the bubble point the last bubble of vapor disappears, and only liquid remains. Obviously, we could run the process in reverse, from high pressure to low, passing through the same states in reverse order. We can also place a sample in the cell and slowly raise the temperature, simultaneously removing mercury at a suitable rate to maintain constant pressure; that process is sketched for benzene–toluene in Figure 8.7d. The typical result of such experiments for a binary mixture is sketched in Figure 10.4.

Table 10.1 Changes in Saturated Vapor and Liquid Water from 2000 psia to the Critical Point

P (psia)	2000	3203.5
T (°R)	1096.67	1165.11
z_{liquid}	0.078	0.23
z_{vapor}	0.576	0.23

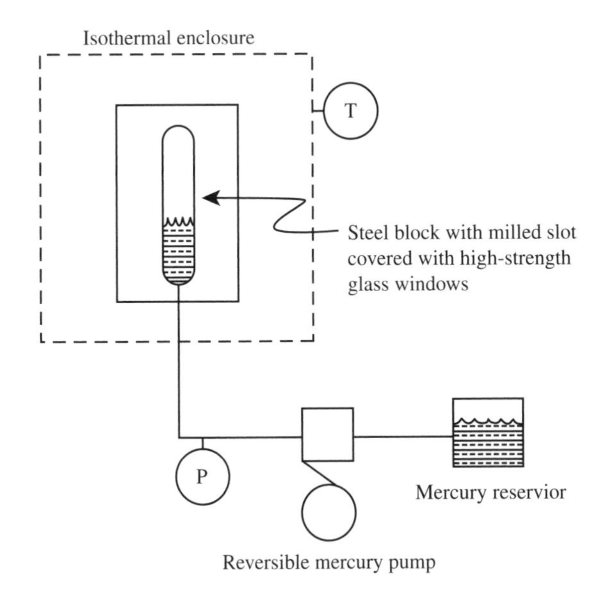

FIGURE 10.3 Schematic of a high-pressure phase-behavior cell. The sample is contained in a stainless steel block with a slot milled in it, which is covered front and rear with high-strength glass windows, held in place with bolted retainers and high-pressure gaskets. The volume, and thus the pressure, in the cell is controlled by pumping mercury from a reservoir into or out of the cell, thus compressing or expanding the sample. The pressure in the cell is practically equivalent to that of the mercury. The whole apparatus is inside a temperature-controlled environment, which can be set for a variety of temperatures. A light shines through the cell from the rear. The cell is viewed from the front through a mirror, so that if the glass windows fail, the fragments will hit the mirror, not the investigator. This sketch does not show the piping and valves for getting the sample into and out of the cell, nor the provisions for rocking the cell to promote equilibrium between gas and liquid [2].

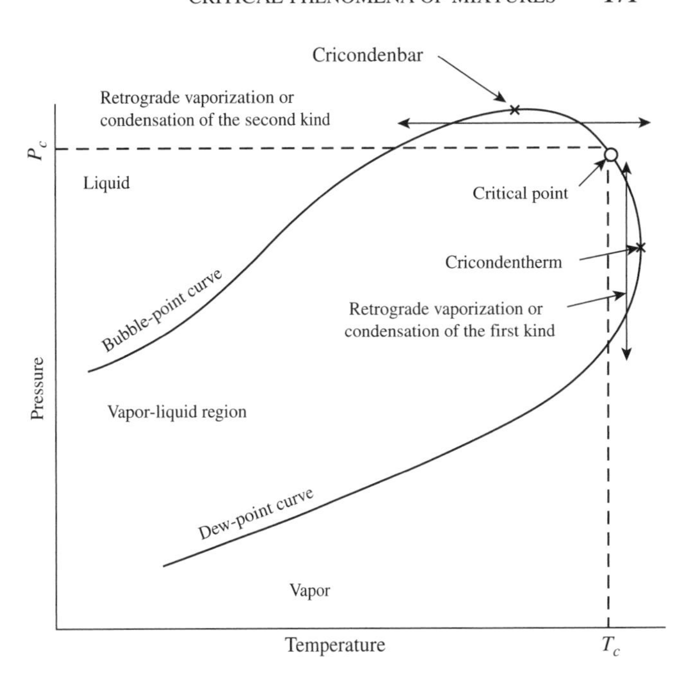

FIGURE 10.4 Typical experimental result for one specific mixture of two species, showing bubble-point and dew-point curves, critical point, cricondentherm and cricondenbar, and the paths for retrograde condensation or vaporization of the first and second kinds.

In Figure 10.4 we see that for a temperature well below T_c the behavior is what we expect at low pressure; increasing or decreasing pressure at constant temperature (moving vertically on the figure) or increasing the temperature at constant pressure (moving horizontally on the figure) produces the set of results described above. Normally, we will not have data on the compositions of the coexisting vapor and liquid, which are difficult to measure with this kind of device, but will simply note the pressure and temperature corresponding to the bubble and dew points. (At the bubble point almost all the material is in the liquid, so that the liquid-phase composition is almost the same as that of the material originally placed in the cell and at the dew point it is almost all in the vapor, so that the vapor-phase composition is almost the same as that of the material originally placed in the cell.)

If we run the same experiment, going up in pressure, at T_c we find that at P_c the two phases become identical and the interface between them vanishes. This is the critical point of this particular mixture and the observed behavior is practically the same as we see in the same experiment for a pure species (except that for a pure species the whole process is colorless, while for mixtures, as we pass through the critical point the system forms fine mists that refract light producing strong colors that disappear with time). However, if we go to some temperature slightly above the critical temperature of the mixture we see behavior that is never seen in a pure species. For a pure species no liquid can exist at a temperature above the critical temperature. For many (but not all) mixtures, there is some range of temperatures above the critical temperature in which if we begin with vapor and compress at constant temperature we will see that a liquid forms at the bottom of our cell (a dew point). As we continue to increase the pressure, the volume of liquid increases, then decreases, and finally the last drop of liquid disappears (another dew point!). Thus, we have crossed the dew-point curve twice. We can see in Figure 10.4 that this is possible. This kind of behavior is called *retrograde vaporization of the first kind.* If, at the same constant temperature, we proceed from high pressure to low, we pass through the same states, in reverse order, and observe *retrograde condensation of the first kind.* As the temperature is increased, the range of pressures over which two phases exist becomes less and less, and as sketched in Figure 10.4 there is some maximum temperature (called the *critical condensation temperature* or *cricondentherm)* at which we barely observe the existence of the liquid. At still higher temperatures no liquid can exist.

Figure 10.4 also shows that for temperatures slightly below the critical temperature the bubble-point curve is at a higher pressure than the critical pressure. The maximum pressure point on the bubble-point curve is called the *critical condensation pressure* or *cricondenbar*. For pressures between the critical pressure and the cricondenbar it is possible to start with a liquid, heat it at constant pressure, and observe that when the bubble point is reached gas begins to appear, the volume of the gas increases, then decreases, and finally the gas disappears and we have all liquid again. We crossed the bubble-point curve twice! This behavior is called *retrograde vaporization of the second kind,* and if, at the same fixed pressure, we proceed from high temperature to low, we see the same states in reverse order, called *retrograde condensation of the second kind.*

Retrograde condensation of the first kind occurs in petroleum reservoirs of the "gas-condensate" type. It has been intensely studied, because the recovery of valuable hydrocarbons is significantly increased if we avoid such retrograde condensation as the pressure in such fields is reduced (at nearly constant temperature) [3]. The economic benefits of understanding and avoiding this behavior are very large. As far as I know. retrograde condensation/vaporization of the second kind is a laboratory curiosity with no industrial or economic significance.

Figure 10.5 shows the results of a set of experiments like that sketched in Figure 10.4 for a variety of mixtures of ethane and *n*-heptane [4]. In this figure we see that curves 1 and 10 represent the vapor pressure curves for pure ethane and pure *n*-heptane, analogous to the vapor–pressure curve shown for water in Figure 1.8 (and represented throughout this book by the Antoine equation, see Chapter 5). Curves 2 through 6 are all of the type sketched in Figure 10.4, each representing some fixed-composition mixture, for various concentrations of ethane. Curves 7 through 9 show only the dew points of those mixtures, which are mostly *n*-heptane. The dotted curve running from the end of curve 1 to the end of curve 10 is the *critical locus*, which simply connects the critical points of each of the pure species and the various mixtures.

The information in Figure 10.5 can be represented in an intuitively useful three-dimensional diagram, sketched below as Figure 10.6. In this figure the axes are *P, T*, and $x_{\text{higher boiling species}}$. The near and far faces of the figure are the simple pure-species vapor–pressure curves (like curves 1 and 10 in Figure 10.5 or Figure 1.8). Between those two faces two surfaces are sketched: the upper is the bubble-point surface, the lower the dew-point surface. For pressures above the bubble-point surface, only liquid exists. For pressures below the dew-point surface only vapor exists. Between the two surfaces both vapor and liquid exist. The two surfaces join along the critical locus and on the front and back (pure species) faces. We may think of this figure as showing the equivalent of a mitten (with no thumb) made by stitching two

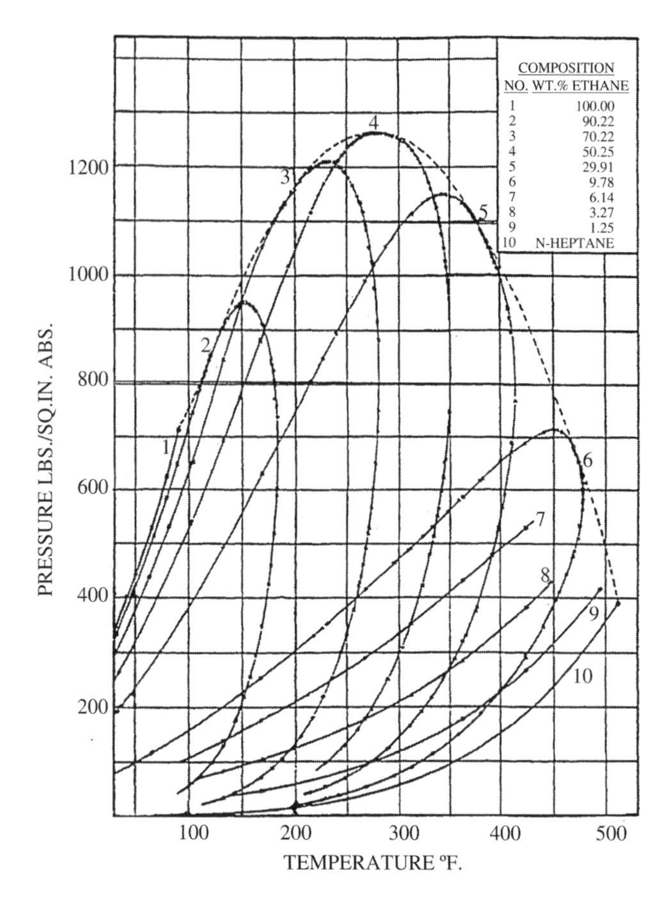

FIGURE 10.5 Bubble- and dew-point curves for a variety of mixtures of ethane and *n*-heptane. See the text for a description of this figure. (Reprinted with permission from Kay, W. B. Liquid–vapor phase equilibrium relations in the ethane-*n*-heptane system, *Ind. Eng. Chem.* 30: 459–465. Copyright (1938) American Chemical Society.)

pieces of fabric together along the two pure-species vapor-pressure curves and along the critical locus. Along the critical locus the junction is quite rounded, as we would have if we stitched the two surfaces together, and then turned them inside out (using a somewhat stiff fabric). Along the vapor-pressure edges the junction is sharp, as if we laid the two surfaces fiat, stitched them together, and then trimmed off the excess beyond the sewn seam. If we make a constant-composition slice through the figure, the result will be like Figure 10.4. A constant pressure slice is a *T*-*x* diagram, like those shown in part d of Figures 8.7, 8.8, 8.9, and 8.12. Figure 10.7 shows such slices for the same system shown in Figure 10.5, at various pressures.

The 400-psia slice produces a curve very much like the 1-atm benzene–toluene curve shown in Figure 8.7d, but with a wider spacing between the curves because of the greater difference in pure-species boiling points. The higher pressure (e.g., the 1200 psia) slice does not reach either of the pure species axes. These high-pressure slices correspond to

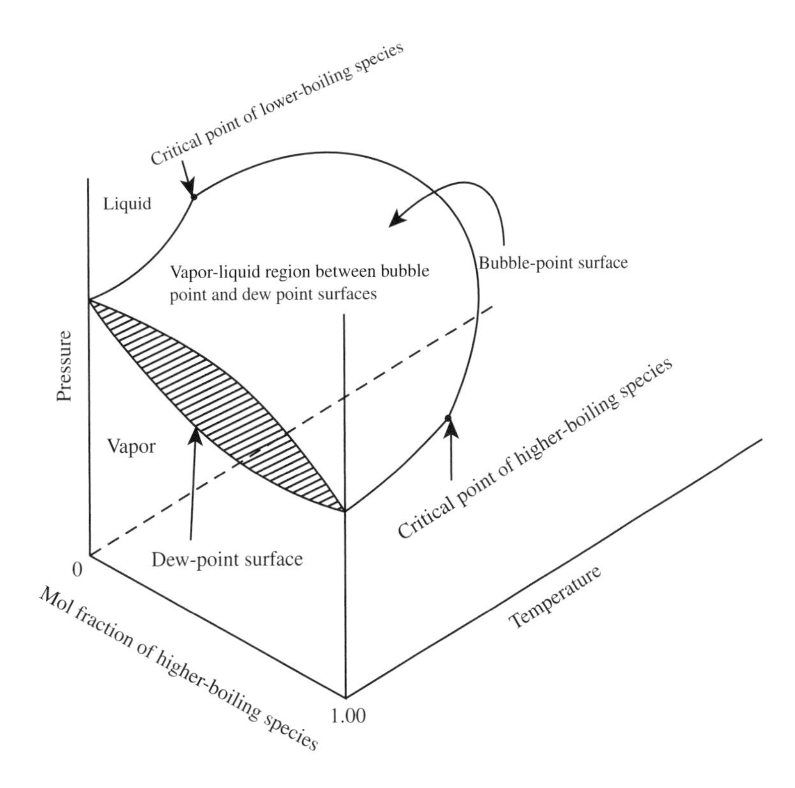

FIGURE 10.6 Three-dimensional representation of the phase behavior of a binary mixture like that in Figure 10.5. This figure is not to scale, nor specific for any pair of species.

the "fingertips of the mitten" in Figure 10.6. A similar plot could be produced by a constant-temperature slice, which would look similar to Figure 10.7, but with axes P and x_{ethane}. The space between a pair of curves would still represent a vapor–liquid mixture, but the space below both curves would represent vapor and that above the curves would represent liquid. (In Figure 10.7 at a fixed pressure the upper curve is the dew point, the lower a bubble point; on a constant temperature slice the upper curve is a bubble point and the lower a dew point.)

Example 10.1 Estimate the bubble-point temperature at 100 psia of a liquid with 10 mol% ethane, balance n-heptane and the mol fraction ethane in the equilibrium vapor (a) from Figure 10.7, (b) from Raoult's law, and (c) from Raoult's law, using the Lewis–Randall (L-R) rule for the vapor.

a. By direct chart reading, tne (bubble-point curve for 100 psia crosses the 10-mol% ethane line at about 165°F. We find the corresponding mol fraction ethane in the equilibrium vapor phase by reading horizontally from that intersection to the dew-point curve, finding $y_{\text{ethane}} \approx 92\%$. (These values are a mixture of chart reading and interpolation in the experimental data tables [4] from which Figure 10.7 was produced.)
b. By Raoult's law this is a trial-and-error procedure on the temperature, identical to Example 3.5. Using the Antoine constants from Table A.2, we find a bubble-point temperature of 133°F, and $y_{\text{ethane}} \approx 96.8\%$.
c. To add the L-R rule, we observe that instead of the simple Raoult's law, we now must use

$$y_i = \frac{x_i p_i}{\phi_i P} \tag{10.1}$$

We calculate the ϕ_i from the little EOS (see Example 7.1(c)). The results are summarized in Table 10.A. ■

This example shows the following:

1. Both Raoult's law and Raoult's law plus the L-R rule substantially underestimate the equilibrium temperature. This is mostly because the extrapolated vapor pressure of ethane is too high to match the experimental values in Figure 10.7. Looking back to Figure 10.1 we see that the Raoult's law values for ethane on that figure lead to a higher K (and thus a lower equilibrium temperature) than the experimental values.
2. The L-R rule makes little change for the ethane, which practically behaves as an ideal gas. But for n-heptane the calculated value is low enough that it makes a

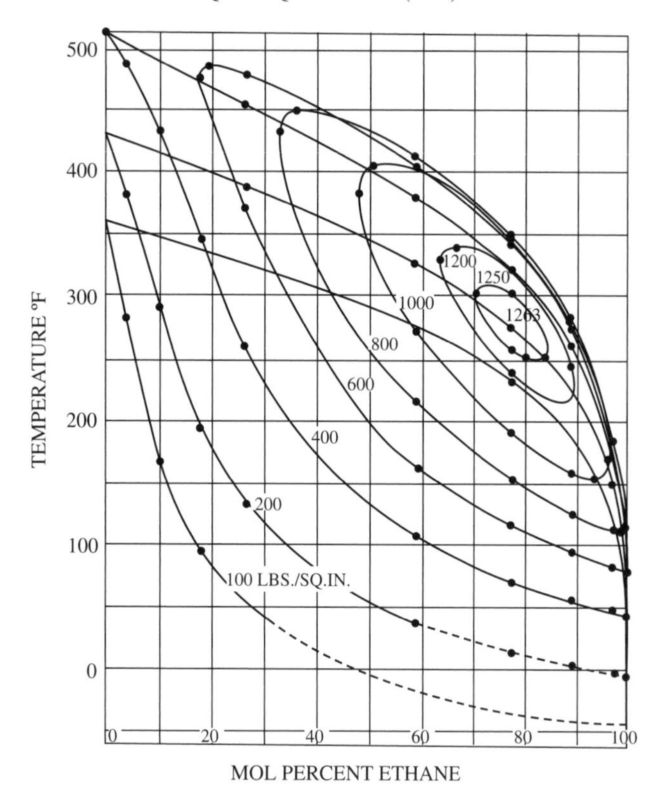

FIGURE 10.7 A series of constant-pressure slices through the equivalent of Figure 10.6, for the ethane–*n*-heptane system. Each pair of curves (e.g., the 400-psia curves) divides the space into liquid (below both curves), vapor-liquid mix (between the curves), and vapor (above the upper curve). (Reprinted with permission from Kay, W. B. Liquid–vapor phase equilibrium relations in the ethane–*n*-heptane system. *Ind. Eng. Chem.* 30: 459–465. Copyright (1938), American Chemical Society.)

substantial difference. Adding the L-R rule makes the error in the calculated temperature greater, but makes the calculated $y_{n\text{-heptane}}$ much closer to the observed value.

3. Ethane is above its critical temperature; it could not exist as a pure liquid at this temperature. The temperature is low enough and the pressure high enough that *n*-heptane could not exist as a vapor at this temperature and pressure. That combination of values makes this a severe test of Raoult's law and the L-R rule. The results shown here demonstrate fair, but not good, agreement between calculation and experiment.
4. The calculated temperature is largely determined by the estimate of K_{ethane}, which must be ≈ 9 because the vapor, experimental or calculated, is mostly ethane, and its mol fraction in the liquid is 0.1. The differences in calculated temperatures are mostly due to different estimates of K_{ethane} as a function of temperature.
5. The calculated mol fractions in the vapor phase are mostly determined by $K_{n\text{-heptane}}$. Both parts (b) and (c) underestimate this value, leading to low estimates of the $y_{n\text{-heptane}}$. Adding the L-R rule to the calculation substantially improves this estimate. The estimate in part (b) is low by a factor of 2.6, that in part (c) by 1.4.

Example 10.2 Repeat Example 10.1 for 800 psia and $x_{\text{ethane}} = 0.6$.

a. The 800-psia bubble-point curve line crosses the $x_{\text{ethane}} = 0.6$ line at $T \approx 210°\text{F}$. Reading horizontally from that intersection to the dew-point curve we, read $y_{\text{ethane}} \approx 95\%$. By linear interpolation in the experimental data tables on which Figure 10.7 is based we make a slightly more reliable estimate, $T \approx 209°\text{F}$ and $y_{\text{ethane}} \approx 94.5\%$.
b. The Raoult's law calculation, similar to that in Example 10.1, leads to a bubble-point temperature of 172°F and $y_{\text{ethane}} \approx 99.6\%$.
c. The Raoult's law plus L-R rule calculation (using the little EOS) leads to a bubble-point temperature of 72°F, and $y_{\text{ethane}} \approx 64\%$. ■

Raoult's law makes a very poor estimate of these data at 800 psia and $x_{\text{ethane}} = 0.6$; the calculated bubble-point temperature is low by about 37°F, and the calculated $y_{n\text{-heptane}}$ is low by a factor of 20. Adding the L-R rule makes the calculated values much worse, because the $\phi_{n\text{-heptane}}$ calculated by the little EOS = 0.001. Clearly, we should not use Raoult's law or Raoult's law plus L-R rule and the little EOS for mixtures like this one!

The treatment in Figures 10.1 to 10.7 is all for straight-chain aliphatic hydrocarbon (HC) mixtures for which we would expect the liquid behavior to be close to ideal solution behavior. These mixtures have been studied in detail because they occur at temperatures and pressures near the critical regions in some petroleum reservoirs and in high-pressure petroleum processing. We have less information about non-hydrocarbon mixtures, because they seldom occur near their critical temperatures and pressures in industrial equipment. (Prudent engineers avoid near-critical states in ordinary process equipment, because gas–liquid behavior becomes very difficult to predict. Some processes, such as *supercritical extraction,* intentionally use such states, taking advantage of the unusual solvent properties of fluids in such states.) We would expect mixtures whose liquids form non-ideal solutions and perhaps azeotropes to have more complex high-pressure VLE behavior than that shown here for straight-chain aliphatic HCs. Some of this is reviewed in [5].

10.3 ESTIMATING HIGH-PRESSURE VLE

The methods used in Chapter 8 and 9 are rarely used for VLE near the critical region of the mixture (the right side of

Table 10.A Results of Example 10.1

Variable	Values Read or Calculated from Figure 10.7	Values Calculated from Raoult's Law	Values Calculated from Raoult's Law with the L-R Rule
P (psia)	100	100	100
x_{ethane}	0.1	0.1	0.1
T (°F)	165	133	124
y_{ethane}	0.92	0.968	0.943
$y_{n\text{-heptane}}$ by difference	0.08	0.031	0.056
$p_{\text{ethane, extrapolated}}$ (psia)		968	896
$p_{n\text{-heptane}}$ (psia)		3.5	2.9
$P_{r,\text{ethane}}$			0.141
$P_{r,n\text{-heptane}}$			0.251
$T_{r,\text{ethane}}$			1.062
$T_{r,n\text{-heptane}}$			0.600
ϕ_{ethane} (little EOS)			0.950
$\phi_{n\text{-heptane}}$ (little EOS)			0.459
K_{ethane}	9.2	9.68	9.43
$K_{n\text{-heptane}}$	0.089	0.035	0.056

Figure 10.1, or the "fingertips of the mitten" in Figure 10.6.) Instead we use one of the following:

1. Empirical K-value correlations
2. Separate correlations for vapor and liquid, not starting from Raoult's law
3. Equations of state that represent both vapor and liquid

10.3.1 Empirical K-Value Correlations

Before we had computers, methods were devised for estimating high-pressure VLE of hydrocarbon mixtures, using graphs, slide rules, and mechanical hand calculators. These are now only of historical interest, but today's student should know about them to understand the older literature. The most widely used method started with figures like Figure 10.1. In that figure the solid lines represent empirical curve fits for mixtures of these hydrocarbons for which the K_i values all *converged* to 1.00 at a pressure of 3000 psia. Data for other mixtures showed that the *convergence pressure* for such systems depended on temperature and composition. Correlations were advanced for estimating at which pressure such convergence would occur, and a library [6] of plots like Figure 10.1 was produced for various convergence pressures, based on experimental data where available and by analogy to Figure 10.1 where data were not available. With the correlations one could estimate the proper convergence pressure, and then estimate K_i values using the proper charts. Then those values were used to estimate VLE, exactly as in Example 8.14. In this method the dependence of the individual K_i on composition appears only in the choice of convergence pressure. For a specified convergence pressure, the K_i are independent of composition. This method is discussed in [3, Chapter 6].

10.3.2 Estimation Methods for Each Phase Separately, Not Based on Raoult's Law

The most widely used method of this type is that of Chao and Seader [7]. This estimation method (and several variants of it proposed by other authors, particularly Grayson and Streed [8]) is offered as a calculation method in many current process-design computer packages and is widely recommended as being very good with hydrocarbon systems at high pressure (like Figure 10.1).

To see how it works we rewrite the general equation for equilibrium between vapor and liquid (Eq. 8.3) as

$$K_i = \frac{y_i}{x_i} = \frac{\gamma_i p_i}{\hat{\phi}_i P} \tag{10.2}$$

In Raoult's law-type formulations, we normally take $\hat{\phi}_i = \phi_i$, (the L-R rule, and ideal solution of nonideal gases), but Chao and Seader retained $\hat{\phi}_i$ and computed it from the Redlich and Kwong (RK) EOS, as described in detail in the next section and in Appendix F. For the liquid they departed radically from Raoult's law by replacing p_i with $(\phi_i)_{\text{pure liquid } i} \cdot P$. Substituting this value in Equation 10.2 produces

$$K_i = \frac{y_i}{x_i} = \frac{\gamma_i (\phi_i)_{\text{pure liquid } i}}{\hat{\phi}_i} \tag{10.3}$$

They calculated γ_i, by a slightly modified version of Equation 9.24, based on Scatchard and Hildebrand's regular solution theory, and found $(\phi_i)_{\text{pure liquid } i}$ by a Pitzer-type equation

$$(\phi_i)_{\text{pure liquid } i} = (\phi_i)^{(0)}_{\text{pure liquid } i} + \omega (\phi_i)^{(1)}_{\text{pure liquid } i} \tag{10.4}$$

where $(\phi_i)^{(0)}_{\text{pure liquid } i}$ and $(\phi_i)^{(1)}_{\text{pure liquid } i}$ are empirical functions of T_r and P_r.

This does not divorce these values from the measured vapor-pressure curve of the individual species, because ω depends on the vapor pressure. (If the curves in Figure 5.4 were totally straight instead of being gently curved, then if we knew ω_i we could write out the whole vapor-pressure curve for species i with complete accuracy. In Figure 5.4 none of the experimental data curves cross one another, and the curvature is more or less proportional to ω, so it makes sense that there could be a universal $P_r = f(T_r, \omega)$ function. Chao and Seader proposed such a function for hydrocarbons; the methods described in the next section also do this, but in a somewhat different way.)

This method and its modifications will not be discussed further here because it has largely been supplanted by the methods discussed next. However, its approach was an important step toward the development of those methods, by showing a logical way to separate the liquid-phase behavior from the vapor-pressure (Raoult's law) data, and instead use an empirical equation for the liquid-phase fugacities.

10.3.3 Estimation Methods Based on Cubic EOSs

Figure 10.8 shows the calculated 100°F isotherm of propane on *P-v* coordinates, calculated by Starling's modification [9] of the Benedict–Webb–Rubin equation of state (BWR EOS). On it the saturation curves are drawn as heavy lines, as is the line of constant pressure in the two-phase region. The predictions of the BWR EOS are shown as a lighter-weight curve. We see, starting at the right (from A to B), that in the vapor phase the BWR curve for 100°F is slightly above the saturation curve and shows the *P-v* behavior of the vapor. At 188 psia (B) the BWR EOS curve crosses the saturation curve

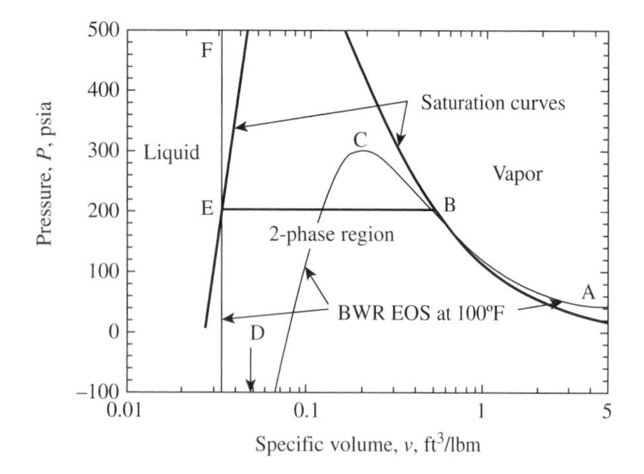

FIGURE 10.8 Pressure-specific volume plot for propane, showing the vapor, liquid, and two-phase regions, and the 100°F equilibrium line (at 188.32 psia), as well as the 100°F isotherm calculated by the BWR EOS. (Observe the logarithmic scale for specific volume and the arithmetic scale for pressure). See the text for a discussion of this figure.

and enters the two-phase region. The BWR EOS, like almost all EOSs, is a one-phase EOS, which should not be expected to directly show the behavior in the two-phase region. Instead, it forms a maximum at 297 psi (C) and then falls to a minimum at – 799 psia (D). After this minimum the curve rises again, crossing the saturation curve at 188 psia (E) and entering the liquid region. From there to F it represents the *P-v* behavior of the liquid at 100°F. We may verify that both the vapor and the liquid *P-v* values for 100°F in [9] correspond to the one-phase curves shown in Figure 10.8 (they were calculated from the same equation!). But what are we to make of the minimum and maximum in the two-phase region? Do those values have any physical meaning?

The part of the curve from B to C represents subcooled vapors, which will condense to form a vapor–liquid mixture if a suitable condensation nucleus is supplied. This is the basis of "cloud seeding," discussed in Chapter 14. The part of the curve from E to D represents a superheated liquid, which will turn to a vapor–liquid mixture if a suitable boiling nucleus is supplied [10], also discussed in Chapter 14. Both of these parts of the curve have real physical meaning and can be demonstrated in careful laboratory experiments. All of the states shown by the curve between B and E are *thermodynamically* unstable. They can lower their Gibbs energies by converting from a single-phase (gas at the right, liquid at the left) to a two-phase mixture. The part of the curve between C and D has no physical meaning and cannot be demonstrated in the laboratory. It represents states that, in addition to their thermodynamic instability (which may persist as a metastable equilibrium for a long period), are also *mechanically* unstable, in which state they cannot persist very long. If we had such a substance in one of our many piston-and-cylinder arrangements, and pushed the piston in slightly, thus lowering the volume, if the substance followed the curve between C and D it would fall in pressure, and thus suck the piston in behind it.

In Example 7.1 we showed that for any pure species, we can calculate the fugacity from an EOS. In Chapters 7 and 9 we used only the little EOS, for which that calculation was easy. For more complex EOSs the mathematics become more complex, as shown in Appendix F, but in principle the procedure is the same. In Figure 10.8 at any value of P we could read the values of the BWR EOS specific volume corresponding to the vapor and the liquid. At the pressure corresponding to the vapor pressure, the calculated fugacities of these two should be equal. That is the method actually used to calculate the values of the saturation vapor pressure in most modern tables of thermodynamic properties. It has the merit that the *PvT* behavior of the liquid, and that of the vapor, and the vapor-pressure curve are all calculated from the same EOS, with the consequence that at the phase boundaries all the values are internally consistent. We might think that this divorces the calculated vapor pressures from the experimental vapor-pressure measurements, but it does not. The adjustable constants in the EOS are chosen to make the

Table 10.2 Summary of Results of Examples 10.2 and 10.3, VLE of Ethane-n-Heptune at 800 psia, with Liquid Containing 60 mol% Ethane

	Experimental Values, Shown in Figure 10.7 and the Tables in [4], Example 10.2	Raoult's Law, Example 10.2	SRK EOS for Both Vapor and Liquid, Example F.3/10.3
Bubble-point temperature T(°F)	209	172.2	210
y_{ethane}	0.945	0.996	0.952
$y_{n\text{-heptane}}$	0.055	0.004	0.048
K_{ethane}	1.52	1.66	1.59
$K_{n\text{-heptane}}$	0.14	0.0096	0.12
f_{ethane} (psia)		796.8	620
$f_{n\text{-heptane}}$ (psia)		3.1	8.9

vapor-pressure curve calculated this way match the experimental vapor-pressure data.

Once the idea that we could calculate pure species vapor pressures (i.e., pure species VLE) from an EOS was widely accepted, it was a fairly short step to attempting to calculate multispecies VLE from a single EOS [11, 12]. Most modern VLE calculation programs do that, using *cubic EOSs*. As discussed in detail in Appendix F, these are EOSs that can be written in the form $z=f(T_r, P_r$. and $\omega)$ in fairly simple algebraic expressions. with z appearing only as z^3, z^2, and z. Cubic equations can be solved algebraically; numerical solutions on spreadsheets are much quicker and easier.

Example 10.3 Repeat Example 10.2, using the Soave–Redlich–Kwong (SRK) EOS.

The necessary background for this calculation and the details of it are shown in Appendix F (Example F.5/10.3). The results are summarized and compared with those of Example 10.2 in Table 10.2. ■

In this example, the SRK EOS was used to compute the fugacities of each species in each of the phases; it estimates the VLE very well. We also see that the departures from ideality are in opposite directions for the two components: Ethane has a slightly lower K value than we would estimate from Raoult's law, and n-heptane has a much higher one. Looking back at Figure 10.1 (which is not directly applicable here because it is only for 150°F), we see that at 800 psia the K value for ethane is just about equal to the Raoult's law value, while that for n-heptane is ≈4 times the Raoult's law value. The K values in Table 10.2 show the same behavior. Table 10.3 shows some more of the details of the SRK calculation. We see the following:

1. At this T and P pure ethane would be a gas and pure n-heptane a liquid. Neither could exist in the other state.
2. The calculated pure component $\phi_i=f/P$ for ethane works well with the $(\hat{\phi}_i/\phi_i)$ formulation (see Table 7.D); the calculated $(\hat{\phi}_i/\phi_i)$ for both phases are quite plausible.
3. The calculated pure component $\phi_i=f/P$ for n-heptane works well with the $(\hat{\phi}_i/\phi_i)$ formulation in the liquid phase; the calculated $(\hat{\phi}_i/\phi_i)$is quite plausible. But in the vapor phase it does not work well at all; the calculated $(\hat{\phi}_i/\phi_i)$makes little sense, mostly because the computed pure species $\phi_i=f/P$ is for a liquid. If we wished to treat this mixture in that way, we would have to estimate a hypothetical $\phi_i=f/P$ for n-heptane vapor (see Fig. 8.14), which would be perhaps 10 times the value shown in Table 10.3.
4. The liquid is behaving more or less as we would expect from Raoult's law, but the vapor is not. The n-heptane content of the vapor is ≈15 times what we would calculate that way. Mostly what we see here is that high-pressure gases, close to their critical states, are much better solvents for other liquids (and solids) than we would expect for an ordinary gas. Most of the unusual behavior encountered in high-pressure VLE is due to the high-pressure vapors behaving in a liquid-like way, which includes becoming very good solvents.
5. The description in terms of the $(\hat{\phi}_i/\phi_i)$ notation shown here is rarely used for high-pressure VLE, because the values of ϕ_i calculated as shown above for n-heptane are unreasonable.
6. This is a difficult mixture for Raoult's law and the L-R rule to estimate, because it consists of only two species with very different normal boiling points (NBPs) (−88.6°C and 98.4°C). Example 10.3 and Problem 10.11 show that it is reasonably well

Table 10.3 More Details from Example E5/10.3, at 800 psia and 210°F, Based on the SRK EOS

	Ethane		n-Heptane	
	Vapor	Liquid	Vapor	Liquid
Pure species z	0.789			0.322
Pure species $\phi_i=f/P$	0.811			0.025
Mixture z	0.723	0.246	0.723	0.246
Mixture $\hat{\phi}_i$	0.815	1.293	0.231	0.027
Calculated $(\hat{\phi}_i/\phi_i)$	1.005	1.593	9.344	1.110

estimated by the SRK equation, both in Example F.3/10.3 and in the process-design programs that carry out that calculation.

10.4 COMPUTER SOLUTIONS

Most chemical engineers in industry have ready access to programs that carry out the VLE calculations in Chapters 8, 9 and 10, often as part of more complex process design calculations. Many university students also have such access. The internal workings of these programs are similar, but not necessarily the same as are shown in this book. Problems 10.11 to 10.18 suggest comparisons between the hand or spreadsheet calculations in Chapters 8, 9 and 10 and these computer packages.

10.5 SUMMARY

1. The behavior of pure species near their critical states is different from what we expect well away from those states. The same is true for mixtures, whose high-pressure VLE behavior is quite different from that at low pressures.
2. Mostly this is due to the high-pressure vapors behaving in a "liquid-like" way, in particular by becoming much better solvents for all sorts of things (leading to processes like "supercritical extraction").
3. The examples in this chapter are all for straight-chain aliphatic hydrocarbons, whose behavior is the simplest possible, because they form close to ideal solutions with each other. For more complex mixtures the behavior is more complex.
4. Current process design computer programs mostly calculate high-pressure VLE using cubic EOSs, of which the SRK is one of the most popular. The procedure is as illustrated in Example F.5/10.3.

PROBLEMS

See common units and values for problems and examples. An asterisk (*) on a problem number indicates that the answer is in Appendix H.

10.1* **a.** Which of curves 2 through 6 of Figure 10.5 show retrograde condensation–vaporization of the first kind?
b. Which of curves 2 through 6 of Figure 10.5 show retrograde condensation–vaporization of the second kind?

10.2 Repeat the chart reading shown in Examples 10.1 and 10.2 to show that it is correct.

Table 10.B Molar Volumes and Solubility Parameters for Problem 10.5

Substance	Molar Volume (mL/mol)	Solubility Parameter $(\text{cal/mL})^{0.5}$
Ethane	68	6.05
n-Heptane	147.5	7.43

10.3 Show the Raoult's law and Raoult's law plus L-R rule calculations in Examples 10.1 and 10.2.

10.4* Solve Example 10.1 using the high-temperature version of Figure 8.20, which is not shown in this text, but is in [13] and [14, pp. 13–18], as well as many other thermodynamics textbooks. This is a repeat of Example 8.14. Is this a better fit of the experimental data than Raoult's law? Or worse? Comment on the probable reasons.

10.5 The calculations in parts (b) and (c) of Example 10.1 assume that the liquid-phase activity coefficients of both species are 1.00. Chao and Seader [7] compute the activity coefficients of such systems by Eq. 9.24, using the values in Table 10.B. Using these, estimate the two liquid-phase activity coefficients in Example 10.1 (b). If we had included these estimated activity coefficients in parts (b) and (c) would they have made the agreement with the experimental results better? Worse? No change?

10.6 Repeat Example 10.1 using the SRK EOS, as shown in Example F.5/10.3E.

10.7* Repeat Example 10.1 (all three parts) for 100 psia and $x_{\text{ethane}} = 0.20$.

10.8 Based on Figure 10.1, estimate the bubble-point and dew-point compositions for mixtures of ethane and n-heptane at 150°F and 200 psia. Compare the results to the values we would read from Figure 10.7.

10.9 In Figure 10.7 for the 1200 psia curve:
a. How many critical points are there?
b. What T and x_{ethane} do they correspond to?
c. Do these values agree with Figure 10.5?

10.10 In Figure 10.8, if the gas and liquid at B and E are in equilibrium, then their fugacities must be equal. From Eq. 7.6 we know that this means that

$$\int_B^E d\ln f = \frac{1}{RT}\int_B^E v\,dP = 0 \tag{10.7}$$

For the horizontal line B-E this is obviously true, because $dP = 0$. But what about for the curve B-C-D-E? Is the integral $= 0$ for this curve as well? The answer is "yes"; this topic is discussed in [15, p. 14].

Computer Problems The following problems require the use of a computer program with an internal VLE package, such as Aspen Plus or ProMax. Such packages offer a wide variety of calculation modules. These problems mostly call for using the NRTL-IG module, which uses Raoult's Law with the NRTL activity coefficient for the liquid and ideal gas in the vapor, the Chao-Seader module, described in Section 10.3.2 and the Soave–Redlich–Kwong (SRK) module, described in Appendix F. The student is encouraged to explore the consequences of rerunning the same problems using some of the many other modules offered.

10.11 Repeat Example 8.9 and compare the results with that example and the experimental values in it:
a. Using the NRTL/Ideal gas thermodynamic module, and
b. Using the NRTL/SRK (or RK) gas thermodynamic module.
Compare and discuss the results in parts (a) and (b)

10.12 Repeat the following examples using the NRTL/Ideal gas thermodynamic modules. In each case compare the calculated results to the spreadsheet solutions in those examples, and the experimental results shown in the examples
a. Example 8.10
b. Example 8.11
c. Example 8.12
d. Example 8.13

10.13 Repeat Example 8.13, using the NRTL/Ideal gas thermodynamic modules, with the change that the feed is 50 mol% ethanol. Explain the results.

10.14 Repeat Example 8.13, using the NRTL/Ideal gas thermodynamic module, with the feed at 1 atm, and the outlets at 0.5 atm, with the flash adiabatic. Compare the results to those for Example 8.13 with feed and outlets all at 1 atm.

10.15 Repeat Example 8.14, using the NRTL/Ideal gas thermodynamic module, which works poorly on this problem and with the Chao-Seader module and SRK module, which do much better. Compare these results with the calculated and experimental values in Example 8.14.

10.16 Repeat Example 9.8, using the NRTL/Ideal gas thermodynamic module. Compare these results with the calculated and experimental values in Example 9.8.

10.17 Estimate liquid composition (i.e., gas solubility) at 1 atm and 25°C for pure oxygen in *n*-hexane and in water using:
a. The NRTL/Ideal gas thermodynamic modules.
b. The Chao-Seader module.
c. The SRK module.
Compare the results to Table 9.2 and discuss them.

10.18 Run the following flashes, using NRTL-IG and/or Chao-Seader and/or SRK modules. Compare the results to the hand calculations.
a. Example 10.1
b. Example 10.2
c. Problem 10.7
d. Problem 10.8

REFERENCES

1. Keenan, J. H., F. G. Keyes, P. G. Hill, and J. G. Moore. *Steam Tables: Thermodynamic Properties of Water Including Vapor, Liquid and Solid Phases.* New York: Wiley (1969).
2. de Nevers, N. *Phase Behavior 1 and II.* Educational videos published by the University of Utah (1968).
3. Katz, D. L., D. Cornell, R. Kobayashi. F. H. Poettmann, J. A. Vary, J. R. Elenbaas, and C. F. Weinaug. *Handbook of Natural Gas Engineering.* New York: McGraw-Hill (1959).
4. Kay, W. B. Liquid–vapor phase equilibrium relations in the ethane–*n*-heptane system. *Ind. Eng. Chem.* 30:459–465 (1938).
5. Prausnitz, J. M., R. N. Lichtenthaler, and E. Gomez de Azevedo. *Molecular Thermodynamics of Fluid-Phase Equilibria*, ed. 3. Upper Saddle River, NJ: Prentice Hall, Chapter 12 (1999).
6. *Equilibrium Ratio Data Book.* Tulsa, OK: Natural Gasoline Association of America (1957).
7. Chao, K. C., and J. D. Seader. A general correlation of vapor–liquid equilibria in hydrocarbon mixtures. *AIChE J.* 7:598–605 (1961).
8. Grayson, H. G., and C. W. Streed. Vapor–liquid equilibrium for high temperature, high pressure hydrogen–hydrocarbon mixtures. *Sixth World Petroleum Congress Proceedings*, Sect. III, pp. 233–245 (1963).
9. Starling, K. E. *Fluid Thermodynamic Properties for Light Petroleum Systems.* Houston, TX: Gulf (1973).
10. Blander, M., and J. L. Katz. Bubble nucleation in liquids. *AIChE J.* 21:833–848 (1975).
11. Wilson, G. M. Vapor–liquid equilibria, correlation by means of a modified Redlich–Kister equation of state. *Adv. Cryog. Eng.* 9:168–176 (1964).
12. Soave, G. Equilibrium constants from a modified Redlich–Kwong equation of state. *Chem. Eng. Sci.* 27:1197–1203 (1972).
13. Smith, J. M., H. C. Van Ness, and M. M. Abbott. *Introduction to Chemical Engineering, Thermodynamics*, ed. 7, New York: McGraw-Hill, p. 366 (2005).
14. Seader, J. D., J. J. Siirola, and S. D. Bamicki. Distillation. In *Perry's Chemical Engineers Handbook*, ed. 7, Green, D. W., and J. O. Maloney, eds. New York: McGraw-Hill (1997).
15. Walas, S. M. *Phase Equilibria in Chemical Engineering.* Boston: Butterworth, p. 535 (1985).

11

LIQUID–LIQUID, LIQUID–SOLID, AND GAS–SOLID EQUILIBRIUM

The major application of vapor–liquid equilibrium (VLE) is distillation. The uses of liquid–liquid equilibrium (LLE), liquid–solid equilibrium (LSE), and gas–solid equilibrium (GSE) are much more diverse. They include extraction (both solid and liquid), decantation as a phase separation, vapor-phase deposition (the heart of the semiconductor business), and a host of environmental applications. In all of these applications the equilibrium state and the rate of approaching it are both important. This book discusses only the equilibrium state, normally asking what are the compositions of the equilibrium phases when the system has minimized its Gibbs energy, subject to the external constraints and the starting conditions. As with VLE, the working criterion for LLE, LSE, and GSE is that the fugacity of any individual species must be the same in all the phases at equilibrium (Eq. 7.4).

11.1 LIQUID–LIQUID EQUILIBRIUM (LLE)

Most chemical species can exist as a liquid at some combination of temperature and pressure. The liquids we are most familiar with are the substances and mixtures that are liquid at 1 atm and about 20°C. Most of these are water and solutions or suspensions of other materials in it (our foods and beverages, the fluids in our bodies, oceans, rivers, rain). We are also familiar with many organic liquids (gasoline, lube oil, paint thinner, cooking oil). Mercury is the only metal that is liquid at room temperature. At cryogenic temperatures gases like oxygen, nitrogen, hydrogen, and helium become liquids. At high temperatures metals like copper, tin, lead, and steel become liquids. Many salts like sodium chloride melt at modest temperatures (801°C for NaCl).

The number of possible LLE is nearly infinite, because there are so many known pure species and we can select any two from the list to test. For most of the rest of this chapter we will consider LLE between water and those organic compounds that are liquids at or near room temperature. This is the most thoroughly studied class of LLE, because of its great industrial, biological, and environmental importance. (It is also the cheapest and easiest type of LLE to test in the laboratory.) We will make occasional reference to other types of LLE, but mostly will talk about water and organics that are liquid at or near room temperature. All of the types of phenomena in LLE not involving water are illustrated in the water-LLE examples we discuss here.

11.2 THE EXPERIMENTAL DETERMINATION OF LLE

If we place samples of two pure liquid species that do not react chemically in a container, shake them to bring them to equilibrium, and then observe the result, there are three possibilities:

1. The two liquids may be totally miscible, forming one liquid phase, for example, water and ethanol.
2. The two liquids may be practically immiscible, forming two liquids phases, each of which contains only a small amount (e.g., less than 1 mol%) of the other, for example, water and most organic compounds ("water and oil don't mix").
3. The two liquids are partly miscible with each other, forming two separate phases, each of which contains

Physical and Chemical Equilibrium for Chemical Engineers, Second Edition. Noel de Nevers.

substantial amounts (more than 1 mol%) of the other, for example, water and some organics like *n*-butanol (see Figure 8.12d).

Which of these results occurs is largely dependent on the molecular interactions between the two pure species.

In the first case, totally miscible behavior, there is no LLE, because there is only one liquid phase. In the other two cases we can mix the two liquids vigorously to obtain equilibrium, then separate the two resulting liquids (which will generally have different densities) by gravity or in a centrifuge, sample each phase, and analyze the samples to find the chemical composition of that phase. This is the basic LLE experiment. The first part of this chapter reports on the results of such experiments, with a little intuitive commentary. Later sections discuss the thermodynamics of LLE and how we can use that thermodynamics to predict, interpolate, or extrapolate LLE data.

11.2.1 Reporting and Presenting LLE Data

The most common way of reporting and presenting two-species LLE data [1] is as a table like Table 11.1. Although [1] presents solubilities in mol percent, which we mostly use in this chapter; many sources [2] present them as weight percent. From Table 11.1 we see that both solubilities increase slowly with increasing temperature. The effect of temperature on solubility is discussed in Section 11.2.6. We also see that the solubility of benzene in water at 25°C is 0.0405 mol% = 0.000,405 mol fraction = 405 ppm by mol. This is "practically insoluble."

Example 11.1 A common way to measure small solubilities like those shown in Table 11.1 is to add one species from a buret, one drop at a time, to a large mass of the other species. After each drop is added the mixture is shaken and observed to see if it has turned cloudy, indicating that the solubility limit has been reached. If we begin with 1 L of pure water and add benzene one drop at a time from a buret at 25°C, how many drops should it take to saturate the water?

Table 11.1 Solubility Data for Benzene (species 1) and Water (species 2) [1]

Temperature (°C)	Solubility of (1) in (2) (mol%)	Solubility of (2) in (1) (mol%)
0	0.04	0.133
5	0.04	0.156
10	0.04	0.180
15	0.04	0.211
20	0.04	0.252
25	0.0405	0.300
30	0.0411	0.356
40	0.0437	0.491
50	0.0474	0.664
60	0.0531	0.895
70	0.0615	1.19

Typically a buret will deliver about 20 drops of liquid per milliliter, so for benzene

$$1 \text{ drop} \approx \frac{\text{mL}}{20} \cdot 0.88 \frac{\text{g}}{\text{mL}} \cdot \frac{\text{mol}}{78\text{g}} = 5.64 \times 10^{-4} \text{mol} \quad (11.A)$$

and 1 L of water ≈ 1000 g/(18 g/mol) ≈ 55.6 mol, so

$$n_{\text{benzene to saturate}} = 405 \times 10^{-6} \cdot 55.6 \text{ mol} = 0.0225 \text{ mol} \quad (11.B)$$

and

$$\text{drops of benzene} \approx \frac{0.0225 \text{ mol}}{0.000{,}564 \text{ mol/drop}} = 40 \text{ drops} \quad (11.C)$$

■

We are all familiar with the expression "water and oil don't mix." From this example and from Table 11.1 we would conclude that since benzene is an "oil," the expression should really be "water and oil don't mix *very much*."

Example 11.2 One thousand pounds of benzene have been leaked into the soil, in contact with the groundwater. The benzene slowly dissolves in the groundwater. How many pounds of groundwater will become saturated with benzene?

The total mols of benzene are

$$n_{\text{benzene}} = \frac{m_{\text{benzene}}}{M_{\text{benzene}}} = \frac{1000 \text{ lbm}}{78 \text{ lbm/lbmol}} = 12.82 \text{ lbmol} \quad (11.D)$$

If this is to be 0.000,405 mol fraction in water then

$$\begin{aligned} n_{\text{benzene}} &= x_b n_T \approx x_b\, n_{\text{water}} \\ n_{\text{water}} &= \frac{n_{\text{benzene}}}{x_b} = \frac{12.82 \text{ lbmol}}{0.000{,}405} = 0.32 \times 10^5 \text{ lbmol} \\ &= 5.5 \times 10^5 \text{ lbm} \end{aligned} \quad (11.E)$$

The federal drinking water standard for benzene [3] is 5 ppb by weight or ≈ (5 · 18/78) = 1.15 ppb by mol (1.15×10^{-9} mol fraction). Thus, to make this benzene-saturated water acceptable as drinking water we must remove (1 – 1.15 ppb/405 ppm) = 99.9997% of the dissolved benzene. ■

This example shows that for benzene the solubility in water is small enough that one pound of benzene will saturate 550 pounds of water, but large enough that to clean this water to meet the federal drinking water standards will require nearly "six nines" removal efficiency. Don't let benzene get into the ground, unless you have a very large cleanup budget!

Table 11.2 Solubility of Hydrocarbons in Water, and Water in Hydrocarbons at 25°C [2]

Hydrocarbon	Formula	Solubility in Water (mol fraction × 10^6)	Solubility of Water in (mol fraction × 10^6)
n-Pentane	C_5H_{12}	9.5	480
n-Hexane	C_6H_{14}	2	520
n-Heptane	C_7H_{16}	9	280
Cyclohexane	C_6H_{12}	12	460
Benzene	C_6H_6	420	2,700
Toluene	C_7H_8	100	1,700
Mixed xylenes	C_8H_{10}	34	2,900
Methylene dichloride	CH_2Cl_2	4000	7,000
Chloroform	$CHCl_3$	1250	13,000
Carbon tetrachloride	CCl_4	90	700
Monochlorobenzene	C_6H_5Cl	78	2,000

11.2.2 Practically Insoluble Liquid Pairs at 25°C

Water is practically insoluble in all hydrocarbons that do not contain oxygen or nitrogen, for example, simple aliphatic or aromatic hydrocarbons and chlorinated hydrocarbons; these same hydrocarbons are practically insoluble in water. Table 11.2 shows some examples of the practical insolubility of water with some of these materials. The values for benzene in this table do not agree exactly with those in Table 11.1; this slight disagreement between reported solubility data is, alas, common. For most of the materials in this table the solubilities are small, but not zero. The solubility of water in these materials is about 5 to 10 times that of these materials in water on a mol basis, but typically only about 2 to 4 times as much on a weight basis.

11.2.3 Partially Soluble Liquid Pairs at 25°C

The species in Table 11.2 are all compounds containing only C, H, and Cl. The only solubility (water-in-organic or organic-in water) greater than 1 mol% is water in chloroform, 13,000 ppm by mol = 1.3 mol%. Compounds containing C, H, and O are much more soluble in water, and water is much more soluble in them. The oxygen-containing groups (hydroxyl, carboxyl, ketone, ether, ester) are hydrophilic, while the remaining hydrocarbon (HC) parts of the molecules are hydrophobic. Such molecules may be thought of as having a water-soluble part and a part soluble in organic liquids. Table 11.3 shows some examples of such solubilities. Comparing Table 11.3 with Table 11.2, we see that the values are of the order of 100 to 10,000 times as large. Within each family, increasing the number of carbon atoms makes the hydrophobic part of each molecule larger, thus decreasing both the organic-in-water and the water-in-organic solubilities.

Table 11.3 Solubility of Oxygen-Containing Organic Compounds in Water, and Water in Oxygen-Containing Organic Compounds at 25°C [2]

Compound	Formula	Solubility in Water (mol%)	Solubility of Water in (mol%)
n-Butanol	C_4H_9OH	1.9	51.3
Cyclohexanol	$C_6H_{11}OH$	0.8	42.6
Methyl ethyl ketone	CH_3-CO-C_2H_5	8.1	35.3
Methyl isobutyl ketone	CH_3-CO-C_4H_9	0.3	9.7
Diethyl ether	C_2H_5-O-C_2H_5	1.8	5.1
Diisopropyl ether	C_3H_7-O-C_3H_7	0.2	3.4
Methyl acetate	CH_3-O-CO-CH_3	7.3	26.9
Ethyl acetate	C_2H_5-O-CO-CH_3	1.7	14.3
n-Butyl acetate	C_4H_9-O-CO-CH_3	0.11	7.8

11.2.4 Miscible Liquid Pairs at 25°C

If we lower the number of carbon atoms within one organic family, thus making the hydrophobic part of the molecules smaller, we would expect the solubilities to increase, which they do. The lower molecular weight alcohols (methanol, ethanol, and isopropanol) and the lowest molecular weight ketone (acetone) are completely miscible with water. The next larger alcohol (*n*-butanol) and the next larger ketone (methyl ethyl ketone) are partly miscible at 25°C, as shown in Table 11.3. Adding a second hydroxyl group to an alcohol to make a glycol increases the water solubility; all the glycols with up to six C atoms are miscible with water at 25°C.

Organic nitrogen compounds are even more water soluble than organic oxygen compounds. If we replace the oxygen in an alcohol with a nitrogen and a hydrogen, forming an amine, the solubility is greater than that of the corresponding alcohol; amines with up to six C atoms are completely miscible with water (at 25°C). If we replace one CH group in practically water-insoluble benzene with an N, we form totally water-soluble pyridine.

Digressing briefly from our water–organic examples, we can say that if any two species are enough alike chemically, the liquids will mix in all proportions. Examples of totally miscible liquid groups are

1. All the liquids (at room temperature) of the families of straight-chain or branched hydrocarbons, cycloparaffins, and aromatics, with each other, for example, *n*-hexane, 2-methyl-pentane, cyclohexane, benzene.
2. Most chlorinated members of the above families, and the low molecular weight alcohol, ketone, or ester derivatives of them with each other and with the above-listed hydrocarbons.
3. Most molten metals, such as copper, zinc, lead, and tin with each other.
4. Many molten salts with each other.

These mixtures need not be ideal solutions (e.g., acetone–water, see Chapters 8 and 9), but they will be closer to ideal solutions than those species pairs that do not form a single phase. In most (but not all!) of these cases the liquids form type II (see Section 8.4.2) interactions with each other (positive deviations from ideal solution, $\gamma_i > 1.00$), but the mutual repulsion is not strong enough to form two phases.

11.2.5 Ternary LLE at 25°C

If, instead of mixing two pure species, we mix three, we find more complex behavior:

1. If we mix three members of one homologous series (such as pentane, heptane, hexane or benzene, toluene, xylene) we would find that they are miscible, forming one liquid phase.
2. If we mix three members of very different chemical types (such as water, benzene, mercury) we would find three separate liquid phases, each containing one of the species practically pure, containing less (often much less!) than 1 mol% of each of the other two.
3. If we mix water, ethanol, and benzene, we would expect more complex behavior, because water and ethanol are totally miscible, and benzene and ethanol are totally miscible, but water and benzene, as discussed above, are practically immiscible.

The most common way to present the experimental data for such systems [4] is shown in Table 11.4. The data in this table are redundant; if we know two of the mol fractions we can calculate the third. But this is the common way of presenting such data.

This table shows the result of 12 experiments, in each of which some amounts of the three species were placed in a container and agitated. Then the two liquid phases were allowed to settle by gravity or centrifuged, and samples of each liquid phase were withdrawn and analyzed. The first row tells us that a liquid with 96.134 mol% water, 3.817 mol% ethanol, balance benzene, is in equilibrium with another liquid that is 98.56 mol% benzene, 1.010 mol% ethanol, balance water. The terms left phase and right phase refer to the common plotting convention shown below; the terms water-rich and benzene-rich are more descriptive. Such a table has less intuitive content than a plot of the same data. The traditional presentation has been on an equilateral triangle plot, as is discussed in most books on unit operations or material balances [5, 6]. Figure 11.1 shows such a plot of the data in Table 11.4 (omitting the sixth row from the top, which practically duplicates the row above it).

Each of the data sets in Table 11.4 (except the one that practically repeats the one above) is plotted as two data points, connected by a *tie line*. Any overall mixture whose composition lies on a tie line will spontaneously separate into the two phases whose compositions are shown at the ends of that tie line. The lowest tie line corresponds to the first row in the table. The ends of the tie lines are connected in two curves (called *binodal curves*), which show the compositions of the two equilibrium phases: the left or water-rich phase and the right or benzene-rich phase. These two curves meet at the *plait point*, shown as a small square in the diagram. For high ethanol concentrations only one phase exists. The single-phase region extends down to the two lower vertices, forming two narrow strips at the edges of the diagram (ending at 2700 ppm on the left and 420 ppm on the right) (see Table 11.2).

Example 11.3 We place into our 25°C mixing container 3.75 mol of water, 2.5 mol of ethanol, and 3.75 mol of benzene, and shake well. When the phases have separated, what will be the composition of the two equilibrium phases?

Table 11.4 Equilibrium Data for the System Water (species 1), Ethanol (species 2), and Benzene (species 3) at 25°C [4]

Left (water-rich) Phase			Right (benzene-rich) Phase		
Mol% 1	Mol% 2	Mol% 3	Mol% 1	Mol% 2	Mol% 3
96.134	3.817	0.049	0.430	1.010	98.560
91.980	7.968	0.052	0.850	3.323	95.827
86.884	12.977	0.139	2.081	5.860	92.059
81.478	18.134	0.388	2.455	9.121	88.424
75.459	23.540	1.001	3.588	12.939	83.474
74.824	24.069	1.107	3.967	13.340	82.694
70.281	27.892	1.828	5.046	16.090	78.864
64.904	31.725	3.371	6.434	18.943	74.623
59.095	35.510	5.395	7.727	22.444	69.829
51.033	39.382	9.584	10.233	26.216	63.551
45.629	41.062	13.309	12.562	29.341	58.096
37.176	41.771	21.053	16.607	33.093	50.300

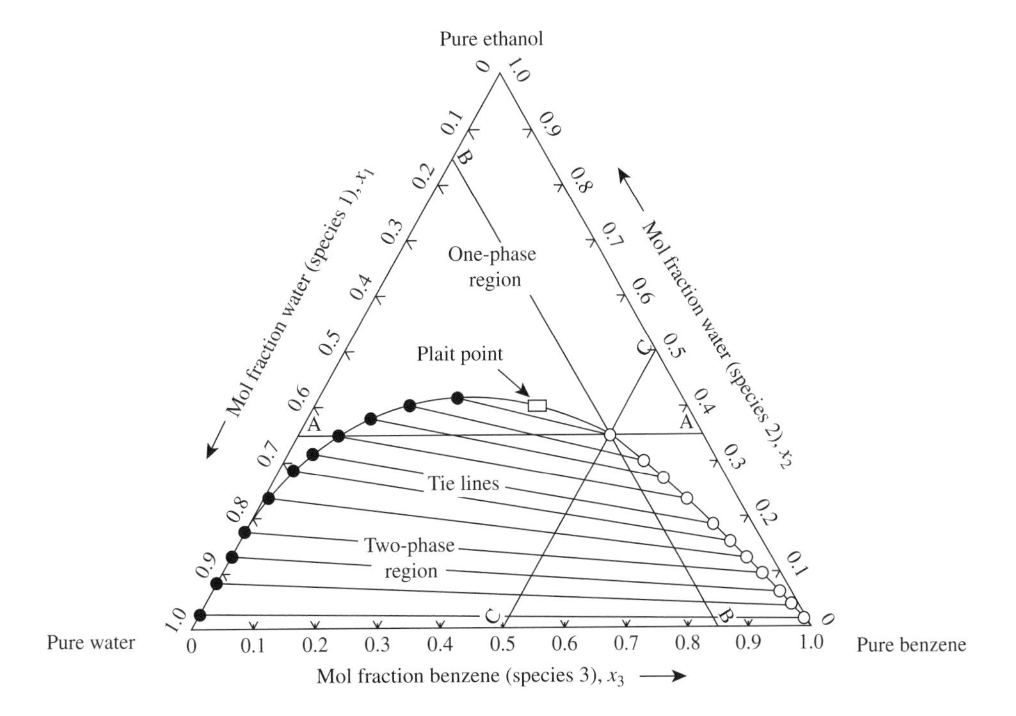

FIGURE 11.1 Equilateral triangular diagram showing the LLE data in Table 11.4 for water–ethanol–benzene at 25°C. On this type of diagram as shown, each vertex represents a pure species. Each of the sides of the triangle represents a two-species mixture; for example the right side represents mixtures of ethanol and benzene. Each point within the figure represents a three-species mixture. To locate the lowest point on the right side of Table 11.4, we draw line AA, parallel to the base, at value 0.3309 on the right hand scale; every point on this line represents a mixture with 33.09 mol% ethanol. Next we draw line BB, parallel to the right border passing through 0.16607 on the left border; every point on this line represents a mixture with 16.607 mol% water. These two lines determine the point. However, just to check, we draw line CC, parallel to the left border passing through 0.5030 on the base; every point on this line represents a mixture with 50.33 mol% benzene. If we have drawn the lines correctly they should cross at one point, representing the lowest entry for the right phase in Table 11.4.

By simple stoichiometry the overall mol fractions (1, 2, 3) are 0.375, 0.25, and 0.375. We locate the point corresponding to this concentration on the diagram, by drawing any two of the three straight lines corresponding to those mol fractions, finding that the point falls almost exactly on the fifth tie line from the top, for which the end-point values (read from Table 11.4, fifth row from the bottom) are water-rich phase 64.9, 31.75, and 3.37 mol%, and benzene-rich phase, 6.43, 18.94, and 74.62 mol%. ■

We could use a similar procedure for any point in the two-phase region, interpolating between tie lines as needed. We could also compute, by material balance, how many mols of each of the phases is present (see Problem 11.5). Although Figure 11.1, the equilateral triangle diagram, is the traditional way of representing these data, we can also replot the same data on a right triangle, as shown in Figure 11.2. This is currently a more popular representation because it is easier to produce by common computer graphics programs. Both Figures 11.1 and 11.2 present exactly the same information.

Figures 11.1 and 11.2 show the simplest case of a triangular diagram representing three-species LLE (called type I). The two-liquid-phase region forms a dome, which contacts only the lower edge of the triangle. This type makes up about 75% of the cases for which such ternary data are available [4]. This does not mean that this is the most common in nature; it is the type of greatest industrial, biological, and environmental importance, hence the most-studied type. The other common type (called type II), accounting for ≈20% of the examples in [4] has the two-phase region touch two of the edges of the triangle; an example is shown in Figure 11.6, below. The remaining few percent are several other uncommon types. If the two-liquid-phase region contacts only one edge of the triangle, it is the common convention to draw the diagram so that the dome rises from the lower edge.

In Figures 11.1 and 11.2 we see that the ethanol acts as a *cosolvent* for benzene and water. (Benzene is a very poor solvent for water, and water a poor solvent for benzene. Ethanol increases the solubility of each in the other, making them effectively better solvents for each other. Hence the

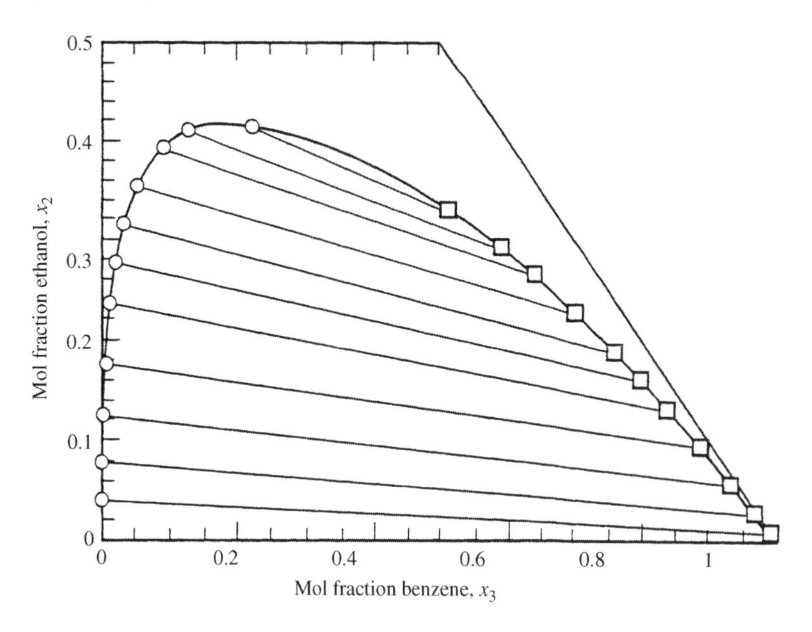

FIGURE 11.2 Replot of Figure 11.1, on rectangular coordinates. The top part of the triangle, above $x_2 = 0.5$ is omitted because it is blank. The scales on the two axes are not the same; the vertical scale is expanded compared to the horizontal scale.

name "cosolvent.") For example, if we have 50 mol% ethanol, then the remaining 50 mol% can be any mixture of benzene and water, and still we will have only one phase. But if we remove the ethanol (move down a straight line from the top vertex through the original composition point on Figure 11.1 or 11.2) then we will pass into the two-phase region, eventually reaching the base of the figure, where the two liquid phases have the compositions shown in Table 11.2. Such cosolvents are widely used industrially to make mutually immiscible liquids miscible, or slightly soluble liquids more soluble. Figures 11.1 and 11.2 also describe LLE suitable for liquid–liquid extraction (see [7]).

11.2.6 LLE at Temperatures Other Than 25°C

As Table 11.1 makes clear, LLE values change with changes in temperature. For practically insoluble materials like the benzene–water binary in Table 11.1 both solubilities increase slightly with increasing temperature. (The solubility of benzene in water increases by a factor of 1.5 and that of water in benzene by a factor of 9 as the temperature increases from 0 to 70°C.) This behavior is observed for most practically insoluble liquid pairs.

For pairs with significant solubility, the behavior is more complex. Figure 11.3 shows the *solubility diagram* for *n*-butanol and water for pressures above the boiling points of the mixtures. From Figure 8.12d we see that at 1 atm pressure the boiling temperature is ≈ 92°C and the two liquid phases have about 65 and 98 mol% water. From Figure 11.3 we read that at 92°C, the two liquid phases in equilibrium have these same values. This should not surprise us; Figure 8.12d is almost certainly based on the same set of experimental data as Figure 11.3. We also see in Figure 11.3 that as the temperature is increased the phases become more and more alike, at ≈ 125°C, the *critical solution temperature* (also called *consolute temperature*), the two become the same, and for higher temperatures there is only one liquid phase. This critical solution temperature is in some ways like the critical temperature at which liquid and gas become identical; in each case the number of phases decreases by one. But 125°C is far below the critical temperatures of water and *n*-butanol

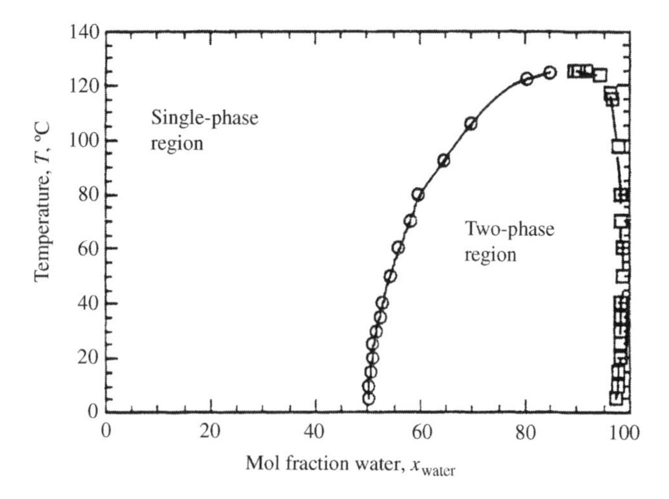

FIGURE 11.3 Solubility–temperature diagram for 1-butanol and water. The circles and squares represent the experimental data [4]. The single-phase region extends as a thin strip down the right side of the figure.

(374 and 290°C), so we should not confuse this critical solution state with the gas–liquid critical state. (There are other events or states in engineering and science that are called "critical"; the word is popular for describing any singular state or occurrence.) On this figure we could determine the composition of the equilibrium phases at any temperature, simply by drawing a horizontal line at that temperature and reading the mol fractions corresponding to the two boundaries between the two-phase region and the single-phase region (which extends as a thin strip along the right axis of the figure).

Figure 11.3 is the most common type of two-species solubility diagram. However, nature presents some other versions of this figure. Figures 11.4 and 11.5 show two of these. In Figure 11.3 there is one point at which increasing the temperature causes the two phases to become identical, which is called an upper critical solution temperature (UCST). In Figure 11.4 there is also one such point, but it is a lower critical solution temperature (LCST). Above that temperature two phases exist; below it only one phase exists. In Figure 11.5 there are both UCST and LCST. Between ≈ 72°C and 135°C two phases exist, while below 72°C or above 135°C only one phase can exist.

Figure 11.1 shows the most common behavior for partly soluble ternary mixtures (called type I). Figure 11.6, for *n*-hexane, methylcyclopentane, and analine, shows the next most common type (called type II) at 25°C. Two of the three binary mixtures form two liquid phases, so that the two-phase region touches two of the edges of the triangle. At 34.5°C the methylcyclopentane–analine binary is at its UCST (the same as 125°C in Figure 11.3). At that temperature the two-phase region has shrunk, compared to its size at 25°C, and its plait point barely touches the M-A edge. At 45°C the two-phase region has shrunk even further and only

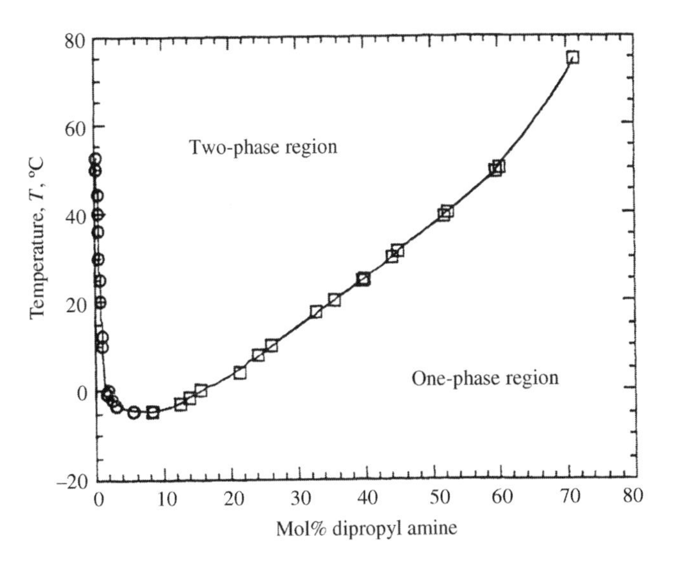

FIGURE 11.4 Solubility–temperature diagram for dipropyl amine and water [1, p. 434]. The squares and circles are the experimental data points. The horizontal axis extends only to 80%.

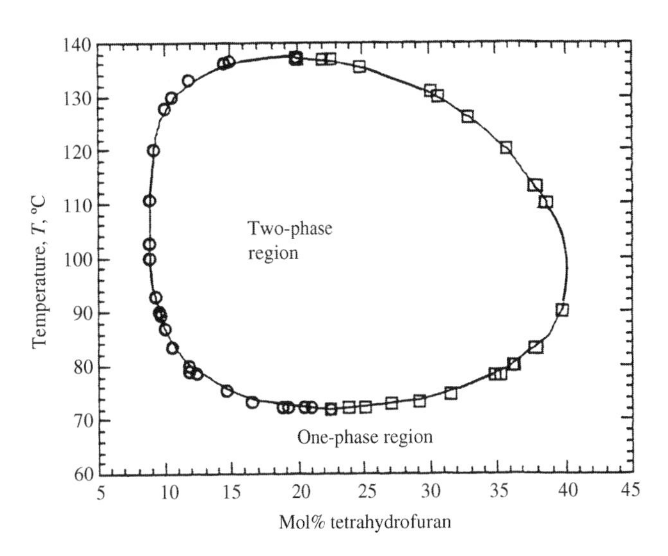

FIGURE 11.5 Solubility–temperature diagram for tetrahydrofuran, THF (1,4-epoxy butane) and water [1, p. 221]. The squares and circles are the experimental data points. The horizontal axis extends only from 5 to 45%. The one-phase region surrounds the two-phase region.

touches one of the three edges, so the diagram has the same form (type 1) as Figure 11.1.

11.3 THE ELEMENTARY THEORY OF LLE

So far this chapter has presented a description of the most common (and interesting and important) types of LLE. There are ample LLE data in the literature, and the experiments to determine LLE (of nontoxic liquids at modest T and P) need not be terribly expensive or difficult. So we could simply assume that we will look up or measure any LLE data we need. Current process design computer programs can estimate such LLE data from theory with fair-to-good accuracy much more quickly and cheaply than we could find it in the library or the laboratory. This section shows the fundamentals of that theory and some simplified (hand or spreadsheet calculable) versions of the more complex algorithms in those computer programs.

The fundamental relation for any LLE is the same as that for VLE, or for any phase equilibrium, that for any of the species present at equilibrium the fugacity must be the same for that species in all of the phases present,

$$f_i^{(1)} = f_i^{(2)} \tag{11.1}$$

where $f_i^{(1)}$ is the fugacity of species i in phase 1, and $f_i^{(2)}$ is the fugacity of species i in phase 2. (Alas, there is no agreement on subscripts and superscripts to represent species and phases. This is one of the most commonly used conventions.) Equation 11.1 is separately obeyed for each of the

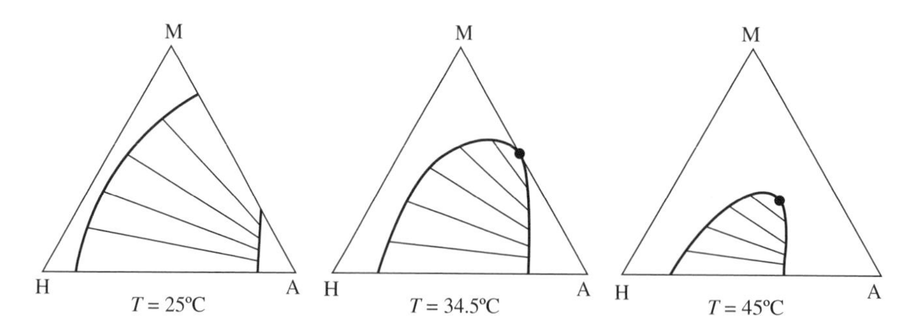

FIGURE 11.6 Effect of temperature on solubility for the system *n*-hexane (H), methylcyclopentane (M), and analine (A). Increasing the temperature reduces the size of the two-phase region. (From Seader, J. D., and E. J. Henley. *Separation Process Principles.* © 1998, New York: Wiley, p. 438. Reprinted by permission of John Wiley & Sons, Inc.)

species present. In almost all LLE we represent the fugacities by the "Raoult's law with activity coefficient" form (see Table 8.2). That changes Eq. 11.2 to

$$x_i^{(1)}\gamma_i^{(1)}p_i^{(1)} = x_i^{(2)}\gamma_i^{(2)}p_i^{(2)} \tag{11.2}$$

here $p_i^{(1)} = p_i^{(2)}$ is the vapor pressure of i at the system temperature, which is the same on both sides of Eq. 11.2, and is normally canceled. Solving for $x_i^{(1)}$ we find

$$x_i^{(1)} = \frac{x_i^{(2)}\gamma_i^{(2)}}{\gamma_i^{(1)}} \tag{11.3}$$

which is true for any number of species and any number of liquid phases. If we consider only binary systems, and practically insoluble liquid pairs, then if species i is the species present to a few parts per million in phase 1, it will be practically pure in the other liquid, phase 2, so that both $x_i^{(2)}$ and $\gamma_i^{(2)}$ will be $\approx$ 1.00 (see Section 8.5). That simplifies Eq. 11.3 to

$$x_i^{(1)} \approx \frac{1}{\gamma_i^{(1)}} \quad \begin{array}{l}\text{practically insoluble}\\ \text{species in binary mixtures}\end{array} \tag{11.4}$$

Example 11.4 Estimate the activity coefficients of benzene in water and water in benzene at 25°C.

The mol fractions of benzene in water and water in benzene (from Table 11.1) are 405 and 3000 ppm by mol, so that the mol fractions of water in the water-rich phase and benzene in the benzene-rich phase are 0.999595 and 0.997, both $\approx$ 1.00. Thus, the assumptions leading to Eq. 11.4 are certainly suitable here. Solving Eq. 11.4 for $\gamma_i^{(1)}$ and inserting the values from Table 11.1, we find

$$\gamma_{\text{benzene}}^{(\text{water-rich phase})} = -\frac{1}{x_{\text{benzene}}^{(\text{water-rich phase})}} = \frac{1}{405 \times 10^{-6}} = 2500 \tag{11.F}$$

and

$$\gamma_{\text{water}}^{(\text{benzene-rich phase})} = \frac{1}{x_{\text{water}}^{(\text{benzene-rich phase})}} = \frac{1}{3000 \times 10^{-6}} = 333 \blacksquare \tag{11.G}$$

For practically insoluble liquids, the activity coefficient values calculated this way, which are of the same type we used in Chapters 8 and 9, are always very large, much larger than any we saw in those chapters. For partly soluble liquid pairs the calculation is more difficult, because in Eq. 11.3 neither of the liquids is practically pure, so we may not assume that any of the terms in the equation is $\approx$1.00.

In low-pressure VLE (see Chapters 8 and 9) we normally begin with experimental data, calculate liquid-phase activity coefficients, use those to estimate the appropriate constants in a suitable liquid-phase activity coefficient equation, and then use that plus a suitable estimate of the vapor-phase nonideality (often the ideal gas law or the L-R rule for low-pressure VLE) to calculate equilibrium phase concentrations. In LLE we most often begin with some kind of liquid-phase activity coefficient equation, use it to calculate the composition of the equilibrium phases (without going through the intermediate step of calculating activity coefficients), and then compare the predicted to the experimental equilibrium concentrations, adjusting our equations as needed to get agreement. Then we use the equation to estimate other data points, the values at other temperatures, and so on.

To do this we return to the equivalent of Figure 6.7, where we showed that a liquid mixture will form two liquid phases only if its $g - x_a$ plot has an internal maximum, and that the equilibrium concentrations are those corresponding to the two points of tangency of a single straight line.

Example 11.5 Species a and b have pure-species Gibbs energies of $g_a^o = 2$ kJ/mol and $g_b^o = 1$ kJ/mol. Their liquid-phase activity coefficients are represented adequately by

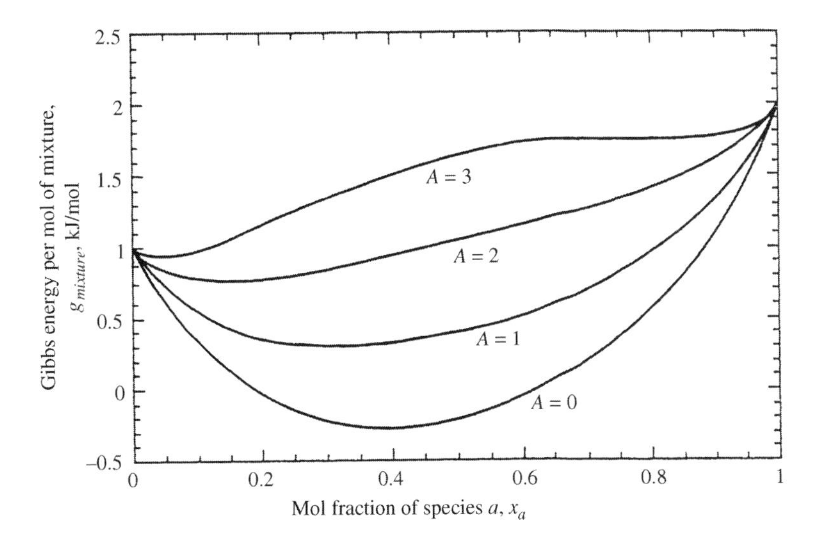

FIGURE 11.7 Calculated values of the molar Gibbs energy of a binary mixture at 25°C, assuming the symmetrical activity coefficient equation, with various values of A.

Eq. 9.G (which leads to the symmetric or 2-suffix Margules equation, Eq. 9.1). Prepare a plot, similar to Figure 6.7 for this mixture at 298.15 K, for values of the constant in Eq. 9.I of $A = 0$, 1, 2, and 3.

Starting with Eq. 9.5, we have

$$g_{\text{mixture}} = g_{\text{solution of the same composition, but ideal}} + g^E$$

$$= \sum x_i g_i^{\text{o}} + \sum x_i (RT \ln x_i) + \sum x_i (RT \ln \gamma_i) \quad (11.5)$$

which is a general expression for the molar Gibbs energy of any mixture. For the symmetric equation the individual $\ln \gamma_i = A(1 - x_i)^2$ so that

$$\sum x_i (RT \ln \gamma_i) = \sum x_i RTA(1 - x_i)^2 \quad (11.\text{H})$$

and

$$g_{\text{mixture}} = \sum x_i g_i^{\text{o}} + \sum x_i (RT \ln x_i) + \sum x_i RTA(1 - x_i)^2 \quad (11.\text{I})$$

As an example point, for $x_a = 0.3$ and $A = 2$, this becomes

$$g_{\text{mixture}} = \left(0.3 \cdot 2 \frac{\text{kJ}}{\text{mol}} + 0.7 \cdot 1 \frac{\text{kJ}}{\text{mol}}\right) + RT(0.3 \ln 0.3 + 0.7 \ln 0.7) + 2RT(0.3 \cdot 0.7^2 + 0.7 \cdot 0.3^2) \quad (11.\text{J})$$

$$g_{\text{mixture}} = 1.3 \frac{\text{kJ}}{\text{mol}} + 8.314 \frac{\text{J}}{\text{mol K}} \cdot 298.15K \cdot (-0.611 + 0.42)$$
$$= 1.3 \frac{\text{kJ}}{\text{mol}} + 2.480 \frac{\text{kJ}}{\text{mol}} \cdot (-0.1909) = 0.827 \frac{\text{kJ}}{\text{mol}} \quad (11.\text{K})$$

Similar calculations (easy on a spreadsheet!) produce curves shown in Figure 11.7.

From Figure 11.7 (and Eq. 11.I) we see that $A = 0$ makes all the activity coefficients = 100, and corresponds to an ideal solution. For $A = 1.0$ the activity coefficients are greater (than 1.00, but not large enough to cause liquid-phase separation (see Figure 6.7!). For $A = 2$ the center of the plot is practically straight. We may show (see Problem 11.12) that for this equation, and $A = 2.00$, at the middle of the plot ($x_a = 0.5$) the second derivative $d^2 g_{\text{mixture}}/dx_a^2 = 0$. This corresponds to the beginning of a local maximum in the curve. For all values $A < 2.00$ there is no local maximum; $d^2 g_{\text{mixture}}/dx_a^2 > 0$ for all x_a. For all values of $A > 2.00$ there is a local maximum; $d^2 g_{\text{mixture}}/dx_a^2 < 0$ for $x_a = 0.5$. So (for this assumed simplest possible activity coefficient relation!) if $A = 2.00$, then the system is at the boundary between one-liquid phase systems and two-liquid-phase systems. For $A = 3$ the curve shows a distinct local maximum. By drawing the line that is tangent to the $A = 3$ curve twice we see that the calculated mol fractions of species 1 in the two phases are approximately 0.1 and 0.9 (see Problem 11.14). ■

We may show (Problem 11.13) that the values of g_a^{o} and g_b^{o} in the above example do not influence whether there is one liquid phase or two or the composition of the two liquids. For that reason the quantity normally plotted in such figures is

$$g_{\text{mixture}} - \sum x_i g_i^{\text{o}} = \sum x_i (RT \ln x_i) + \sum x_i (RT \ln \gamma_i) \quad (11.\text{L})$$

which is the vertical distance in Figure 11.7 (and Figure 6.7) between the local value of g_{mixture} and the straight line connecting the two g_i^{o} values. This quantity has all the useful properties of the quantity plotted in Figures 11.7 and 6.7.

Example 11.6 Estimate whether water and *n*-butanol form two phases as 92°C, and, if so, what the phase compositions are, based on the assumption that the liquid-phase activity coefficients are described by the van Laar equation (Eq. 8.9), in the ln γ form, with constants $A = 1.2739$ and $B = 3.9771$, which are based on VLE measurements, (Table A.7).

As an example point, at $x_a = 0.3$, for water

$$\ln \gamma_a = \frac{B^2 A x_b^2}{(Ax_a + Bx_b)^2} = \frac{3.9771^2 \cdot 1.2739 \cdot (1.-0.3)^2}{(1.2739 \cdot 0.3 + 3.9771 \cdot 0.7)^2}$$
$$= 0.985 \tag{11.M}$$

and correspondingly for *n*-butanol, $\ln \gamma_b = 0.0579$, so

$$g_{\text{mixture}} - \sum x_i g_i^{\text{o}} = 3.037 \frac{\text{kJ}}{\text{mol}} \cdot (0.3 \ln 0.3 + 0.7 \ln 0.7$$
$$+ 0.3 + 0.985 + 0.7 \cdot 0.0579)$$
$$= 3.037 \frac{\text{kJ}}{\text{mol}} \cdot (-0.2748) = -0.835 \frac{\text{kJ}}{\text{mol}} \tag{11.N}$$

Figure 11.8 shows the result of similar calculations, for the whole range of x_a and also the straight line tangent to the curve twice.

Figure 11.8 shows the same type of behavior as the $A = 3$ curve in Figures 11.7 and 6.7 (except that the van Laar equation is not symmetrical, so the internal maximum is not in the center as it was in those figures). The straight line, drawn to be tangent to the curve twice, touches the curve at about $x_a \approx 0.47$ and 0.97. Thus, based on the severe assumption that the van Laar equation (with constants based on VLE measurements) is an accurate representation of the LLE, we would conclude that at 92°C water–*n*-butanol does form two liquid phases, which is correct, and that the two $x_a \approx 0.47$ and 0.97, while the experimental values are 0.65 and 0.98. ■

This disagreement between the experimental equilibrium phase compositions and those calculated from VLE activity coefficient equations is common [8, Chapter 7], The calculated results are quite sensitive to the equation parameters, and the values like those in Table A.7 are the best average values for the whole range of VLE data. Process-design computer programs estimate LLE phase compositions by the same method as Example 11.6, but generally using more complex (and more accurate) liquid-phase activity coefficient relationships. Several of these methods are discussed and compared in [9]. For three species the line is replaced by a plane that must be tangent three times to a three-dimensional surface instead of twice to the two-dimensional curve shown in Figures 6.7 and 11.8. That leads to more complex numerical solutions, which are hard to show by hand, but easily done in large computer programs. In principle, they are the same as Example 11.6.

11.4 THE EFFECT OF PRESSURE ON LLE

All of the information presented so far in this chapter is practically independent of the system pressure, as long as the pressure is above the boiling-point pressure of the mixture. The reason, as shown in Chapter 7, is that change in fugacity with pressure is given by Eqs. 7.6 and 7.14, which show that

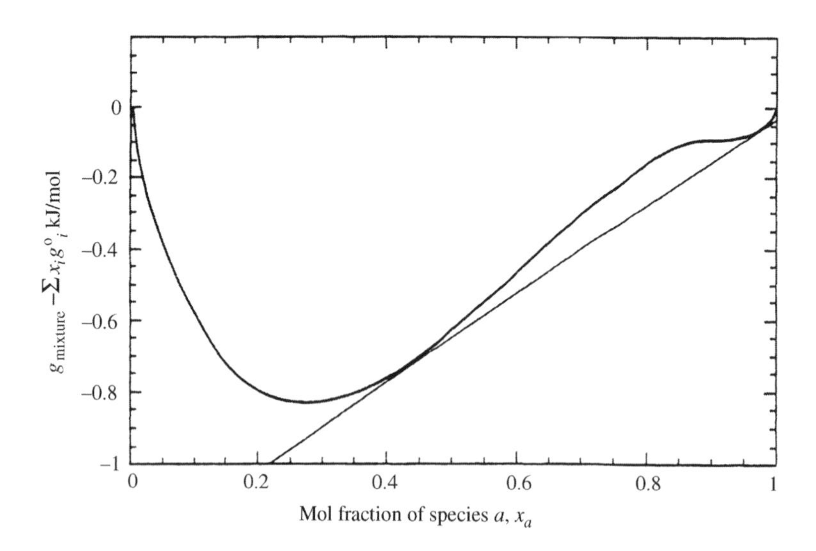

FIGURE 11.8 Calculated values of $g_{\text{mixture}} - \sum x_i g_i^{\text{o}} = (RT \ln x_i) + \sum x_i (RT \ln \gamma_i)$ for water and *n*-butanol at 92 °C, based on the van Laar equation, with constants based on VLE measurements, and the straight line tangent to the curve in two places.

that change is proportional to the exponential of the molar volume for a pure species or the partial molar volume for one species in a mixture. The molar volumes of pure liquids and the partial molar volumes of species in liquid mixtures are small enough that this effect is generally negligible (see Example 7.3). The sources for the data in Tables 11.1–11.4 do not report the pressures for those equilibria, assuming that the readers know that the effect of pressure on LLE is negligible (as long as the pressure is high enough to prevent boiling). At very high pressures (hundreds or thousands of atmospheres) these effects can become significant and must be considered.

11.5 EFFECT OF TEMPERATURE ON LLE

For practically insoluble mixtures we can estimate the effect of temperature change on LLE with some confidence. As shown in Eq. 11.4, the solubility is inversely proportional to the activity coefficient. If we take the ln of both sides of Eq. 11.4, differentiate with respect to temperature, and then substitute from Eq. 7.32, we find.

$$\frac{d\ln x_i^{(1)}}{dT} = -\frac{d\ln\gamma_i^{(1)}}{dT} = -\frac{(h_i^{\circ}-\bar{h}_i)}{RT^2} \quad \text{practically insoluble species in binary mixtures} \tag{11.6}$$

If we assume that $(h_i^{\circ}-\bar{h}_i)$ (minus the differential heat of mixing, see Eq. 6.11) is a constant, independent of temperature, then we can integrate Eq. 11.6, finding

$$\ln x_i^{(1)} = \frac{(h_i^{\circ}-\bar{h}_i)}{RT} + \text{constant of integration} \tag{11.7}$$

which suggests that a plot of ln $x_i^{(1)}$ vs. (1/*T*) should form a straight line. Figure 11.9 shows the data from Table 11.1, plotted as ln $x_i^{(1)}$ vs. (1000 K/*T*). For water in benzene the data do form practically a straight line, while for benzene in water there is slight curvature.

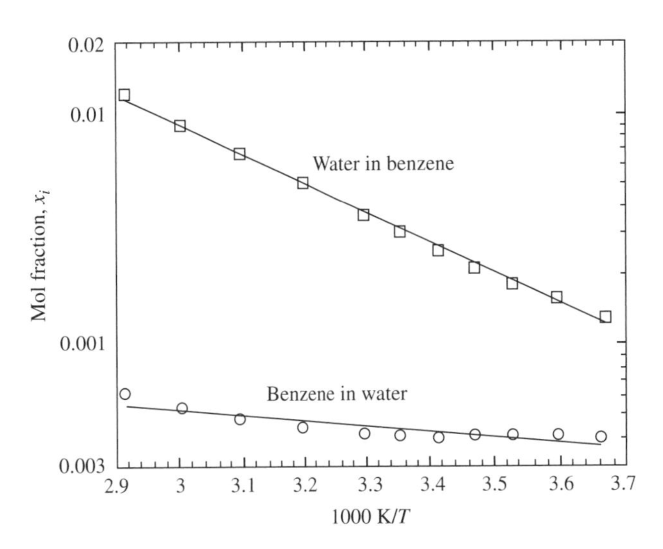

FIGURE 11.9 Benzene–water solubility data, in the form suggested by Eq. 11.7. The points are the values from Table 11.1. The lines are least-squares fits of the data. In this formulation temperature increases from right to left!

Example 11.7 Estimate the heat of mixing of water in benzene and of benzene in water, from the straight lines in Figure 11.9. We find that the water-in-benzene least-squares-fit line has the equation

$$\ln x_{\text{water}}^{(\text{benzene-rich phase})} = 4.175 - \frac{2967.7\text{K}}{T} \tag{11.O}$$

Comparing this term-by-term with Eq. 11.7, we conclude that

$$\frac{(h_i^{\circ}-\bar{h}_i)}{RT} = -\frac{2967.7\text{K}}{T} \tag{11.P}$$

so that

$$(h_i^{\circ}-\bar{h}_i) = -2967.7\text{K}\cdot 8.314\frac{\text{J}}{\text{mol K}} = -24.67\frac{\text{kJ}}{\text{mol}} \tag{11.Q}$$

for benzene-in-water the constant in Eq. 11.P is −522.9 K, leading to a calculated value of $(h_i^{\circ}-\bar{h}_i) = 4.347$ kJ/mol. ■

What physical meaning should we attach to these values? $(h_i^{\circ}-\bar{h}_i)$ is the enthalpy of the pure liquid minus the partial molar enthalpy of that species in solution, both at the solution temperature. Thus, we would have to *add* 24.67 kJ for each mol of water we dissolve in benzene, and *add* 4.37 kJ for each mol of benzene we dissolve in water. These values are 55 and 12% of the heats of vaporization of water and benzene at room temperature. For practically insoluble pairs, the energy required to get the molecules of solute away from each other and to insert them in the solvent is of the same order of magnitude as that needed to vaporize the solute. The type of behavior described here is very common; Figures 11.10 and 11.11 [10] show the plots corresponding to Figure 11.9 for a variety of hydrocarbon–water pairs. Observe that the horizontal axis is reversed, (1/*T*) is plotted from right to left, and the corresponding temperatures shown from left to right.

Example 11.8 Gasoline is put into the tank of an auto, saturated with water at 50°F. A sudden cold wave cools the auto and its fuel system to 20°F. How much water would we expect to come out of solution in the gasoline? What would its effect likely be?

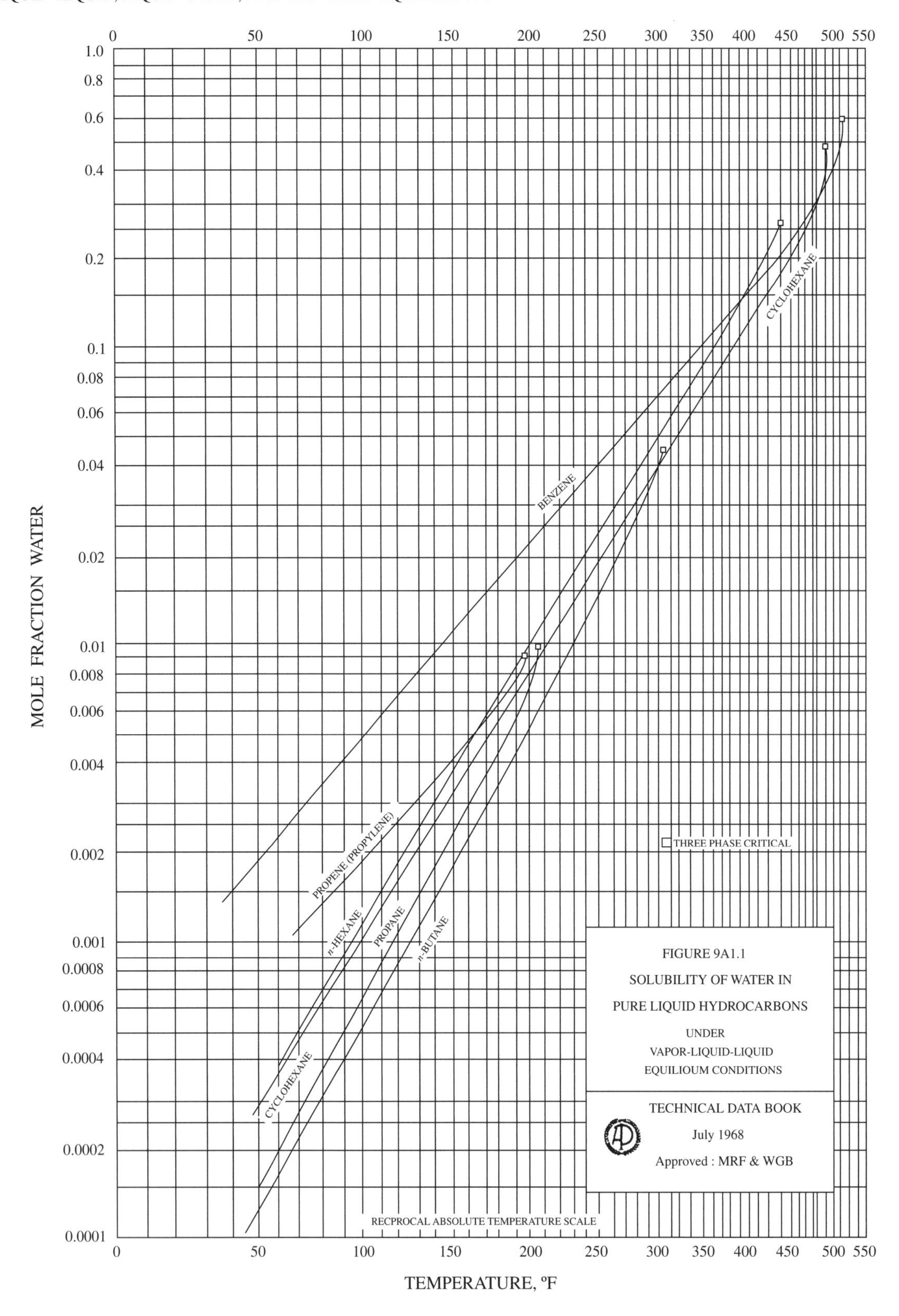

FIGURE 11.10 Solubility of water in pure liquid hydrocarbons. The horizontal scale is $(1/T)$, plotted from right to left, with the corresponding values of T in °F shown. This plot is a summary of all the available experimental data as of 1968. (From Daubert, T. E., and R. P. Danner. Phase equilibria in water–hydrocarbon systems. In *Technical Data Book, Petroleum Refining*, Vol. 2, Chapter 9, Figure 9A1.1. Washington, DC: American Petroleum Institute (1978). Reproduced by permission of the American Petroleum Institute.)

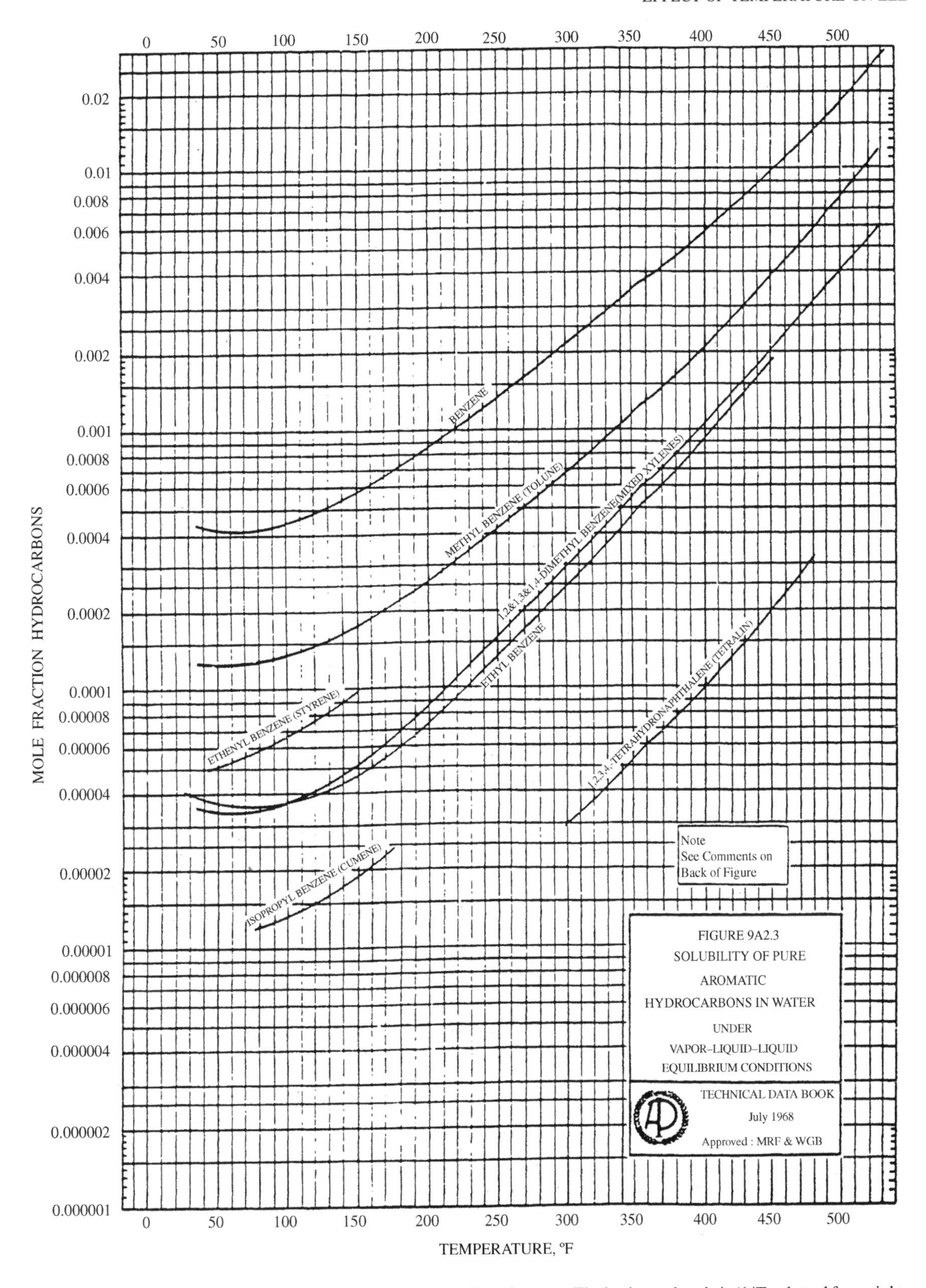

FIGURE 11.11 Solubility of pure liquid aromatic hydrocarbons in water. The horizontal scale is $(1/T)$, plotted from right to left, with the corresponding values of T in °F shown. This plot is a summary of all the available experimental data as of 1968. (From Daubert, T. E., and R. P. Danner. Phase equilibria in water-hydrocarbon systems. In *Technical Data Book, Petroleum Refining*, Vol. 2, Chapter 9, Figure 9A2.3. Washington, DC: American Petroleum Institute (1978). Reproduced by permission of the American Petroleum Institute.)

Another plot in [10], not shown here, suggests that the solubility of water in typical gasoline is very similar to that in cyclohexane, shown in Figure 11.10. From that figure we read the solubility of water in cyclohexane at 50°F as 0.00026 mol fraction. The cyclohexane curve in Figure 11.10, is totally straight, so we linearly extrapolate it to 20°F, finding a water solubility of ≈0.0001 mol fraction. Thus, we would expect the gasoline to reject (0.00026 − 0.0001) = 0.00016 mol of water per mol of gasoline. The molecular weight of gasolines varies somewhat, but a typical value is 115 g/mol; the density of gasoline also varies, but a typical value is 720 g/L at room temperature. Thus, we would expect

$$n_{\text{water rejected}} = 0.00016\frac{\text{mol water}}{\text{mol gasoline}}\cdot\frac{\text{mol gasoline}}{115\,\text{g}}\cdot\frac{720\text{g}}{\text{L}} = 0.001\frac{\text{mol water}}{\text{L gasoline}} = 0.018\frac{\text{g water}}{\text{L gasoline}} \tag{11.R}$$

If the fuel tank contains 10 gallons ≈ 40 L, then we would expect $0.018 \cdot 40 = 0.72\,\text{g} \approx 0.72\,\text{mL}$ of water to come out of solution in the gasoline.

At 20°F we would expect this water to become solid ice, forming a piece large enough to plug the fuel line of a parked auto. This is a disaster for the driver, who has to find a way to thaw the fuel line without freezing her/his fingers or nose! Petroleum refiners try to keep the water content of their gasolines as low as possible. After they have made the final water removal, many add small amounts of cosolvent, typically methanol or ethanol, to the gasoline to increase the water solubility (in winter and in cold climates). You can also purchase gasoline fuel-line deicers (cosolvents) to add to the tank of an auto driven from a warm climate to a cold one; these are practically pure methanol. They work. ■

So far we have discussed only the temperature effect on practically insoluble liquid pairs, based on the assumption (Eq. 11.4) that the solvent was practically pure. That assumption is not available for partly soluble pairs like water–*n*-butanol (Figure 11.3). The logic shown above applies in a *qualitative* way to the upper critical solution curves shown in Figures 11.3 and 11.5, but not in a quantitative way. Clearly, it cannot explain the lower critical solution temperatures on Figures 11.4 and 11.5, even qualitatively. Equation 11.8 shows that if $(h_i^{\text{o}} - \bar{h}_i)$ is negative, as in Figure 11.9, then the solubility must increase with increasing temperature, while if it is positive, the reverse must be true. So the pairs that form lower critical solution temperatures would be expected to evolve heat on mixing. This is not common behavior, but it is observed in the mixtures that form LCSTs [11].

11.6 DISTRIBUTION COEFFICIENTS

For extraction process and some others, we wish to know the relative solubility of one species in two insoluble solutions.

Example 11.9 At 25°C we place 5 mol of water, 5 mol of benzene, and 0.1 mol of ethanol in a container and agitate them to produce equilibrium. At this low an ethanol content we are sure to have two phases present (see Figures 11.1 and 11.2). What will the distribution of ethanol between the two phases be?

The fugacity of ethanol must be the same in each of the two phases, so that from Eq. 11.3 we have

$$K = \begin{pmatrix}\text{distribution}\\ \text{coefficient}\end{pmatrix} = \frac{x_{\text{ethanol}}^{(\text{water-rich phase})}}{x_{\text{ethanol}}^{(\text{benzene-rich phase})}} = \frac{\gamma_{\text{ethanol}}^{(\text{benzene-rich phase})}}{\gamma_{\text{ethanol}}^{(\text{water-rich phase})}} \tag{11.8}$$

Here we have defined the *distribution coefficient* K as the ratio of mol fractions of the distributed solute between the two phases, thus introducing another quantity normally written as K. We observe that for the first experimental data set in Table 11.4

$$K = \begin{pmatrix}\text{distribution}\\ \text{coefficient}\end{pmatrix} = \frac{x_{\text{ethanol}}^{(\text{water-rich phase})}}{x_{\text{ethanol}}^{(\text{benzene-rich phase})}} = \frac{3.817\%}{1.010\%} = 3.78 \tag{11.S}$$

Then we repeat the calculations for all the data points in that table and plot them as shown in Figure 11.12.

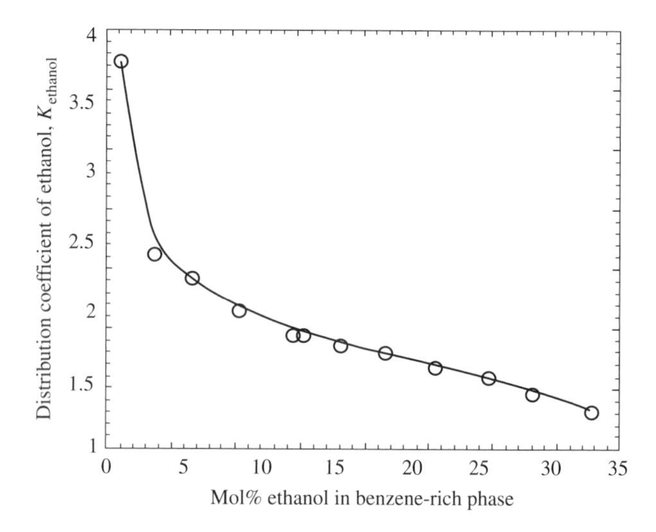

FIGURE 11.12 Distribution coefficient of ethanol between water-rich and benzene-rich phases, based on Table 11.4.

We see that at the low mole percent ethanol in this example, the distribution coefficient will be ≈ 4, so that the total 0.1 mol of ethanol will distribute itself about 80% in the water-rich phase and 20% in the benzene-rich phase (if each phase has the same number of moles). ■

This distribution coefficient > 1.00 indicates that, for example, we could extract ethanol from benzene with water fairly easily, but that extracting ethanol from water with benzene would work poorly. If we consider only concentrations more than about 5 mol% ethanol in the benzene-rich phase, then we see that the ethanol distribution coefficient is nearly a constant ≈ 1.75. If the distribution coefficient is nearly constant, some shortcut calculations can be used for liquid–liquid extraction, so this term is widely used.

11.7 LIQUID–SOLID EQUILIBRIUM (LSE)

A solid is a material at a temperature below its melting point. Almost all solids will melt, but some like CO_2 can exist as liquids only at pressures above atmospheric, and some solids like $CaCO_3$ and wood decompose on heating before they melt. The pure-species solids we are familiar with are the common metals (aluminum, copper, tin,...), salts (sodium chloride, calcium carbonate,...), and some crystalline organic materials (sucrose, naphthalene, waxes), ice (frozen water), and dry ice (frozen CO_2). Most of the solids we encounter in daily life are not pure species, for example, wood, steel (mostly iron and carbon), plastics (mixtures of polymers of various chain lengths with plasticizers and other additives), fabrics (blends of various natural and synthetic polymers with dyes, and surface treatment chemicals), concrete, asphalt, rubbers, and apples. Some apparently solid materials, like glass, are polymers, which are believed to be fluids at a temperature low enough that their viscosity is practically infinite. (Glass has no simple melting-point temperature like ice does; on heating it becomes less and less viscous, thus changing from a material that shatters if struck to a runny fluid, with no sharp boundary between the two.)

11.7.1 One-Species LSE

The simplest LSE is that of a single pure species, represented by the L-S or melting-point curve in Figure 1.9. We see that the curve is practically vertical, indicating that the melting temperature is practically independent of pressure, 0°C for water. Figure 1.10 shows that for pressures in thousands of atmospheres the curve for water has a slightly negative slope. This is true only for the few substances like water that expand on freezing; most substances shrink on freezing, and their equivalents of Figure 1.10 have a freezing curve with a slight positive slope. For one species there is no "solubility" as we would understand the term. At any fixed T and P not exactly on the L-S equilibrium curve any pure species can exist as solid or liquid, but not as one dissolved in the other. So the term LSE refers to systems with more than one species.

11.7.2 The Experimental Determination of LSE

The experimental determination of LSE is similar to that of LLE. We place a sample of the pure-species solid in a container with a sample of pure-species liquid, shake until the liquid has become saturated with the solid, separate by gravity or filtration, and analyze the liquid to determine how much solid is dissolved in it. The solubility of liquids in most pure-species solids is small enough that we ignore it. (Polymers, natural and synthetic, often absorb solvents, and increase in volume, called "swelling" in the polymer field. Metals and salts generally do not *measurably* absorb liquids and swell.) This means that the equivalent of Table 11.1 shows the solubility of the solid in the liquid, but almost never the solubility of the liquid in the solid.

Compared to the three possible outcomes of the LLE experiment (Section 11.2) we have only two here. There is no equivalent of total miscibility; that would require that mixing an infinitesimal amount of liquid with a large amount of solid produced all liquid. The other two possibilities are

1. Practically insoluble solids (less than a fraction of a mol percent dissolved), such as iron, copper, wax, or rubber in water.
2. Substantial solubility (more than a few mol percent dissolved), such as table salt or sugar in water.

11.7.3 Presenting LSE Data

The most common presentation of LSE data is a table like Table 11.5. We see that the solubility of NaCI in water increases very slowly with increases in temperature.

Example 11.10 Table 11.5 shows the most commonly reported unit for reporting such solubility. Show the corresponding values of the weight fraction and mol fraction of NaCl in a saturated solution in water at 20°C:

Table 11.5 Solubility of NaCl in Water [12]

Temperature (°C)	Solubility of NaCl (g per 100 g of water)
0	35.7
20	36.0
40	36.6
60	37.3
80	38.4
100	39.8

$$\text{weight fraction} = \frac{36.0\text{ g}}{(100+36)\text{ g}} = 0.265 = 26.5\text{ wt\%} \tag{11.T}$$

$$\text{mol fraction} = \frac{\dfrac{36\text{ g}}{58.5\text{ g/mol}}}{\dfrac{36\text{ g}}{58.5\text{ g/mol}} + \dfrac{100\text{ g}}{18\text{ g/mol}}} = 0.0997 \approx 0.10 = 10\text{ mol\%} \quad \blacksquare \tag{11.U}$$

This simple example is included to show the relations among the three most common ways of expressing LSE data.

For very low solubility salts the results are almost always presented in terms of the *solubility product*, discussed in Section 13.3. For medium-solubility materials they are often presented in plots like Figure 11.13. The solubilities of calcium salts are much lower than those of sodium salts; the top of the diagram is only 3000 ppm = 0.3% by weight. Calcium salts also exhibit *inverted solubility curves*, that is, the solubility decreases as the temperature increases. Most solids, like NaCl show the opposite effect: The solubility increases slightly with increasing temperature (see Table 11.5). This nasty behavior of calcium salts has practical consequences.

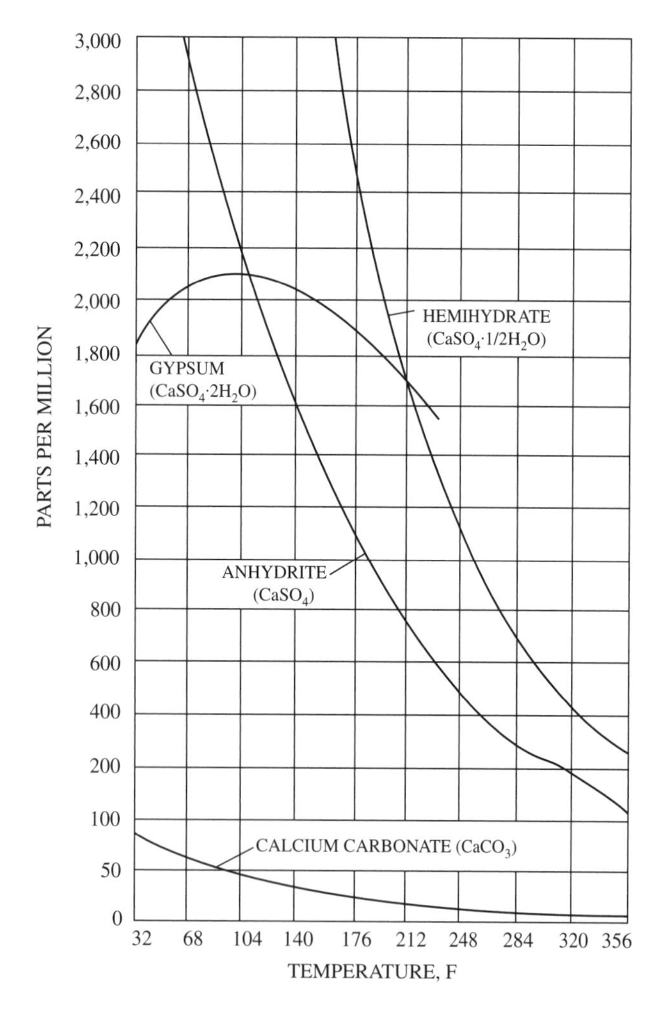

FIGURE 11.13 Solubility of various calcium salts, expressed in ppm by weight, as a function of temperature [13]. Observe the scale changes at 100 and 200 ppm. (Used with permission of Betz Dearborn, Inc., Trevose, PA.)

Example 11.11 Most municipal water systems use surface waters (rivers or lakes) or groundwaters (from wells). The surface waters often have low content of Ca salts, but groundwaters are often saturated at about 20°C with calcium carbonate. Often the water will also contain calcium bicarbonate, $Ca(HCO_3)_2$ which converts to calcium carbonate on heating. Most municipal waters contain much less of the sulfate salts than the carbonate salts, because the world has a lot more limestone than gypsum. If the household water heater takes water saturated with calcium carbonate at 20°C (=68°F) and delivers it to the house at 140°F (a typical household water heater setting) how much solid calcium carbonate will it deposit in the water heater? For this example we ignore the other salts present in the water, such as calcium sulfates and bicarbonates and magnesium salts, all of which contribute to water "hardness."

From Figure 11.13 we read the solubilities of calcium carbonate at 68 and 140°F as about 60 and 30 ppm.

Thus, we would expect approximately (60−30) = 30 ppm of calcium carbonate to come out of solution and remain in the water heater. A typical household water heater heats ≈ 100 gallons/per day (strongly dependent on how much laundry is done and the number of people in the house), so we would expect to deposit

$$\begin{pmatrix}\text{CaCO}_3\\ \text{deposited}\end{pmatrix} = 30 \times 10^{-6} \cdot 100\,\frac{\text{gal}}{\text{day}} \cdot 8.33\,\frac{\text{lb}}{\text{gal}} = 0.025\,\frac{\text{lb}}{\text{day}} \quad \blacksquare \tag{11.V}$$

This is a high estimate of the deposition rate. Most waters do not deposit this much. But this deposition is a serious problem in domestic water heaters; the solids collect in the bottom of the heater, impeding heat transfer and speeding the eventual wear-out of the heater. Water softeners use ion exchange to replace the calcium (and magnesium) ions in water with sodium ions (from NaCl, sending the reject $CaCl_2$ to the sewer). The resulting Na salts remain in solution and do not react with soaps the way Ca salts do, thus producing "soft water." Water softeners lengthen the life of water heaters. Industrial water heaters and boilers use a variety of techniques to prevent this deposition of sulfates and carbonates. (In the early days of the industrial revolution the role of these Ca and Mg salts in boiler feed waters was not understood; they led to some disastrous boiler explosions!)

If we extend the temperature scale of figures like Figure 11.13 down to the freezing temperature of the solvent,

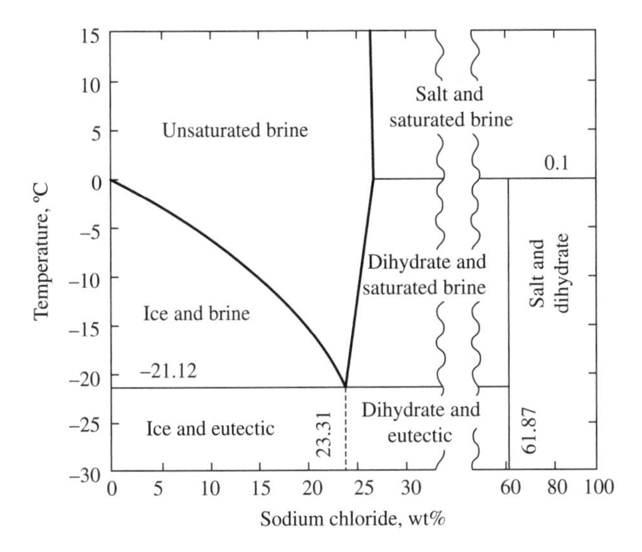

FIGURE 11.14 Phase diagram for water and sodium chloride. If the diagram extended up above 100°C, it would also show the VLE. Observe the contraction of the scale between 30 and 60 wt% NaCl. (From Bertram, B. M. Sodium compounds (sodium halides). In *Kirk-Othmer Encyclopedia of Chemical Technology*, ed. 4, Vol. 22, Kroschwitz, J. I., and M. Howe-Grant, eds. Copyright © 1997, New York: Wiley, p. 360. Reproduced by permission of John Wiley and Sons, Inc.)

then the behavior becomes more complex (and much more interesting). Figure 11.14 shows such a diagram for the NaCl–water system at some pressure above the boiling pressure of the liquid. The effect of changing pressure on such diagrams is so small that the pressure is normally not specified.

Figure 11.14 is a phase diagram; with it for any temperature and weight percent salt we may determine by inspection which phases are present. We may illustrate this complex diagram by mentally following several paths on it. If we begin at 0 wt% NaCl and 5°C and proceed horizontally to the right (by adding NaCl and stirring), we will be in the unsaturated brine (solutions of ordinary salts are called *brines*) region until we reach 26 wt%. At that point the solution becomes saturated; it can hold no more NaCl. If we continue to add NaCl and stir, the solid NaCl will not dissolve. If we stir hard enough it will remain suspended in the brine; when we stop stirring it settles to the bottom of the container. In Figure 11.14 we will have passed into the region marked "Salt and saturated brine," indicating that the composition of the brine does not change as we add more salt; only the ratio of brine to solid salt changes.

Next we consider starting with a brine that is 5 wt% salt at 15°C, and cooling it. At $\approx -3°$ we encounter the phase boundary between "unsaturated brine" and "Ice and brine." At this boundary ice begins to form. This ice contains no salt (if we proceed slowly enough so that none of the brine is mechanically trapped in the ice). As the ice forms it removes water from the brine, so that the salt concentration in the remaining brine increases. Further cooling causes the temperature to fall, more ice to form, and the composition of the remaining brine to follow the phase boundary line, until we reach −21.12°C and a salt content of 23.31 wt%. At that temperature the solution begins to freeze, forming a *eutectic*, which is a mixture of crystals of ice and dihydrate ($NaCl \cdot 2H_2O$). It is not a solid solution, rather it is an intimate mixture of the two crystal types, which form simultaneously. From the phase diagram we see that since the overall composition is still 5 wt% salt, we will have crystals of pure ice and crystals of the eutectic.

11.7.4 Eutectics

Most systems involving water and solid salts like NaCl have phase diagrams like Figure 11.14, in which the liquid has negligible solubility in the solid phases (ice or dihydrate). That is not the case in the many technically important metallurgical eutectics. Figure 11.15 shows the equivalent of Figure 11.14, for the important Sn-Pb (tin-lead) system. This figure shows no intermediate compound like $NaCl \cdot 2H_2O$. It does show, as α and β, two solid solutions, one with a little Pb dissolved in almost pure Sn and the other with a substantial amount of Sn dissolved in Pb. No such solid solutions occur in Figure 11.14 because the three solids on Figure 11.14 (ice, salt, dihydrate) are crystalline solids that do not dissolve measurable amounts of water or the other solids. (Solubilities can never be truly zero, so if we had suitable instruments, we could detect some very, very small amount of water dissolved in these solids.) At the eutectic

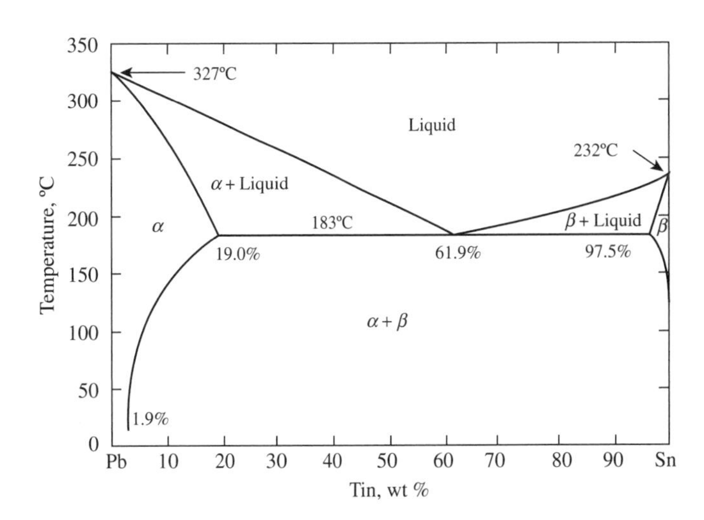

FIGURE 11.15 The phase diagram for Sn-Pb. This differs from that for $NaCl\text{-}H_2O$, because there is no intermediate compound like $NaCl \cdot 2H_2O$, and because the two equilibrium solids are both solid solutions, α, mostly Pb with up to 19% Sn, and β, mostly Sn with up to 2.5% Pb.

composition (183°C, 61.9 wt% Sn) we have equilibrium between a 61.9 wt% Sn liquid, solid α (19.0 wt% Sn) and solid β (2.5 wt% Sn). Industry uses a variety of solders for special purposes; the common solder sold in your hardware store (approximately 61.9 wt% Sn), melts at 183°C, a temperature easily reached by ordinary soldering irons and propane torches. It wets and bonds easily to clean copper, so that it is regularly used to join copper wires and copper plumbing tubing and pipes, making joints that are good electrical conductors and that seal well against leaks. Its melting point is high enough that it does not appreciably soften on hot summer days, nor with hot water flowing in the pipes. This combination of virtues made its use practically universal, until concerns about it transferring Pb into drinking water led many to stop using it in drinking water plumbing. With a good microscope you can see that the solid solder is an intimate mixture of the two different solid solutions (α and β).

Repeating the discussion of Figure 11.14, suppose we cool a liquid that is 40 wt% Sn. When we reach about 240°C, a solid begins to form. Unlike the pure ice that forms in the same situation in Figure 11.14, the solid that forms here is the α solid, with about 12% Sn, 88% Pb. As we continue cooling, the liquid follows the line sloping to the right, as the solid depletes Pb in the liquid. However the deposited solid's tin content increases from about 12% to about 19%. For this to happen we must cool slowly enough for the solid already precipitated to remain in equilibrium with the liquid, whose Sn content is continually increasing as we cool and deposit a solid that is mostly Pb. Finally at 183°C, we reach the eutectic composition, with two intimately mixed solids, one with about 2.5% Sn, the other with about 19% Sn. If we could maintain equilibrium (difficult and slow with mixed solids) then on continued cooling the compositions of the two solids would change, with the α solid reaching 1.9% Sn at about 20°C, and the β solid becoming practically pure tin at about 150°C. That requires diffusion rates in solids larger than we observe in nature, so that practically the compositions at the eutectic are "frozen" in place.

The type of eutectic shown in Figure 11.15 is very common in metallurgy. The Fe-C diagram that directs the science of steelmaking has two eutectics like Figure 11.15 (a liquid and two solids) and one with three solids, and one intermediate compound Fe_3C, like the $NaCl \cdot 2H_2O$ in Figure 11.14. Metallurgy books are full of such diagrams.

In Section 11.3 (LLE) we used equality of the individual species' fugacities as our working equilibrium criterion. That is rarely done with solids, or with systems like those shown in Figures 11.14 and 11.15, because the vapor pressures of most solids are low. Ice has a substantial vapor pressure at its melting point or slightly below (Example 8.16), but the vapor pressure of NaCl at 0°C (extrapolated by the C-C equation from much higher temperatures where it can be measured) is 10^{-30} atm., much too low to measure. The vapor pressures of Sn and Pb at their melting points on Figure 11.15 (extrapolated the same way) are 10^{-27} and 10^{-11} atm. (If we wanted to represent the fugacity of such solids using a Raoult's-law-type formulation, we would find that the pure solid had $f_i \approx 0$!) Instead, metallurgists normally think of such equilibria in terms of Gibbs energy-composition diagrams like Figures 11.7 and 11.8. Figure 11.16 shows an example of that.

The lower right of the five figures that make up Figure 11.16 is a generic representation of a temperature-

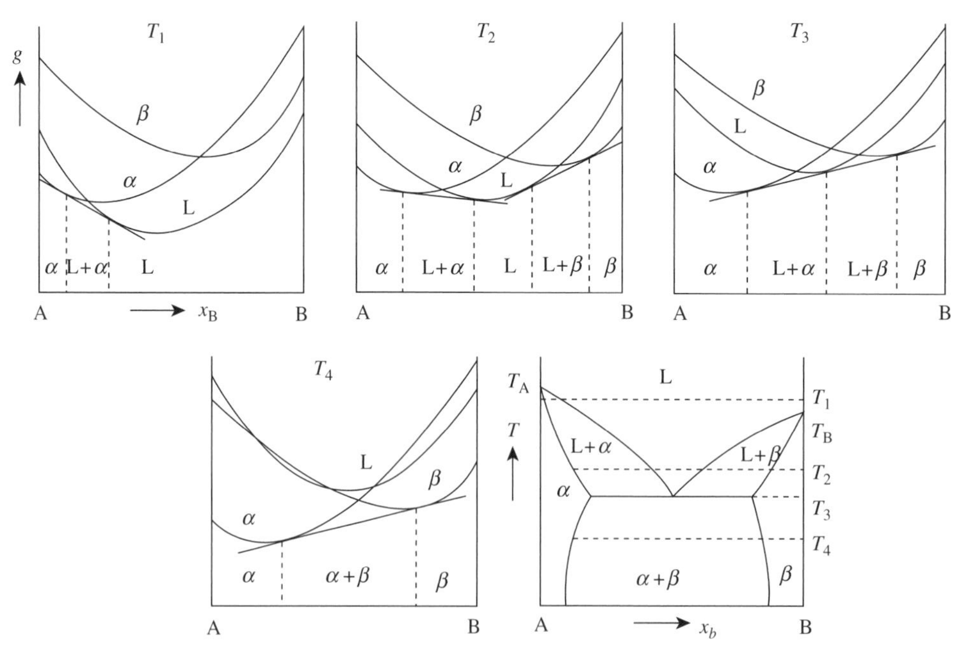

FIGURE 11.16 Gibbs-energy-composition-temperature relations for a system like that in Figure 11.15. See the text for a description.

composition phase diagram like the Sn-Pb on Figure 11.15. It shows the four temperatures, ($T_1 \ldots T_4$) at which the preceding four g-x_b diagrams are made. On each of the constant-temperature diagrams the three curves represent the Gibbs energy of the α solid, the β solid and the liquid (L). In each of these figures the compositions of the equilibrium phases are found by drawing a line tangent to the lowest 2 or 3 of these curves, (see the discussion of this procedure in terms of the chemical potentials in section 6.6). At T_1 there is only one possible such tangent line; the one we could draw between curves α and β lies above the L curve, and thus does not represent an equilibrium situation. At T_2 there are two such tangents showing the four possible equilibrium phases at this temperature. T_3 represents the eutectic, where one line is tangent to all three curves; two solids and a liquid are in equilibrium. At T_4 the line touches only the two solids; no liquid exists at this temperature. I hope this short digression helps chemical engineering students see the direct connection between our way of describing the thermodynamics of multiphase equilibrium, and the same phenomena as described by the metallurgists.

11.7.5 Gas Hydrates (Clathrates)

Figures 11.14 and 11.15 show the common L-S equilibrium situation (see Section 8.10.2), in which a liquid exists at a temperature lower than the freezing temperature of either pure solid; we see this often in metallurgy and in aqueous solutions of mineral salts. In a much less common G-L-S situation, a solid called a gas hydrate can exist at a temperature above the freezing temperature of both of the compounds that make it up; most of these are made by water using other ("guest") molecules with diameters between about 4 and 7Å (roughly argon through *n*-butane) as templates to form solid ice-like structures with internal cages that hold the guest molecules. Several other compounds besides water (e.g., urea) also form these cage compounds (named *clathrate* from Latin words approximately meaning cage; some are also called *adducts*). If the cage-former is water, they are almost always spoken of as hydrates.

The water-NaCl dihydrate in Figure 11.14 has a simple molecular formula and forms regular-structure crystals, as do most such compounds. The cage-structure hydrates discussed here do not, because the guest molecules do not bind to host molecules in a stoichiometric way, like water and NaCl [14, p. 313]. The S-L equilibria in Figures 11.13 and 11.14 are practically independent of pressure. Because the G in G-L-S has a much larger specific volume than liquids or solids, gas hydrates phase diagrams are pressure-dependent.

Figure 1.17 shows a summary of the experimental data for hydrate formation between water and methane [14]. This figure is entirely analogous to Figure 1.9 for water alone, except that the vertical scale is logarithmic because of the range of pressures covered. In Figure 1.9 an area represents

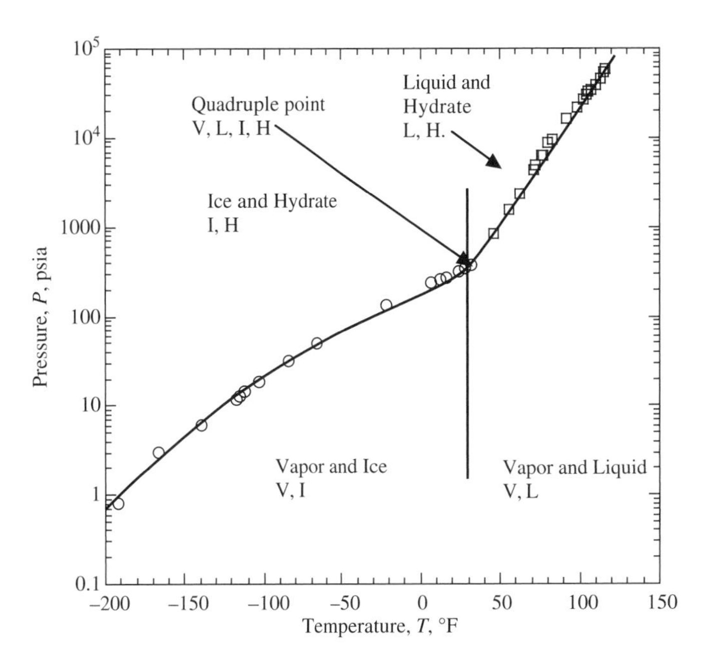

FIGURE 11.17 Phase behavior of the water-methane system [14, p 164] in the hydrate-forming region. Here V stands for vapor, H for hydrate, I for ice, and L for liquid. See the text for a discussion of this figure.

one phase, a curve represent two phases (the two in the areas it divides) and the point of three meeting curves represents all three phases in equilibrium (the triple point). Figure 11.17 represents two species; the phase rule (Chapter 15) tells us that in a figure like Figure 1.9, adding a new species increases all of the numbers of phases by one, so that on Figure 11.17 an area represent two phases, the curve dividing two areas represents three phases and the point where the four curves meet represents four phases in equilibrium (the quadruple point). The four phases on this figure are vapor (V), solid hydrate (H), ice (I) a liquid that is mostly water (L).

The areas are marked on the figure. The curves, (clockwise from the upper right corner) represent the (V, L, H) equilibrium, the (V, L, I) equilibrium, the (V, I, H) equilibrium and the (I, H, L) equilibrium. The quadruple point represents the (V, L, I, H) equilibrium.

On this figure two of the phase boundaries have been carefully measured, as the data points show. The two nearly-vertical boundaries at 32°F represent the ordinary freezing curve for water, drawn on the assumption that the small amount of methane that dissolves in water does not influence the shape of this curve much. At high pressures this curve bends to the left (Problem 11.34) but over the range shown it is practically straight and vertical.

This is one of the simplest phase diagrams for hydrate formation, many others are more complex [14], mostly because the hydrates have at least three different crystalline structures, leading to additional phases. Water is known to form such hydrates with Ar, Kr, N_2, O_2, CH_4, Xe, H_2S, CO_2, C_2H_6, $(CH_3)_2O$, C_3H_8 and i-C_4H_{10}. In Figure 11.17 methane is far below its T_c so that it cannot exist as a liquid. Guest molecules like propane can exist as a liquid at hydrate temperatures forming an additional phase, mostly liquid propane, producing at least two quadruple points on the phase diagram [14, p 164].

We introduce this topic and Figure 11.17 because:

1. The behavior in Figure 11.17 is quite remarkable. Water is shown forming an ice-like solid at temperatures up to 120°F. This requires pressures of 30,000 psia, which are uncommon, but at 1000 psia (a common pressure in deep-sea ocean bottoms and in some natural gas pipelines) the solid can exist up to 50°F. This hydrate looks like compacted snow, with a density practically the same as ice and a methane content of 10 to 13 wt%. When brought to atmospheric pressure it evaporates; the methane gas emitted burns in a typical natural gas flame [15].
2. Hydrates are a major problem for the natural gas industry. The natural gas (and associated natural gas liquids) come out of the ground fairly warm, but in surface (or underwater) gathering pipes they cool enough to form hydrates that plug up pipes and valves at high pressure and/or low temperatures. The defenses against them are rigorous water removal, and/or addition of hydrate-preventing diluents (often glycols or methanol). The deep oceans have temperatures and pressures in the hydrate-forming range for methane and for its mixtures with other natural gas liquids, which was publicly demonstrated in the unsuccessful efforts to capture the flow of the uncontrolled Macondo deep Gulf of Mexico oil well in 2010.
3. The deep ocean bottoms and the arctic permafrost contain buried methane hydrate. The estimated amounts are large enough [16, 17] that some consider the prospect of mining them to capture the methane for fuel. Others fear that their inadvertent release would put enough methane – a potent greenhouse gas - into the atmosphere to seriously increase global warming.

11.8 THE ELEMENTARY THERMODYNAMICS OF LSE

At every point in a phase diagram like Figure 11.14, if more than one phase is present and we have true thermodynamic equilibrium (systems like that in Figure 11.14 often display metastable equilibria), we can say that the system has taken up the state with the lowest Gibbs energy consistent with the external constraints and the initial state. That means that for every individual species present the fugacity must be the same in all of the phases at equilibrium. In Table 11.5 and Figures 11.13 and 11.14 the pressure was not specified. As long as it is greater than the system's vapor pressure and less than hundreds or thousands of atmospheres, it has practically no effect on the values shown. We normally observe these systems at 1 atm in contact with air. That means that there will be some oxygen, nitrogen, CO_2, and so forth dissolved in the liquids. Those dissolved amounts are normally small enough (see Chapter 3) that they make no measurable change in the values shown. If we have one of our many piston-and-cylinder devices containing salt and water, at any point in Figure 11.14, and we use a vacuum pump to remove all the air, then the remaining vapor will be practically pure water at the equilibrium vapor pressure of the phase or phases in the container. At these temperatures salt (melting point 801°C, boiling point 1413°C) has a negligible vapor pressure, so the vapor will be practically pure water.

Equation 11.1, which we applied to LLE, applies to any phase equilibrium, so it must apply to LSE, with phase 1 being the liquid and phase 2 being the solid. Substituting the definition of the fugacity in terms of the standard state fugacity and solving for the mol fraction of dissolved solid, we find

$$x_i^{(1)} = \frac{x_1^{(2)} \gamma_i^{(2)} f_i^{(2)^\circ}}{\gamma_i^{(1)} f_i^{(1)^\circ}} \tag{11.9}$$

which is also true for any two-phase equilibrium. In LLE we made the Raoult's law type of standard state definition, $f_i^{(1)^\circ} = p_i$ for each species in each phase, which allowed us to cancel the two standard state values to find Eq. 11.3. Here we can certainly do that for the pure solid phase (if we have some reasonable way of estimating its vapor pressure), but what are we to use for the standard state of the dissolved solid in the liquid phase? We immediately think we should use the same standard state in both phases, which worked well for LLE. However, in VLE (Chapter 8) we saw that when the phases were of different types we normally had to choose different standard states for each phase, such as Raoult's law and Henry's law. Here a plausible guess for a standard state for the solid dissolved in the liquid is the vapor pressure that the solid would have if it existed as a subcooled liquid below the triple point (Figure 5.7). This is a guess, to be tested for usefulness, not a piece of rigorous thermodynamics.

If we make that guess, we can also assert that for the pure solid phase $x_i^{(2)} \gamma_i^{(2)} = 1$, so Eq. 11.9 becomes

$$x_i^{(1)} = \frac{p_{i,\text{ solid phase}}}{\gamma_i^{(1)} p_{i,\text{ subcooled liquid}}} \tag{11.10}$$

At the triple point $p_{i,\text{ solid phase}} = p_{i,\text{ liquid}}$. If we then assume that we are dealing with low pressures (the vapor pressures of most substances at the triple point are quite low) we may represent each of these vapor pressures by the Clausius–Clapyron equation (Eq. 5.10)

$$\ln\left(\frac{p_{i,\text{ solid phase}}}{p_{i,\text{ solid phase triple point T}}}\right) = \frac{\Delta h_{\text{solid to vapor}}}{R}\cdot\left(\frac{1}{T_{\text{triple point}}}-\frac{1}{T}\right) \tag{11.11}$$

and an almost identical equation for the liquid, with $\Delta h_{\text{solid to vapor}}$ replaced by $\Delta h_{\text{liquid to vapor}}$. We then subtract this second equation from Eq. 11.11 and rearrange to

$$\ln\left(\frac{p_{i,\text{ solid phase}}}{p_{i,\text{ subcooled liquid}}}\right) = \frac{\Delta h_{\text{solid to liquid}}}{R}\cdot\left(\frac{1}{T_{\text{triple point}}}-\frac{1}{T}\right) \tag{11.12}$$

We then take the ln of Eq 11.10, rearrange, and substitute Eq. 11.12, finding

$$\ln(x_i^{(1)}\gamma_i^{(1)}) = \ln\left(\frac{p_{i,\text{ solid phase}}}{p_{i,\text{ subcooled liquid}}}\right) = \frac{\Delta h_{\text{solid to liquid}}}{R}\cdot\left(\frac{1}{T_{\text{triple point}}}-\frac{1}{T}\right) \tag{11.13}$$

Melting-point data are much more widely available than triple-point data, and for most substances the 1-atm melting-point temperature ≈ the triple-point temperature, so it is common to substitute T_{melting} for $T_{\text{triple point}}$ in Eq. 11.13, and also to multiply both sides by minus 1, finding

$$\ln\left(\frac{1}{x_i^{(1)}\gamma_i^{(1)}}\right) = \ln\left(\frac{p_{i,\text{ subcooled liquid}}}{p_{i,\text{ solid phase}}}\right) = \frac{\Delta h_{\text{solid to liquid}}}{R}\cdot\left(\frac{1}{T}-\frac{1}{T_{\text{melting point}}}\right) \tag{11.14}$$

or

$$\ln\left(\frac{1}{x_i^{(1)}\gamma_i^{(1)}}\right) = \ln\left(\frac{p_{i,\text{ subcooled liquid}}}{p_{i,\text{ solid phase}}}\right) = \frac{\Delta h_{\text{solid to liquid}}}{RT_{\text{melting point}}}\cdot\left(\frac{T_{\text{melting point}}}{T}-1\right) \tag{11.15}$$

Equation 11.14 has the Clausius–Clapyron assumptions built into it; the more complex version without those assumptions is in [18, p. 640] (See Problem 11.36).

If we further assume that the dissolved solid forms an ideal solution in the solvent, we may drop the $\gamma_i^{(1)}$ from Eq. 11.15, and find an "ideal solubility curve." If this long and somewhat speculative derivation is correct, then if we had a family of compounds (e.g., aromatic compounds that are solid at room temperature), we would expect them to form ideal solutions in benzene. If they had equal values of $\Delta h_{\text{solid to liquid}}/T_{\text{melting point}} = \Delta s_{\text{fusion at melting point}}$, then we could calculate such an ideal solubility curve for all of them on $\ln x_i^{(1)}$ vs. $(T_{\text{melting point}}/T)$ coordinates. Figure 11.18 shows such a plot. The ideal solubility curve is calculated on the assumption that for all aromatics that melt somewhat above room temperature, $\Delta s_{\text{fusion at melting point}} = 54.5$ J/mol K. The comparison with the experimental data is very good, indicating that, at least for this particularly simple solubility problem, Eq. 11.14 is a very good approximation.

If we wish to extend this idea to nonideal solutions, we must retain the $\gamma_i^{(1)}$ in Eq. 11.15. We may test this idea by dissolving the aromatic compounds in Figure 11.18 in a nonaromatic solvent, in which they would be expected to show $\gamma_i^{(1)} > 1.00$. Figure 11.19 shows a similar plot for the solubilities of these same aromatic compounds in nonaromatic CCl_4. The "ideal solubility curve" is the same as that in Figure 11.18: From Figure 11.19 we see that the solubilities in CCl_4 are less than those in benzene.

Example 11.12 Estimate the activity coefficients at $x_2 = 0.1$ for the aromatic solutes in Figures 11.18 and 11.19.

Those figures do not indicate a constant activity coefficient independent of temperature or solubility; that would require parallel lines on the figure. The two lines meeting in the upper left corner indicate an activity coefficient of 1.00 at the melting point and activity coefficients that increase with

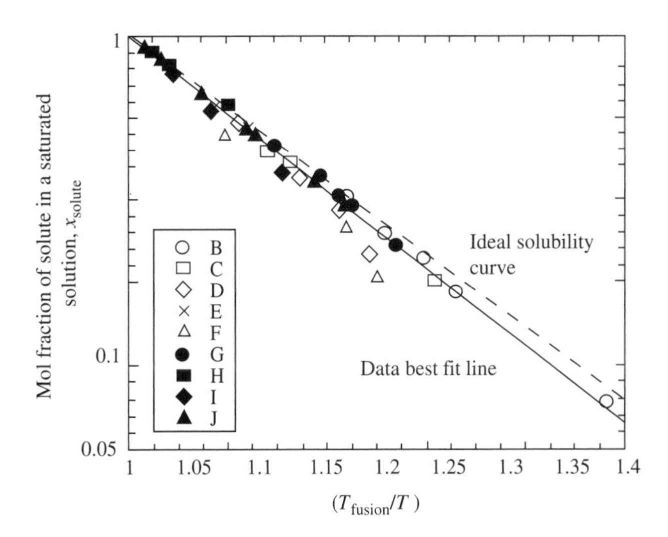

FIGURE 11.18 Solubility of aromatic solids in benzene [19]. The solid line is a best linear fit of the experimental data. The compounds listed from B to J in the figure are pyrene, fluorene, fluoranthene, biphenyl, acenaphthene, phenanthrene, *o*-terphenyl, *m*-terphenyl, and anthracene. The authors also show tabular solubility values for triphenylene, *p*-terphenyl, and 1,3,5-triphenyl benzene, which are not shown in the above plot (or the plot in the original article). Those values differ from the data fit line by up to a factor of 3. The authors cite the fact that these three compounds have fusion temperatures much higher than those of the compounds shown as the probable reason for this disagreement.

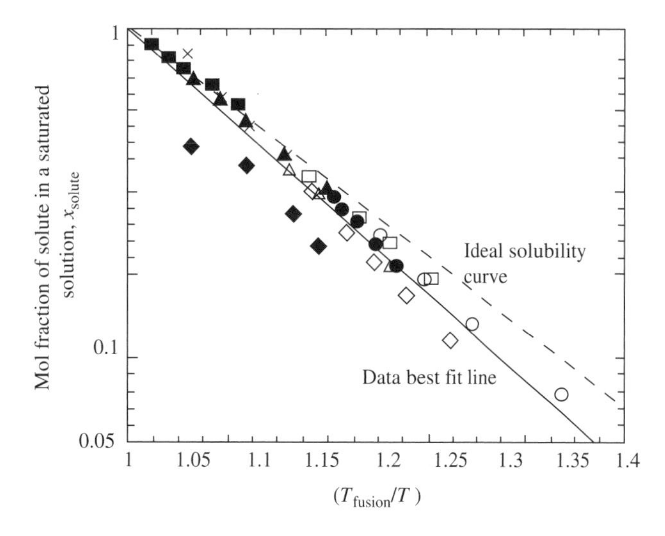

FIGURE 11.19 Solubility of aromatic solids in carbon tetrachloride [20]. The solid line is a best linear fit of the experimental data. The symbols represent the same compounds as in Figure 11.18.

increasing separation from the melting point. (Remember from Chapter 9 that for type II behavior, the activity coefficients increase as the concentration decreases. That behavior is shown here.) From the definitions we can write that

$$\gamma_i^{(1)} = \left[\frac{\left(x_i^{(1)}\right)_{\text{ideal solution}}}{x_i^{(1)}}\right]_T \tag{11.16}$$

From Figures 11.18 and 11.19 (or the equations of the lines on those figures, presented in [18]) we can compute the values in Table 11.6.

We see that at $x_{\text{solute}} = 10\%$ the average of these compounds has $\gamma_i = 1.14$ in benzene, corresponding to practically ideal solution and $\gamma_i = 1.52$ in CCl_4, corresponding to mild type II behavior (see Section 8.4.2). ■

Example 11.13 Estimate the solubility of NaCl in water at 20°C, based on Eq. 11.15.

Ignoring for the moment the wild extrapolation involved, we simply insert the appropriate values, $T_m \approx 800°\text{C}$, $\Delta h_{\text{fusion}} = 517.1\ \text{J/g} = 30{,}219\ \text{J/mol}$ [21. p. 359], and find

Table 11.6 Values for Example 11.12

	Figure 11.18, Solutions in Benzene	Figure 11.19, Solutions in CCl_4
T_m/T at which $\log x_{\text{experimental}} = -1$	1.332	1.288
x_{ideal} at that value of T_m/T	0.114	0.152
$\gamma_i^{(1)} = \left[\frac{\left(x_i^{(1)}\right)\text{ideal solution}}{x_i^{(1)}}\right]_T$	1.14	1.52

$$\ln\left(\frac{1}{x_i^{(1)}\gamma_i^{(1)}}\right) = \frac{30{,}219\,\dfrac{\text{J}}{\text{mol}}}{8.314\,\dfrac{\text{J}}{\text{mol K}}\cdot 297{\cdot}15\text{K}}\cdot\left(\frac{1073.15\ \text{K}}{297.15\ \text{K}} - 1\right)$$

$$= 8.845 \tag{11.W}$$

and

$$x_{\text{NaCl}}^{(\text{acueous solution})}\,\gamma_i^{(1)} = \frac{1}{\exp 8.845} = 0.000{,}144 \tag{11.X}$$

If we make the plausible (??) assumption that $\gamma_i^{(1)} = 1.00$, then $x_{\text{NaCl}}^{(\text{aqueous solution})} = 0.000144$, which is clearly very far wrong. From Example 11.10 we know that the experimental value is 0.10. ■

What does this mean? First, observe that this simple theory of solubility uses no information about the solvent. The above theoretical calculation is for any solvent. If the solute (NaCl in this case) does not interact with the solvent, then this may be a fair estimate. The reported solubility of NaCl in ethanol at 25°C is 0.00025 mol fraction, 1.7 times the value calculated above. If we wanted to know the solubility of NaCl in gasoline or diesel fuel, with which it would not be expected to interact much, the 0.00014 mol fraction computed in Example 11.13 would be a fair estimate. However, we know that water and NaCl interact strongly. The salt ionizes, the ions solvate with the water molecules. We may think of their interaction as a strong example of type III (Section 8.4.3) with a calculated activity coefficient of 0.000,14. All of the salts that dissolve to high concentrations in water are similar to NaCl in this behavior.

The other class of materials that dissolve in water to substantial degrees, well below their melting points, are carbohydrates like sucrose. They have multiple –OH groups exposed to water, with which they form hydrogen bonds. The above type of calculation for the solubility of sucrose in some nonhydrogen-bonding solvent would give a fair estimate, but would not give a reasonable estimate for any solvent, like water, in which sucrose forms hydrogen bonds. At the end of this discussion of LSE we see that if the solute (solid) species does not form quasi-chemical bonds (ionization, solvation, hydrogen bonding) with the solvent, then the simple theory advanced above has some predictive power. It predicts that the solubility falls as $\exp(T_{\text{melting}}/T)$, and allows an estimate of the numerical relationship. For solutes that do form quasi-chemical bonds with the solvent, we need experimental data, molecular-interaction theory, or both.

11.9 GAS–SOLID EQUILIBRIUM (GSE) AT LOW PRESSURES

The simplest gas–solid equilibrium is that between a the solid and gas of a pure species, such as the gas-solid curve for water

Table 11.A Vapor Pressure of Elemental Iodine [12, p. 2.58]

Pressure, mm Hg										Melting Point °C
1	5	10	20	40	60	100	200	400	760	
Temperature, °C										
38.7	62.2	73.2	84.7	97.5	105.4	116.5	137.3	159.8	183.0	112.9

vapor and ice in Figure 1.9. This curve is also a vapor-pressure curve, although we most often see that term applied to VLE, not GSE. For a pure species, GSE and VLE are practical identical, except that the highest point on a pure-species VLE curve is the critical point and on a GSE curve it is the triple point. If the fugacity of the solid is greater than that in the surrounding gas, then the solid at the surface evaporates or *sublimes* into the gas, and if the fugacity of the solid is less, then the gas either crystallizes onto the solid surface or condenses onto it as an amorphous solid. For gas mixtures at low pressures we can treat the gas phase in GSE the same way we did in Chapter 8, assuming most often that the gas is an ideal solution of ideal gases, or, if the pressures are too high for that, assuming that it is adequately described by the L-R rule.

Example 11.14 Solid water (ice) is in equilibrium with air at 1 atm and 30°F. What is the equilibrium concentration of water vapor in the air?

This is practically the same problem we addressed in Example 3.1. The vapor pressure of ice at 30°F is 0.0808 psia. As in that problem we may assume that the solubility of nitrogen and oxygen in solid ice is negligible, and thus use Raoult's law, finding

$$y_{\text{water vapor}} = \frac{x_{\text{water in ice}}\, p_{\text{ice}}}{P} = \frac{1.00 \cdot 0.0808\ \text{psia}}{14.7\ \text{psia}} = 0.0055 \quad \blacksquare \tag{11.Y}$$

We are not used to thinking about the vapor pressure of ice, but Figure 1.9 clearly shows that it has one, whose value we can look up in the steam tables. Residents of cold climates know that ice disappears slowly from our streets (and our laundry if we hang it out to dry frozen) even if the temperature never goes above freezing. This example shows why; the atmospheric air in cold climates rarely contains this much water vapor, so the ice vaporizes into it. If we were to repeat all the parts of Examples 3.1–3.3., we would find that, as in that example, the solubility of nitrogen and oxygen in the ice would not change our answer significantly.

The inverse of this example is *vapor deposition.* If we place a cool object in a space in which the vapor's pressure is higher than the vapor pressure of pure solid at the temperature of the cool object, then the vapor will condense on the cool object. Many variants of this process are used in the production of computer chips. (Some are simple physical deposition as described here; others are *chemical vapor deposition* in which the vapor reacts chemically with the surface.)

The vapor pressure data for solids is often shown inconspicuously in handbooks. For example, in the vapor pressure tables in [12, p. 2.58] the values for elemental iodine are shown as a row in the table (see Table 11.A). From the value in the rightmost column we see that the four rightmost values are the vapor pressure of pure liquid iodine (above the melting point) and the five leftmost values are the vapor pressures of pure solid iodine.

11.10 GSE AT HIGH PRESSURES

As with VLE, increasing the pressure does not change the properties of the solid very much (see Example 7.3), but it does change the properties of the gas. As we saw in Chapter 10, for gas pressures above the critical pressure and temperatures close to the critical temperature, the gaseous material (often called a "dense fluid" because its properties are different from those we expect from a gas) can become a good solvent for solids and liquids (see Example 10.3). Figure 11.20, shows a dramatic example of this effect.

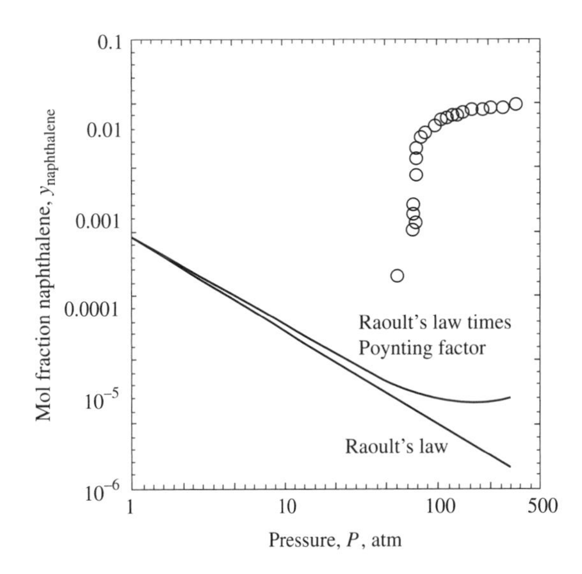

FIGURE 11.20 The measured mol fraction of naphthalene in CO_2 at 35°C as a function of pressure, compared to Raoult's law, and Raoult's law times the Poynting factor. Observe that the scales are different; if the same scale had been used on both axes the Raoult's law line would have a slope of −1. The circles are the experimental data of [22].

Example 11.15 Estimate the solubility of naphthalene in CO_2 at 35°C and 100 atm, by Raoult's law and by Raoult's law times the Poynting correction factor.

The calculated vapor pressure of naphthalene at 35°C is 0.000,65 atm. The solid is practically pure naphthalene, so substituting the appropriate values in Eq. 11.Y, we find the calculated $y_{\text{naphthalene}}$ by Raoult's law is 6.5×10^{-6}. At this high a pressure the volume of solid naphthalene is $\approx 132\,\text{cm}^3/\text{mol}$, so that, from Eq. 7.N,

$$\frac{f^{\circ}_{i,\,100\text{ atm}}}{f^{\circ}_{i,\,1\text{ atm}}} = \exp\frac{132\dfrac{\text{cm}^3}{\text{mol}} \cdot (100-1)\text{ atm}}{82.06\dfrac{\text{cm}^3\text{ atm}}{\text{mol K}} \cdot 308.15\text{ K}} = 1.68 \qquad (11.\text{Z})$$

and the estimated $y_{\text{naphthalene}} = 6.5 \times 10^{-6} \cdot 1.68 = 10.9 \times 10^{-6}$. ■

These are the values plotted in Figure 11.20. From the same figure we read the experimental value as ≈ 001, which is ≈1500 times the Raoult's law value and ≈915 times the value from Raoult's law times the Poynting correction. (This latter ratio is often called the *enhancement factor* for GSE. It shows how much the solubility is increased by the nonideal gas behavior of the solvent.) What are we to make of this huge difference between the experimental values and what we would compute from Example 11.15? At this high a pressure the Poynting factor is significant, but it is only a factor of 1.68, while the solubility is ≈915 times what we would expect. The reason is that in this situation the gas phase is not behaving as a gas at all, but as a dense fluid. The critical temperature and pressure of CO_2 are 304.2 K and 72.9 atm, so at this T and P, the CO_2 is clearly not a gas. The measured density of "gaseous" CO_2 is $\approx 0.73\,\text{g/cm}^3$, comparable to that of liquid gasoline at room temperature.

We may also look back to Figures 11.18 and 11.19, observe that for naphthalene the 1-atm melting point is 353.5 K, so that

$$\frac{T_m}{T} = \frac{353.5}{308.15} = 1.147 \qquad (11.\text{AA})$$

From Figure 11.18 at this value of T_m/T we read $\log x_2 \approx -0.42$, or $x_2 \approx 0.38$. Thus, the solubility of naphthalene in the dense CO_2 in Example 11.15 is about 0.01/0.38 ≈ 3% of its solubility in benzene at the same temperature. Compared to benzene, dense fluid CO_2 is not a very good solvent, but compared to an ideal gas at the same T and P it is an amazingly good solvent. This solvent power has led to its use, and that of other gases at comparable T_r and P_r, for *supercritical extraction* [7, p. 641]. Remarkable results are often obtained, but the high pressures used make the processes expensive.

11.11 GAS–SOLID ADSORPTION, VAPOR–SOLID ADSORPTION

So far we have dealt with well-defined, intuitively obvious phases: gas, liquid solid. But chemical engineers deal with other, less intuitive phases of considerable technical interest and economic importance, for example gases and vapors adsorbed onto solids. Figure 11.21 shows, as an example the equilibrium curve (always called an "adsorption isotherm" or "isotherm" in the adsorption literature) for nitrogen adsorbed on zeolite at a temperature well below room temperature, but far above the critical temperature of nitrogen (126.2 K).

Such data are measured in the apparatus shown schematically in Figure 11.22. About a gram of adsorbent it placed on the sample hanger of an electronic microbalance and inserted in a closed container, surrounded by a constant temperature bath. The sample container is evacuated, and then the gas (or vapor) to be adsorbed is admitted. The pressure and the mass adsorbed are recorded, and then the pressure is changed, producing data points like those in Figure 11.21. The amounts adsorbed are small enough that the standard measure of amount adsorbed is (0.001 mol adsorbed)/(g of adsorbent) = mmol/g. Measurements are often made going up in pressure and then down, to be sure that equilibrium was truly reached. The experimental details are more complex than this simple sketch shows. The electronic microbalance as made this measurement much easier than it was with older volumetric methods.

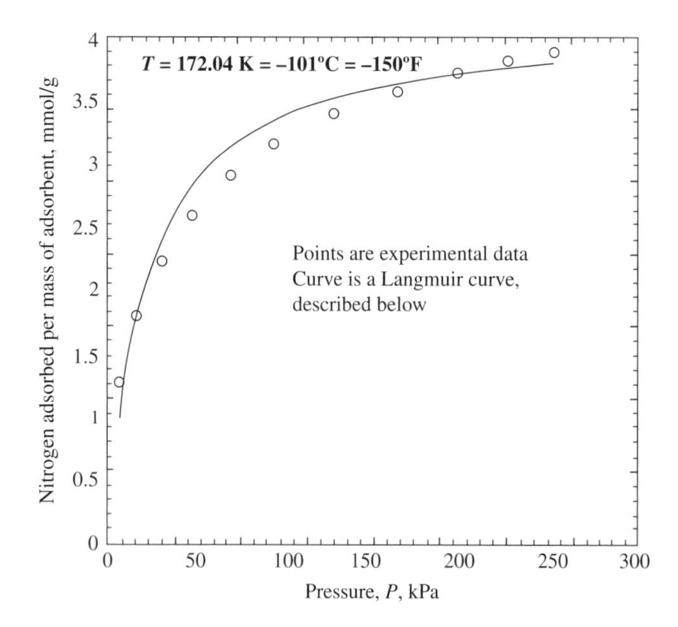

FIGURE 11.21 Experimental measurements of the adsorption of nitrogen on solid zeolite at 172.04 K [23]. The curve is a best fit of Langmuir's absorption equation, described below.

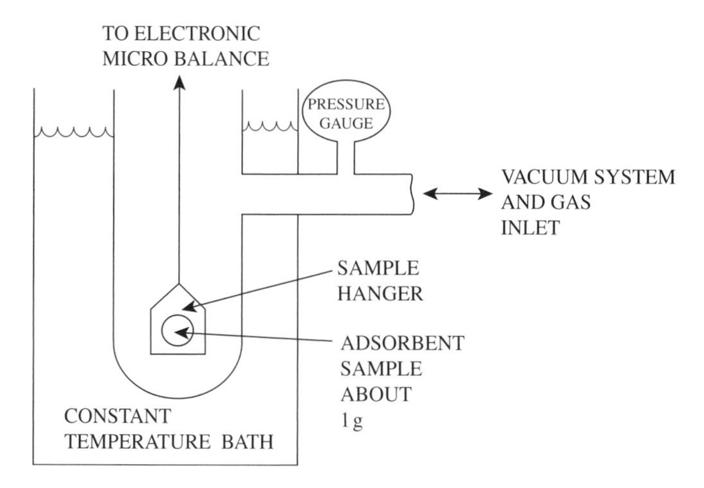

FIGURE 11.22 Simplified schematic of a device for measuring gas-solid or vapor-solid adsorption.

From Figure 11.21 we see that

1. The curve, based on Langmuir's absorption equation discussed below, is only a fair fit of the data.
2. At the atmospheric pressure (101.3 kPa) the adsorbed amount is 3.2 mmol/g $\approx$ 0.09 g adsorbed/g of adsorbent. This is certainly measurable, but not overwhelming.
3. The same source shows this zeolite has an estimated surface area (mostly internal surface) of 672 m^2/g. Based on this value, at one atmosphere the adsorbed amount is = 0.00013 g/m^2. Thus, the surface of an ordinary desk (about 2 m^2) would have about 0.00026 g of adsorbed nitrogen (facing pure nitrogen, not air). This is certainly a small enough value that we have not made serious errors by ignoring gases adsorbed on surfaces in the previous parts of this book.
4. The same source shows that the amount absorbed (at constant pressure) decreases with increasing temperature, falling by about a factor of 7 in going from −101°C to 0°C. The curve for a higher temperature would lie below and to the right of the curve shown on Figure 11.21.
5. At this point you are certainly wondering why we bother with this small an effect. The surface area of adsorbent materials like the zeolite in this example is truly remarkable. A gram of this material is about a cubic centimeter, the size of a standard sugar cube. Its surface area is about 1/8 that of a football field! The large internal surface area of industrial (and laboratory) adsorbents makes this small amount adsorbed per unit area significant.
6. The curve shown is truly an equilibrium curve. One can increase or decrease the pressure (at constant T) and find the same amount adsorbed at any pressure (but see below!). It closely resembles Figure 3.9, but with the axes interchanged. (Adsorption isotherms are always plotted with adsorbed amount – concentration – on the vertical axis and pressure on the horizontal. VLE curves almost always have the concentration variable – most often mol fraction – on the horizontal axis and P or T on the vertical axis.) Making the conceptual interchange of the axes, we see that the curve in this figure is like Figure 3.9 with the difference that instead of mol fraction of nitrogen dissolved in water we have its adsorbed concentration in mmol/g of adsorbent. This similarity leads writers on adsorption to refer to the low-pressure part of the Figure 11.21 as the "Henry's law region," directly analogous to Figure 3.9.
7. This similarity would lead us to believe that this is some kind of a VLE curve, with the adsorbed nitrogen on the solid surface behaving as if it were a liquid. This is an intuitively satisfying idea; it is supported by the fact that when such an adsorption occurs there is always a heat release, generally of the order of the heat of condensation (the heat release on converting from vapor to liquid). However the temperature here is well above the critical temperature of nitrogen, so that liquid nitrogen cannot exist at this temperature. For temperatures below the critical temperature (discussed below) this works better. To maintain this intuitively satisfying idea, we must consider nitrogen-adsorbed-on-zeolite as a kind of liquid, in some ways similar to ordinary liquid, but different. Its "apparent vapor pressure" depends not only on the temperature (as an ordinary liquid's would) but also on the amount adsorbed, which is quite different from the behavior of an ordinary liquid.
8. If the whole adsorbing surface were the same, then we would expect the isotherm to be vertical; once the pressure for adsorption were reached, the pressure would continue constant until the surface was covered. The shape of the curve (and supporting theory) suggests that instead of a simple, uniform plane, the surface is irregular, with some sites having much more attraction for the gas molecules than others. As we introduce gas or vapor to be adsorbed the molecules attach first to the most attractive sites, then as those are taken up it requires more pressure to attach a molecule to the next most attractive site, and so on leading to the shape of the curve shown. Heat of condensation data (discussed below) strongly support this idea.

11.11.1 Langmuir's Adsorption Theory

Langmuir's adsorption theory, like the EOS of van der Walls, was one of the first to set out a scientific model for describing gas adsorption on a solid. Like the vdW EOS it is *not very good* at predicting or correlating experimental data. Compare the curve and the data in Figure 11.21; that is about as good as it ever gets, and this data set was chosen because it gives a better-than-average fit to the Langmuir equation. Also like the vdW EOS, Langmuir's theory has had a profound effect on all subsequent work in the field. Much of subsequent work builds on it. It is simple, intuitively satisfying, and helps us form a mental picture of gas–solid and vapor–solid adsorption, with the knowledge that this picture is a gross simplification of a complex reality and the resulting equation is only a fair representation of experimental data.

Langmuir's equation [24] is regularly derived in a variety of ways, of which the most intuitive is based on chemical equilibrium (Chapter 12) rather than phase equilibrium. He originally based it only on the external surface of crystalline solids, not on the internal surface of the pores in crystalline or amorphous solids – by far the most industrially important adsorption application; it is most often applied to these internal surfaces. The discussion here is a significant simplification of that he presented. He assumed that at equilibrium the rates of adsorption and desorption were equal and that

$$\begin{pmatrix}\text{rate of}\\ \text{adsorption}\end{pmatrix} = \text{pressure}\cdot\begin{pmatrix}\text{area available}\\ \text{for a molecule}\\ \text{to adsorb onto}\end{pmatrix}\begin{pmatrix}\text{adsorption}\\ \text{rate}\\ \text{constant}\end{pmatrix}$$

$$= PA_{\text{available}}K_{\text{forward}}$$

$$= PA_{\text{total}}\begin{pmatrix}\text{fraction}\\ \text{unoccupied}\end{pmatrix}k_{\text{forward}} \quad (11.17)$$

and

$$\begin{pmatrix}\text{rate of}\\ \text{desorption}\end{pmatrix} = \begin{pmatrix}\text{area covered with}\\ \text{adsorbed molecules}\end{pmatrix}\cdot\begin{pmatrix}\text{desorption}\\ \text{rate constant}\end{pmatrix}$$

$$= A_{\text{occupied}}k_{\text{reverse}} = A_{\text{total}}\begin{pmatrix}\text{fraction of}\\ \text{area occupied}\end{pmatrix}k_{\text{reverse}} \quad (11.18)$$

Setting these equal, letting the fraction of the area occupied be θ and the fraction unoccupied be $(1-\theta)$ we find

$$Pk_{\text{forward}}A_{\text{total}}(1-\theta) = A_{\text{total}}\theta k_{\text{reverse}} \quad (11.19)$$

which simplifies to

$$P\frac{k_{\text{forward}}}{k_{\text{reverse}}} = \frac{\theta}{(1-\theta)} \quad \text{or} \quad Pb = \frac{\theta}{(1-\theta)}\left(\text{where } b = \frac{k_{\text{forward}}}{k_{\text{reverse}}}\right)$$

$$\text{or } \theta = \frac{Pb}{(1+Pb)} \quad (11.20)$$

We have no direct measurement of θ, so we assume that there is some number of mols of absorbate (n_{max}) that corresponds to covering the whole surface with an adsorbed layer one molecule thick – a monolayer – and thus that $\theta = n_{\text{adsorbed}}/n_{\text{max}}$. In Figure 11.21 the vertical axis is n_{adsorbed}; the asymptotic value of the curve if we extended the pressure to infinity would presumably be n_{max}. We will see below other ways to estimate that value. Substituting this value of θ and multiplying out we find

$$\frac{n_{\text{adsorbed}}}{n_{\text{max}}} + \frac{n_{\text{adsorbed}}}{n_{\text{max}}}Pb = Pb \quad \text{or} \quad \frac{P}{n_{\text{adsorbed}}} = \frac{1}{bn_{\text{max}}} + \frac{P}{n_{\text{max}}} \quad (11.21)$$

This suggests that if we plot P/n_{adsorbed} vs P the plot should be linear and we should be able to determine n_{max} and b from the slope and intercept. Figure 11.23 shows the data in Figure 11.21 replotted this way.

$$\text{slope} = \frac{1}{n_{\text{max}}} = 0.2372\frac{\text{kPa}/(\text{mmol/g})}{\text{kPa}};$$

$$n_{\text{max}} = \frac{1}{0.2372} = 4.216\frac{\text{mmol}}{\text{g}} \quad (11.\text{AB})$$

$$\text{intercept} = \frac{1}{bn_{\text{max}}}: \quad b = \frac{1}{4.216\dfrac{\text{mmol}}{\text{g}}\cdot 6.33\dfrac{\text{k Pa}}{\text{mmol/g}}}$$

$$= 0.0375\frac{1}{\text{kPa}} \quad (11.\text{AC})$$

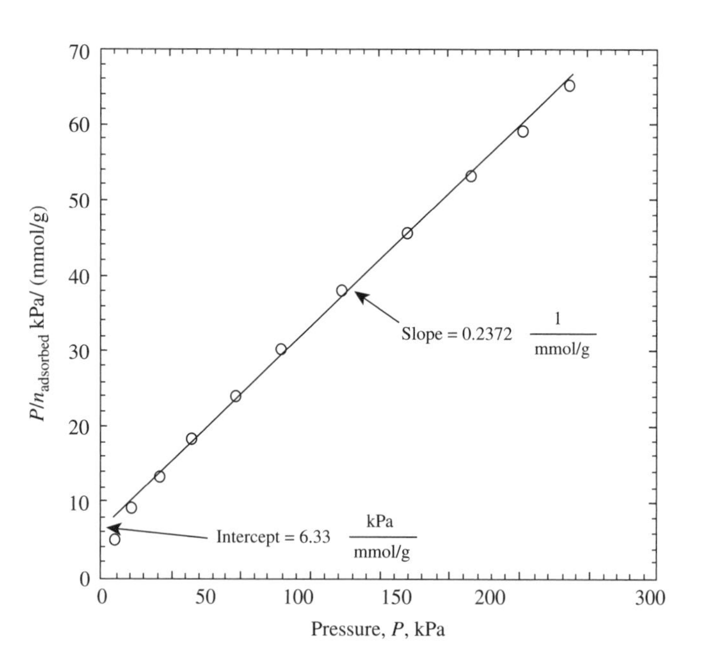

FIGURE 11.23 The same data as shown in Figure 11.21, on the coordinates suggested by Eq. 11.21. The curve on Figure 11.21 is the same as the line on this figure, with the coordinates transformed.

The best fit of Eq. 11.21 to the data on these two figures is

$$n_{\text{absorbed}} = 4.216 \frac{\text{mmol}}{\text{g}} \cdot \frac{\frac{0.375}{\text{kPa}} P}{\left(1 + \frac{0.0375}{\text{kPa}} P\right)} \quad \text{(11.AD)}$$

11.11.2 Vapor-solid Adsorption, BET Theory

If we choose a temperature below the critical temperature of the gas adsorbed, (thus creating a vapor-solid situation), the results are different from (and technically more interesting) than the simpler gas-solid case. Figure 11.24 shows a sample of that case, again nitrogen, but on a different adsorbent (silica gel), at 77 K, nitrogen's normal boiling point.

From this figure we see that

1. The true curve must pass through the origin, so the leftmost part of the curve cannot be completely correct. I have drawn it the way it appears in the original papers.
2. The pressures shown in atmospheres (not kPa as in Figure 11.21) are lower than in that figure; 0.6 atm $\approx$ 62 kPa. Because this plot is at the NBP of nitrogen, the pressure (in atm.) is equal to P/P_0, the ratio of the actual vapor pressure to the boiling pressure at this temperature. Many plots of this type choose P/P_0 as the horizontal axis, to facilitate comparison with adsorption isotherms of other gases with different NBPs.

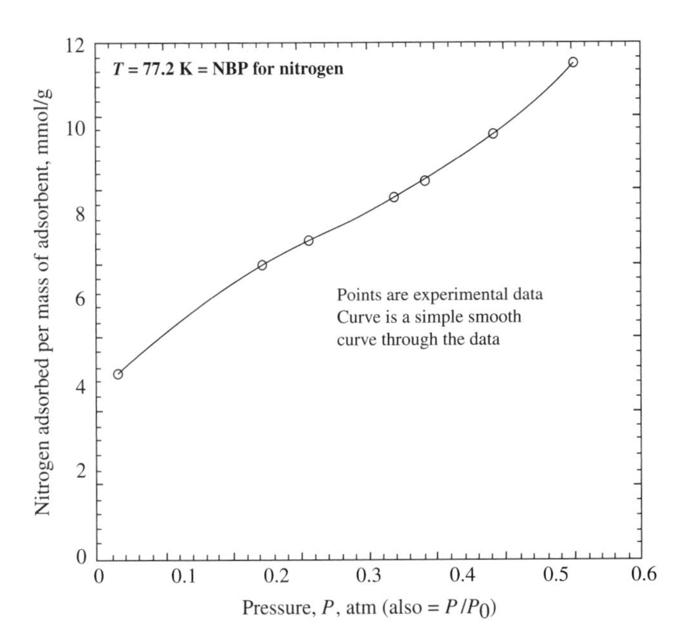

FIGURE 11.24 Experimental measurements of the adsorption of nitrogen on solid silica gel at 77 K, the NBP of nitrogen [25]. The curve is a simple smooth interpolation.

3. The original data source shows that this material has a calculated surface area of 595 m^2/g. Using that value and the calculated area of a single nitrogen molecule on a surface (at the NBP, see Problem 11.41) we compute that about 6 mmol/g should cover the surface with a monolayer of adsorbed nitrogen, (at about $P/P_0 \approx$ 0.15). The part of the curve to the left of this value, with less than a monolayer has the same curvature as Figure 11.21 (and the Langmuir theory) but to the right of $P/P_0 \approx 0.30$, (about 1.3 monolayers) the experimental data curves upward to the right, the opposite of the curvature in Figure 11.21.
4. At any of the pressures shown on this figure, pure liquid nitrogen cannot exist at equilibrium; it would evaporate. So the adsorbed layer is not simply liquid nitrogen. The common, and intuitively satisfying explanation is that the molecules of the solid surface attract nitrogen molecules more strongly than molecules in an ordinary liquid do. In this case we would say that the fugacity of this first layer is about 0.15 as much as it would have if it were the surface layer of a pool of liquid nitrogen at this temperature. The second layer of molecules should not be held on as strongly as the first, but apparently are held on more strongly than they would be as the surface layer of a pool of liquid. At 12 mmol/g, from the observed $P/P_0 \approx 0.5$ we would infer that the fugacity of the surface layer of the adsorbed film was about 0.5 as much as in such a pool of liquid.
5. The above explanation is simple and intuitively satisfying, but is certainly a great simplification of what nature is really doing.

Faced with this type of adsorption, Brunauer, Emmett and Teller (BET) [26] worked out the mathematics of multiple layers of adsorbed molecules, subject to some simplifying assumptions, including that there were multiple one-molecule-thick layers, and that each layer covers a smaller part of the surface than the one below it. Their resulting equation (rewritten to use the notation in this chapter) is

$$\frac{n_{\text{adsorbed}}}{n_{\text{monolayer}}} = \frac{b(P/P_0)}{(1 - P/P_0) \cdot [1 + (b-1) \cdot (P/P_0)]} \quad (11.22)$$

which contains the terms in Eq. 11.20, with some others. There seems to be no simple derivation for Eq. 11.22, as there is for Eq. 11.20, Like that equation, this one can be rewritten

$$\frac{P}{n_{\text{adsorbed}}(P_0 - P)} = \frac{1}{bn_{\text{monolayer}}} + \frac{(b+1)}{bn_{\text{monolayer}}} \cdot \frac{P}{P_0} \quad (11.23)$$

showing that a plot of $P/n_{\text{adsorbed}}(P_0 - P)$ vs P/P_0 should produce a straight line, from whose slope and intercept we should be able to determine b and $n_{\text{monolayer}}$. If one plots all

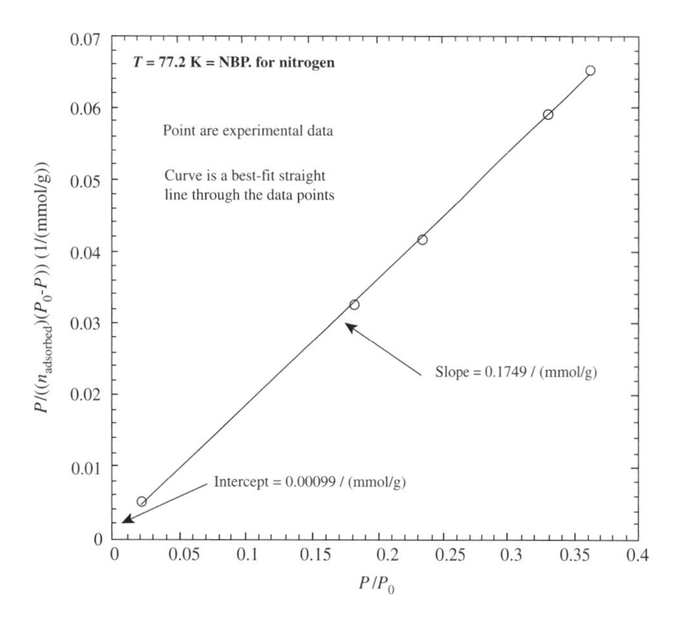

FIGURE 11.25 The data in Figure 11.24, replotted on the coordinates suggested by Eq. 11.23, omitting the two highest-pressure data points.

the data points shown in Figure 11.24 that way and draws the best-fit straight line, one finds that the intercept is negative, leading to the embarrassing conclusion that either b or $n_{\text{monolayer}}$ must be negative! (See Problem 11.44).

Faced with this annoying disagreement with their theory, BET pointed out that if one restricted oneself to data points with P/P_0 less than 0.35 the data points formed a well-behaved straight line with a positive intercept. Figure 11.25 shows such a plot of the data in Figure 11.24, showing only points for P/P_0 less than 0.37.

Example 11.16 Using Figure 11.25 estimate the surface area (m^2/g) for the solid adsorbent in that example. By rearranging Eq. 11.23 (See Problem 11.42) we find that $n_{\text{monolayer}} = 1/(\text{slope} + \text{intercept}) = (\text{mmol/g})/(0.1749 + 0.00099) = 5.685$ mmol/g. Using Avogadro's number we find that this is $3.52 \cdot 10^{21}$ molecules/monolayer. Then assigning each of those molecules a surface area of 16.2 $(\text{Å})^2$. (See Problem 11.41 and Table 1 of [27]) we find a surface area of 571 m^2/g. The authors of the data report 595 m^2/g. Given the fact that the data points for figures 11.24 and 11.25 were obtained by measuring off a graph in [25], this is excellent agreement! ■

The calculation in this example, based on measured adsorption of nitrogen at its NBP, is the standard "BET Method" of determining surface area of porous solids. If the BET theory were perfect then the areas determined for a single solid would be independent of which gas were used. Alas, the values are similar but not identical; the users have agreed to accept the surface areas values based on nitrogen at its boiling point as industry standard. Instrument companies sell equipment specifically designed for this measurement. From this discussion it should be clear that

1. The BET method builds on and extends Langmuir's theory.
2. Its applicability to only P/P_0 less than 0.35 shows that although it is based on multiple-layer adsorption, it is only reliable to adsorption values less than a few complete monolayers.
3. The BET theory, involving the possibility of multimolecular layers, must be closer to physical reality for vapors below their critical temperatures than the single-layer Langmuir theory. However it gains scientific credibility at the expense of loss of intuitive content.

11.11.3 Adsorption from Mixtures

The Langmuir and BET models described above apply to adsorption of pure gases and vapors on solid surfaces. Industrially we are much more interested in adsorption of one or more components of a mixture onto such a surface. The typical example is the facemask worn by painters to prevent inhalation of the evaporating solvent from the paint they spray or brush onto a surface. The concentration of the evaporated solvent in the air must be less than the vapor pressure of the solvent in the paint, or it would not evaporate. The facemask normally has throwaway canisters filled with charcoal adsorbent, whose behavior with the evaporated solvent is like the nitrogen-silica gel system in Figure 11.24. In this case instead of total pressure on the horizontal axis we would place partial pressure, typically less than 0.01 atm, far to the left on Figure 11.24. Here the purpose is not separation but capture. For larger-scale applications the throwaway cartridges are replaced with adsorbent beds, which are regenerated (desorbing the adsorbed material) by raising the temperature, or lowering the pressure, or flowing air or another gas through the filter. The latter approach is used in the carbon canister that adsorbs the evaporative emissions from the fuel tanks on all modern autos.

In the above cases adsorption was used to separate mixed hydrocarbon vapors from air, without influencing the distribution among the various hydrocarbon species in those mixtures. That separation is easy because the NBPs of all the hydrocarbons in gasoline or paint solvents are far above those of air. If the NBPs of the species in a mixed vapor are close to each other, adsorption can be used to separate one from the other. Figure 11.26 shows two equilibrium curves. The one with the circles is a simple curve fit of the experimental data for the adsorption of an

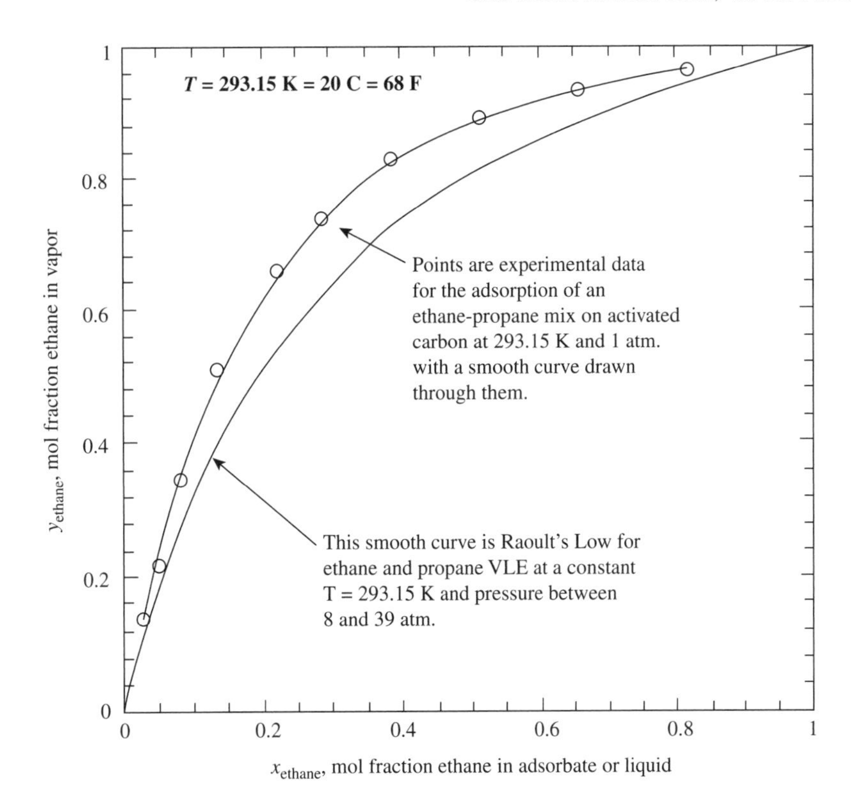

FIGURE 11.26 Comparison of experimental adsorption equilibrium data for an ethane-propane mix ([23] p. 242), with the calculated VLE data for ethane-propane, according to Raoult's law, both at 293.15 K. The adsorption data are all at $P = 1$ atm, while the calculated VLE is at a pressure varying from 38 atm (pure ethane) to 8 atm (pure propane).

ethane-propane mixture on activated carbon. The other is a simple VLE curve for the same mixture, calculated from Raoult's law.

Comparing the two curves in this figure we see

1. The composition of the adsorbed phase (x_{ethane} and ($x_{propane} = 1 - x_{ethane}$)) depends on the composition of the vapor (y_{ethane}, and ($y_{propane} = 1 - y_{ethane}$)) pretty much as it would if this were a vapor-liquid example. In both cases at equilibrium the individual fugacities of each species must be equal in the two equilibrium phases. This data suggests that the ratio [(fugacity of ethane)/(fugacity of propane)] in the adsorbed phase must be quite similar to that in an ordinary, ideal solution liquid containing ethane and propane at the same concentration.
2. Experimental data not shown here [28] indicate that the same is true even for quite nonideal solutions, with some adjustments of activity coefficients.
3. For the VLE example we can specify only two of the three variables (T, P, and one mol fraction in one of the phases) then all the others are fixed. For the adsorption example we can specify three from this list. This is discussed as a phase-rule problem in Chapter 15.

In addition to the adsorptive separation based on the different fugacities of the adsorbed species shown in Figure 11.26 separations are made industrially based on synthetic zeolites ("molecular sieves") whose pore openings are chosen so that the smallest members of the mixture can enter and larger ones cannot. The separation here is more physical (like a sieve with holes about equal to the size of the smaller molecules) than thermodynamic (like VLE).

11.11.4 Heat of Adsorption

The Clapeyron equation (Eq. 5.5) applies to equilibrium between any two phases of a pure substance. If we wish to apply it to the adsorbed phase and the gas or vapor in equilibrium with it, we see that because the vapor pressure of the adsorbed phase depends not only on the temperature but also on the amount adsorbed per mass of adsorbent, we must specify the amount adsorbed as well. Thus, for equilibrium between the adsorbed phase and gas or vapor, Eq. 5.5 becomes

$$\left(\frac{\partial P}{\partial T}\right)_{n_{\text{adsorbed}}} = \frac{\Delta h}{T \Delta v} \tag{11.24}$$

For the low pressures normally involved in adsorption, the Clausius-Clapeyron simplifications (Eqs. 5.8, 5.9, and 5.10) seem reasonable, so that we have

$$\left(\ln\frac{P_2}{P_1} = \frac{\Delta h_{\text{average}}}{R}\cdot\left(\frac{1}{T_1}-\frac{1}{T_2}\right)\right)_{n_{\text{adsorbed}}} \tag{11.25}$$

Where $\Delta h_{\text{average}}$ is the average heat of adsorption over the temperature range (assumed to be a constant in the integration leading to Eq. 5.10.)

Example 11.17 The source of the data in Figure 11.24 also reports the isotherm for 90.15 K, from which we find that at $n_{\text{adsorbed}} = 6$ mmol/g the pressure is 0.566 atm while for 77.2 K we read the table leading to Figure 11.24 as 0.118 atm at the same amount adsorbed. Substituting these values in Eq. 11.25, we find

$$\left(\ln\frac{0.566\text{ atm}}{0.118\text{ atm}} = \frac{\Delta h_{\text{average}}}{R}\cdot\left(\frac{1}{77.2\text{K}}-\frac{1}{90.15\text{ K}}\right)\right)_{n_{\text{adsorbed}}} \tag{11.AE}$$

$$\Delta h_{\text{average}} = R\frac{1.556}{0.001875/\text{K}} = 835\text{ K}\cdot R = 1659\frac{\text{cal}}{\text{mol}}$$

$$= 2986\frac{\text{Btu}}{\text{lbmol}} = 106.7\frac{\text{Btu}}{\text{lbm}} \quad ■ \tag{11.AF}$$

This is somewhat larger than the 85 Btu/lbm at the average of the two temperatures for simple VLE, [29]. If we repeat this calculation at $n_{\text{adsorbed}} = 4.5$ mmol/g we find $\Delta H_{\text{average}} = 131$ Btu/lbm. These two values illustrate the common observation that heats of adsorption are somewhat larger than heats of condensation, and that they are larger at low values of n_{adsorbed} than at higher ones. The explanation normally given is that at the start of adsorption the molecules attach to the strongest adsorption sites, with the largest heat effect, and so on. Heats of adsorption can also be directly measured calorimetrically. The resulting values are in at least fair agreement with those computed in Example 11.17 (with exceptions) [30].

11.11.5 Hysteresis

Alas, adsorption is more complex than VLE or than the simple picture show above. Figure 11.27 shows experimental data for the adsorption and desorption of water on silica gel [31].

From this figure we see that

1. The first adsorption curve starts like Figure 11.24, but then levels out as the pressure approaches the saturation pressure (P/P_0 approaches 1.0). The maximum adsorbed value (0.4 g/g) corresponds roughly to the amount of water that would fill all the pores in the silica gel (porosity about 40%).
2. The first desorption curve shows two kinds of hysteresis. From $P/P_0 \approx 1.0$ to 0.3 the observed values are much larger than the values observed during adsorption. Below $P/P_0 \approx 0.3$ the curve shows a dramatic change in slope, and then parallels the first adsorption curve.
3. For the second adsorption the curve lies somewhat above the first adsorption curve and practically retraces the lower part of the desorption curve to $P/P_0 \approx 0.3$, but then does not follow it upward, but rather practically parallels the first adsorption curve.

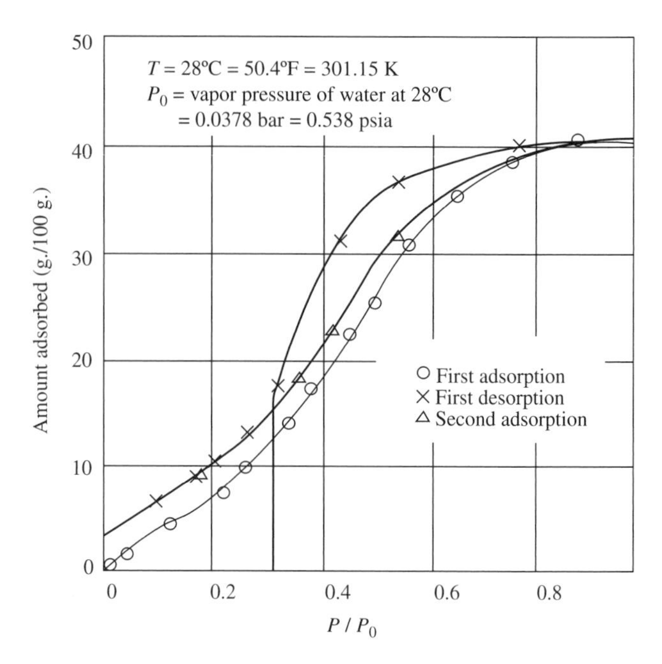

FIGURE 11.27 Adsorption of water on silica gel [31], showing hysteresis. The vertical axis shows $g_{\text{adsorbed}}/100\ g_{\text{adsorbent}}$, rather than the conventional mmol/g. The three curves correspond to the first adsorption of water, the first desorption and then the second adsorption.

Such curves in which the values going up and down are not the same are called *hysteresis* curves. Their explanation is far from agreed upon among experts, but the following, advanced in [31] is widely accepted. On the first desorption, liquid adsorbed into small pores is held in place by surface tension and does not evaporate at the same external pressure as it would if it were a flat sheet. This is called *capillary condensation*, discussed in Chapter 14. As the pressure falls, more and more of the liquid bound that way desorbs, and at about $P/P_0 \approx 0.3$ those pores are empty, and the behavior parallels that of the first adsorption curve. On the second adsorption those pores are filling again but capillary condensation does not affect their adsorption behavior.

The first desorption and second absorption curves are reported as equilibrium curves, meaning that one could go up and down on them, finding the same values. The offset of the second adsorption curve from the first is explained by the authors as an "irreversible type of hysteresis possibly due to imperfect rigidity in the gel structure."

At the end of this discussion of this figure, we can say that thermodynamics (Chapter 14) suggests an explanation of the hysteresis loop in the adsorption-desorption curves, but gives no suggestions about the offset between the first and second adsorption curves. I end this section on adsorption with this unexplained result and the lame explanation of it to remind the reader that the interactions between gas and liquid molecules and heterogeneous solid surfaces is not nearly as simple or as well understood as the interactions between liquids, gases and homogeneous solids.

11.12 SUMMARY

1. LLE. LSE, and GSE all obey the same basic thermodynamics as VLE. At equilibrium the system has taken up the lowest Gibbs energy consistent with the external constraints. The working form of this statement is that for any species in any such equilibrium the fugacity of any species is the same in all the phases at equilibrium.
2. LLE and LSE show extreme nonideal behavior, such as that between water and hydrocarbons. The calculated activity coefficients are much larger than are observed in common VLE situations.
3. For a liquid mixture to form two phases, its isothermal plot of g_{mixture} vs. mol fraction must show an internal maximum, as in Figure 11.8. The calculation procedures for estimating whether two phases exist and the compositions of the two phases are the equivalent of computing such a curve and computing the two points of tangency of a straight line that touches the curve twice. This is done with activity coefficient correlations more powerful (and complex) than the simple ones in the examples presented here.
4. For most LLE and LSE increasing the temperature increases the solubility. There are important exceptions.
5. Eutectics, LSE that occurs most often in metallurgy and also in water-mineral salt systems, involve one liquid and two solid phases at temperatures *below* the melting point of both of the solids. Hydrates (clathrates), GSL that occurs most often with hydrocarbons and water at temperatures *above* the melting points of the chemical species involved, involve a solid, sometimes a gas, and one or more liquid phases. For both the behavior follows the same thermodynamic rules as the other equilibria in this book, but with more complexity.
6. For solids that dissolve without molecular interactions with their solvents, we can make tolerable estimates of the solubility from the extrapolated liquid-phase vapor-pressure curve. Many common solids, such as NaCl and sucrose, do interact with water as a solvent, thus producing much greater solubility than we would estimate this way.
7. For low-pressure GSE we can estimate the vapor-phase composition from Raoult's law with fair certainty.
8. For high-pressure GSE, at pressures above the critical pressure, the solid solubility is many times the value we would compute from Raoult's law (with the Poynting correction). The solvents are acting as dense fluids, whose behavior is more like that of a liquid than that of a gas.
9. Gas-solid adsorption follows the same thermodynamic rules as the rest of this book, but the solid surface is more complex and heterogeneous than the simple solid, liquid and gas phases in the rest of the book, leading to more complex and poorly understood behavior.

PROBLEMS

See the Common Units and Values for Problems and Examples. An asterisk (*) on a problem number indicates that the answer is in Appendix H.

11.1* Repeat Example 11.1 for determining the solubility of water in benzene. How many drops of water must we add to saturate 1.00 kg of benzene?

11.2* At 25°C, what is the solubility of benzene in water and water in benzene, expressed as wt%?

11.3 Repeat Example 11.2 for toluene in water. The Federal drinking water standard for toluene is 1 ppm by weight [3].

11.4 Suggest a molecular explanation for the fact that in Table 11.2 the solubilities of methylene dichloride and chloroform are substantially greater than those of carbon tetrachloride and most hydrocarbons.

11.5* In Example 11.3 find the number of mols of each of the two liquid phases present, by material balance on one of the species.

11.6 Repeat Example 11.3 for 5 mols of benzene, 1 mol of ethanol, and 4 mols of water.

11.7* In Example 11.3 if the total number of mols (n_T) ≈ 10 and me the mols of benzene $= 5$, what number of mols of ethanol correspond to the boundary between the one-phase and two-phase regions?

11.8 Repeat Example 11.3 using Figure 11.2 instead of Figure 11.1. Are the results the same? Should they be?

11.9 The Greek liquor ouzo is a clear liquid (≈ 17 mol% ethanol, 82 mol% water, 1 mol% other substances). When it is diluted with water it turns cloudy. Suggest an explanation, based on this chapter. *Hint*: Sketch a plausible triangular diagram, and sketch the dilution process on it.

11.10* Repeat Example 11.4 for *n*-heptane and water.

11.11 Repeat the numerical part of Example 11.5 for $x_1 = 0.5$, $A = 3$. Compare the result to the value shown in Figure 11.7.

11.12 Show that for the symmetrical equation.

a. If $A = 2.00$, and $x_a = 0.5$ the second derivative $d^2 g_{\text{mixture}}/dx_a^2 = 0$.

b. For all values of $A < 2.00$ there is no local maximum; $d^2 g_{\text{mixture}}/dx_a^2 > 0$ for all x_a.

c. For all values of $A > 2.00$ there is a local maximum; $d^2 g_{\text{mixture}}/dx_a^2 < 0$ for $x_a = 0.5$.

11.13 Show that Eq. 11.L and Figure 11.8 are the equivalent of Eq. 11.5 and Figure 11.7. *Hint*: The requirement of the tangent line is that it have the same slope at both places. Show how that slope, as a function of composition, is influenced by changing from one formulation to the other.

11.14* **a.** Show that if the symmetrical equation is plotted as shown in Figure 11.8, the curve is symmetrical about $x_a = 0.5$; either side is the mirror image of the other.

b. Then show that this requires that the straight line tangent to the curve in two places must have slope $(dg/dx_a) = 0$.

c. Then show that in Figure 11.7 the corresponding requirement is that $dg/dx_a = g_a^{\text{o}} - g_b^{\text{o}}$, which is equal to 1.0 kJ/mol in Figure 11.7.

d. Then use this information to find the two x_a's for which $(dg/dx_a) = 1$ kJ/mol, using the derivative calculated from Eq. 11.L.

e. Visually compare the values calculated in part (d) with those we would read from Figure 11.7.

11.15 The graphical procedure shown in Example 11.6, and Figure 11.8 has a high intuitive content, but is not suitable for computer solution. Solve that same example numerically, as follows:

a. Show that substitution of the van Laar equation into Eq. 11.L leads to

$$\frac{g_{\text{mixture}} - \sum x_i g_i^{\text{o}}}{RT} = \sum x_i (\ln x_i) + x_a \frac{B^2 A x_b^2}{(Ax_a + Bx_b)^2} + x_b \frac{BA^2 x_a^2}{(Ax_a + Bx_b)^2} \tag{11.AG}$$

which can be simplified to

$$\Gamma = \frac{g_{\text{mixture}} - \sum x_i g_i^{\text{o}}}{RT} = \sum x_i (\ln x_i) + \frac{BAx_a x_b}{(Ax_a + Bx_b)} \tag{11.AH}$$

where we have introduced the dummy variable Γ to simplify the following equations.

b. Show that the derivative of Eq. 11.AH is

$$\frac{d\Gamma}{dx_a} = \ln \frac{x_a}{x_b} + BA \frac{(Ax_a + Bx_b)\cdot(1 - 2x_a) - (A - B)\cdot(x_a - x_a^2)}{(Ax_a + Bx_b)^2} \tag{11.AI}$$

c. Show that the conditions at equilibrium are that the values of the derivative in Eq. 11.AI must be the same at the two points of tangency and that the slope of the line between the two points must have the same value. If we let the two points of tangency be α and β, then this slope is $(\Gamma_\beta - \Gamma_a)/(x_a^{(\beta)} - x_a^{(\alpha)})$. Thus, we have two algebraic equations to solve numerically,

$$\frac{\Gamma_\beta - \Gamma_\alpha}{x_a^{(\beta)} - x_a^{(\alpha)}} = \left(\ln \frac{x_a^{(\alpha)}}{x_b^{(\alpha)}} + BA \frac{\left(Ax_a^{(\alpha)} + Bx_b^{(\alpha)}\right)\cdot\left(1 - 2x_a^{(\alpha)}\right) - (A - B)\cdot\left[x_a^{(\alpha)} - \left(x_a^{(\alpha)}\right)^2\right]}{\left(Ax_a^{(a)} + Bx_b^{(a)}\right)^2} \right) \tag{11.AJ}$$

and

$$\frac{\Gamma_\beta - \Gamma_\alpha}{x_a^{(\beta)} - x_a^{(\alpha)}} = \left(\ln \frac{x_a^{(\beta)}}{x_b^{(\beta)}} + BA \frac{\left(Ax_a^{(\beta)} + Bx_b^{(\beta)}\right)\cdot\left(1 - 2x_a^{(\beta)}\right) - (A - B)\cdot\left[x_a^{(\beta)} - \left(x_a^{(\beta)}\right)^2\right]}{\left(Ax_a^{(\beta)} + Bx_b^{(\beta)}\right)^2} \right) \tag{11.AK}$$

in which the Γ's are evaluated from Eq. 11.AH for $x_a^{(\alpha)}$ and $x_a^{(\beta)}$. The x_bs are evaluated by $x_b = (1 - x_a)$. When all of these are combined, we have two algebraic equations with two unknowns, $x_a^{(\alpha)}$ and $x_a^{(\beta)}$, which any competent nonlinear simultaneous algebraic equation solver can handle.

d. Show that solution for Example 11.6.

11.16 Repeat the preceding problem using the Margules equation instead of the van Laar equation. The Margules constants are shown in Problem 9.22.

11.17 Figure 11.8 (like Figure 6.7) shows the Gibbs energy relations for equilibrium between two liquids. Figure 8.12(d) shows a three-phase equilibrium between two liquids and a vapor. Sketch a copy of Figure 11.8 and then sketch on it what the curve for the Gibbs energy of the vapor must be at that temperature and pressure. *Hint*; the vapor curve must be tangent to the L-L tangent line. Where? What does it look like for other compositions?

11.18* Example 11.4 shows the activity coefficients of water in benzene and benzene in water at 25°C. What values of A and B in the van Laar equation reproduce these values?

11.19 Sketch the probable triangular diagram for water–gasoline–ethanol, treating gasoline as if it were a single pure species. Indicate what adding a small amount of ethanol to the gasoline would do to the solubility of water in the gasoline.

11.20 Repeat Example 11.11 for heating the water to its NBP, 212°F. This was the situation in the earliest steam power plant boilers. Suggest ways in which the problem of accumulation of solids in the boilers could be dealt with.

11.21 At 140°F, the solubilities of gypsum and anhydrite, read from Figure 11.13 are approximately 2020 and 1600 ppm. Taking into account the water of hydration in gypsum, which of the two is more soluble?

11.22 Water softeners replace the Ca^{2+} (and Mg^{2+}) ions in the water with equivalent amounts of Na^+ rejecting the calcium as $CaCl_2$ in the waste stream.

a. For the water in Example 11.11, how many pounds of NaCl will be needed per pound of water to replace the Ca^{2+} only?

b. How many pounds of NaCl will be needed per month for the typical water demand in that example?

c. How does this value compare with your personal experience with salt demand if you have a residential water softener? Comment?

11.23 On Figure 11.14, if we begin with pure solid ice at −5°C, and slowly add NaCl and mix, sketch and describe the path of states that we will follow. Assume that we add or subtract heat as needed to hold the temperature at −5°C.

11.24 On Figure 11.14, if we begin with a mixture of ice and eutectic with overall NaCl weight fraction = 10% at −30°C and heat slowly, sketch and describe the path of states that we will follow.

11.25* Calculate the weight percent NaCl in $NaCl{\cdot}2H_2O$.

11.26* Show the details of proceeding from Eq. 11.11 to Eq. 11.13.

11.27 A liquid solution of sodium chloride in water with 25 wt% sodium chloride originally at 0°C is slowly cooled.

a. At what temperature will the first particle of solid appear?

b. What will be the weight percent sodium chloride in this first particle?

11.28 On Figure 11.15, if we begin with pure solid Pb at 200°C, and slowly add Sn and mix, sketch and describe the path of states that we will follow. Assume that we add or subtract heat as needed to hold the temperature at 200°C.

11.29 On Figure 11.15, if we begin with a mixture of Sn and Pb with overall Pb wt fraction = 10%, at 150°C and heat slowly, sketch and describe the path of states that we will follow.

11.30* Calculate the wt ratio of phases α and β at the eutectic on Figure 11.15.

11.31 If we start with a liquid, 10% Sn at 350°C and cool slowly, sketch and describe the path of states that we will follow. Explain the meaning of the changes that occur between 150°C and 100°C. How rapidly would we expect those changes to occur?

11.32 Sketch the equivalent of Figure 11.16 for the water-n-butanol VLE shown in Figure 8.12(d).

11.33 On Figure 11.17 the quadruple point is at 32°F and approximately 383 psia.

a. Estimate the mol fraction of methane dissolved in water at this condition.

b. Estimate the mol fraction of water in the equilibrium vapor.

c. Comment on the reliability of the assumption that these two small values are effectively zero.

11.34 Figure 11.17 shows the melting temperature of water as constant at 32°F ignoring the fact that as pressure increases the melting temperature of water decreases.

a. Estimate the melting temperature of water at the quadruple point on that figure, $P = 383$ psia, using the values from part (b).

b. From Figure 1.10 (actually the table from which that figure was made) we have the following values for points on that curve; *P*, psia: 0, 2900, 7,250, 14,500, 21,750 and 29,000, *T*, °F: 32, 29.6, 23.3, 14.2, 4.94, −6.4. Plot these values on a copy of Figure 11.17 and sketch the curve through them.

c. How bad is the straight, vertical approximation we made in Figure 11.17?

11.35 Sketch a plausible analog to Figure 11.17 for water and propane. As in Figure 11.17 an area represents two phases in equilibrium, a curve three and four curves meeting at a point four phases. The new phase that appears here is a liquid phase, mostly propane. The books on this topic use the symbols L_W and L_{HC} for the mostly-water phase and the mostly-hydrocarbon phase. On this diagram the V-L_W-L_{HC} curve is practically the same as the vapor pressure curve for pure propane; it forms a second quadruple point above and to the right of one shown on Figure 11.17. See [14, Fig 4.2b] for confirmation (?) of your sketch.

11.36 The worst simplification in Eq. 11.15 is the built-in assumption that $\Delta h_{\text{solid to liquid}}$ is a constant, independent of temperature. Reference [18, p. 640] shows that if we do not make that simplification, but do assume that the heat capacities of solid and liquid are constant, independent of temperature, Eq. 11.15 becomes

$$\ln\left(\frac{1}{x_i^{(1)}\gamma_i^{(1)}}\right) = \frac{\Delta h_{\text{solid to liquid}}}{RT_{\text{melting point}}} \cdot \left(\frac{T_{\text{melting point}}}{T} - 1\right) + \frac{\Delta C_P}{R} \cdot \left(\ln\frac{T_{\text{melting point}}}{T} + 1 - \frac{T_{\text{melting point}}}{T}\right) \quad (11.26)$$

Show the derivation of Eq. 11.26.

11.37 Sketch what Figure 11.19 would look like if the solid aromatics dissolved in CCl_4 all had $\gamma_i^{(\text{liquid})} = 2.00$ independent of liquid composition.

11.38* Repeat Example 11.13 at $\log x_2 = -0.6$

11.39 Figure 11.20 indicates that at about 200 atm the "Raoult's law times Poynting factor" curve has a minimum, and then turns upward with increasing pressure. Can that be right? Write the equation for that curve, solve it for the minimum, then insert values to see at what pressure the minimum should occur.

11.40 Typical porous adsorbents have surface areas of up to 1000 m^2/g.

a. If we had a simple sheet of the material, like a sheet of paper, and its density were 2 g/cm^3, how thick would the sheet be to have this high a surface area (both sides of the sheet)?

b. How would one make a porous solid with this much internal surface area, and still some mechanical strength? *Hint*; consider how you make charcoal.

11.41 In the Langmuir curve example we made an estimate of $n_{\text{monolayer}}$, the total amount of nitrogen that would form a complete monolayer on the internal surface of the adsorbent. If we knew the surface area occupied by one molecule of adsorbed nitrogen, we could use it to estimate the total surface area of the solid, Make such an estimate as follows:

a. Assume that the adsorbed nitrogen has the same density as liquid nitrogen at its NBP, 0.808 g/cm^3, = 808 kg/m^3.

b. Determine the number of nitrogen molecules in this many kg. Then divide that number into the above mass to find the average volume occupied by one nitrogen molecule.

c. Then make the strong simplifying assumption that each molecule occupies a cubical volume, and compute the length of the edge of that cube.

d. From that compute the area of one face of the cube. Compare this with the commonly used value of 16.2 (Å)^2 [25].

11.42 Show the algebra to get from Eq. 11.22 to Eq. 11.23.

11.43 Show that if one prepares a plot like Figure 11.25 based on Eq. 11.23, then from the straight line shown, $b = 1 +$ slope/intercept, and $n_{\text{monolayer}} = 1/(\text{slope} + \text{intercept})$.

11.44 **a.** Prepare the equivalent of Figure 11.25, using all the data points in Figure 11.24, and show that the best-fit straight line has a negative intercept. Table 11.B shows the data (captured by

Table 11.B Data points for Figures 11.24 and 11.25

Pressure, atm	n_{adsorbed}, mmol/(g of adsorbent)
0.021	4.337
0.182	6.831
0.235	7.376
0.331	8.395
0.364	8.795
0.441	9.865
0.529	11.509

computer scanning the data points in Figure 6 of [25]. And converting to the consistent dimensions shown,)

b. The negative intercept (with a value close to zero) makes b a large negative number (see Problem 11.43). But does it lead to a very different value of $n_{\text{monolayer}}$ than Example 11.16?

c. Plot these data in the Langmuir form (Eq. 11.21, Figure 11.23). Does it form a straight line?

REFERENCES

1. Sorensen, J. M., and W. Arlt. *Liquid-Liquid Equilibrium Data Collection*, Vol. 1: *Binary Systems*. Frankfurt, Germany: Dechema, p. 341 (1979).
2. Smallwood, I. *Solvent Recovery Handbook*. New York: McGraw-Hill, Appendix I (1993).
3. National Primary Drinking Water Regulations. *40CFR 141.60, Subpart G* (1998).
4. Sorensen, J. M., and W. Arlt. *Liquid-Liquid Equilibrium Data Collection*, Vol. 2, *Ternary Systems*. Frankfurt, Germany: Dechema, 353 (1980).
5. McCabe, W. L., and J. C. Smith. *Unit Operations of Chemical Engineering*, ed. 3. New York: McGraw-Hill, p. 495 (1976).
6. Felder, R. M., and R. W. Rousseau. *Elementary Principles of Chemical Processes*, ed. 3. New York: Wiley, p. 273 (2015).
7. Seader, J. D., and E. J. Henley. *Separation Process Principles*. New York: Wiley, Chapter 8 (1998).
8. Walas, S. M. *Phase Equilibria in Chemical Engineering*. Boston: Butterworth, p. 535 (1985).
9. Poling, B. E., J. M. Prausnitz, and J. P. O'Connell. *The Properties of Liquids and Gases*, ed. 5. New York, McGraw-Hill. pp. 8–157 (2000).
10. Daubert, T. E., and R.P. Danner. Phase equilibria in water-hydrocarbon systems. In *Technical Data Book, Petroleum Refining*, Vol. 2. New York: American Petroleum Institute, Chapter 9 (1978).
11. Stoye, D. *Paints, Coatings and Solvents*. Weinheim, Germany: VCH Verlagsgesellschaft mbH, p. 278 (1993).
12. Liley, P. E., G. H. Thomson, D. G. Friend, T. E. Daubert, and E. Buck. Physical and chemical data. In *Perry's Chemical Engineers' Handbook*, ed. 8. Green, D. W. ed., New York: McGraw-Hill, pp. 2–123 (2008).
13. *Betz Handbook of Industrial Water Conditioning*, ed. 8. Trevose, PA: Betz Laboratories, p. 204 (1980).
14. Sloan, E. D. Jr. *Clathrate Hydrates of Natural Gases*, ed. 2. New York: Marcel Dekker Chapter 2 (1998).
15. Carroll, J. J. *Natural Gas Hydrates: A Guide for Engineers*, ed. 2, Amsterdam: Elsevier, 18 (2009).
16. Kvenvolden, K.A. Gas hydrates—geological perspective and global change. *Rev. Geophysics*, 31(2):173–187 (1993).
17. Milkov, A.V., Global estimates of hydrate-bound gas in methane sediments: how much is really out there? *Earth Science Reviews* 66:183–197 (2004).
18. Prausnitz, J. M., R. N. Lichtenthaler, and E. Gomez de Azevedo. *Molecular Thermodynamics of Fluid-Phase Equilibria*. Upper Saddle River, NJ: Prentice Hall, Chapter 12 (1999).
19. McLaughlin, E., and H. A. Zainal. The solubility behavior of aromatic hydrocarbons in benzene. *J. Chem. Soc.* 863–867 (1959).
20. McLaughlin, E., and H. A. Zainal. The solubility behavior of aromatic hydrocarbons, Part II: solubilities in carbon tetrachloride. *J. Chem. Soc.* 2485–2488 (1960).
21. Bertram, B. M. Sodium compounds (sodium halides). In *Kirk–Othmer Encyclopedia of Chemical Technology*, ed. 4, Vol. 22, Kroschwitz, J. I., and M. Howe-Grant, eds. New York: Wiley, p. 360 (1997).
22. Tsekhanskaya, Y. V., M. B. Iomtev, and E. V. Mushkina. Solubility of naphthalene in ethylene and carbon dioxide under pressure. *Russ. J. Phys. Chem.* 38:1173–1176 (1964).
23. Valenzuela, D. P., and A. L. Myers. *Adsorption Equilibrium Data Handbook*. Englewood Cliffs, NJ: Prentice Hall, p. 153 (1989).
24. Langmuir, I. The adsorption of gases on plane surfaces of glass, mica and platinum. *JACS* 40:1361–1403 (1918).
25. Brunauer, S., and P. H. Emmett. The use of low temperature van der Walls adsorption isotherms in determining the surface areas of various adsorbents. *JACS* 59:2682–2689 (1937).
26. Brunauer, S., P. H. Emmett, and E. Teller. Adsorption of gases in multimolecular layers. *JACS* 60:309–319 (1938).
27. Emmett, P. H. and S. Brunauer, The use of low temperature van der Waals adsorption isotherms in determining the surface area of iron synthetic ammonia catalysts. *JACS* 59:1553–1564 (1937).
28. Talu, O., and I. Zwiebel. Multicomponent adsorption equilibria of nonideal mixtures. *AIChE Journal* 32 (8):1263–1276 (1996).
29. Starling, K. E. *Fluid Thermodynamic Properties for Light Petroleum Systems*. Houston, TY: Gulf (1973).
30. Adamson, A. W. *Physical Chemistry of Surfaces*. 6 ed. New York: Wiley-Interscience, p. 647 (1997).
31. Cohan, L. H. Hysteresis and the capillary theory of adsorption of vapors. *JACS* 66:98–105 (1944).

12

CHEMICAL EQUILIBRIUM

12.1 INTRODUCTION TO CHEMICAL REACTIONS AND CHEMICAL EQUILIBRIUM

Chemical reactions transform one chemical species or set of species to another species or set of species: reactant(s) $\rightarrow$ product(s). Sometimes we do this because the products (e.g., pharmaceuticals) are much more valuable than the reactants from which they are made. Sometimes we want the heat released by the chemical reaction of the materials (e.g., fuels) with air (burning) either to heat some material (cooking our food or heating our homes) or to burn the fuels inside the engines that propel our vehicles or generate electricity. Sometimes we destroy harmful materials by chemical reaction (incineration of hazardous hydrocarbons, destruction of bacteria, protozoa, and viruses in drinking water with chlorine or ozone). The most important chemical reactions are those within our bodies. Every second millions of chemical reactions are occurring in our bodies, accomplishing all the things we call life. The chemical reactions in our nervous system control our muscular movements and our thoughts; the *nerve gases* that interfere with those chemical reactions can kill in seconds.

In this chapter we consider only single chemical reactions, all occurring in one phase. In the next chapter we consider multiple reactions, in series and in parallel, occurring in one or more phases. The next chapter introduces no new principles, only more complex and interesting applications of the ideas of this chapter.

12.2 FORMAL DESCRIPTION OF CHEMICAL REACTIONS

All chemical reactions can be described in general as

$$a\text{Reactant 1} + b\text{Reactant 2} + \cdots \Leftrightarrow r\text{Product 1} + s\text{Product 2} + \cdots \tag{12.1}$$

where $a, b, \ldots, r, s$ are the number of molecules consumed or produced, and the $\Leftrightarrow$ sign indicates that this reaction can go in either direction, and that at some set of concentrations there is an equilibrium in which the rate of the forward (reactant to product) reaction is equal and opposite to that of the backward (product to reactant) reaction so that there is no net change with time, and we have reached *chemical equilibrium*. If the reaction goes only one way we often replace the $\Leftrightarrow$ with a $\rightarrow$ to indicate that. Some books replace both of these symbols with an $=$.

As an example of this kind of reaction, which will appear often in this chapter, consider the reaction of hydrogen and nitrogen to form ammonia

$$N_2 + 3H_2 \Leftrightarrow 2NH_3 \tag{12.A}$$

in which one nitrogen and 3 hydrogen molecules are the reactants and 2 ammonia molecules are the products. (This reaction is the basis of the production of nitrogen fertilizers and explosives; hundreds of factories carry out this reaction

Physical and Chemical Equilibrium for Chemical Engineers, Second Edition. Noel de Nevers.

on a large scale, see Section 1.1.) Equation 12.1 is often written as

$$aA + bB + \cdots \Leftrightarrow pP + rR + \cdots \tag{12.2}$$

where A, B, ... are the reactants and P, R, ... are the products. The numbers of each molecule involved in the reaction, $a, b, \ldots, p, r, \ldots$ are called the *stoichiometric coefficients* of the reaction (see Section 12.6.1). We often also see Reaction 12.A written in the form

$$0.5N_2 + 1.5H_2 \Leftrightarrow NH_3 \tag{12.B}$$

which is obviously Reaction 12.A divided by 2. In the ammonia literature the reaction is almost always expressed as Reaction 12.B, not Reaction 12.A, so in the rest of this chapter we will follow that practice, using Reaction 12.B.

12.3 MINIMIZING GIBBS ENERGY

In principle, all chemical reactions are equilibrium reactions, but, as we shall see later, in practice, many are not. In all chemical reactions, at equilibrium the reacting system has taken up the set of compositions that minimizes the Gibbs energy of the mixture of reactants and products, as described in Section 4.5.3 and Figure 4.12. Thus, all that we do in this chapter is apply Eq. 4.6, minimization of Gibbs energy, to chemical reaction equilibria.

Figure 12.1 (a repeat of Figure 4.11 without the chemical potentials) shows the calculated Gibbs energy for mixtures of normal and isobutane. If there were no change of Gibbs energy on mixing, then the curve would be replaced by a straight line connecting the two pure-component values of the Gibbs energy of normal and isobutane. But from Chapters 7, 8, and 9 we know that for an ideal binary solution the Gibbs energy is given by

$$g_{\text{mix}} = \sum x_i g_i^{\text{o}} + RT \sum x_i \ln x_i \quad \text{[binary ideal solution]} \tag{12.3}$$

where the first term on the right is a straight line between the pure-component values, and the second is the Gibbs energy increase on mixing, which is always negative because the ln of numbers less than 1.00 is always negative.

Example 12.1 Estimate the chemical equilibrium composition of a gaseous mixture of n-butane and isobutane at 298.15 K and 1 bar, based on direct minimization of Gibbs energy. Assume that n-butane and isobutane form an ideal solution of ideal gases at this T and P.

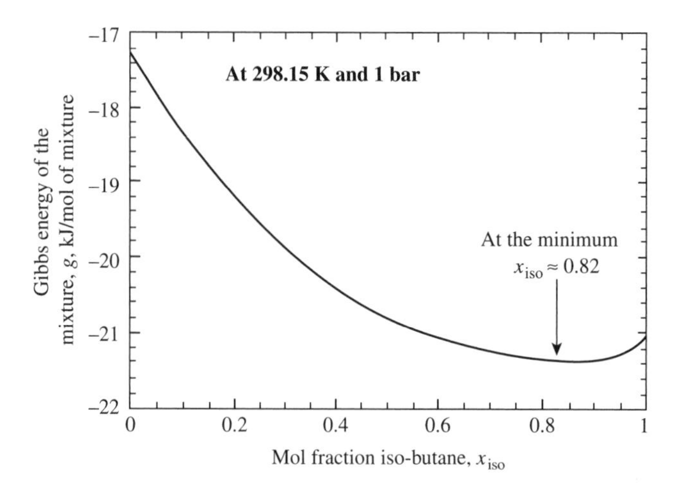

FIGURE 12.1 Calculated Gibbs energy for a mixture of normal and isobutane at 25°C = 298.15 K and 1.00 bar. The pure component values are from Table A.8 and the intermediate values based on an assumed ideal solution. The minimum value, which corresponds to the chemical equilibrium between the two isomeric forms, occurs at $x_{\text{iso}} \approx 0.82$.

We differentiate Eq. 12.3 with respect to x_a, remembering that for a binary mixture $dx_b = -dx_a$, finding

$$\begin{aligned}\frac{dg_{\text{mix}}}{dx_a} &= g_a^{\text{o}} - g_b^{\text{o}} + RT[\ln x_a + 1 - (\ln x_b + 1)] \\ &= g_a^{\text{o}} - g_b^{\text{o}} + RT[\ln x_a - \ln(1 - x_a)] \\ &= g_a^{\text{o}} - g_b^{\text{o}} + RT \ln\left(\frac{x}{1 - x_a}\right)\end{aligned} \tag{12.4}$$

Setting this equal to zero (to find the minimum on the g-x curve) and solving gives

$$\left(\frac{x_a}{1 - x_a}\right) = \exp\left(\frac{g_b^{\text{o}} - g_a^{\text{o}}}{RT}\right) \qquad x_a = \frac{\exp\left(\frac{g_b^{\text{o}} - g_a^{\text{o}}}{RT}\right)}{1 + \exp\left(\frac{g_b^{\text{o}} - g_a^{\text{o}}}{RT}\right)} \tag{12.5}$$

Inserting the values of the pure component Gibbs energies from Table A.8 (discussed below) for isobutane, a, and n-butane, b, we find

$$\exp\left(\frac{g_b^{\text{o}} - g_a^{\text{o}}}{RT}\right) = \exp\left[\frac{-17.2\dfrac{\text{kJ}}{\text{mol}} - \left(-20.9\dfrac{\text{kJ}}{\text{mol}}\right)}{0.008314\dfrac{\text{J}}{\text{mol}\cdot\text{K}} \cdot 298.15\ \text{K}}\right]$$

$$= 4.49 \quad x_a = \frac{4.49}{1 + 4.49} = 0.818 \tag{12.C}$$

The same result is shown graphically in Figure 12.1. ■

If there were no Gibbs energy change on mixing, then the lowest point on the curve (now a straight line) would be at pure isobutane, 100% conversion. The Gibbs energy decrease on mixing guarantees that the minimum of the curve will never be at 100% conversion, but rather will always fall at some incomplete conversion (see Section 12.6.2). We may think of the same fact another way, in terms of the forward and reverse rates of the reaction. Chemical reaction rates almost always increase with an increase in reactant concentration, most often to the first power of the concentration, sometimes to other powers. If we had 100% conversion then the forward rate would become zero because the concentration of the reactant became zero, while the backward rate was still finite, so we would move back away from 100% conversion. By minimizing the Gibbs energy the system chooses the concentrations of reactants and products at which the two rates are equal and opposite.

If the change of Gibbs energy going from reactants to products is small, like most chemical reactions in our bodies, or the isomerization reaction in Figure 12.1, then at equilibrium there will be substantial amounts of reactants in equilibrium with the products of the reaction. If the Gibbs energy change is large, as in combustion or explosive reactions, then at equilibrium there will be practically no reactants left. We commonly refer to such reactions (combustion, explosions, cooking of foods, pyrolysis) as irreversible, even though in principle all chemical reactions are reversible (see Section 12.6.2).

The rest of this chapter and all of the next chapter are devoted to mathematical methods to estimate the chemical composition that corresponds to the minimum in Figure 12.1. As the details become complex, remember that we are simply seeking the minimum value of g on the (often multidimensional) equivalent of Figure 12.1. Many computer programs for chemical equilibrium do exactly what we did in Example 12.1, including the suitable activity and/or fugacity coefficients in Eq. 12.3, instead of making the ideal solution assumption. However, a traditional hand calculation approach, called *the law of mass action*, has historically been more widely used, is almost always used in the literature, and is perhaps easier to understand. It always leads to the same result as this *direct minimization of Gibbs energy* approach. The rest of this chapter and the next chapter present all of chemical reaction equilibrium calculations in the law of mass action form. Remember that it is merely a computationally satisfying way of finding the compositions that correspond to a minimum in the Gibbs energy.

12.4 REACTION RATES, ENERGY BARRIERS, CATALYSIS, AND EQUILIBRIUM

In principle, the equilibrium state is independent of the rate of the reaction, but in practice, many reactions reach a metastable equilibrium, dependent on the reaction rate. For example, at room temperature at equilibrium the combustion reaction

$$2H_2 + O_2 \Leftrightarrow 2H_2O \tag{12.D}$$

is almost complete, with practically zero hydrogen and oxygen remaining. However, at room temperature, without a catalyst, the rate of the reaction is practically zero, so that there is a metastable equilibrium, in which none of the hydrogen and oxygen reacts.

This reaction is typical of combustion reactions; the rate is $\approx$0.00 at room temperature and very fast at high temperatures. If we initiate the reaction in a small part of the reactant mixture, for example, with an automobile spark plug, then the heat liberated by the reaction is enough to raise the temperature of the remaining reactants enough that the reaction propagates, forming a flame or an explosion. The situation can be intuitively visualized in Figure 12.2.

We see that the energy of the products is less than that of the reactants, so that if the reaction proceeds from reactants to products (e.g., fuel and oxygen to combustion products) then the system's energy will be reduced and heat will be given off, as it is for all combustion reactions. But we also see that there is an energy barrier between reactants and products. For the reaction to proceed, the reactants (or the products for the reverse reaction) must go through some intermediate states with a higher energy than either the reactants or products. For combustion at room temperature this energy barrier is high enough that the rate of both the forward and the backward reaction is practically zero. So there exists a state of metastable equilibrium. If we raise the temperature, we raise the energy of both reactants and products but do not raise

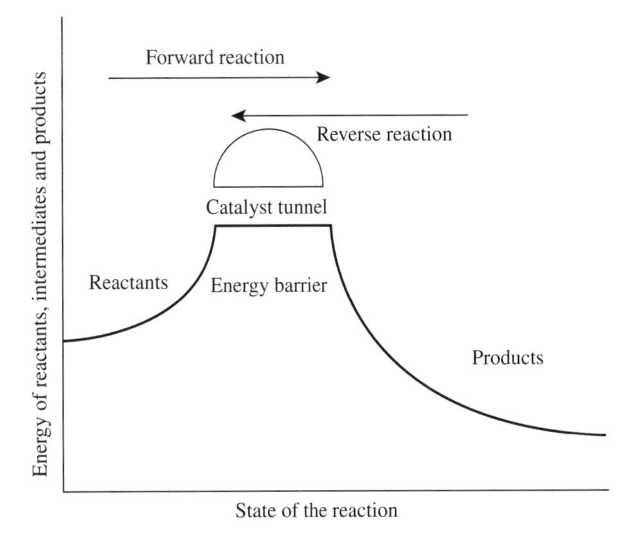

FIGURE 12.2 Intuitive picture of the energy relations for an exothermic chemical reaction.

the height of the energy barrier. If we raise the temperature enough that some reactant molecules can get over the energy barrier, then the reaction will proceed.

In Figure 12.2 we should not think of all the molecules lying on the curve. In any chemical at any temperature there is a distribution of energies, with some minimum. We might think that on either side of the energy barrier in Figure 12.2 the molecules are like a flock of birds, with none flying below the curve shown, most flying near it, and some flying high above it. If the barrier is high enough, none can get over. As the difference between the baseline on either side and the height of the barrier becomes smaller and smaller, more and more of the most energetic molecules in the mix can cross.

Figure 12.2 shows a "catalyst tunnel." Catalysts are substances that allow the reaction to proceed by a route that does not have as high an energy barrier as does the uncatalyzed reaction. With catalysts we can cause the reaction to proceed at temperatures far lower than those needed for the uncatalyzed reaction. At equilibrium the rates of the forward and backward reactions are equal; the presence or absence of the catalyst does not influence the concentration at which these rates are equal, so that while a catalyst does make a reaction go faster (or go at a temperature at which the rate without the catalyst is ≈ 0), the catalyst does not, *in principle*, influence the equilibrium concentrations.

In practice, catalysts can influence the position of metastable equilibrium, which is often what we seek industrially. Carbon monoxide and hydrogen can react in many ways; for example,

$$CO + 2H_2 \Leftrightarrow CH_3OH \tag{12.E}$$

and

$$CO + 3H_2 \Leftrightarrow CH_4 + H_2O \tag{12.F}$$

At a high temperature both of these reactions come to equilibrium; we would expect CO, H_2, CH_3OH, CH_4, and H_2O all to be in chemical equilibrium. However, it is possible to find *selective catalysts* that promote only Reaction 12.E without promoting 12.F, or the reverse, so that we can produce practically pure methanol or practically pure methane; both of these reactions are carried out on a large scale industrially, using such selective catalysts. We may think of this as a three-dimensional equivalent of Figure 12.2; there are two energy barriers, one for each reaction. A selective catalyst provides a tunnel for one reaction, but not the other. In modern industrial catalytic research the goal is almost always to find a catalyst that is extremely selective, facilitating *only* the desired reaction.

In principle, there is no such thing as a negative catalyst, one that prevents a reaction. In practice, there are materials that act *as if* they were negative catalysts. The selective catalyst that makes Reaction 12.E proceed to equilibrium at a temperature low enough that Reaction 12.F does not occur acts as if it were a negative catalyst for Reaction 12.F. The real negative influence is the low temperature, but it seems as if the catalyst is acting as a negative catalyst for the non-selected reaction. We also know that at the molecular level many chemical reactions are chain reactions, in which small amounts of materials that do not appear in the balanced chemical reaction equation play the role of chain carriers, being used over and over again. A famous example of this process is Reaction 12.G:

$$2CO + O_2 \Leftrightarrow 2CO_2 \tag{12.G}$$

Without the hydroxyl radical, OH·, the rate of the reaction is ≈ 0. This means that if there is a little water present CO burns easily in O_2 or air, but that if there is no water present the reaction rate is ≈ 0, even at flame temperatures. Many successful fire-extinguishing agents and some automotive antiknock agents act by capturing or destroying these free radicals and thus stopping chain reactions. Many food preservatives and oxidation inhibitors work this way. Some vitamins and food chemicals are believed to increase human longevity by capturing or destroying free radicals in our cells before they can participate in aging reactions. These *reaction inhibitors* act *as if* they were negative catalysts. But neither negative nor positive catalysts influence the composition of a mixture at true thermodynamic equilibrium unless there is a metastable equilibrium involved, like hydrogen and oxygen at room temperature, or whichever of Reactions I2.E and 12.F is not catalyzed while the other is.

In the rest of this chapter we assume that we have reached equilibrium, by using a high enough temperature for the reaction to proceed without a catalyst, by using a catalyst, by waiting long enough, or by all three. We ask only what is the chemical composition of the mixture of reactants and products at that equilibrium state. That information alone is quite useful for estimating the limits of chemical reactions. Taken together with information about the rates (kinetics) of the reactions it allows us to successfully design chemical reactors.

12.5 THE BASIC THERMODYNAMICS OF CHEMICAL REACTIONS AND ITS CONVENIENT FORMULATIONS

The basic statement for all equilibrium is Eq. 4.6: Nature minimizes Gibbs energy. At any state of chemical or phase equilibrium the Gibbs energy is the lowest value consistent with the external constraints (Figure 12.1). In phase equilibrium we found it convenient to work with the fugacity instead of the Gibbs energy, and we will do the same here. Returning to the definition of the fugacity in Section 7.2, we can ask what it tells us about chemical equilibrium. Consider the reaction of hydrogen and nitrogen to form ammonia

$$0.5\mathrm{N_2} + 1.5\mathrm{H_2} \Leftrightarrow \mathrm{NH_3} \tag{12.B}$$

all in the gas phase. If we apply the same arguments here that we applied in Chapters 4 and 7 for phase equilibrium, we can say that if this reaction is at equilibrium, then a differential conversion of dn_i of nitrogen and hydrogen to ammonia or the reverse at constant T and P must cause zero change in the Gibbs energy. However, we can see that if a differential number of mols of nitrogen disappears by this reaction, three times as many mols of hydrogen must also disappear and two times as many mols of ammonia must be produced.

From Eq. 6.4 we know that for any single-phase system

$$G = n_a\bar{g}_a + n_b\bar{g}_b + n_c\bar{g}_c + \cdots \tag{12.6}$$

and

$$g = \frac{G}{n_T} = x_a\bar{g}_a + x_b\bar{g}_b + x_c\bar{g}_c + \cdots \tag{12.7}$$

This chapter concerns reactions in only one phase; we will not need the phase superscript in this chapter. The next chapter considers reactions in more than one phase, where we will need it. For the net change in the Gibbs energy of the system to be zero, as we convert $dn_{\mathrm{N_2}}$ to ammonia, we must have

$$dG = 0 = (-0.5\bar{g}_{\mathrm{N_2}} - 1.5\bar{g}_{\mathrm{H_2}} + \bar{g}_{\mathrm{NH_3}})dn_{\mathrm{N_2}}$$

or

$$\frac{dg}{dn_{\mathrm{N_2}}} = 0 = -0.5\bar{g}_{\mathrm{N_2}} - 1.5\bar{g}_{\mathrm{H_2}} + \bar{g}_{\mathrm{NH_3}} = -0.5\mu_{\mathrm{N_2}} - 1.5\mu_{\mathrm{H_2}} + \mu_{\mathrm{NH_3}} \tag{12.8}$$

where these are the same partial molar Gibbs energies (chemical potentials) that we saw played the dominant role in phase equilibrium. *The partial molar Gibbs energies control not only phase equilibrium, but also chemical equilibrium.* Here we show both the $\bar{g}_{\mathrm{N_2}}$ and the $\mu_{\mathrm{N_2}}$ forms of this equation. In phase equilibrium we regularly see both forms, but in chemical equilibrium we mostly see the $\bar{g}_{\mathrm{N_2}}$ form and seldom the $\mu_{\mathrm{N_2}}$ form.

Substituting the definition of fugacity (Eq. 7.1) for $\bar{g}_i$, three times we find

$$\frac{dg}{dn_{\mathrm{N_2}}} = 0 = -0.5\left(RT\ \ln f_{\mathrm{N_2}} + g^{\mathrm{o}}_{\mathrm{N_2}}\right) - 1.5\left(RT\ \ln f_{\mathrm{H_2}} + g^{\mathrm{o}}_{\mathrm{H_2}}\right) + \left(RT\ \ln f_{\mathrm{NH_3}} + g^{\mathrm{o}}_{\mathrm{NH_3}}\right) \tag{12.9}$$

Rearranging, we have

$$RT(0.5\ \ln f_{\mathrm{N_2}} + 1.5\ \ln f_{\mathrm{H_2}} - \ln f_{\mathrm{NH_3}}) = g^{\mathrm{o}}_{\mathrm{NH_3}} - 0.5g^{\mathrm{o}}_{\mathrm{N_2}} - 1.5g^{\mathrm{o}}_{\mathrm{H_2}} \tag{12.10}$$

or

$$\ln\left(\frac{f_{\mathrm{NH_3}}}{f^{0.5}_{\mathrm{N_2}} f^{1.5}_{\mathrm{H_2}}}\right) = -\frac{\left(g^{\mathrm{o}}_{\mathrm{NH_3}} - 0.5g^{\mathrm{o}}_{\mathrm{N_2}} - 1.5g^{\mathrm{o}}_{\mathrm{H_2}}\right)}{RT} \tag{12.11}$$

Previously we have said little about g^{o}_i other than that it was a function of temperature only and that each species had its own value of g^{o}_i. We discussed standard states in Chapters 7, 8, and 9. A standard state is some state of matter that we will all agree upon as a suitable basis for constructing tables of properties. For most chemical reaction purposes we choose the standard state of some substance as the pure substance in its normal state (solid, liquid, or gas) at $P = 1$ atm or 1 bar, and an arbitrarily chosen T, normally $= 25°\mathrm{C} = 298.15\ \mathrm{K}$ for the tables of interest in this chapter. Alas, there are other standard states that are much more convenient for some problems, as discussed previously for vapor–liquid equilibrium calculations. However, if we put off for the moment saying what our standard state is, we can use the symbol ° to indicate a property in the standard state, and then say that, for any pure chemical element or compound (pure species) the partial molar Gibbs energy is the same as the pure species Gibbs energy, and in its standard state

$$(\bar{g}_i)^{\mathrm{o}} = g^{\mathrm{o}}_i = RT\ \ln f^{\mathrm{o}}_i + g^{\mathrm{o}}_i \quad \text{[pure species]} \tag{12.12}$$

If we solve this for g^{o}_i and substitute the value three times in Eq. 12.11 and rearrange, we find

$$\frac{f_{\mathrm{NH_3}}}{f^{0.5}_{\mathrm{N_2}} \cdot f^{1.5}_{\mathrm{H_2}}} = \exp\left[\frac{-1}{RT}\left(g^{\mathrm{o}}_{\mathrm{NH_3}} - 0.5g^{\mathrm{o}}_{\mathrm{N_2}} - 1.5g^{\mathrm{o}}_{\mathrm{H_2}}\right)\right]\left[\frac{f^{\mathrm{o}}_{\mathrm{NH_3}}}{\left(f^{\mathrm{o}}_{\mathrm{N_2}}\right)^{0.5} \cdot \left(f^{\mathrm{o}}_{\mathrm{H_2}}\right)^{1.5}}\right] \tag{12.13}$$

or

$$\frac{\left[\dfrac{f_{\mathrm{NH_3}}}{f^{\mathrm{o}}_{\mathrm{NH_3}}}\right]}{\left[\dfrac{f_{\mathrm{N_2}}}{f^{\mathrm{o}}_{\mathrm{N_2}}}\right]^{0.5} \cdot \left[\dfrac{f_{\mathrm{H_2}}}{f^{\mathrm{o}}_{\mathrm{H_2}}}\right]^{1.5}} = \exp\left(\frac{-\Delta g^{\mathrm{o}}}{RT}\right) = K \tag{12.14}$$

where

$$\Delta g^{o} = \left(g^{o}_{NH_3} - 0.5 g^{o}_{N_2} - 1.5 g^{o}_{H_2}\right) \tag{12.H}$$

12.5.1 The Law of Mass Action and Equilibrium Constants

Equation 12.14 is the "law of mass action," which appears in elementary chemistry books, written for Reaction 12.B. The term at the right is the familiar "equilibrium constant," K, which has the following characteristics:

1. Its form follows directly from the definition of the fugacity and the observation that at equilibrium a differential chemical change produces zero change in Gibbs energy.
2. Δg° is the Gibbs energy change in going from *pure reactants to pure products*, with all reactants and products in their standard states (normally 1 atm or 1 bar and 25°C). This is often called the "standard Gibbs energy change of the reaction." Δg° is not the Gibbs energy change in going from the starting to the equilibrium state. In Figure 12.1, Δg° is the decrease in Gibbs energy from pure *n*-butane to pure isobutane. The Gibbs energy change for going from pure *n*-butane to the equilibrium state is greater than Δg° (see Problem 12.3).
3. The individual species' fugacities appear only in the form of the activity

$$\text{activity} = \frac{f_i}{f_i^{\circ}} = a_i \tag{7.26}$$

so that Eq. 12.14 is often written as

$$\frac{[a_{NH_3}]}{[a_{N_2}]^{0.5} \cdot [a_{H_2}]^{1.5}} = \exp\left(\frac{-\Delta g^{o}}{RT}\right) = K \tag{12.15}$$

In Chapter 7, where we introduced the activity, we said little about it; here we say a little more. Equation 7.26 and the definition of the fugacity can be combined to

$$RT \ln a_i = \bar{g}_i - g_i^{o} \tag{12.16}$$

which shows that the value of the activity depends on our choice f_i^{o} or g_i^{o}. In Chapters 7, 8, and 9 we saw that several choices were regularly made. For chemical reactions we almost always use the choices shown in Table A.8, which are all at 1 bar pressure and 25°C. We discuss that table in the next section.

4. This equilibrium constant is dimensionless, as are all equilibrium constants expressed in terms of activities, because the equilibrium constant is a ratio of products of activities to various powers (+ and –), and activities are all dimensionless. That was one of the reasons for defining the activity, to make all equilibrium constants dimensionless. However, the numerical value of the activity depends on the choice of standard states. If we change standard states we will change the computed numerical value of the equilibrium constant, but in a way that will not change the computed concentrations at equilibrium (if we pay careful attention to standard states!). As we will see later, we often use modified equilibrium constants that have built-in dimensions; however, the basic equilibrium constant, which is defined by Eqs. 12.14 and 12.15, is always dimensionless.
5. The equilibrium constant does not depend in any way on the pressure at which we conduct the reaction. The individual activities in Eqs. 12.14 and 12.15 do depend on pressure, as we will discuss in Section 12.8, but their ratio in Eq. 12.15 does not.
6. The equilibrium constant does depend on what pressure we choose for our values of Δg°. If we always choose these as states at which the pressure = 1.00 bar—which most current tables of properties do—then there is little opportunity for confusion here. Unfortunately, some tables of properties also have other choices, so that the values of K calculated from them are different from those calculated for $P = 1.00$ bar. If we are careful to leam what values of Δg° are used, we will always get the correct calculated concentrations. (Older tables mostly used $P = 1.00$ atm; newer ones mostly use $P = 1.00$ bar. The differences are small, but not zero, see Problem 12.15.)
7. The equilibrium constant does depend on the temperature of the reaction, both because a T appears in its defining equation (Eq. 12.14) and because the Δg^{o} of almost all reactions changes with changes in temperature. We will explore this in Section 12.7.
8. The equilibrium constant depends on how we write the reaction. If we had done all of this section in terms of Reaction 12.A instead of Reaction 12.B, then the Δg^{o} would have been exactly twice as large, and according to Eq. 12.14

$$K_{\text{Reaction 12.A}} = \left(K_{\text{Reaction 12.B}}\right)^2 \tag{12.17}$$

We will see that this does not lead to any uncertainty about calculated concentrations, as long as we are careful to observe which form of the reaction is associated with the K for which data are available.

9. Unfortunately, the same symbol, K, is used for other kinds of equilibrium besides chemical equilibrium, for example, VLE (see Section 8.2) and liquid–liquid or liquid–solid phase equilibrium (see Section 11.6). There aren't enough letters in the alphabet to prevent this. From the context we can normally determine which K is meant.

12.6 CALCULATING EQUILIBRIUM CONSTANTS FROM GIBBS ENERGY TABLES AND THEN USING EQUILIBRIUM CONSTANTS TO CALCULATE EQUILIBRIUM CONCENTRATIONS

Equation 12.14 also shows that if we had a table of the values of the standard state pure species Gibbs energy g_i° for all chemical elements and compounds, we could use those tables to calculate the equilibrium constant for any chemical reaction. We do have such tables for a very large number of compounds, widely available in handbooks [1–4]; a sample is shown in Table A.8. How they are actually constructed is discussed in Appendix E. In making them up the physical chemists needed to agree upon some datum values for the Gibbs energy, because we know of no way to calculate an absolute value for it. We know from experiments (see Appendix E), as sketched in Figure 12.3, that in going from gaseous ethanol to the elements C, H_2, and O_2, the Gibbs energy increases by 168.5 kJ/mol of ethanol, and that in going from the elements to ethylene the Gibbs energy increases by 68.5 kJ/mol. If we assign all chemical elements a datum value of $g^o = 0$, at some temperature and pressure, then we can assign any compound a "Gibbs energy of formation from the elements," and use the values to compute the Gibbs energy change for any reaction. The resulting values are listed in Table A.8 as the "Standard Gibbs Energy of Formation from the Elements." This choice of datum values is convenient, because no element can be made from another element, and every chemical compound can, in principle, be made from the elements. To form some element "from the elements" means to make no change, so the Gibbs energy of formation of the elements themselves "from the elements" in those tables is zero. From Figure 12.3 we see that on this basis, the g^o of formation from the elements of ethylene is +68.5 kJ/mol, while that of ethanol is −168.5 kJ/mol.

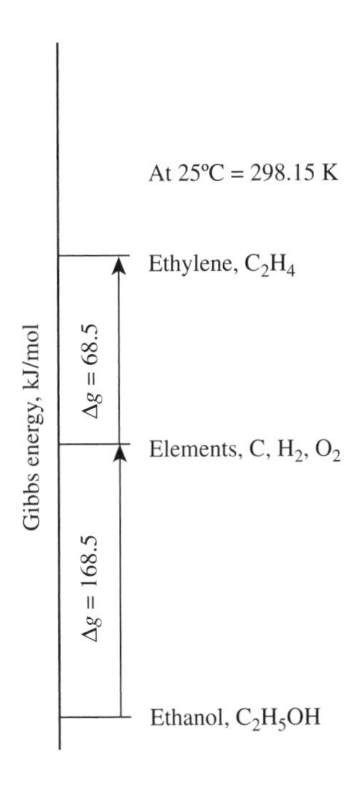

FIGURE 12.3 Gibbs energy changes for two chemical reactions.

The older tables were mostly in kcal/mol, while the newer ones are mostly in kJ/mol; (cal = 4.184 J). Examples and problems in this text are based on Table A.8, which is in kJ/mol, and values in other units are used only where needed (e.g., when quoting a source that uses kcal/mol).

We could have made other choices of datum values; we sometimes see tables that have other choices. However, the only use of these tables is to calculate Δg^o of some chemical reaction, and if we were to choose the Gibbs energy of all the elements as some value X per atom (instead of zero), the calculated values of Δg^o for all reactions would not change, because we would add and subtract the same amount to the Gibbs energy of the reactants and the products, for a net change of zero.

The same tables normally also show the enthalpy change of formation from the elements at 25°C and 1 atm. With that value we can also compute the enthalpy change of reactions ("heat of reaction"). In Section 12.8 we will see the use of that heat of reaction in calculating the change in K with temperature. Some elements have multiple forms in the pure state; for example, diamond and graphite are both pure carbon. For that reason Table A.8 lists the Gibbs energies of formation for both, showing that graphite is chosen as having 0.00 enthalpy and Gibbs energy change of formation and diamond as having enthalpy and Gibbs energy changes of formation—from graphite—of 1.9 and 2.9 kJ/mol (see also Example 4.9).

Example 12.2 Table A.8 shows that at 298.15 K the Gibbs energy of formation from the elements at 1 bar (0.987 atm) of NO (nitric oxide) is 86.6 kJ/mol. (The corresponding values for the elements N_2 and O_2, not shown in the table, are both 0.00.) Using these values estimate the equilibrium constant for the reaction of nitrogen and oxygen to form nitric oxide at 298.15 K by

$$N_2 + O_2 \Leftrightarrow 2NO \tag{12.I}$$

and estimate the equilibrium concentration of NO in air at 1 atm and at this 298.15 and at 2000 K.

Here

$$\Delta g^{\circ} = 2g^{\circ}_{NO} - g^{\circ}_{O_2} - g^{\circ}_{N_2} = 2(86.6) - 0 - 0 = 173.2\frac{\text{kJ}}{\text{mol}} \quad (12.J)$$

$$K_{298.15} = \exp\left(\frac{-\Delta g^{\circ}}{RT}\right) = \exp\left(\frac{-173{,}200\dfrac{\text{J}}{\text{mol}}}{8.314\dfrac{\text{J}}{\text{mol}\,K}\cdot 298.15\,\text{K}}\right)$$

$$= 4.5 \times 10^{-31} = \frac{[a_{NO}]^2}{[a_{N_2}]\cdot[a_{O_2}]} \quad (12.K)$$

The activities are all at $a_i = f_i/f_i^{\circ}$. The f_i° all correspond to the standard states in Table A.8, which for gases are the ideal gas state at 1 bar, so that $f^{\circ}_{N_2} = f^{\circ}_{O_2} = f^{\circ}_{NO} = 1$ bar. If we make the most general statement of the activities (for gases) we would have

$$a_i = \frac{y_i \hat{\phi}_i P}{f_i^{\circ}} \quad (12.18)$$

At this low a pressure we may safely assume that the NO, O_2, and N_2 behave as ideal gases for which $\hat{\phi}_i = 1.00$ and we may substitute for the activities, finding

$$K_{298.15} = \frac{[a_{NO}]^2}{[a_{N_2}]\cdot[a_{O_2}]} = \frac{\left[\dfrac{y_{NO}P}{1\text{ bar}}\right]^2}{\left[\dfrac{y_{N_2}P}{1\text{ bar}}\right]\cdot\left[\dfrac{y_{O_2}P}{1\text{ bar}}\right]} = \frac{[y_{NO}]^2}{[y_{N_2}]\cdot[y_{O_2}]} \quad (12.19)$$

For ideal gas reactions *in which the number of mols does not change* (like this one) the P and the pressure dimension in the mass action equation always cancel, and we find the expression in terms of mol fractions. We will see that for reactions in which the number of mols changes, this cancellation does not occur, and the formulae are more complex.

We may use this constant to compute the equilibrium concentration of NO to be expected in a sample of air ($\approx$21% oxygen, $\approx$78% nitrogen, $\approx$1% argon) at 298.15 K as

$$[y_{NO}]^2 = K_{298.15}\cdot[y_{N_2}]\cdot[y_{O_2}] = 4.5 \times 10^{-31}\cdot 0.78\cdot 0.21$$
$$= 7.4 \times 10^{-32}$$
$$[y_{NO}] = \sqrt{7.4 \times 10^{-32}} = 2.7 \times 10^{-16} \quad (12.L)$$

which shows that at 25°C = 298.15 K the equilibrium NO concentration is negligible. If we repeat the calculation for 2000 K (see Problem 12.6), we will find $K_{2000} = 4.0 \times 10^{-4}$ and

$$[y_{NO}]^2 = K_{2000}\cdot[y_{N_2}]\cdot[y_{O_2}] = 4.0 \times 10^{-4}\cdot 0.78\cdot 0.21$$
$$= 6.55 \times 10^{-5}$$
$$[y_{NO}] = \sqrt{6.55 \times 10^{-5}} = 8.09 \times 10^{-3} \approx 8100\,\text{ppm} \quad (12.M)$$

See also Problem 12.5. ■

The 8100 ppm is a small number, but this reaction plays a significant role in air pollution, where the NO from high-temperature flames contributes to the formation of photochemical oxidant (mostly ozone), by the reaction, *expressed in simplified form* [5] as

$$NO + \text{hydrocarbons} + O_2 + \text{sunlight} \rightarrow NO_2 + O_3 \quad (12.N)$$

and contributes to acid rain by producing nitric acid in the atmospheric reaction

$$2NO + 1.5O_2 + H_2O \Leftrightarrow 2HNO_3 \quad (12.O)$$

It also plays a major role in the creation and sustenance of living things. Most plants cannot use the abundant N_2 from the atmosphere; they need *fixed nitrogen* in which the extremely strong N–N bond has been broken and replaced by a weaker N–O or N–H bond. The above reaction, occurring in the heated air of lightning strikes provides much of the world's naturally occurring fixed nitrogen. It was also the basis for the several fertilizer processes, which were replaced by the more economical ammonia process described in Section 1.1.

12.6.1 Change of Reactant Concentration, Reaction Coordinate

In Example 12.2 we assumed, without stating the assumption, that the change in mol fraction of N_2 and O_2 was so small that we could ignore it and use the starting values in our equilibrium calculation.

Example 12.3 How large an error did we make in Example 12.2, by ignoring the changes in N_2 and O_2 concentrations at 2000 K?

Here we start with 1.00 mol of air and let the mols of NO formed be $2x$. Then the remaining unreacted mols of N_2 and O_2 at equilibrium will be $(0.78 - x)$ and $(0.21 - x)$ (see Eq. 12.I). Substituting these values in Eq. 12.L we have

$$[2x]^2 = 4.0 \times 10^{-4}\cdot[0.78 - x]\cdot[0.21 - x] \quad (12.P)$$

This quadratic equation has roots $x = 0.0040$ and $x = -0.0041$. The negative root is meaningless. The calculated NO concentration is $2x = 0.0080 = 8000$ ppm, 99% of the value computed in Example 12.2. ■

The source of the Gibbs energy values used to compute $K_{2000\,K}$ in Examples 12.2 and 12.3 [6] claims an accuracy of no more than ±5%, so for this problem, taking the change in reactant concentration into account makes a smaller percentage change in the answer than the uncertainty in the equilibrium constant. However, in principle, this should be a better estimate, and for other reactions in which there are major changes from the original reactant concentration, we must take such changes into account. The most easily understood approach is that shown above, in which we implicitly chose one mol of initially reacting mixture, and allowed x mols of one of the reactants to be formed or to disappear, and then by material balance wrote the concentration of the other species present. Then all the concentrations were substituted in the equilibrium relation, which was solved for x.

A more formal approach defines two new, widely used, terms

$$e = \begin{pmatrix}\text{reaction}\\ \text{cordinate}\end{pmatrix} = \begin{pmatrix}\text{some other}\\ \text{names}\end{pmatrix} = \begin{pmatrix}\text{mols of one selected}\\ \text{reactant consumed or}\\ \text{product produced}\end{pmatrix} \tag{12.20}$$

$$v = \begin{pmatrix}\text{stoichiometric}\\ \text{coefficient}\end{pmatrix} = \begin{pmatrix}\text{mols of one species consumed or}\\ \text{produced per mol of selected species}\\ \text{consumed or produced}\end{pmatrix} \tag{12.21}$$

For Reaction 12.I let us arbitrarily choose oxygen as the selected reactant. Then we can write

$$e = n_{\text{oxygen originally present}} - n_{\text{oxygen}} = n_{O_2,0} - n_{O_2} \tag{12.Q}$$

or

$$n_{O_2} = n_{O_2,0} - e \tag{12.R}$$

and correspondingly

$$n_{N_2} = n_{N_2,0} - e \tag{12.S}$$

$$n_{NO} = n_{NO,0} + 2e \tag{12.T}$$

Looking back to Example 12.3, we see that the x in that solution is the same as e. Next we observe that Eqs. 12.R, 12.S, and 12.T can each be written as

$$n_i = n_{i,0} + v_i e \tag{12.U}$$

where the v_i are the stoichiometric coefficients, −1 for O_2, −1 for N_2, and +2 for NO. This makes the meaning of the stoichiometric coefficient clearer; if Reaction 12.I proceeds by one mol of the chosen reactant (O_2 in this example), then the number of mols that appear are −1 for O_2, −1 for N_2, and +2 for NO. The sign convention is that if a mol appears then the stoichiometric coefficient is positive, and if one disappears it is negative. In many reactions, as in this one, there are several choices for the specified reactant or product on which to base e. Normally we choose the limiting reactant, the one that runs out first, which is O_2 in this case. We should choose a reactant or product whose stoichiometric coefficient is +1 or −1. Other choices are not wrong, but the following mathematics is correct only for that choice, and other choices lead to more complex mathematics.

In this example the number of mols does not change, so the advantage of the formulation in terms of reaction coordinate and stoichiometric coefficients is small. Consider again the reaction for the formation of ammonia:

$$0.5N_2 + 1.5H_2 \Leftrightarrow NH_3 \tag{12.B}$$

Here we choose NH_3 as the specified species, because its stoichiometric coefficient is 1. Then e is the number of mols of NH_3 produced, and the three stoichiometric coefficients are −0.5, −1.5, and +1. If we now write the equation for the mol fraction of N_2 in the mixture at equilibrium, we will have

$$\begin{aligned} y_{N_2} &= \frac{n_{N_2}}{n_{N_2} + n_{H_2} + n_{NH_3}} \\ &= \frac{(n_{N_2,0} - 0.5e)}{(n_{N_2,0} - 0.5e) + (n_{H_2,0} - 1.5e) + (n_{NH_3,0} + e)} \\ &= \frac{(n_{N_2,0} + v_{N_2}e)}{(n_{N_2,0} + v_{N_2}e) + (n_{H_2,0} + v_{H_2}e) + (n_{NH_3,0} + v_{NH_3}e)} \\ &= \frac{(n_{N_2,0} + v_{N_2}e)}{(n_{T,0} + e\sum v_i)} \end{aligned} \tag{12.V}$$

where $n_{T,0}$ is the initial total number of mols, and $\sum v_i = -0.5 - 1.5 + 1 = -1$ (for every mol of NH_3 produced, the total number of mols decreases by 1.00). If we assume, for example, that the initial feed was 0.5 mols of N_2

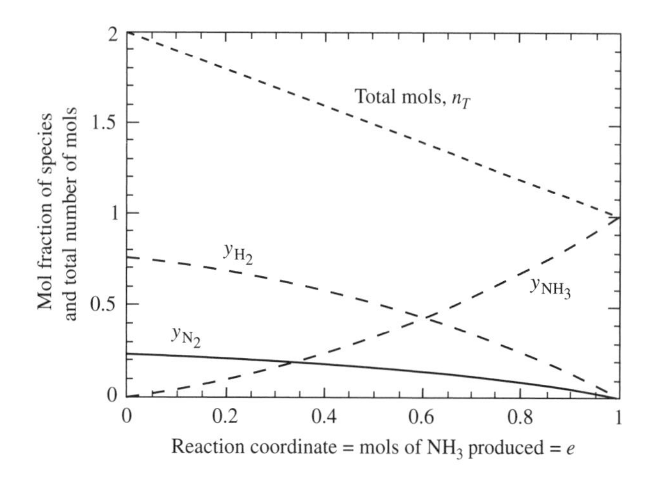

FIGURE 12.4 Changes in mol fractions and total number of mols with increase in the reaction coordinate.

and 1.5 mols of H_2, with no NH_3 ($n_{NH_3,0} = 0, n_{T,0} = 2$), then, at any value of e we will have

$$y_{N_2} = \frac{(n_{N_2,0} + \nu_{N_2} e)}{(n_{T,0} + e\sum \nu_i)} = \frac{0.5 - 0.5e}{2 - e} \tag{12.W}$$

$$y_{H_2} = \frac{(n_{H_2,0} + \nu_{H_2} e)}{(n_{T,0} + e\sum \nu_i)} = \frac{0.5 - 1.5e}{2 - e} \tag{12.X}$$

$$y_{NH_3} = \frac{(n_{NH_3,0} + \nu_{NH_3} e)}{(n_{T,0} + e\sum \nu_i)} = \frac{0 + e}{2 - e} \tag{12.Y}$$

Figure 12.4 shows how these three mol fractions and the total number of mols change with increasing reaction. Observe that we went from a problem with three unknowns, y_{NH_3}, y_{N_2} and y_{H_2} to one with only one unknown, e. That is a vast computational improvement.

This is more mathematics than this situation demands, but for reactions in which the number of mols changes, ($\sum \nu_i \neq 0$), and for multiple reactions the simple, intuitive approaches may lead us astray and this formal approach always works. Furthermore, computer process-design programs formulate their reactor modules in terms of reaction coordinate and stoichiometric coefficients, so we must understand this terminology to use those programs.

Example 12.4 Ethanol is made commercially by the gas-phase hydration of ethylene:

$$C_2H_4 + H_2O \Leftrightarrow C_2H_5OH \tag{12.Z}$$

For this reaction, write the "mass action law" statement on the assumption that the initial reactor feed has no ethanol and 0.833 mols of water per mol of ethylene ($n_{water,0} = 0.833$, $n_{ethylene,0} = 1$, $n_{ethanol,0} = 0$, $n_{T,0} = 1.833$).

In general, we must have

$$\frac{[a_{C_2H_5OH}]}{[a_{C_2H_4}] \cdot [a_{H_2O}]} = K \tag{12.AA}$$

Here again we substitute Eq. 12.18 and assume that we have an ideal solution of ideal gases for which $\hat{\phi}_i = 1.00$ and that for each reactant or product $f_i^o = 1$ bar so that

$$\frac{[a_{C_2H_5OH}]}{[a_{C_2H_4}] \cdot [a_{H_2O}]} = K = \frac{\left[\dfrac{x_{C_2H_5OH} P}{1 \text{ bar}}\right]}{\left[\dfrac{x_{C_2H_4} P}{1 \text{ bar}}\right] \cdot \left[\dfrac{x_{H_2O} P}{1 \text{ bar}}\right]} = \frac{[x_{C_2H_5OH}]}{[x_{C_2H_4}] \cdot [x_{H_2O}]} \cdot \frac{1 \text{ bar}}{P} \tag{12.AB}$$

Here the stoichiometric coefficients are -1, -1, and $+1$, so that $\sum_{\nu_i} = -1$ and

$$\frac{\left[\dfrac{0 + e}{1.833 - e}\right]}{\left[\dfrac{1 - e}{1.833 - e}\right] \cdot \left[\dfrac{0.833 - e}{1.833 - e}\right]} = K \cdot \frac{P}{1 \text{ bar}} \quad ■ \tag{12.AC}$$

Industrially, the reactor feeds are thousands of mols per hour. If we multiply the 0, 1, 0.833, 1.833, and e in this equation by any number, we find this ratio unchanged. The common approach of choosing one mol of one reactant as the basis for equilibrium calculations gives the right equilibrium mol fractions, independent of the actual flow rates of reactants (at the same ratios, one to another!).

Example 12.5 For ideal gases at 25°C = 298 K, the calculated equilibrium constant (based on Table A.8) for Reaction 12.Z is $K = 29.6$. If water, ethylene, and ethanol are in equilibrium at 1 bar and 298 K with the same feed ratios as in Example 12.4, what are the concentrations of reactants and products?

Equation 12.AB, with $K = 29.6$ and (P/1 bar) = 1.00 could be solved analytically as a quadratic equation, but most of us can solve it faster (and more reliably) numerically, using our computers, finding $e = 0.732$ (and a meaningless root, $e = 1.101$). Then

$$y_{ethanol} = \left[\frac{0 + e}{1.833 - e}\right] = \left[\frac{0 + 0.732}{1.833 - 0.732}\right] = 0.664 \tag{12.AD}$$

and similarly, $y_{ethylene} = 0.244$ and $y_{water} = 0.092$. ■

This reaction does not proceed at commercially useful rates at these low temperatures. Commercially, it is carried out over a catalyst at ≈265°C and 70 atm [7] (see Problem 12.26).

12.6.2 Reversible and Irreversible Reactions

In principle, all chemical reactions are reversible; if we wait long enough they come to some equilibrium state at which some amount of reactants are in equilibrium with some amount of products. However, as the following example shows, in practice, some reactions seem irreversible, with practically complete consumption of the reactants.

Example 12.6 Compute the equilibrium concentration of hydrogen and oxygen to be expected when hydrogen and oxygen react to form water, at 25°C and 1 bar by

$$H_2 + 0.5O_2 \Leftrightarrow H_2O \tag{12.AE}$$

Using values of the Gibbs energies of formation in Table A.8 we have

$$\Delta g^o = g^o_{H_2O} - 0.5 g^o_{O_2} - g^o_{H_2} = (-237.1) - 0.5 \cdot 0 - 0 = -237.1 \frac{\text{kJ}}{\text{mol}}$$

and

$$K = \exp\left(\frac{-\Delta g^\circ}{RT}\right) = \exp\left[\frac{-\left(-237,100 \dfrac{\text{J}}{\text{mol}}\right)}{8.314 \dfrac{\text{J}}{\text{mol } K} \cdot 298.15\ \text{K}}\right]$$

$$= 3.5 \times 10^{41} = \frac{[a_{H_2O}]}{[a_{H_2}] \cdot [a_{O_2}]^{0.5}} \tag{12.A.G}$$

Substituting for the activities from Eq. 12.18, and simplifying we find

$$K = 3.8 \times 10^{41} = \frac{[y_{H_2O}]}{[y_{H_2}] \cdot [y_{O_2}]^{0.5}} \cdot \left(\frac{1\ \text{bar}}{P}\right)^{0.5} \tag{12.AH}$$

Choosing oxygen as the selected reactant, and assuming that we begin with 0.5 mols of oxygen and 1.0 mol of hydrogen, we have stoichiometric coefficients of −1, −0.5, and +1, $n_{T,0} = 1.5$, and $\sum v_i = -0.5$. Thus,

$$K = 3.8 \times 10^{41} = \frac{\left[\dfrac{e}{1.5 - 0.5e}\right]}{\left[\dfrac{1 - e}{1.5 - 0.5e}\right] \cdot \left[\dfrac{0.5 - 0.5e}{1.5 - 0.5e}\right]^{0.5}} \tag{12.AI}$$

The numerical solution of Eq. 12.AI is $e \approx (1 - 2.4 \times 10^{-28})$, so the final mol fraction of H_2 is 2.4×10^{-28} and that of oxygen 0.5 times that value (see Problem 12.13). ■

The calculated equilibrium concentration of unreacted hydrogen is far below the limit of detection by any known analytical method, so we are safe in calling it zero. If a reaction consumes all the starting product, we consider it an irreversible reaction. This reaction, if carried to thermodynamic equilibrium would be such a reaction. Many combustion reactions have such high values of $(-\Delta g^o)$ and thus such high values of K that at equilibrium they would be nearly complete and be considered irreversible. Real combustion reactions come close to equilibrium, but are limited by kinetics. So, in principle, there are no irreversible reactions, reactions in which *none* of the reactants remain at equilibrium. But, in practice, many reactions, mostly combustion reactions, have such low equilibrium concentrations of the original reactants that they appear to be irreversible.

Figure 12.5 is the equivalent of Figure 12.1, but for a practically irreversible reaction (Example 12.6). In Figure 12.1 the Δg° of the reaction was small, and the minimum in the curve fell at $x \approx 0.82$, well away from either axis. With the Δg° of this combustion reaction, Figure 12.5 shows that the equivalent of the curved g line is practically straight, except very close to the 1.00 axis, where there exists a minimum, corresponding to the calculated equilibrium state.

12.7 MORE ON STANDARD STATES

In some other life, we will all agree on standard states and on definitions of symbols, and avoid the endless confusion on these topics. In this life we will have to deal with widely varying definitions. Table 12.1 shows the most common definitions of the standards states.

Unfortunately, there is some disagreement among various published tables of Gibbs energies. Table 12.2 compares the published values of the Gibbs energy of formation from the elements, copied from three highly respected sources, for isobutane, *n*-butane, and their computed difference. (Note that these values are in kcal, while the values in Table A.8 are in kJ; both sets of units are widely used.) The differences are small, but troublesome. If the problem is important enough, we can trace the values to their original sources, to see which is most credible; most often that much time and effort is not justified (see Problem 12.9).

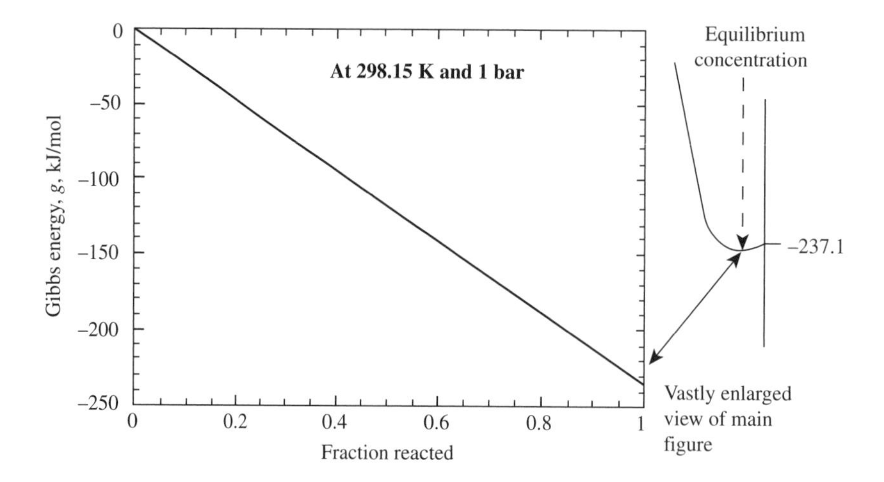

FIGURE 12.5 g-Fraction reacted diagram for a reaction with a very large, negative, Δg°, which makes the reaction practically irreversible, in this case the combustion of hydrogen (see Example 12.6). The main plot is practically but not exactly a straight line, but the vast enlargement at the right shows that in the region very close to 100% reaction, there is a minimum, corresponding to the equilibrium concentration. Compare this figure to Figure 12.1, which is of the same type, but for a small value of Δg^o of the reaction.

Some tables show values for both gas and liquid for substances that regularly occur as gas and liquid (e.g., water). If the NPB is higher than 25°C, then the gas is the hypothetical ideal gas at 1 atm and 25°. Water cannot exist as a pure gas at that T and P, but we can start with its Gibbs energy as a liquid and find the Gibbs energy it would have at that temperature and pressure if it were an ideal gas.

Example 12.7 Estimate the change in g_i° for water, going from liquid at 298.15 K and 1 bar to the hypothetical ideal gas

Table 12.1 Most Commonly Used Standard State Definitions

State	State at Which g_i^o is Measured, Normally in kcal/mol or kJ/mol	Activity $a_i = f_i/f_i^\text{o}$
g (gaseous)	Pure gas at that pressure at which the fugacity is 1 atm or as a ideal gas at 1 atm or either of these at 1 bar	$\frac{f_i}{f_i^\circ} = \frac{y_i\hat{\phi}_i P}{1 \text{ atm or } 1 \text{ bar}}$
1 (liquid)	Pure liquid at a pressure of 1 atm or 1 bar	$\frac{f_i}{f_i^\text{o}} = \frac{x_i \gamma_i p_i}{p_i \text{ at } 25°\text{C} \cdot \text{PF}}$ (PF = Poynting factor ≈ 1.00)
c (crystalline) (used for most solids) or s (solid)	Pure solid at a pressure of 1 atm or 1 bar	$\frac{f_i}{f_i^\circ}$ = same as liquids if $p_{i,\text{solid}}$ is measurable, otherwise $\frac{f_i}{f_i^\circ} = x_i$ or 1.00 for pure solid
aq (aqueous)	The hypothetical aqueous solution with concentration of 1.00 molal of the substance. This applies both to dissolved substances and to ions, see Chapter 13.	$\frac{f_i}{f_i^\circ} = \frac{m_i \gamma_i}{1 \text{ molal}}$
amorph (solids in a noncrystalline state)	Pure solid at a pressure of 1 atm or 1 bar	Same as for crystalline solids

Note. Most sources give these values at 25°C = 298.15 K ≈ 298 K [1, 3]. Same give them at exactly 298.00 K. Some use 1 atm pressure, some use 1 bar (≈0.987 atm). Some show the Gibbs energy change of formation from the elements as Δg_i^o, but the Δ here leads to confusion with the $\Delta g^\text{o}_\text{reaction}$.

Table 12.2 Published Values of the Gibbs Energies of Formation of *n*-Bntane and Isobutane at 25°C = 298.15 K from Three Highly Respected Sources

Source	g^{o} Isobutane (kcal/mol)	$g^{o}n$-Butane (kcal/mol)	$\Delta g^{o} = g^{o}_{iso} - g^{o}_{normal}$ (kcal/mol)
Reid R. C. et al., *The Properties of Liquids and Gases*, ed. 3, p. 643	−4.99	−4.10	−0.89
Lide, D. R. *CRC Handbook of Chemistry and Physics*, ed. 71, pp. 5–74	−5.00	−4.02	−0.98
Liley, P. E. *Perry's Chemical Engineer's Handbook*, ed. 7, pp. 3–148	−4.296	−3.754	−0.542

as the same T and P. Compare this change to the values from Table A.8.

We calculate the Gibbs energy change from liquid to hypothetical gas in three steps:

1. The liquid is reduced in pressure from the standard pressure of 1 bar to its vapor pressure at 298.15 K.
2. The liquid is vaporized at that pressure, for which $\Delta g = 0$ because this is an equilibrium vaporization (see Chapter 4).
3. The vapor is replaced by an ideal gas, which will not condense, and compressed from the vapor pressure at 298.15 K to 1 bar.

Using the equations from Chapter 7, we find

$$\Delta g^{o} = g^{o}_{\text{ideal gas}} - g^{o}_{\text{liquid}} = \int_{1\text{ bar}}^{p_{\text{vapor at }25°\text{C}}} v_{\text{liquid}}\, dP + 0 + \int_{p_{\text{vapor at }25°\text{C}}}^{1\text{ bar}} v_{\text{ideal gas}}\, dP \tag{12.22}$$

Treating the liquid specific volume as a constant, and replacing the gas specific volume by the ideal gas law and integrating, we have

$$\Delta g^{o} = v_{\text{liquid}} \cdot \left(p_{\text{vapor at }25°\text{C}} - 1\text{ bar}\right) + 0 + RT\ \ln \frac{1\text{ bar}}{p_{\text{vapor at }25°\text{C}}} \tag{12.23}$$

$$\Delta g^{o} = 1.805 \times 10^{-5}\frac{\text{m}^3}{\text{mol}} \cdot (0.0317\text{ bar} - 1\text{ bar}) \cdot \frac{10^5\text{J}}{\text{m}^3 \cdot \text{bar}} + 8.314\frac{\text{J}}{\text{mol }K} \cdot 298.15\text{ K} \cdot \ln\frac{1\text{ bar}}{0.0317\text{ bar}}$$
$$= -1.75 + 8556\frac{\text{J}}{\text{mol}} = 8.55\frac{\text{kJ}}{\text{mol}} \tag{12.AJ}$$

From Table A.8 we find

$$\Delta g^{o} = -228.6 - (-237.1) = 8.5\frac{\text{kJ}}{\text{mol}} \quad \blacksquare \tag{12.AK}$$

In Eq. 12.AJ the −1.75 J/mol is the Poynting factor, which is normally small enough to be ignored here (but not in Chapter 14). Students will rarely need to make this kind of calculation, but they need to understand it, because some tables of the type of Table A.8 show the Gibbs energy of formation of all substances, even solids, as that of the hypothetical ideal gas. That appears odd, but has the merit that we can insert such a table into a general-purpose computer program and read values for any compound from it, without any uncertainty as to what standard state is represented. It might appear that the calculated equilibrium compositions would depend on which of the standard states (g) or (l) we choose, but it does not. Changing from one to the other changes the value of K, but, as Table 12.1 shows, it also changes the value of f_i° (see Problem 12.17).

While most tables like A.8 are at 25 °C = 298.15 K, the widely used JANAF tables for combustion calculations [6] give Gibbs energy of formation values from the elements at a wide range of temperatures, in effect defining a new datum state for each temperature. This works perfectly well if we understand what the datum state is (see Problem 12.6).

12.8 THE EFFECT OF TEMPERATURE ON CHEMICAL REACTION EQUILIBRIUM

As discussed in Chapter 7, the values of g_i^{o} are functions of temperature alone. Since the equilibrium constants depend on these values, we would expect the value of the equilibrium constant to depend on temperature, and it does. If we take the ln of both sides of Eq. 12.14 and differentiate with respect to T, we find

$$\frac{d\ \ln K}{dT} = \frac{d}{dT}\left(\frac{-\Delta g^{o}}{RT}\right) = \frac{-1}{R}\frac{d}{dT}\left(\frac{\Delta h^{o}}{T} - \Delta s^{o}\right) = \frac{-1}{R}\left(\frac{-\Delta h^{o}}{T^2} + \frac{1}{T}\frac{d\Delta h^{o}}{dT} - \frac{d\Delta s^{o}}{dT}\right) \tag{12.24}$$

but the two rightmost terms are equal (because of the relation between enthalpy change and entropy change) so they cancel, and we find the *van't Hoff* equation:

$$\frac{d\ \ln K}{dT} = \frac{\Delta h^{o}}{RT^2} \tag{12.25}$$

The Δh^{o} here, like the Δg^{o} we have been using, is that for the reaction going 100% to completion, using the h_i^{o} (enthalpy of formation from the elements) from Table A. 8.

If Δh^o is independent, or practically independent of temperature, then we can integrate Eq. 12.25, finding

$$\int_{T_1}^{T_2} d\ln K = \ln\frac{K_{T_2}}{K_{T_1}} = \frac{\Delta h^o}{R}\int_{T_1}^{T_2}\frac{dT}{T^2} = \frac{\Delta h^o}{R}\left(\frac{1}{T_1}-\frac{1}{T_2}\right) \quad (12.26)$$

which can be rewritten as

$$\ln K_{T_2} = \ln K_{T_1} + \frac{\Delta h^o}{R}\cdot\frac{1}{T_1} - \frac{\Delta h^o}{R}\cdot\frac{1}{T_2} \quad (12.AL)$$

which we often see as

$$\ln K = A + \frac{B}{T} \quad (12.27)$$

which is the same as Eq. 12.AL if

$$A = \ln K_{T_1} + \frac{\Delta h^o}{RT_1} \text{ and } B = -\frac{\Delta h^o}{R} \quad (12.AM)$$

and T equals the T_2 in Eq. 12.AL. Equation 12.27 is the normal formula for specifying $K=f(T)$ in process-design computer programs; it is widely used and widely seen in the literature.

However, it is uncommon for Δh^o to be completely independent of temperature, so that while Eq. 12.27 is useful for quick estimates, for the most accurate work we keep Δh^o inside the integral, and write

$$\ln\frac{K_{T_2}}{K_{T_1}} = \frac{1}{R}\int\frac{\Delta h^o}{T^2}dT \quad (12.28)$$

The relation between Δh^o and temperature is sketched in Figure 12.6. From this figure we see that the enthalpy change of the reaction at T_1 is

$$\Delta h^o_{\text{at } T_1} = h^o_B - h^o_A \quad (12.AN)$$

where the A and B subscripts refer to two corners of the figure. Similarly, at T_2

$$\Delta h^o_{\text{at } T_2} = h^o_D - h^o_C \quad (12.AO)$$

By simple energy relations we have

$$h^o_C - h^o_A = \int_{T_1}^{T_2} C_{P,\text{ reactants}}\, dT \quad (12.AP)$$

and

$$h^o_D - h^o_B = \int_{T_1}^{T_2} C_{P,\text{ products}}\, dT \quad (12.AQ)$$

Summing these four equations we find

$$\begin{aligned}\Delta h^o_{\text{at } T_2} &= \Delta h^o_{\text{at } T_1} + \int_{T_1}^{T_2} C_{P,\text{ products}}\, dT - \int_{T_1}^{T_2} C_{P,\text{ reactants}}\, dT \\ &= \Delta h^o_{\text{at } T_1} + \int_{T_1}^{T_2}(C_{P,\text{ products}} - C_{P,\text{ reactants}})dT\end{aligned} \quad (12.AR)$$

For most common chemicals the experimental values of C_P have been satisfactorily represented by simple polynomial data-fitting equations of the form

$$C_P = a + bT + cT^2 + dT^3 \quad (12.AS)$$

or

$$\frac{C_P}{R} = a + bT + cT^2 + \frac{d}{T^2} \quad (12.AT)$$

Equations 12.AS and 12.AT have no theoretical basis; they are simple data-fitting equations. Such equations reproduce the

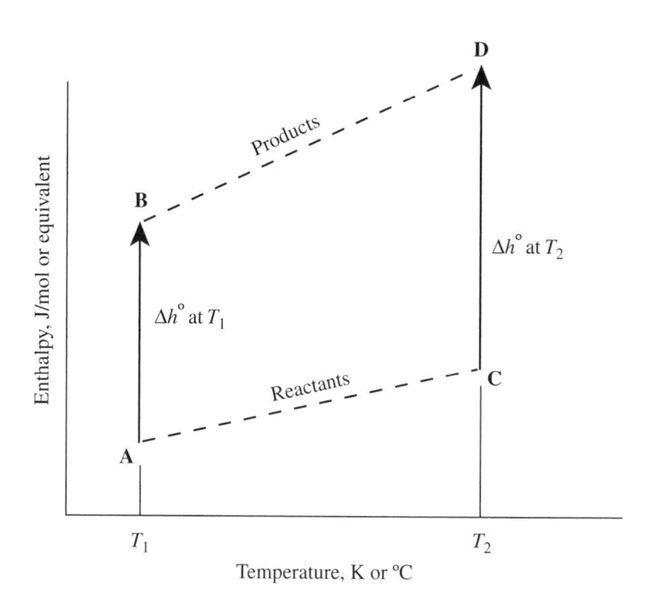

FIGURE 12.6 Relation between enthalpy and temperature for a chemical reaction. Here the reaction (from A to B at T_1 or from C to D at T_2) is endothermic; the enthalpy of the products is greater than that of the reactants. For an exothermic reaction the enthalpy at B and D would be less than that at A and C, and the arrows would point down instead of up. The slopes of the two dotted lines are the C_Ps of the reactants or of the products. For the slopes shown the enthalpy change of the reaction is greater at T_2 than at T_1. If the slopes were equal, then the enthalpy change of the reaction would be independent of temperature. The straight dashed lines correspond to constant C_Ps, independent of T. If C_P varies with T, then they would be curved.

experimental C_P measurements with accuracy satisfactory for most calculations. In using tables of the constants in such equations we must be careful to observe whether the T in these equations is in K or °C; it makes a big difference. Table A.9 shows the values of the constants for some common substances, in the form of Eq. 12. AT for temperatures expressed in K. (Equation 12.AS is normally written as shown, with, for example, the first constant a having the same dimensions as C_P, which are the same as those of R (J/mol K) or equivalent. Equation 12.AT is normally presented as the ratio C_P/R, which is dimensionless; in Eq. 12.AT the first constant a is dimensionless. We may switch from one form to the other by simply multiplying or dividing both sides of the equation by R.)

We use Eq. 12.AT for all the following examples. In Table A.9 all of the entries have only 3 constants, setting either c or $d=0$. The tables in [1, 2] use more complex equations with 5 constants; these are presumably slightly more accurate than the three-constant equations in Table A.9.

Example 12.8 Estimate the value of the equilibrium constant K for the formation of ammonia from hydrogen and nitrogen using Reaction 12.B, at 25°C = 298.15 K and at 400°C = 673.15 K.

Using the values from Table A.8 we may compute

$$\Delta g^{\text{o}}_{298\text{ K}} = g^{\text{o}}_{\text{NH}_3} - 0.5g^{\text{o}}_{\text{N}_2} - 1.5g^{\text{o}}_{\text{H}_2} = -16.5\frac{\text{kJ}}{\text{mol}} - 0.5\cdot 0 - 1.5\cdot 0 = -16.5\frac{\text{kJ}}{\text{mol}} \tag{12.AU}$$

and

$$K_{298\text{ K}} = \exp\left(\frac{-\Delta g^{\text{o}}}{RT}\right) = \exp\left[\frac{-\left(-16,500\dfrac{\text{J}}{\text{mol}}\right)}{8.314\dfrac{\text{J}}{\text{mol }K}\cdot 298.15\text{ K}}\right] = \exp 6.656 = 778 \tag{12.AV}$$

Then to find the value at any other temperature we must use Eq. 12.28. From Table A.8 we compute

$$\Delta h^{\text{o}}_{298\text{ K}} = h^{\text{o}}_{\text{NH}_3} - 0.5h^{\text{o}}_{\text{N}_2} - 1.5h^{\text{o}}_{\text{H}_2} = -46.1\frac{\text{kJ}}{\text{mol}} - 0.5\cdot 0 - 1.5\cdot 0 = -46.1\frac{\text{kJ}}{\text{mol}} \tag{12.AW}$$

and using the constants from Table A.9 to evaluate $\Delta h^{\text{o}}_{\text{at some temperature other than 298.15 K}}$ by Eq. 12.AT, we compute

$$\left(C_{P,\text{products}} - C_{P,\text{reactants}}\right) = R\left(\Delta a + \Delta bT + cT^2 + \frac{\Delta d}{T^2}\right) \tag{12.AX}$$

where

$$\begin{aligned}
\Delta a &= (a_{\text{NH}_3} - 0.5a_{\text{N}_2} - 1.5a_{\text{H}_2}) \\
\Delta b &= (b_{\text{NH}_3} - 0.5b_{\text{N}_2} - 1.5b_{\text{H}_2}) \\
\Delta c &= (c_{\text{NH}_3} - 0.5c_{\text{N}_2} - 1.5c_{\text{H}_2}) \\
\Delta d &= (d_{\text{NH}_3} - 0.5d_{\text{N}_2} - 1.5d_{\text{H}_2})
\end{aligned} \tag{12.AY}$$

or, in general,

$$\Delta a \text{ or } b,c,d = (v_k a_k - v_i a_i - v_j a_j), \text{etc.} \tag{12.AZ}$$

where the v_i are the stoichiometric coefficients of the reaction.

Combining Eqs. 12.AR, 12.AX, and 12.AY, we find

$$\begin{aligned}
\Delta h^{\text{o}}_{\text{at }T_2} &= \Delta h^{\text{o}}_{\text{at }T_1} + R\int_{T_1}^{T_2}\left(\Delta a + \Delta bT + \Delta cT^2 + \frac{\Delta d}{T^2}\right)dT \\
&= \Delta h^{\text{o}}_{\text{at }T_1} + R\left(\Delta aT + \frac{\Delta bT^2}{2} + \frac{\Delta cT^3}{3} - \frac{\Delta d}{T}\right)\Bigg|_{T_1}^{T_2}
\end{aligned} \tag{12.29}$$

Before we substitute this into Eq. 12.28 it is customary to insert the upper and lower limits of this integration, and to group the terms as follows

$$\begin{aligned}
\Delta h^{\text{o}}_{\text{at }T_2} = \Delta h^{\text{o}}_{\text{at }T_1} &+ R\left(\Delta aT_2 + \frac{\Delta bT_2^2}{2} + \frac{\Delta cT_2^3}{3} - \frac{\Delta d}{T_2}\right) \\
&- R\left(\Delta aT_1 + \frac{\Delta bT_1^2}{2} + \frac{\Delta cT_1^3}{3} - \frac{\Delta d}{T_1}\right)
\end{aligned} \tag{12.BA}$$

We observe that the terms involving T_2 are variable, but that those involving T_1 are not, because for the integration of Eq. 12.28 we almost always take T_1 as the temperature of the Gibbs energy tables, normally (but not always) 298.15 K. Thus, we rewrite this as

$$\Delta h^{\text{o}}_{\text{at }T_2} = \Delta h^{\text{o}}_{\text{at }T_1} - I + R\left(\Delta aT_2 + \frac{\Delta bT_2^2}{2} + \frac{\Delta cT_2^3}{3} - \frac{\Delta d}{T_2}\right) \tag{12.BB}$$

where

$$I = R\left(\Delta aT_1 + \frac{\Delta bT_1^2}{2} + \frac{\Delta cT_1^3}{3} - \frac{\Delta d}{T_1}\right) \tag{12.BC}$$

Then we substitute Eq. 12.BB in Eq. 12.28, finding

$$\ln\frac{K_{T_2}}{K_{T_1}} = \frac{1}{R}\int \frac{\Delta h^{\circ}_{\text{at }T_1} - I + R\left(\Delta aT + \frac{\Delta bT^2}{2} + \frac{\Delta cT^3}{3} - \frac{\Delta d}{T}\right)}{T^2} dT$$

$$= \int_{T_1}^{T_2}\left(\frac{\Delta h^{\circ}_{\text{at }T_1} - I}{RT^2} + \frac{\Delta a}{T} + \frac{\Delta b}{2} + \frac{\Delta cT}{3} - \frac{\Delta d}{T^3}\right) dT$$

$$= \left(-\frac{(\Delta h^{\circ}_{\text{at }T_1} - I)}{RT} + \Delta a \ln T + \frac{\Delta bT}{2} + \frac{\Delta cT^2}{6} + \frac{\Delta d}{2T^2}\right)_{T_1}^{T_2} \quad (12.BD)$$

(Here we dropped the subscripts on T_2 because it is in the upper limit of the integration.) Remember that this is a combination of the rigorous van't Hoff equation (Eq. 12.25) with the approximate data fitting equation, Eq. 12.AT, taken three times.

We next look up the appropriate values from Table A.9 and make up Table 12.3. With these values we can evaluate I:

$$I = 8.314\frac{\text{J}}{\text{mol K}}\left[(-2.9355)\cdot 298.15\text{ K} + \frac{2.0905\times 10^{-3}\times 298.15^2}{2}\text{K} + \frac{0\times 298.15^3}{2}\text{K} - \frac{(-0.3305\times 10^5)}{298.15}\text{K}\right]$$

$$= -5582\frac{\text{J}}{\text{mol}} \quad (12.BE)$$

and then

$$\ln\frac{K_{673.15\text{ K}}}{K_{298.15\text{ K}}} = \left[-\frac{-46,100 - (-5582)\frac{\text{J}}{\text{mol}}}{8.314\frac{\text{J}}{\text{mol K}}T} + (-2.967)\ln T + \frac{2.0905\times 10^{-3}T}{2} + 0 + \frac{(-0.3305\times 10^5)}{2T^2}\right]_{298.15\text{ K}}^{673.15\text{ K}}$$

$$= -11.209 - (-0.260) = -10.95 \quad (12.BF)$$

$$K_{673.15\text{ K}} = K_{298.15\text{ K}}\exp(-10.95) = 778\cdot 1.76\times 10^{-5} = 0.0137 \quad ■ \quad (12.BG)$$

Table 12.3 Values from Table A.9 for This Example

	a	$10^3 b$ (1/K)	$10^6 c$ (1/K^2)	$10^{-5} d$(K^2)
NH_3	3.578	3.020	0	−0.186
H_2	3.249	0.422	0	0.083
N_2	3.280	0.593	0	0.040
$\Delta = (NH_3 - 0.5N_2 - 1.5H_2)$	−2.935	2.0905	0	−0.3305

This calculation is tedious, but once it is programmed into a computer, we can easily run out the values for any temperature. That has been done for a variety of reactions, with the results presented in Figure 12.7. The reader may verify that, within chart reading accuracy, it shows the values of K computed in the above example at 25 and 400°C. If we have all the necessary data (which we do for all reactions involving all common chemicals) and have the time to program them (or have a preprogrammed computer package) then this is the most reliable method of calculating the effect of temperature on equilibrium constants. The procedure is rigorous and the results are as reliable as are the input data (and the accuracy of the empirical data-fitting heat capacity equations).

We can gain some insight into the previous example and into Figure 12.7 by reconsidering Eq. 12.25. The curvature on Figure 12.7 is due to the change in $\Delta h^{\circ}/T^2$ with temperature. If the effects of temperature on Δh° are zero or negligible, then the integration of Eq. 12.25 becomes simple, with Eq. 12.26 as the result. For this assumption, a plot of ln K vs. (1/T) should form a straight line. Figure 12.8 shows a plot like Figure 12.7 for a variety of reactions on ln K vs. 1/T coordinates. The resulting curves are much closer to straight than those in Figure 12.7, but most show some modest curvature, indicating that Δh° does change modestly with temperature for most chemical reactions. But this plot shows that the curvature of the lines in Figure 12.7 is mostly due to the T^2 in the denominator of the integrand, and much less caused by the change in Δh° with temperature. Figure 12.8 does not show the ammonia reaction, but (see Problem 12.19) we can show that if we estimate $K_{673.15\text{ K}}$ from Eq. 12.26, starting from $K_{298.15\text{ K}}$ the resulting estimate is about 2.1 times the value calculated above; for some problems this would be a satisfactory estimate.

As Figures 12.7 and 12.8 make clear, for some reactions, increasing T increases K while for others it reduces K. Equation 12.25 shows that which of these occurs depends on the sign of the heat of the reaction Δh°. If this is positive (an endothermic reaction), then increasing T increases K; if it is negative (exothermic reaction), then increasing T reduces K. The larger the absolute value of Δh°, the larger the rate of increase or decrease of K with a change in T. If there were a reaction with $\Delta h^{\circ} = 0$, its value of K would be independent of T. This can be seen in terms of Le Chatelier's principle that nature opposes what we want to do. If we heat

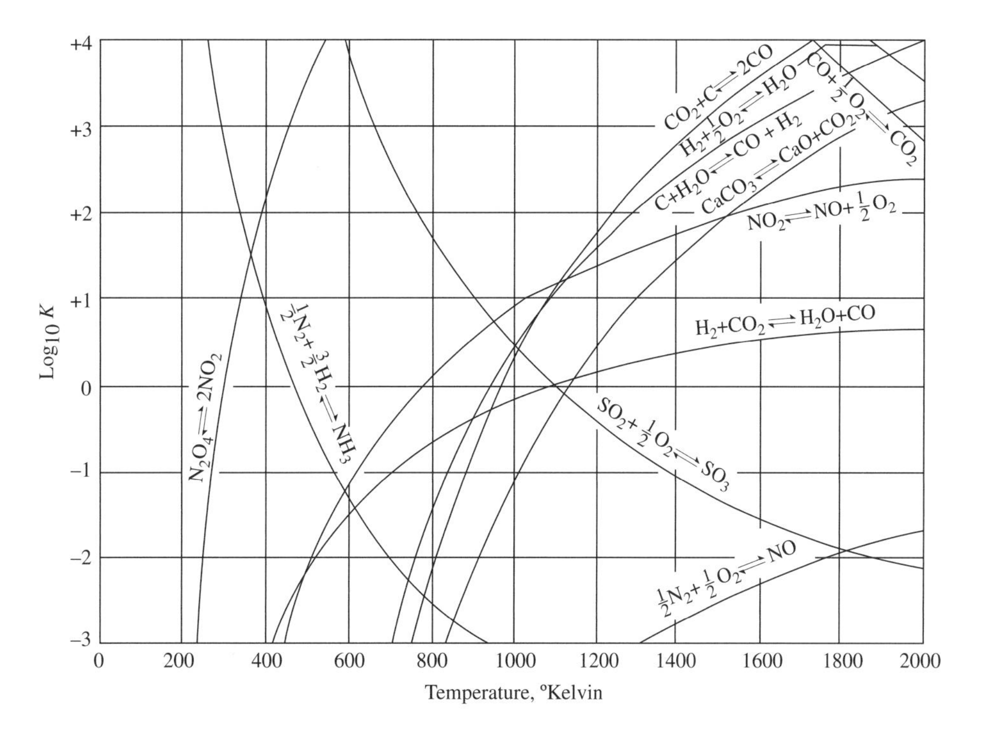

FIGURE 12.7 Calculated values of the equilibrium constants K for a variety of reactions. (From Hougen, O. A., K. M. Watson, and R. A. Ragatz, *Chemical Process Principles*, Part II: *Thermodynamics*, ed. 2. © 1959, New York: Wiley, p. 1003. Reprinted by permission of the estate of O. A. Hougen.)

any reaction, then the equilibrium will move the system in the heat-absorbing direction making it harder to raise the system temperature, and conversely a heat-producing reaction will move in the heat-releasing direction, making it harder to lower the system temperature.

Using the values of K from Examples 12.8 and Table A.9, we may estimate the effect of changes in temperature on the composition at equilibrium.

Example 12.9 Estimate the equilibrium conversion of a mixture of 1.5 mol H_2 and 0.5 mol of N_2 to NH_3 at 1 bar pressure at temperatures of 298.15 K and 400°C = 673.15 K.

We start with Eq. 12.14. For a pressure of 1 bar, with the assumption of ideal solution of ideal gases and standard state fugacities of 1 bar,

$$a_i = \left[\frac{f_i}{f_i^\circ}\right] \approx \left[\frac{Py_i}{1\ \text{bar}}\right] \approx y_i \tag{12.BH}$$

Then we substitute Eqs. 12.W through 12.Y in the equilibrium relation, finding

$$\frac{[y_{NH_3}]}{[y_{N_2}]^{0.5}\cdot[y_{H_2}]^{1.5}} = K = \frac{\dfrac{0+e}{2-e}}{\left[\dfrac{0.5-0.5\cdot e}{2-e}\right]^{0.5}\cdot\left[\dfrac{1.5-1.5\cdot e}{2-e}\right]^{1.5}} \tag{12.BI}$$

From Example 12.8 we know that $K_{25°C=298.15\ K} = 778$ and $K_{400°C=673.15\ K} = 0.0137$. The numerical solutions to Eq. 12.BI for these two values of K are $e_{298.15} = 0.97$ and $e_{673.15} = 0.0088$. The corresponding mol fractions of NH_3 in the equilibrium mixture (see Eq. 12.Y) are 94 and 0.44%. ■

From these values it is clear that if we could reach equilibrium at 1 atm and room temperature, we would have almost complete conversion to NH_3; the equilibrium is very favorable for the reaction. Alas, the reaction rate at this temperature is ≈0, and no catalyst is known that will make the reaction go at temperatures below about 300°C. (Fame and fortune await the student who can find one!) Most industrial ammonia synthesis reactors operate in the temperature range 350–520°C, at which the rates of the reaction are satisfactorily rapid (in the presence of a catalyst). At this temperature and 1 atm the calculated equilibrium concentration of ammonia is small enough to make the reaction completely impractical. Industrially, the reaction is conducted at high pressures, as discussed below.

This is an exothermic (heat-producing) reaction (Δh° is negative). From Eq. 12.25 it is clear that for any exothermic reaction increasing the temperature lowers the value of K. Many industrially significant chemical reactions are exothermic, so this means that much of the time, when we raise the temperature to make the reaction go at

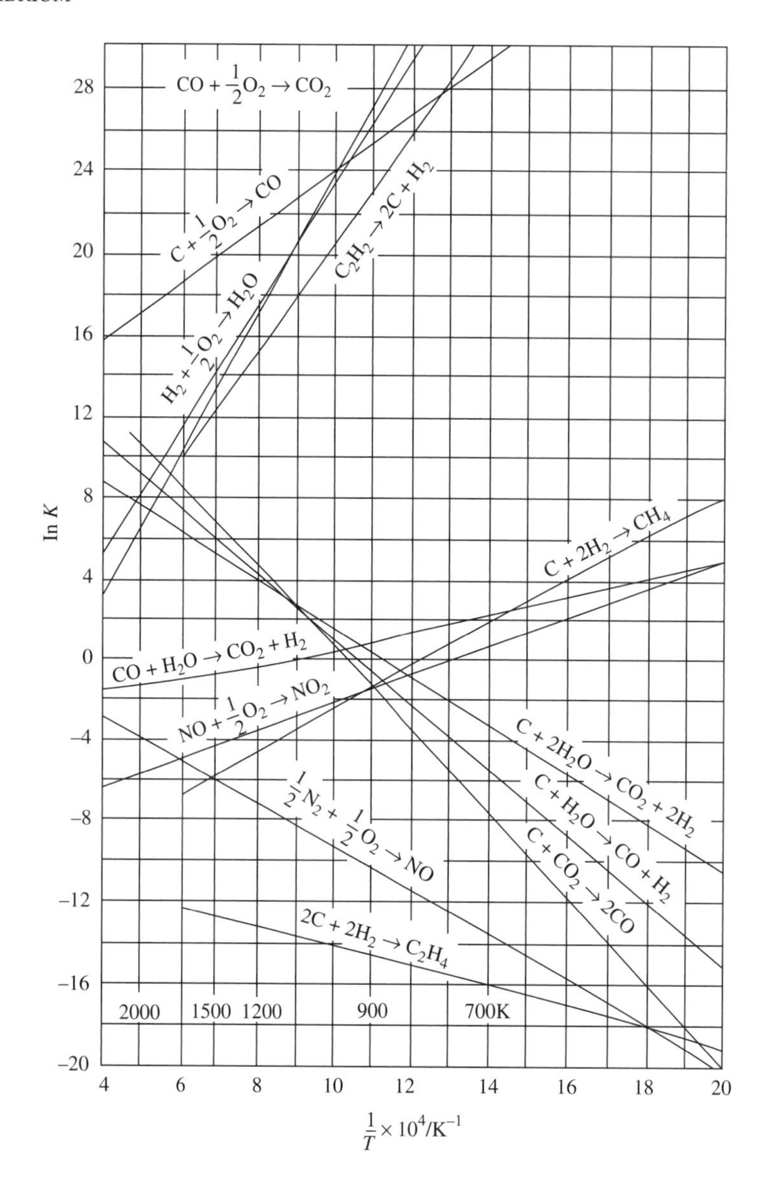

FIGURE 12.8 Calculated values of the equilibrium constants K for a variety of reactions, plotted on $\ln K$ vs. $(10^4/T)$ coordinates. Taking $(10^4/T)$ as the abscissa makes T increase from right to left, making the sign of the slopes opposite to those shown in Figure 12.7. Some values of temperature are shown above the abscissa. (From Smith, J. M., H. C. Van Ness, and M. M. Abbott, *Introduction to Chemical Engineering Thermodynamics*, ed. 5. New York: McGraw-Hill, p. 569 (1996), reproduced with permission of the McGraw-Hill Companies).

industrially useful rates, we lower the equilibrium conversion. It would make you believe in the perversity of nature! The effect of changes in temperature on equilibrium concentration discussed in this section are the same, both in theory and in practice, whether the reactions involve solids, liquids, or gases. The same is not true for the effect of pressure on equilibrium, discussed in the next section.

12.9 THE EFFECT OF PRESSURE ON CHEMICAL REACTION EQUILIBRIUM

The value of the equilibrium constant K does not depend on the pressure, because the standard state Gibbs energies do not. However, the fugacities that appear in the law of mass action do depend on pressure. We show this effect by returning to Eq. 12.14, writing out the individual fugacities in the form used in Chapter 8 (Eq. 8.9), and writing the f_i^o as 1 bar. We find

$$K = \frac{\left[\dfrac{f_{NH_3}}{f_{NH_3}^o}\right]}{\left[\dfrac{f_{N_2}}{f_{N_2}^o}\right]^{0.5} \cdot \left[\dfrac{f_{H_2}}{f_{H_2}^o}\right]^{1.5}} = \frac{\left[\dfrac{(y\hat{\phi})_{NH_3}P}{1 \text{ bar}}\right]}{\left[\dfrac{(y\hat{\phi})_{N_2}P}{1 \text{ bar}}\right]^{0.5} \cdot \left[\dfrac{(y\hat{\phi})_{H_2}P}{1 \text{ bar}}\right]^{1.5}} \tag{12.30}$$

We next factor the term on the right to have

$$K = (\text{bar})^{1.5+0.5-1} \cdot \frac{y_{NH_3}P}{(y_{N_2}P)^{0.5} \cdot (y_{H_2}P)^{1.5}} \cdot \frac{\hat{\phi}_{NH_3}}{(\hat{\phi}_{N_2})^{0.5} \cdot (\hat{\phi}_{H_2})^{1.5}} \tag{12.31}$$

Then we define two terms,

$$K_p = \frac{y_{NH_3}P}{(y_{N_2}P)^{0.5} \cdot (y_{H_2}P)^{1.5}} \tag{12.32}$$

and

$$K_{\hat{\phi}} = \frac{\hat{\phi}_{NH_3}}{(\hat{\phi}_{N_2})^{0.5} \cdot (\hat{\phi}_{H_2})^{1.5}} \tag{12.33}$$

so that

$$K = (\text{bar})^{(1.5+0.5-1)} \cdot K_p \cdot K_{\hat{\phi}} \tag{12.34}$$

What did this buy us? We know that K is independent of pressure. K_p and $K_{\hat{\phi}}$ are both pressure dependent in some cases. The $(\text{bar})^{(1.5+0.5-1)}$ term is simply accounting for the dimensions of our answers. By writing out K in this way we can consider various cases easily.

12.9.1 Ideal Solution of Ideal Gases

If all the reactants and products behave as ideal gases, then $K_{\hat{\phi}} = 1.00$, and Eq. 12.30 simplifies to

$$K = (\text{bar})^{(1.5+0.5-1)} \cdot \frac{y_{NH_3}P}{(y_{N_2}P)^{0.5} \cdot (y_{H_2}P)^{1.5}} \quad \text{[ideal solution of ideal gases]} \tag{12.BJ}$$

Logic suggests that we should factor this to the form shown in Eq. 12.BK, but the form show in Eq. 12.BJ, in terms of partial pressures, is the form most often seen.

If the number of mols does not change in the reaction, then changing the pressure will not change the equilibrium concentrations. However, if the number of mols changes, then the reaction can be driven in one direction or the other by changing the pressure. We can see this by rearranging Eq. 12.BJ to

$$K\left[\frac{P}{1 \text{ bar}}\right]^{(1.5+0.5-1)} = K\left[\frac{P}{1 \text{ bar}}\right]^{(-\sum \nu_i)} = \frac{y_{NH_3}}{y_{N_2}^{0.5} \cdot y_{H_2}^{1.5}} \tag{12.BK}$$

We see that if $\sum \nu_i$ is zero (no change in number of mols), then the P term is raised to the zero power, $= 1.00$, and changing the pressures does not change the equilibrium concentrations. But if, as in this case, $\sum \nu_i = -1$, then the term on the left is proportional to the pressure to the $+1$ power. Thus, if the number of mols decreases as the reaction proceeds, then we can drive the reaction in the forward direction by increasing the pressure.

Example 12.10 Modern large ammonia plants mostly carry out the ammonia production reaction at *about* 400°C and 150 atm pressure. Estimate the equilibrium conversion at these conditions, assuming that all reactants and products behave as ideal gases.

Comparing this example with Example 12.9, we see that we can use the same equation, but that in place of $K_{673.15 \text{ K}}$ we use

$$K\left[\frac{P}{1 \text{ bar}}\right]^{(1.5+0.5-1)} = 0.0137 \cdot \left[\frac{150 \text{ atm}}{1 \text{ bar}} \cdot \frac{1.013 \text{ bar}}{\text{atm}}\right] = 2.08 \tag{12.BL}$$

Solving Eq. 12.BI with this value of the constant we find $e_{673.15 \text{ K}, 150 \text{ atm}} = 0.48$ and an NH_3 mol fraction in the gas of ≈ 0.316. ■

In this reaction two mols of reactants combine to form one mol of product. That leads to the (P/bar) term. Physically, it means that as the pressure is increased, the system can lower its Gibbs energy by decreasing the number of mols of gas, so the equilibrium is shifted in the direction of the smaller number of mols. If we wished to make the reaction operate in the opposite direction, we would choose as low a pressure as practical, because that would shift the equilibrium in the opposite direction. That is exactly what is done in some fuel-cell power plants. The fuel is stored as liquid ammonia and then brought to equilibrium over a catalyst at high temperature and 1 atm pressure. The H_2–N_2 mixture (plus small amounts of NH_3) passes to the fuel cell in which the H_2 reacts with O_2 to produce electricity. (It is often cheaper and safer to store and transport hydrogen as liquid ammonia and dissociate it when needed than to store and transport gaseous or liquid hydrogen.)

12.9.2 Nonideal Solution, Nonideal Gases

For high-pressure gas reactions, like the ammonia synthesis reaction, the $K_{\hat{\phi}}$ term is often important. If we retain it, then we would rewrite Eq. 12.34 as

$$\frac{K}{K_{\hat{\phi}}}\left[\frac{P}{1\text{ bar}}\right]^{(-\sum \nu_i)} = \frac{y_{NH_3}}{y_{N_2}^{0.5} \cdot y_{H_2}^{1.5}} \quad \text{(12.BM)}$$

We rarely have data on gas-phase nonideality, so we normally make the L-R assumption (an ideal solution of nonideal gases, Section 8.6.2), $\hat{\phi}_i = \phi_i$, changing $K_{\hat{\phi}}$ to K_ϕ. Gillespie and Beattie [8] computed this parameter for the ammonia synthesis reaction, representing the specific volume of the various gases by the Beattie–Bridgeman EOS. They presented their results in the form

$$\log\left(\frac{1}{K_\phi}\right) = \left(\frac{0.1191849}{T} + \frac{91.87212}{T^2} + \frac{25122730}{T^4}\right) \cdot P \quad \text{(12.BN)}$$

for T in K and P in atm. Figure 12.9 shows this function.

We see that at low pressures, as the behavior of the gases becomes practically ideal, $K_\phi \to 1.00$. K_ϕ becomes smaller with increasing pressure and decreasing temperature, as the compressibility factor of ammonia becomes smaller and smaller. These values reflect both the fact that the fugacity coefficient of ammonia is decreasing with increasing P and decreasing T and the fact (see Figure 7.1) that at these temperatures the fugacity coefficients of N_2 and H_2 are >1.00.

Example 12.11 Repeat Example 12.10, taking into account the nonideal gas behavior of the reactants and products.

Combining Eqs. 12.30 and 12.BM, we find

$$\frac{K}{K_\phi} \cdot \left[\frac{P}{1\text{ bar}}\right]^{(1.5+0.5-1)} = \frac{y_{NH_3}}{y_{N_2}^{0.5} \cdot y_{H_2}^{1.5}}$$

$$= \frac{\dfrac{0+e}{2-e}}{\left[\dfrac{0.5-0.5\cdot e}{2-e}\right]^{0.5} \cdot \left[\dfrac{1.5-1.5\cdot e}{2-e}\right]^{1.5}} \quad \text{(12.BO)}$$

Then, from Eq. 12.BN (or Figure 12.9 if we can read it well enough), we find $K_\phi = 0.84$ and

$$\frac{K}{K_\phi} \cdot \left[\frac{P}{1\text{ bar}}\right]^{(1.5+0.5-1)} = \frac{0.0137}{0.84}[150 \cdot 1.013] = 2.48 \quad \text{(12.BP)}$$

Inserting this value in Eq. 12.BO and solving numerically, we find $e_{673.15\text{ K, 150 atm, nonideal gas}} = 0.51$ and an NH_3 mol fraction in the gas of ≈ 0.345 ■

We see that in this case the nonideal gas behavior increases the fractional conversion, because the product behaves less like an ideal gas than do the reactants. The effect is small, but significant (34.5 vs. 31.6 mol% NH_3 in the equilibrium gas). This same calculation has been carried out for a variety of temperatures and pressures; the results are summarized in Figure 12.10. The reader may verify that (within chart-

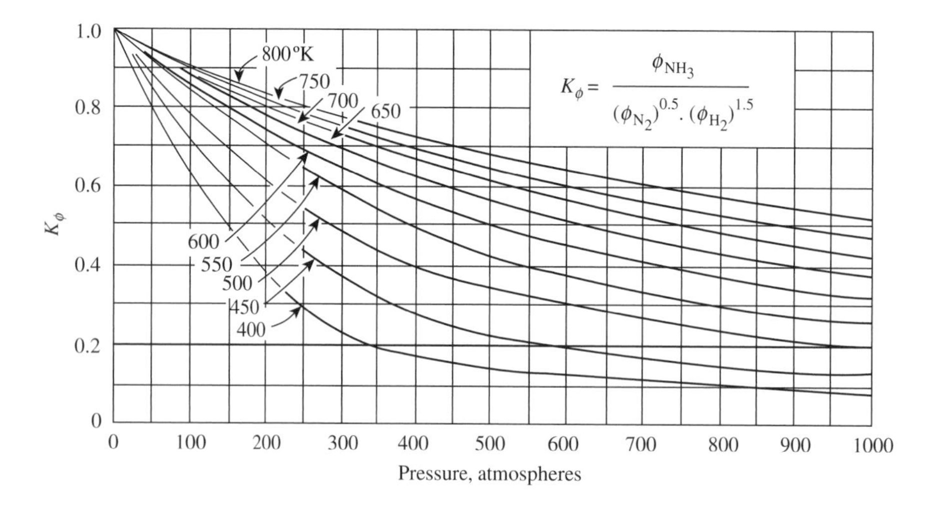

FIGURE 12.9 K_ϕ for the ammonia synthesis reaction, calculated from Eq. 12.BN. (From Hougen, O. A., K. M. Watson, and R. A. Ragatz. *Chemical Process Principles*, Part II: *Thermodynamics*, ed. 2. © 1959, New York: Wiley, p. 1009. Reprinted by permission of the estate of O. A. Hougen.)

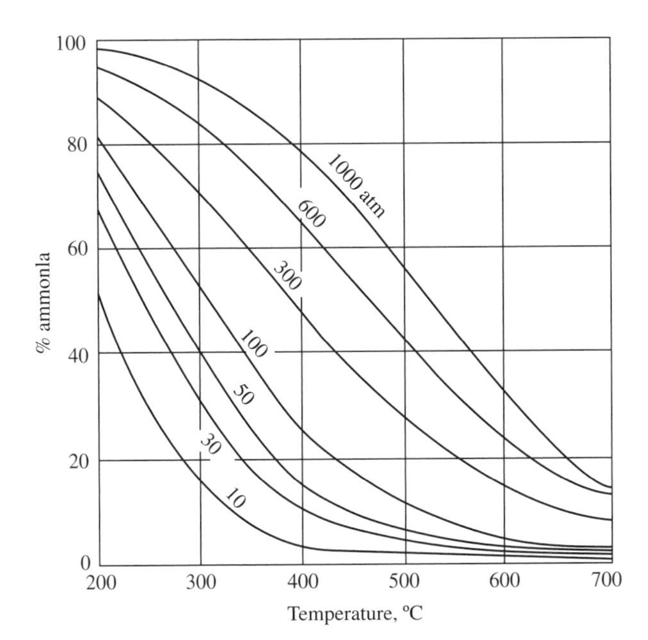

FIGURE 12.10 Calculated equilibrium concentration of NH_3 for a variety of temperatures and pressures, starting with a feed that is 75 mol% H_2 and 25 mol% N_2. (From Comings, E. W. *High Pressure Technology*. New York: McGraw–Hill, p. 410, 1956. Reproduced with permission of the McGraw-Hill Companies).

reading accuracy) this calculation and that figure agree. Such plots are available for a variety of gas-phase reactions.

12.9.3 Liquids and Solids

The effect of pressure on equilibrium concentration is large for gases, if the number of mols changes in the reaction or if one of the reactants or products behaves much more non-ideally than the others. The effect of pressure on equilibrium concentration for liquids and solids is small. We may think about this by remembering the derivative

$$\left(\frac{dg}{dP}\right)_T = v \tag{4.32}$$

which shows that for gases that have large values of the molar (or specific) volume, changing P makes a large change in g, while for liquids and solids, which have small values of the molar volume, the effect is much smaller (typically smaller by a factor of 500 to 1000).

Example 12.12 Estimate the effect of a pressure increase from 1 to 150 atm at 25°C on the equilibrium concentration of the liquid-phase production of ethyl acetate

$$\underset{\text{etahnol}}{C_2H_5OH} + \underset{\text{acetic acid}}{CH_3COOH} \Leftrightarrow \underset{\text{ethyl acetate}}{C_2H_5OOCCH_3} + \underset{\text{water}}{H_2O} \tag{12.BQ}$$

Table 12.4 Data for Example 12.12

	Molecular Weight (g/mol)	Density at 20°C (g/mL)	Molar Volume (mL/mol)
Ethanol	46	0.789	58.30
Acetic acid	60	1.049	57.20
Ethyl acetate	88	0.901	97.67
Water	18	0.9985	18.03

The physical properties for these chemicals at 20°C [1, pp. 2–28] are shown in Table 12.4. Ignoring the difference between 20 and 25°C, we can use the values in Table A.8 to compute that $\Delta g^{\text{o}} = 10.54\,\text{kJ/mol}$ and thus $K = 0.0142$. Choosing ethanol as the selected reactant, and assuming that we start with one mol each of ethanol and acetic acid, we have

$$K = \frac{a_{\text{ethyl acetate}} \cdot a_{\text{water}}}{a_{\text{ethanol}} \cdot a_{\text{acetic acid}}} = \frac{x_{\text{ethyl acetate}} \cdot x_{\text{water}}}{x_{\text{ethanol}} \cdot x_{\text{acetic acid}}} = \frac{e \cdot e}{(1-e) \cdot (1-e)} \tag{12.BR}$$

where we have assumed that at the same conditions as the standard states (1 bar, 25°C) the activities are equal to the mol fractions. (This assumes ideal solution behavior, which is not likely to be exactly correct, but is not important for this example.) The solution of Eq. 12.BR, for $K = 0.0142$ is $e = 0.106$ and the equilibrium mol fractions of ethyl acetate and water are each $x = e/2 = 0.053$.

To see the effect of changing the pressure we first compute from the above table that the volume increase of the reaction is

$$\Delta v = 97.67 + 18.03 - 58.30 - 57.20 = 0.20 \frac{\text{cm}^3}{\text{mol}} \tag{12.BS}$$

which is about 0.15% of the volume of the reacting mixture. Thus, from Le Chatelier's principle, we would expect increasing the pressure to lower the fraction converted, forcing the reaction in the direction of the reactants, away from the products. But how much? First write the activity of each of the four species as

$$a_i = \frac{f_i}{f_i^{\circ}} = \frac{x_i \gamma_i p_i}{p_i} \cdot \exp\left[\frac{v}{RT} \cdot (P - p_i)\right] \tag{12.BT}$$

where we have switched standard states to the Raoult's law type, so that we can simply follow Example 7.3. The rightmost term is the Poynting factor. Substituting Eq. 12.BT four times in Eq. 12.BR, again assuming ideal solution so that all the γ_is = 1.00, and simplifying algebraically, we have

$$K = \frac{a_{\text{ethyl acetate}} \cdot a_{\text{water}}}{a_{\text{ethanol}} \cdot a_{\text{acetic acid}}} = \frac{x_{\text{ethyl acetate}} \cdot x_{\text{water}}}{x_{\text{ethanol}} \cdot x_{\text{acetic acid}}}$$

$$\cdot \exp\left[\frac{\left(\sum_{\text{products}} v - \sum_{\text{reactants}} v\right)}{RT} \cdot (P - p_i)\right]$$

$$= \frac{e \cdot e}{(1-e) \cdot (1-e)}$$

$$\cdot \exp\left[\frac{\left(\sum_{\text{products}} v - \sum_{\text{reactants}} v\right)}{RT} \cdot (P - p_i)\right] \quad (12.BU)$$

where the exp term is the Poynting factor for the reaction. Its numerical value is

$$\exp\left[\frac{\left(\sum_{\text{products}} v - \sum_{\text{reactants}} v\right)}{RT} \cdot (P - p_i)\right]$$

$$= \exp\left[\frac{0.20\,\text{cm}^3/\text{mol}}{82.06\dfrac{\text{cm}^3\,\text{atm}}{\text{mol}\,\text{K}} \cdot 298\,\text{K}} \cdot (150 - 1)\text{atm}\right]$$

$$= \exp[0.0012] = 1.0012 \quad (12.BV)$$

So that

$$\frac{e \cdot e}{(1-e) \cdot (1-e)} = \frac{K}{\exp[1.0012]} = \frac{0.0142}{1.0012} = 0.01418 \quad (12.BW)$$

We do not know the value of K to that many significant figures, but ignoring that for the moment and solving we find that e is still 0.106, and that if we carry out the calculation to enough figures, the new value of e is 0.9993 times the previously calculated value of e. ■

Here we have ignored the effect of the pressure change on the activity coefficients, but from Chapter 9 we can show that the expected value of that effect is at least an order of magnitude less than the effect shown here. Comparing this result to Examples 12.10 and 12.11, we see that for the gaseous ammonia synthesis reaction raising the pressure from 1 to 150 atm increased the mol fraction of ammonia in the exit stream by a factor of $0.345/0.0044 \approx 78$. Here it multiplies it by a factor of 0.9993. The reason, as stated before, is that for gases the values of v are large, and thus the effect of changes in P on Gibbs energies are significant, while for liquids and solids the values of v are small, with the result shown. Only for the highest pressure reactions (e.g., inside explosions) need we consider the effect of pressure changes on chemical reaction equilibrium of solids or liquids.

12.10 THE EFFECT OF NONIDEAL SOLUTION BEHAVIOR

We rarely have data on gas phase nonideality, and normally simply use the L-R rule. However modern process design computer programs often compute $\hat{\phi}_i$ using equations of state for gas mixtures (Chapter 10 and Appendix F), which include gas-phase nonideality (and thus allow us to compare $\hat{\phi}_i$ to ϕ_i). Only at pressures of hundreds of atmospheres or more do their values differ substantially.

12.10.1 Liquid-Phase Nonideality

For liquids the situation is different. If we rewrite Eq. 12.30 using the normal definition of activity for the liquid phase, and considering the reaction n-butane ⇔ isobutane, we will have

$$K = \frac{\left[\dfrac{f_{\text{isobutane}}}{f^{\text{o}}_{\text{isobutane}}}\right]}{\left[\dfrac{f_{n\text{-butane}}}{f^{\text{o}}_{n\text{-butane}}}\right]} = \frac{\left[\dfrac{(x\gamma p)_{\text{isobutane}}}{p_{\text{isobutane at } T^{\text{o}}}}\right]}{\left[\dfrac{(x\gamma p)_{n\text{-butane}}}{p_{n\text{-butane at } T^{\text{o}}}}\right]} \quad (12.35)$$

in which the K is based on Δg^{o} for the conversion of liquid n-butane to liquid isobutane at ΔT^{o}. This would logically be factored to

$$K = \frac{x_{\text{isobutane}}}{x_{n\text{-butane}}} \cdot \frac{\gamma_{\text{isobutane}}}{\gamma_{n\text{-butane}}} \cdot \frac{(p/p_{\text{at } T^{\text{o}}})_{\text{isobutane}}}{(p/p_{\text{at } T^{\text{o}}})_{n\text{-butane}}} \quad (12.36)$$

For this reaction the two rightmost ratios (normally called K_γ and the pressure term, which has no common name or symbol) are both practically 1.00, so this is almost the same result as in Example 12.1. However, if the reaction of interest were one in which the liquid-phase nonideality were substantial, Eq. 12.36 shows how that would be accounted for (see Problems 12.30 and 12.31).

12.11 OTHER FORMS OF K

The K shown so far is the basic dimensionless K, formulated in terms of the activities and calculated by Eq. 12.14. We very often see in its place K_p, defined by Eq. 12.32.

Example 12.13 Estimate the value of K_p for the formation of ammonia at 150 atm and 400°C.

Using the values from Example 12.11, we know that $K = 0.013$, $K_\phi = 0.84$ so that

$$K_p = \frac{K}{K_\phi} \cdot \left[\frac{1}{\text{bar}}\right]^{(-\sum v_i)} = \frac{0.0137}{0.84} \cdot \left[\frac{1}{\text{bar}}\right]^{+1} = \frac{0.0163}{\text{bar}} \quad (12.BX)$$

The basic K is dimensionless, but K_p has dimensions of pressure to the $\sum v_i$ power. ■

In Chapter 16 we will see that the biochemical engineers and biologists use several other definitions of K that seem useful in their work.

12.12 SUMMARY

1. Chemical reactions are of paramount importance in chemical and in environmental engineering, and in every aspect of our daily life.
2. This chapter and this book deal only with equilibrium, not with the rate of chemical reactions. By using selective catalysts we can influence the rate of reactions so that we produce mixtures that are at or near equilibrium for one reaction, but not for others. Thermodynamics tells us little about that.
3. At true thermodynamic equilibrium, any chemically reacting mass of matter has taken up the chemical composition that has the minimum possible Gibbs energy for that T, P, and starting composition. All that we do in this chapter is work out the consequences of this statement.
4. The most common formulation is in terms of the law of mass action, which leads to the equilibrium constant K, stated in terms of the activities and stoichiometric coefficients of the reactants and products.
5. K is dimensionless, and can be calculated from Δg^{o} and T. Other forms of K, such as K_p, are widely used; they are often not dimensionless.
6. For almost all reactions, changing T changes K and thus changes the equilibrium concentrations.
7. Changing P does not change the value of K. It does change the values of K_p and K_ϕ.
8. For reactions involving only liquids and solids, changing P has negligible effect on the equilibrium concentrations.
9. For gases, if the number of mols changes in the reaction, then raising the pressure will drive the reaction in the direction of the decreasing number of mols, and lowering the pressure will do the reverse.
10. For nonideal gases, raising the pressure will drive the reaction in the direction of the reactant(s) or product(s) with the lowest values of the fugacity coefficient $\hat{\phi}_i$.
11. For reactions in the liquid phase, K_ϕ may be important.

PROBLEMS

See the Common Units and Values for Problems and Examples. An asterisk (*) on the problem number indicates that the answer is in Appendix H.

12.1 Show the construction of Figure 12.1, using the values in Table A.8 and the ideal solution of ideal gases assumption.

12.2 If there were no Gibbs energy change on mixing (which there actually always is!)

a. What would Figure 12.1 look like? Show a simple sketch.

b. What would the equilibrium composition be?

c. How would this be described in terms of the equilibrium state being the one at which the forward and backward chemical reaction rates are equal and opposite?

12.3* In Example 12.1,

a. What is the value of Δg°?

b. What is the value of g at the equilibrium state?

c. What is the change in g in going from pure n-butane to the equilibrium state?

d. What is the change in g in going from pure isobutane to the equilibrium state?

12.4 Some authors write Reaction 12.I as

$$0.5N_2 + 0.5O_2 \Leftrightarrow NO \quad (12.BY)$$

Repeat Example 12.2 (for 298.15 K only) for this reaction instead of Reaction 12.I. Clearly, the final concentrations cannot depend on which of these forms of expressing the reaction we choose, but what do the details of the calculation look like?

12.5* Repeat Example 12.2 (at 2000 K only) for a typical combustion gas, in which the N_2 content is the same as in Example 12.2, but the O_2 content is typically 4%.

12.6 For reactions at temperatures other than 298.15 K we cannot directly use the $g^{\text{o}}_{\text{formation from the elements at 298.15}}$ values from Table A.8, but must use the methods shown in Section 12.8. This is not particularly difficult with our computers, but before the computer age it was a giant pain. To simplify this problem, the

JANAF tables [6] computed those values for the common gases, and then presented them as $g^{o}_{\text{formation from the elements at temperature } T}$. This is equivalent to taking the Gibbs energy of the elements as zero at all temperatures, which is equivalent to using a different standard state for each temperature. Using this basis, at 2000 K, we have the values in Table 12.5. These values were used in Example 12.2 and should be used for the high-temperature gas reaction problems in this chapter. See also Problem 12.28. Observe that these values are in kcal, not kJ. Using Table 12.5,

a. Show the calculation of the equilibrium constant in Example 12.2.

b. Estimate the equilibrium concentration of NO_2 in air at 2000 K. The overall reaction is

$$N_2 + 2O_2 \Leftrightarrow 2NO_2 \qquad (12.BZ)$$

(b-1) Assume that it proceeds as shown in Eq. 12.BZ.

(b-2) Assume that it actually proceeds by Eq. 12.I, followed by

$$NO + 0.5O_2 \Leftrightarrow NO_2 \qquad (12.CA)$$

and that the equilibrium NO concentration is that found in Example 12.2. This latter is the principal mechanism for forming NO_2.

Table 12.5 Gibbs Energy of Formation of Some Gases at 2000 K, Based on the JANAF Tables [6]

Element or Compound	h^{o} of Formation from the Elements, at this T (kcal/mol)	g^{o} of Formation from the Elements, at this T (kcal/mol)
N_2	0	0
NO	21.626	15.548
NO_2	7.908	38.002
O_2	0	0

12.7* Estimate the values of the equilibrium constant K at 25°C for the following reactions, assuming that all the reactants and products are ideal gases:

a. $C_2H_6 \Leftrightarrow C_2H_4 + H_2$ (12.CB)

b. $C_2H_4 + H_2O \Leftrightarrow C_2H_5OH$ (12.CC)

c. Discuss the economic significance of these reactions.

12.8 In the preceding problem you estimated the value of K for Reaction 12.CB at 25°C. Estimate the value of K for that reaction at 1100 K. For this problem, but not in real life, you may assume that Δh^{o} for this reaction is independent of temperature. (This reaction, one of the principal industrial sources of ethylene, is normally conducted at about 1100 K.)

12.9 Table 12.2 shows the reported values of the Gibbs energies of isobutane and n-butane from three sources. Based on that table, estimate the value of K and the mol fraction of isobutane in the equilibrium mix, both at 298.15 K and 1 bar, for each of these three sets of values. Assume that at this condition isobutane and n-butane form an ideal solution of ideal gases.

12.10 Repeat Example 12.3 for a typical combustion gas, in which $x_{\text{oxygen}} \approx 0.04 = 4\%$. See Problem 12.5.

12.11 Repeat Example 12.4 for the following reactions:

a. Reaction 12.G

b. Reaction 12.BZ

c. 12.CA.

In each case assume that the initial reacting mix is stoiciometrically balanced, contains none of the final products, and that the reactants and products are all ideal gases. Do not assume that $P = 1.00$ bar.

12.12 Show the solution (both roots) for e in Example 12.5. Does the other root have any significance?

12.13 Show the numerical solution of Eq. 12.AI in Example 12.6. Most spreadsheets will not solve this directly because the exponent is beyond their permitted range. If we substitute $e \approx 1.00$ in that equation and simplify, we find an equation with an analytical solution, leading to the stated result.

12.14 Figure 12.5 shows that as the fraction reacted approaches 1.00 there is a minimum in the g-fraction reacted plot. If, as shown, the value of g for 100% completion is −237.1 kJ/kg, what is the value of g at the minimum?

12.15 Estimate the change in the Gibbs energy of formation from the elements when we switched from tables with standard state of $P = 1$ atm to $P = 1$ bar. Show the results for

a. A solid like sodium chloride.

b. A liquid like water.

c. An ideal gas in which the number of mols does not change on formation (e.g., NO).

d. An ideal gas in which the number of mols does change on formation (e.g., NO_2).

12.16* Repeat Example 12.7 for ethanol. Compare your result to the values in Table A.8.

12.17 In Problem 12.7 you computed the equilibrium constant K for Reaction 12.CC on the assumption that all reactants and products were ideal gases.

a. Repeat that calculation using the value of g^{o}_{water} for *liquid* water from Table A.8. Is K the same as that calculated in Problem 12.7? Should it be?

b. Show the equation equivalent to Eq. 12.AB that goes with this value of K. Is it the same as Eq. 12.AB? Should it be?

c. Discuss the real? or apparent? difficulty caused by this change in standard states.

12.18 The table of thermodynamic properties [2] presents the standard enthalpies and Gibbs energies of formation for all materials as that of an ideal gas at 298.15 K and 1 atm. That table does not include common solid minerals like NaCl, $CaCO_3$, diamonds, or graphite. What difficulties might we encounter while attempting to insert these materials in that table?

12.19 Example 12.8 estimates the effect of temperature change on K for the ammonia synthesis reaction, taking the effect of change in Δh^o into account.

a. Estimate $K_{673.15K}$ for this reaction, assuming that Δh^o is a constant (i.e., using Eq. 12.26) and the Δh^o for this reaction at 293.15 K.

b. Estimate the value of Δh^o for this reaction at 673.15 K.

c. Repeat part (a) using the average of the Δh^o at 298.15 K and 673.15 K.

d. Compare the results of parts (a) and (c) with the result in Example 12.8.

e. Sketch the equivalent of Figure 12.6 for this reaction.

12.20 Repeat Example 12.8, also at 673.15 K, for Reaction 12.I and compare the results with Figure 12.8.

12.21* In Figure 12.8 the curve for Reaction 12.I is practically straight.

a. Based on that fact, estimate the value of Δh^o for this reaction from Figure 12.8.

b. Estimate the value of Δh^o for this reaction using the values in Table A.8. Compare the results to those from part (a).

c. Based on elementary thermochemistry, explain why the Δh^o for this reaction should be practically independent of temperature.

12.22 For the chemical reaction $C_3H_8 \Leftrightarrow C_3H_6 + H_2$(propane $\Leftrightarrow$ propylene + hydrogen)

a. Estimate the equilibrium constant K at 298.15 K.

b. Estimate the temperature at which the equilirium constant $K = 1.00$, on the assumption (only a fair assumption, but suitable for this problem) that ΔH^o of this reaction is totally independent of temperature.

12.23 The kinetic theory of gases shows that if we ignore internal molecular vibrations, (a good assumption at low temperatures, but not at high) then for ideal gases, $C_P/R = 2.5$ for monatomic gases, 3.5 for diatomic and 4 for gases with 3 or more atoms, independent of temperature.

a. If this were rigorously true for real gases, then in Table A.9, what would the values of a, b, c and d be? How closely do the values shown there match these values? Explain why the match is much better for some classes of molecules than others.

b. Repeat example 12.8 using these kinetic theory values of C_P/R. How much difference does it make?

12.24 Examples 12.9, 12.10, and 12.11 all assume that the starting material is 3 mols of H_2 and 1 mol of N_2. In real ammonia plants the reactor feed always contains some argon and some CH_4, which are practically inert. Figure 1.1 shows that a small bleed stream must be taken to remove these. Repeat Example 12.11, assuming that the feed consists of 3 mols of H_2, 1 mol of N_2, 0.22 mols of Ar, and 0.22 mols of CH_4.

12.25 We regularly see equations for K of the form of Eq. 12.27. For example, for Reaction 12.B, Frear and Baber [9] show that for 700 to 1000°F

$$\log K_p = -5.963 + \frac{2740\,K}{T} \tag{12.CD}$$

with K_p having dimension atm.

a. Using this equation estimate the values of K_p at 700°F (371°C), 400°C, and 1000°F (537.8°C).

b. Compare them to the values computed in Example 12.9 for 400°C, and to values computed the same way as Example 12.9 for the other two temperatures.

12.26 Repeat Example 12.5 for the reaction temperature 265°C. At that temperature $K \approx 0.005$. As in Example 12.5, assume that the feed is 0.833 mol of water per mol of ethylene and that all the reactants and products are ideal gases. Give the value of e and the mol fraction of ethanol at equilibrium for

a. $P = 1.00$ atm.

b. $P = 70$ atm.

(The values computed here for ideal gases are similar to but not identical to those for real gases [7].)

12.27 In Example 12.12, show the mol fraction of each of the reactants and products at equilibrium.

12.28 Estimate the Gibbs energy of formation of NO from the elements at 2000 K, based on Tables A.8, and A.9. Compare that value to the one from the JANAF tables, shown in Problem 12.6 Are they identical? Close to one another? Discuss.

12.29 Many tables, like Table A.8, also show a function $\log K_f$, with K_f defined by

$$K_f = \exp \frac{-g^{\text{o}}_{\text{formation from the elements}}}{R \cdot 298.15\ \text{K}} \quad (12.37)$$

a. Estimate the values of K_f for ammonia, nitrogen, and hydrogen.

b. Write the equation for K for the ammonia synthesis reaction (12.B) in terms of K_{f_i} for ammonia, nitrogen, and hydrogen.

c. Using the numerical values from part (a), show the numerical value of K for the equilibrium between ammonia, nitrogen, and hydrogen.

12.30 In Eq. 12.36:

a. If the activity coefficient of the product is greater than the activity coefficient of the reactant, does that lead to greater conversion of reactant to product than for ideal solution? Or to lesser conversion than for ideal solution?

b. If the rightmost term is to be independent of temperature, what must the two vapor pressure curves look like on Figure 5.2?

12.31* For the reaction A ⇔ B, in the liquid phase at 298.15 K, for which the rightmost term in Eq. 12.36 = 1.00, what is the mol fraction of A in the equilibrium liquid:

a. If the liquid forms an ideal solution and $K = 0.1$? 1.00? 10?

b. Same as part (a), all three parts, but the liquids form a type II solution which obeys the symmetrical activity coefficient equation (Eq. 9.I), with $A = 1.5$.

REFERENCES

1. Liley, P. E., G. H. Thompson, D. G. Friend, T. E. Daubert, and E. Buck. Physical and chemical data. In *Perry's Chemical Engineer's Handbook*, ed. 8, D. W. Green, ed. New York: McGraw-Hill, pp. 2–187 (2008).
2. Poling, B. E., J. M. Prasunitz, and J. P. O'Connell. *The Properties of Liquids and Gases*, ed. 5, New York: McGraw-Hill, Appendix A (2001).
3. Lide, D. R. *Handbook of Chemistry and Physics*, ed. 71, Boca Raton, FL: CRC Press, pp. 10–29 (1990). The 91st edition (2010) does not contain Gibbs energy of formation values.
4. Speight, J. G. *Lange's Handbook of Chemistry*, ed. 16, New York: McGraw-Hill, p. 1.237 (2005).
5. de Nevers, N. *Air Pollution Control Engineering*, ed. 2. Long Grove, IL: Waveland, Appendix D (2000, reissued 2010).
6. *JANAF Thermochemical Tables, PB 168 370*. Washington, DC: ARPA, DOE (1964).
7. Logsdon, J. E. Ethanol. In *Kirk–Othmer Encyclopedia of Chemical Technology*, Vol. 9, ed. 4, Howe-Grant, M., ed. New York: Wiley, pp. 812–860 (1994).
8. Gillespie, L. J., and J. A. Beattie. The thermodynamic treatment of chemical equilibria in systems composed of real gases, I: an approximate equation for the mass action function applied to the existing data on the Haber equilibrium. *Phys. Rev.* 36:743–753 (1930).
9. Frear, G. L., and R. L. Baber. Ammonia. In *Kirk-Othmer Encyclopedia of Chemical Technology*, Vol. 2, ed. 2, Mark, H. F., J. J. McKetta, and D. F. Othmer, eds. New York: Interscience–Wiley, p. 269 (1963).

13

EQUILIBRIUM IN COMPLEX CHEMICAL REACTIONS

Chapter 12 discussed only single reactions in one phase, involving only nonionized species. In this chapter we apply the results of that chapter to more complex equilibria. There are no new principles or new ideas in this chapter, only applications of the principles and ideas from previous chapters to a new set of more complex problems. Again, nature minimizes Gibbs energy; all we do here is estimate the concentrations at which that minimum occurs.

13.1 REACTIONS INVOLVING IONS

All of the discussion in Chapter 12 applies to reactions involving ions.

Example 13.1 Estimate the equilibrium concentration of hydrogen ions (H^+) and hydroxyl ions (OH^-) in equilibrium with pure water at 25 °C.

Here the reaction is

$$H_2O \Leftrightarrow H^+ + OH^- \tag{13.A}$$

Reading the necessary values from Table A.8, we find

$$\Delta g^o = g^o_{H^+} + g^o_{OH^-} - g^o_{H_2O} = 0 + (-157.29) - (-237.1)$$
$$= 79.81 \frac{\text{kJ}}{\text{mol}} \tag{13.B}$$

$$K = \exp\left(\frac{-\Delta g^o}{RT}\right) = \exp\left[\frac{-\left(79,810\dfrac{\text{J}}{\text{mol}}\right)}{8.314\dfrac{\text{J}}{\text{mol K}}\cdot 298.15\text{K}}\right]$$
$$= 1.04 \times 10^{-14}$$
$$= \frac{\left[\dfrac{[H^+]}{1 \text{ molal}}\right]\cdot\left[\dfrac{[OH^-]}{1 \text{ molal}}\right]}{[a_{\text{water}}]} \tag{13.C}$$

The activity of any pure liquid at its standard state is 1.00, and here the water is practically pure and at its standard T and P, so

$$\left[\frac{[H^+]}{1 \text{ molal}}\right]\cdot\left[\frac{[OH^-]}{1 \text{ molal}}\right] = 1.04 \times 10^{-14} \approx 10^{-14} \quad \blacksquare \tag{13.D}$$

This is the result given in all elementary chemistry books, that in water at 25 °C, the product of the hydrogen ion and hydroxyl ion concentrations, both expressed in molality (= mol/kg of solvent) = 10^{-14}. For dilute solutions, molality ≈ mol/L = molarity. In tables like Table A.8, the standard states of ions are all based on electrochemical measurements, which can only give values for ion pairs (e.g., H^+ and OH^-), never a direct measurement for one ion alone. The convention adopted is to take the standard state Gibbs energy of H^+, $g^o_{H^+}$, as 0.0000 at a concentration of 1 molal at 25 °C and 1.00 bar. With that convention, the

Physical and Chemical Equilibrium for Chemical Engineers, Second Edition. Noel de Nevers.

Gibbs energies of all ions can be measured, taking the standard state for each ion as $m_i = 1$ molal, $T = 25\,°C$, $P = 1$ bar. This standard-state definition is different from that for Raoult's or Henry's laws (see Tables 8.4 and 12.1). With this convention, the common definition of the activity coefficient is that the activity coefficient of any ion at infinite dilution $\gamma_i^\infty = 1.00$, which has the unusual property that in the standard state, $m_i = 1.00$, the activity coefficient is not necessarily = 1.00. The observational fact is that for dilute solutions (including those up to $m_i = 1.00$) the $\gamma_i \approx 1.00$, so it is common to leave the activity coefficients out of expressions for the activity of ions in dilute solutions. However, at high concentrations the activity coefficients of ions can be significantly different from 1.00. For the rest of this chapter we will express the activities of dissolved or ionic species simply as [species] with the understanding that this is the ((concentration · activity coefficient)/1 molal) and that for dilute solutions the activity coefficient ≈ 1.00. For dimension-checking purposes this means that we have multiplied K by (1 molal) to the $\sum v_i$ power.

13.2 MULTIPLE REACTIONS

Multiple reactions, in series, in parallel, or both, introduce no new concepts. Nature minimizes Gibbs energy, regardless of how many reactions are occurring. As the number or reactions increases, the mathematics and the number of variables to be accounted for increase.

13.2.1 Sequential Reactions

For sequential reactions we have one or more of the products of the first reaction serving as a reactant in the second reaction, and so on. Most dibasic acids dissociate in two steps:

$$H_2SO_4(aq) \Leftrightarrow HSO_4^- + H^+ \tag{13.E}$$

and

$$HSO_4^- \Leftrightarrow SO_4^{2-} + H^+ \tag{13.F}$$

Figure 13.1 illustrates the two reactions, showing that a product from the first reaction (bisulfate ion, HSO_4^-) is a reactant for the second, and that both reactions produce H^+. The Gibbs energies of formation for all the species in Eqs. 13.E and 13.F are shown in Table A.8. Using these values in Eq. 12.14 we find that at 25 °C, $K_{\text{Reaction 13.E}} = 104$ and $K_{\text{Reaction 13.F}} = 0.010$ (Problem 13.1). These are normally described as the *first ionization constant* or *first dissociation constant*, K_1, and the *second ionization constant* or *second dissociation constant*, K_2.

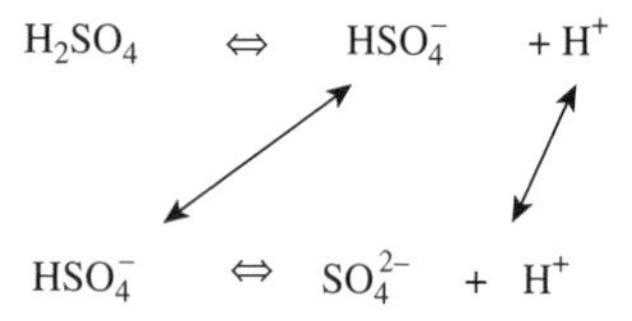

FIGURE 13.1 Schematic of Reactions 13.E and 13.F.

Example 13.2 If we add one mol of H_2SO_4 to 1000 g of water at 25 °C, at equilibrium what will be the concentrations of H_2SO_4, HSO_4^-, SO_4^{2-} and H^+?

Writing Eq. 12.14 twice, inserting the calculated dissociation constants, we find that

$$K_1 = 104 = \frac{[HSO_4^-]\cdot[H^+]}{[H_2SO_4]} \tag{13.G}$$

and

$$K_2 = 0.010 = \frac{[SO_4^{2-}]\cdot[H^+]}{[HSO_4^-]} \tag{13.H}$$

We use the scheme outlined in Section 12.6.1, letting e_1 be the mols of H^+ produced by the first ionization reaction and e_2 that by the second. Then we can write out Table 13.A, which shows that for H^+ and for HSO_4^- both reactions must be considered simultaneously to compute the equilibrium. This formulation in terms of e_1 and e_2 reduces the number of unknowns from four to two!

Substituting these values into Eqs. 13.G and 13.H, we have

$$K_1 = 104 = \frac{(e_1 - e_2)\cdot(e_1 + e_2)}{(1 - e_1)} \tag{13.I}$$

and

$$K_2 = 0.010 = \frac{(e_2)\cdot(e_1 + e_2)}{(e_1 - e_2)} \tag{13.J}$$

This set of equations is easily solved numerically, finding $e_1 = 0.9906$ mol/kg of solvent, $e_2 = 0.00988$ mol/kg of solvent. The corresponding equilibrium concentrations are shown in Table 13.B. The reader may substitute these values in Eqs. 13.I and 13.J and verify that these equations are

Table 13.A Concentrations in Sequential Ionization

	Mols = Molality in This Example
H_2SO_4	$1 - e_1$
HSO_4^-	$e_1 - e_2$
SO_4^{2-}	e_2
H^+	$e_1 + e_2$

Table 13.B Solutions to Eqs. 13.I and 13.J

	Mols = Molality in This Example
H_2SO_4	0.0094
HSO_4^-	0.9807
SO_4^{2-}	0.00988
H^+	1.0004

satisfied. Because this is a quadratic, it has two roots. The other set of values that satisfy this equation set, $e_1 = 1.000386$, $e_2 = -1.0202$, has no physical meaning. ■

We see that at this concentration the first ionization is practically complete, but that the second is only about 1 % complete. Most of the sulfuric acid placed in this solution is in the form of HSO_4^- (bisulfate) ion. In this example we computed the ionization constants from Table A.8 (see Problem 13.1). Most handbooks list the ionization constants directly, saving us this minor effort (see Problem 13.3 (b)).

13.2.2 Simultaneous Reactions

Example 13.3 Methanol is mostly produced [1,2] by the simultaneous gas-phase reactions

$$CO + 2H_2 \Leftrightarrow CH_3OH \tag{12.E}$$

and

$$CO_2 + 3H_2 \Leftrightarrow CH_3OH + H_2O \tag{13.K}$$

at pressures of about 10 MPa ($\approx$ 98.9 atm) and $\approx$250 °C, using a copper oxide–zinc oxide catalyst. Although we could analyze the reaction with these two equations, it is common to subtract Eq. 12.D from Eq. 13.K, finding

$$CO_2 + H_2 \Leftrightarrow CO + H_2O \tag{13.L}$$

At equilibrium all three of these reactions must be at equilibrium, but they are not independent of one another (see Problem 13.4). For 10 MPa and 250 °C (= 523.15 K) we have the values in Table 13.C [1, 2] (see Examples 12.9–12.11).

Table 13.C Values for Example 13.3 at 10 MPa and 250 °C

	Reaction 12.E	Reaction 13.L
K	0.00164	0.0117
$\sum \nu_i$	−2	0
K_ϕ	0.32	0.36
$\frac{K}{K_\phi}\left[\frac{P}{1\text{ atm}}\right]^{(-\sum \nu_i)}$	49.9	0.032

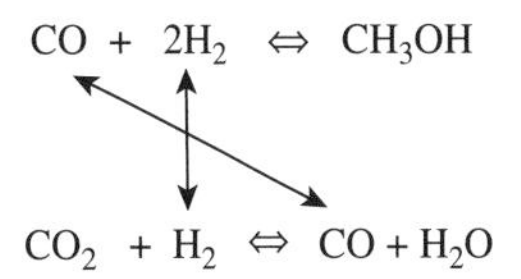

FIGURE 13.2 Schematic of Reactions 12.E and 13.L.

Figure 13.2 shows schematically Reactions 12.D and 13.L, and the linkage between the two via the CO and H_2 molecules.

The typical reactor feed (synthesis gas) is 15 mol% CO, 8 mol% CO_2, 74 mol% H_2, and 3 mol% CH_4, which latter is assumed to be inert in this reaction. Estimate the percent conversion in each of the two reactions, and the mol percent methanol at equilibrium.

Let e_1 = the mols of CO reacted and e_2 = the mols of CO_2 reacted. Then

$$y_{CO} = \frac{(n_{CO,0} + \nu_{CO,1}e_1 + \nu_{CO_2}e_2)}{(n_{T,0} + e_1(\sum \nu_i)_1 + e_2(\sum \nu_i)_2)} = \frac{0.15 - 1\cdot e_1 + 1\cdot e_2}{1 + e_1(-2) + e_2(0)} = \frac{0.15 - e_1 + e_2}{1 - 2e_1} \tag{13.M}$$

and correspondingly

$$y_{H_2} = \frac{(n_{H_2,0} + \nu_{H_2,1}e_1 + \nu_{H_2,2}e_2)}{(n_{T,0} + e_1(\sum \nu_i)_1 + e_2(\sum \nu_i)_2)} = \frac{0.74 - 2e_1 - e_2}{1 - 2e_1} \tag{13.N}$$

$$y_{CH_3OH} = \frac{(n_{CH_3OH,0} + \nu_{CH_3OH,1}e_1 + \nu_{CH_3OH,2}e_2)}{(n_{T,0} + e_1(\sum \nu_i)_1 + e_2(\sum \nu_i)_2)} = \frac{0 + e_1}{1 - 2e_1} \tag{13.O}$$

$$y_{CO_2} = \frac{(n_{CO_2,0} + \nu_{CO_2,1}e_1 + \nu_{CO_2,2}e_2)}{(n_{T,0} + e_1(\sum \nu_i)_1 + e_2(\sum \nu_i)_2)} = \frac{0.08 - e_2}{1 - 2e_1} \tag{13.P}$$

$$y_{H_2O} = \frac{(n_{H_2O,0} + \nu_{H_2O,1}e_1 + \nu_{H_2O,2}e_2)}{(n_{T,0} + e_1(\sum \nu_i)_1 + e_2(\sum \nu_i)_2)} = \frac{0 + e_2}{1 - 2e_1} \tag{13.Q}$$

and

$$\left(\frac{K}{K_\phi}\left[\frac{P}{1\text{ atm}}\right]^{(-\sum \nu_i)}\right)_{12.D} = \frac{\dfrac{0 + e_1}{1 - 2e_1}}{\left(\dfrac{0.15 - e_1 + e_2}{1 - 2e_1}\right)\cdot\left(\dfrac{0.74 - 2e_1 - e_2}{1 - 2e_1}\right)^2} = 49.9 \tag{13.R}$$

$$\left(\frac{K}{K_\phi}\left[\frac{P}{1\ \text{atm}}\right]^{\left(-\sum v_i\right)}\right)_{13.L} = \frac{\left(\frac{0.15 - e_1 + e_2}{1 - 2e_1}\right)\cdot\left(\frac{0 + e_2}{1 - 2e_1}\right)}{\left(\frac{0.08 - e_2}{1 - 2e_1}\right)\cdot\left(\frac{0.74 - 2e_1 - e_2}{1 - 2e_1}\right)} = 0.032 \tag{13.S}$$

These substitutions in terms of e_1 and e_2 reduce the number of unknowns from five to two. The resulting equations can be simplified algebraically and then solved numerically, giving $e_1 = 0.176$, $e_2 = 0.038$, and $y_{CH_3OH} = 0.272$. ■

From this calculation we see the following:

1. The conversion of the CO is almost complete:

$$\left(\frac{\text{mols CO reacted}}{\text{mols CO fed or produced}}\right) = \frac{0.176}{0.15 + 0.032} = 0.967 \tag{13.T}$$

2. The fractional conversion of the CO_2 is less, 0.038/0.08 = 0.475.
3. The hydrogen in the feed gas could react with CO and CO_2 not only to form methanol, but also to form methane (see Section 12.4). If the catalyst facilitated those reactions in addition to the methanol reactions, then most of the product would be methane. This synthesis of methanol (which is widely used industrially) is possible only because materials have been found that catalyze the methanol reactions at a temperature low enough that the methane reactions do not occur.
4. The mathematics of Example 13.3 are longer and more involved than that of the examples in Chapter 12. There is no new principle involved, and the equations are each similar to those in Chapter 12. But as the number of parallel reactions increases, the amount of stoichiometric accounting, the number of equations, and their order in the independent variables increase. (Eq. 13.S is cubic in both independent variables!) For this reason systems with large numbers of reactions in parallel lead to complex computational problems. Smith and Missen [3] devote a whole book to the mathematics of solving for the equilibrium state in systems with many reactions (mostly with many parallel reactions). For combustion, in which there are dozens or hundreds of parallel reactions, the simple hand calculation methods shown here are practically never used; [3] is mostly devoted to solving for the equilibrium concentrations in that type of reaction. Either way, the goal of the calculation is to find the compositions which corresponds to the minimum in the Gibbs energy.

13.2.3 The Charge Balance Calculation Method and Buffers

In the previous calculations we used e (the extent of the reaction) as a calculation tool. However in aqueous chemistry calculations [4] a different calculation method, "charge balance," is more convenient and widely used. We introduce it here and illustrate its use with examples of buffer solutions.

Example 13.4 To show the effects of a buffer solution, we begin with asking the pH at 25 °C of an aqueous solution of 0.1 molal acetic acid (HAc, where Ac stands for the acetate ion, CH_3COO^-). HAc is a weak acid, whose ionization (HAc) ⇔ $[H^+] + [Ac^-]$ at 25 °C is described by

$$K_{HAc} = 1.76 \times 10^{-5} = \frac{\left[\frac{H^+}{1\ \text{molal}}\right]\cdot\left[\frac{Ac^-}{1\ \text{molal}}\right]}{\left[\frac{HAc}{1\ \text{molal}}\right]} \tag{13.U}$$

For the rest of this section we drop the [1 molal] denominators, remembering that concentrations are all expressed in molality (practically equal to molarity for dilute solutions). Instead of using the extent of reaction, we ask what species will be present at equilibrium, excluding H_2O, which is present in excess, finding, for this reaction the species H^+, Ac^-, HAc and OH^-. By electroneutrality, we can write

$$[H^+] = [Ac^-] + [OH^-]. \tag{13.V}$$

We will replace $[OH^-]$ by

$$K_W = [H^+]\cdot[OH^-] \quad \text{or} \quad [OH^-] = \frac{K_W}{[H^+]} \tag{13.W}$$

Where K_w is the ionization constant for water = 10^{-14} at room temperature. By conservation of Ac we can say that

$$(HAc)_0 = (HAc) + [Ac^-] \tag{13.X}$$

Where $(HAc)_0$ is the amount of HAc introduced into the solution. Combining this with Eq. 13.U

$$K_{HAc} = \frac{[H^+]\cdot[Ac^-]}{(HAc)} = \frac{[H^+]\cdot[Ac^-]}{(HAc)_0 - [Ac^-]} \quad \text{or}$$

$$[Ac^-] = \frac{K_{HAc}(HAc)_0}{K_{HAc} + [H^+]} \tag{13.Y}$$

We now use Eqs. 13.W and 13.Y to eliminate $[Ac^-]$ and $[OH^-]$ from Eq. 13.V, finding

$$[H^+] = \frac{K_{HAc}(HAc)_0}{K_{HAc} + [H^+]} + \frac{K_W}{[H^+]} \tag{13.Z}$$

For acid solutions like this one the rightmost term in this equation is certainly negligible, but for other cases it is not, so we retain it for now. If we clear fractions and group terms we find that this is a cubic equation in $[H^+]$ with the only other quantities being the two Ks and $(HAc)_0$. With any of several equation solvers (Goal Seek on Excell) we find $[H^+] = 0.00133$ molal, $pH = -\log(0.00133) = 2.88$. Only about 1.3% of the HAc is ionized. ■

We could certainly have done this more easily using *e* methods, but as the complexity of the problems grows you will come to appreciate the benefits of this method. We may also observe that the term involving K_w contributes less than 10^{-11} molal to the answer, and could have been safely ignored in this problem.

Example 13.5 Estimate the pH of the above solution if we add to it (a) 0.03 mols of HC1, or (b) 0.03 of NaOH.

(a) The charge balance equation becomes

$$[H^+] = [Ac^-] + [OH^-] + [Cl^-] \tag{13.AA}$$

HC1 is almost totally ionized in dilute solution (K_{HCl} is some large number, often written as 1000) so that $[Cl^-] \approx (HCl)_0$. Inserting this value in the charge balance and rearranging we find

$$[H^+] = \frac{K_{HAc}(HAc)_0}{K_{HAc} + [H^+]} + \frac{K_w}{[H^+]} + (HCl)_0 \tag{13.AB}$$

The numerical solution to this cubic equation is $[H^+] = 0.0309$ molal, $pH = 1.52$.

(b) Adding NaOH instead of HCl makes charge balance

$$[H^+] + [Na^+] = [Ac^-] + [OH^-] \tag{13.AC}$$

As with HC1 so also with NaOH the ionization is practically complete, so that we may write $[Na^+] \approx (NaOH)_0$. Inserting this value in the charge balance and rearranging, we find

$$[H^+] = \frac{K_{HAc}(HAc)_0}{K_{HAc} + [H^+]} + \frac{K_w}{[H^+]} - (NaOH)_0 \tag{13.AD}$$

The numerical solution to this cubic equation is $[H^+] = 0.000041$ molal, $pH = 4.38$. ■

Figure 13.3 shows the three values from these two examples as part of the curve marked "0.1 mol/L HAc." The reader may verify that those three values fall on that curve, which was made by repeating Example 13.5 for a variety of values, plotting them and drawing a smooth curve through them. The other curve will be discussed below.

Now we start with the 0.1 mol/L solution of HAc and add 0.08 mols of NaAc, creating a buffer solution, whose properties are also sketched on Figure 13.3.

Example 13.6 Repeat the calculations in Examples 13.4 and 13.5 for the buffer solution described above.

The equilibrium constants are unchanged, and the charge balance is the same as Eq. 13.AB. NaAc is practically totally ionized, so that $[Na^+] \approx (NaAc)_0$. The balance on Ac^- becomes

$$(HAc)_0 + (NaAc)_0 = (HAc) + [Ac^-] \tag{13.AE}$$

the analog of Eq. 13.Y becomes

$$K_{HAc} = \frac{[H^+] \cdot [Ac^-]}{(HAc)_0 + (NaAc)_0 - [Ac^-]} \quad \text{or}$$

$$[Ac^-] = \frac{K_{HAc}\left((HAc)_0 + (NaAc)_0\right)}{K_{HAc} + [H^+]} \tag{13.AF}$$

and the analog of Eq. 13.Z becomes

$$[H^+] = \frac{K_{HAc}\left((HAc)_0 + (NaAc)_0\right)}{K_{HAc} + [H^+]} + \frac{K_w}{[H^+]} \tag{13.AG}$$

The numerical solution to this cubic equation is $[H^+] = 2.18 \times 10^{-5}$, $pH = 4.66$. The reader may verify that this is the value for 0 added HCl or NaOH on the upper curve on Figure 13.3.

Then, if we add 0.03 mols of HCl, the charge balance becomes

$$[H^+] + [Na^+] = [Ac^-] + [OH^-] + [Cl^-] \tag{13.AH}$$

As in the previous example, HC1 is almost totally ionized, so that $[Cl^-] \approx (HCl)_0$ and Eq. 13.AF becomes

$$[H^+] = \frac{K_{HAc}\left((HAc)_0 + (NaAc)_0\right)}{K_{HAc} + [H^+]} + \frac{K_w}{[H^+]} + (HCl)_0 \tag{13.AI}$$

The numerical solution to this cubic equation is $[H^+] = 4.56 \times 10^{-5}$, $pH = 4.34$. The reader may verify that this is the value for 0.03 mol added HCl on the upper curve on Figure 13.3.

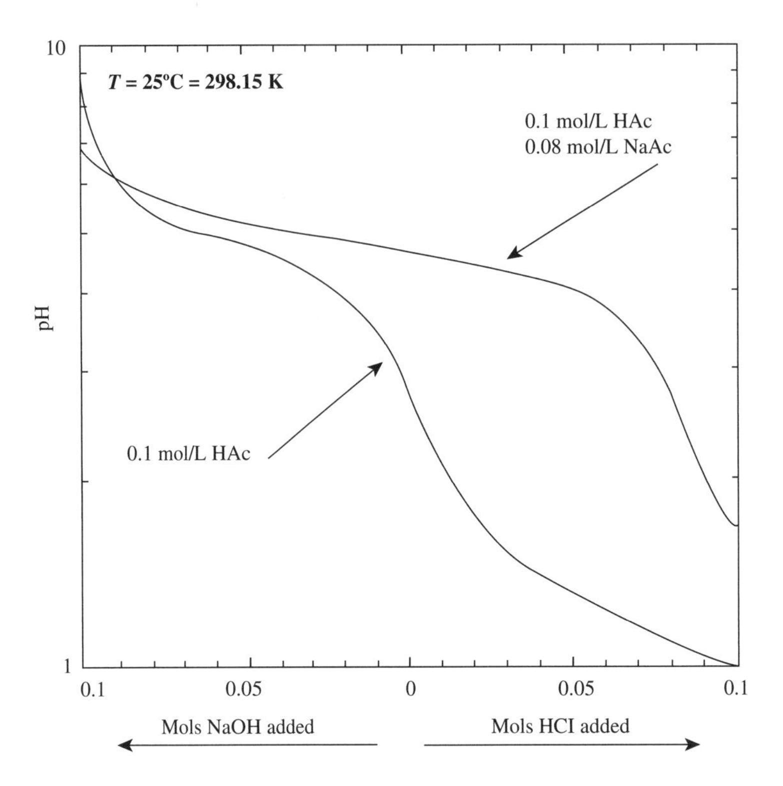

FIGURE 13.3 Response of an ordinary solution of a weak acid (HAc) and of a buffer solution of HAc and NaAc to additions of HC1 and of NaOH. Observe the log scale for pH, and the unusual horizontal coordinate with the values plotted increasing in each direction from zero.

Finally, if we add 0.03 mols of NaOH instead of HCl, then by the same procedure we find

$$[H^+] = \frac{K_{HAc}\left((HAc)_0 + (NaAc)_0\right)}{K_{HAc} + [H^+]} + \frac{K_w}{[H^+]} - (NaOH)_0 \tag{13.AJ}$$

The numerical solution to this cubic equation is $[H^+] = 1.11 \times 10^{-5}$, pH = 4.95. The reader may verify that this is the value for 0.03 mols of added NaOH on the upper curve on Figure 13.3. ■

From these calculations and their summary on Figure 13.3 we can say

1. The buffer works by shifting the equilibrium $[H^+] + [Ac^-] \Leftrightarrow (HAc)$. If we add HCl or some other acid, the additional H^+ reacts with Ac^- to remove H^+ from the solution, minimizing the effect of the added acid. If we add NaOH or some other base, the OH^- ions take H^+ ions out of solution but the HAc supplies additional H^+ ions, minimizing the effect of the added base.
2. The effect can be dramatic. From the above examples we see that shifting from 0.03 mols of HCl added to 0.03 mols of NaOH added increases the pH of the unbuffered acid by $(4.38 - 1.52 = 2.86$, increasing $[H^+]$ by $10^{2.86} = 724)$ while the same change for the buffered solution changes the pH by $(4.95 - 4.34 = 0.61$, increasing $[H^+]$ by $10^{0.61} = 4)$. On Figure 13.3 the buffer curve between 0.03 HCl and 0.03 NaOH is practically flat, while the corresponding curve for the unbuffered HAc is steep.
3. Figure 13.3 shows that the buffering works well for acid or base additions up to about half of the amount of buffering agent (0.08 mol/L), but becomes weaker as the buffering capacity approaches exhaustion.
4. Buffer solutions play little role in large-scale industrial chemistry; they have some uses in laboratory chemistry. But they play a significant role in biochemistry. The reactions in living organisms are very sensitive to pH; nature provides the buffers to make those reactions proceed; see Chapter 16. Your body is using buffer solutions millions of times a second to regulate the biochemistry that makes your life possible.
5. Although the charge balance is the quickest and easiest way to solve this class of problems, it is not the universal solver. The method shown in Example 13.2 is easiest for that class of problems. (Try that example by charge balance, you will be impressed!)

13.3 REACTIONS WITH MORE THAN ONE PHASE

All of the previous examples have been for reactions completely in one phase. However, there is no restriction in the basic equations in Chapter 12 that says that they must all be in the same phase, as the following examples show.

13.3.1 Solubility Product

The above discussion also applies to reactions involving ions and solids.

Example 13.7 Silver chloride is very slightly soluble in water. The dissolution is believed to follow Eq. 13.AK:

$$\text{AgCl(s)} \Leftrightarrow \text{Ag}^+ + \text{Cl}^- \tag{13.AK}$$

This suggests that there are no molecules of solid AgCl dissolved in the water, but rather the crystalline AgCl that does enter the liquid phase exists only in the form of these ions. The data on solubility of slightly soluble salts suggest that this is the case. Estimate the amount dissolved in equilibrium with solid AgCl at 25 °C.

Using the values from Table A.8, we find

$$\begin{aligned}\Delta g^{\text{o}} &= g^{\text{o}}_{\text{Ag}^+} + g^{\text{o}}_{\text{Cl}^-} - g^{\text{o}}_{\text{AgCl}} \\ &= 77.12 + (-131.26) - (-109.8) \\ &= 55.66\,\frac{\text{kJ}}{\text{mol}}\end{aligned} \tag{13.AL}$$

$$\begin{aligned}K &= \exp\left(\frac{-\Delta g^\circ}{RT}\right) = \exp\left[\frac{-\left(55{,}660\,\dfrac{\text{J}}{\text{mol}}\right)}{8.314\,\dfrac{\text{J}}{\text{mol K}}\cdot 298.15\text{ K}}\right] \\ &= 1.77\times10^{-10} = \frac{\left[\dfrac{[\text{Ag}^+]}{1\text{ molal}}\right]\cdot\left[\dfrac{[\text{Cl}^-]}{1\text{ molal}}\right]}{[a_{\text{AgCl}}]}\end{aligned} \tag{13.AM}$$

For solids $f_i^{\text{o}}$is normally taken as the fugacity of the pure crystalline solid at the temperature of the system (see Table 12.1). If the solid involved in the reaction is that pure crystalline solid (which means that the other species in the system do not dissolve in it to any significant extent at T° and P°), then the activity of the solid = 1.00. This is almost always assumed in this type of calculation, and will be assumed here. Then

$$\left[\frac{[\text{Ag}^+]}{1\text{ molal}}\right]\cdot\left[\frac{[\text{Cl}^-]}{1\text{ molal}}\right] = 1.77\times10^{-10} = K_{\text{sp}} = \begin{pmatrix}\text{solubility}\\ \text{product}\end{pmatrix} \;\blacksquare \tag{13.AN}$$

This shows the definition of the *solubility product.* Extensive tables of solubility products reside in handbooks [5, p. 8–39]. The reported value there for AgCl is 1.77×10^{-10}. The fact that this value and the result of the above example are identical does not prove that they are right. Instead, it shows that a consistent data set of some kind was used to generate both the g° values in Table A.8 and the K_{sp} values in the corresponding table, calculated exactly as shown in this example. The point of this example is to show that solubility product and other quantities regularly used in dilute aqueous chemistry are computed exactly the same way as chemical reaction equilibria. (Here we have assumed ideal solutions of ions, $\gamma_{\text{Cl}^-} = \gamma_{\text{Ag}^+} = 1.00$. For dilute solutions this is a good assumption, but for concentrated solutions—those with a high "ionic strength"—these activity coefficients can take on values much different from 1.00.)

13.3.2 Gas-Liquid Reactions

Several industrially important gases dissolve in liquids and then react chemically in the liquids. For carbon dioxide in water the sequence seems to be

$$CO_2\text{ (in gas)} \Leftrightarrow CO_2\text{ (dissolved in liquid)} \tag{13.AO}$$

$$CO_2\text{ (dissolved in liquid)} + H_2O \Leftrightarrow H_2CO_3 \tag{13.AP}$$

$$H_2CO_3 \Leftrightarrow H^+ + HCO_3^- \tag{13.AQ}$$

$$HCO_3^- \Leftrightarrow H^+ + CO_3^{2-} \tag{13.AR}$$

The sequence for SO_2 (see Problem 13.16) seems to be exactly the same, with all the Cs in the preceding sequence of reactions replaced by Ss. In both cases the pure, undissociated acid, H_2CO_3 or H_2SO_3, does not exist as a pure substance. It exists only is solution, and if we attempt to concentrate it by removing water, Reactions 13.AO and 13.AP (or their analogs for SO_2) operate in the reverse direction, giving back the gas and water and destroying the acid. The existence of undissociated H_2CO_3 has been inferred by spectroscopic measurements, and its existence seems to be real [6]. However, because it does not exist in a pure state, the values for it in Table A.8 are based on the *convention* that

$$H_2CO_3\text{(aq)}\begin{pmatrix}\text{is the same}\\ \text{material as}\end{pmatrix}\text{ or }\begin{pmatrix}\text{has the same}\\ \text{thermodynamic}\\ \text{properties as}\end{pmatrix}(CO_2\text{(aq)} + H_2O\text{(l)}) \tag{13.AS}$$

Table 13.D Gibbs Energy and Enthalpy Changes of the Formation of Carbonic and Sulfurous Acids, Using Values from Table A.8

Gibbs Energies and Enthalpies (kJ/mol)	CO_2	SO_2
g_i^o (aq)	−386.0	−300.7
g_i^o water	−237.1	−237.1
g_i^o H_2CO_3 or H_2SO_3(aq)	−623.1	−537.9
Δg° of Reaction 13.AP or 13.CE	0.0	−0.1≈0.0
h_i^o(aq)	−413.8	−323.0
h_i^owater	−285.8	−285.8
h_i^o H_2CO_3 or H_2SO_3 (aq)	−699.7	−608.8
Δh° of Reaction 13.AP or 13.CE	−0.1 ≈ 0.0	0.0

and correspondingly for SO_2. We may see this from Table 13.D.

We see that the values shown agree with the convention in Eq. 13.AS and correspondingly

$$H_2SO_3(aq)\begin{pmatrix}\text{is the same}\\ \text{material as}\end{pmatrix} \text{ or }$$

$$\begin{pmatrix}\text{has the same}\\ \text{thermodynamic}\\ \text{properties as}\end{pmatrix}(SO_2(aq) + H_2O(l)) \tag{13.AT}$$

The reason for this convention is that when we are thinking about the gas-liquid equilibrium, we treat it by Henry's law, which relates the partial pressure of CO_2 or SO_2 in the gas to the concentration of the same material dissolved in the liquid, while when we are talking about the ionization reactions we think about the dissolved CO_2 or SO_2 as the undissociated carbonic or sulfurous acid.

Figure 13.4 shows the overall scheme of these reactions, without the acid H_2CO_3 being shown. In this formulation, the vertical arrow represents the dissolution of CO_2 in water. This is seen as a physical equilibrium (not a chemical one) and is described by Henry's law. The two horizontal arrows represent the first and the second ionizations of carbonic acid (= dissolved $CO_2 + H_2O$), which are seen as chemical reactions. Ions have almost no vapor pressure; they do not exist in the gas phase (except at flame temperatures). The carbon-containing species can leave the liquid only by forming dissolved CO_2, which has a vapor pressure.

Example 13.8 Estimate the amount of CO_2 dissolved and the concentrations of HCO_3^-, CO_3^{2-}, and H^+ when water is in equilibrium with atmospheric air, which contain 390 ppm of CO_2 at 1 atm and 20 °C. Assume that air behaves as an ideal gas and that the liquid and the ions behave as ideal solutions.

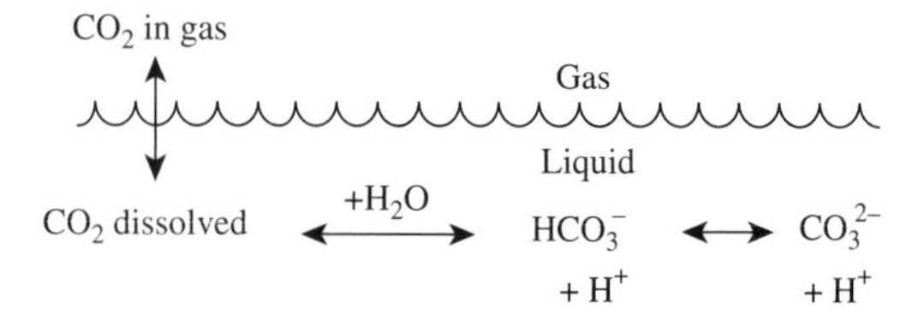

FIGURE 13.4 Typical representation of the CO_2–water equilibrium.

Taking the two ionizations constants from Problem 13.10 and the value of the Henry's law constant from Table A.3 (and ignoring the difference between 20 and 25 °C), we can write

$$\frac{[HCO_3^-]\cdot[H^+]}{[CO_2(aq)]} = \frac{[HCO_3^-]\cdot[H^+]}{[H_2CO_3]} = K_1 = 4.5\times 10^{-7} \tag{13.AU}$$

$$\frac{[CO_3^{2-}]\cdot[H^+]}{[HCO_3^-]} = K_2 = 4.7\times 10^{-11} \tag{13.AV}$$

and

$$x_{CO_2} = \frac{Py_{CO_2}}{H} = \frac{1\text{ atm}\cdot 390\times 10^{-6}}{1480\text{ atm}} \tag{13.AN}$$

We can solve these three equations one at a time, finding

$$x_{CO_2} = \frac{Py_{CO_2}}{H} = \frac{1\text{ atm}\cdot 390\times 10^{-6}}{1480\text{ atm}} = 2.63\times 10^{-7} \tag{13.AX}$$

This gives the mol fraction. The dissociation constants are based on molalities as standard states (see Problem 13.10), so

$$[CO_2(aq)] \approx x_{CO_2}\cdot\frac{55.6\text{ mol}}{1000\text{ g of }H_2O} = 2.63\times 10^{-7}\cdot 55.6$$
$$= 1.47\times 10^{-5}\text{ molal} \tag{13.AY}$$

Then we assume that almost all the H^+ comes from the dissociation of dissolved CO_2, so

$$[HCO_3^-] \approx [H^+] = \sqrt{K_1\cdot[CO_2(aq)]}$$
$$= \sqrt{4.5\times 10^{-7}\cdot 1.47\times 10^{-6}}$$
$$= 2.6\times 10^{-6}\text{ molal} \tag{13.AZ}$$

Then we compute

$$[CO_3^{2-}] = K_2\frac{[HCO_3^-]}{[H^+]} \approx K_2 = 4.7\times 10^{-11}\text{ molal} \quad ■ \tag{13.BA}$$

From this example we see the following:

1. About 85% of the dissolved CO_2 is in the undissociated form ($CO_2(aq)$ or H_2CO_3). The ratio of

$$\frac{[HCO_3^-]}{[HCO_3^-]+[CO_2(aq)]}=\frac{2.6\times10^{-6}}{2.6\times10^{-6}+1.32\times10^{-5}}=0.15 \quad (13.BB)$$

   (See Problem 13.10.)
2. The contribution of the second dissociation to the total H^+ concentration is negligible, ($4.7\times10^{-11} \ll 2.6\times10^{-6}$).
3. We normally see the H^+ concentration expressed as

$$\text{pH}=-\log(a_{H^+})\approx-\log(\text{molality of } H^+) \quad (13.BC)$$

For dilute solutions, molality ≈ molarity, which is the way this is most often presented. In this case it is $-\log(2.6\times10^{-6})=5.59$, which is the observed pH of rainfall (which practically comes to equilibrium with the CO_2 in the atmosphere through which it falls) in unpolluted environments.

We can shift this equilibrium in either direction by making the water acid or alkaline.

Example 13.9 Repeat Example 13.8 for water that contains enough alkali (NaOH) that $[H^+]=10^{-10}$ molal, (i.e., pH = 10).

Our Henry's law calculations (Eqs. 13.AW and 13.AX) are independent of the subsequent fate of the dissolved CO_2. The concentration of dissolved CO_2 in equilibrium with the atomsphere is $[CO_2]=1.47\times10^{-5}$ molal, independent of that acidity or basicity of the water. Then

$$[HCO_3^-]=K_1\frac{[CO_2(aq)]}{[H^+]}=K_1\frac{[H_2CO_3]}{[H^+]}=4.5\times10^{-7}\cdot\frac{1.47\times10^{-5}}{10^{-10}}=0.066\ \text{molal} \quad (13.BD)$$

and

$$[CO_3^{2-}]=K_2\frac{[HCO_3^-]}{[H^+]}=4.7\times10^{-11}\frac{0.066}{10^{-10}}=0.0310\ \text{molal} \quad ■ \quad (13.BE)$$

From this example we see that the dissolved CO_2 is 68% in the form of bicarbonate ion, 32% in the form of carbonate ion, and 0.01% in the form of $CO_2(aq)$. The total amount of dissolved carbonate species is ≈6000 times the amount that exists as dissolved CO_2, which is ≈ the amount that can be dissolved in pure water. This type of equilibrium is very important industrially. In the production of ammonia (and many other processes) it is necessary to remove CO_2 from a gas stream that is mostly H_2 (or a mixture of H_2 and N_2). The general scheme for removing one component from a gas stream is shown in Figure 13.5. The solvent used is normally water containing

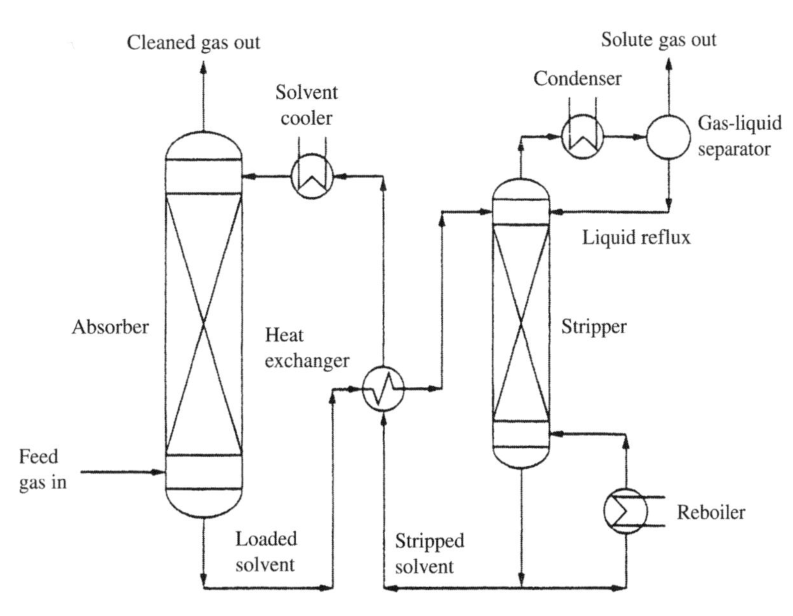

FIGURE 13.5 The general scheme for removing one species from a gas stream. This only works if a solvent can be found that is *selective* for the species to be removed. In the case of removing CO_2 from a stream that contains no other *acid gases* this works quite well with an aqueous solution of any weak base as the absorbent. (From de Nevers, N. *Air Pollution Control Engineering,* ed. 2. New York: McGraw–Hill, p. 362 (2000). Reproduced with permission of the McGraw-Hill Companies.)

a weak base (monoethanol-amine or potassium carbonate), which can form an un-ionized salt with carbonate or bicarbonate ion. That removes the carbonate or bicarbonate ion, and drives the equilibria in the direction of absorption of CO_2. In the absorber the P and T are chosen to have the CO_2 highly soluble in the absorbing solution. The stripper P and T are chosen to greatly reduce the CO_2 solubility, (normally a lower P and/or a higher T than in the absorber) driving CO_2 out of solution, regenerating the solvent and often recovering CO_2 as a practically pure gas.

In the scrubbing of combustion gases to remove SO_2 there are two acid gases present, SO_2 and CO_2. The scrubbing solution is kept acid enough (pH 4 to 6) to prevent the unwanted absorption of CO_2 but not so acid as to exclude SO_2, which forms a stronger acid than does CO_2 [7, Chapter 11]. The equilibria discussed in this section also occur with other gases which react with water. Edwards et al. [8] show the data and calculation methods for aqueous solutions of CO_2, H_2S, HCN, SO_2, and NH_3, both dilute and concentrated, taking into account the nonideal behavior of the solutions, which is ignored in the above examples. The type of equilibrium discussed in the previous two examples is widely presented as a plot like Figure 13.6. Such plots are available for a variety of systems [9, p. 508]. In Example 13.8 we computed that (at pH = 5.59) HCO_3^- was about 15% of the dissolved carbonate species, CO_3^{-2} was negligible, and thus ≈ 85% of the total carbonate species were dissolved CO_2, which Figure 13.6 shows in the alternative form, H_2CO_3; these are the values we read from Figure 13.6.

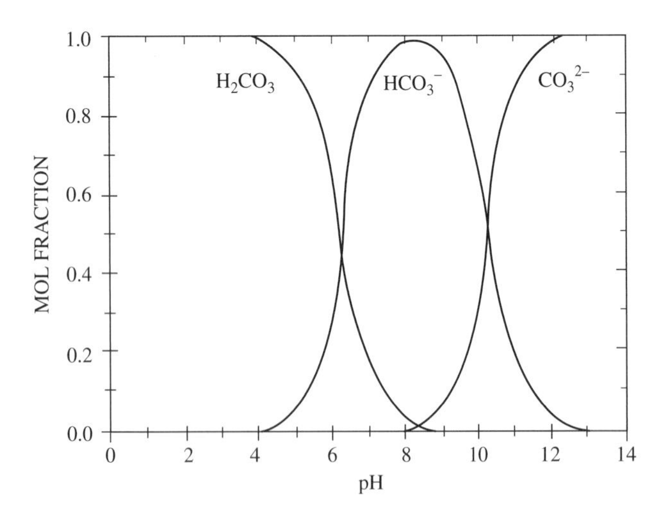

FIGURE 13.6 Distribution among carbonate species in aqueous solution as a function of pH. In this and similar plots, the vertical axis, labeled "mol fraction" is actually the fraction by mol of the total carbonate species, which is in each of the three forms shown. (From Kohl, A., and R. Nielsen. *Gas Purification,* ed. 5. Houston, TX: Gulf, p. 508 (1997)).

13.4 ELECTROCHEMICAL REACTIONS

Many chemical reactions have large positive Gibbs energy changes, and hence very low values of the equilibrium constant; for example,

$$Al_2O_3 + 1.5C \Leftrightarrow 2Al + 1.5CO_2 \tag{13.BF}$$

for which

$$\Delta g^o_{298.15K} = 2 \cdot 0 + 1.5 \cdot (-394.4) - (-1582.3) - 1.5 \cdot 0 = 990.7 \frac{kJ}{mol}$$

$$K_{298.15K} = 2.7 \times 10^{-174} \tag{13.BG}$$

The −174 is not a misprint. All of the aluminum in the world is made by this reaction (the Hall–Heroult process), using electricity to drive it "up a Gibbs energy hill." The opposite side of this coin is the production of electricity by chemical reactions, such as the reactions in dry cells, lead storage batteries, and fuel cells, in which chemical reactions with large decreases in Gibbs energy are used to produce or store electricity.

Figure 13.7 shows a schematic of a steady flow, isothermal electrochemical reactor. The steady flow first law statement (leaving out kinetic, potential, surface, etc. energies) for this reactor is

$$0 = \sum (\bar{h}_i \dot{n}_i)_{in} - \sum (\bar{h}_i \dot{n}_i)_{out} + \dot{Q} - EI \tag{13.1}$$

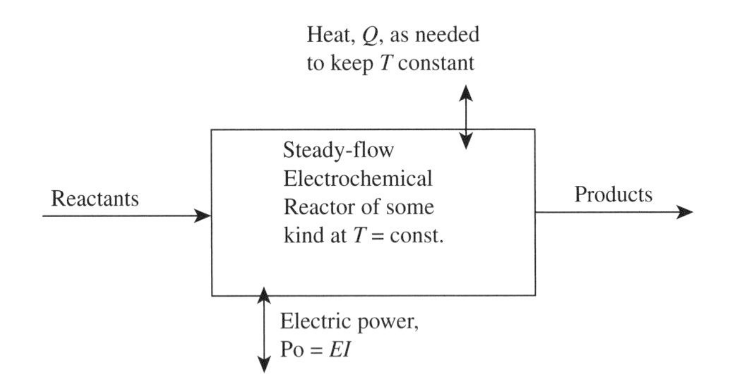

FIGURE 13.7 Schematic of a steady-flow, isothermal, electrochemical reactor. The heat flow arrow is two-headed because heat may flow in or out as needed to hold the temperature constant. The electric power arrow is shown two-headed because if this is an electrochemical cell like those that produce metallic aluminum, then the flow is in, while if it is a fuel cell the electric energy flow is out, and if it is a storage battery the electric energy flow is in while it is charging and out while it is discharging. Only one arrow is shown for reactants or products, but there may be multiple flows in or out, or the flows in and out may be zero, such as for a dry cell battery.

and the steady-flow second law statement is

$$0 = \sum (\bar{s}_i \dot{n}_i)_{\text{in}} - \sum (\bar{s}_i \dot{n}_i)_{\text{out}} + \frac{\dot{Q}}{T} + \frac{dS_{\text{irreversible}}}{dt} \quad (13.2)$$

If we now restrict our attention to reversible reactors (i.e., those that, among other things, operate at chemical equilibrium), then we may drop the rightmost term of Eq. 13.2, and eliminate $\dot{Q}$ between the two equations, finding

$$0 = \sum (\bar{g}_i \dot{n}_i)_{\text{in}} - \sum (\bar{g}_i \dot{n}_i)_{\text{out}} - EI \quad (13.3)$$

Comparing this to ordinary chemical reactions we see that if $EI = 0$, then the Gibbs energy of the outlet streams must be the same as that of the inlet streams, or $\Delta(n_i \bar{g}_i) = 0$, which we have been using as the ordinary criterion for chemical equilibrium. If we solve Eq. 13.3 for E, we find

$$E = \frac{\sum (\bar{g}_i \dot{n}_i)_{\text{in}} - \sum (\bar{g}_i \dot{n}_i)_{\text{out}}}{I} = \frac{-\Delta(\bar{g}_i \dot{n}_i)}{I} = \frac{-\dot{e}\Delta g}{I} \quad (13.4)$$

where $\dot{e}$ is the rate of consumption of the reactant with $v = -1$, or of production of the reactant with $v = +1$.

Next we make use of the experimental fact, demonstrated by Michael Faraday and sometimes called *Faraday's law*, that 1.00 mol of electrons (i.e., 6.02×10^{23} electrons) = 96,500 coulombs, and that I is expressed (in SI) in ampere = coulomb/second:

$$I = n_e \frac{\left(\begin{array}{c}\text{mols electrons}\\ \text{transfered}\end{array}\right)}{\text{mol reacted}} \cdot \dot{e}\frac{(\text{mol reacted})}{\text{s}} \cdot \frac{96{,}500 \text{ coulombs}}{\text{mol electrons}} = n_e \dot{e} F \quad (13.5)$$

F, called *Faraday's constant*, is a simple unit conversion factor, equal to one; we may multiply or divide any quantity by F without changing its value. However, its use in the formulae of electrochemistry is almost universal. It will be used in this book, even though we all know that it is equal to 1.00. n_e is the number of mols of electrons transferred for the reaction as written.

Combining Eqs. 13.4 and 13.5 we find

$$E = \frac{\sum (\bar{g}_i \dot{n}_i)_{\text{in}} - \sum (\bar{g}_i \dot{n}_i)_{\text{out}}}{I} = \frac{-\Delta(\bar{g}_i \dot{n}_i)}{n_e \dot{e} F} = \frac{-\Delta g}{n_e F} \quad (13.6)$$

which is general for any value of Δg. If we let $\Delta g = \Delta g^\circ$ then we calculate $E = E^\circ$, where E° is the "standard state voltage," corresponding to Δg°, the "standard Gibbs energy change." If we write Eq. 13.6 twice, once for any set of reactant and product concentrations, and once for reactants and products all in their standard states, and subtract one from the other, we find

$$E - E^\circ = \frac{-(\Delta g - \Delta g^\circ)}{n_e F} = \frac{-RT \ln \dfrac{\left[a_i^{v_i} \cdot \text{etc}\right]_{\text{Products}}}{\left[a_i^{v_i} \cdot \text{etc}\right]_{\text{Reactants}}}}{n_e F} \quad (13.7)$$

which is the *Nernst equation.* We often see this equation without the minus sign and with the $\left[a_i^{v_i} \cdot \text{etc}\right]_{\text{Products}} / \left[a_i^{v_i} \cdot \text{etc}\right]_{\text{Reactants}}$ term inverted. The standard state voltages are easily calculated (see Example 13.10) and their values are widely published. With Eq. 13.7 we can see the effect of changing concentrations on the reversible (equilibrium) voltage.

Example 13.10 Estimate the standard state cell voltage for the production of aluminum (Eq. 13.BF) at 298.15 K.

For that reaction as written $n_e = 6$, and all the reactants and products enter or leave the reactor as pure species in their standard states (there are almost no solutions, because the solubility of CO_2 gas in molten Al or solid C is ≈ 0, and the vapor pressure of Al at 25 °C is ≈ 0), so $\Delta g = \Delta g^\circ$ and $E = E^\circ$, the "standard state voltage," so

$$E^\circ_{298.15\,\text{K}} = \frac{-\Delta g^\circ}{n_e F} = \frac{-990{,}700 \dfrac{\text{J}}{\text{mol}}}{6 \dfrac{\text{mol electrons}}{\text{mol}} \cdot 96{,}500 \dfrac{\text{coulomb}}{\text{mol electrons}}} \cdot \frac{\text{volt} \cdot \text{coulomb}}{\text{J}} = -1.71\text{V} \quad ■ \quad (13.\text{BH})$$

Industrially this reaction is carried out at 920–980 °C at which the equilibrium voltage is −2.23 V. To produce aluminum at commercially useful rates, voltages about twice this equilibrium value are used. In addition to Reaction 13.BF, there is also the reaction

$$Al_2O_3 + 3C \Leftrightarrow 2Al + 3CO \quad (13.\text{BI})$$

which uses up twice as much carbon and has a higher equilibrium voltage (see Problem 13.22). We would prefer that this reaction not occur, but industrially 10 to 50% of the gas produced is CO, balance CO_2 [10]. Shreve [11] presents a list of 35 electrochemical processes in current industrial use; many of them are like this example, using electrical work to overcome a large positive Gibbs energy change.

Example 13.11 Calculate the reversible voltage for some kind of electrochemical device that would react pure lithium with pure fluorine, producing solid LiF.

From Table A.8 we find that $\Delta g^\circ = -587.7$ kJ/mol. We also know that $n_e = 1$, because the valence state of Li and of F change by 1, so one electron is transferred per molecule of LiF. Thus,

$$E^\circ_{298.15\,\mathrm{K}} = \frac{-\Delta g^\circ}{n_e F}$$

$$= \frac{-\left(-587{,}700\,\dfrac{\mathrm{J}}{\mathrm{mol}}\right)}{1\,\dfrac{\text{mol electrons}}{\mathrm{mol}} \cdot 96{,}500\,\dfrac{\text{coulomb}}{\text{mol electrons}}} \cdot \frac{\text{volt} \cdot \text{coulomb}}{\mathrm{J}}$$

$$= 6.09\ \mathrm{V} \quad \blacksquare \qquad (13.BJ)$$

The normal sign convention has E negative for power-consuming cells and E positive for power-producing cells, as shown in these two examples. This example shows the voltage for the reaction between the pure materials to produce a pure product. In most electrochemical cells the reactants are dissolved in an electrolyte, normally water, and the resulting voltage is slightly different. The Li-F cell shown here has one of the highest theoretical voltages of any simple electrochemical cell. The ordinary 12-V storage battery in our cars has six 2-V cells in series. If we could make this theoretical lithium-fluorine cell a practical cell, we could use two in series instead. The practical difficulties of making a lithium-fluorine cell are serious.

Here we computed the voltage by computing the standard Gibbs energy change. Handbooks [5, p. 8–20] present extensive tables of "half-cell" voltages. From those tables we may read

$$\begin{array}{ll} F_2 + 2e \Leftrightarrow 2F^- & E^\circ = 2.866\ \mathrm{V} \\ Li^+ + e \Leftrightarrow Li & E^\circ = -3.0401\ \mathrm{V} \end{array} \qquad (13.BK)$$

Subtracting the second from the first we find a cell voltage of + 5.907 V. The difference between this result and the 6.09 V in Example 13.11 is that 6.09 V corresponds to starting with pure materials and producing pure LiF, while this example produces a 1 molal solution of LiF. As shown in Problem 13.26, if we compute the Gibbs energy change for this cell, ending with a 1 molal solution of LiF (aq), the computed value is very close to the 5.907 V shown above.

These examples were chosen to have $\Delta g = \Delta g^\circ$, making them very simple. We can also consider cells in which the reactants or products do not enter or leave at their standard states, finding that they are not much more difficult. The general approach is to use the Nernst equation (Eq. 13.7), but the approach shown in the next example gives the same result in a more intuitively satisfying way.

Example 13.12 A steady-flow electrolytic cell produces hydrogen and oxygen by the reaction,

$$H_2O\ (I) \rightarrow H_2(g) + \frac{1}{2}O_2(g) \qquad (13.BL)$$

The gases are discharged to their respective storage reservoirs at pressure P. When $P = 1$ atm and $T = 25\,^\circ$C, the required ("standard") equilibrium cell voltage is $E^\circ = -1.229$ V. Estimate the equilibrium cell voltage if feed and products are all at 100 atm. Assume that O_2 and H_2 are ideal gases and that the specific volume of water is negligible.

Equation 13.6 applies to this reaction, but we must compute for each material

$$g = g^\circ + \int_{P^\circ}^{P} \left(\frac{\partial g}{\partial P}\right)_T dP = g^\circ + \int_{P^\circ}^{P} v_T dP = g^\circ + v_T \ln\frac{P}{P^\circ} \qquad (13.8)$$

In the rightmost term we have replaced v by RT/P, which is correct only for ideal gases. According to the assumption, we can ignore the change in Gibbs energy with pressure of the liquid water, so that

$$\Delta g = \Delta g^\circ + 1.5RT \ln\frac{P}{P^\circ} \qquad (13.BM)$$

and

$$E = \frac{-\Delta g}{n_e F} = \frac{-\left(\Delta g^\circ + 1.5RT\ln\dfrac{P}{P^\circ}\right)}{n_e F} = E^\circ + \frac{-1.5RT\ln\dfrac{P}{P^\circ}}{n_e F} \qquad (13.BN)$$

$$E - E^\circ = \frac{-1.5RT\ln\dfrac{P}{P^\circ}}{n_e F}$$

$$= \frac{-1.5 \cdot 8.314\,\dfrac{\mathrm{J}}{\mathrm{mol\ K}} \cdot 298.15\,\mathrm{K} \cdot \ln\dfrac{P}{P^\circ}}{2\,\dfrac{\text{mol electrons}}{\mathrm{mol}} \cdot 96{,}500\,\dfrac{\text{coulomb}}{\text{mol electrons}}} \cdot \frac{\text{volt} \cdot \text{coulomb}}{\mathrm{J}}$$

$$= -0.01926\ \mathrm{V} \cdot \ln\frac{100\ \mathrm{atm}}{1\ \mathrm{atm}} = -0.0887\ \mathrm{V} \qquad (13.BO)$$

and

$$E = -1.229 - 0.0887 = -1.318\ \mathrm{V} \quad \blacksquare$$

To find the equilibrium voltage at some other temperature, we compute the change in Δg° with temperature the same way we did in Chapter 12 (see Problem 13.30).

We can readily show that, for any kind of constant-pressure cell, the reversible voltage depends only on the concentrations and pressures of the chemicals involved and not on whether the cell is flow or batch. In practice, all real cells operated at commercially useful rates have voltages different than these equilibrium voltages, lower voltages when power is being withdrawn, and higher voltages when power is being supplied. Normally, cell efficiencies are defined in terms of the ratio of actual to equilibrium voltages.

13.5 CHEMICAL AND PHYSICAL EQUILIBRIUM IN TWO PHASES

In the previous sections we have considered chemical equilibrium in one phase, or chemical equilibrium in one phase with simultaneous physical equilibrium with another phase, for example, gas-liquid or liquid-solid equilibrium, with chemical equilibrium in the liquid phase. If we had the universal catalyst, which made all chemical reactions come to equilibrium in all phases, what would the equilibrium situation be? The situation is sketched in Figure 13.8.

Here for the reaction A ⇔ B (e.g., *n*-butane ⇔ isobutane), we show chemical equilibrium between A and B in each of the two phases, plus physical equilibrium between A in the gas and A in the liquid and similarly for B. For chemical equilibrium in any reaction in which the number of moles does not change, we know that $\bar{g}_a = \bar{g}_b$. This must be true in each phase. Similarly, for physical equilibrium we know that

$$\bar{g}_a^{(\text{liquid})} = \bar{g}_a^{(\text{gas})} \tag{13.9}$$

and that a similar relation exists for B. Thus, for the particularly simple case sketched in Figure 13.8, we would have

$$\bar{g}_a^{(\text{liquid})} = \bar{g}_a^{(\text{gas})} = \bar{g}_b^{(\text{liquid})} = \bar{g}_b^{(\text{gas})} \tag{13.10}$$

This does not mean that the compositions are the same. For the *n*-butane-isobutane example shown in Figure 12.1, in the gas phase (assuming ideal gases and ideal solutions) we calculated in Example 12.1 that at 25 °C, $y_{\text{isobutane}} \approx 0.82$. That was for an assumed pressure of 1 atm, at which the system would be all gas. If we were to isothermally compress the gas until an equilibrium liquid formed, we would calculate its composition by the temperature-specified dew-point calculation shown in Example 8.10. If we assume that the liquid is an ideal solution (a good assumption for this system), then using Raoult's law we would find

$$x_{\text{isobutane}} = \frac{y_{\text{isobutane}} P}{p_{\text{isobutane}}} \tag{13.BP}$$

and similarly for n-butane. We would also find that

$$P = x_{\text{isobutane}} \cdot p_{\text{isobutane}} + x_{n-\text{butane}} \cdot p_{n-\text{butane}} \tag{13.BQ}$$

If the two vapor pressures were equal, then Eqs. 13.9 and 13.BP would predict the same mol fractions in the liquid as in the gas. In reality, the two vapor pressures are similar, but not the same (NBP's are −0.6 and −10 °C); however, Eq. 13.10 is still obeyed. We may see this by replacing Eq. 13.10 by the same equation in terms of the fugacities,

$$\begin{aligned} \bar{g}_{\text{isobutane}} &= RT \ln f_{\text{isobutane}} + \overset{\circ}{g}_{\text{isobutane}} \\ &= y_{\text{isobutane}} P + \overset{\circ}{g}_{\text{isobutane}} \\ &= x_{\text{isobutane}} p_{\text{isobutane}} + g^{\circ}{}_{\text{isobutane}} \end{aligned} \tag{13.BR}$$

which can be solved to give Eq. 13.BP.

Why bother with this calculation? There are very few industrial examples of reactions that come to equilibrium simultaneously in two phases, but

1. There are some, of which one is discussed below.
2. When we discuss the phase rule (Chapter 15) we will refer back to this example.

13.5.1 Dimerization (Association)

Several industrially important materials form dimers (or trimers, or tetramers,...) reversibly, in both gas and liquid phases. Important examples are sulfur [12]

$$S_6 \Leftrightarrow 3S_2 \qquad S_8 \Leftrightarrow 4S_2 \tag{13.BS}$$

and acetic acid, CH_3COOH (abbreviated HAc), which can form hydrogen-bonded dimers of the type shown in Figure 13.9. For HAc, the R is the methyl group. This same type structure has been observed in all the lower molecular weight carboxylic acids, for example, formic, propionic, and butyric acids [13]. This behavior is called dimerization or association. It occurs in many hydrogen-bonding species,

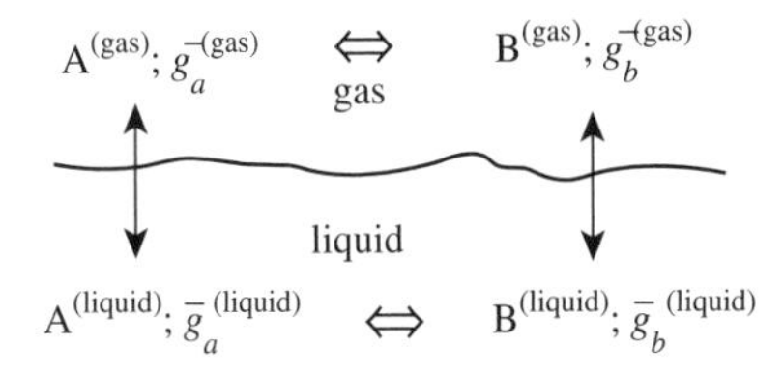

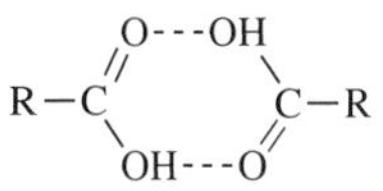

FIGURE 13.8 Gas–liquid equilibrium with chemical equilibrium in both phases.

```
      O---OH
     //      \
R—C           C—R
     \       //
      OH---O
```

FIGURE 13.9 Structure of dimmers formed by carboxylic acids. For HAc, R is the methyl group, CH_3.

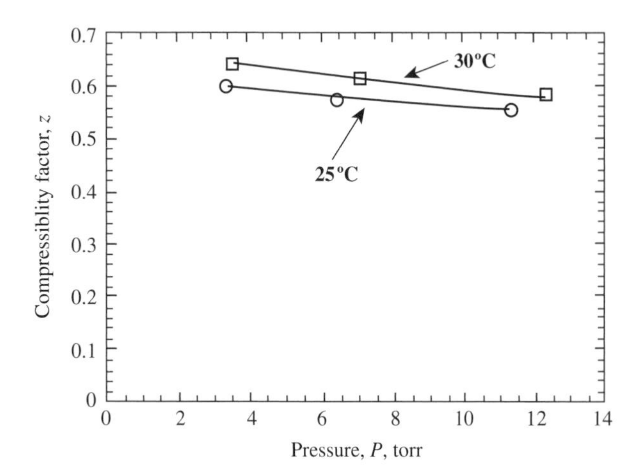

FIGURE 13.10 Experimental compressibility factors of HAc as a function of pressure at very low pressures [14].

such as water, but much more strongly in HAc and its near chemical relatives than with most other species.

Example 13.13 Estimate the value of the compressibility factor z for HAc at 25 °C and 11.38 torr from the measured value of the specific volume (0.6525 L/0.04346 g of HAc) [14] with the assumption that the HAc is all in the form of monomer with molecular weight 60.05 g/mol.

$$z = \frac{Pv}{RT} = \frac{\left(\dfrac{11.38}{760}\text{atm}\right)\cdot\left(\dfrac{0.6525\,\text{L}}{0.04345\,\text{g}}\right)\cdot\left(60.05\dfrac{g}{\text{mol}}\right)}{0.08206\dfrac{\text{L}\cdot\text{atm}}{\text{mol}\cdot\text{K}}\cdot 298.15\,\text{K}}$$

$$= 0.552 \quad \blacksquare \tag{13.BT}$$

This value, plus those at lower pressures, at 25 and at 30 °C are shown in Figure 13.10.

The pressures in Example 13.13 and Figure 13.10 are low enough that we would expect HAc vapor to behave practically as an ideal gas, so that if there were no dimerization, we would expect $z \approx 1.00$. That is clearly not the case if, as assumed in this calculation, the molecular weight is that of the monomer. If, on the other hand, there was complete dimerization, so that the true molecular weight of the vapor were twice that of HAc ($2 \cdot 60.05 = 120.1$ g/mol) then we would calculate a z twice as large, or 1.104. This is too large for an ideal gas, and, furthermore, Figure 13.10 shows that the computed values of z are increasing as the pressure falls, so according to the assumption of complete dimerization the value of z is going away from ideal gas behavior as the pressure falls!

This strange behavior is explained [15] by assuming that the formation of the dimer occurs in the gas phase by

$$2\text{HAc} \Leftrightarrow (\text{HAc})_2 \tag{13.BU}$$

for which

$$K = \frac{(Py_{(\text{HAc})_2})}{(Py_{\text{HAc}})^2} \qquad \log(\text{K}\cdot\text{torr}) = -10.4184 + \frac{3164\,\text{K}}{T} \tag{13.BV}$$

The numerical values in Eq. 13.BV are based on fitting experimental data like those in Figure 13.10.

Example 13.14 Estimate the mol fractions of monomer and dimer HAc in the vapor at 25 °C and 11.38 torr, based on Eqs. 13.BU and 13.BV.

From Eq. 13.BV,

$$K = \frac{Py_{(\text{HAc})_2}}{(Py_{\text{HAc}})^2} = \frac{1}{\text{torr}} \times 10\left(-10.4184 + \frac{3164\,\text{K}}{298.15\,\text{K}}\right) = \frac{1.56}{\text{torr}} \tag{13.BW}$$

$$y_{(\text{HAc})_2} = 11.38\ \text{torr}\cdot\frac{1.56}{\text{torr}}\cdot(y_{\text{HAc}})^2 = 17.75\cdot(1 - y_{(\text{HAc})_2})^2 \tag{13.BX}$$

which is readily solved, showing $y_{(\text{HAc})_2} = 0.789$, and, correspondingly, $y_{\text{HAc}} = 0.211$. ■

Example 13.15 Estimate the value of z for the above example, assuming that the vapor has a molecular weight corresponding to the fractional dimerization shown in Example 13.14.

The average molecular weight of the vapor is

$$M_{\text{avg}} = M_{(\text{HAc})_2}\cdot y_{(\text{HAc})_2} + M_{\text{HAc}}\cdot y_{\text{HAc}}$$

$$= 120.1\frac{\text{g}}{\text{mol}}\cdot 0.789 + 60.05\frac{\text{g}}{\text{mol}}\cdot 0.211 = 107.5\frac{\text{g}}{\text{mol}} \tag{13.BY}$$

and the computed value of z is the value found in Example 13.13 (0.552), which was based on a molecular weight of 60.05, times the ratio (107.5/60.05) = 0.988. ■

This value (≈1.00) shows that the vapor is indeed practically an ideal gas, but that its true molecular weight is not that of the monomer, but that of the above-calculated mixture of monomer and dimer.

This dimerization complicates the interpretation of VLE data for associating species. Figure 13.11 shows the values of the activity coefficients for water and HAc, calculated from the published VLE data [15] based on the assumption that the HAc is present in both phases entirely as the monomer. Clearly, the values in Figure 13.11 cannot be right: γ_{HAc} does not approach 1.00 as x_{HAc} approaches 1.00, nor do the slopes of the curves agree with the Gibbs–Duhem equation. The scatter in the data seems large, but observe the small values of the activity coefficients: All are between 0.7 and 1.2.

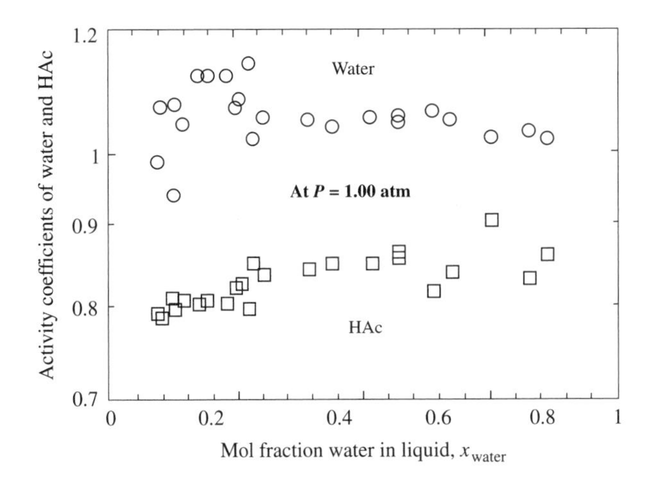

FIGURE 13.11 Activity coefficients for water and HAc in liquid solution at 1 atm pressure, based on the measured vapor and liquid compositions [15] and the *clearly incorrect* assumption that the HAc in both phases is present as all monomer.

This industrially important VLE system has been thoroughly analyzed [15]. The reversible dimerization is believed to be rapid in both phases, so that equilibrium by Reaction 13. BV is believed to exist in both phases. When we apply the logic shown in Figure 13.9 separately to the monomer and the dimer, taking into account that the measured vapor pressure of the liquid is that of a mixture of monomer and dimer, whose composition changes with temperature, and take the dimerization equilibrium into account in each phase, we find that the activity coefficients are of type II (Figure 8.8), with values similar to those shown in Figure 8.8b.

13.6 SUMMARY

1. There are no new principles in this chapter, merely the application of previously discussed principles to more complex reactions.
2. In all chemical reactions, at equilibrium nature has selected those chemical compositions that correspond to the minimum Gibbs energy consistent with the starting composition and the external constraints.
3. The basic calculation schemes introduced in Chapter 12 apply as well to sequential reactions, simultaneous reactions, reactions involving ions, and reactions in more than one phase.
4. Most of the examples in this chapter use extent of reaction *(e)* as a computation tool. For aquatic chemistry, the charge-balance computation tool is often more satisfactory.
5. Electrochemical reactions require us to expand our definition of the equilibrium state, to take into account the possibility of doing electrical work.
6. Association and dimerization occur often in liquids, sometimes in gases. When they do occur (sulfur and carboxylic acids) we would make serious errors by ignoring them. They are best treated as chemical reactions, by the methods shown here.

PROBLEMS

See the Common Units and Values for Problems and Examples. An asterisk (*) on the problem number indicates that the answers is in Appendix H.

13.1 Show the calculation of the two dissociation constants for H_2SO_4 in Example 13.2, using the values from Table A.8.

13.2* Repeat Example 13.2 for 0.1 mol of H_2SO_4 dissolved in 1000 g of water at 25 °C.

13.3 **a.** Repeat Example 13.2 for dissolving 1.00 mol SO_2 (=sulfurous acid=H_2SO_3, see Eq. 13.AD) in 1000 g of water at 25 °C, using the values from Table A.8.

b. The values of K_1 and K_2 that you calculate from Table A.8 do not always agree exactly with values from other handbooks. The same handbook from which the Gibbs energy values were taken shows in [5, p. 8–37] a table of ionization constants. For sulfurous acid that book shows K_1 and K_2 at 18 °C as 0.0154 and 1.02 E-7. Compare those with the values you must calculate in part (a). How much difference do these different ionization constants make in the calculated equilibrium concentrations?

c. Reference [16] shows equations for K_1 and K_2 as a function of temperature. Based on them, the two values at 18 °C are 0.0164 and 7.02×10^{-8} and at 25 °C are 0.0140 and 6.83×10^{-8}. Are these in good or fair agreement with the values shown above? Do they show the same trends with temperature? At the end of this set of comparisons are you convinced that all sources are in fair agreement on these values, but that there is still some disagreement? Do you think the same is likely to be the case for other dissociation constant data?

13.4* Show the relation between the *K*s for reactions 12.E, 13.K, and 13.L. Does this result prove that these three reactions are not independent of each other?

13.5 Show the calculation of the two values of K and K_ϕ at 10 MPa and 523.15 K in Example 13.3.

a. The K_ϕ are computed from Figure A.5 or its equivalent.

b. For reaction 12.E, [2] shows

$$K_{12.D} = 9.740 \times 10^{-5} \cdot \exp\left[21.225 + \frac{9143.6}{T} - 7.492 \ln T + 4.076 \times 10^{-3} T - 7.161 \times 10^{-8} T^2\right] \quad (13.BZ)$$

with T expressed in K, and for Reaction 13.L

$$K_{13.M} = \exp\left[13.148 - \frac{5639.5}{T} - 1.077 \ln T - 5.44 \times 10^{-4} T + 1.125 \times 10^{-7} T^2 + \frac{49,170}{T^2}\right] \quad (13.CA)$$

with T expressed in K. Equation 13.BZ certainly contains some error, because the values calculated from it are not consistent with the other information in [2]. They are consistent with the following equation from [1]:

$$\Delta g^\circ = -17,835 - 16.08 T \ln T - 0.01119 T^2 + 1.018 \times 10^{-6} T^3 + 0.0811 \times 10^{-9} T^4 - 48.3 T \quad (13.CB)$$

with T expressed in K and Δg° in cal/mol.

This is mentioned here to make clear to the reader that such equations are regularly presented in literature with copying or typographic errors.

c. Estimate the value of $K_{12.E}$ at 523.15 K, using the values in Tables A.8 and A.9, and the method in Example 12.7.

13.6 What would be the outcome in Example 13.3 if the catalyst facilitated the conversion of CO and H_2 not only to methanol, but also to methane?

a. Estimate the value of K and K_ϕ for Reaction 12.F, using Tables A.8 and A.9 and Figure 7.1.

b. Write the equivalent of Eq. 13.R for Reaction 12.F.

c. Extend Example 13.3 to include Reaction 12.F.

13.7 Show the solution to Eqs. 13.Z, 13.AA, 13.AC, 13.AF, 13.AH, and 13.AI (or as many of these as interest you). Suggested procedure; Rewrite the equations moving all the terms to the left of the equals sign, so that the right hand term = 0. On a spreadsheet in one column enter: A guess of the final $[H^+]$, then the formulae for the each of the individual terms, using the Ks, the amounts added and the guessed value of $[H^+]$, and then the sum of the terms. Then use the spreadsheet's equation solver (Goal Seek on Excell) to make that sum = 0 by changing the values of the guess of the final $[H^+]$. Once you have this debugged, you can use subsequent columns to solve the rest of the equations.

Do not be surprised if your solver reduces the sum to a small value, but not to zero. The problem is round-off error; in the solution to Eq. 13.Z the largest term is 0.0013 and the smallest is 7×10^{-12}. Spreadsheets have a hard time with equations with that much disparity in size. Also relax if your spreadsheet reports a small negative value for $[H^+]$. Manually change that to +, and observe that it doesn't change the other values materially. Again, the difference in size of the terms is the cause of the difficulty.

13.8 Repeat Examples 13.4, 13.5, and 13.6 (as many parts as you like) for the buffer solution made from ammonia and ammonium chloride. For ammonia $NH_3 + H_2O \Leftrightarrow NH_4^+ + OH^-$; $K = 1.78$ E–5. See the suggestions for how to do this in Problem 13.7.

13.9 In Example 13.4, Eq. 13.Z, the K_w term is certainly much smaller than the others. Set that equal to zero, and determine how much it changes your answer. If you have set up the spreadsheet recommended in Problem 13.7, you can manually set $K_w = 0$ and your spreadsheet will solve if for you. Alternatively, you can note that this change makes the equation a quadratic, easy to solve numerically or analytically.

13.10 Using the values from Table A.8,

a. Estimate the values of the two ionization constants of H_2CO_3.

b. Estimate the Henry's law constant for CO_2 dissolved in water, and compare it to the value given in Table A.3. *Hint:* Pay attention to units; molal is not the same as mol fraction!

c. The experimental determination of Henry's law constants is normally made by measuring the amount of gas that goes into solution. But the equation is written in terms of $[CO_2\ (aq)]$, not the sum of $([CO_2\ (aq)] + [HCO_3^-] + [CO_3^{2-}])$. For the values shown in Example 13.5, estimate how different the Henry's, law constant would be if we based it entirely on $[CO_2\ (aq)]$, and on $([CO_2(aq)] + [HCO_3^-] + [CO_3^{2-}])$.

13.11 Repeat Example 13.8, not making the assumption that the H^+ comes all from Reaction 13.AU, but including that from Reaction 13.AV in the calculations. How much difference does it make?

13.12* In Example 13.9, as we add more and more NaOH to the solution, we eventually reach the solubility limit of Na_2CO_3, which is 17.7 wt% in water at 20 °C. Estimate the highest pH we can have in that example as we continue adding NaOH, always waiting for the solution to come to equilibrium with the CO_2 in the atmosphere.

13.13 The scrubbers that are widely used to remove sulfur dioxide from the exhaust gases from coal-fired power plants use a scrubbing solution with pH ≈ 6, which is maintained at that value by the addition of solid lime or limestone. The exhaust gas from such a power plant, which passes through the scrubber, has ≈ 12 mol% CO_2. Assume that the pH = 6 solution leaving the scrubber is in equilibrium with a vapor containing 12 mol % CO_2. For this solution, estimate the concentrations of: H_2CO_3 (or CO_2(aq)), HCO_3^- and CO_3^{2-}.

(The values in Table A.8 are for 25 °C and these scrubbers operate at about 50 °C, but we will ignore that difference for this problem!)

13.14* Estimate the amount of CO_2 dissolved, and the concentrations of carbonic acid, its two ions and the hydrogen ion, and the pH, when water is in equilibrium with pure carbon dioxide at 1 atm pressure and 20 °C.

13.15 Table A.3 shows the Henry's law constants for CO_2 in water at several temperatures, based on *ICT*, (Vol. 3, p. 260). Edwards et al. [8] represent the Henry's law constant for this system by

$$\ln H = \frac{B_1}{T} + B_2 \ln T + B_3 T + B_4$$

$$B_1 = -6789.04, \quad B_2 = -11.4518,$$

$$B_3 = -0.010454, \quad B_4 = 94.4914$$

H expressed in kg · atm/mol, T in K (13.CC)

a. Compute the value according to Eq. 13.CC at 0 °C and compare it to the value in Table A.3. (Observe the difference in units!)

b. Edwards et al. [8] show the literature sources for their equation. As part (a) shows, the values computed from Eq. 13.CC are about 4 to 9% less than those in Table A.3. As a research project, consult the original sources cited by Edwards et al. and those cited in *ICT* (Vol. 3 p. 260) and determine which is more likely to be correct and whether the constants are defined differently, in addition to being expressed in different units.

13.16 Examples 13.8 and 13.9 treat the carbon dioxide water equilibrium. The corresponding problem for SO_2 is of paramount importance in air pollution control [16]. The corresponding reactions seem to be

$$SO_2(g) \Leftrightarrow SO_2(aq) \tag{13.CD}$$

$$SO_2(aq) + H_2O(l) \Leftrightarrow H^+ + HSO_3^- \tag{13.CE}$$

$$HSO_3^- \Leftrightarrow H^+ + SO_3^{2-} \tag{13.CF}$$

for which the equilibrium relationships for dilute solutions are

$$H_{SO_2} = \frac{P y_{SO_2}}{[SO_2(aq)]} \tag{13.CG}$$

$$K_1 = \frac{[HSO_3^-] \cdot [H^+]}{[SO_2(aq)]} \tag{13.CH}$$

$$K_2 = \frac{[SO_3^{2-}] \cdot [H^+]}{[HSO_3^-]} \tag{13.CI}$$

where, H_{SO_2}, K_1, and K_2 are the Henry's law constant and the first and second ionization constants, with dimensions atm/(mol/L), (mol/L), and (mol/L). Reference [16] shows equations for these three equilibrium constants (as well as others important in air pollution control), as a function of temperature. The values calculated from their equations for two temperatures are shown in Table 13.E.

a. Henry's law (Eq. 3.6) is most often written using the mol fraction x_i, of the dissolved gas as the liquid concentration variable. In the above table and in much of the air pollution scrubber literature, Henry's law is written with the molarity (mol/L) or mass concentration (mol/kg or lbmol/ft^3) as the liquid–phase concentration variable. These concentration values can all be converted

Table 13.E Equilibrium Data for SO_2 in Water [16]

	T = 68°F = 20°C	T = 125°F ≈ 52°C
H_{SO_2} [atm/(L/mol)]	0.679	1.870
K_1 (mol/L)	0.0156	0.00678
K_2 (mol/L)	6.86×10^{-8}	5.25×10^{-8}

to the others, so the H for one can be converted to the H for any of the others. Estimate the value of the Henry's law constant at 20 °C in the mol fraction form (Eq. 3.6) based on the value from the above table.

b. Estimate the values of the concentrations at 20 °C of $SO_2(aq)$, H^+, HSO_3^-, and SO_3^{2-} in a solution that is in equilibrium with a gas at 1 atm, with $y_{SO_2} = 0.001$. The procedure to follow is the same as in Example 13.8, assuming first that the SO_3^{2-} concentration is negligible, and so on.

c. Estimate the pH of the solution in (b).

d. The Henry's law constant shown in Eq. 13.CG is based on the dissolved SO_2 only, not its ionization products. If we wrote it to take the ionization products into account we would have

$$H_{SO_2} = \frac{Py_{SO_2}}{[SO_2(aq)] + [HSO_3^-] + [SO_3^{2-}]} \quad (13.CJ)$$

For the situation in part (b), what would be the value of H_{SO_2}, both in atm and in atm/(mol/L)?

13.17 Kohl and Nielson [9, p. 508] present a plot similar to Figure 13.6 for the sulfite equilibrium in water, shown here as Figure 13.12. Compare the values computed in parts (b) and (c) of the preceding problem to those we would estimate from this figure.

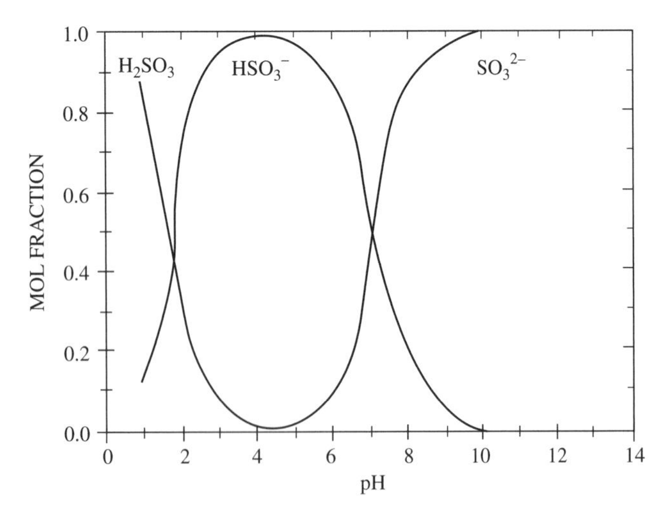

FIGURE 13.12 Distribution among sulfite species in aqueous solution as a function of pH. In this and similar plots, the vertical axis, labeled "mol fraction" is actually the fraction by mol of the total sulfite species, which is in each of the three forms shown. (From Kohl, A., and R. Nielsen. *Gas Purification,* ed. 5. Houston, TX: Gulf, p. 508 (1997)).

13.18 *Perry's* [17] presents a table of the vapor pressures of SO_2 over aqueous solutions. At 20 °C, for dissolved SO_2 concentrations of 0.01, 0.1, 1, and 10 g/(100 g of water), the reported SO_2 vapor pressures are 0.07, 3.03, 58.4, and 714 torr.

a. Can these four values be represented by simple Henry's law, which requires that in dilute solutions the vapor pressure is proportional to the concentration?

b. Estimate the corresponding vapor pressures using the equilibrium data in the preceding problem. Do they agree with the data from *Perry's*?

c. If we wish to fit the water–SO_2 equilibrium into the mold of simple Henry's law, with H in atm and dissolved gas concentration in mol fraction, what values will we use for the Henry's law "constant" at the above 4 concentrations?

13.19 Repeat Example 13.9 for pH = 1.00, provided by adding some strong acid, such as HCl, to the water.

13.20* Figures 13.6 and 13.12 are from the same source.

a. Are their general forms the same?

b. Do they show the changes from un-ionized to first ionization to second ionization at the same values of the pH?

c. Why?

13.21* Estimate the number of kilowatt-hours of electricity required to make a pound of aluminum metal. As described in Section 13.4, the actual cell voltage required is about 4 V. The atomic weight of aluminum is 23.

13.22* **a.** Repeat Example 13.10, assuming that the reaction is

$$Al_2O_3 + 3C \Leftrightarrow 2Al + 3CO \quad (13.BI)$$

b. Based on your result, discuss the role of this reaction in commercial aluminum production.

13.23 **a.** Repeat Example 13.10 assuming that the reaction is

$$Al_2O_2 \Leftrightarrow 2Al + 1.5O_2 \quad (13.CK)$$

b. Based on your result, discuss the feasibility of carrying out this reaction commercially.

13.24* Estimate the required equilibrium cell voltage at 298.15 K for the reaction $MgCl_2 \Leftrightarrow Mg + Cl_2$. This is actually carried out at a temperature high enough for the $MgCl_2$ and Mg to be molten, but for this problem use the values in Table A.8, which correspond to solid $MgCl_2$, solid Mg, and gaseous Cl_2.

13.25 The Gibbs energy change going from solid LiF to a 1 molal solution of LiF in water at 25 °C is estimated to be +13.39 kJ/mol. Using that value, estimate the reversible cell voltage for a cell that begins with pure Li and F and produces a 1 molal solution of LiF. Compare the result to the values shown in Eq. 13.BJ.

13.26 Repeat Example 13.12 starting with the Nernst equation (Eq. 13.7).

13.27 In Example 13.12 we computed the effect of pressure on the equilibrium cell voltage for Reaction 13.BL. Now we wish to conduct the same operation under conditions where the water will enter the cell at 1 atm and the hydrogen will leave the cell at 1 atm but the oxygen will leave the cell at 100 atm. Everything else is exactly the same as in Example 13.12. What is the equilibrium cell voltage for this situation?

13.28 In Example 13.12 we computed the effect of pressure on the reversible equilibrium steady-flow cell voltage for the Reaction 13.BL. We assumed that the specific volume of water was negligible. This is not rigorously true. Calculate the numerical value of the error caused by this assumption. Assume that $v_{\text{water}} = 1.00$ L/kg, independent of pressure. Present your answer as

$$\Delta\left[\frac{\partial E}{\partial P}\right]_T = \left\{\left[\frac{\partial E}{\partial P}\right]_{T,\ \text{taking water volume into account}} - \left[\frac{\partial E}{\partial P}\right]_{T,\ \text{taking water volume}=0}\right\}\frac{\text{volts}}{\text{atm}} \tag{13.CL}$$

We know that the answer is small. How small?

13.29 For the reaction shown in Example 13.10 estimate the derivative of the equilibrium cell voltage with temperature $(\partial E/\partial T)_P$. We know it is a small number. How small?

a. Start with Eq. 13.6, take the derivative with respect to temperature, finding

$$\frac{\partial E}{\partial T} = \frac{-1}{n_e F}\frac{\partial \Delta g}{\partial T} = \frac{-1}{n_e F}\frac{\partial(\Delta h - T\Delta s)}{\partial T} = \frac{\Delta s}{n_e F} = \frac{(\Delta g - \Delta h)}{n_e F T} \tag{13.11}$$

the *Gibbs–Helmhotz equation.*

b. Explain why this comes out differently from the seemingly similar problem of the change of chemical reaction equilibrium constant with temperature (Eq. 12.25).

c. Find the suitable values from Table A.8 and Example 13.10 to compute $(\partial E/\partial T)_P$.

13.30 **a.** In Example 13.12, calculate the cell voltage at 1 atm and 30 °C, using the formula you develop in the preceding problem.

b. The following is an *incorrect* solution to part (a):

Incorrect Solution

$$\Delta E = \frac{\Delta(-\Delta g)}{n_e F}$$
$$\Delta G = \Delta g_2 - \Delta g_1 = \Delta h_2 - \Delta h_1 - (T_2\Delta s_2 - T_1\Delta s_1) \tag{13.CM}$$

For small values of $(T_2 - T_1)$ this is

$$\Delta(\Delta g) = \Delta T(C_{P,\text{ products}} - C_{P,\text{ reactants}}) - T_{\text{avg}}\left(C_{P,\text{ products}} \cdot \frac{\Delta T}{T} - C_{P,\text{ reactants}} \cdot \frac{\Delta T}{T}\right) \tag{13.CN}$$

Take the limit of $\Delta T/T$ as ΔT goes to zero:

$$\Delta(\Delta g) = \Delta T(C_{P,\text{products}} - C_{P,\text{reactants}}) - T_{\text{avg}}\left(C_{P,\text{products}} \cdot \ln\frac{T_2}{T_1} - C_{P,\text{reactants}} \cdot \ln\frac{T_2}{T_1}\right) \tag{13.CO}$$

$$\Delta(\Delta g) = \left(\Delta T - T_{\text{avg}} \cdot \ln\frac{T_2}{T_1}\right) \cdot (C_{P,\text{ products}} - C_{P,\text{ reactants}}) \tag{13.CP}$$

$$\Delta E = \frac{\left(\Delta T - T_{\text{avg}} \cdot \ln\frac{T_2}{T_1}\right) \cdot (C_{P,\text{ products}} - C_{P,\text{ reactants}})}{n_e F} \quad ?? \tag{13.CQ}$$

You can easily show that this solution is incorrect by inserting numerical values into Eq. 13.CQ and comparing the result with the answer to part (a). Here $(C_{P,\text{products}} - C_{P,\text{reactants}}) = 12.0$ J/mol.

What is the fundamental error in this incorrect solution? Explain clearly what error one makes starting with the first equation, which is absolutely correct, and proceeding as shown to the final equation, which is absolutely wrong.

13.31 In Example 13.12, instead of running the cell to produce H_2 and O_2 at 100 atm, we could run the cell at 1 atm, and then compress the gases to 100 atm. If we had reversible, isothermal compressors (which exist only in textbook examples), what would the work required to drive them be, compared to the extra electrical work needed to run the cell with the outlet pressure 100 atm?

13.32 What is the most economical way to supply 1 kg/day of hydrogen? 1000 kg/day? 10^6 kg/day?

13.33 The overall reaction of a lead storage battery of the type used in most automobiles (during discharge, when power is being withdrawn) is

$$\text{Pb(s)} + \text{PbO}_2\text{(s)} + 2\text{H}_2\text{SO}_4\text{(aq)} \Leftrightarrow 2\text{PbSO}_4\text{(s)} + 2\text{H}_2\text{O(l)}. \qquad (13.CR)$$

Here two electrons per mol are transferred for the equation as written. What is the equilibrium voltage of this cell? (Those of you who have ever looked under the hood of your car know the answer, but show the calculations to support it!)

13.34 Using the data in the preceding problem, estimate the amount of heat released or absorbed when 1 g of Pb reacts reversibly and isothermally according the equation shown.

13.35 In Problem 13.33 you computed the equilibrium voltage for a single cell of a lead-acid battery. The calculation there was for E°, the "standard cell voltage," which corresponds to all of the reactants being in their standard states. For H_2SO_4(aq) the standard state is an ideal solution with concentration 1 molal. How much would the calculated voltage change if the concentration of H_2SO_4(aq) was 5 molal?

We know the answer is "not much." But how much? Which way, plus or minus? The expected answer is the value of $(E - E^\circ)$. For this problem you may assume that solutions of sulfuric acid in water are ideal solutions (not a very good assumption, but acceptable for this problem). *Hint:* Start with Eq. 13.7, expand the term in brackets, find its numerical value.

13.36* Repeat Example 13.14 for $P = 1$ atm and $T = 117.9\,^\circ$C (the NBP).

13.37* At 25 °C, what pressure corresponds to a $y_{(\text{HAc})_2} = 90\%? = 10\%?$

13.38 Repeat Examples 13.13 and 13.14 for 25 °C and $P = 3.16$ torr, for which the measured specific volume = (2.315 L/0.04345 g of HAc) [14].

13.39 For sulfur in the vapor phase, the equilibrium reactions are reported [12] to be

$$\text{S}_6 \Leftrightarrow 3\text{S}_2 \quad \ln\frac{K}{\text{atm}^2} = 37.4199 - \frac{33{,}534\ \text{K}}{T} \qquad (13.CS)$$

and

$$\text{S}_8 \Leftrightarrow 4\text{S}_2 \quad \ln\frac{K}{\text{atm}^3} = 57.1441 - \frac{49{,}862\ \text{K}}{T} \qquad (13.CT)$$

a. Using these values, estimate the mol fractions of S_2, S_6, and S_8 in an equilibrium vapor that is made up of pure sulfur at 500 °C and 1 atm. *Hint:* Guess a value of y_{S_2} and use it to compute the values of y_{S_6} and y_{S_8}. Then compare the sum of the vapor-phase mol fractions to 1.00. Adjust your guess of y_{S_2} and repeat until the mol fractions sum to 1.00. This goes very quickly on a spreadsheet.

b. Compare your calculated values to the values shown in Figure 13.13.

REFERENCES

1. Woodward, H. F. J. Methanol. In *Kirk–Othmer Encyclopedia of Chemical Technology*, Vol. 13, ed. 2, Standen, A., ed. New York: Wiley, pp. 370–398 (1967).
2. Wade, L. E., R. B. Gungelbach, J. L. Trumbley, and W. L. Hallbauer. Methanol. In *Kirk–Othmer Encyclopedia of Chemical Technology*, Vol. 15, ed. 3, Grayson, M., ed. New York: Wiley, pp. 398–415 (1981).
3. Smith, W. R., and R. W. Missen. *Chemical Reaction Equilibrium Analysis: Theory and Applications*. New York: Wiley (1982).
4. Stumm, W., and J. J. Morgan, *Aquatic Chemistry: Chemical Equilibria and Rates in Natural Waters*, ed. 3, New York: Wiley, p. 105 (1996).

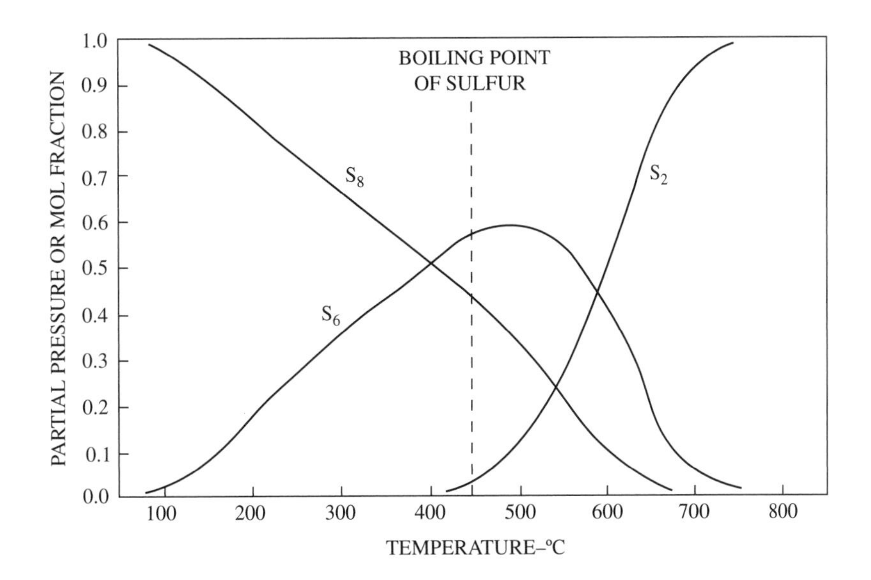

FIGURE 13.13 Distribution among sulfur species in the gas phase as a function of temperature at pressures that increase from left to right. Above the NBP the plot shows the mol fractions at 1 atm pressure. Below the NBP the plot shows that the mol fractions at the vapor pressure of liquid sulfur, given approximately by $\log(p/\text{Pa}) = 12.3256 - 3268.2\ \text{K}/T$ for temperatures between 325 and 550°C. (From Gamson, B. W., and R. H. Elkins. Sulfur from hydrogen sulfide. *Chem. Eng. Prog.* 49: 203–215 (1953). Reproduced with permission of the American Institute of Chemical Engineers.)

5. Haynes, W. M. *Handbook of Chemistry and Physics*, ed. 91, Boca Raton, FL: CRC Press, pp. 8–127 (2010).
6. Butler, J. N. *Carbon Dioxide Equilibria and Their Applications*. Reading, MA: Addison-Wesley (1982).
7. de Nevers, N. *Air Pollution Control Engineering*, ed. 2. New York: McGraw–Hill, Appendix D (2000). (Reissued 2010 by Waveland Press, Long Grove, IL).
8. Edwards, T. J., G. Maurer, J. Newman, and J. M. Prausnitz. Vapor–liquid equilibria in multicomponent aqueous solutions of volatile weak electrolytes. *AIChE J.* 24:966–976 (1978).
9. Kohl, A., and R. Nielsen. *Gas Purification*, ed. 5. Houston: Gulf (1997).
10. Staley, J. T., and W. Haupin. Aluminum and its alloys. In *Kirk–Othmer Encyclopedia of Chemical Technology*, Vol. 2, ed. 4, Howe-Grant, M., ed. New York: Wiley, pp. 184–251 (1992).
11. Austin, G. T. *Shreve's Chemical Process Industries*, ed. 5. New York: McGraw–Hill (1984).
12. Gamson, B. W., and R. H. Elkins. Sulfur from hydrogen sulfide. *Chem. Eng. Prog.* 49:203–215 (1953).
13. Tsonopoulos, C, and J. M. Prausnitz. Fugacity coefficients in vapor-phase mixtures of water and carboxylic acids. *Chem. Eng. J.* 1:273–277 (1970).
14. MacDougall, F. H. The molecular state of the vapor of acetic acid at low pressures at 25, 30, 35, and 40 °. *J. Am. Chem. Soc.* 58:2583–2591 (1936).
15. Sebastiani, E., and L. Lacquaniti. Acetic-acid—water system thermodynamical correlation of vapor–liquid equilibrium data. *Chem. Eng. Sci.* 22:1155–1162 (1967).
16. Pasiuk-Bronikowska, W., and K. J. Rudzinski. Absorption of SO_2 into aqueous systems. *Chem. Eng. Sci.* 46:2281–2291 (1991). Observe that this paper uses a Henry's law definition that is the reciprocal of the common one.
17. Liley, P.E., G.H. Thompson, D.G. Friend, T.E. Daubert, and E. Buck. Physical and chemical data. In *Perry's Chemical Engineer's Handbook*, ed. 8, D.W. Green, ed. New York: McGraw-Hill, pp. 2–128 (2008).

14

EQUILIBRIUM WITH GRAVITY OR CENTRIFUGAL FORCE, OSMOTIC EQUILIBRIUM, EQUILIBRIUM WITH SURFACE TENSION

In the previous chapters we ignored the effect of gravity, surface tension, and electrostatic or magnetic forces, and ignored the possibility that there may be semipermeable membranes in our system at equilibrium. There are important equilibrium situations in which we must take all of these into account. In this chapter we consider three ways in which we can have equilibrium without a uniform pressure, as a result of gravity, semipermeable membranes, or surface forces.

14.1 EQUILIBRIUM WITH OTHER FORMS OF ENERGY

In Section 4.1 we showed that in the absence of gravity, surface, tensile, electrostatic, and electromagnetic forces, the criterion of phase and chemical equilibrium was $dG_{T,P} \geq 0$, with the "=" corresponding to the equilibrium state and the ">" corresponding to moving away from the equilibrium state. Correspondingly, the approach to equilibrium from a nonequilibrium state must be one for which $dG_{T,P} < 0$; nature minimizes Gibbs energy! If we now return to Eq. 4.1, and allow for some of these other kinds of energy, we will have

$$d(U + KE + PE + TE + EE + ME) = dQ + dW \quad (14.1)$$

where *KE*, *PE*, *TE*, *EE*, and *ME* represent the kinetic, potential, tensile, electrostatic, and magnetic energies of the system. In this case we must also expand the work term to

$$dW = -PdV + \begin{pmatrix} \text{terms for other kinds of work,} \\ \text{e.g. electrostatic, magnetic, etc.} \end{pmatrix} \quad (14.2)$$

If we restrict our attention to systems that can only exchange heat and *PdV* work with the surroundings, then we can drop the rightmost term of Eq. 14.2, and substitute Eq. 14.1 for Eq. 4.1. If we then follow the derivation through to the analog of Eq. 4.9, we find

$$d(G + KE + PE + TE + EE + ME)_{\text{system}} \geq 0 \quad (14.3)$$

with the "=" corresponding to the equilibrium state and the ">" corresponding to a change that takes us away from the equilibrium state. Equation 14.3 leads rapidly to the conclusion (see below) that we can have physical or chemical equilibrium in a system that is not at a constant pressure, if that pressure is balanced by gravity, surface forces, semipermeable membranes, and so on. We will also see that omitting that from consideration has not caused significant errors in the calculations in the previous chapters. Equation 14.3 does not in any way change the statement in Chapter 3 that any system at physical or chemical equilibrium is an isothermal system.

Intuitively, we may consider that these other forms of energy can, *in principle*, be reversibly converted one to the other or to Gibbs energy, so that if there are changes in one of them at equilibrium, there must be a corresponding change in one of the others or in the Gibbs energy for our

Physical and Chemical Equilibrium for Chemical Engineers, Second Edition. Noel de Nevers.

system to remain at the minimum, which corresponds to equilibrium.

14.2 EQUILIBRIUM IN THE PRESENCE OF GRAVITY

This subject is of relatively little practical significance. The more practical case is equilibrium in centrifuges, which have significant changes in equilibrium composition over short distances due to strong centrifugal force fields. However, the gravity case is more intuitively comfortable and simpler mathematically. Once we have mastered it, the extension to the centrifugal force fields is fairly straightforward.

Consider some long, thin column of a mixture of species in a container at constant temperature and with the pressure fixed at the top, as sketched in Figure 14.1. A model might be a natural gas well, which had been "shut in" for a long enough time to be at equilibrium. (Real natural gas wells always have a temperature gradient, because the center of the earth is hotter than the surface; but for this problem we ignore that and assume constant temperature.) In this system, we now withdraw dn_i mols of substance i at z_1, and insert it at z_2.

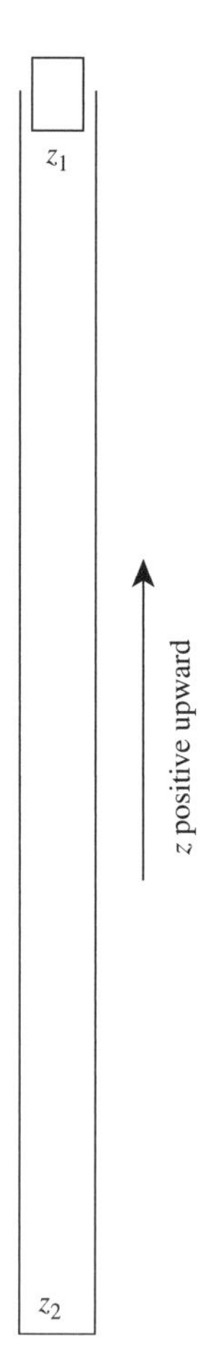

FIGURE 14.1 Deep well at a constant temperature.

The change of Gibbs energy plus potential energy for this transaction must be

$$d(G+PE)_{\text{System}} = \left[-\frac{d}{dn_i}(G+PE)_{\text{at } z_1} + \frac{d}{dn_i}(G+PE)_{\text{at } z_2}\right] dn_i \tag{14.4}$$

If this is to be an equilibrium change then $d(G + PE)_{\text{system}}$ must be zero. We know that the derivatives are partial molar derivatives (see Chapter 6), and that

$$\frac{d}{dn_i} PE = M_i g z \tag{14.5}$$

so that

$$\bar{g}_{i2} - \bar{g}_{i1} = RT \ln \frac{f_{i2}}{f_{i1}} = -M_i g (z_2 - z_1) \tag{14.6}$$

or

$$\frac{f_{i2}}{f_{i1}} = \exp \frac{-M_i g (z_2 - z_1)}{RT} \tag{14.7}$$

The derivation of Eq. 14.7 is straightforward, but it has little intuitive content. We may form an *approximate* intuitive idea of why nature behaves that way by considering the forces acting on a single molecule. The diffusional force, which tries to make the concentration the same everywhere, is *roughly* the same for each molecule, independent of its molecular weight, because in their multiple collisions with all the other molecules in the gas, the individual gas molecules will all take up the same temperature and thus more or less the same velocity. (If gas molecules all had the same size, this would be exactly true; gas molecules don't all have the same sizes, but the differences in sizes is generally less than the difference in molecular weights.) The gravity force on each molecule depends on its molecular weight. So at equilibrium, gravity is trying to arrange the molecules, with a layer of the highest molecular weight species on the bottom, the next higher molecular weight on top of it, and so on. Diffusion is trying to make the concentration uniform throughout the container. Equation 14.7 shows the outcome of these two competing forces. Under most circumstances (with gravity not much different from that at the surface of the earth) diffusion wins almost completely.

All the discussion and equations in this chapter up to this point apply to solids, liquids, or gases. The rest of this section is restricted to ideal solutions of ideal gases for which we can replace the fugacities by partial pressures, $f_i = y_i P$.

Example 14.1 A natural gas well (Figure 14.1) is at constant temperature of 300 K and in thermodynamic equilibrium. At the surface the gas is 85 mol% methane, 10 mol% ethane, balance propane, and the pressure is 2 MPa. At this temperature and pressure the gas mixture may be assumed to be an ideal solution of ideal gases. Estimate the concentrations at a depth of 1000 m.

For methane we solve Eq. 14.7:

$$\frac{f_{\text{methane, 2}}}{f_{\text{methane, 1}}} = \exp - \frac{16\dfrac{\text{g}}{\text{mol}} \cdot 9.81\dfrac{\text{m}}{\text{s}^2} \cdot (-1000\,\text{m})}{8.314\dfrac{\text{J}}{\text{mol K}} \cdot 300\,\text{K}} \cdot \frac{\text{J s}^2}{\text{kg m}^2} \cdot \frac{\text{kg}}{1000\,\text{g}} = \exp 0.0629 = 1.065 \quad (14.\text{A})$$

For ethane $M_i = 30$ g/mol, so

$$\frac{f_{\text{ethane, 2}}}{f_{\text{ethane, 1}}} = \exp\left(0.0629 \cdot \frac{30}{16}\right) = 1.125 \quad (14.\text{B})$$

and for propane, $M = 44$ g/mol, so

$$\frac{f_{\text{propane, 2}}}{f_{\text{propane, 1}}} = \exp\left(0.0629 \cdot \frac{44}{16}\right) = 1.189 \quad (14.\text{C})$$

For the assumption of an ideal solution of ideal gases, the fugacities can be replaced by the partial pressures, so that at the surface

$$f_{\text{methane, 1}} = 0.85 \cdot 2\,\text{MPa} = 1.7\,\text{MPa}$$
$$f_{\text{ethane, 1}} = 0.2\,\text{MPa} \quad f_{\text{propane, 1}} = 0.1\,\text{MPa} \quad (14.\text{D})$$

and at the bottom

$$f_{\text{methane, 2}} = y_{\text{methane, 2}} P_2 = 1.7\,\text{MPa} \cdot 1.065 = 1.811\,\text{MPa} \quad (14.\text{E})$$

and

$$f_{\text{ethane, 2}} = y_{\text{ethane, 2}} P_2 = 0.2\,\text{MPa} \cdot 1.125 = 0.225\,\text{MPa} \quad (14.\text{F})$$

$$f_{\text{propane, 2}} = y_{\text{propane, 2}} P_2 = 0.1\,\text{MPa} \cdot 1.189 = 0.119\,\text{MPa} \quad (14.\text{G})$$

$$P_2 = P_2(y_{\text{methane, 2}} + y_{\text{ethane, 2}} + y_{\text{propane, 2}}) = 1.811 + 0.225 + 0.119 = 2.155\,\text{MPa} \quad (14.\text{H})$$

$$y_{\text{methane, 2}} = \frac{1.811\,\text{MPa}}{2.155\,\text{MPa}} = 0.840$$
$$y_{\text{ethane, 2}} = \frac{0.225\,\text{MPa}}{2.155\,\text{MPa}} = 0.105$$
$$y_{\text{propane, 2}} = \frac{0.119\,\text{MPa}}{2.155\,\text{MPa}} = 0.055 \quad (14.\text{I})$$

The equilibrium mol fractions at a depth of 1000 m are up to 10% different from those at the surface. The mol fraction of the two higher molecular weight species have increased, while that of the lowest molecular weight species has decreased. Gravity has partially sorted by molecular weight, but mostly diffusion has won, and the mol fractions are only slightly different from top to bottom of the well. ■

The calculation scheme shown in Example 14.1 is applicable to any number of species, as long as the ideal-solution-of-ideal-gases assumption is applicable. For the special case of an ideal solution of ideal gases with only two species, this result (see Problem 14.4) can be simplified to

$$\frac{y_{i2}}{y_{i1}} = \frac{1}{y_{i1} + y_{j1}/\alpha} \quad \text{where } \alpha = \exp\left[-(M_i - M_j)\frac{g(z_2 - z_1)}{RT}\right] \quad (14.8)$$

Subsequent examples show the convenience of this formulation.

The calculation method in Example 14.1 is applicable in principle to the atmosphere. However, the real atmosphere is far from equilibrium, because atmospheric winds and turbulence keep the atmosphere well-mixed.

Example 14.2 Estimate the nitrogen concentration that would exist at the top of the atmosphere, if the winds stopped mixing the atmosphere and it came to equilibrium. For this calculation only, replace the real atmosphere, which is quite complex, with an isothermal atmosphere with $T = 288$ K and thickness (from the ground up) of 15 km. Assume that the nitrogen mol fraction at the surface is 0.79 and the rest of the atmosphere is O_2.

From Eq. 14.8, taking 1 at the surface and 2 at $z = 15$ km,

$$\alpha = \exp\left[-(28 - 32)\frac{\text{g}}{\text{mol}} \cdot \frac{9.81\dfrac{\text{m}}{\text{s}^2} \cdot (15{,}000\,\text{m} - 0)}{8.314\dfrac{\text{J}}{\text{mol K}} \cdot 288\,\text{K}} \cdot \frac{\text{J s}^2}{\text{kg m}^2} \cdot \frac{\text{kg}}{1000\,\text{g}}\right] = 1.28 \quad (14.\text{J})$$

$$\frac{y_{i2}}{y_{i1}} = \frac{1}{0.79 + 0.21/1.28} = 1.05 \quad ■ \quad (14.\text{K})$$

Without the winds to mix the nitrogen and oxygen, the concentration of nitrogen at the top of the atmosphere would be ≈5% greater than at the surface. In the real atmosphere the wind mixing is strong enough that this difference is undetectably small.

If Eqs. 14.7 and 14.8 apply over distance of thousands of meters, do they also apply in industrial-sized equipment? Obviously, they must. We have ignored the effect of gravity in all our previous discussions, because we asserted that it was negligible. These equations allow us to see how large an error we made by neglecting them.

Example 14.3 Most industrial reactors have a vertical dimension less than 10 m (32.8 ft). Repeat Example 14.2, for an elevation change from top to bottom of the reactor of 10 m.

$$\alpha = \exp\left[-(28-32)\frac{\text{g}}{\text{mol}} \cdot \frac{9.81\frac{\text{m}}{\text{s}^2}\cdot(10\,\text{m}-0)}{8.314\frac{\text{J}}{\text{mol K}}\cdot 288\,\text{K}} \cdot \frac{\text{J s}^2}{\text{kg m}^2}\cdot\frac{\text{kg}}{1000\,\text{g}}\right] = 1.00016 \tag{14.L}$$

$$\frac{y_{i2}}{y_{i1}} = \frac{1}{0.79+0.21/1.00016} = 1.00003 \quad ■ \tag{14.M}$$

Based on this calculation we see that if there is no mechanical mixing we make an error of 0.003% in the equilibrium concentration of nitrogen at the top of the reactor by ignoring the effect of gravity. We seldom have data precise enough to justify worrying about this small an error, so the calculations we made in the preceding chapters are not seriously in error because we ignored gravity.

14.2.1 Centrifuges

We can make the gravity effects computed above occur over short distances if we replace gravity with centrifugal force. We know from basic mechanics that the centrifugal force is a pseudo-force, which results from Newton's law of motion that all bodies in motion continue to move in a straight line unless acted on by an external force. To use this idea we simply replace the g in the above equations with $-\omega^2 r$, where ω is the angular velocity. (The minus sign appears because gravity points in the $-z$ direction and centrifugal force points in the $+r$ direction.) Integrating from r_1 to r_2 is the equivalent of replacing $g\Delta z$ with $-0.5\omega^2(r_2^2 - r_1^2)$, so the centrifugal equivalent of Eq. 14.8 becomes

$$\frac{y_{i2}}{y_{i1}} = \frac{1}{y_{i1}+y_{j1}/\alpha} \quad \text{where } \alpha = \exp\left[(M_i - M_j)\frac{\omega^2(r_2^2 - r_1^2)}{2RT}\right] \tag{14.9}$$

The only current industrial application of centrifugal separation of gases known to the author is the separation of uranium isotopes. Natural uranium is about 0.7% ^{235}U, balance ^{238}U. The *uranium enrichment* process increases the ^{235}U content to about 2 to 3% for electric power reactors and to about 90% for nuclear weapons, (There are several types of uranium enrichment processes [1], of which only the centrifuge process is discussed here.) All enrichment processes currently in use convert the uranium to uranium hexafluoride, UF_6, which sublimes to a gas (as does solid carbon dioxide, "dry ice") at one atmosphere and 56.5°C. The resulting gas is 0.7 mol% $^{235}UF_6$ ($M = 349$ g/mol) and 99.3 mol% $^{238}UF_6$ ($M = 352$ g/mol). Enrichment processes use this small difference in molecular weights to produce *enriched uranium*, which has more than 0.7 mol% $^{235}UF_6$ rejecting the *depleted uranium*, which has much less than this amount of ^{235}U. Then the enriched hexafluoride gas is converted to uranium metal or uranium oxide for peaceful or military purposes.

Example 14.4 A uranium enrichment centrifuge has external radius 10 cm and internal radius 2 cm, and rotates at 800 revolutions/s (=48,000 rpm). The feed is a natural mixture of uranium hexafluoride with 0.7 mol% $^{235}UF_6$ at 300 K. Estimate the ratio of the mol fraction of $^{235}UF_6$ at the 2-cm radius to that at the 10-cm radius, at a state of thermodynamic equilibrium.

From Eq. 14.9, taking i to be $^{235}UF_6$ and assuming an ideal solution of ideal gases,

$$\alpha = \exp\left[(349-352)\frac{\text{g}}{\text{mol}} \cdot \frac{(2\pi\cdot 800/\text{s})^2\cdot(10^2-2^2)\text{cm}^2}{2\cdot 8.314\frac{\text{J}}{\text{mol K}}\cdot 300\,\text{K}} \cdot \frac{\text{J s}^2}{1000\,\text{g m}^2}\cdot\frac{\text{m}^2}{10^4\,\text{cm}^2}\right] = 0.864 \tag{14.N}$$

$$\frac{y_{i2}}{y_{i1}} = \frac{1}{0.007+0.993/0.864} = 0.865 \quad \text{or} \quad \frac{y_{i1}}{y_{i2}} = 1.156 \quad ■ \tag{14.O}$$

This shows that at equilibrium (for an assumed ideal solution of ideal gases) we would expect the gas at the 2-cm radius would contain about 16 mol% more $^{235}UF_6$ than that at the 10-cm radius. If we put the feed in at 10 cm radius and withdraw the product at 2 cm radius, then we increase the ^{235}U concentration from 0.7 to 0.8 mol%. The commercial plants that perform this operation use about 11 stages of such separation to produce uranium enriched to 3% [2]. There are formidable mechanical engineering challenges in building thousands of small (20-cm-diameter) centrifuges that will

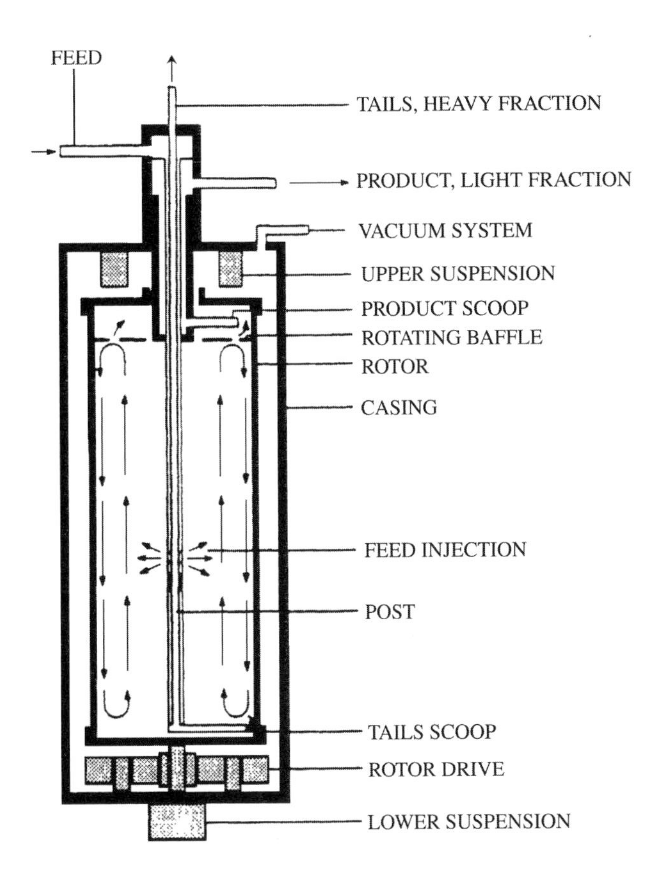

FIGURE 14.2 Cross section of a single-stage gas centrifuge for uranium enrichment. (From Benedict, M., T. H. Pigford, and H. W. Levi. *Nuclear Chemical Engineering*, ed. 2. New York: McGraw–Hill, pp. 847–876 (1981). Reproduced by permission of the McGraw-Hill Companies.)

operate at 48,000 rpm without shutting down for many years. But the economic advantages of this method of uranium enrichment are great enough that there are several plants in the world that enrich uranium this way and this is apparently the least technically-challenging and least-expensive route to a nuclear weapon. Figure 14.2 shows a simplified cross-sectional view of such a centrifuge [1]. In it a countercurrent flow along the axis is induced by the two takeoff scoops, the lower of which is in the main body of the rotor, while the upper is shielded from it by a baffle. This causes the enrichment process to occur as the gas circulates from top to bottom, producing a higher separation of $^{235}UF_6$ at the top from $^{238}UF_6$ at the bottom than could be produced in one equilibrium stage.

14.3 SEMIPERMEABLE MEMBRANES

A semipermeable membrane allows the free passage of one or more species in a solution and forbids the passage of others. The perfect semipermeable membrane would offer no resistance to the transferred species and allow no leakage of any nontransferred species. No perfect semipermeable membranes are known, but there are several examples of industrially useful membranes that are practically semipermeable. These include hot palladium metal, which pass hydrogen but not other gases [3]; some glasses, which pass helium but not other gases; and cellophane, which passes water, urea, and some salts but not proteins in artificial kidneys. Most of the cell walls in living things are semipermeable membranes, such as the membranes in our lungs, which let oxygen and carbon dioxide diffuse across, but do not let blood leak out, or the walls of our red blood corpuscles, which let oxygen pass freely, but hold in the hemoglobin. These membranes are not the "perfect semipermeable membranes" of thermodynamics texts, but they make life possible.

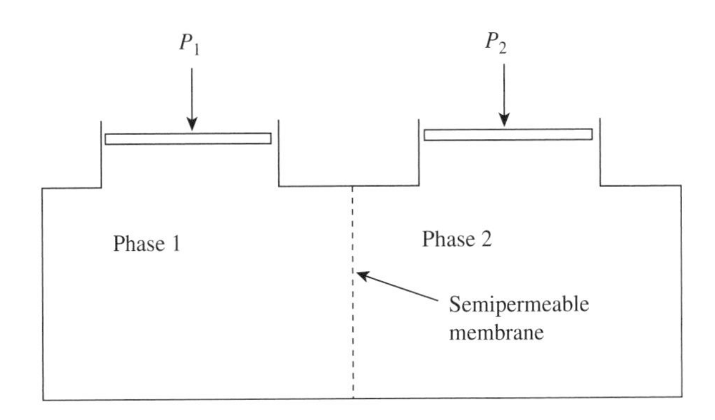

FIGURE 14.3 Two phases in equilibrium, separated by a semipermeable membrane.

Figure 14.3 shows another of our piston-and-cylinder containers, in which two phases are in equilibrium, separated by a semipermeable membrane. Each of these phases is an equilibrium phase, completely internally mixed. The phases could be gas, liquid or solid; most often the two phases are either both liquids or both gases (but not in the case of our lungs). We assume that each of the phases consists of two chemical species, *i* and *j*.

Unlike the other piston-and-cylinders we have used, for example Figure 4.1, this figure shows two pistons and cylinders. The reason is that if phases 1 and 2 are both mixtures and are not chemically identical, and if the semipermeable membrane allows the passage of one or more of the chemical species present but not one or more other species, then, at equilibrium, the two parts of the device will be at different pressures. The pressure difference can be large, as shown in the next example. In industrial devices the membrane is often quite thin; clever mechanical devices support the membrane so that the pressure difference does not rupture it, while blocking as little of the membrane surface as possible.

If the membrane is only permeable to *i* but not to *j*, then at equilibrium

$$\mu_i^{(1)} = \mu_i^{(2)} \tag{14.10}$$

but there is no corresponding relation for *j*. $\mu_j^{(1)}$ and $\mu_j^{(2)}$ can take on any values. We can arbitrarily change one of them, without changing the other. If we restate this in terms of the fugacities

$$f_i^{(1)} = f_i^{(2)} \tag{14.11}$$

but $f_j^{(1)}$ and $f_j^{(2)}$ can take on any values.

14.3.1 Osmotic Pressure

Example 14.5 In Figure 14.3, phase 1 is seawater and phase 2 is freshwater, both at 40°F. The two phases are separated by a membrane permeable to water but not to salts. If they are at equilibrium, what is the pressure difference between them?

Seawater varies from time to time and place to place, and contains at least trace amounts of all the elements of the periodic table. It is approximately 3 wt% dissolved solids, mostly NaCl, Na_2SO_4, $MgCl_2$, and KCl, which, if they are totally ionized, lead to a *water* mol fraction ≈0.98 [4]. To estimate the two fugacities in Eq. 14.11, we return to Example 7.3, which shows the effect of pressure on fugacity of liquids, and combine that with Eq. 7.27 to find

$$f_i = x_i\ \gamma_i\ p \cdot \exp\int_p^P \frac{\bar{v}}{RT}\,dP \tag{14.12}$$

(This a Raoult's law-type fugacity statement with the Poynting factor included.) We write this equation twice, once for pure water and once for the water in the ocean water, and equate the fugacities, finding

$$\left(x_i\ \gamma_i\ p \cdot \exp\int_p^P \frac{\bar{v}}{RT}\,dP\right)_{\text{pure water}} = \left(x_i\ \gamma_i\ p \cdot \exp\int_p^P \frac{\bar{v}}{RT}\,dP\right)_{\text{water in solution}} \tag{14.13}$$

This is very similar to the boiling-point elevation and freezing-point depression cases we considered in Section 8.10. In both of those cases the solute was inactive, either because its vapor pressure was ≈0.00 in those cases or because its permeability through the membrane is ≈0.00 in this case. The behavior of the solvent, which is close to pure, can be estimated by Raoult's law in all three cases. For the pure water, x_i and γ_i are unity, and for the water in the solution, with mol fraction 0.98, Raoult's law is certain to be practically obeyed, so that γ_i is certain to be practically unity. The partial molar volume of water in pure water is practically the same as that in dilute solutions, so we may take the ln of both sides and combine the two integrals, noting that the pressure of the salt water is greater than that of the freshwater, finding

$$-\ln x_{\text{water}}^{(\text{in salt water})} = \int_{P_{\text{pure water}}}^{P_{\text{salt water}}} \frac{\bar{v}_{\text{water}}^{(\text{in salt water})}}{RT}\,dP = \frac{\bar{v}_{\text{water}}^{(\text{in salt water})}}{RT}\left(P_{\text{salt water}} - P_{\text{pure water}}\right) \tag{14.14}$$

$$\left(P_{\text{salt water}} - P_{\text{pure water}}\right) = \Delta P_{\text{osmotic}} = \frac{\left(-RT\ \ln x_{\text{water}}^{(\text{in salt water})}\right)}{\bar{v}_{\text{water}}^{(\text{in salt water})}}$$

$$= \left[\frac{-\left(\dfrac{10.73\ \text{psi ft}^3}{\text{lbmol}^\circ\text{R}}\right)\cdot(500^\circ\text{R})\cdot(\ln 0.98)}{\left(\dfrac{18}{62.4}\right)\dfrac{\text{ft}^3}{\text{lbmol}}}\right]$$

$$= 375\ \text{psi} \qquad \blacksquare \tag{14.15}$$

This quantity, called the *osmotic pressure* $\Delta P_{\text{osmotic}}$, depends on the mol fraction and specific volume of the solvent, but not on the identity of the solute (or mixture of solutes). Referring back to Section 8.10, we recall that boiling-point elevation, freezing-point depression, and osmotic pressure are called the *colligative properties* of solutions. All three describe the situation in which, in a dilute solution, the solute has negligible vapor pressure for boiling-point elevation or freezing-point depression, or negligible permeability through a membrane for osmotic pressure, and the solvent practically obeys Raoult's law.

The experimental value of the osmotic pressure of the ocean ≈340 psi, indicating that some of the simplifications used here are not exactly correct. (See Problem 14.18.) This says that if ocean water and freshwater are separated by a membrane (Figure 14.3), permeable only to water, and if there is to be equilibrium, with no flow of water in either direction, then the ocean water must be at a pressure 340 psi higher than the freshwater. If the ocean pressure is more than 340 psia greater than that of the freshwater, then water will flow through the membrane from the ocean side to the freshwater side, leaving its salts behind. This is the *reverse osmosis* method of water desalination, which is economical for some brackish waters (waters with less salt than the ocean, but too salty for drinking water) in some locations. It is not currently economical for preparing freshwater from ocean water. (The U. S. Army regularly uses it to prepare *emergency* drinking water from seawater when that is needed.) If the pressure difference is less than 340 psi, freshwater will flow through the membrane into the ocean water. Car washes and building humidifiers need water with practically zero dissolved solids; many find that reverse osmosis of city water is the cheapest way to obtain it. See also [5].

As a simple example of the importance of osmotic equilibria, consider the various ways humans have developed to

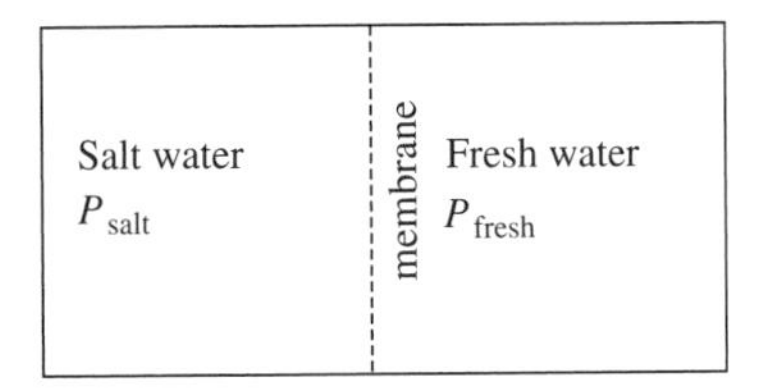

If $(P_{salt} - P_{fresh}) > \Delta P_{osmotic}$ water flows from salt to fresh sides, in Reverse osmosis.

If $(P_{salt} - P_{fresh}) < \Delta P_{osmotic}$ water flows from fresh to salt sides, in salt preservation of foods.

If $(P_{salt} - P_{fresh}) = \Delta P_{osmotic}$ we have osmotic equilibrium and there is no flow.

FIGURE 14.4 Osmotic flow directions as a function of $(P_{salt} - P_{fresh})$ and $\Delta P_{osmotic}$.

preserve food. Before the invention of canning and freezing, we preserved food by salting, with sugar or honey (jams and jellies), in alcohol or vinegar (pickles and sauerkraut), and by air, sun, or oven drying. All of these methods rely on the fact that the bacteria, yeasts, and other organisms that spoil foods need a fairly high water content to live and multiply. That air, sun, or oven drying deprives them of water is fairly obvious. The other methods all work by causing the bacteria and yeasts, whose cell walls are practically semipermeable membranes, to try to come to osmotic equilibrium with their surroundings whose water concentrations are lower than their own, and thus to lose so much water that they die. Our distant ancestors who discovered these food preservation methods did not know about bacteria or osmotic equilibrium, but they found ways to "osmose the little rascals to death." These relations are sketched in Figure 14.4.

14.4 SMALL IS INTERESTING! EQUILIBRIUM WITH SURFACE TENSION

Gas–liquid interfaces exert forces due to the surface tension of the liquid. For most systems of engineering importance, these forces are negligible. However, for small bubbles and drops, surface forces are very important, as shown here. Several other situations in which surface tension is important are not considered here, such as emulsions, coatings, candle wicks, sweat solder fittings, multiphase flow in porous media, and ink-jet printers [6, Chapter 14].

14.4.1 Bubbles, Drops, and Nucleation

A small liquid droplet is surrounded by pure vapor of the same material at the same temperature (e.g., a water droplet

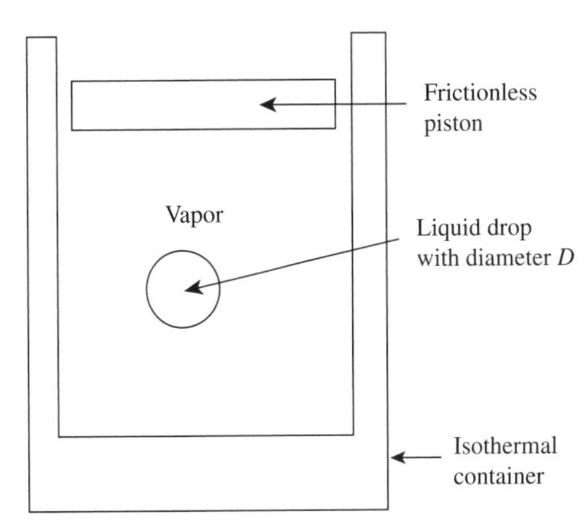

FIGURE 14.5 A small drop of liquid suspended in a vapor of the same chemical species. We ignore the very slow gravity settling of the drop.

surrounded by steam) (see Figure 14.5). The drop falls due to gravity, but very small drops fall very slowly, so we may ignore the gravity settling of the drop.

What is the criterion of equilibrium in this situation? From any elementary fluid mechanics book or from a simple force balance around the droplet (see Figure 14.6) we can find that

$$P_{inside} = P_{outside} + \frac{4\sigma}{D} \tag{14.16}$$

(the *Young–La Place equation*), where σ is the surface tension.

Example 14.6 Estimate the difference in pressure between inside and outside of a droplet of water suspended in steam at 100°C, with $P_{outside} = 1$ atm, for various drop diameters.

The metric steam table [7, p. 267] (see Problem 14.22) gives the surface tension between steam and water at 100°C as 0.05892 N/m, so for a 1-mm drop

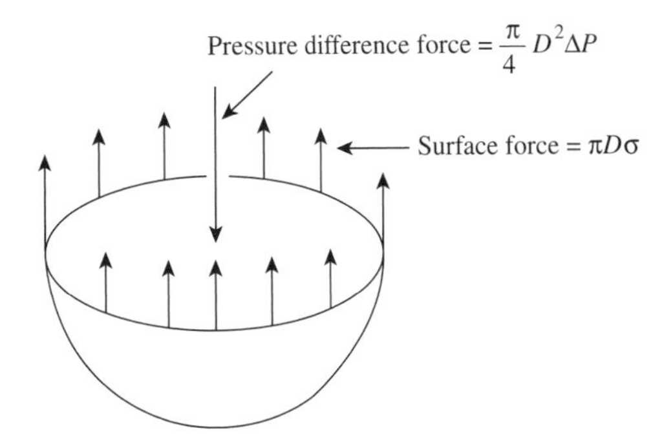

FIGURE 14.6 Sketch for force balance around a small droplet, cut in half. The surface force, acting on the cut surface and the pressure force acting on the cut projected area are equal and opposite, leading to Eq. 14.16. (From de Nevers, N. *Fluid Mechanics for Chemical Engineers*, ed. 2. New York: McGraw–Hill, p. 490 (1991). Reproduced by permission of the McGraw-Hill Companies.)

$$P_{\text{inside}} - P_{\text{outside}} = \frac{4 \cdot 0.05892\,\text{N/m}}{0.001\text{m}} = 235.7\text{Pa} = 0.00233\,\text{atm} \tag{14.P}$$

which is certainly negligible. However, if we reduce the diameter to $10^{-6}\text{m} = 1\,\mu\text{m}$, a common drop size in sprays, then the calculated pressure is 2.33 atm, and if we reduce it to 0.01 μm (about the smallest sized drop that is likely to exist, see Problem 14.28), then the calculated pressure difference is 233 atm = 3240 psig! ■

The surface of the drop acts like a semipermeable membrane, preventing the contained liquid from expanding, but not preventing the liquid in the drop and the gas outside the drop from being in thermodynamic equilibrium. This means that the Gibbs energies (or fugacities) of the liquid inside the drop and the vapor outside it must be equal. In this case (for a pure species instead of a mixture) it seems simpler to work directly with the Gibbs energies, as shown below.

The practical problem of interest is the stability of small drops or bubbles. In these problems the condensation or evaporation normally begins on some small solid particle.

This may be a very small dust particle in the atmosphere, a chemical particle generated in a cloud-seeding operation in the atmosphere, or a very small impurity particle present in gas or liquid. Such particles quickly form equivalent bubbles or drops by adsorbing molecules of the water on their solid surfaces, so that they become "artificial drops" or "artificial bubbles" of practically constant size.

Example 14.7 Steam at 1 atm and 100°C is being slowly compressed isothermally. There is no liquid present and the distance from the nearest solid walls is so large that condensation on them is unimportant. The condensation nuclei present in the steam (due to impurity particles) have diameters of 0.01 μm ($=10^{-8}\,\text{m} = 3.94 \times 10^{-7}$ inches). At what pressure will the steam begin to condense on these nuclei and therefore convert the all-gaseous system to a two-phase mixture?

At equilibrium the Gibbs energy per pound will be the same inside and outside the drops. From Example 14.6 we know that the $(P_{\text{inside}} - P_{\text{outside}}) = 4\sigma/D = 233\,\text{atm} = 235.7$ bar. Taking the Gibbs energy per pound at the normal boiling point (the same in gas and liquid) as g_{NBP} we have

$$\begin{aligned} g_{\text{small drop equilibrium}} &= g_{\text{NBP}} + \int_{P_{\text{NBP}}}^{P_{\text{gas}}} v_{\text{water}}^{(\text{gas})} dP \\ &= g_{\text{NBP}} + \int_{P_{\text{NBP}}}^{P_{\text{gas}} + 4\sigma/D} v_{\text{water}}^{(\text{liquid})} dP \end{aligned} \tag{14.17}$$

If we assume that the specific volume of the liquid is a constant, independent of pressure, and that the volume of the vapor is given by the ideal gas law, then we can perform the integrations and cancel the g_{NBP} terms, finding the *Kelvin equation*:

$$RT\,\ln\frac{P_{\text{gas}}}{P_{\text{NBP}}} = v_{\text{water}}^{(\text{liquid})} \cdot \left(P_{\text{gas}} + \frac{4\sigma}{D} - P_{\text{NBP}}\right) \tag{14.18}$$

It has no analytic solution, but can be solved numerically. However, for very small drops

$$(P_{\text{gas}} - P_{\text{NBP}}) \ll \frac{4\sigma}{D} \tag{14.19}$$

so that we can write it approximately as

$$\frac{P_{\text{gas}}}{P_{\text{NBP}}} \approx \exp\left(\frac{v_{\text{water}}^{(\text{liquid})} \cdot 4\sigma/D}{RT}\right) \tag{14.20}$$

which is also called the Kelvin equation. Inserting the values corresponding to the normal boiling point gives

$$\begin{aligned} \frac{P_{\text{gas}}}{P_{\text{NBP}}} &\approx \exp\left(\frac{\dfrac{\text{m}^3}{958.39\,\text{kg}} \cdot \dfrac{4 \cdot 0.05892\text{N/m}}{10^{-8}\text{m}} \cdot \dfrac{\text{bar}}{10^5\text{Pa}} \cdot \dfrac{0.018\,\text{kg}}{\text{mol}}}{0.08314\dfrac{\text{L}\cdot\text{bar}}{\text{mol}\cdot\text{K}} \cdot \dfrac{\text{m}^3}{1000\,\text{L}} 373.15\text{K}}\right) \\ &= \exp\ 0.1427 = 1.1534 \end{aligned} \tag{14.Q}$$

$$\begin{aligned} P_{\text{gas}} - P_{\text{NBP}} &= 1.1534 - 1 = 0.1534\,\text{atm} = 0.1554\,\text{bar} \\ &= 2.25\,\text{psi} \end{aligned} \tag{14.R}$$

At equilibrium, the pressure in the gas is 2.25 psia above the normal boiling pressure at this temperature, and the pressure inside the drop is 3420 psi greater than the pressure in the gas. ■

This says that if all the foreign particles present have diameters of $\le 0.01\,\mu\text{m}(= 10^{-8}\ \text{m} = 3.94 \times 10^{-7}$ inches), then the steam can be compressed by 2.25 psi above its normal boiling point pressure before it will begin to condense. When it does begin to condense, as each drop grows its D increases so that it no longer requires as high a pressure for moisture to condense on it, and it will grow as rapidly as it can reject the heat of condensation.

Example 14.8 A drop of water with temperature 100°C and a diameter of 0.01 μm($= 10^{-8}$ m $= 3.94 \times 10^{-7}$ inches) is suspended in steam at a temperature of 100°C and gauge pressure 0.15 bar. Will this drop grow by condensation?

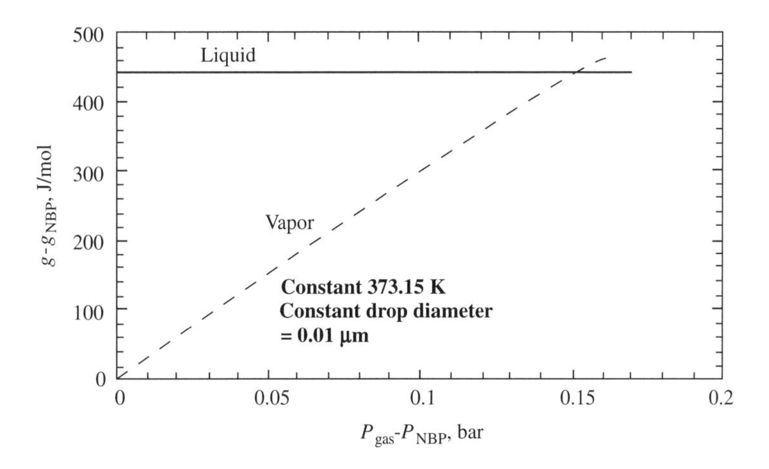

FIGURE 14.7 Computed $(g - g_{\mathrm{NBP}})$ for water as a function of $P_{\mathrm{gas}} - P_{\mathrm{NBP}}$ at 100°C = 373.15 K, for a fixed drop diameter of 0.01 μm.

Shrink by evaporation? Or remain at its present size because it is in equilibrium with its surroundings?

This is the same size drop and the same temperature as in Example 14.7, so the calculation of the pressure difference from inside to outside is the same. We begin with Eq. 14.17 and solve for the specific Gibbs energy of the liquid and then that of the gas. For the liquid

$$
\begin{aligned}
g_{\mathrm{water}}^{(\mathrm{liquid})} - g_{\mathrm{NBP}} &= \int_{P_{\mathrm{NBP}}}^{P_{\mathrm{gas}}+4\sigma/D} v_{\mathrm{water}}^{(\mathrm{liquid})}\, dP \\
&= v_{\mathrm{water}}^{(\mathrm{liquid})} \cdot \left(P - P_{\mathrm{NBP}} + \frac{4\sigma}{D} \right) \\
&= \left(\frac{0.018}{0.95839} \frac{\mathrm{L}}{\mathrm{mol}} \right) \cdot (0.15 + 235.7)\ \mathrm{bar} \\
&\quad \cdot \frac{10^5 \mathrm{N}}{\mathrm{m}^2 \cdot \mathrm{bar}} \cdot \frac{\mathrm{m}^3}{1000\ \mathrm{L}} \cdot \frac{\mathrm{J}}{\mathrm{N} \cdot \mathrm{m}} = 442.64 \frac{\mathrm{J}}{\mathrm{mol}}
\end{aligned} \tag{14.S}
$$

and for the gas, again using Eq. 14.17,

$$
\begin{aligned}
g_{\mathrm{water}}^{(\mathrm{gas})} - g_{\mathrm{NBP}} &= \int_{P_{\mathrm{NBP}}}^{P_{\mathrm{gas}}} v_{\mathrm{water}}^{(\mathrm{gas})}\, dP = RT \ln \frac{P_{\mathrm{gas}}}{P_{\mathrm{NBP}}} \\
&= 8.314 \frac{\mathrm{J}}{\mathrm{mol} \cdot \mathrm{K}} \cdot 373.15\ \mathrm{K} \\
&\quad \cdot \ln \frac{(1.013 + 0.15)\ \mathrm{bar}}{1.013\ \mathrm{bar}} = 428.39 \frac{\mathrm{J}}{\mathrm{mol}}
\end{aligned} \tag{14.T}
$$

$$
g_{\mathrm{water}}^{(\mathrm{liquid})} - g_{\mathrm{water}}^{(\mathrm{gas})} = 442.64 - 428.39 = 14.25 \frac{\mathrm{J}}{\mathrm{mol}} \tag{14.U}
$$

The liquid can lower its Gibbs energy 14.25 J/mol by changing to a gas, so that even at 0.15 bar above the normal boiling point, a drop this small is unstable and will quickly *evaporate*. ■

The relations between the Gibbs energies of the liquid and the gas at a constant 373.15 K and a constant drop diameter of 0.01 μm are sketched in Figure 14.7; the individual values are calculated exactly as in Example 14.8. The liquid line appears to be horizontal, but actually increases by a negligible amount as the gas pressure increases. Because of its much larger specific volume, the Gibbs energy of the gas increases much more rapidly with increasing pressure and, as shown, the two curves cross at the equilibrium gauge pressure, 0.155 bar.

We can see that for any particular relation between pressure and temperature in any gas that is unstable with regard to condensation, there is one size of drop that is stable, and that those smaller than it can lower their Gibbs energy by evaporating, while those larger than that size can lower their Gibbs energy by increasing in size. As a practical matter, we see the terrific instability of a small drop (if it is not fixed in size by forming around a solid particle). If a drop is at equilibrium with its surroundings at the equilibrium Gibbs energy shown in Figure 14.7 and there is an infinitesimal increase in pressure, then $g_{\mathrm{water}}^{(\mathrm{gas})} > g_{\mathrm{water}}^{(\mathrm{liquid})}$ and the gas will spontaneously condense onto the drop. As the drop's size increases its internal pressure falls and so its Gibbs energy falls, making the gas condense onto it more rapidly, increasing its size even more, and so forth. Conversely, if the drop is at equilibrium, as shown on Figure 14.7, and the gas pressure falls slightly, then $g_{\mathrm{water}}^{(\mathrm{gas})} < g_{\mathrm{water}}^{(\mathrm{liquid})}$ and liquid will evaporate off the drop, decreasing its diameter, raising its internal pressure and thus its Gibbs energy, causing it to evaporate more rapidly and lower its size rapidly. Referring to Figure 1.5, this situation is the unstable case, with the ball resting on the top of a very narrow, steep-sided peak.

Thus, this will be a highly irreversible, spontaneous process. This kind of equilibrium, when it exists, is metastable. This is of prime significance in the whole area of nucleation [8]. If we have a superheated liquid or a subcooled vapor and introduce nuclei that are larger than the stability limit calculated this way, they will grow and cause boiling or condensation. We can easily calculate the required nucleus diameter for the nucleus to be effective. That calculation guides the whole field of nucleation. A nucleus larger than the diameter calculated above, if introduced into a supersaturated system, will grow rapidly, causing rapid, sometimes explosive conversion to the stable state. A smaller nucleus will do nothing. The same ideas presented here apply to cloud seeding, crystallization, and other nucleation phenomena.

Example 14.9 Pure liquid water is at 400 psia and at the exact saturation temperature for that pressure (444.70°F = 904.7°R). We now reduce the pressure on the liquid, at constant temperature. There is no free surface, so boiling can begin only around the boiling nuclei in the liquid (generally submicroscopic solid particles), which have a diameter of 10^{-5} inch. At what liquid pressure will these boiling nuclei (which may be considered the equivalent of small bubbles of steam) begin to grow and thus initiate boiling? At 444.70°F, $\sigma_{\text{water}} \approx 1.76 \times 10^{-4}$ lbf/inch.

This is practically the same as the growing drop problem, except that in this case the gas is inside the bubble, at a pressure much higher than that of the surrounding liquid. Again, the criterion of equilibrium is that the Gibbs energy of the gas inside the bubble must be the same as that of the liquid outside the bubble. We can rewrite Eq. 14.17 as

$$g_{\text{small bubble equilibrium}} = g_{\text{NBP}} + \int_{P_{\text{NBP}}}^{P_{\text{liquid}}+4\sigma/D} v_{\text{water}}^{(\text{gas})}\, dP$$

$$= g_{\text{NBP}} + \int_{P_{\text{NBP}}}^{P_{\text{liquid}}} v_{\text{water}}^{(\text{liquid})}\, dP \quad (14.22)$$

and again set these equal and assume that the liquid has practically constant density and that the gas behaves as an ideal gas (which becomes a poor assumption as the gas pressure rises), to find

$$RT \ln \frac{P_{\text{liquid}} + 4\sigma/D}{P_{\text{NBP}}} = v_{\text{water}}^{(\text{liquid})} \left(P_{\text{liquid}} - P_{\text{NBP}}\right) \quad (14.22)$$

This equation has no analytical solution, but can be easily solved numerically for P_{liquid}. First we evaluate

$$\frac{4\sigma}{D} = \frac{4 \cdot 1.76 \times 10^{4} \,\frac{\text{lbf}}{\text{in}}}{1 \times 10^{-5}\text{in}} = 70.4 \frac{\text{lbf}}{\text{in}^2} \quad (14.V)$$

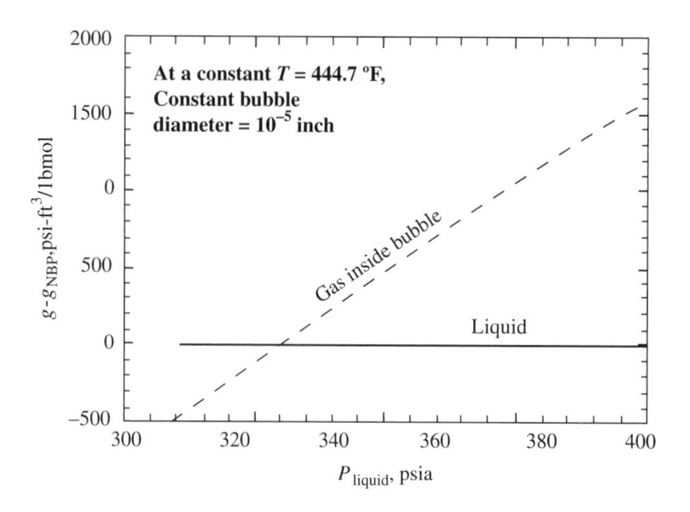

FIGURE 14.8 Calculated values for Example 14.9. The Gibbs energy of the liquid changes only slightly with decreasing liquid pressure (at constant temperature) because of its small specific volume.

then in Eq. 14.22

$$10.73 \frac{\text{psi} \cdot \text{ft}^3}{\text{lbmol} \cdot \text{R}} \cdot 904.7^{\circ}\text{R} \cdot \ln \frac{\left(P_{\text{liquid}} + 70.4 \text{ psia}\right)}{400 \text{ psia}}$$

$$= \left(18 \cdot 0.01934 \frac{\text{ft}^3}{\text{lbmol}}\right) \cdot \left(400 \text{ psia} - P_{\text{liquid}}\right) \quad (14.\text{W})$$

or

$$27{,}885 \text{ psia} \cdot \ln \frac{\left(P_{\text{liquid}} + 70.4 \text{ psia}\right)}{400 \text{ psia}} = \left(400 \text{ psia} - P_{\text{liquid}}\right) \quad (14.\text{X})$$

for which the numerical solution is 328.6 psia. At this external pressure, the pressure inside the bubble is 399.0 psia. ■

The pressure-Gibbs energy relations in Example 14.9 are sketched in Figure 14.8, which is very similar to Figure 14.7, but has the equilibrium P below the ordinary boiling-point pressure instead of above because the gas is at $P_{\text{liquid}} + 4\sigma/D$, while in Figure 14.7 the liquid was at $P_{\text{gas}} + 4\sigma/D$. We may easily show that for smaller bubbles the gas curve is shifted to the left, and for larger bubbles it is shifted to the right.

Again, we see the extreme instability of small bubbles and drops. If the liquid pressure falls slightly below the equilibrium pressure, then $g_{\text{water}}^{(\text{liquid})} > g_{\text{water}}^{(\text{gas})}$, so the bubble will expand, increasing its diameter and shifting the vapor curve to the right, causing rapid bubble growth. If the liquid pressure rises slightly from the equilibrium value, then $g_{\text{water}}^{(\text{liquid})} < g_{\text{water}}^{(\text{gas})}$, so the babble will contract, shifting the

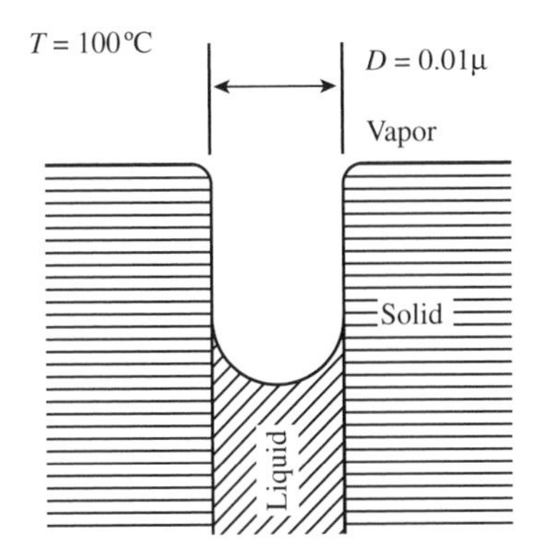

FIGURE 14.9 A practically-cylindrical pore of diameter D in a solid. The pore is partially filled with liquid, which wets the sides of the pore, so that surface tension pulls the interface into a hemisphere.

vapor curve to the left and leading to rapid bubble contraction and collapse. If the bubble is stabilized against such collapse by being formed around some kind of solid particle, then it will remain in place and maintain its diameter as long as the pressure is higher than the equilibrium pressure.

14.4.2 Capillary Condensation

Porous solids often contain pores of the same sizes as the bubbles and drops in the previous section, or smaller. Figure 14.9 shows a simplified view of such a pore, which extends a long way into the solid.

The hemispherical-shaped-interface-in-a-cylindrical-tube is the simplest mathematically of the possible geometries. For it we can use Equation 14.16 to find that the pressure inside the fluid is less than the pressure of the gas or vapor at the mouth of the pore by $4\sigma/D$. From Example 14.6 we know that for ordinary sized holes this difference is negligible, but for microscopic ones it becomes quite large. If we repeat Example 14.7 for this geometry ($D = 0.01\ \mu$ case only), we find almost the same numbers, but with different meanings. As before the pressure difference across the hemispherical 0.01 μ gas-liquid interface must be 3420 psi, but in this case that makes the pressure in the column of fluid inside the pore = −3240 psig!

This is certainly a surprise, but in this situation the negative pressure simply indicates that the trapped liquid is in tension. An ordinary liquid under tension will boil and convert to a gas-liquid mixture, but at this extremely small size the needed bubble cannot form, so the liquid under tension is quite stable.

Example 14.10 Estimate the vapor pressure of the liquid water held in place in the $D = 0.01\ \mu$ pore in Figure 14.9, if, as in Example 14.7 the vapor above the pore is pure steam at 100°C, and the liquid in the pore is at 100°C.

Returning to Equation 14.17 we see that the pressure we are seeking is that at which the liquid in the pore is in equilibrium with the external gas, so we seek the surrounding steam's pressure. The only change is in the upper integration limit on the right term, which changes from $P_{gas} + 4\sigma/D$ to $P_{gas} - 4\sigma/D$. If we then follow the rest of that example we see that the sign change affects all the calculations in a minor way, and finally $P_{gas} = 0.867$ atm ■

From this example we see:

1. Liquid water can exist at its NPB and a pressure of 0.867 atm if it is constrained by the surface forces in an 0.01 μ pore. This behavior, called *capillary condensation* plays a significant role in adsorption.
2. The simple cylindrical pore sketched in Figure 14.9 would not produce the bizarre hysteresis behavior shown in Figure 11.27. But a pore shaped like a bottle with a narrow neck (called an *ink bottle pore* in the adsorption literature) would [10]. If such a pore was filled up to the top of its neck with water at 100°C and the external steam pressure were gradually lowered, at a pressure of 0.867 atm the liquid would be in equilibrium with the vapor, and begin to evaporate. As the interface receded through the neck, the liquid's vapor pressure would remain constant. But when the neck emptied and the interface moved into the larger part of the pore, D would increase, $4\sigma/D$ would decrease, and the vapor pressure of the remaining liquid would *increase*, producing the behavior shown in Figure 11.27.
3. Here we have used the Poynting Factor (Section 7.4.2) to show the change in vapor pressure for *decrease* in system pressure. That is uncommon, but correct. The result is the Kelvin equation (14.20) with a minus sign inserted in the argument of the exponent (also called the Kelvin equation.)
4. This simple explanation shows how the hysteresis loop in Figure 11.27 *could* occur. The adsorption literature [10] shows that real experimental systems are more complex than what is described here.

14.5 SUMMARY

1. If we take into account the effects of gravity, magnetic, tensile, or electrostatic energies we find that the criterion of equilibrium is expanded, as shown in Eq. 14.3.
2. This equation shows that we can have systems in equilibrium with different pressures in different parts of the system, if those pressure differences are balanced by gravity, and so forth. That equation does not change our previous statement that any system at physical or chemical equilibrium is an isothermal system.

3. A mixture of two substances with different molecular weights will have a change in chemical composition with depth at equilibrium in a gravity field or with radius in a centrifugal force field. For ordinary situations this is negligible, but not for high speed centrifuges.
4. An ideal semipermeable membrane allows equality of the chemical potentials for one or more species of a mixture, but not for others. Many real membranes, such as those in our lungs, come close to this ideal.
5. A semipermeable membrane with solutions of different compositions on either side at equilibrium produces an osmotic pressure. By adjusting the real pressure difference we can cause the flow of the transferred species in either direction across such a membrane.
6. In very small droplets and bubbles there is a substantial pressure difference between the inside and the outside at equilibrium, due to surface tension. This plays an important role in small bubble and drop behavior. The same kind of forces, acting on liquids inside of very small pores in solids can produce capillary condensation, which complicates adsorption processes.

PROBLEMS

See the Common Units and Values for Problems and Examples. An asterisk (*) on the problem number indicates that the answer is in Appendix H.

14.1 In Example 14.1, the calculated pressure at the bottom of the well is 2.155 MPa. What pressure would we calculate for that depth using the basic equation of fluid statics (barometric equation) and assuming that the average molecular weight of the gas in the well was the same as at that the top? Assume isothermal, ideal gas behavior.

14.2 Repeat Example 14.1 for a difference in elevation from top to bottom of 1 m. Are we justified in normally ignoring such differences?

14.3* Repeat Example 14.1 for a well of 5000 m depth.

14.4 Show the derivation of Eq. 14.8. Write Eq. 14.7 twice, once for species *i* and once for species *j*. Divide one equation by the other (thus eliminating the pressure at the bottom of the well). Then replace the y_{j2} with $(1 - y_{i2})$, and perform the necessary algebra.

14.5* There is a slow but steady leakage of helium (produced by radioactive decay of uranium) out of the ground into our atmosphere. But the concentration in the atmosphere is only $\approx$5 ppm. With this steady inflow from the earth, over geologic time we would expect a large value. It is suggested that over geological time the atmosphere loses light gases (hydrogen and helium) from its top and that, although the winds stir the atmosphere, there is a concentration gradient that carries helium up to the top, where it can escape. To test this idea, compute the equilibrium ratio of helium concentration at the top of the atmosphere to that at the bottom. For this calculation only, replace the real atmosphere, which is quite complex, with an isothermal atmosphere with $T = 288$ K, average molecular weight 29g/mol, and thickness (from the ground up) of 15 km. Assume that the helium mol fraction at the surface is 5×10^{-6}.

This calculated result only partly explains the scarcity of helium. It is easier for a light molecule, like helium to get enough kinetic energy to escape from the earth's gravity than for a heavier molecule like nitrogen. That combines with this effect to remove light gases from the atmosphere.

14.6 Equations 14.7 and 14.8 seem to be simply a balance of mechanical effects, but the temperature appears in these equations. Why? Give a physical, intuitively satisfying answer. No calculations are needed.

14.7* In Example 14.4 the outlet concentration of $^{235}UF_6$ is 1.16 times the inlet value, which is assumed to be 0.007 mol fraction.

a. If we repeated the process 11 times, what would the concentration be?

b. How many times would we have to repeat it to get weapons-grade (90 mol%) $^{235}UF_6$?

14.8 The feed to a uranium enrichment plant is contaminated with nitrogen, at a concentration of 0.001 mol fraction. What will be the ratio at equilibrium of nitrogen at a radius of 2 cm to that at a radius of 10 cm, in the centrifuge discussed in Example 14.4?

14.9 For a gas centrifuge with the dimensions and angular velocity shown in Example 14.4, estimate the ratio of the pressure at the outer wall to that at the inner wall.

14.10 Explain how the scoop arrangement shown in Figure 14.2 induces the circulating flow shown in that figure. *Hint*: Observe that the baffle between the upper scoop and the main chamber has holes both at the periphery and near the center.

14.11 It is often asked how it is possible for a tall tree to get liquid (sap) to the top. One answer regularly given is that it is done by osmotic pressure. Is that possible? The tallest trees in the world are the redwood trees in California of which the tallest one measured to date is 367 feet tall.

a. Assuming that the fluid inside the tree has the density of pure water, what is the pressure difference that must be overcome to get water from its roots (assumed at atmospheric pressure at the base) to the top?

b. Assuming that this is done by osmotic pressure using plant cell walls that are totally permeable to water but not permeable to salts and sugars, what mol fraction of dissolved salts and sugars would have to exist in the sap?

c. Another explanation that is sometimes advanced is capillarity. Is that plausible? Here assume that inside the trees are small, cylindrical, open tubes of wood, in which the water rises by capillary action. Assume that there is a hemispherical air–water interface inside these tubes at the top, and that $\sigma_{\text{water}} = 4.2 \times 10^{-4}$ lbf/in. Assume that at the top of the tree the pressure of the air opposite the hemispherical interface is 1 atm. Estimate the tube diameter that would be needed to get the water to the top of a 367-ft-high tree.

d. Comment on the feasibility of this approach for bringing water to the top of the trees.

14.12 We wish to design a reverse osmosis plant to prepare drinking water from water with too high a salt content to drink. If the maximum pressure we can afford is 100 psig, what is the highest salt content in the water we can tolerate and still produce pure drinking water? Assume that the salt is all sodium chloride, that it is 100% ionized, and that it forms an ideal solution.

14.13 The army uses portable reverse osmosis equipment driven by diesel engines to make emergency freshwater from seawater. Estimate how many pounds of freshwater we can make per pound of diesel fuel consumed in such a device. Assume the following:

- The required pressure difference is as shown in Example 14.5.
- The power required to run a reverse osmosis device is given by $\text{Po} = Q\Delta P$, where Po is the power, Q is the volumetric flow rate of fluid pumped to the pressure of the reverse osmosis cell, and ΔP is the pressure increase from ambient to the inlet of that cell.
- We must pump twice as much saltwater into the cell as we get freshwater out, so that the other half can carry away all the salts (at a concentration ≈ twice that at the inlet).
- Diesel fuel has a heat of combustion of 19,000 Btu/lb.
- The overall efficiency of the motor-pump combination is 30%, that is, 30% of the combustion energy of the fuel goes into the work of compressing the fluid.

14.14 Show the details of the derivation of Eq. 14.14 from Eq. 14.13.

14.15 Pure hydrogen is produced commercially from impure hydrogen streams using palladium metal at 575 to 750°F as a semipermeable membrane [3]. If we have a gas stream that is 10 mol% hydrogen, and the hydrogen activity coefficient in this stream is $\gamma_{\text{hydrogen}} = 0.95$, what total pressure must we supply on the impure hydrogen side so that there will be physical equilibrium with pure hydrogen at 1 atm on the other side?

14.16 Why does the temperature appear in the equation for the osmotic pressure? This would seem to be simply a balance of mechanical effects. But the temperature appears in the equations. Give a physical, intuitively satisfying answer. No calculations are needed.

14.17* One hundred grams of a substance of unknown molecular weight was dissolved in 1 kg of pure water at 20°C and the osmotic pressure of the resulting solution, relative to pure water, was measured as 100 psi. Estimate the molecular weight of this substance on the assumption that this substance does not ionize in solution in water.

14.18 Example 14.15 computes the osmotic pressure of seawater as 375 psia, based on an estimated water mol fraction $x_{\text{water}}^{(\text{in salt water})} = 0.98$. What value must we use for this mol fraction to have the computed pressure match the observed value of 340 psia at 40°F?

14.19 Equation 14.15 for the osmotic pressure is written on the assumption that the dilute solvent has practically ideal solution behavior. As the concentration of solute increases (and that of solvent decreases) this assumption becomes unreliable.

a. Rewrite Eq. 14.15, taking liquid phase nonideality into account.

b. Solve the resulting equation for the activity coefficient of the solvent.

c. Discuss the pros and cons of using this method to measure the activity coefficient of solvents with various solutes [9,p. 173].

14.20 In Example 14.7 we assumed that $P_{\text{gas}} - P_{\text{NBP}}$ was negligible compared to the pressure difference from inside to outside of the drop (Eq. 14.19). How much difference does that simplification make in our answer? Repeat that example, not making that simplification.

14.21 Estimate the pressure at which steam will be in equilibrium with water droplets with diameters of 10^{-5}, 10^{-6}, and 10^{-7} inches, at 212°F, for which $\sigma_{\text{water}} \approx 3.3 \times 10^{-4}$ lbf/inch.

14.22* In the cgs system of units, surface tensions are expressed in dyn/cm, and in the SI system they are expressed in N/m. If a fluid has a surface tension of 60 dyn/cm (a typical value), what is its surface tension in N/m? The reported values of the surface tension in [7] are high by a factor of 1000. Comment on the probable origin of the values shown there.

14.23 Steam (gaseous water) is at 212.000°F = 100.00°C and a pressure of 14.8 psia = 0.10 psig. What is the diameter of the smallest drop of liquid water that will spontaneously grow when placed in contact with this steam? At this temperature $\sigma_{\text{water}} \approx 3.3 \times 10^{-4}$ lbf/inch.

14.24 Liquid water at 1 atm and 212°F = 100°C is being slowly heated at constant pressure in a closed, constant pressure container with no free surface. There are tiny impurities in the water with diameter 10 μm (= 3.97×10^{-4} inch), which are coated with gas, so they behave as tiny gas bubbles of that diameter. How much must the temperature of the water be raised before these bubbles begin to grow? Assume that steam is an ideal gas, and that at 212°F $\sigma_{\text{water}} \approx 3.3 \times 10^{-4}$ lbf/in.

14.25 When cold beer is placed in a clean, clear drinking glass, we observe that streams of bubbles form as certain sites on the wall of the glass and flow steadily upward, while practically no other bubbles appear. Explain this observation in terms of the ideas in this chapter.

14.26 When air bubbles with diameters smaller than a few microns are introduced into pure water, they collapse rapidly. Explain this observation in terms of the ideas in this chapter.

14.27 Sketch a desorption curve like the one in Figure 11.27 for the assumption that all the adsorbed material was inside a single pore, whose opening to the surroundings had a diameter of 0.01 μ and whose vapor liquid interface was at the end of the pore, for assumed pore geometries of: (a) cylinder, (b) a cone open at the end with the interface, and (c) an "ink bottle pore" with an 0.01 μ neck and an 0.1 μ main body.

14.28 **a.** Several places in this chapter it says that the smallest bubble or drop is likely to have a diameter of about 0.01 μ. If such a drop is a sphere, containing water at its NPB ($\rho = 958$ kg/m^3), what are the volume, mass and number of water molecules in the drop?

b. Repeat part (a) for a drop diameter of 0.001 μ. Would such a drop have enough water molecules to hold together?

REFERENCES

1. Benedict, M., T. H. Pigford, and H. W. Levi. *Nuclear Chemical Engineering*, ed. 2. New York: McGraw-Hill, pp. 847–876 (1981).
2. Krass, A. S., P. Boskma, B. Elzen, and W. A. Smit. *Uranium Enrichment and Nuclear Weapon Proliferation*. London: Stockholm Int. Peace Res. Inst., Taylor & Francis, Chapter 6 (1983).
3. Palladium diffusion yields high-volume, hydrogen. *Chem. Eng.* 72(5):36 (1965).
4. Haynes, W. M., *Handbook of Chemistry and Physics*, ed. 91. Boca Raton, FL: CRC Press, pp. 14–16 (2010).
5. Levenspiel, O., and N. de Nevers. The osmotic pump. *Science* 183:157–160 (1974).
6. de Nevers, N. *Fluid Mechanics for Chemical Engineers*. ed. 3. New York: McGraw–Hill. Chapter 14 (2005).
7. Haar, L., J. S. Gallagher, and G. S. Kell. *NBS/NRC Steam Tables*. New York: Hemisphere (1984).
8. Blander, M., and J. L. Katz. Bubble nucleation in liquids. *AIChE J.* 21:833–848 (1975).
9. Walas, S. M. *Phase Equilibria in Chemical Engineering*. Boston: Butterworth, p. 535 (1985).
10. Grossman, A., and C. Ortega. Capillary condensation in porous materials, hysteresis and interaction mechanism without pore blocking/percolation process. *Langmuir* 24:3977–3986 (2008).

15

THE PHASE RULE

15.1 HOW MANY PHASES CAN COEXIST IN A GIVEN EQUILIBRIUM SITUATION?

We are all aware that water can exist in the form of gas, liquid, and solid (called steam, water, and ice). We also know that water and steam can coexist over a finite range of pressures and temperatures (as described by the vapor-pressure curve, Figure 1.8). Most of us are less aware that ice and water can coexist over a range of temperatures and pressures (the "freezing-point curve"), because we have less experience with that. Most of us have heard that solid, liquid, and vapor water can coexist only at one specific temperature and pressure, called the triple point (Figure 1.9).

Could all three phases of water coexist over some finite range of temperatures? Could the vapor–liquid equilibrium exist over a range of pressures at one temperature, instead of at just one pressure for any given temperature? We have all been told, in previous courses, that the answer is no. But how would you prove that? The answer is that we would use the *phase rule*, often called *Gibbs' phase rule* after Josiah Willard Gibbs (1790–1861).

For the simple case of asking whether three phases of a single pure substance can coexist over a finite range of pressures, it is instructive to study the problem by Clapeyron's equation before we take up the phase rule. Consider the possibility of three phases of some single pure substance (like water or propane) coexisting over some finite range of temperature and pressure. For the three combinations of two phases each we can write Clapeyron's equation (Eq. 5.5) three times for the three binary equilibria (1 = solid, 2 = liquid, 3 = gas)

$$\left(\frac{dP}{dT}\right)_{1-2} = \frac{\Delta s_{1-2}}{\Delta v_{1-2}} \tag{15.1}$$

$$\left(\frac{dP}{dT}\right)_{1-3} = \frac{\Delta s_{1-3}}{\Delta v_{1-3}} \tag{15.2}$$

$$\left(\frac{dP}{dT}\right)_{2-3} = \frac{\Delta s_{2-3}}{\Delta v_{2-3}} \tag{15.3}$$

But if the three phases coexist over some finite range of temperatures, then dP/dT must be the same for the three combinations indicated above, or

$$\frac{(s_2 - s_1)}{(v_2 - v_1)} = \frac{(s_3 - s_1)}{(v_3 - v_1)} = \frac{(s_3 - s_2)}{(v_3 - v_2)} \tag{15.4}$$

which can be satisfied only if the three phases lie on a straight line on an entropy–volume diagram, as sketched on Figure 15.1.

This is certainly conceivable, but very unlikely. No substance is known for which this is true. For example, the entropies and volumes per pound for the three phases of water at a triple point, 32.018°F, 0.08866 psia, from [1] are shown in Table 15.1 and sketched in Figure 15.2. We might think that we need to use absolute entropies here, instead of steam table entropies. However, that would merely shift all the values on Figure 15.2 upward by an equal amount, and not change the shape of the figure.

Physical and Chemical Equilibrium for Chemical Engineers, Second Edition. Noel de Nevers.

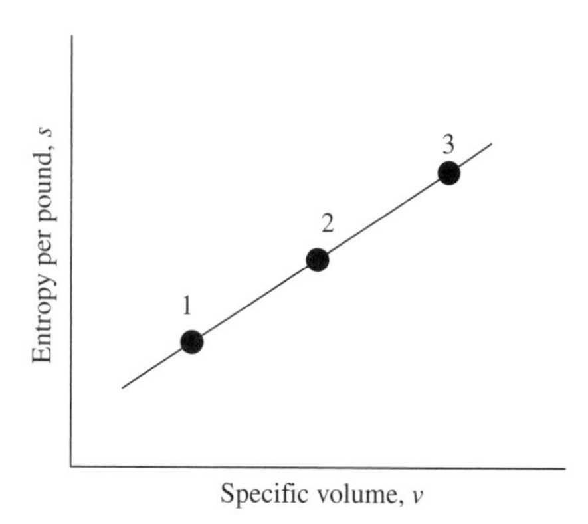

FIGURE 15.1 Required relation between entropies and volumes of the three individual phases for three phases of one pure substance to coexist over some finite range of T and P.

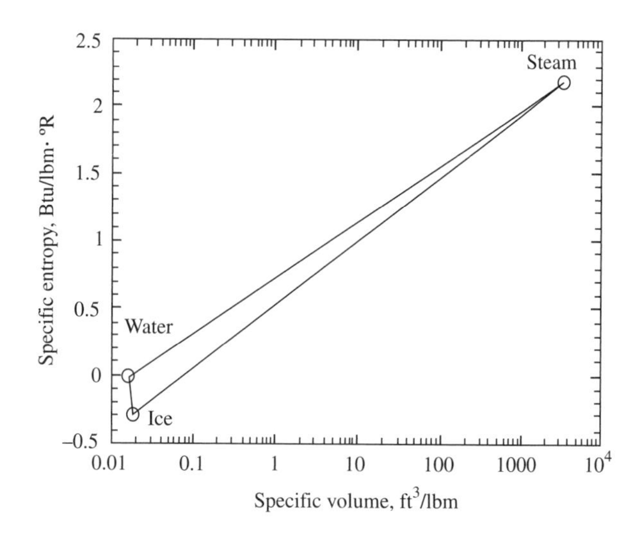

FIGURE 15.2 Properties of the three coexisting phases at the triple point of water, shown on semilogarithmic s-v coordinates.

Table 15.1 Properties at the Water Triple Point

Phase	v (ft^3/lbm)	s [Btu/(lbm · °R)]
Solid (ice)	0.01747	−0.292
Liquid	0.016022	0 (definition)
Gas	3302	2.1869

From this figure and the arguments in Eqs. 15.1 to 15.3, we can show from Clapeyron's equation alone that, for the system water–ice–steam, the three phases *cannot* coexist over any finite range of pressures and temperatures. In this case we did not need Gibbs' phase rule. But for more complicated cases it is much harder to settle such questions by Clapeyron's equation alone; the more complex cases are easier to settle by Gibbs' phase rule.

15.2 WHAT DOES THE PHASE RULE TELL US? WHAT DOES IT NOT TELL US?

The phase rule provides no numerical values of temperatures, pressures, mol fractions, relative masses of phases, or other quantitative values. The variables it discusses are all intensive variables, normally T, P, and mol fractions. It does not deal with extensive variables like system volume, mass, enthalpy, or entropy, or with the number of mols of any species in any phase of any system. It only answers questions of a yes or no variety, or questions whose answers are dimensionless integers. We will illustrate its application by starting an example here and finishing it later.

Example 15.1 A constant-volume, isothermal, transparent container contains a gas phase (mostly carbon dioxide, some water vapor), one liquid phase (mostly water, some dissolved carbon dioxide, some ions), and two solid phases, at some specified temperature T. The two solid phases are practically pure NaOH and practically pure solid H_2O (ice). In this container there is a good mixer, and diffusion and chemical reactions are so fast that we always have thermodynamic equilibrium. We now begin to steadily introduce CO_2 gas into the container. This leads to the conversion of NaOH to $NaHCO_3$. Eventually, all the solid NaOH will be gone and there will be solid $NaHCO_3$. The piece of solid ice is large enough that it may shrink, but it cannot disappear before the piece of NaOH does.

If we watch the container we can determine whether the solid NaOH disappears completely before any solid $NaHCO_3$ appears, whether one solid phase disappears at exactly the same moment the other solid phase begins to appear, or whether there is some significant period when only one solid phase (mostly ice) is present in the container. During the entire process the liquid will contain dissolved NaOH, $NaHCO_3$, CO_2, and various ions. These three possibilities are sketched in Figure 15.3.

We could answer this question by conducting the experiment, but we don't have to. Based on the phase rule alone we can find the answer. We will reopen Example 15.1 later in this chapter and show how that answer is found. ■

15.3 WHAT IS A PHASE?

The nature of a phase is discussed in detail in Section 1.8. There can be only one gas phase, because all gases are miscible with each other. There can be several liquid phases in equilibrium, although more than three is uncommon. There can be any number of solid phases in equilibrium. There is seldom any problem in identifying the proper number of phases or telling one from the other.

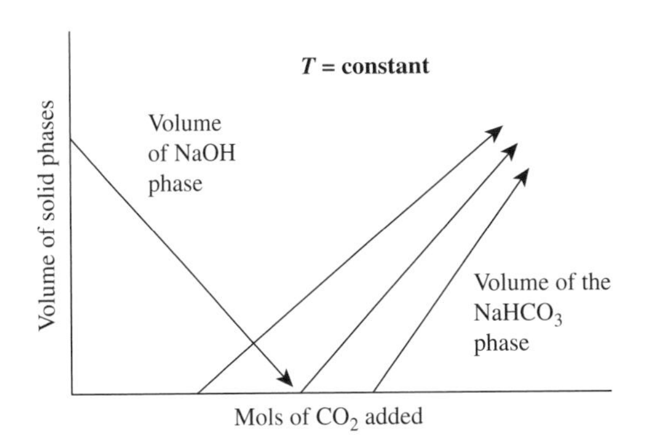

FIGURE 15.3 Possible outcomes of the experiment in Example 15.1. The $NaHCO_3$ solid phase may begin to form before the NaOH phase disappears, exactly when it disappears, or after it disappears. This figure intentionally shows no numerical values of any kind. The curves are drawn as straight lines, but that is a simple admission of ignorance of the true behavior. The phase rule tells us nothing about numerical values on this figure or about the shape of the curves. But it does tell us which is the correct choice among the three "volume of $NaHCO_3$" curves shown.

15.4 THE PHASE RULE IS SIMPLY COUNTING VARIABLES

The phase rule is obtained by counting the variables that nature allows us to arbitrarily specify. We begin by imagining that we wish to confine within a piston and cylinder (Figure 15.4) a single phase of some mixture of N identifiable chemical species (e.g., N_2, H_2, NH_3; in this case, $N = 3$). We can arbitarily set the temperture of the system by placing the whole piston and cylinder in a thermostated oven, and within limits (e.g., no condensation or vaporization) we can set the pressure by moving the piston in or out. (To avoid forming a second phase we must stay above the condensation temperature if the phase is a gas, or below the melting point if the phase is a solid, etc.)

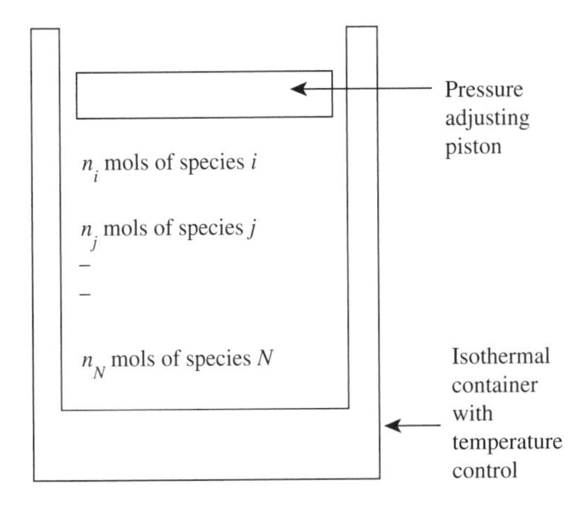

FIGURE 15.4 Another of our piston and cylinder arrangements, showing how we make up some phase.

The phase rule does not deal with the amounts of each of the materials we put into the piston and cylinder, but it does deal with *concentrations.* For the purposes of this discussion we will assume that all concentrations are specified in mol fractions. The same arguments can be made for molalities (lbm/ft^3 of species i, etc.) But it is easiest to see for mol fractions. In making up this piston and cylinder with N species, we can set $(N - 1)$ mol fractions. We know that the sum of the mol fractions must be one, so after we set $(N - 1)$ of the mol fractions the remaining one is fixed. Since we can arbitrarily set T and P, we can arbitrarily set $(2 + (N - 1))$ variables in making up such a piston and cylinder. (Here we have assumed that there are no magnetic, surface, electrostatic, or gravity effects.) Try the "thought experiment" of finding any way to modify the properties of a phase without changing the values of T, P or the number of mols of each species. Once you agree that you cannot, then the rest of this argument will feel comfortable.

Having made up one such phase in a piston and cylinder, we now make up additional phases, each in its own cylinder, until we have a total of M such piston and cylinders, each containing one phase with N identifiable chemical species. We further specify that the M piston and cylinders each contain different phases; for example one contains a gas, another one liquid, another a solid, another some other solid, and so on.

Next we place all the pistons and cylinders together in a pile and surround them by an adiabatic, constant-volume cover. In making up this pile of pistons and cylinders we have been able to arbitrarily set $M[2 + (N - 1)]$ variables.

Now suppose that we wish to make up this pile in such a way that, after we have assembled it, we can remove the pistons and cylinders and find that the phases are all already in a state of chemical and physical equilibrium with all the other phases they will contact when the pistons and cylinders are removed. In this case, how many of these $M[2 + (N - 1)]$ variables can we arbitrarily set? The answer must be $M\,[2 + (N - 1)]$ minus the number of relations that must exist among the phases if they are to be at equilibrium. We know that for these phases to be in physical equilibrium

$$T^{(1)} = T^{(2)} = T^{(3)} = \cdots = T^{(M)} \tag{15.5}$$

$$P^{(1)} = P^{(2)} = P^{(3)} = \cdots = P^{(M)} \tag{15.6}$$

$$\mu_i^{(1)} = \mu_i^{(2)} = \mu_i^{(3)} = \cdots = \mu_i^{(M)} \tag{15.7}$$

$$\mu_j^{(1)} = \mu_j^{(2)} = \mu_j^{(3)} = \cdots = \mu_j^{(M)} \tag{15.8}$$

$$\mu_N^{(1)} = \mu_N^{(2)} = \mu_N^{(3)} = \cdots = \mu_N^{(M)} \tag{15.9}$$

Each of these rows is $(M - 1)$ equations, and there are $(2 + N)$ rows. Thus, there are $(M - 1)\cdot(2 + N)$ independent relations among the $(M - 1)[2 + (N - 1)]$ variables, which must be satisfied if we are to have *physical* equilibrium among all the phases. Therefore, we could set

$$M[2+(N-1)]-(M-1)(2+N)$$
$$= 2M+MN-M-2M-MN+2+N$$
$$= 2+N-M \quad (15.10)$$

variables. But so far we have not taken into account the possibility of chemical reactions among the N chemical species present. Suppose, for example, we have in the gas phase NH_3, N_2, and H_2. Then we can have the reaction

$$N_2 + 3H_2 \Leftrightarrow 2NH_3 \quad (12.A)$$

If there is such an equilibrium, then we know from Chapter 12 that

$$-\mu_{N_2} - 3\mu_{H_2} + 2\mu_{NH_3} = 0 \quad (12.6)$$

We know that the μs are functions of mol fraction, temperature, and pressure, so this is a restriction on our ability to arbitrarily set those values. Thus, this is an additional restriction to the compositions we can arbitrarily set, shown above. There is one such restriction for each balanced, independent equation we can write among the species present. Is there a separate one for each phase? Here, if these three kinds of molecules are in equilibrium in the gas phase, they are certainly in equilibrium in the liquid phase also, if one exists. However, from Eqs. 15.8 to 15.9 we know that by physical equilibrium the μs are the same for any species in all phases present, so although Eq. 12.6 is obeyed independently in each of the phases, it represents only one restriction (see Section 13.5).

Let us call the number of such balanced independent equations we can write Q. Then the remaining number of variables V under our independent control (if we want physical and chemical equilibrium) is

$$V = 2+N-Q-M \quad (15.11)$$

Now let $(N - Q) = C$, the number of *components*, and $M = P$, the number of phases (not to be confused with P, the pressure), so this is

$$V = C+2-P \quad (15.12)$$

which is *Gibbs' phase rule.* Remember what it tells us: V is the number of variables that we can arbitrarily select (also called *degrees of freedom*) when we assemble P phases of C components and still guarantee that there will be physical and chemical equilibrium.

15.5 MORE ON COMPONENTS

In applying the phase rule, there is seldom any argument about the number of phases or what we mean by degrees of freedom; we can generally agree on these with little trouble. The difficulty almost always comes with finding the right number of components. Above we said that the number of components is the number of identifiable chemical species minus the number of independent balanced chemical equations among them. We will see below that there are two additional items. However, let us begin by calculating the number of components for a few examples.

Example 15.2 Suppose we have a system that contains C, O_2, CO_2, and CO. This is a system with four identifiable chemical species. The balanced equations we can write among them are

$$C + 0.5O_2 \Leftrightarrow CO \quad (15.A)$$

$$C + O_2 \Leftrightarrow CO_2 \quad (15.B)$$

$$CO + 0.5O_2 \Leftrightarrow CO_2 \quad (15.C)$$

$$CO_2 + C \Leftrightarrow 2CO \quad (15.D)$$

Thus, there are four relations, but they are not independent. If we add Eqs. 15.A and 15.C and cancel like terms, we obtain Eq. 15.B. Thus, if this is to be a list of *independent* chemical equilibria we must delete Eq. 15.C above. If it bothers you to delete one, remember that each chemical equilibrium is really a relation among the μs and if we already have an equilibrium relation among the some subset of the μs we cannot have an additional independent one among the same μs.

Now, if we reverse the direction of Eq. 15.B and add it to Eq. 15.A, we see that Eq. 15.D is also not independent. Thus, there are only two independent relations among these four species and

$$C = 4-2 = 2$$

Thus, this is a two-component system. ■

Example 15.3 Suppose that the system we are considering has three species: H_2, N_2, and NH_3 ($N = 3$). From Eq. 12.A we know that there is one balanced chemical reaction among these species so

$$C = N - Q = 3 - 1 = 2 \tag{15.13}$$

Now consider the possibility that we made up the system by starting with pure ammonia, and dissociating it over a catalyst. Further, assume that all of the species are in the gas phase. In this case all the hydrogen and all the nitrogen in the system have come from the ammonia that was dissociated. Their molar ratio must be 3 : 1. We can write an equation among their mol fractions, viz:

$$y_{H_2} = 3y_{N_2} \tag{15.E}$$

This is another restriction on the number of variables we can arbitrarily set, independent of all the ones introduced so far. It is called a *stoichiometric restriction*, and it reduces the number of degrees of freedom by one. We might modify the phase rule to put in another symbol for stoichiometric restrictions, but the common usage is to write that

$$\text{Components} = \text{species} - \begin{pmatrix}\text{independent}\\ \text{reactions}\end{pmatrix} - \begin{pmatrix}\text{stoichiometric}\\ \text{restrictions}\end{pmatrix}$$

$$(\textit{preliminary}) \tag{15.F}$$

$$C = N - Q - SR = 3 - 1 - 1 = 1 \ ■$$

Stoichiometric restrictions seem to be a permanent problem for students (and for textbook authors, as one of the problems shows), so some more discussion on them seems warranted. Figure 15.5 shows a triangular diagram, which is a common way of representing compositions of ternary mixtures. The three pure components are shown at the three vertices. Any point on the surface of the triangle represents some composition (either in weight or mol fractions); the sum of the weight or mol fractions is always unity.

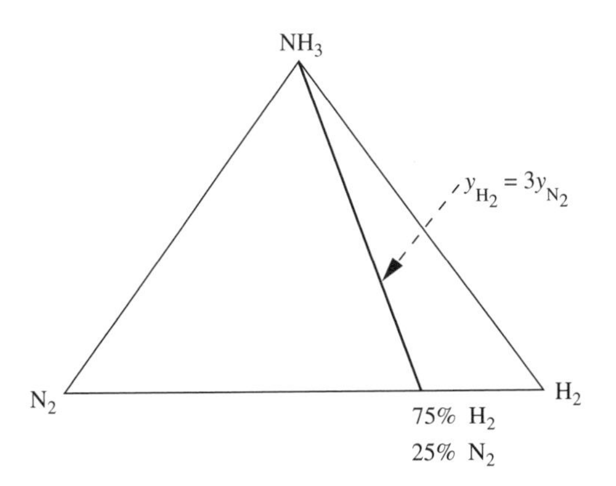

FIGURE 15.5 A ternary mol fraction diagram for Eq. 12.A, showing the stoichiometric restriction that occurs if the mixture is made up by dissociating pure ammonia.

If we are making up a sample of H_2, N_2, and NH_3 in the laboratory, using lecture bottles of the pure species, we can obviously make up any possible combination of weight or mol fractions (which sums to unity), so we can get to any point on this diagram. This is true independent of the temperature and pressure. Now suppose we introduce a catalyst, which causes the Reaction 12.A to go to equilibrium. If we assume ideal gas behavior (not really true at the high pressures normally used for this reaction, but a satisfactory assumption for this example), then we will have

$$\frac{[y_{NH_3}P]^2}{[y_{N_2}P] \cdot [y_{H_2}P]^3} = K \tag{12.BH}$$

where K is an equilibrium constant that depends on temperature only. This sets some limits on where on Figure 15.5 we can be at equilibrium. For example, we can never have zero values of y_{H_2} or y_{N_2}. But we can make one of those values arbitrarily small, by making the other large (i.e., if we use a 100 : 1 hydrogen : nitrogen ratio, y_{N_2} at equilibrium will be very small). We can make y_{NH_3} large by increasing K or small by lowering K, so the chemical equilibrium relation places little restriction of where we can go in Figure 15.5 (if we are free to manipulate T, P, and the ratio of the three species introduced).

However, for the case in Example 15.3, the stoichiometric restriction says that we can never get to any point in Figure 15.5 that is not on the line $y_{H_2} = 3y_{N_2}$, which is sketched in Figure 15.5. So we see that the stoichiometric restriction converts our range of possible compositions from an area (two dimensional) to a line (one dimensional). This is a graphical confirmation of the loss of one degree of freedom, due to the stoichiometric restriction.

However, there is more to it than that. Let us suppose that we have the system in Example 15.3, all made up by dissociating ammonia, and that we have a catalyst that will maintain chemical equilibrium at all temperatures. (Real catalysts won't carry out this reaction at low temperatures; commercially, the reaction is carried out at 350 to 520°C, but in thermodynamics texts we assume the existence of such catalysts. If you can invent and patent a low-temperature catalyst for this reaction you will become rich beyond your wildest dreams!) Now we begin to cool the system, holding the pressure constant. The first species to condense is ammonia. That does not change the ratio of $y_{H_2} = 3y_{N_2}$ until it becomes cold enough that some of the hydrogen or nitrogen dissolves in the liquid. If they dissolved in the ratio 3 : 1 that they have in the gas phase, then Eq. 15.E will still be correct, and we will still have a one component system. If they do not dissolve in this ratio, then Eq. 15.E is no longer observed, and this has become a two-component system. In all the stoichiometric restriction examples I know of, the restriction applies to the ratio of mol fractions *in one phase*. If there is a stoichiometric restriction, then it is possible to write an equation among mol fractions in one phase, like $y_{H_2} = 3y_{N_2}$. If we cannot write such an equation, then there is no stoichiometric restriction.

Adding another phase (into which one of the species can dissolve) removes the restriction, and thus increases the number of components. This is perhaps made clear by the next example.

Example 15.4 A sample of pure solid $CaCO_3$ is placed in an evacuated test tube and heated. It dissociates according to the reaction

$$CaCO_3(s) \Leftrightarrow CaO(s) + CO_2(g) \tag{15.G}$$

How many components are present?

Here we have three species and one balanced chemical reaction between them. One might be tempted to look for a stoichiometric restriction based on the ratio of CaO to CO_2; but since CO_2 will mostly be in the gas phase and $CaCO_3$ and CaO will each form separate solid phases, there is no equation we can write among the mol fractions in any of the phases. Hence, there is no stoichiometric restriction, and the number of components is $C = 3 - 1 = 2$ ■

Example 15.5 In the preceding example it is asserted that the thermal decomposition of $CaCO_3$ (Eq. 15.G) leads to a two-component system. If this is a two-component system, and if solid CaO and $CaCO_3$ form two separate phases, then

(a) How many phases are present?
(b) How many degrees of freedom are there?
(c) If we place a sample of pure $CaCO_3$ in an evacuated container and heat it, will we find a unique *P-T* curve?

The $CaCO_3$ and CaO form separate solid phases, so we have three phases, two solid and one gas. From the phase rule

$$V = C + 2 - P = 2 + 2 - 3 = 1 \tag{15.H}$$

If there is only one degree of freedom, then the system should have a unique *P-T* curve. Findlay et al. [2, p. 214] shows the data to draw such a curve, which can be well represented by

$$\ln\left(\frac{p}{\text{torr}}\right) = 23.6193 - \frac{19{,}827\ \text{K}}{T} \quad ■ \tag{15.I}$$

In this example the gas phase is certain to be practically pure CO_2, $y_{CO_2} \approx 1.00$. This is not a stoichiometric restriction; all of the stoichiometric restriction examples known to the author involve *ratios* of mol fractions. This example also shows that the number of components may be *more* than the number of pure species originally introduced into the system. If more than one phase is present, this is a common occurrence. Some textbooks give the "rule" that the number of components is the minimum number of pure species that must be introduced to make up the system. This example shows that "rule" is not correct.

Another complicating factor in computing the number of components is illustrated by the next example.

Example 15.6 Suppose our system consists of H_2O, HCl, H^+, OH^-, and Cl^-. Here we have five species and two chemical relations:

$$H_2O \Leftrightarrow H^+ + OH^- \tag{15.J}$$

$$HCl \Leftrightarrow H^+ + Cl^- \tag{15.K}$$

In addition we have *electroneutrality*, which says that at equilibrium the total number of positive charges on ions in the solution must be the same as the total number of negative charges on ions, or

$$[H^+] = [OH^-] + [Cl^-] \tag{15.L}$$

Here $[H^+]$ stands for the molality of hydrogen ion. This is convertible (at least in principle) to a relation among the μs; hence, it is an additional restriction and the number of components is

$$C = 5 - 2 - 1 = 2 \quad ■$$

This does not include the possibility that the system we are considering is one plate of a charged capacitor, or a highly charged part of a thundercloud, neither of which are electroneutral. If we are dealing with charged systems, we must reconsider this restriction; for most systems of interest it applies, because most systems of interest are eiectroneutral.

Sometimes we do not know whether some compound really exists. This causes no problem in selecting the number of components.

Example 15.7 Our system consists of Au and H_2O. If no compounds are formed, then we have $C = 2 - 0 = 2$. However, if there is also the chemical reaction

$$Au + H_2O \Leftrightarrow AuH_2O \quad (??) \tag{15.M}$$

we have $C = 3 - 1 = 2$. The number of components is independent of the existence or nonexistence of such compounds of questionable existence. ■

Hence, our final working rule is

$$\text{Components} = \text{species} - \begin{pmatrix}\text{independent}\\ \text{reactions}\end{pmatrix} - \begin{pmatrix}\text{stoichiometric}\\ \text{restrictions}\end{pmatrix} - \begin{pmatrix}\text{1, if ionic}\\ \text{species are}\\ \text{present}\end{pmatrix} \tag{15.14}$$

15.5.1 A Formal Way to Find the Number of Independent Equations

In the previous examples we have found the number of independent chemical equations by intuition or knowledge of the system. This works well for simple systems, but is harder and more unreliable for complex ones. Of the various algorithms for doing this, the simplest reliable one seems to be the following:

1. Write the formulas for the formation of all the compounds in the species list from the elements.
2. Algebraically eliminate those elements that do not appear in the species list.

The following example shows how this is done.

Example 15.8 Determine the maximum number of balanced chemical equations that exist among $CaCO_3$, CaO, and CO_2.

This set of species, the subject of Examples 15.4 and 15.5 is chosen here to illustrate the process. First we write

$$Ca + C + 1.5O_2 \Leftrightarrow CaCO_3 \tag{15.N}$$

$$Ca + 0.5O_2 \Leftrightarrow CaO \tag{15.O}$$

$$C + O_2 \Leftrightarrow CO_2 \tag{15.P}$$

Then we observe that Ca, C, and O_2 do not appear on the species list, so we must eliminate them from these three equations. If we solve Eq. 15.P for C and substitute that in Eq. 15.N, we find

$$Ca + CO_2 - O_2 + 1.5O_2 \Leftrightarrow CaCO_3 \tag{15.Q}$$

We then add Eq. 15.Q to Eq. 15.O, finding

$$CO_2 \Leftrightarrow -CaO + CaCO_3 \tag{15.R}$$

which we can rearrange to Eq. 15.G. This is the only independent balanced chemical reaction between the species on the species list. It includes all of them and does not include any of the elements. The same procedure is reliable for more complex systems. The final list of equations need not contain all the materials on the species list. If we added N_2 to the species list in this example the number of balanced equations among the species would still be one. ■

15.6 THE PHASE RULE FOR ONE- AND TWO-COMPONENT SYSTEMS

For a one-component system, $V = C + 2 - P = 3 - P$. So if

P	V
1	2
2	1
3	0

Thus, for one phase (e.g., gas) we can arbitrarily set two variables, such as T and P, T and h, or u and s. For two phases we can arbitrarily select one variable, such as T, or the value of h or s in one of the two phases, but then everything else is set for us. For three phases we can select none, since that will be a fixed point. We can only observe under what conditions this set of phases coexist. Four phases cannot coexist. This is easily seen in Figure 1.9, the P-T diagram for water, reprinted here as Figure 15.6. We see in it that one phase (G, L, or S) corresponds to an area, two phases (G-L, S-L, or S-G) correspond to a curve, and three phases (G-L-S) correspond to a point. In an area we can move in two perpendicular directions without leaving the area, hence $V = 2$. On a curve we can move in one direction (up and down along the curve, but not off it), hence $V = 1$. At a point we cannot move in any direction without leaving the point, so $V = 0$.

We normally think of a substance having only one triple point. Figure 1.10, the very high-pressure P-T diagram for water, shows six triple points, in addition to the one we are familiar with, which is lost into the horizontal axis. If we measured the very high-pressure diagram for other substances, we would probably observe similar behavior. Normally when we use the term "triple point" we are speaking about the G-L-S triple point, of which any pure substance can have only one. But we must keep our minds open to the other possibilities shown in Figure 1.10.

For a two-component mixture we can repeat the above table, finding

P	V
1	3
2	2
3	1
4	0

Consider a binary like benzene-toluene (Figure 8.7d). For one phase (e.g., gas) we can set three independent variables, such as, P, T, and one mol fraction (but not the other because for a binary solution, setting one x_i sets the other). For two phases (e.g., gas and liquid) we can set two independent

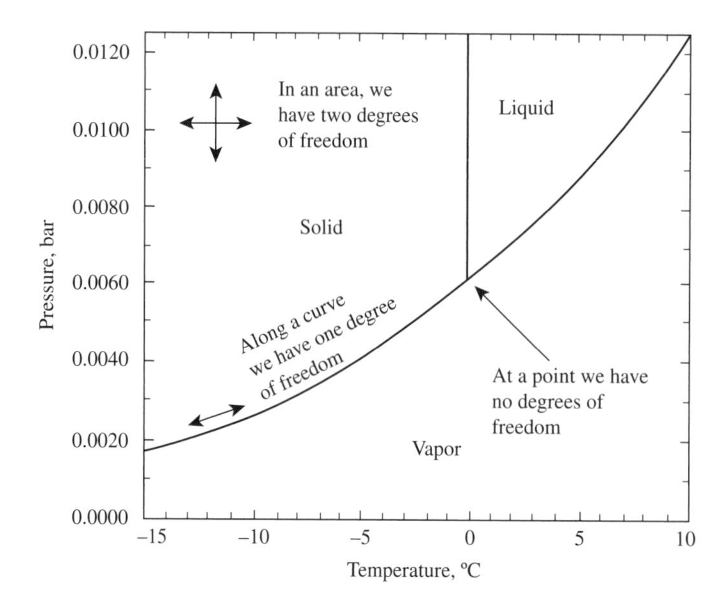

FIGURE 15.6 This is a repeat of Figure 1.9, showing that in the area corresponding to the solid (or the liquid or the vapor) we can move in two perpendicular directions and stay within the area, so we have two degrees of freedom. Along the solid–vapor curve (or the solid—liquid or vapor—liquid curves) we must stay on the curve, so that we have one degree of freedom. At the triple point, we have no degrees of freedom.

variables, such as, P and T, but then the mol fractions of both phases are fixed. For three phases (e.g., gas, liquid, and solid) we can set one variable, such as T, and then the corresponding P and mol fractions are fixed. For four phases in equilibrium we can set none of the variables. Five phases cannot coexist. If we fix the pressure, then we have given up one degree of freedom and, as shown in Figure 15.7, in a one-phase area

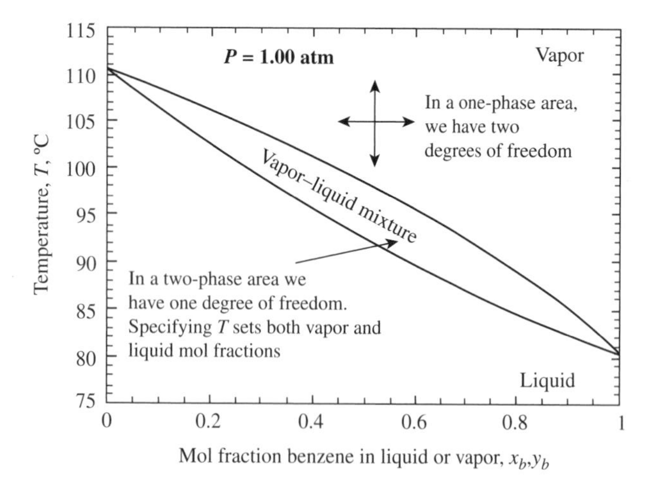

FIGURE 15.7 This is a replot of Figure 3.12, a T-x diagram for benzene-toluene at 1 atm. As shown, in the single-phase regions (vapor and liquid) we have two degrees of freedom, while in the two-phase (vapor–liquid) region we have only one.

(either vapor or liquid) we have two degrees of freedom; we can independently change both T and the mol fraction of benzene. In the two-phase region we have only one degree of freedom; at any specified T, both phase mol fractions are fixed.

One more example and the completion of Example 15.1 will show how the phase rule answers some engineering questions.

Example 15.9 In systems involving the oxidation and reduction of iron ores, the following species may occur: Fe, O_2, FeO, Fe_2O_3, Fe_3O_4. We now place pure solid iron (Fe) and pure gaseous oxygen in a piston and cylinder arrangement with a pressure gage and slowly decrease the volume, at constant temperature.

Sketch a P-V diagram for this system under the following assumptions:

1. Isothermal
2. Physical and chemical equilibrium at all times
3. Gas phase is always present
4. No liquid phase is ever present
5. Fe, FeO, Fe_2O_3, and Fe_3O_4 do not form solid solutions with each other.

In each region of the P-V diagram (see Fig. 15.8), indicate which phases are present.

Here we have five species and three equations among them

$$Fe + 0.5O_2 \Leftrightarrow FeO \tag{15.S}$$

$$2Fe + 1.5O_2 \Leftrightarrow Fe_2O_3 \tag{15.T}$$

$$3Fe + 2O_2 \Leftrightarrow Fe_3O_4 \tag{15.U}$$

so that the number of components is $C = 5 - 3 = 2$. We have used up one degree of freedom by specifying the temperature so that the remaining degrees of freedom are $V = 2 + 2 - P - 1 = 3 - P$. Thus, when two phases are present we will have one degree of freedom, when three phases are present we will have no degrees of freedom, and there can never be four phases at this fixed T. If we start with pure Fe and pure O_2, we have two phases and one degree of freedom, so lowering the volume will cause the pressure to increase. When we reach the pressure at which Reaction 15.S will occur, FeO will begin to form, so three phases will be present until all the Fe is consumed. Thus, the pressure must remain constant, and we must have a horizontal part of the P-V diagram. Then when all the Fe is gone, we will again have only two phases present. The pressure will rise until we reach the pressure at which the reaction

$$3FeO + \tfrac{1}{2}O_2 \Leftrightarrow Fe_3O_4 \tag{15.V}$$

occurs. Again we will have a constant pressure until all the FeO disappears. The same situation will occur for the other transition, so the whole diagram is as sketched in Figure 15.8. ■

This example illustrates the great utility of the phase rule. Here, *without benefit of any experimental data,* we can consider an equilibrium process and know for certain what combinations of phases can be present at any time. We learn from the phase rule alone. (In deciding in what order the phases appear, we used either our intuition or Le Chatelier's rule, which could be restated to say that the iron phases appear in order of increasing O/Fe ratio.) The phase rule does not tell us any of the pressures or volumes in Figure 15.8, only the correct shape of the curve.

Example 15.1 continued. In the first part of Example 15.1 we posed the question of when the $NaHCO_3$ phase would appear. For this system the species are H_2O, OH^-, Na^+, NaOH, CO_2, $HCO_3{}^-$, $NaHCO_3$, $N = 7$. The balanced chemical reactions are

$$NaOH \Leftrightarrow Na^+ + OH^- \tag{15.W}$$

$$CO_2 + OH^- \Leftrightarrow HCO_3^- \tag{15.X}$$

$$Na^+ + HCO_3^- \Leftrightarrow NaHCO_3 \tag{15.Y}$$

We also have electroneutrality, so the number of components is $C = 7 - 3 - 1 = 3$. The maximum number of phases that can coexist in equilibrium (for $V = 0$) is $C + 2 = 5$. Here we have used up one degree of freedom by specifying the temperature, so the maximum number of phases that can coexist is four.

The container originally contains four phases, so the system is invariant. At equilibrium, adding CO_2 cannot change the pressure or the chemical composition of any phase. As we add CO_2 it will mostly dissolve in the liquid. Some of the NaOH must also dissolve to keep the ratios of the mol fractions of all dissolved species constant. Some of the ice must also melt to keep the concentrations of dissolved species constant. As long as both solids are present they can keep the liquid composition constant while we add CO_2 with vigorous mixing. Eventually one of the solids is used up, and

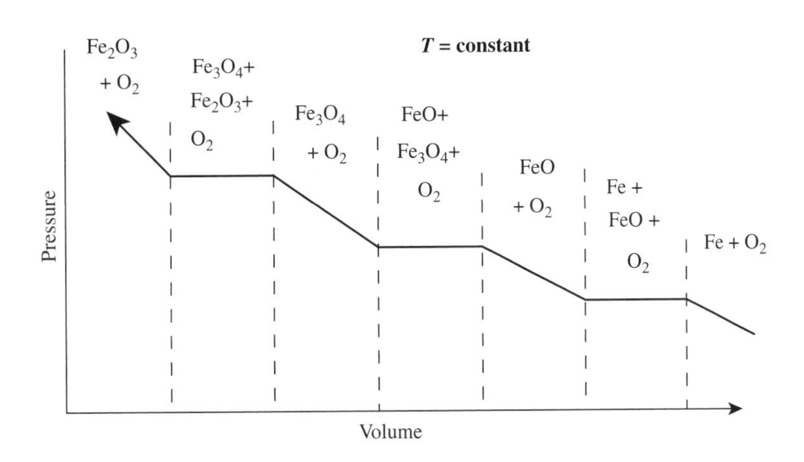

FIGURE 15.8 Qualitative P-V diagram for Example 15.9.

for this example we have specified that there is much more ice than NaOH, so the NaOH will disappear first.

When that happens, we have only three phases, so we have one degree of freedom. The pressure can rise, and the gas and liquid mol fractions can change. Eventually we reach a high enough pressure (and/or a high enough concentration of $NaHCO_3$ in the liquid) that solid $NaHCO_3$ will begin to appear. Then the system becomes invariant again until some phase (probably the ice) disappears. So the answer to the question raised in the first part of this example is that the two curves in Figure 15.3 do not cross, but that there is a gap between them, and the rightmost line on Figure 15.3 is correct. ■

This example shows the power and the limitations of the phase rule. It does not tell us how many mols of CO_2 must be admitted, the pressure or any of the phase compositions. It does not tell us if the solid that actually appears is $NaHCO_3$ or one of its hydrates, such as sodium sesquicarbonate ($Na_2CO_3 \cdot NaHCO_3 \cdot 2H_2O$). But it does tell us, without benefit of any experiment, what is possible in this system and what is not. Often that is of great value.

15.7 HARDER PHASE RULE PROBLEMS

In the previous parts of this chapter the phases have been solid, liquid or gas, and perfectly intuitive. But nature is more complex than that, as illustrated by applying the phase rule to simple adsorption of a pure gas onto an adsorbent solid, for example the adsorption of N_2 on silica gel in Figure 11.24.

1. How shall we count components? Are there 2, N_2 and silica gel? Or does the silica gel play no more role than the walls of the glassware in which we conduct other experiments, in which we do not count the surface of the glass as participating in the reaction? In that case there is only one component, N_2.
2. How many phases are there? There is certainly a gas phase, and some other phase. Normally people see the second phase as consisting of the N_2 adsorbed on the solid so $P = 2$. If $C = 1$ and $P = 2$, then by Eq. 15.12, $V = 1$. This is the same answer we would get for vapor and liquid ($P = 2$) in equilibrium for one component (e.g., N_2). But we see that in Figure 11.24 one must set two variables (T and P) to uniquely determine the absorbed concentration, meaning that experimentally $V = 2$.
3. If we decide that the silica gel is a component and also a separate phase, that increases C and P each by 1, so V remains unchanged at 1.
4. If we decide that the silica gel is a component, but that the adsorbed N_2 and the silica gel on which it is adsorbed are together only one phase, then $C = 2$ and $P = 2$, and by Eq. 15.12, $V = 2$, which matches what we see on Figure 11.24.
5. Most of those who have analyzed this problem [6, 7] have rejected the idea [4] and insisted that the adsorbed material is itself a phase. If it is a separate phase, then how can the amount of it adsorbed on the solid (mmol/g on Figure 11.24) be a phase rule variable? Phase rule variable are always intensive properties, (T, P and mol fractions or their equivalent). At the low-pressure end of Figure 11.24 the coverage is less than a monolayer, so that we could consider the fractional coverage θ as equivalent to a mol fraction. But as we move to higher values of P/P_0, multiple layers occur ($\theta > 1$??). As P/P_0 approaches 1.0, the uppermost layers change from adsorbed to liquid-like (see section 14.4.2), requiring some other description. If we double the thickness of the solid layers making up the adsorbent, without changing the surface in any way, that will divide the values of (mmol adsorbed/gm adsorbent) by a factor of 2, making it harder to think of (mmol adsorbed/gm adsorbent) as an intensive phase rule variable.

Those who have analyzed this problem [6, 7] have not been looking to the phase rule for guidance as to what nature does, but rather trying to adjust the phase rule to agree with the experimental results. You are invited to read their explanations and judge for yourself how plausible they are.

One rarely sees the phase rule even mentioned in books and articles on biochemistry. The main reason is that the phase rule deals with systems at physical and chemical equilibrium, and biological systems are almost never at or very near to physical and chemical equilibrium, (See Figure 1.4). In addition, the phases in biological systems are mostly even more complex physically and chemically than the adsorption case considered here. For clear, simple and well-defined phases and ordinary chemical reactions, the phase rule has explanatory and predictive power. For complex phases and complex biochemical reactions (see Chapter 16) it apparently has less such power.

15.8 SUMMARY

1. The phase rule relates only to intensive variables; it makes no statements about the relative amounts of individual phases. It makes no statements about the values of the intensive variables, but only uses them to count phases.
2. The phase rule is derived by counting the variables we can specify for an individual phase, and then subtracting the number of relations that must exist between those

variables at equilibrium. The algebraic sum of these lists leads directly to the phase rule.

3. There is seldom much difficulty in counting phases or degrees of freedom. Determining the number of components seems to cause the most trouble.
4. The only quantitative information supplied by the phase rule is the number of phases or the number of degrees of freedom at equilibrium.

PROBLEMS

See the Common Units and Values for Problems and Examples. An asterisk (*) on the problem number indicates that the answers is in Appendix H.

15.1 We place 3 lbm of water at the triple-point temperature and pressure (32.018°F, 0.08866 psia) in a piston and cylinder container. We adjust the proportions so that there is 1 lbm of solid (ice), 1 lbm of liquid (water), and 1 lbm of gas (steam).

a. Now we reversibly, adiabatically expand the mixture. Which phase disappears first? Which phase disappears second? What would the answers be if we adiabatically compressed instead of expanding the mixture?

b. Instead of adiabatically expanding the mixture we cool it at constant volume. Which phase disappears first? Which phase disappears second? What would the answers be if we heated instead of cooling the mixture at constant volume?

c. In part (a) there is some initial relation of the masses of ice, water, and steam (different from 1 : 1 : 1) for which adiabatic compression will lead to the simultaneous disappearance of two phases. What is that relation?

15.2* An all-gas system consists of CH_4, H_2O, CO, CO_2, and H_2. It is made up by mixing 2 mols of CH_4 and 3 mols of H_2O, and then bringing the system to thermodynaic equilibrium over a catalyst at about 1500°F. (This is the actual current industrial process used to make hydrogen in large quantities, called "methane reforming.")

a. How many phase rule components C are there for this system, made up this way?

b. Now consider the possibility that we could find a negative catalyst (perfect reaction inhibitor) that would prevent all reactions except

$$2CH_4 + 3H_2O \Leftrightarrow CO_2 + 7H_2 + CO \quad (15.Z)$$

No other reactions occur. The entire system is in the gas phase. How many phase rule components C are in this system?

There is no known "negative catalyst," and theory suggests that there cannot be. However, if all possible reactions are very slow at low temperatures and we find a catalyst that is active at low temperatures for just one reaction, then the effect is the same as if there were a negative catalyst for all the others. Many industrial processes are of this type.

15.3* For a system containing the following identifiable chemical species, H_2O, H_2S, NH_3, H^+, OH^-, $NH_4{}^+$, HS^-, NH_4OH,

c. What is the maximum number of phases that can coexist at equilibrium?

d. If you wished to determine some kind of P-T curve for gas–liquid equilibrium, how many concentration variables would you have to hold constant?

15.4* **a.** What is the number of components in the system containing CO_2, CaO, $CaCO_3$, H_2O, Ca^{2+}, $CO_3{}^{2-}$, $HCO_3{}^-$?

b. What is the number of components in the above system if we add to the list OH^-, H^+?

15.5* Repeat Example 15.8 for the system consisting of CO, CO_2, H_2, H_2O, and CH_4 (a system of great industrial interest in the production of hydrogen). How many independent balanced chemical equations exist between these 5 species?

15.6 For the system consisting of two liquid phases and one gas phase and containing H_2O, H_2S, NH_3, NH_4OH, NH_4SH, and C_2H_6, how many independent variables must be specified before the state of the system is specified? Give one possible such list of fixed variables.

15.7 At 1 atm pressure and 70°F, how many phases can coexist at equilibrium in a system containing the following identifiable chemical species? H_2O, H^+, OH^-, $BaCl_2$, Ba^{2+}, Cl^-, $Ba(OH)_2$, HCl.

15.8 At 1 atm pressure, how many phases can coexist in equilibrium in a system that contains the following identifiable chemical species? H_2O, HCl, H^+, Cl^-, NH_4^+, NH_3, NH_4Cl.

15.9* Iron pyrite, FeS_2, is roasted with pure oxygen in a furnace. The plausible materials that might be present in the equilibrium system are the feed materials plus SO_2, SO_3, FeO, Fe_2O_3, Fe_3O_4, $FeSO_4$. Assuming that no liquid phases are present, that we have physical and chemical equilibrium at all times, that none of the iron compounds are present as a gas, and that they do not form solid solutions in each other, how many of the above list of iron compounds would we expect to find present at an arbitrarily selected temperature, pressure, and mol fraction of SO_2 in the gas phase?

15.10 One mole each of CO_2 and O_2 are placed in an evacuated container. A catalyst causes equilibrium in Reaction 15.C to occur. Only the gas phase is present. Is there a stoichiometric restriction in the application of the phase rule to this system? Are there more than one? If there is (are) write the appropriate equation(s) of the restriction(s), and sketch it (them) on an appropriate triangular composition diagram.

15.11 Ten grams of pure, solid ammonium carbonate $[(NH_4)_2CO_3 \cdot H_2O]$ are placed in an evacuated 20-mL flask, and the flask is sealed. A little of the solid vaporizes and forms a gas phase according to the reaction

$$[(NH_4)_2CO_3 \cdot H_2O](s) \Leftrightarrow 2NH_3(g) + 2H_2O(g) + CO_2(g) \tag{15.AA}$$

a. Is there a stoichiometric restriction in the application of the phase rule to this system? Are there more than one? If there is (are) write the appropriate equation(s) of the restriction(s), and sketch it (them) on an appropriate triangular composition diagram.

b. Is this system univariant, that is, does fixing the temperature fix the pressure? Give a complete phase rule analysis to support your answer.

c. Same as (b) except that the flask instead of being initially evacuated is initially full of pure nitrogen gas.

15.12 A small sample of sodium sesquicarbonate ($Na_2CO_3 \cdot NaHCO_3 \cdot 2H_2O$) is placed in an evacuated container and heated. When a high enough temperature is reached, this can be expected to decompose. The species that may possibly be present in addition to the starting material are Na_2CO_3, $NaHCO_3$, NaOH, H_2O, and CO_2. What is the maximum number of phases that can ever be present in this system?

15.13 In one process for the production of zinc, pure ZnO and pure C are placed in an evacuated reactor and heated. The species present are ZnO, C, Zn, CO, and CO_2.

a. Below 1292 K no liquid is zinc present. How many components are there? What is the maximum number of phases possible? How many degrees of freedom are there?

b. Above 1293 K liquid zinc is present. How many components are there? What is the maximum number of phases possible? How many degrees of freedom are there?

This is a worked example in [3, p. 1054]. The phase rule discussion there is puzzling. The discussion in [4] is much clearer.

15.14* At a constant temperature of 20°C, we drop a piece of pure solid tin into a closed container full of an aqueous solution of HC1. The reaction that occurs is

$$2HCl + Sn \Leftrightarrow H_2 + SnCl_2 \tag{15.AB}$$

The hydrogen comes out of solution, forming a separate gas phase. By moving a piston up or down, we can adjust the system pressure. Will solid tin and solid $SnCl_2$ (as well as gas and liquid phases) be present at equilibrium at this temperature,

c. Not at any pressure?

d. At only one pressure?

e. Over some range of pressures?

15.15 The following quote is from a thermodynamics book [5]. Is the application of the phase rule there correct? If not, what errors exist, and what is the correct analysis?

> **ILLUSTRATION 7-2**
> What are the degrees of freedom for the reaction of tin and aqueous hydrochloric acid to produce solid stannous chloride and hydrogen?
>
> ***Solution***
> There are five chemical species, that is, Sn, HCl, H_2O, H_2, $SnCl_2$
>
> $$C = 5$$
>
> There is one independent reaction:
>
> $$Sn_{(s)} + 2HCl_{(aq)} = SnCl_{2(s)} + H_{2(g)}$$
>
> therefore, $R = 1$.
> But the total pressure on the system is a function of the partial pressures, and
>
> $$P_{H_2} = f(P_{HCl})$$
>
> so that
>
> $$R' = 1$$

and

$$N = 5 - 1 - 1 = 3$$

The number of phases is three, so that

$$\pi = 3$$

and

$$D = 3 + 2 - 3 = 2$$

Or two degrees of freedom are available, that is, we could select, say, temperature and total pressure to fix the equilibrium condition.

15.16 For a one-component system, on a *P-T* plot, zero degrees of freedom corresponds to a triple point, from which three curves leave. Each of them represents the equilibrium of two phases. The space between the curves represent single phases.

What is the corresponding situation for two- and three-component systems? If we make a *P-T* plot for them, how many lines join at an invariant point? What do the lines represent? What do the spaces between them represent?

15.17 Example 15.1 assumes that all the CO_2 introduced into the system is present as either gaseous or dissolved CO_2 or as HCO_3^- or $NaHCO_3$. As discussed in Chapter 13, there will most likely be some CO_3^{2-} and some Na_2CO_3 also present, and we do not know (without some study) whether the first new solid that appears is the $NaCHCO_3$ assumed there or is sodium sesquicarbonate ($Na_2CO_3 \cdot NaHCO_3 \cdot 2H_2O$) or perhaps Na_2CO_3 $7H_2O$.

Repeat that example, with the assumption that CO_3^{2-} and Na_2CO_3 are also present. How much is the answer changed?

15.18 Example 15.1 starts with CO_2 already present and at equilibrium in the system. Suppose that instead we started with only H_2O and NaOH, and then added CO_2? Initially the container would still have four phases: gas, liquid, and two solids. How many components are there? How many degrees of freedom? Can we arbitrarily specify the temperature for this equilibrium? When we add some CO_2, how do the answers to the above questions change?

REFERENCES

1. Keenan, J. H., F. G. Keyes, P. G. Hill, and J. G. Moore. *Steam Tables: Thermodynamic Properties of Water Including Vapor, Liquid and Solid Phases*. New York: Wiley (1969).
2. Findlay, A., A. N. Campbell, and N. O. Smith. *The Phase Rule*, ed. 9. New York: Dover, p. 47 (1951).
3. Hougen, O. A., K. M. Watson, and R. A. Ragatz. *Chemical Process Principles*, Part II: *Thermodynamics*, ed. 2. New York: Wiley (1959).
4. Denbigh, K. G. *The Principles of Chemical Equilibrium*. Cambridge, UK: Cambridge University Press, p. 190 (1955).
5. Contact the author if you wish to know from which textbook this example was copied. Textbook authors are loath to point the finger at each other. I hope there are no errors that obvious in this book.
6. Langmuir, I. An extension of the phase rule for adsorption under equilibrium and non-equilibrium conditions, *J. Chem. Phys.* 1: 3–12 (1933).
7. Smith, J. M., H. C. Van Ness, and M. M. Abbott. *Introduction to Chemical Engineering Thermodynamics*, ed. 7. New York: McGraw-Hill, p. 607 (2005).

16

EQUILBRIUM IN BIOCHEMICAL REACTIONS

Everything we said about equilibrium in chemical reactions in Chapters 12 and 13 is true for biochemical reactions, those that occur in living things. The following example suggests that while this is true, it is generally not very useful.

16.1 AN EXAMPLE, THE PRODUCTION OF ETHANOL FROM SUGAR

Example 16.1 Estimate the equilibrium constant for the conversion of sugar (sucrose) to ethanol at 25°C= 298.15 K.

The overall reaction is

$$\underset{\text{Sucrose}}{C_{12}H_{22}O_{11}(s)} + H_2O(1) \rightarrow \underset{\text{Ethanol}}{4C_2H_5OH(1)} + 4CO_2(g) \tag{16.A}$$

$$\begin{aligned}\Delta g^\circ &= 4g^\circ{}_{CO_2} + 4g^\circ{}_{\text{ethanol}} - g^\circ{}_{\text{sucrose}} - g^\circ{}_{\text{water}} \\ &= 4(-394.4) + 4(-178.4) - (-1553.7) - (-237.1) \\ &= -486 \frac{\text{kJ}}{\text{mol}}\end{aligned} \tag{16.B}$$

$$\begin{aligned}K &= \exp\left(\frac{-\Delta g^\circ}{RT}\right) = \exp\left(\frac{-\left(-486{,}000\dfrac{\text{J}}{\text{mol}}\right)}{8.314\dfrac{\text{J}}{\text{mol K}}\cdot 298.15\text{ K}}\right) \\ &= 1.4\cdot 10^{85} = \frac{[a_{\text{ethanol}}]^4\cdot[a_{CO_2}]^4}{[a_{\text{sucrose}}]\cdot[a_{H_2O}]} \quad \blacksquare\end{aligned} \tag{16.C}$$

The extreme value ($K = 10^{85}$) suggests that if you dropped some water in your sugar bowl and equilibrium occurred in this reaction, you would be left with a bowl of almost pure ethanol and a cloud of CO_2! You can try that experiment at home and discover that this is not a description of what nature does. Why?

Eq.16.A is the correct overall chemical description of a reaction that is carried out on a massive industrial scale to make ethanol for motor fuels (Brazil produces about 25 billion pounds a year), and on a small scale in thousand of breweries, wineries, moonshine operations and home beer and wine operations (most often using cheaper impure sugar or starch sources, like grapes, barley or corn). But we only know how to carry it out in aqueous solution, with the aid of living cells (mostly yeasts in commercial practice, some others in laboratory settings). Any reaction with $K = 10^{85}$ is practically irreversible, so this calculation tells us that it should be possible to conduct the reaction (which most of us already knew) but little else.

16.2 ORGANIC AND BIOCHEMICAL REACTIONS

Most of modern industrial organic chemistry (plastics, fibers, fuels, fertilizers, pharmaceuticals, agricultural chemicals, explosives, paints and coatings) is based on materials derived from petroleum. All the organic compounds listed in Table A.8 meet that description. Petroleum is ultimately derived from the fats in the microscopic bodies of living organisms (biochemical materials), transformed by heat and pressure in the ground over geological time, by what we would call organic reactions, not biochemical reactions, because after the bodies were buried, their further processing did not require living organisms, as do biochemical reactions.

Physical and Chemical Equilibrium for Chemical Engineers, Second Edition. Noel de Nevers.

The same is true of coal (starting with vegetable rather than animal debris).

Comparing biochemical reactions (and equilibrium in them) to the reactions discussed in Chapters 12 and 13, we see the following major differences:

1. The sizes of the molecules are very different. Sucrose has 45 atoms and $M = 342.3$. Ethyl acetate has the most atoms (13) of any of the substances in Chapters 12 and 13; $PbSO_4$ has the highest molecular weight, (303, mostly due to the high atomic weight of lead, 207). Of the organic compounds listed in Table A.8 *methyl*-cyclohexane has the most atoms, 21. The increase in complexity between a 21-atom compound and a 45-atom compound is major. Furthermore, many of the important molecules in biochemistry (enzymes, proteins) have hundreds or thousands of atoms, and molecular weights ranging to millions of Daltons (g/mol).
2. The phases are different. Almost all biochemical reactions occur in aqueous media at temperatures between the freezing and boiling points of water, while many reactions discussed in Chapters 12 and 13 occur in the gas phase, many at high temperatures (e.g., ammonia synthesis). Many industrial chemical reactions (e.g., hydrogenations, most petroleum refining reactions) involve gases dissolved in liquids. Most biochemical reactants and products do not exist as gases because their $T_{\text{decomposition}}$ is lower than their $T_{\text{vaporization}}$. Few exist in nature as solids (although we can crystallize many of them, e.g., sucrose).
3. The catalysts for modern industrial chemical (non-biochemical) reactions are mostly very porous solids with small amounts of metals (e.g., platinum) dispersed on their surfaces, produced by inorganic chemistry. The catalysts for biochemical reactions are mostly enzymes, high-molecular weight, water-soluble proteins produced biochemically by living cells. Much of human, animal and plant DNA directs the synthesis of these enzymes that then regulate the chemistry of life.
4. The equilibrium in industrial reactions involving gases can be shifted dramatically by changing the pressure (NH_3), or in other reactions by applying an external voltage (Al). Almost all biochemical reactions occur at 1 atm, and without external voltages. (At the molecule level charges on one part of a molecule, balanced by opposite charges elsewhere in the same molecule can play a significant role, but that is not the same as the electrochemical production of aluminum.)
5. The living organisms (yeasts, bacteria, molds) that facilitate biochemical reactions have their own energy and nutrient needs, and often consume some of the raw material intended to be converted to product, that is, the yeasts that ferment sugar or starch to alcohol consume about 13% of the sugar or starch. Their reaction mix must include sugar, water, yeast and the other materials that the yeasts need. Some of the nutrients they use are for simply living; others are for growth in numbers. The sugar-to-ethanol process is a net producer of yeast cells.
6. Many compounds can be produced organically from petroleum, or produced biochemically from biological raw materials. For example, ethanol is regularly produced by catalyzed gas-phase hydrogenation of ethylene, Example 12.4, and also produced biochemically from sugar (or starch), Example 16.1. The resulting ethanol is the same from either source (except in the eyes of the US laws that subsidize corn-based ethanol motor fuels but not hydrocarbon-based ethanol.)
7. In most industrial chemistry (e.g., ammonia production) the chemical reaction on the surface of the catalyst probably has several steps, but the intermediate products do not exist separately as pure chemicals, but only as short-lived intermediates (often free radicals or charged complexes). In contrast, biochemical reactions like the fermentation of sucrose to make ethanol go through several steps in which easily–identified chemicals are produced. One suggested list of intermediate, identifiable products for Eq. 16.A [1] is glucose, fructose-biphosphate, triose phosphate, phosphoglyceric acid, pyruvic acid (acetaldehyde and CO_2), and finally ethanol. Each of those steps has its own decrease in Gibbs energy; their sum is the change of Gibbs energy shown above.
8. The organisms that conduct the reaction are sensitive to the concentrations of reactants and products. Most of the yeasts that will conduct Reaction 16.A stop working when the ethanol concentration in the solution reaches 10–14%; this limits the "strength" of beers and wines (but not of distilled spirits, which rely on distillation, Chapter 8 to get around this limitation.)

16.3 TWO MORE SWEET EXAMPLES

Example 16.2 Estimate the equilibrium constant for the hydrolysis of sucrose to make "invert sugar," an equimolar mixture of glucose and fructose.

The reaction is

$$\underset{\text{Sucrose}}{C_{12}H_{22}O_{11}(aq)} + H_2(O)(I) \rightarrow \underset{\text{glucose}}{C_6H_{12}O_6(aq)} + \underset{\text{fructose}}{C_6H_{12}O_6(aq)} \tag{16.D}$$

Glucose and fructose have the same overall formula, but different structures (glucose is an oxygen-containing 6-membered ring with one $-CH_2OH$ group, fructose is a

5-membered oxygen-containing ring with two such groups). The Gibbs energy change of this reaction is given in [2] as −29.64 kJ/mol, so that

$$K = \exp\left(\frac{-\Delta g^\circ}{RT}\right) = \exp\left[\frac{-\left(-29{,}640\dfrac{\text{J}}{\text{mol}}\right)}{8.314\dfrac{\text{J}}{\text{mol K}}\cdot 298.15\,\text{K}}\right]$$

$$= 1.6\cdot 10^5 = \frac{[a_{\text{glucosel}}]\cdot[a_{\text{fructose}}]}{[a_{\text{sucrose}}]\cdot[a_{\text{H}_2\text{O}}]} \quad \blacksquare \tag{16.E}$$

The high value of K shows that this reaction should go to virtual completion, which it does. It is interesting because:

1. The reaction can be performed chemically; adding simple acids to the sugar solution drives the reaction toward equilibrium. Our stomach acids perform it as the first step in our digestion of sugar. It can also be done biologically, using an enzyme catalyst, *invertase.* That is the first of many steps summarized by the overall reaction Eq. 16.A.
2. The product, "invert sugar" is sweeter than its sucrose raw material, because while glucose is only 60% as sweet (per lbm or kg) as sucrose, fructose is roughly 170% as sweet and the mix about 115% as sweet.
3. This reaction is performed on a large scale industrially, mostly by acid hydrolysis. Food processors prefer invert sugar to sucrose, because of its greater sweetness (per pound and thus per dollar) and because it has some other desirable properties in food production.

Example 16.3 Corn syrup, a practically pure solution of glucose (also called dextrose) in water, is made by the enzyme-catalyzed hydrolysis (two step, two different enzymes) of cornstarch. It is then isomerized (with the *glucose isomeraze* enzyme) to make fructose by the reaction

$$\underset{\text{glucose}}{C_6H_{12}O_6(aq)} \Leftrightarrow \underset{\text{fructose}}{C_6H_{12}O_6(aq)} \tag{16.F}$$

The reported K for this reaction [3] is ≈0.865 so that

$$K = \frac{a_{\text{frusctose}}}{a_{\text{glucose}}} \approx \frac{x_{\text{frusctose}}}{x_{\text{glucose}}} = 0.865:$$

$$x_{\text{frusctose}} = \frac{K}{K+1} = \frac{0.865}{1.865} = 0.46. \quad \blacksquare \tag{16.G}$$

The typical industrial reactor output (dry basis) is 42% fructose, 53% glucose and 5% higher sugar byproducts. The fructose is concentrated by chromatography to 90% (dry basis) and then blended with the above 42% mix to make a 55% fructose (dry basis) syrup, sold as "high fructose corn syrup" [4]. It is widely used in soft drinks, and other foods because it is much cheaper than sugar (on an equal sweetness basis).

This is an unusual biochemical reaction, because it truly operates near to equilibrium, catalyzed by a biochemical enzyme that will promote the reaction in either direction.

16.4 THERMOCHEMICAL DATA FOR BIOCHEMICAL REACTIONS

A chemical engineer trying to estimate the equilibrium in a biochemical reaction is normally frustrated to find that books titled "Thermochemistry ... Organic." [5, 6] have large tables of $h^{\text{o}}_{\text{formation}}$ but nothing on the $g^{\text{o}}_{\text{formation}}$ that the engineer needs to calculate the K of the reaction. The handbook tables of thermochemical properties [7] (from which Table A.8 is a brief excerpt) give values of both enthalpy and Gibbs energy of formation for inorganic compounds and simple organics, but only $h^{\text{o}}_{\text{formation}}$ for the compounds of biochemical interest. The reason (suggested by [5], page 18), is that for biochemical reactions $T\Delta s^{\text{o}}_{\text{reaction}}$ is much smaller than $\Delta h^{\text{o}}_{\text{reaction}}$ and the uncertainty in the computed value of K is mostly caused by uncertainty in $\Delta h^{\text{o}}_{\text{reaction}}$ and very little is caused by uncertainty in $\Delta s^{\text{o}}_{\text{reaction}}$. Thus, if one has values of the much more easily-measured $\Delta h^{\text{o}}_{\text{reaction}}$ one can make a good estimate of K without knowing the much more difficult-to-measure value of $\Delta s^{\text{o}}_{\text{reaction}}$ or the resulting $\Delta g^{\text{o}}_{\text{reaction}} = \Delta h^{\text{o}}_{\text{reaction}} - T\Delta s^{\text{o}}_{\text{reaction}}$.

Example 16.4 Estimate the value of K for Examples 16.1, 16.2, and 16.3 from the simplifying assumption that $\Delta g^{\text{o}}_{\text{reaction}} \approx \Delta h^{\text{o}}_{\text{reaction}}$. Some enthalpies of formation from the elements are in Table A.8. The others we need are: sucrose, glucose, fructose (all in kJ/mol): −2225.5, −1266.8, and −1264.4 [8].

For Example 16.1

$$\Delta h^{\text{o}} = 4h^{\text{o}}_{\text{CO}_2} + 4h^{\text{o}}_{\text{ethanol}} - h^{\text{o}}_{\text{sucrose}} - h^{\text{o}}_{\text{water}}$$

$$= 4(-393.5) + 4(-277.7) - (-2225.5) - (-285.5)$$

$$= -173.5\frac{\text{kJ}}{\text{mol}} \tag{16.H}$$

and

$$K \overset{??}{\approx} \exp\left(\frac{-\Delta h^{\text{o}}}{RT}\right) = \exp\left(\frac{-\left(-173{,}500\dfrac{\text{J}}{\text{mol}}\right)}{8.314\dfrac{\text{J}}{\text{mol}\,K}\cdot 298.15\,\text{K}}\right)$$

$$= 2.45\cdot 10^{30} \tag{16.I}$$

For Example 16.2

$$\Delta h^o = h^o_{\text{glucose}} + h^o_{\text{fructose}} - h^o_{\text{sucrose}} - h^o_{\text{water}}$$

$$= (-1266.8) + (-1264.4) - (-2225.5) - (-285.8)$$

$$= -305.7 \frac{\text{kJ}}{\text{mol}} \tag{16.J}$$

and

$$K \overset{??}{\approx} \exp\left(\frac{-\Delta h^o}{RT}\right) = \exp\left(\frac{-\left(-305,700\frac{\text{J}}{\text{mol}}\right)}{8.314\frac{\text{J}}{\text{mol K}} \cdot 298.15\text{ K}}\right)$$

$$= 3.85 \cdot 10^{53} \tag{16.K}$$

For Example 16.3

$$\Delta h^o = h^o_{\text{fructose}} - h^o_{\text{glucose}} = (-1264.4) - (-1266.8)$$

$$= 2.34 \frac{\text{kJ}}{\text{mol}}$$

$$K \overset{??}{\approx} \exp\left(\frac{-\Delta h^o}{RT}\right) = \exp\left(\frac{-\left(-2,340\frac{\text{J}}{\text{mol}}\right)}{8.314\frac{\text{J}}{\text{mol K}} \cdot 298.15\text{ K}}\right) = 0.46 \tag{16.L}$$

The results of all three of these estimates are compared to the values from the examples in Table 16.A.

We see that the values based on $\Delta g^o_{\text{reaction}} \approx \Delta h^o_{\text{reaction}}$ are not very good matches to those based on $\Delta g^o_{\text{reaction}}$. But they do show that the first two reactions have practically infinite values of K, and that for the third the two values K agree within a factor of 2. Thus, this widely-used approximation seems useful, but hardly quantitative. ■

Table 16.A Results of Example 16.4

Example Number	K in example	K based on $\Delta g^o_{\text{reaction}} \approx \Delta h^o_{\text{reaction}}$
16.1	1.4×10^{85}	2.56×10^{30}
16.2	1.6×10^{6}	3.84×10^{53}
16.3	0.865	0.46

In 1957, Krebs and Kornberg [9] published the (apparently) first table of Gibbs energies of formation of biochemical compounds. Others have mostly not followed their example; such tables are rare.

16.5 THERMODYNAMIC EQUILIBRIUM IN LARGE SCALE BIOCHEMISTRY

The three reactions examined above are all large-scale industrial processes; many chemical engineers work in those industries. Austin [10] lists 14 food products, 6 vitamins, 9 industrial enzymes, 28 industrial chemicals and 33 antibiotics produced by biochemical means. In the World War II effort to move penicillin from lab glassware to industrial scale production the technical directors assigned a chemical engineer and a microbiologist to work together on each aspect of the problem, because neither working separately had much chance of getting it right [11]. Engineers who work in this industry today all become pretty good microbiologists, who know how to select, feed and nurture the microorganisms that work for them.

The biochemists who work on the thermodynamics of biochemistry see the world quite differently than chemical engineers do, and would rarely express biochemical equilibrium the way it is shown the first three examples above. The main reasons for this are:

1. For inorganic and organic chemistry we can find the values of g^o_i for most or all of the reactants and products in our reactions from short tables like Table A. 8 or the much more extensive tables in [12] or the NIST web site [13]. But those sites rarely if ever show the pure species g^o_i for biochemical materials, such as sucrose (see Problem 16.1).
2. Because the reactions of interest to biochemists almost all occur in aqueous solutions, they are rarely concerned with pure reactants or products. Their published g^o_i values are for the materials in solution, at specified values of the pH and/or the concentrations of various metallic ions.
3. If the necessary g^o_i values are known, the computation of the equilibrium K is worthwhile, giving some idea of the feasibility of the reaction. But for most industrial biochemistry (e.g., the fermentation route to ethanol) other factors, like the life story of the yeasts, play a much more important role in deciding what can be done and how to do it than does the equilibrium K.

16.6 TRANSLATING BETWEEN BIOCHEMICAL AND CHEMICAL ENGINEERING EQUILIBRIUM EXPRESSIONS

This short chapter will not teach you biochemistry, but I hope it will help you understand the biochemical literature when it talks about biochemical equilibrium. This translation of biochemical to chemical language and nomenclature is based on Lehninger [14].

16.6.1 Chemical and Biochemical Equations

One of the key energy-transfer reactions inside the cells of all living things is described as

$$ATP + H_2O = ADP + P_i \tag{16.M}$$

where ATP stands for adenosine tri phosphate, ADP stands for adenosine di phosphate and P_i stands for a liberated phosphate ion (PO_4^{3-}). This is the recommended *biological equation.* It does not list all the species involved, nor does it balance atoms, nor indicate charged species. In this equation ATP stands for the equilibrium mixture of ATP^{4-}, $HATP^{3-}$, H_2ATP^{2-}, $MgATP^{2-}$, MgHATP and Mg_2ATP all at the specified pH and pMg (which is analogous to pH, $pMg = -\log(Mg^{2+})$ with Mg^{2+} expressed as molarity), and similarly for the other species. The corresponding *chemical equation* is

$$ATP^{4-} + H_2O = ADP^{3-} + HPO_4^{2-} + H^+ \tag{16.N}$$

which does balance atoms and balance charge.

16.6.2 Equilibrium Constants

The standard chemical equilibrium constant (Chapter 12) for reaction 16.N would be

$$K = \exp\left(\frac{-\Delta g^{\circ}}{RT}\right) = \frac{[ADP^{3-}][HPO_4^{2-}][H^+]}{[ATP^{4-}][H_2O]} \tag{16.O}$$

where the terms in brackets [] are the activities of the species. Since activities are dimensionless, this K is always dimensionless; its value depends only on the g_i° of the reactants and products in their pure states at the temperature of the reaction. By biological convention this same chemical equilibrium is written

$$K_c = \frac{[ADP^{3-}][HPO_4^{2-}][H^+]}{[ATP^{4-}](c^{o})^2} \tag{16.P}$$

where now the terms in brackets are concentrations, not activities, expressed as molarities and $(c^{o})^2 = (1\ \text{molar})^2$is inserted to make K_c be dimensionless. K_c depends on the g_i° of the reactants and products not necessarily in their pure states, and also on the pH and the *ionic strength* discussed below.

Biochemists define a *apparent equilibrium constant*, K' for biochemical reactions as (for Equation 16.P)

$$K' = \frac{[ADP][P_i]}{[ATP]c^{o}} \tag{16.Q}$$

where the terms in brackets representing the molar concentrations, and ADP, P_i and ATP are shorthand for the sum of all the ADP, ATP and P compounds present, and c^{o} is inserted to make the dimensions come out right.

To maintain parallelism with Chapter 12, they define

$$\Delta_f g'^{\circ}(i) = \left(\begin{array}{c}\text{transformed standard Gibbs}\\ \text{energy of formation of species } i\end{array}\right) \tag{16.1}$$

with which

$$K' = \exp\left(\frac{-\Delta_r g'^{\circ}}{RT}\right) \tag{16.2}$$

where the r subscript stands for reaction.

The equilibrium constants in common biochemical use have the following symbols, K_x, K_m, and K_c with concentrations expressed in mol fraction, molality and molarity [15]. One also sees K_a representing concentrations shown as activities, such as the K in Chapter 12.

Example 16.5 The hypothetical reaction

$$A(aq) \Leftrightarrow B(aq) + C(aq) \tag{16.R}$$

has the following equilibrium constant we are familiar with from Chapters 12 and 13

$$K_{\text{Chapter12}} = K_{\text{activities}} = 2.00 = \frac{a_b \cdot a_c}{a_a} = \frac{x_b\gamma_b \cdot x_c\gamma_c}{x_a\gamma_a} \tag{16.S}$$

What are the values of K_x, K_m and K_c?

For this reaction, K_x is defined as

$$K_x = \frac{x_b \cdot x_c}{x_a} = K_a\frac{\gamma_a}{\gamma_b \cdot \gamma_c} \tag{16.T}$$

If the activity coefficients are all = 1.00, then the rightmost term is 1.0 and $K_x = K_a$. But suppose that they are all = 1.50? In that case

$$K_x = \frac{x_b \cdot x_c}{x_a} = K_a\frac{\gamma_a}{\gamma_b \cdot \gamma_c} = K_a\frac{1.5}{1.5^2} = \frac{K_a}{1.5} = \frac{2}{1.5} = 1.333 \tag{16.U}$$

Both K_a and K_x are dimensionless, (because x and γ are dimensionless), so these two equilibrium constants are dimensionless, = 2 and 1.333. The equilibrium constant based on molalities

$$K_m = \frac{m_b \cdot m_c}{m_a} \tag{16.V}$$

has the dimension of molality (mols substance/kg solvent). To find its numerical value, we start with K_x and observe that

$$x_a = \frac{\text{mols A}}{\text{mols A} + \text{mols B} + \text{mols C} + \text{mols H}_2\text{O}} = \frac{\text{mols A}}{\sum \text{mols}} \tag{16.W}$$

so that

$$K_x = \frac{(\text{mols B}) \cdot (\text{mols C})}{(\text{mols A})} \cdot \frac{1}{\sum \text{mols}}$$

$$= \frac{(m_b \cdot \text{kg}_{\text{solvent}}) \cdot (m_b \cdot \text{kg}_{\text{solvent}})}{(m_a \cdot \text{kg}_{\text{solvent}})} \cdot \frac{1}{\sum \text{mols}}$$

$$= K_m \cdot \frac{\text{kg}_{\text{solvent}}}{\sum \text{mols}} \tag{16.X}$$

or

$$K_m = K_x \frac{\sum \text{mols}}{\text{kg}_{\text{solvent}}} \tag{16.Y}$$

For dilute solutions in water

$$\sum \text{mols} = \text{mols A} + \text{mols B} + \text{mols C} + \text{mols H}_2\text{O} \approx \text{mols H}_2\text{O} \tag{16.Z}$$

and

$$\frac{\text{mols H}_2\text{O}}{\text{kg}} = \frac{1000\ \text{g/kg}}{18\ \text{g/mol}} = 55.5 \frac{\text{mol}}{\text{kg}} \tag{16.AA}$$

so that

$$K_m = K_x \frac{\sum \text{mols}}{\text{kg}_{\text{solvent}}} \approx 1.333 \cdot 55.5 \frac{\text{mol}}{\text{kg}} = 74.0 \frac{\text{mol}}{\text{kg}} \tag{16.AB}$$

Similar calculations, again restricted to dilute solutions, lead to

$$K_c = 74.0 \frac{\text{mol}}{\text{liter}} \quad \blacksquare \tag{16.AC}$$

Chemical engineers who are comfortable with $K = \exp(-\Delta g^{o}/RT)$ in which K is dimensionless and determined by the Gibbs energies of formation of the pure reactants and products (in their appropriate standard states at temperature T) are uncomfortable (I am!) with Ks that have dimensions and whose values depend on the concentrations in the solution. But that is the normal definition of equilibrium constants in biochemistry.

16.6.3 pH and Buffers

In Chapters 12 and 13 the hydrogen ion (H^+) is taken as a reactant (See Example 13.1). In biochemical equilibrium calculations it is generally not so shown. Instead the pH is specified and the aqueous solution is expected to provide or accept these ions as needed, without appearing in the biochemical equation. Buffers (Section 13.2.3) are very important in biochemistry; almost all biochemical equilibrium calculations specify the pH at which the equilibrium exists.

16.6.4 Ionic Strength

The *ionic strength* is defined by

$$I = \begin{pmatrix} \text{ionic} \\ \text{strength} \end{pmatrix} = \frac{1}{2} \sum_{i=1}^{n} c_i z_i^2 \tag{16.3}$$

where c_i is the concentration of the ion (normally in molarity, sometimes in molalilty) and z_i is charge on that ion.

Example 16.5 Estimate the ionic strength of a 100% ionized, 0.05 molar solution of Na_2SO_4 in water. The two ions are Na^+ and SO_4^{2-}. There must also be H^+ and OH^- but these are normally in small enough concentrations to ignore.

$$I = \tfrac{1}{2}[(2 \times 0.05 \times 1^2) + (1 \times 0.05 \times 2^2)]$$

$$= \tfrac{1}{2}[(0.10) + (0.20)] = 0.15 \frac{\text{mol}}{\text{liter}} \quad \blacksquare \tag{16.AD}$$

The $g^{o}_{\text{formation}}$ of Mg^{2+} is shown in ([14] p 1647) as −455.30 (kJ/mol) at $I = 0$ and −458.54 at $I = 0.25$, Chemical engineers would show that change as a change in activity coefficient, keeping $g^{o}_{\text{formation}}$ and K independent of concentrations and of pH. Biochemists take those changes into the values of $g^{o}_{\text{formation}}$ and K.

16.7 EQUILIBRIUM IN BIOCHEMICAL SEPARATIONS

The fermentation of sugars (and starches) to ethanol is presumably the oldest known example of humans domesticating a micro-organism to produce a product we wanted, known from Egyptian hieroglyphics. (Perhaps the organisms chose us, doing our work in return for us caring for them, as cats and dogs apparently have.) The basic pattern is that some organism grows in a prepared nutrient soup (possibly grape juice was the first) producing a byproduct we like. We then separate the organism from the soup (mostly by sedimentation or filtration in the ethanol case). In the ethanol case, the clear liquid (beer, wine, mead or pulque) is used as is, as a beverage. Later we learned to separate the part we wanted by distillation to produce distilled spirits and even later, industrial alcohol.

Starting in the 19 th century we began to use similar processes with other organisms to produce chemicals we wanted. Starting about 1940 we made the first large-scale pharmaceutical that way, penicillin. Since then we have made dozens of biologically active materials by the same general approach: some organism (bacterium, yeast, mold) is found and/or biochemically engineered to produce (internally or externally) the product we want. (In the recent past we have learned to do the same with mammalian cells growing in similar media.) We then grow the organism in a highly-engineered soup of nutrients, normally in a batch mode until we reach the peak concentration of what we want. Then we stop the reaction, and separate the desired product from the reaction mix.

The production of penicillin shows the classic example of producing a high value biochemical by biochemical means. In a very well-managed soup of nutrients, a mold grows and excretes the antibiotic product we want. When the mold will produce no more, the concentration of penicillin in the soup is about 1% (1 part penicillin to 99 parts of waste products). The two challenges in the process are first, to get the mold to produce as much as possible, as quickly as possible, and second to separate the pure (easily destroyed) penicillin away from the waste products at an acceptable cost.

Chemical equilibrium plays little if any role in the growth step, but physical equilibrium is very important in the separation steps. In industrial organic chemistry the separation tools (distillation, crystallization, extraction) all depend on using differences in equilibrium concentration to enrich one species in a mixture relative to the others. The same is true in industrial biochemistry. The main differences are

1. The starting concentrations of biochemical products are generally much smaller than in industrial organic chemistry. Biochemical processes often start the separation with much less than 1% of the desired product in the mix. Industrial organic chemistry rarely does.
2. The desired product is often very unstable. Penicillin is destroyed by heating, and by long exposure to acids or bases. The same is not true of most industrial organic chemicals.
3. The products purity requirements for materials that are injected into human bodies are much stricter than those for other materials.

The separation scheme for penicillin starts with the broth from the reactor, which is chilled to reduce the rate of natural decomposition of the desired product, and then filtered to remove the solids. From there to the finished product the steps are:

1. Extraction from aqueous to an organic phase, with the pH adjusted to 2.5 to make the penicillin more soluble in organic than in water.
2. Charcoal adsorption of the organic phase (see Chapter 11) to remove those contaminants that will adsorb on charcoal.
3. Extraction from the organic phase back to an aqueous one, with the pH adjusted to 7.5 to make the penicillin more soluble in water than in organic.
4. Precipitation from the aqueous phase as a crude salt by any of several techniques, all designed to shift the equilibrium from soluble penicillin to insoluble, followed by final cleanup.

In each of these steps the product is moved from the phase in which penicillin is present in low concentration to one in which its concentration is much higher, mostly by adjusting pH, sometimes by changing the solvent. For more complex separations various kinds of chromatography are used. All of these steps obey the thermodynamic equilibrium relations shown in the previous chapters.

16.8 SUMMARY

1. All of biochemistry obeys the general principle that nature minimizes Gibbs energy.
2. The application of this principle in biochemistry generally does not take the same forms as in the preceding chapters.
3. The symbols and notation in biochemistry are often quite different from those we are familiar with in chemical engineering.
4. This chapter has said nothing about the biochemical reactions inside individual cells. Those are more complex that the simple examples shown here.
5. The study of the thermodynamics of chemical reactions within cells is now called Bioenergetics; for a lucid introduction, see [14]. For help translating between chemical engineering and biochemical notation see [15].

PROBLEMS

See the Common Units and Values for Problems and Examples. An asterisk (*) on the problem number indicates that the answer is in Appendix H.

16.1 **a.** Estimate g° of formation from the elements of sucrose at 298.15K. The reported h° of formation is -2221.2 kJ/mol and s° (absolute, not of formation) is reported as 392.4 J/mol K [17]. The standard absolute entropies of elemental C, O_2 and H_2 at 298.15 ([18] page 992), all in kJ/mol K, are 0.00569, 0.2050 and 0.1306.

b. Compare this value to the -1551.8 kJ/mol (aq) reported by [9], and the values of -1557.6 kJ/mol (cr) and -1564.7 kJ/mol (aq) reported by [2].

16.2 In example 16.3 replace Equation 16.F with

$$\underset{\text{glucose}}{C_6H_{12}O_6(aq)} \Leftrightarrow \underset{\text{fructose}}{C_6H_{12}O_6(aq)} + \begin{pmatrix}\text{higher molecular} \\ \text{weight sugars (aq)}\end{pmatrix} \tag{16.AE}$$

Assume that we start with pure glucose, and that in the equilibrium mix the mol fraction of dissolved higher-molecular-weight sugars is 0.05. Repeat the calculation in Equation 16.G for this assumption.

16.3 Example 16.3 shows that after glucose and fructose are brought to chemical equilibrium, the fructose can be mostly removed from the glucose-fructose-water solution by chromatography. Sketch a flow diagram for a process that receives a glucose-water solution and delivers a practically pure fructose-water solution, using up practically all the glucose. Then sketch a similar flow diagram for *n*-butane and *i*-butane. See Example 4.3 and the observation that the tallest distillation tower in a typical modern oil refinery is making the separation between *n*-butane and *i*-butane.

16.4* Tewari and Goldberg [3] determined the K for the isomerization of glucose to fructose by measuring the equilibrium concentrations, coming both ways, from pure glucose (aq) to equilibrium and from pure fructose (aq) to equilibrium, finding $K_x = 0.865$. Krebs and Kornberg [9] report the following values for the g°(aq) for glucose and fructose as -917.24 and -915.40 kJ/mol.

a. Does the K for the isomerization computed from these two g°(aq) s agree with the experimental K_x?

b. What would the equilibrium conversion of glucose to fructose be for the value of K_x computed in part (a)?

16.5 For the hypothetical reaction in Example 16.4, assume that we add 0.1 mol of A to one kg of water. The activity coefficients are all = 1.5 as in that example, and the volume of the resulting solution is 1.01 L. Estimate the equilibrium activities, mol fractions, molalities and molarities of A, B and C. Compare those values to the four equilibrium constants in that example.

16.6 Krebs and Kornberg ([9], Table 11) report the Gibbs energy change for the reaction:

$$\text{glucose(aq)} \Leftrightarrow 2(\text{ethanol(aq)} + CO_2(g)) \tag{16.AF}$$

as -234.72 kJ/mol.

a. Reaction 16. A consists of twice this reaction, plus Reaction 16.D. Combining the Δg° s for these reactions, estimate the Δg° for reaction 16.A and compare it with the value shown in Example 16.1.

b. Putnam and Boreiro-Goates [2] show that the publicly-accepted value before 1993 of Δg° for sucrose was 10.2 kJ/mol more negative than their experimental results. Does adding this correction make the calculated value match Example 16.1 better?

REFERENCES

1. Schuler, M. L., and F. Kargi, *Bioprocess Engineering: Basic Concepts*. Upper Saddle River, NJ: Prentice Hall, p. 148 (2002).
2. Putnam, R. L., and J. Boerio-Goates. Heat-capacity measurements and thermodynamic functions of crystalline sucrose from 5 K to 342 K. (Plus additional text). *J. Chem. Thermodynamics*, 25: 607–613 (1993).
3. Tewari, R. B., and R N. Goldberg. Thermodynamics of the conversion of aqueous glucose to fructose, *J. Solution Chemistry*, 11(8): 523–547 (1984).
4. Orthhoefer, F. T., Corn starch modification and uses. In *Corn: Chemistry and Technology*, A. S, Watson, and P. E. Ramstead, eds. St. Paul, MN: Amer. Assn. of Cereal Chemists, Inc. p. 514 (1987).
5. Cox, J. D., and G. Pilcher. *Thermochemistry of Organic and Organometallic Compounds*. London: Academic Press (1970) .
6. Pedley, J. B., R. D. Naylor, and S. P. Kirby. *Thermochemical Data of Organic Compounds*. ed. 2. London: Chapman & Hall (1986).
7. Haynes, W. M. *Handbook of Chemistry and Physics*. ed. 91. Boca Raton, FL: CRC Press (2010).
8. Domalski, E. S. Selected values of heats of combustion and heats of formation of organic compounds containing the elements C. H, N, O, P and S. *J. Phys. Chem. Ref. Data*, 1(2): 221–243 (1972).
9. Krebs, H. A., and H. Kornberg. *Energy Transformations in Living Matter*. Berlin: Springer Verlag (1957).

10. Austin, G. T. *Shreve's Chemical Process Industries*. ed. 5. New York: McGraw-Hill. p. 579 (1984).
11. Schuler, M. L., and F. Kargi. *Bioprocess Engineering: Basic Concepts*. Upper Saddle River, NJ: Prentice Hall, p. 7 (2002).
12. Lide, D. R. *CRC Handbook of Chemistry and Physics*. ed. 89. Boca Raton, FL: CRC Press (2008).
13. *Chemistry WebBook,* National Institute of Science and Technology.
14. Lehninger, A. L. *Bioenergetics: The Molecular Basis of Biological Energy Transformations*, ed. 2. Reading, MA: Addison Wesley (1971).
15. Segel, I. H. *Biochemical Calculations*, ed. 2. New York: Wiley (1976).
16. Alberty, R.A. Recommendations for nomenclature and tables in biological thermodynamics. *Pure and Appl. Chem.* 66(8): 1641–1666 (1994).
17. Goldberg, R. N., and Y. B. Tewari. Thermodynamics of enzyme-catalyzed reactions: Part 5. Isomerases and ligases. *J. Phys. Chem. Ref. Data* 24(6):1765–1798 (1995).
18. Hougen, O.A., K.M. Watson, and R.A. Ragatz, *Chemical Process Principles, Part II, Thermodynamics*, ed. 2. New York: Wiley, p. 600 (1947).

APPENDIX A

USEFUL TABLES AND CHARTS

A.1 USEFUL PROPERTY DATA FOR CORRESPONDING STATES ESTIMATES

The values given in Table A.1 are useful in "corresponding states" estimates of thermodynamic properties.

Bar = 0.987 atm.

Table A.1 Property Data for Corresponding States Estimates

	Molar Mass (= Molecular Weight) M (g/mol)	ω	T_c(K)	P_c (bar)	z_c
Methane	16.043	0.012	190.6	45.99	0.286
Ethane	30.07	0.1	305.3	48.72	0.279
Propane	44.097	0.152	369.8	42.48	0.276
n-Butane	58.123	0.2	425.1	37.96	0.274
n-Pentane	72.15	0.252	469.7	33.7	0.27
n-Hexane	86.177	0.301	507.6	30.25	0.266
n-Heptane	100.204	0.35	540.2	27.4	0.261
n-Octane	114.231	0.4	568.7	24.9	0.256
n-Nonane	128.258	0.444	594.6	22.9	0.252
n-Decane	142.285	0.492	617.7	21.1	0.247
Isobutane	58.123	0.181	408.1	36.48	0.282
Isooctane	114.231	0.302	544	25.68	0.266
Cyclopentane	70.134	0.196	511.8	45.02	0.273
Cyclohexane	84.161	0.21	533.6	40.73	0.273
Methylcyclopentane	84.161	0.23	532.8	37.85	0.272
Methylcyclohexane	98.188	0.235	572.2	34.71	0.269
Ethylene	28.054	0.087	282.3	50.4	0.281
Propylene	42.081	0.14	365.6	46.65	0.289
1-Butene	56.108	0.191	420	40.43	0.277
cis-2-Butene	56.108	0.205	435.6	42.43	0.273
trans-2-Butene	56.108	0.218	428.6	41	0.275
1-Hexene	84.161	0.28	504	31.4	0.265
Isobutylene	56.108	0.194	417.9	40	0.275
1,3-Butadiene	54.092	0.19	425.2	42.77	0.267
Cyclohexene	82.145	0.212	560.4	43.5	0.272

(continued)

Physical and Chemical Equilibrium for Chemical Engineers, Second Edition. Noel de Nevers.

Table A.1 (*Continued*)

	Molar Mass (= Molecular Weight) M (g/mol)	ω	T_c(K)	P_c (bar)	z_c
Acetylene	26.038	0.187	308.3	61.39	0.271
Benzene	78.114	0.21	562.2	48.98	0.271
Toluene	92.141	0.262	591.8	41.06	0.264
Ethylbenzene	106.167	0.303	617.2	36.06	0.263
Cumene	120.194	0.326	631.1	32.09	0.261
o-Xylene	106.167	0.31	630.3	37.34	0.263
m-Xylene	160.167	0.326	617.1	35.36	0.259
p-Xylene	160.167	0.322	616.2	35.11	0.26
Styrene	104.152	0.297	636	38.4	0.256
Naphthalene	128.174	0.302	748.4	40.51	0.269
Biphenyl	154.211	0.365	789.3	38.5	0.295
Formadlehyde	30.026	0.282	408	65.9	0.223
Acetaldehyde	44.053	0.291	466	55.5	0.221
Methyl acetate	74.079	0.331	506.6	47.5	0.257
Ethyl acetate	88.106	0.366	523.3	38.8	0.255
Acetone	58.08	0.307	508.2	47.01	0.233
Methyl ethyl ketone	72.107	0.323	535.5	41.5	0.249
Diethyl ether	74.123	0.281	466.7	36.4	0.263
Methyl *tert*-butyl ether	88.15	0.266	497.1	34.3	0.273
Methanol	32.042	0.564	512.6	80.97	0.224
Ethanol	46.069	0.645	513.9	61.48	0.24
1-Propanol	60.096	0.622	536.8	51.75	0.254
1-Butanol	74.123	0.594	563.1	44.23	0.26
1-Hexanol	102.177	0.579	611.4	35.1	0.263
2-Propanol	60.096	0.668	508.3	47.62	0.248
Phenol	94.113	0.444	694.3	61.3	0.243
Ethylene glycol	62.068	0.487	719.7	77	0.246
Acetic acid	60.053	0.467	592	57.86	0.211
n-Butyric acid	88.106	0.681	615.7	40.64	0.232
Benzoic acid	122.123	0.603	751	44.7	0.246
Acetonitrile	41.053	0.338	545.5	48.3	0.184
Methylamine	31.057	0.281	430.1	74.6	0.321
Ethylamine	45.084	0.285	456.2	56.2	0.307
Nitromethane	61.04	0.348	588.2	63.1	0.223
Carbon tetrachloride	153.822	0.193	556.4	45.6	0.272
Chloroform	119.377	0.222	536.4	54.72	0.293
Dichloromethane	84.932	0.199	510	60.8	0.265
Methyl chloride	50.488	0.153	416.3	66.8	0.276
Ethyl chloride	64.514	0.19	460.4	52.7	0.275
Chlorobenzene	112.558	0.25	632.4	45.2	0.265
Neon	14.0099	0	44.4	27.6	0.311
Argon	39.948	0	150.9	48.98	0.291
Krypton	83.8	0	209.4	55.02	0.288
Xenon	165.03	0	289.7	58.4	0.286
Helium 4	4.003	−0.39	5.2	2.28	0.302
Hydrogen	2.016	−0.216	33.19	13.13	0.305
Oxygen	31.999	0.022	154.6	50.43	0.288
Nitrogen	28.014	0.038	126.2	34	0.289
Chlorine	70.905	0.069	417.2	77.1	0.265
Carbon monoxide	28.01	0.048	132.9	34.99	0.299
Carbon dioxide	44.01	0.224	304.2	73.83	0.274
Carbon disulfide	76.143	0.111	552	79	0.275
Hydrogen sulfide	34.082	0.094	373.5	89.63	0.284
Sulfur dioxide	64.065	0.245	430.8	78.84	0.269

(*continued*)

Table A.1 (*Continued*)

	Molar Mass (= Molecular Weight) M (g/mol)	ω	T_c(K)	P_c (bar)	z_c
Sulfur trioxide	80.064	0.424	490.9	82.1	0.255
Nitric oxide (NO)	30.006	0.583	180.2	64.8	0.251
Nitrous oxide (N_2O)	44.013	0.141	309.6	72.45	0.274
Hydrogen chloride	36.461	0.132	324.7	83.1	0.249
Hydrogen cyanide	27.026	0.41	456.7	53.9	0.197
Water	18.015	0.345	647.1	220.55	0.229
Ammonia	17.031	0.253	405.7	112.8	0.0242
Nitric acid	63.013	0.714	520	68.9	0.231
Sulfuric acid	98.08	****	924	64	0.147

Source: From Smith, J. M., H. C. van Ness, and M. M. Abbot, *Introduction to Chemical Engineering Thermodynamics*, ed. 5. New York: McGraw-Hill (1996). Reproduced with permission of the McGraw-Hill Companies.

A.2 VAPOR-PRESSURE EQUATION CONSTANTS

Table A.2 gives the Antoine equation constants for

$$\log p = A - \frac{B}{T+C}, \qquad p \text{ in mm Hg, } T \text{ in } ^\circ\text{C} \qquad (A.1)$$

We often see this equation as

$$\log p = A - \frac{B}{T-C}, \qquad p \text{ in mm Hg, } T \text{ in K} \qquad (A.2)$$

The C in Eq. A.2 is [273.15 minus the C in Eq. A.1]; the A and B are not changed. We also see it with $\ln p$ instead of $\log p$ and with p expressed in psia, bar, or atm. Changing from one of those forms to the other requires simple multiplication of A and B by suitable constants.

The original sources often give temperature ranges over which these constants should be used. However, we can normally extrapolate beyond those ranges, with only modest loss of accuracy. Most of these constants reproduce the temperature region near the NBP very well, temperatures far above and below the NBP not as well (see Figure 5.5).

Table A.2 Antoine Equation Constants

Substance	Formula	A	B	C
Acetaldehyde	C_2H_4O	7.05648	1070.60	236
Acetic acid	$C_2H_4O_2$	7.29964	1479.02	216.81
Acetone	C_3H_6O	7.02447	1161	224
Acetylene	C_2H_2	7.09990	711.00	253.38
Ammonia	NH_3	7.36048	926.13	240.17
Argon	Ar	6.61562	304.2283	267.31
Benzene	C_6H_6	6.90565	1211.033	220.79
n-Butane	C_4H_{10}	6.80897	935.86	238.73
n-Butanol	$C_4H_{10}O$	7.838	1558.190	196.881
Carbon dioxide, solid	CO_2(s)	9.81064	1347.788	272.99
Carbon dioxide, liquid	CO_2(l)	7.5788	863.35	273.15
Carbon monoxide,	CO	6.24021	230.272	260
Carbon tetrachloride	CCl_4	6.9339	1242.43	230
Chlorine	Cl_2	6.9317	859.17	246.14
Chlorobenzene	C_6H_5Cl	6.9781	1431.06	217.55
Chloroform	$CHCl_3$	6.9547	1170.97	226.23
Cyclohexane	C_6H_{12}	6.84132	1201.53	222.65
Ethane	C_2H_6	6.80267	656.4028	255.99
Ethyl acetate	$C_4H_8O_2$	7.01457	1211.90	216

(*continued*)

Table A.2 (*Continued*)

Substance	Formula	A	B	C
Ethyl alcohol	C_2H_6O	8.04494	1554.3	222.65
Ethylbenzene	C_8H_{10}	6.95719	1424.255	213.206
Ethylene	C_2H_4	6.74756	585.00	255
Fluorine	F_2	6.80540	310.130	267.15
Helium	He	5.32072	14.6500	274.94
Hydrogen	H_2	5.92088	71.6153	276.34
Hydrogen chloride	HCl	7.16761	744.490	258.7
Isopropyl alcohol	C_3H_8O	8.11822	1580.92	219.61
Isopentane	C_5H_{12}	6.78967	1020.012	233.097
Lead	Pb	7.827	9845.4	273.15
Mercury	Hg	7.8887	3148.0	273.15
Methane	CH_4	6.61184	389.9278	265.99
Methyl alcohol	CH_4O	8.07247	1574.99	238.86
Methyl ethyl ketone	C_4H_8O	6.97421	1209.6	216
n-Decane	$C_{10}H_{22}$	6.95367	1501.2724	194.48
n-Heptane	C_7H_{16}	6.9024	1268.115	216.9
n-Hexane	C_6H_{14}	6.87776	1171.53	224.366
n-Pentane	C_5H_{12}	6.85221	1064.63	232
Neon	Ne	6.08443	78.37729	270.54
Nitric oxide	NO	8.74295	682.9382	268.27
Nitrogen	N_2	6.49454	255.6784	266.55
Nitrogen dioxide	NO_2	8.91717	1798.543	276.8
Oxygen	O_2	6.69147	319.0117	266.7
Ozone	O_3	6.83670	552.5020	250.99
Propane	C_3H_8	6.82970	813.2008	247.99
Styrene	C_8H_8	6.92409	1420	206
Toluene	C_7H_8	6.95334	1343.943	219.377
Water	H_2O	7.94917	1657.462	227.02
o-Xylene	C_8H_{10}	6.99893	1474.68	213.69

Source. These values are taken from a variety of sources. Longer lists are in Dean, J. A. *Lange's Handbook of Chemistry,* ed. 12. New York. McGraw–Hill, pp. 10–29 to 10–54 (1979); Reid, R. C., J. M. Prausnitz, and T. K. Sherwood. *The Properties of Liquids and Gases,* ed. 3. New York: McGraw-Hill, Appendix A (1977); and Lide, D. R., ed. *CRC Handbook of Chemistry and Physics,* ed. 71. Boca Raton, FL: CRC Press, pp. 6–70 (1990).

A.3 HENRY'S LAW CONSTANTS

Henry's law

$$y_i = \frac{x_i \cdot H_i}{P} \tag{3.6}$$

is quite useful for gases well above their critical temperatures, dissolved in liquids. It is less applicable for gases at or near their critical temperatures like ethane or CO_2, and more reliable for gases that do not ionize in the liquid, like oxygen in water, than for those that do ionize, like carbon dioxide in water. It is applicable for gases dissolved in any liquid, but most of the published Henry's law constants are for gases in water. A high value of H_i indicates a low solubility of the gas. Henry's law is introduced in Chapter 3 and discussed in more detail in Chapter 9.

Table A.3 gives the reported values of the Henry's law constant H for a variety of gases *dissolved in water* at common temperatures. All values are in atmospheres $\times 10^4$; that is, the Henry's law constant for oxygen at $0°C = 2.55 \times 10^4$ atm = 25,500 atm.

To make life hard for the student and the working engineer, Henry's law is expressed in a variety of ways, with a variety of dimensions. We can rewrite Eq. 3.6 as

$$y_i = \frac{H_i}{P} \cdot \begin{pmatrix} \text{concentration of dissolved} \\ \text{gas, in some set of units} \end{pmatrix} \tag{A.3}$$

If the concentration is expressed in mol fraction, then Eq. A.3 is the same as Eq. 3.6. But we regularly see g/L, mol/L, g/kg, mol/kg, lb/ft^3, and lbmol/ft^3 as the concentration units. For dilute aqueous solutions, for which

$$\rho \approx 1.00 \frac{\text{kg}}{\text{L}} \quad \text{and} \quad \rho_{\text{molar, solvent}} \approx \frac{1000\ g/L}{18.015\ g/mol} = 55.5 \frac{\text{mol}}{\text{L}} \tag{A.A}$$

the concentration of dissolved gas is

$$c_i = x_i \cdot 55.5 \frac{\text{mol}}{\text{L}}$$

so that

$$H_{\text{Eq.A.3}} = 55.5 \frac{\text{mol}}{\text{L}} \cdot H_{\text{Eq.3.6}} \tag{A.B}$$

Chemists regularly write Henry's law as

$$x_i = \begin{pmatrix} \text{a different kind of Henry's} \\ \text{law constant} \end{pmatrix} \cdot Py_i \tag{A.C}$$

This Henry's law constant is the reciprocal of the one used in this book.

Table A.3 Henry's Law Constants for Common Gases in Water

	T (°C)					
	0	10	20	30	40	50
Acetylene	0.072	0.096	0.121	0.146		
Carbon dioxide	0.073	0.105	0.148	0.194	0.251	0.321
Ethane	1.26	1.89	2.63	3.42	4.23	5.00
Helium	12.9	12.6	12.5	12.4	12.1	11.5
Hydrogen	5.79	6.36	6.83	7.29	7.51	7.65
Hydrogen sulfide	0.0268	0.0367	0.0483	0.0609	0.0745	0.0884
Methane	2.24	2.97	3.76	4.49	5.20	5.77
Nitrogen	5.29	6.687	8.04	9.24	10.4	11.3
Oxygen	2.55	3.27	4.01	4.75	5.35	5.88
Ozone	0.194	0.248	0.376	0.598	1.20	2.74

Source: Henry's law values (not all in the same units) are found in various editions of *Perry's Chemical Engineers' Handbook*, McGraw-Hill and *The Handbook of Chemistry and Physics*, CRC Press.

A.4 COMPRESSIBILITY FACTOR CHART (z CHART)

The basic idea of the *theorem of corresponding states* is that this plot (Figure A.4) must be the same for all gases. That is *only approximately true.* Much more accurate ways of estimating fluid densities are available in our computers. But the classic compressibility factor chart (used with the values in Table A.1) allows us to make a very quick estimate of the departure from ideal gas behavior, and it gives some insight into the form of that departure.

Here

$$z = \begin{pmatrix} \text{compressibility} \\ \text{factor} \end{pmatrix} = \frac{Pv}{RT}$$

For a ideal gas, $z = 1$.

A.5 FUGACITY COEFFICIENT CHARTS

According to the *theorem of corresponding states* this plot (Figure A.5, both forms) must be the same for all gases. That is *only approximately true.* Much more accurate ways of estimating pure-component fugacity coefficients are available in our computers. But the classic fugacity coefficient charts (used with the values in Table A.1) allow us to make a very quick estimate of the departure from ideal gas behavior, and give some insight into the form of this function.

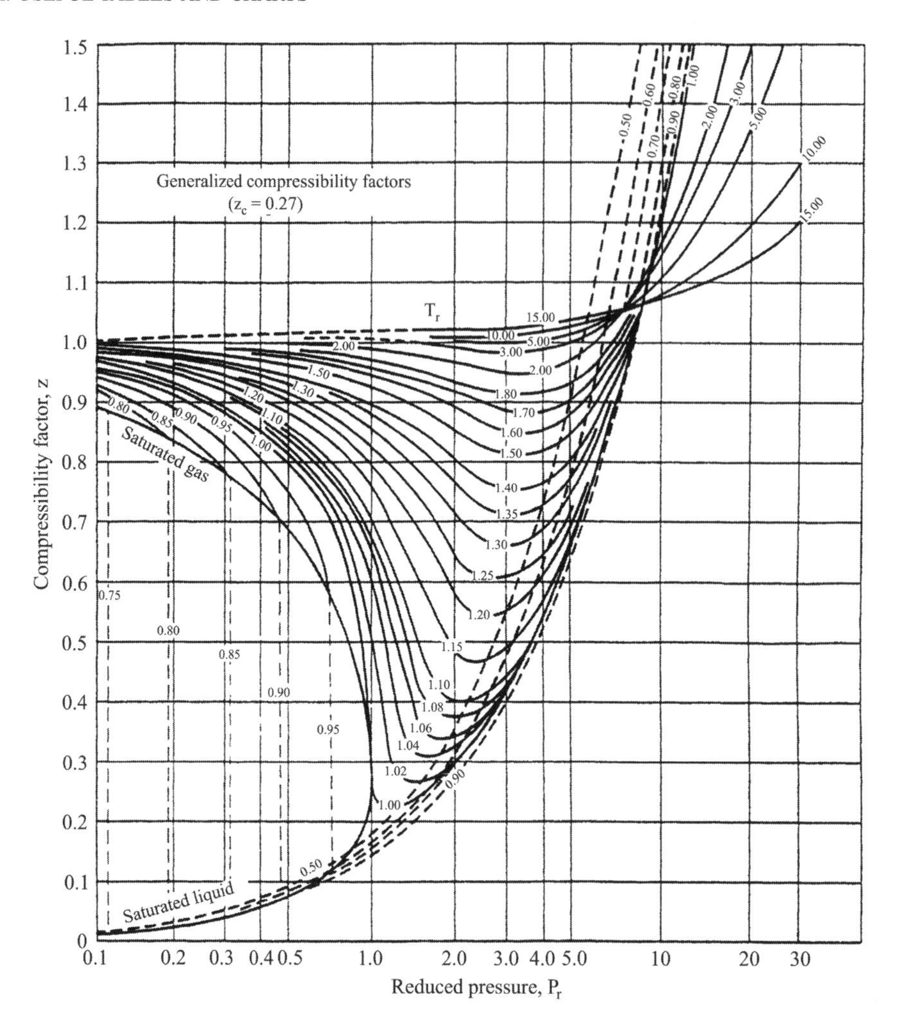

FIGURE A.4 Compressibility factor chart. (From Hougen, O. A., K. M. Watson, and R. A. Ragatz, *Chemical Process Principles*, Part II: *Thermodynamics*, ed. 2. ©1959. New York: Wiley. Reprinted by permission of the estate of O. A. Hougen.)

A.6 AZEOTROPES

Tables A.6.1 and A.6.2 are excerpts from the longer tables in *Perry's Chemical Engineers' Handbook*. More extensive tables can be found in *CRC Handbook of Chemistry and Physics*, and in Horsley, *Advances in Chemistry Series*. ed. 6, Washington, DC: American Chemical Society (1952).

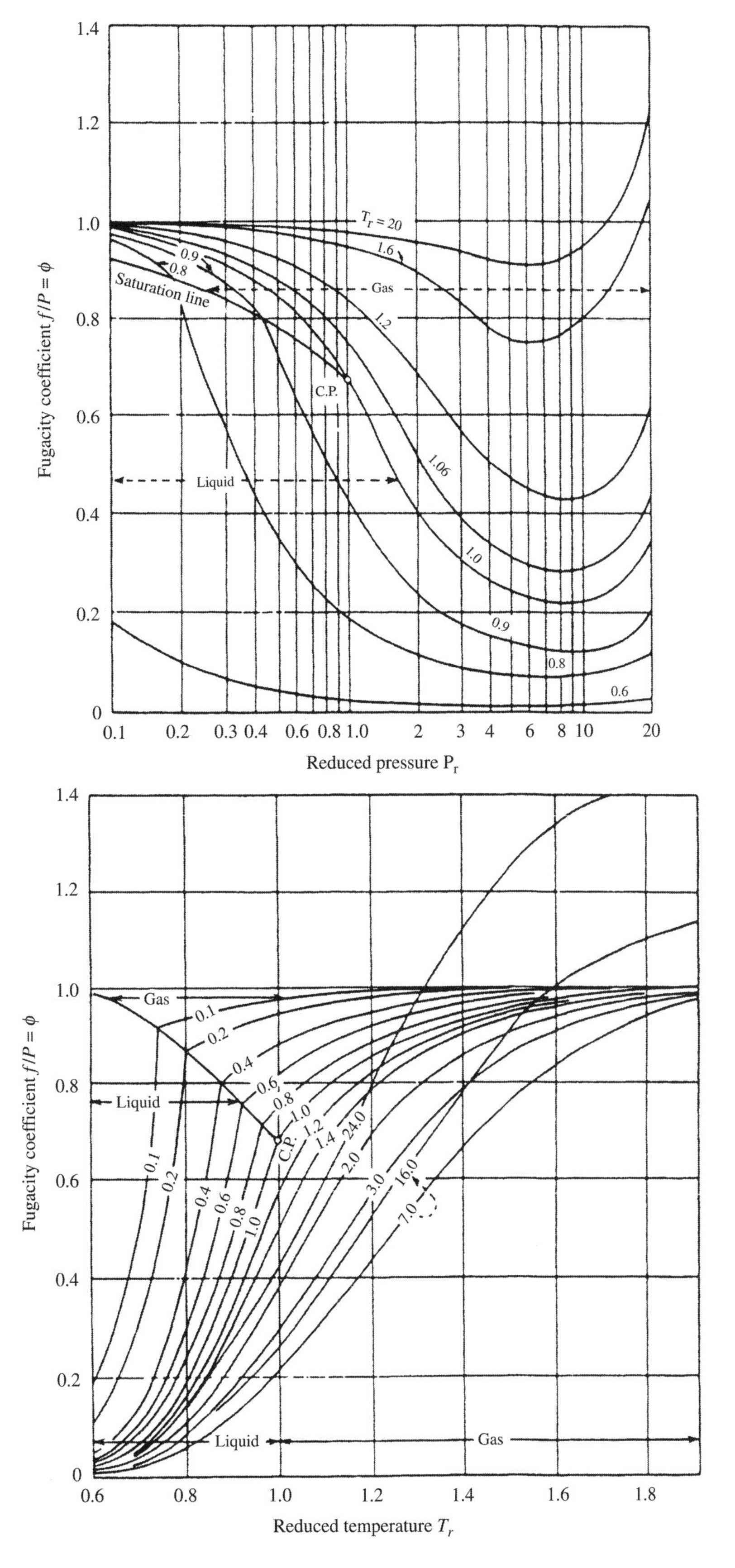

FIGURE A.5 Fugacity coefficient chart. (From Hougen, O. A., K. M. Watson, and R. A. Ragatz, *Chemical Process Principles*, Part II: *Thermodynamics*, ed. 2. © 1959. New York: Wiley. Reprinted by permission of the estate of O. A. Hougen.)

Table A.6.1 Minimum Boiling Binary Azeotropes at 760 torr

System			
A	B	Mol% A	Temperature (°C)
Water	Ethanol	10.57	78.15
	Allyl alcohol	54.5	88.2
	Propionic acid	94.7	99.98
	Propyl alcohol	56.83	87.72
	Isopropyl alcohol	31.46	80.37
	Methyl ethyl ketone	33	73.45
	Isobutyric acid	94.5	99.3
	Ethyl acetate (2 phase)	24	70.4
	Ethyl ether (2 phase)	5	34.15
	n- Butyl alcohol (2 phase)	75	92.25
	Isobutyl alcohol	67.14	89.92
	sec-Butyl alcohol	66	88.5
	tert-Butyl alcohol	35.41	79.91
	Isoamyl alcohol (2 phase)	82.79	95.15
	Amyl alcohol (tert) (2 phase)	65	87
	Benzene (2 phase)	29.6	69.25
	Toluene (2 phase)	55.6	84.1
Carbon tetrachloride	Methanol	44.5	55.7
	Ethanol	61.3	64.95
	Allyl alcohol	73	72.32
	n-Propyl alcohol	75	72.8
	Ethyl acetate	43	74.75
Carbon disulfide	Methanol	72	37.65
	Ethanol	86	42.4
	Acetone	61	39.25
	Methyl acetate	69.5	40.15
Chloroform	Methanol	65	53.5
	Ethanol	84	59.3
	Isopropyl alcohol	92	60.8
n-Butyl alcohol	Cyclohexane	11	79.8
	Toluene	37	105.5
Isobutyl alcohol	Isoamyl bromide	60	103.8
	Benzene	10	79.84
	Toluene	50	101.15
n-Amyl alcohol	*i*-Amyl acetate	96.4	131.3
	i-Butyl propionate	85	130.5
Isoamyl alcohol	Chlorobenzene	42	124.3
	o-Xylene	64	128
	m-Xylene	58	127
	p-Xylene	56	126.8
Nitrobenzene	Benzyl alcohol	39	204.3
Phenol	*p*-Bromotoluene	58	176.2
Acetic acid	Chlorobenzene	72.5	114.65
	Benzene	2.5	80.05
	Toluene	62.7	105.4
	m-Xylene	40	115.38
Ethyl alcohol	Methyl ethyl ketone	45	74.8
	Ethyl acetate	46	71.8
	Methyl propionate	67.5	73.2
	n-Propyl formate	72	73.5
	Benzene	44.8	68.24
	Cyclohexane	44.5	64.9
	n-Hexane	33.2	58.68
	Toluene	81	76.65
	n-Heptane	67	72

(*continued*)

Table A.6.1 (*Continued*)

System			
A	B	Mol% A	Temperature (°C)
Allyl alcohol	Benzene	22.2	76.75
	Cyclohexane	26.6	74
	n- Hexane	6.5	65.5
	Toluene	61.5	92.4
Acetone	Methyl acetate	61	56.1
	Isobutyl chloride	81	55.8
	Diethylamine	43.5	51.5
n-Propyl alcohol	Ethyl propionate	64	93.4
	Benzene	20.9	77.12
	n-Hexane	6	65.65
	Toluene	60	92.6
Isopropyl alcohol	Ethyl acetate	30.5	74.8
	Benzene	39.3	71.92
	n-Hexane	29	61
	Toluene	77	80.6
Tetrachloroethylene	Ethanol	6	77.95
	Allyl alcohol	27	94
	Propionic acid	81	118.95
	n-Propyl alcohol	24	94
	Isopropyl alcohol	8	81.7
	n-Butyl alcohol	47	110
	Isobutyl alcohol	40	103.05

Source: Taken from Perry R. H., and D. W. Green, eds. *Perry's Chemical Engineer's Handbook*, ed. 6, pp. 13–59 and 13–50, McGraw-Hill (2003).

Table A.6.2 Maximum Boiling Binary Azeotropes at 760 torr

System			
A	B	Mol% A	Temperature (°C)
Water	Hydrofluoric acid	65.4	120
	Hydrochloric acid	88.9	110
	Perchloric acid	32	203
	Hydrobromic acid	83.1	126
	Hydriodic acid	84.3	127
	Nitric acid	62.2	120.5
	Formic acid	43.3	107.1
Chloroform	Acetone	65.5	64.5
Formic acid	Diethyl ketone	48	105.4
	Methyl propyl ketone	47	105.3
Phenol	Cyclohexanol	90	182.45
	Benzaldehyde	54	185.6
	Benzyl alcohol	8	206
o- Cresol	Acetophenone	24	203.7
	Phenyl acetate	42.5	198.6
	Methyl hexyl ketone	97	191.5
	Isoamyl butyrate	80	192
m-Cresol	Acetophenone	54	209
	Isoamyl lactate	60	207.6
p–Cresol	Benzyl alcohol	38	207
	Acetophenone	52	208.45

Source: Taken from Perry R. H., and D. W. Green, eds. *Perry's Chemical Engineer's Handbook*, ed. 6, pp. 13–59 and 13–50, McGraw-Hill (2003).

A.7 VAN LAAR EQUATION CONSTANTS

Table A.7 is an excerpt from a much longer table taken from Holmes and Van Winkle, which gives the original literature citations and the corresponding constants for several other equations. The traditional form is

$$\log \gamma_1 = \frac{Ax_2^2}{\left(\frac{A}{B}x_1 + x_2\right)^2} \quad \log \gamma_2 = \frac{Bx_1^2}{\left(x_1 + \frac{B}{A}x_2\right)^2} \tag{8.12}$$

For programming our computers the following form is simpler

$$\log \gamma_1 = \frac{B^2Ax_2^2}{(Ax_1 + Bx_2)^2} \quad \log \gamma_2 = \frac{A^2Bx_1^2}{(Ax_1 + Bx_2)^2} \tag{8.L}$$

These are in the log form. We also often see this equation in the ln form ($\ln \gamma_1 =$ etc.) for which the constants A and B are 2.303 times as large as those shown here. We must always check to see which form the reported constants correspond to.

Most of the values in this table are for data at a constant pressure of 760 torr. Some are constant temperatures, as shown. As discussed in Chapter 9, there should not be much difference between the constants obtained either way, which is observed for most of the pairs in this table for which both forms are shown.

Some pairs, such as acetone–water, are shown twice, once with acetone as component 1 and once with water as component 1. The reader may check to see that this simply interchanges the values of A and B.

Table A.7 Van Laar Equation Constants

Component 1	Component 2	A	B	Pressure (torr)	Temperature (°C)
Acetone	Benzene	0.2039	0.1563	760	
	Carbon tetrachloride	0.3889	0.3301	760	
	Chloroform	−0.3045	−0.2709	760	
	2,3-Dimethylbutane	0.6345	0.6358	760	
	Ethanol	0.2574	0.2879	760	
	Methanol	0.2635	0.2801	760	
		0.2763	0.2878		55
	n-Pentane	0.7403	0.6364	760	
	2-Propanol	0.2186	0.269	760	
		0.3158	0.2495		55
	Water	0.9972	0.6105	760	
Acetonitrile	Water	1.068	0.8207	760	
Benzene	Acetone	0.1563	0.2039	760	
	1-Butanol	0.3594	0.5865	760	
	Carbon tetrachloride	0.036	0.0509	760	
	Chloroform	−0.0858	−0.0556	760	
	Cyclohexane	0.1466	0.1646	760	
	Cyclopentane	0.1655	0.1302	760	
	Ethanol	0.5804	0.7969	760	
	n-Heptane	0.0985	0.2135	760	
		0.1072	0.2361		75
	n-Hexane	0.1457	0.2063	760	
	Methanol	0.7518	0.8975	760	
	Methyl acetate	0.1292	0.0919	760	
	Methylcyclohexane	0.091	0.1901	760	
	Methylcyclopentane	0.136	0.1605	760	
	1-Propanol	0.3772	0.7703	760	
		0.4508	0.7564		75
	2-Propanol	0.4638	0.6723	760	
		0.5455	0.7716	500	
1-Butanol	Benzene	0.5865	0.3594	760	
	Toluene	0.543	0.3841	760	

(continued)

Table A.7 (*Continued*)

Component 1	Component 2	*A*	*B*	Pressure (torr)	Temperature (°C)
Carbon tetrachloride	Acetone	0.3301	0.3889	760	
	Benzene	0.0509	0.036	760	
	2-Propanol	0.4918	0.7868	760	
Chloroform	Acetone	−0.2709	−0.3045	760	
	Benzene	−0.0556	−0.0858		760
	2,3-Dimethylbutane	0.1736	0.279	760	
	Ethyl acetate	−0.2868	−0.4478	760	
	Methanol	0.4104	0.8263	760	
	Methyl acetate	−0.2249	−0.3343	760	
	Methyl ethyl ketone	−0.299	−0.3486	760	
Water	Acetone	0.6105	0.9972	760	
	Acetonitrile	0.8207	1.0680	760	
	Ethanol	0.4104	0.7292	760	
	Methanol	0.2439	0.3861	760	
	1-Propanol	0.5037	1.1433	760	
		0.5305	1.2315		40
		0.5224	1.1879		60
	2-Propanol	0.4750	1.0728	760	
	n-butanol	0.5531	1.7269	760	

Source: Taken from Holmes, M. J., and M. van Winkle, Prediction of ternary vapor-liquid from binary data. *Ind. Eng. Chem.* 62:21–31 (1970).

A.8 ENTHALPIES AND GIBBS ENERGIES OF FORMATION FROM THE ELEMENTS IN THE STANDARD STATES, AT $T = 298.15$ K = 25°C, AND P = 1.00 BAR

The standard states in Table A.8 are (g), pure ideal gas at 1 bar; (l) and (s), the normal state of that substance at 1 bar and 25°C; (aq), an ideal 1-molal solution of that substance in water at 1 bar and 25°C. For several substances, such as water, values are shown for two standard states, (l) and (g). The relation between these is shown in Example 12.7.

Table A.8 Enthalpies and Gibbs Energies of Formation from the Elements

Chemical Species	Formula	State	$h°$ of Formation from the Elements (kJ/mol)	$g°$ of Formation from the Elements (kJ/mol)
PARAFFINIC HYDROCARBONS				
Methane	CH_4	(g)	−74.5	−50.5
Ethane	C_2H_6	(g)	−83.8	−31.9
Propane	C_3H_g	(g)	−104.7	−24.3
n-Butane	C_4H_{10}	(g)	−125.8	−17.2
Isobutane	C_4H_{10}	(g)	−134.5	−20.9
n-Pentane	C_5H_{12}	(g)	−146.8	−8.7
n-Pentane	C_5H_{12}	(1)	−173.1	9.2
n-Hexane	C_6H_{14}	(l)	−166.9	0.2
n-Heptane	C_7H_{16}	(g)	−187.8	8.3
n-Octane	C_8H_{18}	(g)	−208.8	16.3
n-Octane	C_8H_{18}	(I)	−255.1	
UNSATURATED HYDROCARBONS				
Acetylene	C_2H_2	(g)	227.5	210.0
Ethylene	C_2H_4	(g)	52.5	68.5
Propylene	C_3H_6	(g)	19.7	62.2
1-Butene	C_4H_8	(g)	1.2	70.3
1-Pentene	C_5H_{10}	(g)	−21.3	78.4
1-Hexene	C_6H_{12}	(g)	−42.0	86.8
1-Heptene	C_7H_{14}	(g)	−62.3	95.8
1,3-Butadiene	C_4H_6	(g)	109.2	149.8

(*continued*)

Table A.8 (*Continued*)

Chemical Species	Formula	State	h° of Formation from the Elements (kJ/mol)	g° of Formation from the Elements (kJ/mol)
AROMATIC HYDROCARBONS				
Benzene	C_6H_6	(g)	82.9	129.7
Benzene	C_6H_6	(l)	49.1	124.5
Ethylbenzene	C_8H_{10}	(g)	29.9	130.9
Naphthalene	$C_{10}H_8$	(g)	1501.0	223.6
Styrene	C_8H_8	(g)	147.4	213.9
Toluene	C_7H_8	(g)	50.2	122.1
Toluene	C_7H_8	(1)	12.2	113.6
CYCLIC HYDROCARBONS				
Cyclohexane	C_6H_{12}	(g)	−123.1	31.9
Cyclohexane	C_6H_{12}	(l)	−156.2	26.9
Cyclopropane	C_3H_6	(g)	53.3	104.5
methyl-Cyclohexane	C_7H_{14}	(g)	−154.8	27.5
methyl-Cyclohexane	C_7H_{14}	(1)	−190.2	20.6
Cyclohexene	C_6H_{10}	(g)	−5.4	106.9
OXYGENATED HYDROCARBONS				
Acetaldehyde	C_2H_4O	(g)	−166.2	−128.9
Acetic acid	CH_3COOH	(l)	−484.5	−389.9
Acetic acid	CH_3COOH	(aq)	−486.1	−396.5
1,2-Ethanediol (ethylene glycol)	$C_2H_6O_2$	(1)	−454.8	−323.1
Ethanol	C_2H_6O	(g)	−235.1	−168.5
Ethanol	C_2H_6O	(l)	−277.7	−174.8
Ethyl acetate	$CH_3COOC_2H_5$	(l)	−463.3	−318.4
Ethylene oxide	C_2H_4O	(g)	−52.6	−13.0
Formaldehyde	CH_2O	(g)	−108.6	102.5
Formic acid	HCOOH	(l)	−424.7	−361.4
Methanol	CH_4O	(g)	−200.7	−162.0
Methanol	CH_4O	(1)	238.7	−166.3
Phenol	C_6H_5OH	(g)	−165.0	−50.9
INORGANIC COMPOUNDS				
Aluminum oxide	Al_2O_3	(s, α)	−1675.7	−1582.3
Aluminum chloride	$AlCl_3$	(s)	−704.2	−628.8
Ammonia	NH_3	(g)	−46.1	−16.5
Ammonia	NH_3	(aq)	−80.3	−26.6
Ammonium nitrate	NH_4NO_3	(s)	−365.6	−183.9
Ammonium chloride	NH_4Cl	(s)	−314.4	−202.9
Barium oxide	BaO	(s)	−553.5	−525.1
Barium chloride	$BaCl_2$	(s)	−856.6	−810.4
Bromine	Br_2	(l)	0	0
Bromine	Br_2	(g)	30.9	3.1
Calcium carbide	CaC_2	(s)	−59.8	−64.9
Calcium carbonate	$CaCO_3$	(s)	−1206.9	−1128.8
Calcium chloride	$CaCl_2$	(s)	−795.8	−748.1
Calcium chloride	$CaCl_2$	(aq)		−8101.9
Calcium chloride hexahydrate	$CaCl_2 \cdot 6H_2O$	(s)	−2607.9	
Calcium hydroxide	$Ca(OH)_2$	(s)	−986.1	−898.5
Calcium hydroxide	$Ca(OH)_2$	(aq)	−1002.82	−868.1
Calcium oxide	CaO	(s)	−635.1	−604.0
Carbon (graphite)	C	(s)	0	0
Carbon (diamond)	**C**	(s)	1.9	2.9
Carbon dioxide	CO_2	(g)	− 393.5	−394.4
Carbon dioxide	CO_2	(aq)	−413.8	−386.0
Carbon disulfide	CS_2	(l)	89.7	65.3

(*continued*)

Table A.8 (*Continued*)

Chemical Species	Formula	State	h° of Formation from the Elements (kJ/mol)	g° of Formation from the Elements (kJ/mol)
Carbon monoxide	CO	(g)	−110.5	−137.2
Carbon tetrachloride	CCl_4	(l)	−135.4	−65.2
Carbonic acid	H_2CO_3	(aq)	−699.7	623.1
Hydrochloric acid	HCl	(g)	−92.3	−95.3
Hydrochloric acid	HCl	(aq)	−167.2	−131.2
Hydrogen bromide	HBr	(g)	−36.4	−53.5
Hydrogen cyanide	HCN	(g)	135.1	124.7
Hydrogen cyanide	HCN	(l)	108.9	125.0
Hydrogen fluoride	HF	(g)	−271.1	−273.2
Hydrogen iodide	HI	(g)	26.5	1.7
Hydrogen peroxide	H_2O_2	(l)	−187.8	−120.4
Hydrogen sulfide	H_2S	(g)	−20.6	−33.6
Hydrogen sulfide	H_2S	(aq)	−39.7	−27.8
Iodine	I_2	(g)	62.3	19.8
Iodine	I_2	(s)	0	0
Iron (II) oxide	FeO	(s)	−272.0	
Iron (III) oxide (hematite)	Fe_2O_3	(s)	−824.2	−742.2
Iron (II) sulfide	FeS	(s, $\propto$)	−100.0	−100.4
Iron sulfide (pyrite)	FeS_2	(s)	178.2	166.9
Lead oxide	PbO	(s, yellow)	−217.3	−187.9
Lead oxide	PbO	(s, red)	−219.0	−188.9
Lead dioxide	PbO_2	(s)	−277.4	−217.3
Lead sulfate	$PbSO_4$	(s)	−919.9	−813.2
Lithium chloride	$LiCl$	(s)	−408.6	−216.7
Lithium chloride	$LiCl \cdot H_2O$	(s)	−712.6	−631.8
Lithium chloride	$LiCl \cdot 2H_2O$	(s)	−1012.7	
Lithium chloride	$LiCl \cdot 3H_2O$	(s)	−1311.3	
Lithium fluoride	LiF	(s)	−615.96	587.7
Magnesium oxide	MgO	(s)	−601.7	−569.4
Magnesium carbonate	$MgCO_3$	(s)	−1095.8	−1012.1
Magnesium chloride	$MgCl_2$	(s)	−641.3	−591.8
Mercury (I) chloride	Hg_2Cl_2	(s)	−265.2	−210.8
Mercury (II) chloride	$HgCl_2$	(s)	−224.3	−178.6
Nitric Acid	HNO_3	(l)	−174.1	−80.7
Nitric Acid	HNO_3	(aq)	−207.4	−111.3
Nitric oxide	NO	(g)	90.3	86.6
Nitrogen dioxide	NO_2	(g)	33.2	51.3
Nitrous oxide	N_2O	(g)	82.1	104.2
Nitrogen tetroxide	N_2O_4	(g)	9.2	97.9
Potassium chloride	KCl	(s)	−436.8	−409.1
Silicon dioxide	SiO_2	(s, α)	−910.9	−856.6
Silver bromide	$AgBr$	(s)	−100.4	−96.9
Silver chloride	$AgCl$	(s)	−127.1	−109.8
Silver nitrate	$AgNO_3$	(s)	−124.4	−33.4
Sodium bicarbonate	$NaHCO_3$	(s)	−945.6	−847.9
Sodium carbonate	Na_2CO_3	(s)	−1130.7	−1044.4
Sodium carbonate decahydrate	$Na_2CO_3 \cdot 10H_2O$	(s)	−4081.3	−3428,2
Sodium chloride	$NaCl$	(s)	−411.2	−384.1
Sodium chloride	$NaCl$	(aq)		−393.1
Sodium hydroxide	$NaOH$	(s)	−425.6	−379.5
Sodium hydroxide	$NaOH$	(aq)		−419.2
Sodium sulfate	Na_2SO_4	(s)	−1382.8	−1265.2
Sodium sulfate decahydrate	$Na_2SO_4 \cdot 10H_2O$	(s)	−4322.5	−3642.3

(*continued*)

Table A.8 (*Continued*)

Chemical Species	Formula	State	h° of Formation from the Elements (kJ/mol)	g° of Formation from the Elements (kJ/mol)
Sulfur	S_2	(g)	129.8	81.0
Sulfur	S_2	(l)	1.1	0.3
Sulfur	S_2	(s)	0	0
Sulfur dioxide	SO_2	(g)	−296.8	−300.2
Sulfur dioxide	SO_2	(aq)	−323.0	−300.7
Sulfur trioxide	SO_3	(g)	−395.73	−371.1
Sulfur trioxide	SO_3	(1)	−441.0	−368.4
Sulfuric acid	H_2SO_4	(l)	−814.0	−690.0
Sulfuric acid	H_2SO_4	(aq)	−909.3	−744.5
Sulfurous acid	H_2SO_3	(aq)	−608.8	537.9
Water	H_2O	(g)	241.8	−228.6
Water	H_2O	(l)	−285.8	−237.1
Zinc oxide	ZnO	(s)	−348.3	−318.3
IONS				
Hydrogen	H^+	(aq)	0	0
Aluminum	Al^{3+}	(aq)	−531.37	−485.34
Ammonium	NH_4^+	(aq)	−132.51	− 79.37
Calcium	Ca^{2+}	(aq)	−542.83	−553.54
Cupric	Cu^{2+}	(aq)	64.77	65.52
Cuprous	Cu^+	(aq)	71.67	50.00
Ferric	Fe^{3+}	(aq)	−48.53	−4.60
Ferrous	Fe^{2+}	(aq)	−89.12	−78.87
Lead	Pb^{2+}	(aq)	−1.67	− 24.39
Lithium	Li^+	(aq)	−278.49	− 293.3
Magnesium	Mg^{2+}	(aq)	−466.85	−454.80
Potassium	K^+	(aq)	−252.38	−283.26
Silver	Ag^+	(aq)	105.57	77.12
Sodium	Na^+	(aq)	−240.12	−261.66
Zinc	Zn^{2+}	(aq)	−153.89	−147.03
Bicarbonate	HCO_3^-	(aq)	−691.99	−586.85
Bisulfate	HSO_4^-	(aq)	−887.34	−756.01
Bisulfide	HS^-	(aq)	−17.7	12.6
Bisulfite	HSO_3^-	(aq)	626.2	−527.8
Bromide	Br^-	(aq)	−121.54	−103.97
Carbonate	CO_3^{2-}	(aq)	−677.14	−527.89
Chloride	Cl^-	(aq)	−167.16	−131.26
Fluoride	F^-	(aq)	−332.63	−278.82
Hydroxyl	OH^-	(aq)	−229.99	−157.29
Iodide	I^-	(aq)	−55.19	−51.59
Nitrate	NO_3^-	(aq)	−207.36	−111.34
Perchlorate	ClO_4^-	(aq)		−10.8
Sulfate	SO_4^{2-}	(aq)	−909.3	−744.62
Sulfide	S^{2-}	(aq)	+ 30.1	+ 79.5
Sufite	SO_3^{2-}	(aq)	−635.5	−486.6

Sources: Taken from Perry R. H., and D.W. Green, eds. *Perry's Chemical Engineer's Handbook*, ed. 6, pp. 13–59 and 13–50, McGraw-Hill (2003); D. R. Lide, *CRC Handbook of Chemistry and Physics*. Boca Raton, FL: CRC Press.

A.9 HEAT CAPACITIES OF GASES IN THE IDEAL GAS STATE

For ideal gases, C_P is independent of pressure; for real gases at modest pressures it is almost independent of pressure. Table A.9 shows the constants in the equation $C_P/R = a + bT + cT^2 + dT^{-2}$; T in K, up to T_{max}.

Table A.9 Heat Capacity Equation Constants

Chemical Species	Formula	T_{max}	a	$10^3 b$	$10^6 c$	$10^{-5} d$
PARAFFINS						
Methane	CH_4	1500	1.702	9.081	−2.164	
Ethane	C_2H_6	1500	1.131	19.225	−5.561	
Propane	C_3H_8	1500	1.213	28.785	−8.824	
n-Butane	C_4H_{10}	1500	1.935	36.915	−11.402	
Isobutane	C_4H_{10}	1500	1.677	37.853	−11.945	
n-Pentane	C_5H_{12}	1500	2.464	45.351	−14.111	
n-Hexane	C_6H_{14}	1500	3.025	53.722	−16.791	
n-Heptane	C_7H_{16}	1500	3.570	62.127	−19.486	
n-Octane	C_8H_{18}	1500	8.163	70.567	−22.208	
1-ALKENES						
Ethylene	C_2H_4	1500	1.424	14.394	−4.392	
Propylene	C_3H_6	1500	1.637	22.706	6.915	
1-Butene	C_4H_8	1500	1.967	31.630	−9.873	
1-Pentene	C_5H_{10}	1500	2.691	39.753	−12.447	
1-Hexene	C_6H_{12}	1500	3.220	48.189	−15.157	
1-Heptene	C_7H_{14}	1500	3.768	56.588	−17.847	
MISCELLANEOUS ORGANICS						
Acetaldehyde	C_2H_4O	1000	1.693	17.978	−6.158	
Acetylene	C_2H_2	1500	6.132	1.952		−1.299
Benzene	C_6H_6	1500	−0.206	39.064	−13.301	
1,3-Butadiene	C_4H_6	1500	2.734	26.786	−8.882	
Cyclohexane	C_6H_{12}	1500	−3.876	63.249	−20.928	
Ethanol	C_2H_6O	1500	3.518	20.001	−6.002	
Ethylbenzene	C_8H_{10}	1500	1.124	55.380	−18.476	
Formaldehyde	CH_2O	1500	2.264	7.022	−1.877	
Methanol	CH_4O	1500	2.211	12.216	−3.450	
Toluene	C_7H_8	1500	0.290	47.052	−15.716	
Styrene	C_8H_8	1500	2.050	50.192	−16.662	
MISCELLANEOUS INORGANICS						
Air		2000	3.355	0.575		−0.016
Ammonia	NH_3	1800	3.578	3.020		−0.186
Bromine	Br_2	3000	4.493	0.056		−0.154
Carbon monoxide	CO	2500	3.376	0.557		−0.031
Carbon dioxide	CO_2	2000	5.457	1.045		−1.157
Carbon disulfide	CS_2	1800	6.311	0.805		−0.906
Chlorine	Cl_2	3000	4.442	0.089		0.344
Hydrogen	H_2	3000	3.249	0.422		0.083
Hydrogen sulfide	H_2S	2300	3.931	1.490		−0.232
Hydrogen chloride	HCl	2000	3.156	0.623		0.151
Hydrogen cyanide	HCN	2500	4.736	1.359		0.725
Nitrogen	N_2	2000	3.280	0.593		0.040
Nitrous oxide	N_2O	2000	5.328	1.214		−0.928
Nitric oxide	NO	2000	3.387	0.629		0.014
Nitrogen dioxide	NO_2	2000	4.982	1.195		−0.792
Oxygen	O_2	2000	3.639	0.506		−0.227
Sulfur dioxide	SO_2	2000	5.699	0.801		−1.015
Water	H_2O	2000	3.470	1.450		0.121

Sources: Taken from Smith, J. M., H. C. van Ness, and M. M. Abbott, *Introduction to Chemical Engineering Thermodynamics*, ed. 5. New York: McGraw-Hill (1996) based on H.M. Spencer, *Ind. Eng. Chem.* 40: 2152–2154 (1948), K.K. Kelly, *U.S. Bur. Mines Bull. 584* (1960) and L.B. Pankratz, *U.S. Bur. Mines Bull. 672* (1982).

APPENDIX B

EQUILIBRIUM WITH OTHER RESTRAINTS, OTHER APPROACHES TO EQUILIBRIUM

The discussion in Chapter 4 deals with equilibria in which the system was constrained to remain at one temperature and pressure. These were externally specified by placing the system in a constant temperature bath and subjecting it to some fixed external pressure through a moveable piston (see Figure 4.1). However, it is interesting to consider some other systems of restraint. Consider, for example, a mixture of hydrogen and oxygen placed in a perfectly insulated, perfectly rigid container. If we choose the mixture as our system, then any process it undergoes must be an adiabatic, constant-volume one. This system is not at equilibrium (except a metastable equilibrium) because it can undergo a spontaneous chemical reaction. If that reaction is started by introducing an infinitesimal electric spark, what will the situation be when equilibrium is reached?

When equilibrium is reached, no further spontaneous process can occur in the system. Considering the three criteria of equilibrium as stated in Section 3.1, we see first that the temperature must be uniform throughout the system so that there can be no spontaneous heat flow. In the cases we have previously considered, the temperature was externally imposed on the system; here it is a variable that the system will ultimately find for itself. However, when the system has found its final temperature, the temperature will be uniform throughout the system. By entirely similar arguments, the final equilibrium pressure must be the same throughout the system unless there are counteracting forces like gravity, surface, or tensile forces. For this system there is no possibility of electrical potential differences but, in general, the criteria about equality of electrical potentials or their balancing by other potentials or by perfect resistors must apply, regardless of the direction from which we approach equilibrium.

Consider the problem of phase equilibrium as shown in Figure 3.5. For the hydrogen-oxygen system described above, we probably have only one phase, but we can consider other systems in which more phases are present in which we have replaced the constant temperature and pressure constraints of Figure 3.5 with adiabatic, constant-volume restrains. The analogous situation would be Figure 3.5 with the piston firmly fixed in place and the walls of the container changed to perfect insulators.

We first choose as our system the entire contents of the container. The first law statement for this system reduces to

$$dU = 0 \tag{B.1}$$

because there can be no flow in or out, no heat transfer, and no work. The second law statement at equilibrium reduces to

$$dS = 0 \quad \text{[at equilibrium]} \tag{B.2}$$

just as it did in Chapter 3. There is a difference between Eqs. B.1 and B.2: Equation B.1 applies to the whole process of going from the starting state to the equilibrium state, whereas Eq. B.2 applies only at the equilibrium state. The process of going from the starting state to equilibrium state is a spontaneous process for which

$$dS \geq 0 \quad \text{[for the process of coming to equilibrium]} \tag{B.3}$$

Physical and Chemical Equilibrium for Chemical Engineers, Second Edition. Noel de Nevers.

But at the equilibrium state, Eq. B.3 becomes Eq. B.2. The constant-volume restriction is that

$$dV = 0 \tag{B.4}$$

Now, if we write the total differential of the energy of each of the two phases for a small transfer of material by diffusion from phase 1 to phase 2, taking V, S, and the n_i as the independent variables, we find

$$dU^{(1)} = \left(\frac{\partial U}{\partial V}\right)^{(1)} dV^{(1)} + \left(\frac{\partial U}{\partial S}\right)^{(1)} dS^{(1)} + \left(\frac{\partial U}{\partial n_a}\right)^{(1)} dn_a^{(1)} + \left(\frac{\partial U}{\partial n_b}\right)^{(1)} dn_b^{(1)} + \cdots \tag{B.5}$$

$$dU^{(2)} = \left(\frac{\partial U}{\partial V}\right)^{(2)} dV^{(2)} + \left(\frac{\partial U}{\partial S}\right)^{(2)} dS^{(2)} + \left(\frac{\partial U}{\partial n_a}\right)^{(2)} dn_a^{(2)} + \left(\frac{\partial U}{\partial n_b}\right)^{(2)} dn_b^{(2)} + \cdots \tag{B.6}$$

We note as we did in Section 4.2 that the material balance allows us to replace $dn_a^{(2)}$ and $dn_b^{(2)}$ with minus $dn_a^{(1)}$ and $dn_b^{(1)}$. We can also note that $(\partial U/\partial V)_{S,n_a,\text{ etc.}} = -P$, and $(\partial U/\partial S)_{V,n_a,\text{ etc.}} = T$; making these substitutions in Eqs. B.5 and B.6, noting that at equilibrium $P^{(1)} = P^{(2)}$ and $T^{(1)} = T^{(2)}$ we add these two equations and find

$$\begin{aligned} dU = 0 &= dU^{(1)} + dU^{(2)} \\ &= -P(dV^{(1)} + dV^{(2)}) + T(dS^{(1)} + dS^{(2)}) \\ &\quad + \left[\left(\frac{\partial U}{\partial n_a}\right)^{(1)} - \left(\frac{\partial U}{\partial n_a}\right)^{(2)}\right] dn_a^{(1)} \\ &\quad + \left[\left(\frac{\partial U}{\partial n_b}\right)^{(1)} - \left(\frac{\partial U}{\partial n_b}\right)^{(2)}\right] dn_b^{(1)} + \cdots \end{aligned} \tag{B.7}$$

Since we know that dU, dV, and dS are all zero for this change, we can see that this equation can only be satisfied for

$$\left(\frac{\partial U}{\partial n_a}\right)^{(1)} = \left(\frac{\partial U}{\partial n_a}\right)^{(2)} \tag{B.8}$$

$$\left(\frac{\partial U}{\partial n_b}\right)^{(1)} = \left(\frac{\partial U}{\partial n_b}\right)^{(2)}, \text{ etc.} \tag{B.9}$$

Equations B.8 and B.9 are obviously similar to Eqs. 4.15 and 4.16. However, here we have $(\partial U/\partial n_a)_{S,V,n_b,\text{etc.}}$, whereas before we had $(\partial G/\partial n_a)_{T,P,n_b,\text{etc.}}$. What is the relation between these sets of quantities?

We know in general that

$$dG = dU + d(PV) - d(TS) \tag{B.10}$$

If we now write out the values of dU from Eq. B.5 (substituting T and P where needed) and expand the two derivatives on the right, we find

$$\begin{aligned} dG = -P\,dV + T\,dS &+ \left(\frac{\partial U}{\partial n_a}\right)_{S,V,n_b,\text{etc.}} dn_a \\ &+ \left(\frac{\partial U}{\partial n_b}\right)_{S,V,n_a,\text{etc.}} dn_b \\ &+ P\,dV + V\,dP - S\,dT - T\,dS \end{aligned} \tag{B.11}$$

Four of the terms of the right cancel each other. Dropping those and dividing by dn_a while holding P, T, and n_b constant (which makes three more terms on the right zero) leads to

$$\left(\frac{\partial G}{\partial n_a}\right)_{T,P,n_b} = \left(\frac{\partial U}{\partial n_a}\right)_{S,V,n_b} \tag{B.12}$$

which indicates that the criterion for equilibrium in this system at constant volume and entropy is exactly the same as the criterion for equilibrium at constant temperature and pressure, namely equality of the chemical potentials between phases, and, by analogy, equality of the chemical potentials for any possible chemical reaction. If we repeat the same kind of derivation for an isothermal constant-volume process and an adiabatic constant-pressure process, we will find that Eq. B.12 becomes

$$\left(\frac{\partial G}{\partial n_a}\right)_{T,P,n_b} = \left(\frac{\partial U}{\partial n_a}\right)_{S,V,n_b} = \left(\frac{\partial H}{\partial n_a}\right)_{S,P,n_b} = \left(\frac{\partial A}{\partial n_a}\right)_{T,V,n_b} \tag{B.13}$$

where H is the enthalpy and A is the Helmholz energy (U–TS). Thus, for any combination of conditions of restraint, the criteria of equilibrium are the same, namely equality of the chemical potentials between phases and no change of Gibbs energy for a differential chemical reaction.

At first this may appear startling, but on reflection it is obvious. For the various kinds of restraints, the direction of approach to equilibrium will be different; but when we arrive at an equilibrium state, that state will be the same whether the state was arrived at along a constant pressure path or a constant volume path, etc. Climbers have reached the top of Mt. Everest from the west, the east, and the north. The summit is the same, regardless of the route taken to reach it.

Thus, we may conclude that for interphase equilibrium under any condition of restraints, each species will have the same chemical potential in each phase, and the potentials will be balanced for any possible chemical reaction. We can also see from Eq. B.13 that the chemical potential is equal to any of the four derivatives shown in that equation, since they are all the same.

APPENDIX C

THE MATHEMATICS OF FUGACITY, IDEAL SOLUTIONS, ACTIVITY AND ACTIVITY COEFFICIENTS

The derivation of the various relations involving fugacity, ideal solutions, activity, and activity coefficients covers several pages with mathematics. Professors and graduate students enjoy that, but most undergraduates do not. For that reason most of that mathematics is placed in this appendix, so that the discussion in Chapter 7 can flow more easily. The pertinent results of that mathematics are transferred from here to Chapter 7.

C.1 THE FUGACITY OF PURE SUBSTANCES

For a pure substance Eq. 7.1 reduces to

$$g = RT \ln f + g^{\circ}(T) \tag{7.5}$$

If we now take the total differential of Eq. 7.5 and substitute the total differential of the Gibbs energy of a pure substance (Eq. 4.30) we find

$$dg = -s dT + v dP = R(\ln f \, dT + T d \ln f) + dg^{\circ} \tag{C.1}$$

Now if we restrict ourselves for the moment to constant temperatures (for which $dg^{\circ}(T)_T = 0$) we can rearrange Eq. C.1 to

$$\left(\frac{\partial \ln f}{\partial P}\right)_T = \frac{v}{RT} \tag{7.6}$$

For a ideal gas, we know that

$$\left(\frac{v}{RT}\right) = \frac{1}{P} = \left(\frac{\partial \ln f}{\partial P}\right)_T \qquad \text{(C.2)}^*$$

(Here we use the asterisk (*) to indicate that this equation is limited to ideal gases. In the remainder of this appendix we will use the asterisk (*) on a symbol to denote a ideal gas property; e.g., h^* is the enthalpy of a ideal gas at the T and P in question.) Thus, for an ideal gas

$$\left(\frac{\partial \ln f}{\partial P}\right)_T = \left(\frac{\partial \ln P}{\partial P}\right)_T \qquad \text{(C.3)}^*$$

Multiplying through by dP and integrating we find

$$\ln f = \ln P + \text{some constant of integration, which may depend on } T \qquad \text{(C.4)}^*$$

This constant is not defined so far, and we could really make many choices of its value. However, the most sensible choice is zero, so that for ideal gases we have

$$f = P \qquad \text{(C.5)}^*$$

For any real gas, as P approaches zero the behavior of real gases approaches ideal gas behavior, so we can say that for any material in the gas state

$$\lim_{P \to 0} \left(\frac{f}{P}\right) = 1 \tag{7.7}$$

Equation 7.6 shows the most useful derivative of the fugacity. The other one regularly seen is $(\partial \ln f / \partial T)_P$. In principle, we could begin with Eq. C.1 and divide by dT, holding P constant, but that approach leads to a mathematical mess because the derivative of g° appears. Therefore, the

Physical and Chemical Equilibrium for Chemical Engineers, Second Edition. Noel de Nevers.

more practical approach is to write out the value of g from Eq. 7.5 twice, once for some general state and once for the same temperature, but a pressure so low that the material behaves as a ideal gas. We subtract the second equation from the first and see that the two g° terms (which depend on T only) cancel and

$$g - g^* = RT \ln f - RT \ln f^* = RT \ln \frac{f}{f^*} \tag{C.6}$$

Now we differentiate Eq. C.6 with respect to T at constant P, finding

$$\left(\frac{\partial g}{\partial T}\right)_P - \left(\frac{\partial g^*}{\partial T}\right)_P = R \ln \frac{f}{f^*} + RT\left(\frac{\partial \ln f}{\partial T}\right)_P - RT\left(\frac{\partial \ln f^*}{\partial T}\right)_P \tag{C.7}$$

But for a ideal gas $f^* = P$, so the last term in Eq. C.7 is

$$(\partial \ln f^*)_P = (\partial \ln P)_P = 0 \tag{C.8}$$

and it may be dropped. Then we note that

$$R \ln \frac{f}{f^*} = \frac{(g - g^*)}{T} = \frac{h}{T} - \frac{h^*}{T} - s - s^* \tag{C.9}$$

and

$$\left(\frac{\partial g}{\partial T}\right)_P = -s, \quad \left(\frac{\partial g^*}{\partial T}\right)_P = -s^* \tag{C.10}$$

Substituting Eqs. C.9 and C.10 into Eq. C.7 we find

$$-s + s^* = \frac{(h - h^*)}{T} - s + s^* + RT\left(\frac{\partial \ln f}{\partial T}\right)_P \tag{C.11}$$

or finally

$$\left(\frac{\partial \ln f}{\partial T}\right)_P = \frac{(h^* - h)}{RT^2} \tag{7.8}$$

To make up $f/P = \phi$ charts like Figure 7.1 and Appendix A.5, we define a new convenience property called the "volume residual":

$$\alpha = \text{volume residual} = \frac{RT}{P} - v = \frac{RT}{P}(1 - z) \tag{C.12}$$

Solving this for v we find

$$v = \frac{RT}{P} - \alpha \tag{C.13}$$

Now we substitute this value of v in Eq. 7.6 and find

$$\left(\frac{\partial \ln f}{\partial P}\right)_T = \frac{1}{P} - \frac{\alpha}{RT} = \frac{1}{P} - \frac{(1 - z)}{P} \tag{C.14}$$

Multiplying through by dP and noting that $dP/P = d\ln P$, we get

$$(\partial \ln f)_T = (\partial \ln P)_T - \alpha \frac{dP}{RT} = (\partial \ln P)_T - \frac{(1 - z)dP}{P} \tag{C.15}$$

Moving the $\partial \ln P$ term to the left, reversing terms to eliminate a minus sign, and combining terms, we have

$$\left(\partial \ln \frac{f}{P}\right)_T = -\frac{\alpha dP}{RT} = \frac{(z - 1)dP}{P} \tag{C.16}$$

Integrating this between $P = 0$ and $P = P$, we find

$$\left[\ln \frac{f}{P}\right]_0^P = -\frac{1}{RT}\int_0^P \alpha\, dP = \int_0^P \frac{(z - 1)dP}{P} \tag{C.17}$$

At the lower limit of the term on the left we have $f/P = 1$ because any material with a finite vapor pressure is practically a idea gas at $P = 0$. Hence, the left term is simply $\ln f/P$. Taking exponentials of both sides we have, finally

$$\frac{f}{P} = \phi = \exp \frac{-1}{RT}\int_0^P \alpha dP = \exp \int_0^P \frac{(z - 1)dP}{P} \tag{7.9}$$

which is the formula that is actually used to make up the f/P charts like Figure 7.1. Note that the integral is evaluated at a constant temperature, the temperature of the state for which we wish to know f/P. We show two forms in Eq. 7.9 (and its mixture analog 7.17) because the first integral on the right is most convenient if we are working from a table of PvT data, while the second is most convenient if we are working from an EOS.

Equations 7.5 to 7.9 are the whole story on the fugacity of pure substances. They are derived here and copied into the main text.

C.2 FUGACITIES OF COMPONENTS OF MIXTURES

Now let us return to the more interesting case of mixtures. We need the equivalents of Eqs. 7.5 to 7.9 for individual components of mixtures.

We already have the equivalent of Eq. 7.5 in Eq. 7.1, which defines the fugacity. We can easily obtain the analog of Eq. 7.6 by differentiating Eq. 7.5 at constant T and noting that

$$d\bar{g}_i = -\bar{s}_i dT + \bar{v}_i dP \tag{C.18}$$

so

$$(\partial \bar{g}_i)_T = \bar{v}_i \partial P_T = (RT\, \partial \ln f_i)_T \tag{C.19}$$

from which it follows that for mixtures

$$\left(\frac{\partial \ln f_i}{\partial P}\right)_T = \frac{\bar{v}_i}{RT} \tag{7.14}$$

To find the analog of Eq. 7.7 we multiply both sides of Eq. 7.14 by P:

$$\left(\frac{\partial \ln f_i}{\partial \ln P}\right)_T = \frac{P\bar{v}_i}{RT} \tag{C.20}$$

We wish to find the relation between f_i and P as $P \to 0$. In that case the material must behave as a ideal gas, so we can use the ideal gas value of $\bar{v}_i$.

$$\bar{v}_i^* = \left(\frac{\partial V^*}{\partial n_i}\right)_{T,P,nj} = \left[\frac{\partial (n_T \bar{v}^*)}{\partial n_i}\right]_{T,P,nj} = \frac{RT}{P} \tag{C.21*}$$

Thus,

$$\lim_{P\to 0}\left(\frac{\partial \ln f_i}{\partial \ln P}\right)_T = \frac{P\bar{v}^*}{RT} = 1 \tag{C.22}$$

Separating variables and integrating we have

$$\lim_{P\to 0} \ln f_i = \ln P + \ln(\text{constant of integration}) \tag{C.23}$$

or

$$\lim_{P\to 0} f_i = P \cdot (\text{constant of integration}) \tag{C.24}$$

What value ought we assign to this constant of integration? We know that the constant must be dimensionless and that to make Eq. C.24 agree with Eq. 7.5 for the case of a pure component, the constant must be 1.0 for a pure component. Similarly, we know that if the concentration of component i is 0, then its fugacity must be 0 so that as the concentration of a component approaches 0 this constant of integration must approach 0. The logical choice to make here is the mol fraction, which is dimensionless and has both of the above properties. If we make this choice, it will conveniently lead to the properties of ideal solutions, which we discuss below and which will be consistent with the previous definitions. Recognize, however, that we could have made some other choice, which would have made everything that follows different and much more complicated and difficult.

Thus, we may write the analog of Eq. 7.7 as

$$\lim_{P\to 0}\left(\frac{f_i}{Px_i}\right) = \lim_{P\to 0} \hat{\phi}_i = 1 \tag{7.15}$$

To find the mixture equivalent of Eq. 7.8, we repeat derivation from Eqs. C.7 to C.11, replacing all of the gs with $\bar{g}_i$s and all of the fs with f_is. This is the equivalent of substituting Eq. 7.1 for Eq. 7.5. If we then follow the derivation straight through to Eq. 7.8, we see that the result is

$$\left(\frac{\partial \ln f_i}{\partial T}\right)_P = \frac{(\bar{h}_i^* - \bar{h}_i)}{RT^2} \tag{C.25}$$

However, as will be shown later, for an ideal gas $\bar{h}_i^*$ is the same as h_i^* because for ideal gases partial molar enthalpies are the same as pure component enthalpies. Thus, the final analog of Eq. 7.8 is

$$\left(\frac{\partial \ln f_i}{\partial T}\right)_P = \frac{(h_i^* - \bar{h}_i)}{RT^2} \tag{7.16}$$

To find the analog of Eq. 7.9 we use basically the same procedure we used to get Eq. 7.9. First we define

$$\begin{aligned}\bar{\alpha}_i &= \frac{\partial}{\partial n_{i\,T,\,P,\,nj}}\left(\frac{nRT}{P} - V\right) = \frac{RT}{P} - \bar{v}_i \\ &= \frac{\partial}{\partial n_{i\,T,\,P,\,nj}}\left(\frac{nRT}{P} - z\frac{nRT}{P}\right) = \frac{RT}{P}(1 - \bar{z}_i)\end{aligned} \tag{C.26}$$

This is a rare example of a partial molar derivative of an intensive property $\bar{z}_i$. However, we see that it is the logical result of representing an extensive property V by the number of mols and a set of intensive properties. This causes no difficulties, and the resulting equation is widely used.

Then we substitute from Eq. C.26 in Eq. C.20:

$$\begin{aligned}\left(\frac{\partial \ln f_i}{\partial P}\right)_T &= \frac{1}{RT}\left(\frac{RT}{P} - \bar{\alpha}_i\right) = \frac{1}{RT}\left[\frac{RT}{P} - \frac{RT}{P}(1 - \bar{z}_i)\right] \\ &= \frac{1}{P} - \frac{1}{P}(1 - \bar{z}_i)\end{aligned} \tag{C.27}$$

Multiplying by ∂P and rearranging gives

$$(\partial \ln f_i)_T = \frac{\partial P}{P} - \frac{\bar{\alpha}_i}{RT}\partial P = \frac{\partial P}{P} - \frac{(1 - \bar{z}_i)\partial P}{P} \tag{C.28}$$

Noting that $(\partial P)/P$ is $(\partial \ln P)_T$, we move it to the left, rearrange terms to eliminate a minus sign, and combine terms

$$\left(\partial \ln \frac{f_i}{P}\right)_T = -\frac{\bar{\alpha}_i}{RT}\partial P_T = \frac{\bar{z}_i - 1}{P}\partial P_T \tag{C.29}$$

and integrate

$$\left[\ln\frac{f_i}{P}\right]_{P=0}^{P=P} = \frac{-1}{RT}\int_{P=0}^{P=P}\bar{\alpha}_i dP_T = \int_{P=0}^{P=P}\frac{\bar{z}_i - 1}{P}dP_T \quad \text{(C.30)}$$

In deriving Eq. 7.9 for a pure component we showed that the lower limit on the left was ln 1 = 0. Here, from Eq. 7.14, we see it is $\ln(f_i/P) = \ln x_i$, so

$$\ln\frac{f_i}{P} - \ln x_i = \ln\frac{f_i}{Px_i} = -\frac{1}{RT}\int_{P=0}^{P=P}\bar{\alpha}_i dP_T = \int_{P=0}^{P=P}\frac{\bar{z}_i - 1}{P}dP_T \quad \text{(7.17)}$$

which is the mixture analog of Eq. 7.9.

The equations for calculating the fugacity of any component in a mixture and its derivatives are Eqs. 7.1 and 7.14 to 7.17, derived here and copied to the main text.

C.3 THE CONSEQUENCES OF THE IDEAL SOLUTION DEFINITION

See Sections 7.7 and 7.8 for an introduction to ideal solutions. If we substitute Eq. 7.21, which defines an ideal solution, into the definition of the fugacity, we find

$$\begin{aligned}\bar{g}_i &= RT\ln f_i + g_i^\circ(T) = RT\ln f_i^\circ x_i + g_i^\circ(T)\\ &= RT(\ln f_i^\circ + \ln x_i) = g_i^\circ(T)\end{aligned} \quad \text{(C.31)}^\dagger$$

(Here we put an † on an equation to show that it applies only to ideal solutions.) Now let us write this equation twice, once as shown above and once for pure component i at the same temperature and pressure. In this case it becomes

$$g_i^\circ = RT(\ln f_i^\circ) + g_i^\circ(T) \quad \text{(C.32)}^\dagger$$

Here the $\ln x_i$ term has dropped out, because $x_i = 1$ and $\ln 1 = 0$. The f_i° term in Eq. C.32 is the same as in Eq. C.31, because both equations are written for the same component at the same temperature and pressure. Subtracting Eq. C.32 from Eq. C.31 we find

$$\bar{g}_i - g_i^\circ = RT\ln x_i \quad \text{(7.23)}^\dagger$$

Now we differentiate both sides of Eq. 7.23 with respect to P at constant T and constant x_i, finding

$$\left(\frac{\partial \bar{g}_i}{\partial P}\right)_{T,\,x_i} - \left(\frac{\partial g_i^\circ}{\partial P}\right)_{T,\,x_i} = 0 \quad \text{(C.33)}^\dagger$$

But we know that these two derivatives can be written as v, and v_i°, so it follows that for an ideal solution

$$\bar{v}_i - v_i^\circ = 0 \quad \text{(7.24)}^\dagger$$

This relation is an inescapable consequence of the definition of an ideal solution and depends on no further assumptions.

Returning to Eq. 7.23 we can now differentiate it with respect to temperature at constant pressure and x_i, finding

$$\left(\frac{\partial \bar{g}_i}{\partial T}\right)_{P,\,x_i} - \left(\frac{\partial g_i^\circ}{\partial T}\right)_{P,\,x_i} = R\ln x_i \quad \text{(C.34)}^\dagger$$

But we know the values of these two derivatives, so we may substitute them, finding

$$\bar{s}_i - s_i^\circ = RT\ln x_i \quad \text{(7.25)}^\dagger$$

Now we return to Eq. 7.23 and expand it to

$$\bar{g}_i - g_i^\circ = -RT\ln x_i = \bar{h}_i - h_i^\circ - T(\bar{s}_i - s_i^\circ) \quad \text{(C.35)}^\dagger$$

Now we substitute Eq. 7.25 into Eq. C.35 and simplify, finding

$$\bar{h}_i - h_i^\circ = 0 \quad \text{(7.26)}^\dagger$$

This is also an inescapable consequence of the definition of an ideal solution with no further assumptions.

Thus, ideal solutions are defined by Eq. 7.22; their important properties, derived here and transferred to Chapter 7, are Eqs. 7.23 to 7.26.

C.4 THE MATHEMATICS OF ACTIVITY COEFFICIENTS

See Section 7.10, where activity and activity coefficients are defined. If we take the logarithm of both sides of Eq. 7.27 we find

$$\ln\gamma_i = \ln f_i - \ln f_i^\circ - \ln x_i \quad \text{(C.36)}$$

Now we differentiate this with respect to P at constant T and constant x_i, and find

$$\left(\frac{\partial \ln\gamma_i}{\partial P}\right)_{T,\,x_i} = \left(\frac{\partial \ln f_i}{\partial P}\right)_{T,\,x_i} - \left(\frac{\partial \ln f_i^\circ}{\partial P}\right)_{T,\,x_i} - 0 \quad \text{(C.37)}$$

Substituting the known values of these derivatives we obtain

$$\left(\frac{\partial \ln\gamma_i}{\partial P}\right)_{T,\,x_i} = \frac{\bar{v}_i - v_i^\circ}{RT} \quad \text{(7.31)}$$

Similarly, if we differentiate Eq. C.36 with respect to T at constant P and x_i, we find

$$\left(\frac{\partial \ln \gamma_i}{\partial T}\right)_{P,\,x_i} = \left(\frac{\partial \ln f_i}{\partial T}\right)_{P,\,x_i} - \left(\frac{\partial \ln f_1^\circ}{\partial P}\right)_{T,\,x_i} - 0 \tag{C.38}$$

Substituting the known values of the derivatives in Eq. C.38 we find

$$\left(\frac{\partial \ln \gamma_i}{\partial T}\right)_{P,\,x_i} = \frac{(\bar{h}_i^* - \bar{h}_i)}{RT^2} - \frac{(h_i^{\circ *} - h_i^\circ)}{RT^2} \tag{C.39}$$

But for ideal gases

$$\bar{h}_i^* = (h_i^\circ)^* = h^* \tag{C.40}$$

so we may simplify Eq. C.39 to

$$\left(\frac{\partial \ln \gamma_i}{\partial P}\right)_{P,\,x_i} = \frac{(h_i^\circ - \bar{h}_i)}{RT^2} \tag{7.32}$$

APPENDIX D

EQUATIONS OF STATE FOR LIQUIDS AND SOLIDS WELL BELOW THEIR CRITICAL TEMPERATURES

EOSs for gases begin with the ideal gas law and add terms to correct for nonideal behavior. At low P and high T (relative to the critical values) all EOSs for gases approach the ideal gas law. With the ideal gas law, we need only know the molecular weight of the gas to make quite useful estimates of its density. The same is not true for liquids or solids. There is no corresponding "ideal" EOS for liquids or solids. We have useful theories and correlations, but we do not have any simple EOS for liquids and solids that would allow us to start with the molecular weight, or even with a structural formula, and write out comparably useful estimates of the density. We can surely say that almost all hydrocarbon liquids and solids have densities in the range of 0.7 to 1.0 g/cm^3 at 20°C and make similar statements for minerals of various kinds, but these estimates have a high % uncertainty. Thus, the working equations for the density of liquids and solids begin with some measured value of the density at some reference state, most often 20°C or 25°C and 1 atm, and use the approach shown below to estimate the changes in density from that reference state.

D.1 THE TAYLOR SERIES EOS AND ITS SHORT FORM

For any material (liquid, solid, or gas), the density can be written as a Taylor series:

$$\rho = \rho_0 + \frac{\partial \rho}{\partial T}(T-T_0) + \frac{\partial \rho}{\partial P}(P-P_0) + \frac{1}{2!}\left[\frac{\partial^2 \rho}{\partial T^2}(T-T_0)^2 + 2\frac{\partial^2 \rho}{\partial P \partial T}(P-P_0)(T-T_0) + \frac{\partial^2 \rho}{\partial P^2}(P-P_0)^2\right] + \frac{1}{3!}\left[\frac{\partial^3 \rho}{\partial T^3}(T-T_0)^3 + 3\frac{\partial^3 \rho}{\partial P^2 \partial T}(P-P_0)^2(T-T_0) + \cdots\right] \quad \text{(D.1)}$$

where ρ_0, P_0, and T_0 are the density, pressure, and temperature at a suitable reference state and the derivatives are all taken at that state. This EOS is correct for all conditions, if we use an infinite series of terms. For gases we normally do not use this form; instead, we use those shown in Chapter 2. For liquids near the critical temperature, and in the preparation of tables like the steam tables, we use one EOS for both liquid and gas. However, for liquids at temperatures well below the critical temperature (say, 100°F below the critical), and also for solids, we may generally neglect all but the first three terms on the right and write the equation as

$$\rho \approx \rho_0\left[1 + \frac{1}{\rho_0}\left(\frac{\partial \rho}{\partial T}\right)(T-T_0) + \frac{1}{\rho_0}\left(\frac{\partial \rho}{\partial P}\right)(P-P_0)\right] \approx \rho_0[1 + \alpha(T-T_0) + \beta(P-P_0)] \quad \text{(D.2)}$$

where

$$\alpha = -\frac{1}{\rho_0}\left(\frac{\partial \rho}{\partial T}\right) = \text{coefficient of volume thermal expansion} \tag{D.3}$$

with dimension of (1/*T*), and

$$\beta = \frac{1}{\rho_0}\left(\frac{\partial \rho}{\partial P}\right) = \text{isothermal compressibility} = 1/\text{bulk modulus} \tag{D.4}$$

with dimension (1/*P*).

For solids, handbooks most often present the coefficient of linear thermal expansion. If we write the expression for the volume of any regular solid, take the logarithms, and differentiate, we find

$$\frac{dV}{V} = 3\frac{dL}{L} \tag{D.5}$$

Thus, the coefficient of volume thermal expansion is ≈ three times the linear thermal expansion coefficient. Values of the density, isothermal compressibility, and coefficient of volume thermal expansion for some common fluids at 20°C or 25°C are listed in Table D.1 (plus another column to be explained later).

D.2 EFFECT OF TEMPERATURE ON DENSITY

Figure D.1 shows the experimental values of the changes in density of ethanol with changes of temperature at various pressures. If Eq. D.2 were absolutely correct (instead of being only a *useful approximation*), then the effects of increasing pressure and increasing temperature would be completely independent of each other and the curves for various pressures in Figure D.1 would be parallel straight lines. They are close to being parallel and close to being straight, but are neither exactly parallel nor exactly straight.

Table D.1 Isothermal Compressibilities and Coefficients of Thermal Expansion, for Some Liquids and Solids

Substance	$\frac{\rho_{20°C}}{g/cm^3}$	$\beta_{25°C} \times 10^4$ atm	$\left(\frac{\beta_{1000\ atm}}{\beta_{1\ atm}}\right)_{25°C}$	$\alpha_{20°C} \times 10^3$ °C
LIQUIDS				
Acetic acid	1.049	0.92		1.071
Acetone	0.792	1.26	0.49	1.487
Analine	1.022	0.47	0.69	
Benzene	0.879	0.98	0.52	1.237
Carbon disulfide	1.263			1.218
Carbon tetrachloride	1.595	1.08	0.50	1.236
n-Hexane	0.659	1.69	0.40	
Ethanol	0.789	1.13		1.12
Mercury	13.546	0.04 (20°C)	0.98	
Methanol	0.792	1.23		
Methylene chloride	1.336	0.99	0.55	
n-Octane	0.703	1.40		
Toluene	0.866	0.91 (20°C)		
Water	1.00	0.46	0.76	0.207
m-Xylene	0.861	0.86		
SOLIDS				
Wood	≈0.9 to 1.0			≈0.15
Graphite	2.29	0.0030		
Diamond	3.51	0.00016		0.00354
Glass	≈2.7			0.021 to 0.028
Ice	0.95			0.1125
Rock salt	2.163			0.112
Sulfur	1.96 to 2.07			0.223
Copper	8.92			0.05
Steel	7.85	0.0064		0.011
Aluminum	2.71	0.015		0.022
Limestone	2–7 to 2.9			0.027
Paraffin wax	≈0.9			0.39

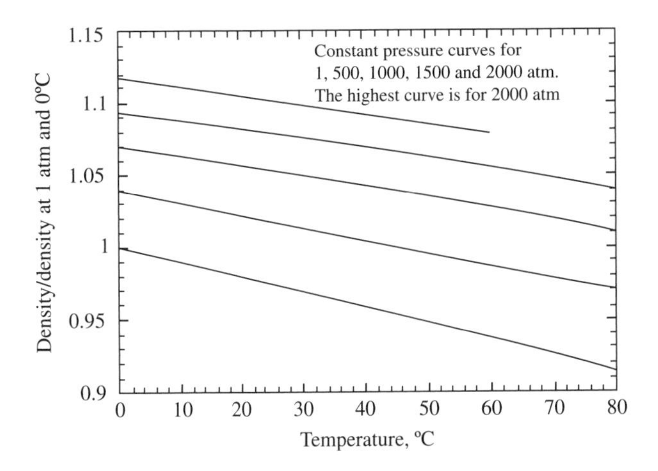

FIGURE D.1 Density–temperature plot for ethanol [1].

Example D.1 Compare the effect of temperature shown in Figure D.1 to Eq. D.2. For the 1-atm curve we can compute the slope of the (almost straight) line from its end values

$$\alpha = -\frac{1}{\rho_0}\left(\frac{\partial \rho}{\partial T}\right) \approx -\frac{1}{\rho_0}\left(\frac{\Delta \rho}{\Delta T}\right) = -\frac{\Delta(\rho/\rho_0)}{\Delta T}$$
$$= -\frac{0.916-1.00}{80°\text{C}} = \frac{0.00105}{°\text{C}} \tag{D.A}$$

This is the average value of α between 0 and 80°C. Table D.1 shows that at 20°C, $\alpha = 0.00112/°\text{C}$, which is in reasonable but far from perfect agreement with the above result. ■

From Figure D.1 we see that the various curves are slightly concave downward, indicating that α increases slightly with increasing temperatures. Many sources [2, p. 2–131] present values of $[-(1/\rho_0)(\partial\rho/\partial T)]$ not as a constant (as in Table D.1) but as a second- or third-order polynomial equation, for example,

$$\rho = \rho_0\left[1 + \alpha(T-T_0) + \beta(T-T_0)^2 + \gamma(T-T_0)^3\right]_{P=\text{const.}} \tag{D.6}$$

where α is minus the α in Eq. D.3, but β and γ are additional terms in this representation of thermal expansion, with dimensions $(1/T^2)$ and $(1/T^3)$.

Comparing Eq. D.6 to Eq. D.1, we see that

$$\beta_{\text{Eq.D.6}} = \left(\frac{1}{2\rho_0} \cdot \frac{\partial^2 \rho}{\partial T^2}\right)_{\text{Eq.D.1}} \tag{D.7}$$

and

$$\gamma_{\text{Eq.D.6}} = \left(\frac{1}{6\rho_0} \cdot \frac{\partial^3 \rho}{\partial T^3}\right)_{\text{Eq.D.1}} \tag{D.8}$$

When data are available in the form of Eq. D.6 they probably lead to better estimates of the effect of temperature change on density than does Eq. D.2.

D.3 EFFECT OF PRESSURE ON DENSITY

Figures D.2 and D.3 show the experimental values of the changes in density of ethanol with changes of pressure at various temperatures.

If Eq. D.2 were absolutely correct (instead of being only a *useful approximation),* then the effects of increasing pressure and temperature would be independent of each other and the curves for various pressures in Figures D.2 and D.3 would be parallel straight lines. They are close to being parallel but show substantial curvature, mostly in the low-pressure range.

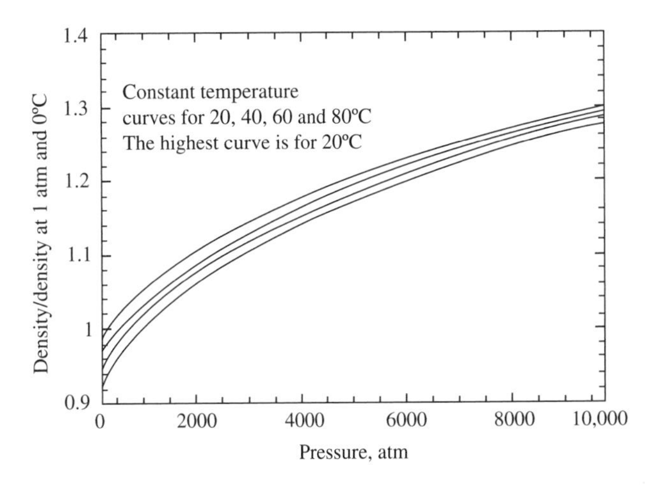

FIGURE D.2 Density–pressure plot for ethanol [1].

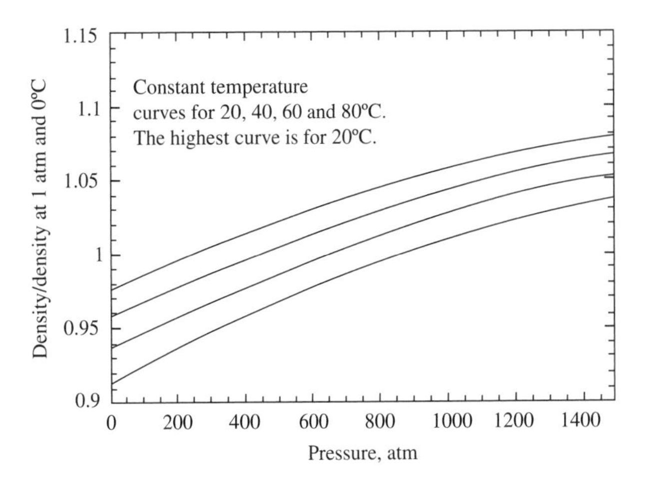

FIGURE D.3 Density–pressure plot for ethanol [1], low-pressure corner.

Example D.2 Compare the effect of pressure change in Figure D.3 to Eq. D.2. As Figure D.2 makes clear, above about 5000 atm the curves are much closer to being straight than below that pressure. Ignoring this for the moment, for the 20°C curve we may estimate from Figure D.3:

$$\beta = \frac{1}{\rho_0}\left(\frac{\partial \rho}{\partial P}\right) \approx \frac{1}{\rho_0}\left(\frac{\Delta \rho}{\Delta P}\right) = -\frac{\Delta(\rho/\rho_0)}{\Delta P} = \frac{1.08-0.98}{1500\text{ atm}} = \frac{0.67\times 10^{-4}}{\text{atm}} \tag{D.B}$$

In Table D.1 this value is reported as 1.13×10^{-4}/atm. Looking back at Figure D.3 we see that if we draw a tangent to the 20°C curve at 1 atm (≈0) we will read its intercept on the right-hand axis as ≈1.16, and compute a value of

$$\beta = \frac{1}{\rho_0}\left(\frac{\partial \rho}{\partial P}\right) \approx \frac{1}{\rho_0}\left(\frac{\Delta \rho}{\Delta P}\right) = -\frac{\Delta(\rho/\rho_0)}{\Delta P} = \frac{1.16-0.98}{1500\text{ atm}} = \frac{1.2\times 10^{-4}}{\text{atm}} \tag{D.C}$$

close to, but not identical to the value in Table D.1 ■

This example (and Figures D.2 and D.3) show that a power series representation of β as a function of P is probably better justified than that for α in Eq. D.6. Common handbooks rarely show this formulation or present the data needed to use it. (In our common experience we encounter changes of hundreds of °C in our kitchens and thousands of °C in our furnaces, but we rarely encounter pressures over a few hundred psia, so this shortage of high-pressure data in handbooks should not surprise us.) The high-pressure physics literature [3, 4] reports several competing EOSs. It also shows that the best first-step improvement over β = constant is

$$\beta = \beta_0 + \left(\frac{d\beta}{dP}\right)_{T,\text{ measured at }P=0} \cdot (P-P_0) \tag{D.9}$$

where

$$\left(\frac{d\beta}{dP}\right)_{T,\text{ measured at }P=0} = \frac{1}{2\rho_0}\left(\frac{\partial^2 \rho}{\partial P^2}\right)_{\text{Eq. D.1}}$$

is always small and negative. This equation shows the density as a parabolic function of P, as do the data shown in Figures D.2 and D.3.

Example D.3 Using the values in Example D.2, estimate the value of $(d\beta/dP)_{T,\text{ measured at }p=0}$ for ethanol at 20°C.

To save writing, we let $(d\beta/dP)_{T,\text{ measured at }p=0} = a$. Then, taking $P_0 \approx 0$, we have

$$\int_{\rho_0}^{\rho} d\rho = \rho_0 \int_0^P (\beta_0 + aP)dP \tag{D.10}$$

$$\rho - \rho_0 = \rho_0 \cdot \left(\beta_0 P + a\frac{P^2}{2}\right) \tag{D.11}$$

$$a = \frac{\frac{\rho}{\rho_0} - 1 - \beta_0 P}{\frac{P^2}{2}} = \frac{\frac{1.08}{0.98} - 1 - \frac{1.2\times 10^{-4}}{\text{atm}}1500\text{ atm}}{\frac{(1500\text{ atm})^2}{2}} = \frac{-6.9\times 10^{-8}}{\text{atm}^2} \quad ■ \tag{D.D}$$

Table D.1 shows the values of $(\beta_{1000\text{ atm}}/\beta_{1\text{ atm}})_{25°\text{C}}$ for many of the substances represented.

Example D.4 Estimate the value of $(\beta_{1000\text{ atm}}/\beta_{1\text{ atm}})_{25°\text{C}}$ for ethanol from the results in the previous example.

From Eq. D.9 we find

$$\frac{\beta_{1000\text{ atm}}}{\beta_0} = 1 + \frac{aP}{\beta_0} = 1 + \frac{\frac{-6.9\times 10^{-8}}{\text{atm}^2}\cdot 1000\text{ atm}}{\frac{1.2\times 10^{-4}}{\text{atm}}} = 1 - 0.575 = 0.425 \quad ■ \tag{D.E}$$

Table D.1 shows that for water this is 0.76 and for most organic compounds it is 0.4 to 0.7. Thus, the value calculated here is plausible.

D.4 SUMMARY

From these examples and an examination of the tables and figures we can see the following:

1. The simple linear formulation in Eq. D.3, with the effects of temperature and pressure independent of each other and of T and P is a reasonable approximation, but should not be used for large changes in T and P. We can make better estimates if we retain more terms in Eq. D.1, at the expense of greater mathematical complexity and greater difficulty in finding published values of the higher derivatives beyond α and β.
2. For most liquids, an increase of 1% in density corresponds to about −10°C or about +150 atm. For solids, the corresponding values are about −200°C or 10^3 to 10^4 atm.

For more complete sets of values that include the effect of increasing temperature on α, see Forsythe [5], Lange [6], Reid et al. [7, Chapter 3], and Lide [8, p. 6–94]. The behavior of liquids near their critical states can be best estimated from an appropriate EOS or Appendix A.4.

REFERENCES

1. Lindsay, R. Density and compressibility of liquids. In *American Institute of Physics Handbook*, Gray, D. E., ed. New York: McGraw-Hill (1957).
2. Liley, P. E., G. H. Thomson, D. G. Friend, T. E. Daubert, and E. Buck. Physical and chemical data. In *Perry's Chemical Engineers' Handbook*, ed. 7, Perry, R. H., D. W. Green, and J. O. Maloney, eds. New York: McGraw-Hill, pp. 2–54 (1997).
3. Anderson, O. L. The use of ultrasonic measurements under modest pressure to estimate compression at high pressure. *J. Phys. Chem. Solids* 27:547–565 (1966).
4. Macdonald, J. R. Some simple isothermal equations of state. *Rev. Mod. Phys.* 38:669–697 (1966).
5. Forsythe, W. E. *Smithsonian Physical Tables*, ed. 9. Washington, DC: Smithsonian Institution (1954).
6. Lange, N. A. *Handbook of Chemistry*. New York: McGraw-Hill, p. 1682 (1982).
7. Reid, R. C., J. M. Prausnitz, and T. K. Sherwood. *The Properties of Liquids and Gases*, ed. 3. New York: McGraw-Hill, Appendix A (1977).
8. Lide, D. R. *CRC Handbook of Chemistry and Physics*, ed. 71. Boca Raton, FL: CRC Press, pp. 6–118 (1990).

APPENDIX E

GIBBS ENERGY OF FORMATION VALUES

In Chapters 12 and 13 we used the Gibbs energy and enthalpy of formation values in Table A.8, without explaining how those values were determined. This appendix discusses that.

E.1 VALUES "FROM THE ELEMENTS"

As discussed in Section 12.6, any such compilation can only show the values of *changes* from one state to another. All energy values are relative to some arbitrary datum (see Section 2.7). For chemical reactions the simplest, most satisfactory, and almost universally used convention is to assign values of

$$(h_{\text{elements}} = g_{\text{elements}} = 0)_{\substack{\text{at the standard state} \\ T \text{ and } P\text{, in the normal} \\ \text{phase at that condition}}} \tag{E.1}$$

at some appropriate standard state, normally 298.15 K and 1.00 bar. Then all values for compounds are the values of the changes in h and g in going from the elements to that compound, at the standard state. This requires us to say which form of the element is the standard one, e.g. graphite (pure carbon), not diamond (also pure carbon) for which $h_{\text{elemental carbon as diamond}} = 1.9$ kJ/mol and $g_{\text{elemental carbon as diamond}} = 2.9$ kJ/mol at the standard state. Since no element can be made from another (by chemical means), we can safely apply Eq. E.1 separately to each element, without any chance of difficulty.

E.2 CHANGES IN ENTHALPY, ENTROPY, AND GIBBS ENERGY

We have no way of determining changes in Gibbs energy directly. Instead, we compute them from

$$(\Delta g = \Delta h - T\Delta s)_{T=\text{constant}} \tag{E.2}$$

finding the values of Δh and Δs separately.

E.2.1 Enthalpy Changes

We can directly measure changes in enthalpy, in a calorimeter (see Section 2.9). For example, if we place one mol of pure H_2 in a calorimeter with a surplus of O_2, and initiate the oxidation reaction with an infinitesimal spark, the temperature of the gas will rise, explosively. If we then cool the mixture to its starting temperature by letting it transfer heat to a large mass of water, we can measure the temperature increase of the water, from which we can compute the amount of heat released by this reaction. The measured quantity is the molar *heat of combustion* of H_2. It is also the negative of the heat of formation (enthalpy change of formation from the elements) of H_2O, which is listed in Table A.8 as −285.8 kJ/mol at 25°C.

We can use simple calorimetry, starting from the elements for many simple compounds, but not for complex ones. For example, we cannot begin with C and Cl_2 and produce pure CCl_4 in a calorimeter, as we did with H_2O. However, if we have some reaction of the form $A + B \rightarrow C + D$, for which we can experimentally measure the enthalpy change and for which we know the enthalpies of formation from the

Physical and Chemical Equilibrium for Chemical Engineers, Second Edition. Noel de Nevers.

elements of all but one of the four species listed, then we can compute the enthalpy of formation of the one unknown species from that measured enthalpy change of the reaction and our table of already-known enthalpies of formation. In this way we can build up a table of enthalpies of formation, one compound at a time, until we have a fairly complete table. The enthalpy of formation values in Table A.8 are a combination of some measured directly, like that for H_2O, and many measured indirectly as shown here.

E.2.2 Entropy Changes

There is no correspondingly direct way of measuring the entropy changes of chemical reactions. There is no experimental device like a calorimeter that provides such information. However, if for some reaction A + B → C + D we can measure the equilibrium constant (by measuring the concentrations of all four species at equilibrium), then from Eq. 12.15 we can compute Δg° for the reaction, at some known temperature T. If we have the enthalpy data to compute Δh° for the reaction, then we can compute

$$\Delta s^\circ = \frac{\Delta h^\circ - \Delta g^\circ}{T} \tag{E.3}$$

If this were a reaction involving the elements and only one compound, then we could use this information to find the entropy change of formation of that compound. However, equilibrium measurements are much more difficult and unreliable than calorimetric measurements, so this is not a very satisfactory method. Nonetheless, it has been carried out enough to discover that for almost all the reactions studied, as the reaction temperature approached 0 K, the numerical value of Δs^o approached 0.00.

This was a new fundamental discovery in thermodynamics, not demonstrable or predictable from previously known thermodynamic principles. It says that as we approach 0 K, the entropies of all known species (elements and compounds) approach the same value. It doesn't make much difference what that value is, as long as it is the same per mol of atoms, because the number of mols of atoms is conserved in any chemical reaction. The universally adopted convention is to make that value zero. (Other choices would have been correct, but would lead to much nastier mathematics.) This is called the *third law of thermodynamics,* stated formally as follows: *The entropy of any pure crystalline substance at* 0 K *is zero*. The "pure crystalline" is explained below. The entropy based on this statement is called the *absolute entropy,* which means entropy relative to that of a pure crystalline substance at 0 K. It is not the same as *steam table entropies,* which are relative to an arbitrarily chosen datum (see Section 2.7). Steam table entropies can be positive or negative. Absolute entropies are always positive or zero.

If we can measure the heat capacity C of some pure species (element or compound) from 0 to 298.15 K, then we can compute its absolute entropy at 298.15 K from the general relationship

$$s_{\text{at } T} = s_{\text{at } T=0\text{ K}} + \int_{T=0\text{ K}}^{T=298.15\text{ K}} C\frac{dT}{T} \tag{E.4}$$

which is true whether the third law is correct or not. If the third law is correct, then the $s_{\text{at } T=0\text{ K}}$ term = 0.00. Here we have not specified whether the heat capacity is at constant pressure, constant volume, or some other condition. For solids and liquids they are practically the same. For gases we must specify whether the integration is at constant pressure or constant volume. Such calorimetric measurements have been carried out for the elements and for many compounds, so from them we can compute the Δs^o of formation, which we can then combine with the Δh^o values to make up our tables of Δg°, which we then use to estimate equilibrium constants. Most of the values in Table A.8 were found that way. If there is a phase change between 0 and 298.15 K (e.g., melting or vaporization or change of crystal form), then we must add to Eq. E.4 a term of the form $\Delta s_{\text{phase change}} = \Delta h_{\text{phase change}}/T$, which uses the experimentally measured enthalpy change of melting, vaporization, or change of crystal form.

When such tables became widely available they were used to compute equilibrium constants for reactions whose equilibrium constants were known experimentally. Most agreed very well, but there were some discrepancies, indicating that for some substances $s_{\text{at } T=0\text{ K}}$ was not 0.00, but some small but significant value. Analysis of this problem led to the conclusion that for the substances that formed perfectly regular crystals at 0 K, $s_{\text{at } T=0\text{ K}}$ was indeed 0.00, but for substances that could have some randomness in the crystal it was not. The easiest example to visualize is CO. This is a linear molecule, with the atoms at either end more or less the same size. In a perfect crystal the molecules would be all lined up head to tail, i.e., CO–CO–CO–. However, the difference between the two ends is small enough that when CO freezes there is some randomness in the orientation, for example, CO–CO–OC–CO–OC–CO–OC. This introduces randomness—imperfection—into the crystal, so that its entropy at $0\text{ K} > 0$.

At the same time that this calorimetric approach was being developed, the theory of statistical thermodynamics showed that for a pure ideal gas, we could calculate the absolute entropy, if we knew some details about the gas molecule's internal vibrations. Entropies calculated that way were compared with those based on calorimetry, generally with agreement within the levels of experimental uncertainty. Based on this comparison, the entropy of CO as a crystal at 0 K $\approx$ 1.1 cal/mol K. If the arrangement of the molecules in the crystal were perfectly random the value would be 1.38 cal/mol K. This indicates that the true situation is somewhere between perfect crystal (no randomness) and perfectly random alignment of the molecules. The other materials for which this measurable entropy at 0 K has been found are H_2, H_2O, and N_2O. For each of these species the randomness in the crystals at 0 K has been explained theoretically. For all the other substances that have been studied, the entropy in the crystalline form at 0 K appears to be 0.00.

E.3 IONS

The calorimetric and/or statistical thermodynamic approach described above cannot be easily applied to ions. However, as shown in Section 13.1, the equilibrium voltage of an electrochemical cell can be directly calculated from its Δg°. The converse is true; from the measured voltage we may calculate the Δg° of the reaction. Most ions can form a cell with a standard hydrogen electrode, so that if we assign a value of $g^\circ = 0$ to a standard hydrogen electrode, we can then use equilibrium voltage measurements to compute g° values for other ions, relative to this standard hydrogen datum. The values of the Δg° of formation from the elements for ions in Table A.8 are all measured that way. The values of Δh° of formation of ions can be determined experimentally, by carrying out electrochemical reactions in a calorimeter.

E.4 PRESENTING THESE DATA

From the above it is clear that the experimental measurements are

1. Calorimetric heats of reaction or combustion
2. Calorimetric heat capacity and enthalpy change of phase change
3. Electrochemical cell voltages

In addition, there are statistical mechanical calculations for gases, for which the molecules' internal vibration frequencies must be determined spectroscopically. Finally, for some substances that exist only at a limited range of conditions, equilibrium constant measurements can be used.

Tables like Table A.8 are produced based on these measurements. The form of Table A.8 is currently the most widely used form. In earlier times, we regularly saw tables of absolute entropies, to be used with tables of heats of reaction [1, p. 992]. That form is now seldom seen. This is an inactive research topic in thermodynamics; we have reliable tables like Table A.8, for all common species. The history and theory of this topic is in [2].

REFERENCES

1. Hougen, O. A., K. M. Watson, and R. A. Ragatz. *Chemical Process Principles*, Part II: *Thermodynamics*, ed. 2. New York: Wiley (1959).
2. Aston, J. G. The third law of thermodynamics and statistical mechanics. In *A Treatise on Physical Chemistry*, ed. 3, Vol. 1, Taylor, H. S., and S. Glasstone, eds. New York: Van Nostrand, Chapter IV (1942).

APPENDIX F

CALCULATION OF FUGACITIES FROM PRESSURE-EXPLICIT EOSs

F.1 PRESSURE-EXPLICIT AND VOLUME-EXPLICIT EOSs

In Examples 7.1 and 9.9 we calculated the fugacities of pure species and of species in mixtures using the little EOS in Eqs. 7.9 and 7.16. That required solving an integral of the form

$$\int\left(\frac{RT}{P}-v\right)dP_T \quad \text{or} \quad \int\left(\frac{RT}{P}-\bar{v}\right)dP_T \tag{F.1}$$

The little EOS is *volume-explicit*, meaning that it can solved for v, in a form like $v=f(P,T)$; that leads to easy solution of the above integrals. Unfortunately, no one has devised a volume-explicit EOS that will represent experimental PvT data well except at low pressures and high temperatures (i.e., near the ideal gas limit). All of the EOSs that will represent the experimental data far from the ideal-gas state are *pressure-explicit*, that is, have the form $P=f(v,T)$. For example, the BWR EOS (Eq. 2.46) shows P as a function of v to powers $1, 2, 3, \ldots, 6$. There is no way it can be easily solved for v for substitution into Eq. F.1. (We can perform the integration numerically, but that is awkward and slow.)

F.2 *f/P* OF PURE SPECIES BASED ON PRESSURE-EXPLICIT EOSs

The general approach to calculating fugacities from pressure-explicit EOSs is to use integration by parts to replace the vdP integral,

$$\int_{P=0}^{P=P} v\,dP = [Pv]_{(Pv)_0}^{Pv} - \int_{v=\infty}^{v} P\,dv \tag{F.2}$$

When this is substituted in Eq. 7.9, and we observe that at the lower limit the ideal gas law applies, then we have

$$RT\ln\frac{f}{P} = \left[Pv - RT - \int_{v=\infty}^{v} Pdv - RT\int_{P=0}^{P=P}\frac{dP}{P}\right] \tag{F.3}$$

To perform the integration (and simplify the problem of lower integration limits of zero and infinity), we add and subtract $\int_{v=\infty}^{v}(RT/v)dv$ to the right-hand side of Eq. F.3 and factor to

$$RT\ln\frac{f}{P} = \left[Pv - RT - \int_{v=\infty}^{v}\left(P-\frac{RT}{v}\right)dv - \int_{v=\infty}^{v}\left(\frac{RT}{v}\right)dv - RT\int_{P=0}^{P=P}\frac{dP}{P}\right] \tag{F.4}$$

Then we note that

$$\begin{aligned}\int_{v=\infty}^{v}\left(\frac{RT}{v}\right)dv - RT\int_{P=0}^{P=P}\frac{dP}{P} &= RT\left[\ln\frac{v}{RT}+\ln P\right]_{P=0,v=\infty}^{P,v} \\ &= RT\left[\ln\frac{Pv}{RT}\right]_{P=0,v=\infty}^{P,v} = RT\ln\frac{z}{1}\end{aligned} \tag{F.5}$$

We then substitute this back into Eq. F.4 and divide by RT to find

$$\ln\frac{f}{P} = \frac{-1}{RT}\int_{P=0}^{P}\left(\frac{RT}{P} - v\right)dP$$
$$= z - 1 - \ln z - \frac{1}{RT}\int_{v=\infty}^{v}\left(P - \frac{RT}{v}\right)dv \quad \text{(F.6)}$$

Equation F.6 is applicable to any EOS. The two left terms are exactly Eq. 7.9, suitable for use with a volume-explicit EOS. The right term is the exact equivalent of the middle term. It shows the same integration as Eq. 7.9, which is a $v\,dP$ integration; by suitable algebra we have found a form of that integration that works with P-explicit EOSs.

Example F.1 Show the form that Eq. F.6 takes for the van der Waals (vdW) EOS

$$P = \frac{RT}{(v-b)} - \frac{a}{v^2} \quad \text{(2.40)}$$

First we solve for z,

$$z = \frac{Pv}{RT} = \frac{v}{v-b} - \frac{a}{RTv} \quad \text{(F.7)}$$

and for

$$\frac{1}{RT}\int_{v=\infty}^{v}\left(P - \frac{RT}{v}\right)dv = \frac{1}{RT}\int_{v=\infty}^{v}\left(\frac{RT}{(v-b)} - \frac{a}{v^2} - \frac{RT}{v}\right)dv$$
$$= \frac{1}{RT}\left[RT\ln(v-b) - \frac{-a}{v} - RT\ln v\right]_{v=\infty}^{v}$$
$$= \frac{1}{RT}\left[RT\ln\frac{(v-b)}{v} + \frac{a}{v}\right]_{v=\infty}^{v}$$
$$= \left[\ln\frac{(v-b)}{v} + \frac{a}{RTv}\right]_{v=\infty}^{v} \quad \text{(F.8)}$$

at the lower limit of integration the first term approaches $\ln(1.00) = 0.00$ and the second term approaches $1/\infty = 0.00$, so we may drop the lower limits and substitute back into Eq. F.6, finding

$$\ln\frac{f}{P} = z - 1 - \ln z - \left(\ln\frac{(v-b)}{v} + \frac{a}{RTv}\right)$$
$$= z - 1 - \frac{a}{RTv} - \ln\left(z \cdot \frac{(v-b)}{v}\right) \quad ■ \quad \text{(F.9)}$$

This shows that if we know the values of a and b we can solve for z and then use a, b, and z to find f/P. We will illustrate this process below.

F.3 CUBIC EQUATIONS OF STATE

Figure 10.8 shows that the BWR EOS, which contains the specific volume to the sixth power, calculates an isotherm on P-v coordinates that has one maximum and one minimum, and that can be used to calculate the PvT behavior of both gas and liquid (but not directly the behavior in the two-phase region). The same result can be obtained with any EOS that has the volume to the third or higher power. Complex EOSs like the BWR can represent experimental PvT data more accurately than *cubic equations*, which have the volume only to the third power, but the latter are much easier to use in high-pressure VLE calculations and are more widely used. The cubic EOSs of interest are as follows:

1. The van der Waals (vdW) EOS

$$P = \frac{RT}{v-b} - \frac{a}{v^2} \quad \text{(2.40)}$$

which is now only of historic interest, but which has played a major role in the development of this field and is used in some examples in this chapter because of its simplicity.

2. The Redlich and Kwong (RK) EOS

$$P = \frac{RT}{v-b} - \frac{a}{\sqrt{T}v(v+b)} \quad \text{(F.10)}$$

which is obviously a fairly minor modification of the vdW EOS, but which fits experimental PvT data much better. It is now principally of historic interest.

3. The Soave or Soave–Redlich–Kwong (SRK) EOS

$$P = \frac{RT}{v-b} - \frac{a\alpha}{v(v+b)} \quad \text{(F.11)}$$

is a modified RK EOS, which has largely replaced the RK EOS and is now probably the most widely used EOS of this type.

4. The Peng–Robinson (PR) EOS

$$P = \frac{RT}{v-b} - \frac{a\alpha}{v^2 + 2bv - b^2} \quad \text{(F.12)}$$

which is similar to the others and is considered superior for some applications. In all of these EOSs the constants a, b, and α are specified in terms of P_r, T_r, and ω. The formulation of these constants are generally a mix of theory and data fitting. (In the vdW EOS they were based on theory alone; the others, which adjusted theory to much experiement, are much more accurate.) The details of these EOSs are summarized in various texts. For the remainder of this appendix we will consider only the vdW, which is the simplest, and the SRK, which seems to be the most widely used.

Example F.2 Using the SRK EOS, estimate $f/P = \phi$ for pure propane at 100°F and 188.32 psia (see Figure 10.8).

As shown in Problems F.3, F.4, and F.5, it is common to rewrite all the cubic EOSs in z form, and then to redefine the constants, as shown below. For the SRK EOS the forms [1] are

$$z^3 - z^2 + z(A - B - B^2) - AB = 0 \qquad (F.13)$$

$$A = \frac{a\alpha P}{R^2T^2} = 0.42747 \cdot [1 + (0.480 + 1.574\omega - 0.176\omega^2) \cdot (1 - T_r^{0.5})]^2 \frac{P_r}{T_r^2} \qquad (F.14)$$

and

$$B = \frac{bP}{RT} = 0.08664 \frac{P_r}{T_r} \qquad (F.15)$$

in which a, b, and α are those in Eq. F.11. For this EOS the equivalent of Eq. F.9 is

$$\ln\frac{f}{P} = z - 1 - \ln(z - B) - \frac{A}{B}\ln\left(\frac{z - B}{z}\right) \qquad (F.16)$$

Table F.1 shows the preliminary steps for this solution. From Figure 10.8 we see that this temperature and pressure correspond to a point on the vapor–liquid equilibrium curve for propane, so we should look for three solutions to Eq. F.13. Table F.2 shows all of those. The procedure is to guess a value of z, and then numerically solve Eq. F.13 (using "goal seek" on a spreadsheet or any other suitable numerical method) for the value of z that makes the sum of the terms on the left of Eq. F.13 equal 0.00. To find all three solutions, we begin with an initial guess larger than the expected vapor z, one smaller than the expected liquid z, and then one about halfway between the two results found for the liquid and the vapor. Convergence is rapid. Simple insertion of the values in Table F.2 into Eq. F.16 leads to the values of f/P shown in Table F.2. The lower lines in Table F.2 are discussed below. ■

Table F.1 Preliminary Steps in Example F.2

T (°F)	100
T (K)	311.11
P (psia)	188.32
P (bar)	12.977
R (L bar/mol K)	0.08314
T_c (K, Table A.1)	369.8
P_c (bar. Table A.1)	42.48
ω (Table A.1)	0.152
T_r	0.841
P_r	0.3055
A (Eq. F.14)	0.2071
B (Eq. F.15)	0.3146

Table F.2 The Three Solutions to Example F.2

		Unstable Liquid–Vapor	
	Liquid	Mixture	Vapor
z (Eq. F.13)	0.0519	0.159	0.789
f/P (Eq. F.16)	0.838	1.030	0.826
v (ft^3/lbm)	0.0377	0.115	0.572
v (ft^3/lbm from [2])	0.0339		0.559
f/P (from [2])	0.8152		0.8152

At the end of this example we observe the following:

1. There is only little interest in computing pure species f/P from cubic EOSs. As in Chapter 7, we make this calculation only as a preliminary step to the more interesting and useful f_i/y_iP for individual species in mixtures and for textbook illustrations.
2. As discussed in Chapter 10 (see Figure 10.8) we expected and found three solutions, corresponding to liquid, vapor, and unstable two-phase mixture.
3. We could compare the values found here with those we would find in Figure 10.8, if we could read that figure accurately enough. Instead, we read the values from [2], which are identical to those in Figure 10.8 and much easier to read. We see that the calculated vapor volume in this example is 2.3% more than that in [2], and the calculated liquid volume is 11% more. The additional terms in the BWR EOS used in [2] are mostly there to give a more accurate estimate of the liquid specific volume.
4. We see that the calculated values of f/P for vapor and liquid, which should be equal for two phases in equilibrium, differ from each other by 0.8% ≈ 0.00, indicating that in this case the SRK EOS does a good job of representing the values of f/P. The calculated values are ≈2% more than those calculated by the BWR EOS in [2].

We may show that the calculated values of z and f/P agree reasonably with Figures A.4 and 7.1. Thus, if we were writing a computer program, for which we wished the computer equivalents of those plots, we could program the results of Example F.2, and find those equivalent results.

F.4 f_i/Py_i FOR INDIVIDUAL SPECIES IN MIXTURES, BASED ON PRESSURE-EXPLICIT EOSs

In principle, we should be able to begin with Eq. F.6, take the partial derivatives with respect to y_i, and use the method of tangent intercepts to find f_i/y_iP. In practice, this is very difficult, because it is very difficult to write out the derivatives of z from a pressure-explicit EOS. Instead, we begin with Eq. 7.15, rewritten as

$$RT\,d\ln f_i = \bar{v}_i dP_T = \left[\frac{\partial(n_T v)}{\partial n_i}\right] dP_T \tag{F.17}$$

all at a constant T (this is the method of tangent slopes, see Section 6.3). Next we eliminate the dP by noting that (at constant T) $P = f(n_i, n_T v)$, where $(n_T v) = V$, the total volume. We retain the $n_T v$ form because it makes the subsequent mathematics easier. Next we write the "chain-rule" derivatives as

$$\left(\frac{\partial P}{\partial(n_T v)}\right)\cdot\left(\frac{\partial n_i}{\partial P}\right)\cdot\left[\frac{\partial(n_T v)}{\partial n_i}\right] = -1 \tag{F.18}$$

from which

$$\left(\frac{\partial(n_T v)}{\partial n_i}\right) dP = -\left(\frac{\partial P}{\partial n_i}\right)\cdot\partial(n_T v) \tag{F.19}$$

The $(\partial P/\partial n_i)$ is a strange-looking derivative. Remember that we have taken $n_T v = V$ as one of the independent variables, so this is the increase in P when we add 1 mol of i at constant V, T, and n_j.

Next we substitute Eq. F.19 in Eq. F.17 and add $RT\,d\ln(v/RT)$ to both sides and factor to

$$RT\,d\ln\frac{f_i v}{RT} = -\left(\frac{\partial P}{\partial n_i}\right)\cdot\partial n_T v + RT\,d\ln\frac{v}{RT} \tag{F.20}$$

We multiply the argument of the ln in the rightmost term by n_T/n_T and carry out the differentiation, finding

$$RT\,d\ln\frac{n_T v}{n_T RT} = RT\frac{1}{n_T v/n_T RT}\cdot\frac{d(n_T v)}{n_T RT} = \frac{RT}{n_T v}d(n_T v) \tag{F.21}$$

We substitute Eq. F.21 into Eq. F.20 and factor to

$$RT\,d\ln\frac{f_i v}{RT} = \left[-\left(\frac{\partial P}{\partial n_i}\right) + \frac{RT}{n_T v}\right]\cdot\partial(n_T v) \tag{F.22}$$

Next we integrate both sides from $v = \infty$ to $v = v$, finding

$$RT\ln\frac{f_i v}{RT}\Big|_{v=\infty}^{v} = \int_{v=\infty}^{v}\left[-\left(\frac{\partial P}{\partial n_i}\right) + \frac{RT}{n_T v}\right]\cdot\partial(n_T v) \tag{F.23}$$

At the lower limit on the left we have ideal gas behavior for which $f_i = y_iP$ and $RT/v = P$ so that

$$RT\ln\frac{f_i v}{RT}\Big|_{v=\infty}^{v} = RT\ln\frac{f_i v/RT}{y_iP/P} = RT\ln\frac{f_i}{y_i} + RT\ln\frac{v}{RT} \tag{F.24}$$

Substituting Eq. F.24 in Eq. F.23 and subtracting $RT\ln P$ from both sides, we rearrange to

$$RT\ln\frac{f_i}{y_iP} = RT\ln\hat{\phi}_i = \int_{v=\infty}^{v}\left[-\left(\frac{\partial P}{\partial n_i}\right) + \frac{RT}{n_T v}\right]\cdot\partial(n_T v) - RT\ln z \tag{F.25}$$

This is the final form we are seeking. It shows $RT\ln(f_i/y_iP) = RT\ln\hat{\phi}_i$ in terms of P derivatives, z, and $n_T v$ but not its derivative! This is exactly the same as Eq. 7.18, but written out (with considerable effort!) in terms suitable for using with any pressure-explicit EOS.

Example F.3 Evaluate Eq. F.25 for the vdW EOS.

We observe first that we must evaluate $(\partial P/\partial n_i)_{T,V,n_j}$ for the vdW EOS (Eq. 2.40). We write the EOS and replace v (the volume per mol) wherever it appears with V/n_T, where V is the total volume, and n_T is the total number of mols. So we need

$$\left(\frac{\partial P}{\partial n_i}\right)_{T,V,n_j} = \left(\frac{\partial}{\partial n_i}\right)_{T,V,n_j}\left(\frac{n_T RT}{(V-n_T b)} - \frac{n_T^2 a}{V^2}\right) = \frac{RT}{(V-n_T b)} + \frac{n_T RT}{(V-n_T b)^2}\cdot\left[\frac{\partial(n_T b)}{\partial n_i}\right] - \frac{1}{V_2}\cdot\left[\frac{\partial(n_T^2 a)}{\partial n_i}\right] \tag{F.26}$$

Here we have taken advantage of the fact that $(\partial n_T/\partial n_i)_{n_j} = 1.00$. We then substitute Eq. F.26 in Eq. F.25, finding

$$RT\ln\frac{f_i}{y_iP} = \int_{v=\infty}^{v}\left(-\left\{\frac{RT}{(V-n_T b)} + \frac{n_T RT}{(V-n_T b)^2}\cdot\left[\frac{\partial(n_T b)}{\partial n_i}\right] - \frac{1}{V^2}\cdot\left[\frac{\partial(n_T^2 a)}{\partial n_i}\right]\right\} + \frac{RT}{V}\right)dV - RT\ln z \tag{F.27}$$

This formidable-looking integral is quite easy, because the two derivatives in the integrand are not functions of V, so we may integrate, finding

$$RT\ln\frac{f_i}{y_iP} = \left(-\left\{ RT\ln(V - n_Tb) + \frac{n_TRT}{(V - n_Tb)^2} \cdot \left[\frac{\partial(n_Tb)}{\partial n_i}\right] + \frac{1}{V} \cdot \left[\frac{\partial(n_T^2a)}{\partial n_i}\right] \right\} + RT\ln V \right)_{v=\infty}^{v} - RT\ln z \tag{F.28}$$

which simplifies to

$$RT\ln\frac{f_i}{y_iP} = RT\ln\frac{V}{V - n_Tb} - RT\ln z - \left\{ \frac{n_TRT}{(V - n_Tb)} \cdot \left[\frac{\partial(n_Tb)}{\partial n_i}\right] + \frac{1}{V} \cdot \left[\frac{\partial(n_T^2a)}{\partial n_i}\right] \right\} \tag{F.29}$$

Here we have dropped the lower integration limit, because all of terms on the right in Eq. F.29 are zero at $v = \infty$. Equation F.29 is correct for the vdW EOS for any set of mixing rules for determining the mixture values of a and b. ■

F.5 MIXING RULES FOR CUBIC EOSs

Please review Sections 7.12 and 9.7, which present the idea of mixing rules, and show a simple example. To use any cubic EOS for mixtures, we must use some set of mixing rules. The study of mixing rules for such EOSs is an active area of research [3]; formulating a set of rules that are simpler and more accurate than those in common use will bring the author (temporary) fame!

Example F.4 Show the form of Eq. F.29 for the following commonly used mixing rules for the vdW EOS (similar to those in Section 9.7):

$$a = \left(\sum y_i\sqrt{a_i}\right)^2 = \left(\sum \frac{n_i}{n_T}\sqrt{a_i}\right)^2 \tag{F.30}$$

and

$$b = \sum y_ib_i = \sum \frac{n_i}{n_T}b_i \tag{F.31}$$

so that

$$\left(\frac{\partial(n_T^2a)}{\partial n_i}\right) = \left(\frac{\partial\left\{n_T^2\left[\left(\sum\frac{n_i}{n_T}\sqrt{a_i}\right)^2\right]\right\}}{\partial n_i}\right) = \frac{\partial\left(\sum n_i\sqrt{a_i}\right)^2}{\partial n_i} = 2\left(\sum n_i\sqrt{a_i}\right)\sqrt{a_i} = 2n_T\sqrt{aa_i} \tag{F.32}$$

and

$$\left(\frac{\partial(n_Tb)}{\partial n_i}\right) = \left(\frac{\partial\left(n_T\sum\frac{n_i}{n_T}b_i\right)}{\partial n_i}\right) = \frac{\partial(n_ib_i)}{\partial n_i} = b_i \tag{F.33}$$

Substituting these two derivatives into Eq. F.29, we find

$$RT\ln\frac{f_i}{y_iP} = RT\ln\frac{V}{V - n_Tb} - RT\ln z - \left(\frac{n_TRT}{(V - n_Tb)} \cdot b_i + \frac{2n_T\sqrt{aa_i}}{V}\right) \tag{F.34}$$

Dividing out the n_Ts to get back to the specific volume v, we find

$$RT\ln\frac{f_i}{y_iP} = RT\ln\hat{\phi}_i = RT\ln\frac{v}{v - b} - RT\ln z - \left(\frac{RTb_i}{(v - b)} + \frac{2\sqrt{aa_i}}{v}\right) \tag{F.35}$$

This form is correct only for the arbitrary mixing rules shown in Eqs. F.30 and F.31 (which are the most widely used rules for the vdW EOS). ■

These two examples were shown for the vdW EOS, because it is the simplest of the cubic EOSs and leads to the simplest algebra. We now return to the SRK EOS, for which the "standard" mixing rule for b is the same as Eq. F.31 above, but for which the other mixing rule is

$$A = \sum\sum y_iy_jA_{ij} \tag{F.36}$$

where the expansion is exactly the same as in Section 9.7 and the cross-coefficient

$$A_{ij} = (1 - k_{ij})\sqrt{A_i \cdot A_j} \tag{F.37}$$

This is the same k_{ij} defined in Section 9.7. (If $k_{ij} = 0$, which is often assumed, then this mixing rule is the same as Eq. F.30, see Problem F.10.)

Making these substitutions into the SRK EOS, for $k_{ij} = 0$ we find

$$\ln\frac{f_i}{y_iP} = \ln\hat{\phi}_i = \frac{B_i}{B}(z - 1) - \ln(z - B) - \frac{A}{B} \cdot \left(2\sqrt{\frac{A_i}{A}} - \frac{B_i}{B}\right)\ln\left(1 - \frac{B}{Z}\right) \tag{F.38}$$

where A and B are the values for the mixture, and A_i and B_i are those for the pure individual species in the mixture.

F.6 VLE CALCULATIONS WITH A CUBIC EOS

We now have all the tools to make VLE calculations using a cubic EOS to represent both vapor and liquid (see Section 10.3.3).

Example F.5/10.3 Repeat Example 10.2, using the SKR EOS. This example is started in Chapter 10 as Example 10.3 and the results are discussed there. However, the solution requires the application of Eq. F.38, four times, so the solution is shown here.

Here we wish to find the bubble-point temperature and the dew-point composition for a liquid mixture with 60 mol% ethane, balance n-heptane, at 800 psia, to compare with the experimental values in Figure 10.7. This problem involves nested trial and error solutions. We begin by assuming that the bubble-point temperature is 200° F and that the vapor is 90 mol% ethane. (We could make better starting guesses based on reading Figure 10.7, but these guesses are more illustrative of the process.) The preliminary calculations are shown in Table F.3, with the assumption that $k_{ij} = 0$.

Next we compute the A and B of the mixture for each of the two phases, using the liquid composition in the problem statement and the assumed composition of the equilibrium vapor. The results are shown in Table F.4.

The only trial-and-error procedure involved in Table F.4 is the solution of the Eq. F.13 for z, twice, one for liquid and one for vapor. The many blanks in the table correspond to values that we need not calculate.

From this result we see that the calculated y_is do not correspond to the assumed values and do not sum to 1.00. The next step is to substitute the calculated y_{ethane} from the next to last row into the top row as the assumed y_{ethane}. This changes the calculated values of A_{mix} and B_{mix}, and leads to a new trial-and-error solution for z_{vapor}. Then that value is substituted into Eq. F.38, leading to new estimates of all the values below it in the table. This is repeated until the new computed y_{ethane} at the bottom of the table is practically the same as the assumed value at the top of the table. This trial-and-error procedure leads to a converged value of $y_{\text{ethane}} = 0.906$, and of $(y_{\text{ethane}} + y_{n\text{-heptane}} = 0.967)$, which indicates that we have guessed too low a temperature. A new temperature must be selected and the whole process repeated, converging on a temperature of 210°F for which the calculated $(y_{\text{ethane}} + y_{n\text{-heptane}}) = 1.00$. The summary of the results is shown in Chapter 10, Example 10.3. ■

This example is slow and tedious with a spreadsheet, because the multiple trial and error calculations must be done one at a time. It is shown that way because it is clear what is happening. The calculation is much easier in a Fortran program with nested DO loops.

Table F.3 Preliminary Steps in Example F.5

T (°F)	200	
T (K)	366.67	
P (psia)	800	
P (bar)	55.13	
R (L bar/mol K)	0.08314	
	FOR ETHANE	FOR n-HEPTANE
T_c (K, Table A.1)	305.3	540.2
P_c (bar, Table A.1)	48.72	27.4
ω (Table A.1)	0.1	0.35
T_r	1.201	0.679
P_r	1.315	2.012
A (Eq. F.14)	0.2957	2.5895
B (Eq. F.15)	0.0816	0.2568

Table F.4 Calculated Values of Liquid and Vapor Properties for the First assumed Temperature and Vapor Composition in Example F.5

	Ethane in Liquid	n-Heptane in Liquid	Ethane in Vapor	n-Heptane in Vapor
Assumed mol fraction, x_i or y_i	$x_{\text{ethane}} = 0.6$	$x_n\text{-heptane} = 1 - x_{\text{ethane}} = 0.4$	assumed $y_{\text{ethane}} = 0.9$	$y_n\text{-heptane} = 1 - y_{\text{ethane}} = 0.1$
A_{mixture} (Eq. F.30)	0.9408		0.4229	
B_{mixture} (Eq. F.31)	0.1517		0.09915	
z_{mixture} (Eq. F.13)	$z_{\text{liquid}} = 0.2458$		$z_{\text{vapor}} = 0.5867$	
$\frac{f_i}{y_i P} = \hat{\phi}_i$ (Eq. F.38)	1.24451	0.0233	0.8291	0.1461
$K_i = \frac{\hat{\phi}_i, \text{liquid}}{\hat{\phi}_i, \text{vapor}}$	1.502	0.1595		
$y_i = K_i x_i$			0.901	0.0638
$y_1 + y_2$			0.965	

F.7 SUMMARY

1. Volume-explicit EOSs, like the little EOS, are useful for vapors whose behavior is close to ideal gas behavior. They are not useful for gases far from the ideal gas state.
2. Pressure-explicit EOSs can represent both liquid and vapor and behavior near the critical state with fair accuracy for simple EOSs (like cubic EOSs) and very good accuracy for more complex EOSs, like the BWR EOS.
3. With suitable mixing rules, these pressure-explicit EOSs can make good-to-excellent estimates of high-pressure VLE. They are very widely used for that purpose.
4. This appendix has more mathematics than seems appropriate in the main text.

PROBLEMS

See the Common Units and Values for Problems and Examples.

F.1 Sketch on P-v coordinates an arbitrary smooth function ($P = f(v)$) from $P_1 v_1$ to $P_2 v_2$. Then show on the sketch which areas correspond to the three terms in Eq. F.2.

F.2 In going from Eq. F.3 to F.4 we wrote

$$\int RT\frac{dP}{P} = RT\ln P \text{ and } \int RT\frac{dv}{v} = RT\ln\frac{v}{RT}$$

Is this correct? Show the appropriate mathematics to support these choices.

F.3 The vdW EOS (Eq. 2.40) may be cleared of fractions by multiplying by $v^2 \cdot (v - b)$ to find

$$P \cdot v^2 \cdot (v - b) = RTv^2 - a(v - b) \tag{F.39}$$

This may be multiplied out and grouped as

$$v^3 \cdot P + v^2 \cdot (-Pb - RT) + va - b = 0 \tag{F.40}$$

and put in the standard cubic form

$$v^3 + v^2 \cdot \left(-b - \frac{RT}{P}\right) + v\frac{a}{P} - \frac{b}{P} = 0 \tag{F.41}$$

Show the equivalent derivations for
a. The RK EOS.
b. The SRK EOS.
c. The PR EOS.

F.4 Equation F.41 may be multiplied through by $(P/RT)^3$ to find

$$z^3 - z^2\left(\frac{bP}{RT} + 1\right) + z\frac{aP}{(RT)^2} - \frac{bP^2}{(RT)^3} = 0 \tag{F.42}$$

in which all of the terms are dimensionless. Show the equivalent derivations for
a. The RK EOS.
b. The SRK EOS.
c. The PR EOS.

F.5 If we define

$$B = \frac{bP}{RT} \quad \text{and} \quad A = \frac{aP}{(RT)^2} \tag{F.43}$$

which are both dimensionless, and substitute into Eq. F.42, we find

$$z^3 - z^2(B + 1) + zA - AB = 0 \tag{F.44}$$

which is the form most often used for computing z in the vdW EOS. Show the equivalent derivations for
a. The RK EOS.
b. The SRK EOS.
c. The PR EOS.
Observe that the definitions of A and B for the RK, SRK, and PR EOSs are similar to those in Eq. F.43, but not identical to them. From the results of Problem F.4, the choice of values of A and B will seem obvious.

F.6 Show the pure species equations for f/P for the following EOSs:
d. Beattie–Bridgeman (Eqs. 2.42–2.45).
e. BWR (Eq. 2.46).

F.7 Show the numerical solutions for the 3 values of z in Table F.2.

F.8 Compare the values of z and f/P found in Example F.2 to those we would read from Appendixes A.4 and A.5. Do they agree to within chart-reading accuracy? Should they?

F.9 Show the difficulty of attempting the integration of Eq. F.1 for the vdW EOS, which is one of the simplest pressure-explicit EOSs.

F.10 Show that if $k_{ij} = 0$, then Eq. F.37 leads to the same mixing rule as Eq. F.30.

F.11 Show the calculations leading to Tables F.3 and F.4.

F.12 Show the calculations that change the final value of y_{ethane} in Table F.4 to match its assumed initial value.

F.13 Show the calculations leading to $y_1 + y_2 = 1.00$ and a bubble point of 210°F in Example F.5/10.3.

REFERENCES

1. Soave, G. Equilibrium constants from a modified Redlich–Kwong equation of state. *Chem. Eng. Sci.* 27:1197–1203 (1972).
2. Starling, K. E. *Fluid Thermodynamic Properties for Light Petroleum Systems*. Houston, TX: Gulf (1973).
3. Orbey, H., and S. I. Sandler. *Modeling Vapor–Liquid Equilibria: Cubic Equations of State and Their Mixing Rules*. Cambridge, UK: Cambridge University Press (1998).

APPENDIX G

THERMODYNAMIC PROPERTY DERIVATIVES AND THE BRIDGMAN TABLE

In Section 2.10 and Table 2.2 we showed the five useful equations for calculating the changes in common thermodynamic properties with changes in T and P. Those five satisfy the needs of most undergraduates and most working engineers. However some uncommon problems require other mathematical relations among thermodynamic properties; those can be found using the methods in this appendix.

These relations can all be derived starting with the property equation (Eq. 2.32), and the definitions of h, g, a, C_P and C_V. The derivations are shown in many thermodynamics books and form a favorite exercise in differential calculus for graduate students. All 168 of the possible relations between the variables u, h, s, g, a, v, P, and T can be worked out quickly and easily using a Bridgman table, Table G.1 (thus missing out on all that fun calculus and algebra).

For any of the properties u, h, s, g, and a we can write a two-term Taylor series expansion of the derivative. For example, for s as a function of T and P,

$$ds = \left(\frac{\partial s}{\partial T}\right)_P dT + \left(\frac{\partial s}{\partial P}\right)_T dP \tag{G.1}$$

Comparing this to Eq. 2.35

$$ds = \frac{C_P}{T} dT - \left(\frac{dv}{dT}\right)_P dP \tag{2.35}$$

We see that these are the same if

$$\left(\frac{\partial s}{\partial T}\right)_P = \frac{C_P}{T} \text{ and } \left(\frac{\partial s}{\partial P}\right)_T = -\left(\frac{dv}{dT}\right)_P \tag{G.2}$$

The first of these comes from the definition of the entropy; the second comes from one of the Maxwell relations. Many thermodynamics texts spend considerable effort showing how these come about and how to derive any of the other derivatives of this type we might need. These derivatives are all of the form $(\partial a/\partial b)_c$ where a, b, and c are any of the following variables, T, P, v, u, h, s, g, and a. Taking 3 variables at a time from a list of 8 allows for 336 combinations, and thus 336 such derivatives, but half of those are the reciprocals of others so there are only 168 such derivatives among this list of variables. Of these 168 the most useful 10 are shown in Table 2.2. But some of the others are sometimes useful; they are easily found from Table G.1

Example G.1 Show the construction of the first six derivatives in Table 2.2 from the Bridgman table.

$$\left(\frac{\partial u}{\partial T}\right)_P = \frac{(\partial u)_P}{(\partial T)_P} = \frac{[\text{Eq.BT.4}]}{[\text{Eq.BT.2}]} = \frac{C_P - P(\partial v/\partial T)_P}{1} = C_P - P(\partial v/\partial T)_P \tag{G.3}$$

$$\left(\frac{\partial u}{\partial P}\right)_T = \frac{(\partial u)_T}{(\partial P)_T} = \frac{[\text{Eq.BT.101}]}{-[\text{Eq.BT.2}]} = \frac{T(\partial v/\partial T)_P + P(\partial v/\partial P)_T}{-1} = -[T(\partial v/\partial T)_P + P(\partial v/\partial P)_T] \tag{G.4}$$

$$\left(\frac{\partial h}{\partial T}\right)_P = \frac{(\partial h)_P}{(\partial T)_P} = \frac{[\text{Eq.BT.5}]}{[\text{Eq.BT.2}]} = \frac{C_P}{1} = C_P \tag{G.5}$$

Physical and Chemical Equilibrium for Chemical Engineers, Second Edition. Noel de Nevers.

$$\left(\frac{\partial h}{\partial P}\right)_T = \frac{(\partial h)_T}{(\partial P)_T} = \frac{[\text{Eq.BT.11}]}{-[\text{Eq.BT.2}]}$$

$$= \frac{-v + T(\partial v/\partial T)_P}{-1} = v - T(\partial v/\partial T)_P \tag{G.6}$$

$$\left(\frac{\partial s}{\partial T}\right)_P = \frac{(\partial s)_P}{(\partial T)_P} = \frac{[\text{Eq.BT.3}]}{[\text{Eq.BT.2}]} = \frac{C_P/T}{1} = C_P/T \tag{G.7}$$

$$\left(\frac{\partial s}{\partial P}\right)_T = \frac{(\partial s)_T}{(\partial P)_T} = \frac{[\text{Eq.BT.9}]}{-[\text{Eq.BT.2}]} = \frac{(\partial v/\partial T)_P}{-1} = -(\partial v/\partial T)_P \tag{G.8}$$

These were easy, because the denominators were all $= \pm 1$. The following example shows a more complex derivative taken from a practical problem.

Example G.2 A rigid container is filled completely with saturated liquid propane at 100 psia. We now transfer heat to it, allowing time for perfect thermal mixing, and ask how fast does the pressure rise as we introduce heat. From the first law we know that for a closed system at constant volume $dU = m du = dQ$, so we are asking for $(\partial P/\partial u)_V$. This is one of the 168 derivatives derivable from Table G.1 but not one of the 10 most often used. This example is a very simplified version of the problem addressed in [1].

From Table G.1

$$\left(\frac{\partial P}{\partial u}\right)_V = \frac{(\partial P)_V}{(\partial u)_V} = \frac{-[\text{Eq.BT.1}]}{[\text{Eq.BT.15}]} = \frac{-(\partial v/\partial T)_P}{C_P(\partial v/\partial P)_T + T(\partial v/\partial T)_P^2}$$

$$= \frac{-1}{C_P\dfrac{(\partial v/\partial P)_T}{(\partial v/\partial T)_P} + T(\partial v/\partial T)_P} \tag{G.9}$$

Taking values from [2],

$$T \approx 55°\text{F} = 515°\text{R}; \quad C_P \approx 0.6\frac{\text{Btu}}{\text{lbm°F}};$$
$$\left(\frac{\partial v}{\partial P}\right)_T \approx -10^{-6}\frac{\text{ft}^3/\text{lbm}}{\text{psi}}$$

and

$$\left(\frac{\partial v}{\partial T}\right)_P \approx 3.9 \cdot 10^{-5}\frac{\text{ft}^3/\text{lbm}}{°\text{F}}.$$

Thus

$$\left(\frac{\partial P}{\partial u}\right)_V = $$
$$= \frac{-1}{0.6\dfrac{\text{Btu}}{\text{lbm°F}}\dfrac{-10^{-6}\dfrac{\text{ft}^3/\text{lbm}}{\text{psi}}}{3.9 \cdot 10^{-5}\dfrac{\text{ft}^3/\text{lbm}}{°\text{F}}} + 515°\text{R}\left(3.9 \cdot 10^{-5}\dfrac{\text{ft}^3/\text{lbm}}{°\text{F}}\right)} \tag{G.A}$$

The first term in the denominator, after simple cancellation of units becomes

$$0.6\frac{\text{Btu}}{\text{lbm psi}}\frac{-10^{-6}}{3.9 \cdot 10^{-5}} = -0.01538\frac{\text{Btu}}{\text{lbm psi}},$$

while the second becomes

$$515°\text{R}\left(3.9 \cdot 10^{-5}\frac{\text{ft}^3/\text{lbm}}{°\text{F}}\right) \cdot \left(\frac{\text{Btu}}{778\text{ ft lbf}}\right) \cdot \left(\frac{144\text{ lbf/ft}^2}{\text{psi}}\right)$$
$$= 0.00372\frac{\text{Btu}}{\text{lbm psi}}$$

and

$$\left(\frac{\partial P}{\partial u}\right)_V = \frac{-1}{(-0.01538 + 0.00371)\dfrac{\text{Btu/lbm}}{\text{psi}}}$$
$$= 85.7\frac{\text{psi}}{\text{Btu/lbm}} = 253\frac{\text{kPa}}{\text{kJ/kg}} \tag{G.B}$$

which shows that heating liquids in closed containers leads to rapid pressure rises. ■

PROBLEMS

See the Common Units and Values for Problems and Examples.

G.1 Show the derivation of the P and T derivatives of g and a using Table G.1 and compare them to the values in Table 2.2.

G.2 Estimate the change in enthalpy, h, of liquid propane at 55°F, as it is isothermally compressed from 100 psia to 1000 psia, using values from Example G.2. Over this pressure range for liquid propane, v is practically constant $\approx 0.0304\text{ ft}^3/\text{lbm}$. Compare the result with the interpolated value of 1.75 Btu/lbm from [2].

G.3 Show the forms that the five equations in Table 2.2 take for an ideal gas.

G.4 To convince yourself of the utility of Table G.1, derive the formula for $(\partial P/\partial u)_V$ without using Table G.1.

TABLE G.1 BRIDGMAN TABLE

This version, presented by Hougen et al. [3], is much more compact than the original by Bridgman [4]. Its use is illustrated in Examples G.1 and G.2.

1. Pressure Constant and Pressure Variable

$$(\partial v)_P = -(\partial P)_V = (\partial v/\partial T)_P \quad \text{(BT.1)}$$

$$(\partial T)_P = -(\partial P)_T = 1 \quad \text{(BT.2)}$$

$$(\partial s)_P = -(\partial P)_S = C_P/T \quad \text{(BT.3)}$$

$$(\partial u)_P = -(\partial P)_U = C_P - P(\partial v/\partial T)_P \quad \text{(BT.4)}$$

$$(\partial h)_P = -(\partial P)_H = C_P \quad \text{(BT.5)}$$

$$(\partial a)_P = -(\partial P)_A = -[s + P(\partial v/\partial T)_P] \quad \text{(BT.6)}$$

$$(\partial g)_P = -(\partial P)_G = -s \quad \text{(BT.7)}$$

2. Temperature Constant and Temperature Variable

$$(\partial v)_T = -(\partial T)_V = -(\partial v/\partial P)_T \quad \text{(BT.8)}$$

$$(\partial s)_T = -(\partial T)_S = (\partial v/\partial T)_P \quad \text{(BT.9)}$$

$$(\partial u)_T = -(\partial T)_U = T(\partial v/\partial T)_P + P(\partial v/\partial P)_T \quad \text{(BT.10)}$$

$$(\partial h)_T = -(\partial T)_H = -v + T(\partial v/\partial T)_P \quad \text{(BT.11)}$$

$$(\partial a)_T = -(\partial T)_A = P(\partial v/\partial P)_T \quad \text{(BT.12)}$$

$$(\partial g)_T = -(\partial T)_G = -v \quad \text{(BT.13)}$$

3. Volume Constant and Volume Variable

$$(\partial s)_V = -(\partial v)_S = (1/T)[C_P(\partial v/\partial P)_T + T(\partial v/\partial T)_P^2] \quad \text{(BT.14)}$$

$$(\partial u)_V = -(\partial v)_U = C_P(\partial v/\partial P)_T + T(\partial v/\partial T)_P^2 \quad \text{(BT.15)}$$

$$(\partial h)_V = -(\partial v)_H = C_P(\partial v/\partial P)_T + T(\partial v/\partial T)_P^2 - v(\partial v/\partial T)_P \quad \text{(BT.16)}$$

$$(\partial a)_V = -(\partial v)_A = -s(\partial v/\partial P)_T \quad \text{(BT.17)}$$

$$(\partial g)_V = -(\partial v)_G = -[v(\partial v/\partial T)_P + s(\partial v/\partial P)_T] \quad \text{(BT.18)}$$

4. Entropy Constant and Entropy Variable

$$(\partial u)_S = -(\partial s)_U = (P/T)[C_P(\partial v/\partial P)_T + T(\partial v/\partial T)_P^2] \quad \text{(BT.19)}$$

$$(\partial h)_S = -(\partial s)_H = -(vC_P/T) \quad \text{(BT.20)}$$

$$(\partial a)_S = -(\partial s)_A = (1/T)[P(\partial v/\partial P)_T + T(\partial v/\partial T)_P^2] + sT(\partial v/\partial T)_P\} \quad \text{(BT.21)}$$

$$(\partial g)_S = -(\partial s)_G = -(1/T)[vC_P - sT(\partial v/\partial T)_P] \quad \text{(BT.22)}$$

5. Internal Energy Constant and Internal Energy Variable

$$(\partial h)_U = -(\partial u)_H = v[C_P - P(\partial v/\partial T)_P] - P[C_P(\partial v/\partial P)_T + T(\partial v/\partial T)_P^2] \quad \text{(BT.23)}$$

$$(\partial a)_U = -(\partial u)_A = P[C_P(\partial v/\partial P)_T + T(\partial v/\partial T)_P^2] + s[T(\partial v/\partial T)_P + P(\partial v/\partial P)_T] \quad \text{(BT.24)}$$

$$(\partial g)_U = -(\partial u)_G = v[C_P - P(\partial v/\partial P)_T] + s[T(\partial v/\partial T)_P + P(\partial v/\partial P)_T] \quad \text{(BT.25)}$$

6. Enthalpy Constant and Enthalpy Variable

$$(\partial a)_H = -(\partial h)_A = -[s + P(\partial v/\partial T)_P] \cdot [v - T(\partial v/\partial T)_P] + PC_P(\partial v/\partial T)_P \quad \text{(BT.26)}$$

$$(\partial g)_H = -(\partial h)_G = -v(C_P + s) + Ts(\partial v/\partial T)_P \quad \text{(BT.27)}$$

7. Helmholz Energy Constant and Helmholz Energy Variable

$$(\partial a)_G = -(\partial g)_A = -s[v + P(\partial v/\partial P)_T] - Pv(\partial v/\partial T)_P \tag{BT.28}$$

Comments on the Bridgman Table

1. Making up your own Bridgman table is harder than it looks (and it looks pretty hard!). According to Hougen et al. [3] Nobel Prize physicist Percy Bridgman, who invented it, had 2 errors in the first one he published in [5].
2. The properties v, u, h, s, g, and a are all shown lower case, indicating that they apply to one lbm or one kg or one mol or lbmol. One can convert them to properties for some specified mass or number of mots by multiplying them by m or n.
3. These use only the constant-pressure heat capacity, the most commonly-used heat capacity, (see Table A.9).
4. The derivative $(\partial v/\partial P)_T$ can be derived in algebraic form with either a v-explicit EOS or (as its reciprocal) from a P- explicit EOS. But $(\partial v/\partial T)_P$ cannot be easily derived algebraically with a P-explicit EOS. All the commonly used EOSs (see section 2.11 and Appendix F) are P-explicit, and cannot be solved to give simple algebraic expressions for $(\partial v/\partial T)_P$. For liquids and solids these two derivatives are equal to the coefficient of thermal expansion and the isothermal compressibility, (see Appendix D). Various numerical techniques approximate $(\partial v/\partial T)_P$; if all else fails, one can evaluate it by

$$\left(\frac{\partial v}{\partial T}\right)_P = -\frac{(\partial P/\partial T)_V}{(\partial P/\partial v)_T} \tag{BT.30}$$

which can be computed algebraically from a P- explicit EOS.
5. The derivatives that incorporate v cannot be easily programmed using P-explicit equations of state.
6. If you must derive thermodynamic relations without the Bridgman table, you will use the historically important *Maxwell Relations*; $(\frac{\partial T}{\partial v})_s = (\frac{\partial P}{\partial s})_v : (\frac{\partial T}{\partial P})_s = -(\frac{\partial v}{\partial S})_P : (\frac{\partial S}{\partial v})_T = (\frac{\partial P}{\partial T})_v$ and $(\frac{\partial S}{\partial P})_T = -(\frac{\partial v}{\partial T})_P$. If you have a Bridgman table you need never use these, but as a student of thermodynamic history you should know about them.

REFERENCES

1. de Nevers, N., Pipe rupture by sudden fluid heating. *Ind. Eng. Chem. Process Des. Dev.* 23:669–674 (1984).
2. Starling, K. E. *Fluid Thermodynamic Properties for Light Petroleum Systems*. Houston: Gulf. (1973).
3. Hougen, O. A., K. M. Watson, and R. A. Ragatz, *Chemical Process Principles,* Part II *Thermodynamics*, ed. 2. New York: Wiley, p. 545 (1947).
4. Bridgman, P. W. *A Condensed Collection of Thermodynamic Formulas.* Cambridge: Cambridge University Press (1925).
5. Bridgman, P. W. *Phys. Rev.* (2) 3:273 (1914).

APPENDIX H

ANSWERS TO SELECTED PROBLEMS

The answers shown here are for problems marked by an asterisk (*) in the individual chapters.

1.3 0.026 mol%, 0.01468 molality, 5.089 g/L, 0.0149 mol/L

2.2 1.12×10^{14} Btu/h; 2.9 lbm/h

2.5 9500 ft/s = 6500 mi/h. Collision of the earth with a stationary object in space.

2.12 -1.69×10^5 kJ

3.1 26.1 times, 29/1, 0.0018

3.2 1.58 ppm, 0.0132 g/m^3, 528, 0.056 g/(8 h)

3.6 mol fraction = 37×10^{-6}

3.12 0.00134

4.2 $\Delta h = 889.2$ Btu/lbm, $\Delta s = 1.1290$ Btu/lbm·°R, $\Delta g \approx 0.00$

5.2 2.23 psia, 384°F

5.4 ≈ 60 psia

5.8 1.14×10^{-18} psia

5.15 1085 psia

5.21 -211 J/mol

5.24 (a) -3.37×10^3 kPa/K, (b) $+1.73 \times 10^3$ kPa/K, (c) b is plausible, a is not.

6.2 -2.28, 0.79, and 3.50 cm^3/mol

6.5 (a) 1.197,1.180, 1.166, 1.154, 1.144, 1.135, 1.114 cm^3/g, (b) 0.052 L/mol

6.8 19.0 cm^3/mol

6.10 -18, -43, -376 Btu/lbm

7.1 $\alpha = -RT\left(\dfrac{dz}{dP}\right)_T$

7.8 1.54 psia

8.1 $K_{\text{acetone}} = 1.013$, $K_{\text{water}} = 0.746$, $\alpha = 1.358$

8.2 (a) $\gamma_{\text{acetone}} = 4.881$, $\gamma_{\text{water}} = 1\text{–}039$, (b) $T = 93.0$°C, $y_{\text{acetone}} = 0.303$

8.6 $y_a = 0.270$, $P = 445$ torr

8.9 $\gamma_a = 0.889$, $\gamma_b = 0.715$

8.11 $x_a = 0.2373$, $P = 12.8$ atm

8.15 $T \approx 117$°C, $x_{\text{water}} \approx 0.02$

8.30 $x_{\text{ethane}} \approx 0.8 \times 10^{-4}$, up to the vapor pressure of ethane at this temperature, 23 atm.

8.33 (a) 0.00223074 (b) 0.0023071 (c) 0.0023244. These show more digits than the data justifies, in order to show how large the changes are between the three parts.

8.34 (a) practically pure benzene and practically pure toluene, (b) ≈ 70 mol% isopropanol, practically pure water, (c) practically pure acetone, ≈ 34.5 mol % acetone, (d) ≈ 73 mol% water, practically pure n-butanol

8.47 From the van Laar equation, the γs are 2.2951 and 1.0066.

8.48 $C = 2.00$

8.55 $T_{\text{boiling}} = 100.255$°C

8.61 $\gamma_{\text{water}} = 0.92$

9.7 (a) 1.024 and 2.43, (b) 1.025 and 2.44

9.9 (a) 1.00, (b) 1.155

9.10 (a) 9.93, (b) 4.08

9.11 (a) 11.92, (b) 1.127

Physical and Chemical Equilibrium for Chemical Engineers, Second Edition. Noel de Nevers.

9.38 vs; 0.982, 0.988, γs; 0.9984, 0.99997

9.42 6×10^{10} and 152

10.1 (a) Curves 2, 3, 4, 5, and barely 6. (b) Curves 5 and 6, and possibly 4. Curves 2 and 3 also show another kind of retrograde behavior that does not have a common name. If we heat a sample with the composition of curve 2 at a constant 900 psia, we cross the dew-point curve twice, which is clearly retrograde behavior of some kind.

10.4 $T \approx 190°F$, $y_{ethanc} \approx 0.90$, somewhat better than Raoult's law.

10.7 T: (a) 80°F, (b) 65 °F, (c) 59°F, y_{ethane}; (a) 0.99, (b) 0.995, (c) 0.989

11.1 14 drops

11.2 b in w, 0.00176, w in b 0.00074

11.5 $n_{left} = 0.473$, $n_{right'} = 0.527$

11.7 $x_{ethanol} = 0.34$, $n_{ethanol} = 3.4$ mol

11.10 $\gamma_{heptane} = 1.11 \times 10^5$; $\gamma_{water} = 3570$

11.14 (d), 0.0707 and 0.9293

11.18 $A = 3.40$; $B\ 5 = 2.52$

11.25 61.9%

11.38 The two γs are 1.08 and 1.29.

12.3 −21.4, −4.2, −0.52, all kJ/mol

12.5 ≈ 3500 ppm

12.7 2.57×10^{-18}, 29.6

12.16 $g_i^\circ(\text{vap}) - g_i^\circ(\text{liq}) = 6.3$ kJ/mol

12.21 (a) 42.5 kJ/mol, (b) 43.2 kJ/mol

12.25 (a) 0.0195, 0.0128, 0.0026

12.31 (a) 0.0909, 0.5, 0.909, (b) 0.023, 0.5, 0.976

13.2 Molalities of H_2SO_4, HSO_4^-, SO_4^{2-}, H^+; 9.5×10^{-5}, 0.091, 0.0085, 0.108,

13.4 $K_{13.L} = K_{13.K}/K_{12.E}$

13.12 pH ≈ 10.9

13.14 Molalities of H_2CO_3, HCO_3^-, CO_3^{2-}, H^+; 0.0376, 1.27×10^{-4}, 4.7×10^{-11}, 1.27×10^{-4}, pH = 3.895.

13.19 Molalities of H_2CO_3, HCO_3^-, CO_3^{2-}; 1.32×10^{-5}, 5.9×10^{-11}, 2.8×10^{-20}

13.21 6.3 kWh/lbm

13.24 −3.07 V

13.36 $y_{dimer} = 0.972$

13.37 57.7 torr, 6226 torr

14.3 $P = 2.9270$ MPa, x_i; 0.7956, 0.1233, 0.0811

14.7 (a) 3.41%, (b) 47 stages

14.9 $P_2/P_1 \approx 27$ million

14.12 0.8 wt% NaCl

14.13 ≈ 2800 lb water/lb diesel fuel

14.17 354 g/mol (≈ that of sucrose)

15.2 (a) 1 component, (b) 1 component

15.3 (a) 5 phases, (b) 2 concentrations

15.4 3, 3

15.5 2 independent equations

15.9 1 gas phase and 2 solid phases

15.14 (c), there is a range of pressures at which four phases exist at equilibrium.

16.4 (a) $K = 0.476$ (b) conversion = 0.323

INDEX

Physical and Chemical Equilibrium for Chemical Engineers, Second Edition. Noel de Nevers.